Springer Monographs in Mathematics

Akihiro Kanamori

The Higher Infinite

Large Cardinals in Set Theory from Their Beginnings

Second Edition

Springer

Akihiro Kanamori
Department of Mathematics
111 Cummington Street
Boston, MA 02215
USA
aki@math.bu.edu

The first edition was published in 1994 by Springer-Verlag under the same title in the series *Perspectives in Mathematical* Logic

First softcover printing 2009

ISBN 978-3-540-88866-6 e-ISBN 978-3-540-88867-3

DOI 10.1007/978-3-540-88867-3

Springer Monographs in Mathematics ISSN 1439-7382

Library of Congress Control Number: 2008940025

Mathematics Subject Classification (2000): 03E05, 03E15, 03E35, 03E55, 03E60

Coverdesign: WMXDesign GmbH, Heidelberg

Printed on acid-free paper

9 8 7 6 5 4 3 2 1

springer.com

Dedicated to my parents Kiyoo and Sakiko
and to Tamara, Danny, and Ari

Acknowledgements

My first thanks goes to Gert Müller who initially suggested this project and persisted in its encouragement. Thanks also to Thomas Orowan who went through many iterations of the difficult typing in the early stages. James Baumgartner, Howard Becker, and Jose Ruiz read through large portions of the text and offered extensive suggestions. The host of people who over the years provided suggestions, information, or encouragement is too long to enumerate, but let me mention Matthew Foreman, Thomas Jech, Alexander Kechris, Menachem Magidor, Tony Martin, Marion Scheepers, Stevo Todorčević, and Hugh Woodin. Of course, the usual exculpatory remarks are very much in order. My final thanks to Burton Dreben and Gerald Sacks. *Honi soit qui mal y pense.*

Note to the Corrected First Edition

This printing mainly incorporates corrections of typographical and grammatical errors, changes in four minor proofs, and updated publication details of cited papers. It also introduces, on page 248, a new embedding concept $\prec^-$ subsequently used in place of several unwarranted uses of $\prec$, elementary embedding. My thanks to Kai Hauser for pointing out these unwarranted uses, as well as to Sakae Fuchino, Richard Laver, Jose Ruiz, and several other people who have sent me corrections and suggestions. Inevitably, several advances in a variety of directions have been made since the first printing, but no major attempt has been made to incorporate these.

Brookline, Massachusetts
20 January 1997

Note to the Second Edition

This edition incorporates further corrections and improvements as well as updating remarks together with new bibliographical citations. The proofs of 23.6, 24.4, 28.12, and 32.7 have been significantly changed. My particular thanks to Alessandro Andretta and Benedikt Löwe for providing corrections and suggestions, and to the Dibner Institute for the History of Science and Technology for providing support and encouragement.

Since the previous printing of the book, Sakae Fuchino has provided a Japanese translation incorporating most of the new corrections and published by Springer Tokyo in 1998. I wish to express my particular gratitude to Fuchino-san for his interest, industry, and efforts to publicize large cardinals in Japan.

The new remarks and citations have to do with advances directly pertinent to the topics elaborated in the book. The projected volume II will provide expositions of many other topics, particular in combinatorics and forcing, and the many new advances made in these directions will be fully explored there.

Brookline, Massachusetts
2 January 2003

Note to the Corrected Second Edition

This printing incorporates corrections of typographical and grammatical errors and cites a few advances. In the Chart of Cardinals there is now an arrow from 'strongly compact' to 'Woodin' because of advances in core model theory (cf. Schimmerling-Steel [96]). Also, p. 350 notes several advances, with 25.20 having become a theorem. My thanks to Masahiro Shioya and to Andrew Brooke-Taylor for their corrections and suggestions.

Brookline, Massachusetts
13 December 2004

Table of Contents

Introduction XI

§0. Preliminaries 1

Chapter 1. Beginnings

§1. Inaccessibility 16
§2. Measurability 22
§3. Constructibility 28
§4. Compactness 36
§5. Elementary Embeddings 44
§6. Indescribability 57

Chapter 2. Partition Properties

§7. Partitions and Trees 70
§8. Partitions and Structures 85
§9. Indiscernibles and $0^{\#}$ 99

Chapter 3. Forcing and Sets of Reals

§10. Development of Forcing 114
§11. Lebesgue Measurability 132
§12. Descriptive Set Theory 145
§13. $\mathbf{\Pi}^1_1$ Sets and $\mathbf{\Sigma}^1_2$ Sets 162
§14. $\mathbf{\Sigma}^1_2$ Sets and Sharps 178
§15. Sharps and $\mathbf{\Sigma}^1_3$ Sets 192

Chapter 4. Aspects of Measurability

§16. Saturated Ideals I 210
§17. Saturated Ideals II 220
§18. Prikry Forcing 234
§19. Iterated Ultrapowers 244

§20. Inner Models of Measurability 261
§21. Embeddings, $0^{\#}$, and $0^{\dagger}$ 277

Chapter 5. Strong Hypotheses

§22. Supercompactness 298
§23. Extendibility to Inconsistency 311
§24. The Strongest Hypotheses 325
§25. Combinatorics of $\mathcal{P}_{\kappa}\gamma$ 340
§26. Extenders 352

Chapter 6. Determinacy

§27. Infinite Games 368
§28. AD and Combinatorics 383
§29. Prewellorderings 403
§30. Scales and Projective Ordinals 417
§31. Det(α-$\mathbf{\Pi}_1^1$) 437
§32. Consistency of AD 450

Chart of Cardinals 472

Appendix 473

Indexed References 483

Subject Index 531

Introduction

The higher infinite refers to the lofty reaches of the infinite cardinalities of set theory as charted out by *large cardinal hypotheses*. These hypotheses posit cardinals that prescribe their own transcendence over smaller cardinals and provide a superstructure for the analysis of strong propositions. As such they are the rightful heirs to the two main legacies of Georg Cantor, founder of set theory: the extension of number into the infinite and the investigation of definable sets of reals. The investigation of large cardinal hypotheses is indeed a mainstream of modern set theory, and they have been found to play a crucial role in the study of definable sets of reals, in particular their Lebesgue measurability. Although formulated at various stages in the development of set theory and with different incentives, the hypotheses were found to form a *linear* hierarchy reaching up to an inconsistent extension of motivating concepts. All known set-theoretic propositions have been gauged in this hierarchy in terms of consistency strength, and the emerging structure of implications provides a remarkably rich, detailed and coherent picture of the strongest propositions of mathematics as embedded in set theory.

The first of a projected multi-volume series, this text provides a comprehensive account of the theory of large cardinals from its beginnings through the developments of the early 1970's and several of the direct outgrowths leading to the frontiers of current research. A further volume will round out the picture of those frontiers with a wide range of forcing consistency results and aspects of inner model theory. A genetic account through historical progression is adopted, both because it provides the most coherent exposition of the mathematics and because it holds the key to any epistemological concerns. With hindsight however the exposition is inevitably Whiggish, in that the consequential avenues are pursued and the most elegant or accessible expositions given. Each section is a modular unit, and later sections often describe how concepts discussed in earlier sections inspired the next advance. With speculations and open questions provided throughout, the reader should not only come to appreciate the scope and significance of the overall enterprise but also become prepared to pursue research in several specific areas.

In what follows a historical and conceptual overview is given, one that serves to embed the sections of the text into a larger framework. In an appendix larger and more discursive issues that may be raised by the investigation of large cardinals are taken up. See Hallett [84], Lavine [94], Moore [82], and Fraenkel–Bar-Hillel–Levy [73] for more on the development of set theory; several themes that are only broached here are substantiated in at least one of these sources.

The Beginnings of Set Theory

Set theory had its beginnings in the great 19th Century transformation of mathematics that featured the arithmetization of analysis and a new engagement with abstraction and generalization. Very much new mathematics growing out of old, the subject did not spring Athena-like from the head of Cantor but in a gradual process out of problems in mathematical analysis. In the wake of the founding of the calculus by Isaac Newton and Gottfried Leibniz the function concept had been steadily extended from analytic expressions toward arbitrary correspondences, in the course of which the emphasis had shifted away from the continuum taken as a whole to its construal as a collection of points, the real numbers. The first major expansion had been inspired by the explorations of Leonhard Euler and featured the infusion of infinite series methods and the analysis of physical phenomena, particularly the vibrating string.

Working out of this tradition the young Cantor in the early 1870's established uniqueness theorems for trigonometric series in terms of their points of convergence, theorems based on collections of reals *defined* through a limit operation iterable into the infinite. In a crucial conceptual move Cantor began to investigate such collections and infinitary enumerations for their own sake, and this led first to basic concepts in the study of sets of reals and then to the formulation of the transfinite numbers. Set theory was born on that December 1873 day when Cantor established that the reals are uncountable, i.e. there is no one-to-one correspondence between the reals and the natural numbers, and in the next decades was to blossom through the prodigious progress made by him in the theory of ordinal and cardinal numbers. But a synthesis of the reals as representing the continuum and the new numbers as representing well-orderings eluded him: Cantor could not establish the Continuum Hypothesis, that the cardinality $2^{\aleph_0}$ of the set of reals is the least uncountable cardinality $\aleph_1$, part of his problem being that he could not define a well-ordering of the reals.

Cantor came to view the finite and the transfinite as all of a piece, similarly comprehendable within mathematics, and delimited by what he termed the "Absolute" which he associated mathematically with the class of all ordinals and metaphysically with God. As part of this realist picture Cantor viewed sets, at least until the early 1890's, as inherently structured with a well-ordering of their members. Ordinal and cardinal numbers resulted from successive abstraction, from a set x to its ordertype $\overline{x}$ and then to its cardinality $\overline{\overline{x}}$.

But such a structured view served to accentuate a growing stress among mathematicians, who were already exercised by two related issues: whether infinite collections can be investigated within mathematics at all and how far the function concept is to be extended. The positive use of an arbitrary function having been made explicit, there was open controversy after Ernst Zermelo [04] formulated what he soon called the Axiom of Choice and established his Well-Ordering Theorem, that the axiom implies every set can be well-ordered.

With axiomatization assuming a general methodological role in mathematics Zermelo [08a] soon published the first axiomatization of set theory. But as with Cantor's work the move was in response to mathematical pressure for a new context: Beyond the stated purpose of securing set theory from paradox Zermelo's main motive was apparently to buttress his Well-Ordering Theorem by making explicit its underlying set existence assumptions. In the process, he shifted the focus away from the transfinite numbers to an abstract view of sets structured solely by $\in$ and simple operations. Extracted from a specific proof (for the Well-Ordering Theorem in his [08]) Zermelo's axioms had the advantages of simplicity and open-endedness. The generative set formation axioms, especially Power Set, were to lead to Zermelo's later adumbration [30] of the cumulative hierarchy view of sets, and the vagueness of the *definit* property in the Separation Axiom was to invite Thoralf Skolem's [23] proposal to base it on first-order logic.

Skolem's move was in the wake of a mounting initiative, one that was to expand set theory with new viewpoints and techniques as well as to invest it with a larger foundational significance. Gottlob Frege is regarded as the greatest philosopher of logic since Aristotle for developing his 1879 *Begriffsschrift* (quantificational logic), establishing a logical foundation for arithmetic, and generally stimulating the *analytic tradition* in philosophy. The architect of that tradition is Bertrand Russell who in his early years, influenced by Frege and Giuseppe Peano, wanted to found all of mathematics on the certainty of logic. The vaulting expression of that ambition was the 1910-3 three volume *Principia Mathematica* by Alfred Whitehead and Russell. But Russell was exercised by his well-known paradox, one which led to the tottering of Frege's mature formal system. As a result *Principia* was encased in a complex logical system of different types and intensional predications ultimately breaking under his Axiom of Reducibility, a fearful symmetry imposed by an artful dodger.

It remained for David Hilbert to shift the ground and establish mathematical logic as a field of mathematics. Russell's philosophical disposition precluded his axiomatizing logic, but Hilbert brought it under scrutiny as he did Euclidean geometry by establishing an axiomatic context and raising the crucial questions of consistency and later, completeness. This largely syntactic approach was soon given a superstructure when in response to intuitionistic criticism by Luitzen Brouwer and Hermann Weyl, Hilbert developed proof theory and proposed his program of establishing the consistency of classical mathematics with his metamathematics. These issues gained currency because of Hilbert's preeminence, just as mathematics in the large was expanded by his reliance on non-constructive proofs and transcendental methods. Through this expansion the Axiom of Choice became a mathematical necessity, particularly because of maximality arguments in algebra, and arbitrary functions became implicitly accepted in the growing investigation of higher function spaces. With the increasing emphasis on frameworks and structures, the power set operation became incorporated into mathematics.

Throughout, Zermelian set theory grew as the mathematical repository of foundational concerns and initiatives. As the first result of his axiomatic set theory Zermelo [08a] himself put the Russell paradox argument to use to show that for any set x there is a set $y \subseteq x$ such that $y \notin x$ (so that there is no universal set). Friedrich Hartogs [15] in effect converted the Burali-Forti paradox into the existence for any set x of a well-orderable set y not injectible into x. Analyzing the Zermelo [08] proof Kazimierz Kuratowski [21] provided that definition of the ordered pair, antithetical to Russell's type-ridden theory, which became the standard way to reduce the theory of relations to sets. And then Skolem [23] made his proposal of rendering Zermelo's Separation Axiom in terms of properties expressible in first-order logic.

More than that, Skolem intended for set theory to be based on first-order logic with $\in$ construed syntactically and without a privileged interpretation. This becomes clear in his application of the Löwenheim-Skolem theorem to get the Skolem paradox: the existence of countable models of set theory although it entails the existence of uncountable sets. Ironically, Skolem intended by this means to deflate the possibility of set theory becoming a foundation for mathematics, but following Kurt Gödel's work Skolem's syntactical approach to set theory came to be accepted. And again the ways of paradox were absorbed into set theory, as the Löwenheim-Skolem theorem came to play an important internal role when semantic methods were ushered in by Alfred Tarski.

Skolem [23] also and Abraham Fraenkel [21, 22] independently proposed the addition of the Replacement Axiom to Zermelo's list, and this axiom soon figured in a counter-reformation of sorts. John von Neumann [23] introduced the ordinals (transitive sets well-ordered by $\in$) and showed that every well-ordering is isomorphic to an ordinal, thereby restoring Cantor's transfinite numbers *as* sets. No longer were the numbers abstractions, but in the new formulation became incorporated into the Zermelian framework of sets built up by $\in$ and simple operations. Von Neumann's particular approach to axiomatization fostered the liberal use of proper classes in set theory and brought Replacement into prominence through its role in definitions by transfinite recursion.

With these developments before him Zermelo [30] presented his final axiomatization of set theory, incorporating Replacement and also Foundation. This axiomatization was in second-order terms, allowed urelements, and eschewed the Axiom of Infinity, but shorn of these features it became the standard Zermelo-Fraenkel (ZFC) one when recast in the soon to emerge terms of first-order logic. The Foundation Axiom had been prefigured as a restricting possibility by Dmitry Mirimanov [17], Skolem [23], and von Neumann [25]. Zermelo offered a synthetic view of a succession of natural models for set theory, each a member of a next, essentially realizing that Foundation ranks the sets in these models into a cumulative hierarchy. In current terms the axiom stratifies the formal universe V of sets as $\bigcup_\alpha V_\alpha$, where V_0 is $\emptyset$, $V_{\alpha+1}$ is the power set of V_α, and V_δ for limit ordinals δ is the union of the V_α's for $\alpha < \delta$. In a notable inversion this *iterative conception* came to be accepted after Gödel's later advocacy as a heuristic for motivating the axioms of set theory generally, its open-endedness moreover promoting a principle of tolerance for motivating new hypotheses mediating toward Cantor's Absolute. Foundation is the one axiom unnecessary for the recasting of mathematics in set-theoretic terms, but it came to be the salient feature that distinguishes structural investigations specific to set theory. Indeed, it can be fairly said that modern set theory is the study of well-foundedness, the Cantorian well-ordering doctrines adapted to the Zermelian combinatorial conception of sets.

In the 1930's Gödel's incisive analyses brought about a transformation of mathematical logic based on new initiatives for mathematical elucidation. The main source was of course his Incompleteness Theorem [31], which led to the undecidability of validity for first-order logic and the development of recursion theory. But starting an undercurrent, the earlier Completeness Theorem [30] clarified the distinction between the semantics and syntax of first-order logic and secured its key instrumental property, compactness. Then Tarski [33, 35] set out his schematic definition of truth in set-theoretic terms, exercising philosophers to a surprising extent ever since. The groundwork had been laid for the development of model theory, and set theory was to be considerably enriched since the 1950's by model-theoretic techniques. First-order logic came to be accepted as the canonical language because of its mathematical possibilities, Skolem's earlier suggestion for set theory taken up generally, and higher-order logics became downgraded as the workings of the power set operation in disguise.

So enriched and fortified by axioms, results, and techniques axiomatic set theory was launched on its independent course by Gödel's construction of L [38, 39] leading to the relative consistency of the Axiom of Choice and the Continuum Hypothesis. Synthesizing what came before, Gödel built on the von Neumann ordinals as sustained by Replacement to formulate a relative Zermelian universe of sets based on logical definability, a universe imbued with a Cantorian sense of order.

Large Cardinals

If the foregoing in brief (and with interpretative twists) is the high tradition of set theory from Cantor to Gödel, large cardinals are the trustees of older traditions in direct line from Cantor's original investigations of definable sets of reals and of the transfinite numbers. Before taking up the more continuous tradition having to do directly with the transfinite the other tradition is described, one that was to be revitalized in the 1960's by major new initiatives.

Descriptive set theory is the definability theory of the continuum, the study of the structural properties of definable sets of reals. In his most substantive approach to the Continuum Hypothesis Cantor had structured the problem via perfect sets and established that the closed sets have the *perfect set property* (11.3). Related were his contributions to measure theory, a theory that led to the Borel sets and of course to Lebesgue measure. The major incentives of descriptive set theory have been to approach sets of reals through definability as Cantor had done, and to investigate the extent of the *regularity properties*, of which Lebesgue measurability and Cantor's perfect set property are two. In a seminal paper Henri Lebesgue [05] provided the first hierarchy for the Borel sets and applied Cantor's diagonalization argument to show that the hierarchy is both proper and does not exhaust the definable sets of reals. The subject really began with Mikhail Suslin's discovery [17] of the *analytic* sets and fundamental results about this first level of the later *projective hierarchy*. The subsequent development by Nikolai Luzin, Wacław Sierpiński, and their collaborators featured *tree representations* of sets of reals, and it was through this opening that well-founded relations entered mathematical practice, the later tradition leading to Foundation and the iterative conception being quite separate and motivated by heuristics. The transfinite numbers, at least the countable ones, gained a further legitimacy through their necessary involvement in this work, contributing to the mathematical pressure for their general acceptance. Pressing upward in the projective hierarchy, by the early 1930's the descriptive set theorists had reached an impasse, one that was to be later explained by Gödel's delimitative results with L. (These matters are taken up in §§12, 13.)

The other, more primal Cantorian initiative, the mathematical investigation of the transfinite, was vigorously advanced into the higher infinite by Felix Hausdorff [08]. Dismissive of foundational issues, he pursued the structure of transfinite ordertypes for its own sake and was first to consider a large cardinal, a *weakly inaccessible* cardinal, as a natural limit point. Paul Mahlo [11, 12, 13] then studied stronger limit points, the *Mahlo* cardinals. Closure under the power set operation, intrinsic to the Zermelian set concept, was later incorporated in the concept of a *(strongly) inaccessible* cardinal by Sierpiński-Tarski [30] and Zermelo [30]. In the early semantic investigations before the general acceptance of first-order logic these cardinals provided the natural models for set theory, i.e. the corresponding initial segments of the cumulative hierarchy. (These topics are developed in §1.)

Measurability, the most prominent of all large cardinal hypotheses, embodied the first confluence of the Cantorian initiatives. Isolated by Stanisław Ulam [30] from measure-theoretic considerations related to Lebesgue measure, the concept *also* entailed inaccessibility in the transfinite. Moreover, the initial airing generated an open problem that was to keep the spark of large cardinals alight for the next three decades: *Can the least inaccessible cardinal be measurable?* (Measurability is discussed in §2.)

The further development of the higher infinite was to depend on model-theoretic techniques brought into set theory in the course of its larger development. Gödel's L was the first example of an *inner model*, a class (definable by a formula of first-order logic) including all the ordinals, which with $\in$ restricted to it is a model of the axioms. Gödel with L had in fact established the minimum possibility for the set-theoretic universe, and large cardinals were to provide the counterweight first in reaction and then for generalization. Gödel's realist speculations, especially about Cantor's Continuum Problem, contained the seeds of later heuristic arguments for large cardinal hypotheses:

The set-theoretic universe V viewed as the cumulative hierarchy $\bigcup_\alpha V_\alpha$ is open-ended and under-determined by the set-theoretic axioms, and invites further postulations based on *reflection* and *generalization*. In 1946 remarks Gödel [90: 151] suggested reflection in terms of a set-theoretic proposition being provable in "the next higher system above set theory", which proof being replaceable by one from "an axiom of infinity". This ties in with V cast as Cantor's Absolute being mathematically incomprehendable, so that any property ascribable to it must already hold in some sufficiently large V_α, some properties leading directly to large cardinal hypotheses. In a 1966 footnote Gödel [90: 260ff] acknowledged "strong axioms of infinity of an entirely new kind", generalizations of properties of $\aleph_0$ "supported by strong arguments from analogy". This ties in with Cantor's unitary view of the finite and transfinite, with properties like inaccessibility and measurability technically satisfied by $\aleph_0$ being too accidental were they not also ascribable to higher cardinals. Both reflection and generalization are latent in the eternal return of successive domains as envisioned by Zermelo [30]. Whatever the heuristics, the theory of large cardinals like other mathematical investigations was to be driven by open problems and growing structural elucidations. (These matters are taken up in §3. Other heuristic arguments are described in Maddy [88, 88a].)

The generalization of first-order logic allowing infinitary logical operations was to lead to the solution of that problem of whether the least inaccessible cardinal can be measurable. Tarski [62] defined the *strongly compact* and *weakly compact* cardinals by ascribing natural generalizations of the key compactness property of first-order logic to the corresponding infinitary languages. A strongly compact cardinal is measurable, and a measurable cardinal is weakly compact. Tarski's student William Hanf [64] then established (4.7) that *there are many inaccessible cardinals (and Mahlo cardinals) below a weakly compact cardinal.* In particular, *the least inaccessible cardinal is not measurable.* Hanf's work radically altered size intuitions about properties coming to be understood in terms of large cardinals. (These topics are developed in §4.)

In the early 1960's set theory was veritably transformed by structural initiatives based on new possibilities for constructing well-founded models and establishing relative consistency results. This was due largely to the creation of *forcing* by Paul Cohen [63, 64], who happened upon a remarkably fertile technique for producing extensions of models of set theory. In a different vein, a seminal result of Dana Scott [61] stimulated the investigation of *elementary embeddings* of inner models. The ultraproduct construction of model theory was just gaining currency when Scott took an ultrapower of V itself to establish (5.5) that *if there is a measurable cardinal, then* $V \neq L$. Large cardinal hypotheses thus assumed a new significance as a means for maximizing possibilities away from Gödel's delimitative construction. And Cantor's Absolute notwithstanding, Scott's construction began the liberal use of manipulative inner model constructions in set theory. It was in this richer setting that measurable cardinals came to play a central structural role, being necessary for securing well-founded ultrapowers (see 5.6 and before): *There is an elementary embedding* $j\colon V \to M$ *for some inner model* M *iff there is a measurable cardinal.* (These matters are taken up in §5.)

With reflection arguments emerging in the model-theoretic approaches taken in set theory, Azriel Levy [60a] established their broader significance and the close involvement of Mahlo cardinals. Then Hanf-Scott [61] formulated the *indescribable* cardinals, directly positing reflection properties in terms of higher-order languages, and showed that these cardinals provide a schematic approach to comparing large cardinals by size. Levy [71] then provided a systematic analysis, features of which were to occur in later contexts. (Indescribability is described in §6.)

Scott's result that if there is a measurable cardinal then $V \neq L$ naturally led to refinements both weakening the hypothesis and strengthening the conclusion. Notably, the first moves were made in the context of the infinitary combinatorics then being developed by Paul Erdős and his collaborators, the study of *partition* properties, which are transfinite generalizations of a result of Frank Ramsey [30]. Frederick Rowbottom [64, 71] then established a partition property for measurable cardinals (7.17), and using model-theoretic methods showed that such properties already imply that there are only countably many reals in L (8.3). This blending of model theory and infinitary combinatorics led to a spectrum of large cardinals positing strong versions of the Lowënheim-Skolem theorem, the *Rowbottom* and *Jónsson* cardinals in particular generating intriguing questions. Weaving in the crucial model-theoretic concept of a *set of indiscernibles* Jack Silver [66, 71] then analyzed what came to be regarded as the essence of transcendence over L, encapsulated by him and Robert Solovay [67] as a set $0^{\#}$ of integers coding a collection of sentences uniquely specified by indiscernibility conditions. Beyond a web of implications encircling the merely negative conclusion $V \neq L$, the existence of $0^{\#}$ is a strikingly informative assertion about just how starkly L is generated in a transcendent V. Subsequent results have buttressed the existence of $0^{\#}$ as a pivotal hypothesis, and its isolation is the first real triumph for large cardinals in the elucidation of set-theoretic structure. (These matters are taken up in Chapter 2.)

Returning to the early 1960's, if Gödel's construction of L had launched axiomatic set theory as a distinctive field of mathematics, then Cohen's technique of forcing began its transformation into a modern, sophisticated one. Starting with his work on the Continuum Hypothesis many problems that had been left unresolved were shown to be independent, as set theorists were presented a remarkably general and flexible scheme with intuitive underpinnings for constructing models of set theory. The thrust of research gradually deflated the Cantor-Gödel realist view with an onrush of new models, and shedding some of its foundational burden set theory became an intriguing mathematical subject where formalized versions of truth and consistency became matters for combinatorial manipulation as in algebra. From Skolem relativism to Cohen relativism the role of set theory for mathematics became even more evidently one of an open-ended framework rather than an elucidating foundation. From this point of view, that the ZFC axioms do not determine the cardinality $2^{\aleph_0}$ of the set of reals seems an entirely satisfactory state of affairs. With the richness of possibility for arbitrary reals and mappings, no axioms that do not directly impose structure from above should constrain a set as open-ended as the collection of reals or its various possibilities for well-ordering.

Inaccessible cardinals figured from the beginning in this sea-change, first in the concept of the *Levy collapse* and then in its use in Solovay's inspiring result [65b, 70] that *if there is an inaccessible cardinal, then in a submodel of a forcing extension every set of reals is Lebesgue measurable and has the perfect set property.* (The Axiom of Choice necessarily fails in this submodel.) As Cohen's independence of the Continuum Hypothesis did for the transfinite, this result on the regularity of sets of reals not only resolved old axiomatic issues but reinvigorated the Cantorian initiatives by suggesting new mathematical possibilities. Solovay [69] soon applied the ideas of his proof to show that measurable cardinals directly imply the regularity properties at the level of Gödel's delimitative results with L, revitalizing the classical program of descriptive set theory. Then Donald Martin and Solovay (cf. their [69]) applied large cardinal hypotheses at the level of $0^{\#}$ to push forward the old tree representation ideas, with the hypotheses cast in the new role of securing well-foundedness in this context. (These matters are taken up in Chapter 3.)

The perfect set property led to the first instance of a new phenomenon in set theory: the derivation of *equiconsistency* results based on the complementary methods of forcing and inner models. A large cardinal hypothesis is typically transformed into a proposition about sets of reals by forcing that "collapses" the cardinal to $\aleph_1$ or "enlarges" the power of the continuum to the cardinal. Conversely, the proposition entails the same large cardinal hypothesis in the clarity of an inner model. Solovay's result provided the forcing direction from an inaccessible cardinal to the proposition that every set of reals has the perfect set property. But Ernst Specker [57] had in effect established that if every set of reals has the perfect set property (and $\aleph_1$ is regular), then $\aleph_1$ is inaccessible in L (11.6). Thus, Solovay's use of an inaccessible cardinal was necessary, and its collapse to $\aleph_1$ complemented Specker's observation. Years later, Saharon Shelah [84] was able to establish the necessity of Solovay's inaccessible also for the proposition that every set of reals is Lebesgue measurable.

The emergence of such equiconsistency results is a subtle transformation of earlier hopes of Gödel: Propositions can indeed be resolved if there are enough ordinals, how many being specified by positing a large cardinal. But the resolution is in terms of the Hilbertian concept of consistency, the methods of forcing and inner models being the operative modes of argument. In a new synthesis of the two Cantorian initiatives, hypotheses of length concerning the extent of the transfinite are correlated with hypotheses of width concerning sets of reals. There is a telling antecedent in the result of Gerhard Gentzen [36, 43] that the consistency strength of arithmetic can be exactly gauged by an ordinal ε_0, i.e. transfinite induction up to that ordinal in a formal system of notations. Although Hilbert's program of establishing consistency by finitary means cannot be realized, Gentzen provided an exact analysis in terms of ordinal length. Proof theory blossomed in the 1960's with the analysis of other theories in terms of such lengths, the proof theoretic ordinals.

In the late 1960's a wide-ranging investigation of measurability was carried out with forcing and inner models. These developments not only provided an illuminating structural analysis, but suggested new questions and provided paradigms for the subsequent investigation of stronger hypotheses. Solovay [66, 71] brought the concept of *saturated ideal* to the forefront, establishing an equiconsistency result about *real-valued measurability*. Subsequent work showed that saturated ideals are a flexible generalization of measurability that can occur low in the cumulative hierarchy. Exploiting the technique of *iterated ultrapowers* developed by Haim Gaifman [64], Kenneth Kunen [70] established the main structure theorems for *inner models of measurability*. Not only do these models have the minimal structure of Gödel's L, but they turn out to be exactly the ultrapowers of each other, and such coherence amounts to strong evidence for the consistency of the concept of measurability. Kunen also established a characterization of the existence of $0^{\#}$ in terms of the non-rigidity of L: $0^{\#}$ *exists iff there is an elementary embedding* $j\colon L \to L$. Solovay isolated a set $0^{\dagger}$ that plays an analogous role for inner models of measurability that $0^{\#}$ does for L, and its existence has a similar characterization in terms of non-rigidity. (These topics are developed in Chapter 4.)

Even as measurability was being methodically investigated, Solovay and William Reinhardt were charting out stronger hypotheses. Taking the concept of elementary embedding as basic they independently formulated the concept of *supercompact* cardinal as a generalization of both measurability and strong compactness, and Reinhardt formulated the stronger concept of *extendible* cardinal with motivating ideas based directly on reflection. Reinhardt briefly considered an ultimate reflection property along these lines, but in a dramatic turn of events Kunen [71b] established that this *prima facie* extension is inconsistent: *There is no elementary embedding* $j\colon V \to V$. Kunen's argument turned on what seemed to be a combinatorial contingency, but his particular formulation has stood as the upper bound for large cardinal hypotheses. The initial guiding ideas shaped and delimited by a mathematical result, hypotheses just on the verge of this inconsistency were subsequently analyzed, as well as the weaker *n-huge* cardinals and *Vopěnka's Principle*, to chart the terrain down to the extendible cardinals. The supercompact cardinals in particular became prominent as a source of new combinatorics and relative consistency results. Also, when refinements of elementary embedding in the form of *extenders* were formulated, weakenings of supercompactness in the form of *strong, Woodin,* and *superstrong* cardinals came to play crucial roles in later developments. (These topics are developed in Chapter 5.)

With this charting out of the higher infinite, the extensive research through the 1970's and 1980's considerably strengthened the view that the emerging hierarchy of large cardinals provides *the* measuring rod of exhaustive principles against which all possible consistency strengths can be gauged. First, the various hypotheses though arising from diverse motivations and historical happenstance nonetheless form a *linear* hierarchy, one neatly delimited by Kunen's inconsistency result. Typically for two large cardinal hypotheses, below a cardinal satisfying one there are many cardinals satisfying the other, in a sense prescribed by the first. Moreover, the weaker hypotheses through strong forms of measurability have been bolstered by a variety of equiconsistency results involving combinatorial propositions low in the cumulative hierarchy. In this respect, particularly intriguing is the work on the *Singular Cardinals Problem*, which showed that something as basic as rendering 2^κ large for singular strong limit cardinals κ essentially requires large cardinals. Finally, a variety of strong propositions have been informatively bracketed in consistency strength between two large cardinal hypotheses: The stronger hypothesis implies that there is a forcing extension in which the proposition obtains; and if the proposition obtains, there is an inner model satisfying the weaker hypothesis. Supercompactness has often figured as the upper bound, but sometimes n-hugeness and even the hypotheses just short of Kunen's inconsistency have played this role. (This wide-ranging exploration is the subject of volume II.)

If set theory serves as an open-ended framework *for* mathematics, as an autonomous field *of* mathematics it has become a remarkably successful investigation of well-foundedness, in large measure because large cardinals have been found to provide an elegant and fully sufficient superstructure for the study of consistency strength.

Determinacy

One of the great successes for large cardinals has to do with perhaps the most distinctive and intriguing development in modern set theory. Although the *determinacy of games* has roots as far back as Zermelo [13], the concept for infinite games only began to be seriously explored in the 1960's when it was realized that it led to the regularity properties for sets of reals. Jan Mycielski and Hugo Steinhaus in their [62] proposed the *Axiom of Determinacy*, at least for some inner model since it contradicts the Axiom of Choice. Then in 1967 Solovay made an initial connection with large cardinals and David Blackwell [67] with methods of descriptive set theory. Investigating further consequences of determinacy, fine mathematicians like Solovay, Martin, Yiannis Moschovakis, Kunen, and Alexander Kechris soon established an elaborate web of connections in the unabashed pursuit of structure for its own sake. Determinacy hypotheses seemed to settle many questions and provide new modes of argument, leading to an opaque realization of the old Cantorian initiatives concerning sets of reals and the transfinite with determinacy replacing well-ordering as the animating principle. By the late 1970's a more or less complete theory for the projective sets was in place, and with this completion of a main project of descriptive set theory attention began to shift to questions of overall consistency.

Martin [70] had early on shown that the existence of a measurable cardinal implies the determinacy of games for analytic sets, and through the 1970's he established results equating many measurable cardinals with levels of a difference hierarchy for analytic sets and then showed that a large cardinal hypothesis near Kunen's inconsistency implied determinacy at the next projective level. Then in the mid-1980's Matthew Foreman, Menachem Magidor, and Shelah made a major breakthrough about strong large cardinal hypotheses, and although not directly involving determinacy Martin, John Steel, and Hugh Woodin were able to build on this to establish the consistency of the Axiom of Determinacy relative to large cardinals. Woodin in fact established that *the Axiom of Determinacy is equiconsistent with the existence of infinitely many Woodin cardinals*, pinpointing the axiom in consistency strength above measurable cardinals but far below supercompact cardinals. This unifying result was a resounding triumph for the modern methods of set theory and an unexpected affirmation of the relevance of large cardinals. Woodin's subsequent results about other determinacy hypotheses and infinite combinatorics speak to the great progress that has been made and the promise of deeper insights to come. (These matters are taken up in Chapter 6.)

0. Preliminaries

This section sets out the necessary mathematical preliminaries for the text. Generally speaking, familiarity is assumed with the development of set theory through the basics about the constructible hierarchy L and the method of forcing, and with the basic concepts and constructions of model theory. Nevertheless, in taking a historical approach well-known concepts are often formulated anew, their basic facts reviewed, and references provided as part of the development. In particular, a discussion of the forcing formalism is deferred until results achieved by that method are dealt with squarely. In what follows, some basic terminology and concepts are affirmed whose contextual review would break the pace of exposition, and some standing conventions established.

Set-Theoretic Notation

For the set theory, the texts Jech [03], Kunen [80], Drake [74] and Levy [79] each provide the basic development of the subject and more. The first three contain the necessary preliminaries about L, the first two such preliminaries about forcing, and the first a good deal of information about large cardinals. $\mathcal{L}_\in$ denotes the language of set theory: first-order predicate calculus with equality and the binary predicate symbol $\in$. In this language AC denotes the Axiom of Choice, CH the Continuum Hypothesis, and GCH the Generalized Continuum Hypothesis. ZF denotes Zermelo-Fraenkel set theory in $\mathcal{L}_\in$, ZFC that theory with AC adjoined, and ZF^- and ZFC^- these theories with the Power Set Axiom deleted.

The results in this book are theorems of ZFC ,

unless a different theory is specified either at the beginning of a section or statement of a result. Thus, by "class" is meant definable class, and although ordered pairs or even transfinite sequences of classes may be used, they will be definable as single classes. The set-theoretic notation used in the text is generally standard, with the possible deviations stipulated in the following précis:

Unless otherwise specified the first lower case Greek letters $\alpha, \beta, \gamma, \delta, \ldots$ denote ordinals, whereas the middle letters $\kappa, \lambda, \mu, \nu, \ldots$ are reserved for *infinite* cardinals. This convention is sometimes extended to allow these middle letters to denote infinite cardinals only in the sense of some model, and this should be clear from the context. Concerning the αth uncountable cardinal, tradition dictates that an intensional distinction be maintained by referring to its ordertype by ω_α and its cardinality by $\aleph_\alpha$, although this distinction is not always sharp or illuminating. $\text{cf}(\gamma)$ denotes the cofinality of γ, and γ^+ denotes the least cardinal greater than γ. A cardinal κ is a strong limit *iff* for every $\lambda < \kappa$, $2^\lambda < \kappa$.

On denotes the class of ordinals, V the universe of sets, V_α the set of sets of rank less than α, and H_κ the set of sets hereditarily of cardinality less than κ. The cumulative hierarchy $\bigcup_\alpha V_\alpha$ provides the basic stratification of V through the full exercise of the Power Set Axiom, but the H_κ's are often more suitable as approximations. Not only is their use more parsimonious since H_κ is usually much smaller than V_κ, but for regular κ, H_κ models Replacement.

For a set x, $|x|$ denotes its cardinality, $\mathcal{P}(x)$ its power set, and $\mathcal{P}_\kappa x = \{y \subseteq x \mid |y| < \kappa\}$. α^β denotes *cardinal* exponentiation for cardinals α and β, and $\alpha^\beta = |\alpha|^{|\beta|}$ for arbitrary α and β unless it is contextually clear that ordinal exponentiation is meant. Also, $\alpha^{<\kappa} = \bigcup\{\alpha^\beta \mid \beta < \kappa\}$, so that $\lambda^{<\kappa}$ is the cardinality of $\mathcal{P}_\kappa\lambda$.

For a function f, dom(f) denotes its domain, ran(f) its range, $f``x = \{f(y) \mid y \in x\}$, the image of $x$ under $f$ (after Whitehead and Russell), and $f|x = f \cap (x \times V)$, the restriction of $f$ to $x$. More generally, for a binary relation $R$, $R``x = \{b \mid \exists a \in x(\langle a, b\rangle \in R)\}$. ${}^y x$ denotes the collection of functions: $y \to x$, so that λ^κ is the cardinality of ${}^\kappa\lambda$, and ${}^{<\alpha}x = \bigcup_{\beta<\alpha} {}^\beta x$. For $1 \leq n < \omega$, an n-tuple is regarded alternately as a member of some n-fold Cartesian product $X_1 \times \ldots \times X_n$ (defined recursively as $(X_1 \times \ldots \times X_{n-1}) \times X_n$ for $n > 1$), or as a sequence, i.e. a function with domain n. For example, ${}^k X \times Y$ for $0 < k < \omega$ is viewed as consisting of $\langle u_0, \ldots, u_{k-1}, u_k\rangle$ such that $u_i \in X$ for $i < k$ and $u_k \in Y$. Concatenation $\frown$ and initial segment have the expected meaning for n-tuples. For finite sequences s, $|s|$ is a convenient way of denoting its length. For a function f and an n-tuple $\langle x_1, \ldots, x_n\rangle$ in its domain, $f(x_1, \ldots, x_n)$ is written for $f(\langle x_1, \ldots, x_n\rangle)$.

$x \subseteq y$ denotes inclusion and $x \subset y$ proper inclusion, with $\supseteq$ and $\supset$ having the derived meanings. $x - y$ denotes set subtraction, $x \triangle y$ the symmetric difference $(x-y)\cup(y-x)$, and tc(x) the transitive closure of x, the smallest transitive set $\supseteq x$. For $x \subseteq$ On, ot(x) denotes its ordertype; $\min(x) = \bigcap x$ when $x \neq \emptyset$; $\sup(x) = \bigcup x$; $\max(x) = \bigcup x$ when $\bigcup x \in x$; $[x]^\alpha = \{y \subseteq x \mid \mathrm{ot}(y) = \alpha\}$; $[x]^{<\alpha} = \bigcup_{\beta<\alpha}[x]^\beta$ so that $\mathcal{P}_\kappa\lambda = [\lambda]^{<\kappa}$; $[x]^{\leq\alpha} = \bigcup_{\beta\leq\alpha}[x]^\beta$; and an initial segment of x is a set of form $x \cap \alpha$ for some α. Furthermore, if $\{\alpha_1, \ldots, \alpha_n\}$ is used to denote a member of $[x]^{<\omega}$, it is with the understanding that $\alpha_1 < \ldots < \alpha_n$, and for f a function with domain $\subseteq [x]^{<\omega}$, $f(\alpha_1, \ldots, \alpha_n)$ is written for $f(\{\alpha_1, \ldots, \alpha_n\})$. In all of these notations proper classes are allowed when the resulting expression is formalizable, i.e. it again denotes a class.

Closed Unbounded and Stationary Sets

For $X \subseteq$ On, γ is a *limit point of X iff* $\bigcup(X \cap \gamma) = \gamma > 0$. C is *closed unbounded in δ iff* C is an unbounded subset of δ containing all its limit points less than δ. For regular $\nu < \delta$, C is *ν-closed unbounded in δ iff* C is an unbounded subset of δ containing all its limit points less than δ of cofinality ν. For limit ordinals δ, S is *stationary in δ iff* $S \subseteq \delta$ and $S \cap C \neq \emptyset$ for any C closed unbounded in δ. The "in δ" is deleted when clear from the context. If $\langle X_\alpha \mid \alpha < \delta\rangle \in {}^\delta\mathcal{P}(\delta)$, then its *diagonal intersection* is $\{\xi < \delta \mid \xi \in \bigcap_{\alpha<\xi} X_\alpha\}$, denoted by $\triangle_{\alpha<\delta}X_\alpha$. For $X \subseteq$ On and $f\colon X \to$ On, f is *regressive iff* $f(\alpha) < \alpha$ for every $\alpha \in X - \{0\}$. The following is basic:

0.1 Proposition. *Suppose that $\lambda > \omega$ is regular.*

(a) If $\gamma < \lambda$ and $\langle C_\alpha \mid \alpha < \gamma \rangle$ is a sequence of sets closed unbounded in λ, then $\bigcap_{\alpha<\gamma} C_\alpha$ is closed unbounded in λ.

(b) If $\langle C_\alpha \mid \alpha < \lambda \rangle$ is a sequence of sets closed unbounded in λ, then its diagonal intersection $\Delta_{\alpha<\lambda} C_\alpha$ is closed unbounded in λ.

(c) (Fodor [56]) *If S is stationary in λ and $f\colon S \to \lambda$ is regressive, then there is an $\alpha < \lambda$ such that $f^{-1}(\{\alpha\})$ is stationary in λ.*

(d) If $\nu < \lambda$ is regular, $S \subseteq \{\xi < \lambda \mid \mathrm{cf}(\xi) = \nu\}$ is stationary in λ, and C is ν-closed unbounded in λ, then $S \cap C \neq \emptyset$. ⊣

Filters and Ideals

For a non-empty set S, $F \subseteq \mathcal{P}(S)$ is a *filter (*resp. *ideal) over S iff* F is a proper, non-principal filter (resp. ideal) *on* the Boolean algebra $\mathcal{P}(S)$. The previous sentence exemplifies the distinction intended between "over" and "on". For $E \subseteq \mathcal{P}(S)$, *the filter (*resp. *ideal) over S generated by E* is $\{X \subseteq S \mid \exists Y \in E(Y \subseteq X)\}$ (resp. $\{X \subseteq S \mid \exists Y \in E(X \subseteq Y)\}$), provided that it is proper and non-principal. A filter (resp. ideal) F over S is *λ-complete iff* for any $\gamma < \lambda$ and $\{X_\alpha \mid \alpha < \gamma\} \subseteq F$, $\bigcap_{\alpha<\gamma} X_\alpha \in F$ (resp. $\bigcup_{\alpha<\gamma} X_\alpha \in F$). U is an *ultrafilter over S iff* it is a maximal filter over S, i.e. for any $X \subseteq S$ either $X \in U$ or else $S - X \in U$. A filter F over S is *uniform iff* $|X| = |S|$ for any $X \in F$.

As a convenient convention it is assumed that *for a filter F over a cardinal λ the final segments $\lambda - \alpha = \{\xi \mid \alpha \leq \xi < \lambda\} \in F$ for every $\alpha < \lambda$.* Concomitantly it is assumed that *for an ideal I over a cardinal λ the initial segments $\alpha = \{\xi \mid \xi < \alpha\} \in I$ for every $\alpha < \lambda$.* For a regular cardinal $\lambda > \omega$, $\mathcal{C}_\lambda$ denotes the *closed unbounded filter over* λ, the filter generated by the closed unbounded subsets of λ, which is λ-complete by 0.1(a). Corresponding to it is NS_λ, *the non-stationary ideal over* λ, the λ-complete ideal of non-stationary subsets of λ.

Model-Theoretic Notation

For the model theory, familiarity is assumed with the basic notions of structures and satisfaction, elementary equivalence, substructure and embeddability, compactness, Skolem hull arguments, and the ultraproduct construction (although these last two are reviewed below), as well as with the formalizability of the satisfaction relation for set structures. The standard text for model theory is Chang-Keisler [90], and the texts Devlin [84], Kunen [80], Moschovakis [80], and Drake [74] work out set-theoretic formalizations.

For a formula φ of $\mathcal{L}_\in$, φ^M denotes the relativization of the quantifiers to the class M, defined by recursion on the complexity of φ. This then extends to definable terms and classes t to yield t^M. More generally, for $\mathcal{M}$ a structure and t a definable term in the corresponding language, $t^{\mathcal{M}}$ denotes the interpretation of t in $\mathcal{M}$. The notation $t(v_1, \ldots, v_n)$ for a term indicates that its variables are among $v_1, \ldots, v_n$, and $\varphi(v_1, \ldots, v_n)$ for a formula indicates that its free variables are among $v_1, \ldots, v_n$. For convenience, variables are usually indexed as here from

1, but on occasion, from 0. In any case the particular indexing is not restrictive, since just as φ should be regarded as a meta-variable for formulas, so too should the v_i's be regarded as meta-variables for formal variables. For $t(v_1, \ldots, v_n)$ and $\varphi(v_1, \ldots, v_n)$ as above and for terms $t_1, \ldots, t_n$, $t(t_1, \ldots, t_n)$ is the result of replacing each occurrence of v_i by t_i, and $\varphi(t_1, \ldots, t_n)$ the result of replacing each free occurrence of v_i by t_i, this allowed only when no variable in any t_i becomes bound by a quantifier of φ. Finally, $\mathcal{M} \models \varphi[x_1, \ldots, x_n]$ indicates that φ is satisfied in the structure $\mathcal{M}$ with the variable assignment taking v_i to x_i in the domain of $\mathcal{M}$, and $t^{\mathcal{M}}(x_1, \ldots, x_n)$ denotes the interpretation of t under this assignment. $\models$ is extended to collections of formulas as in $\mathcal{M} \models \mathrm{ZFC}$ with the expected meaning.

For a theory T, $\mathrm{Con}(T)$ should be a formal assertion of its consistency in terms of an arithmetization of its language. For theories T_1 and T_2, T_2 is *consistent relative to* T_1 *iff* $\mathrm{Con}(T_1) \to \mathrm{Con}(T_2)$; and T_1 and T_2 are *equiconsistent iff* $\mathrm{Con}(T_1) \leftrightarrow \mathrm{Con}(T_2)$. However, such assertions are better construed as part of the mathematical English, the proofs in each case providing the sense.

Although these conventions are maintained when issues of satisfaction and definability are discussed, specific formulas of $\mathcal{L}_\in$ in our arguments are rendered as usual in a mathematical English with the assignment of sets to variables indicated by a direct substitution into the formula. Even when structures are being discussed with some care, this conflating of syntax and semantics may occur to the right of a $\models$; formulas are then set off from the rest of the text by the Quinean quotes $\ulcorner\ \urcorner$ when confusion is possible. For a formula φ, $\ulcorner\varphi\urcorner$ may also denote its code according to some contextually established arithmetization.

We work most often with structures for $\mathcal{L}_\in$ of form $\langle M, \in \cap (M \times M)\rangle$, consisting of a domain together with the real membership relation restricted to it. For simplicity such a structure is usually denoted by $\langle M, \in\rangle$, or just M when contextually clear. For a theory T of $\mathcal{L}_\in$, M is an $\in$-*model of* T *iff* $\langle M, \in\rangle$ is a model of T. We also work with expanded structures of form $\langle M, \in \cap (M \times M), R\rangle$ where $R \subseteq M$, denoted by $\langle M, \in, R\rangle$. It is then to be understood that R is interpreting a new unary predicate symbol, although in the set-theoretic context $x \in R$ is often written instead of $R(x)$. Similar remarks apply to further expanded structures. When there is need to be explicit, $\mathcal{L}^X_\in(\dot{R}_1, \ldots, \dot{R}_n)$ denotes the language $\mathcal{L}_\in$ of set theory expanded by constants $\dot{x}$ for every $x \in X$ and predicate symbols $\dot{R}_1, \ldots, \dot{R}_n$ of specified arity, with $\mathcal{L}_\in(\dot{R}_1, \ldots, \dot{R}_n)$ being $\mathcal{L}^\emptyset_\in(\dot{R}_1, \ldots, \dot{R}_n)$.

We often consider structures with domains that are proper classes, like V itself. The previous conventions apply, but there is an important caveat: By the Gödel-Tarski undefinability of truth argument (Jech [03: 162]) the satisfaction relation $\langle\ulcorner\varphi\urcorner, \langle x_1, \ldots, x_n\rangle\rangle \in S$ *iff* $V \models \varphi[x_1, \ldots, x_n]$ is formally undefinable in ZF, and so also is the general satisfaction relation for proper classes $\langle\ulcorner\varphi\urcorner, \langle z_1, \ldots, z_k\rangle, \ulcorner\psi\urcorner, \langle x_1, \ldots, x_n\rangle\rangle \in S^+$ *iff* $\{y \mid \varphi[y, z_1, \ldots, z_k]\} \models \psi[x_1, \ldots, x_n]$. This is the source of possible unformalizability in the text, and the issue is discussed as it arises.

Hierarchies of Formulas

Several hierarchies of formulas will figure in analyses of definability, particularly in absoluteness and reflection arguments. For any (first-order) extension $\mathcal{L}$ of $\mathcal{L}_\in$, the *Levy hierarchy* of formulas of $\mathcal{L}$ is formulated as follows: A formula is Σ_0 and Π_0 if its only quantifiers are *bounded*, i.e. can be rendered as $\forall v \in w$ or $\exists v \in w$. Recursively, a formula is Σ_{n+1} if it is of the form $\exists v_1 \ldots \exists v_k \varphi$ where φ is Π_n, and Π_{n+1} if it is of the form $\forall v_1 \ldots \forall v_k \varphi$ where φ is Σ_n. For convenience this terminology is extended by stipulating that every formula is Σ_ω and Π_ω. The classification of definable concepts in this hierarchy depends on the ambient theory. For a theory T of $\mathcal{L}$, a formula φ is Σ_n^T *iff* for some Σ_n formula φ', $T \vdash \varphi \leftrightarrow \varphi'$; and similarly for Π_n^T. A formula is Δ_n^T *iff* it is both Σ_n^T and Π_n^T. This terminology is extended to classes through the analysis of their defining formulas.

Σ_n^{ZF} and Π_n^{ZF} formulas are equivalent to formulas with blocks of like quantifiers contracted into one through applications of the Pairing Axiom. Also, bounded quantification does not add to complexity in ZF:

(i) If φ is Σ_n^{ZF}, so is $\exists v \in w\varphi$ and $\forall v \in w\varphi$.
(ii) If φ is Π_n^{ZF}, so is $\exists v \in w\varphi$ and $\forall v \in w\varphi$.

To show this, assume inductively that (i) and (ii) hold with n replaced by any $k < n$. (ii) follows from (i) by taking negations, and the first part of (i) is immediate. For the second part of (i), φ can be taken to be of form $\exists v_0 \psi$ where ψ is Π_{n-1}. Then by Replacement,

$$\forall v \in w \exists v_0 \psi \;\leftrightarrow\; \exists v_1 \forall v \in w \exists v_0 \in v_1 \psi \;,$$

and the latter formula is Σ_n^{ZF} by induction.

A typical analysis of transfinite recursion leads to a useful observation:

0.2 Lemma. *$\ulcorner \mathrm{rank}(v_0) = v_1 \urcorner$ is Δ_1^{ZF}, and $\ulcorner V_{v_0} = v_1 \urcorner$ is Π_1^{ZF}.*

Proof. Let $\varphi(f)$ be the formula

$$f \text{ is a function } \wedge \forall x \in \mathrm{dom}(f)(x \subseteq \mathrm{dom}(f) \wedge f(x) = \bigcup\{f(y)+1 \mid y \in x\}) \,.$$

This asserts that f must satisfy the usual recursive definition for the rank function on a transitive domain. Hence,

$$\begin{aligned} \mathrm{rank}(v_0) = v_1 \quad & \textit{iff} \;\; \exists f(\varphi(f) \wedge \langle v_0, v_1\rangle \in f) \\ & \textit{iff} \;\; \forall f(\varphi(f) \wedge v_0 \in \mathrm{dom}(f) \to \langle v_0, v_1\rangle \in f) \,. \end{aligned}$$

$\varphi(f)$ is Δ_0^{ZF}, and so the first result follows. For the second, note that

$$V_{v_0} = v_1 \;\; \textit{iff} \;\; \forall v_2(v_2 \in v_1 \leftrightarrow \exists v_3 \in v_0(\mathrm{rank}(v_2) = v_3)) \,.$$

The latter is Π_1^{ZF} by previous remarks about absorbing bounded quantifiers. ⊣

A well-known observation, established by induction on formula complexity, is that Σ_0 *formulas are absolute for transitive structures*, i.e. if $\varphi(v_1, \ldots, v_n)$ is Σ_0, M is transitive, and $x_1, \ldots, x_n \in M$, then

$$\langle M, \in\rangle \models \varphi[x_1, \ldots, x_n] \text{ iff } \langle V, \in\rangle \models \varphi[x_1, \ldots, x_n] .$$

Moreover, Δ_1^{ZF} *formulas are absolute for transitive* $\in$*-models of* ZF, i.e. if $\varphi(v_1, \ldots, v_n)$ is Δ_1^{ZF}, M is such a model, and $x_1, \ldots, x_n \in M$, then

$$\langle M, \in\rangle \models \varphi[x_1, \ldots, x_n] \text{ iff } \langle V, \in\rangle \models \varphi[x_1, \ldots, x_n] .$$

This can be illustrated in terms of the proof of 0.2: If $\langle M, \in\rangle \models \mathrm{rank}(x) = \alpha$, then by the Σ_1 formulation, in M there is an f satisfying φ such that $f(x) = \alpha$, and hence this holds in $V \supseteq M$, so that $\mathrm{rank}(x) = \alpha$ in V, i.e. there is *upward persistence.* If $\mathrm{rank}(x) = \alpha$, then by the Π_1 formulation, for any f satisfying φ with $x \in \mathrm{dom}(f)$ we have $f(x) = \alpha$, and hence this holds in $M \subseteq V$ so that $M \models \mathrm{rank}(x) = \alpha$, i.e. there is *downward persistence.* We often use the basic observation that if a φ is upward persistent, then so is $\exists v\varphi$, and if φ is downward persistent, then so is $\forall v\varphi$.

For more information on the Levy hierarchy see Levy [65], Jech [03: 183ff], Devlin [84: 27ff], or Drake [74: 75ff].

Concerning the satisfaction relation, the latter two texts provide (pp. 31ff and pp. 89ff resp.) formalizations of the relation for set structures that are Δ_1^{ZF}, and hence absolute for transitive $\in$-models of ZF. We will use the following strategem: If M is transitive, $y \in M$ is transitive, $\varphi(v_1, \ldots, v_n)$ is a formula of $\mathcal{L}_\in$, and $x_1, \ldots, x_n \in y$, then

$$\langle y, \in\rangle \models \varphi[x_1, \ldots, x_n] \text{ iff } \langle M, \in\rangle \models \varphi^{(y)}[x_1, \ldots, x_n] .$$

Thus, although $\langle M, \in\rangle$ may not be a model of ZF, it can cast the satisfiability of φ in $\langle y, \in\rangle$ in the absolute terms of the Σ_0 formula $\varphi^{(y)}$.

For proper classes, the unformalizability of the satisfaction relation was already mentioned. It will be useful to know that for any particular n and transitive class M, the satisfaction relation $\models_M^n$ for M restricted to Σ_n formulas *is* formalizable in ZF: For $n = 0$, using the aforementioned absoluteness of Σ_0 formulas for transitive structures,

$$\begin{aligned}\models_M^0 \varphi[x_1, \ldots, x_k] \text{ iff } & \varphi(v_1, \ldots, v_k) \text{ is } \Sigma_0 \text{ and } \exists y \in M(y \text{ is transitive} \\ & \wedge x_1, \ldots, x_k \in y \wedge \langle y, \in\rangle \models \varphi[x_1, \ldots, x_k]) ,\end{aligned}$$

and recursively,

$$\begin{aligned}\models_M^{n+1} \varphi[x_1, \ldots, x_k] \text{ iff } & \varphi(v_1, \ldots, v_k) \text{ is } \Sigma_{n+1}, \text{ say for simplicity} \\ & \exists v_{k+1} \ldots \exists v_{k+r} \neg\psi(v_1, \ldots, v_{k+r}) \text{ where } \psi \text{ is } \Sigma_n, \\ & \text{and } \exists y_1 \ldots \exists y_r \neg(\models_M^n \psi[x_1, \ldots, x_k, y_1, \ldots, y_r]) .\end{aligned}$$

General higher-order languages have typed variables of every finite type (or order), quantifications of these, and beyond the atomic formulas specified by the language, $X \in Y$ and $X = Y$ for any typed variables X and Y. In the intended semantics, if D is the domain of a structure, type 1 variables play the usual role of first-order variables, type 2 variables range over $\mathcal{P}(D)$, and generally, type $i+1$ variables range over $\mathcal{P}^i(D)$ where $\mathcal{P}^i$ denotes i iterations of the power set operation. A formula is Π^m_n *iff* it starts with a block of universal quantifiers of type $m+1$ variables, followed by a block of existential quantifiers of type $m+1$ variables, and so forth with at most n blocks in all, followed afterwards by a formula containing variables of type at most $m+1$ and quantified variables of type at most m. A formula is Σ^m_n *iff* it starts instead with existential quantifiers. Of course, formulas containing only type 1 variables can be construed as the usual first-order formulas. Note that this classification of formulas is cumulative because of the "at most": any Π^m_n or Σ^m_n formula is also Π^r_s and Σ^r_s for any $r > m$, or $r = m$ and $s > n$. Over the structures that will be considered these formulas will always be equivalent to like formulas where blocks of like quantifiers have been contracted into one.

Well-Foundedness

For a binary irreflexive relation R with field a set X,

R is *well-founded iff* any non-empty $Y \subseteq X$ has an R-minimal element .

It is not assumed that R is transitive. The following is basic:

0.3 Lemma. *For a binary relation R with field X, R is well-founded iff there is a $\rho\colon X \to \mathrm{On}$ which is order-preserving, i.e. if $\langle x, y\rangle \in R$, then $\rho(x) < \rho(y)$. In particular, $\ulcorner R$ is well-founded$\urcorner$ is Δ^{ZF}_1 and hence absolute for transitive $\in$-models of* ZF. ⊣

In the substantive direction, define a rank function ρ by recursion:

$$\rho(y) = \sup(\{\rho(x) + 1 \mid \langle x, y\rangle \in R\}) .$$

The definition of well-foundedness is Π^{ZF}_1 and the rank function formulation, Σ^{ZF}_1.

Also basic is the *Collapsing Lemma:*

0.4 Lemma (Mostowski [49], Shepherdson [51]). *Suppose that $\langle M, E, \ldots\rangle$ is a (possibly proper class) structure with E a binary relation on M satisfying:*

(a) E is well-founded;

(b) $\langle M, E\rangle$ is extensional, i.e. if $a, b \in M$ and $x\ E\ a$ iff $x\ E\ b$ for every $x \in M$, then $a = b$; and

(c) E is set-like, i.e. $\{x \mid x\ E\ a\}$ is a set for every $a \in M$.

Then there is a unique isomorphism $\pi\colon \langle M, E, \ldots\rangle \to \langle \overline{M}, \in, \ldots\rangle$ where $\overline{M}$ is transitive. ⊣

$\langle \overline{M}, \in, \ldots\rangle$ is the *transitive collapse* of $\langle M, \in, \ldots\rangle$; π is defined by recursion on E by: $\pi(x) = \{\pi(y) \mid y\ E\ x\}$ for $x \in M$. The ... accommodates further relations and functions. For example, for $\langle M, E, R\rangle$ where $R \subseteq {}^n M$ there is an $S \subseteq {}^n\overline{M}$ such that $\pi\colon \langle M, E, R\rangle \to \langle \overline{M}, \in, S\rangle$ satisfies $x \in R$ *iff* $\pi(x) \in S$ for $x \in {}^n M$.

Conditions (b) and (c) are usually immediate for applications of the lemma in this text, the former since the $\langle M, E, \ldots\rangle$ being considered is usually elementarily equivalent to some $\langle X, \in, \ldots\rangle$ where X is transitive. (a) is more consequential, but the lemma will often be invoked when $\langle M, E, \ldots\rangle$ is an elementary substructure of some $\langle X, \in, \ldots\rangle$ when it too is immediate. As applications of the lemma become routine, these conditions, and indeed the lemma, will be mentioned less and less.

Skolem Hull Arguments

To specify the approach and terminology, by a *Skolem hull argument* is meant an argument based on some version of the following construction, detailed in Chang-Keisler [90: §3.3]. Let $\mathcal{M} = \langle M, \ldots\rangle$ be a structure for a language $\mathcal{L}$. For a formula $\varphi(v_0, \ldots, v_n)$ of $\mathcal{L}$, $f\colon {}^n M \to M$ is a *Skolem function for* φ *iff* for any $x_1, \ldots, x_n \in M$,

$$\mathcal{M} \models \exists v_0 \varphi[x_1, \ldots, x_n] \text{ implies that } \mathcal{M} \models \varphi[f(x_1, \ldots, x_n), x_1, \ldots, x_n]\,.$$

Such functions can be provided through a well-ordering of M by taking least witnesses, and so if $\mathcal{M}$ has a definable well-ordering of M, there are definable Skolem functions for every φ. A *complete set of Skolem functions for* $\mathcal{M}$ is the *closure under functional composition* of some collection $\{f_\varphi \mid \varphi \text{ is a formula}\}$ where f_φ is a Skolem function for φ. Such a collection has cardinality $|\mathcal{L}|$, the number of formulas of $\mathcal{L}$.

Let $\mathcal{M}^*$ be the expansion $\langle \mathcal{M}, f_\alpha\rangle_{\alpha<|\mathcal{L}|}$, where $f_\alpha\colon {}^{k(\alpha)}M \to M$ and $\{f_\alpha \mid \alpha < |\mathcal{L}|\}$ is a complete set of Skolem functions for $\mathcal{M}$. The function symbols corresponding to the f_α's in the expanded language are the *Skolem terms*. For any $X \subseteq M$,

$$\{f_\alpha(x_1, \ldots, x_{k(\alpha)}) \mid \alpha < |\mathcal{L}| \ \wedge\ x_1, \ldots, x_{k(\alpha)} \in X\}$$

is the domain of a substructure of $\mathcal{M}^*$, and its reduct $\mathcal{H}(X)$ to the original language is the *Skolem hull of* X *in* $\mathcal{M}$ *with respect to* $\{f_\alpha \mid \alpha < |\mathcal{L}|\}$. By the usual (Tarski) criterion, $\mathcal{H}(X) \prec \mathcal{M}$, i.e. $\mathcal{H}(X)$ is an elementary substructure of $\mathcal{M}$. This provides a useful version of the *Löwenheim-Skolem Theorem:*

0.5 Theorem (Tarski-Vaught [57: 92]). *Suppose that* $\mathcal{M} = \langle M, \ldots\rangle$ *is a structure for a language* $\mathcal{L}$, *and* $X \subseteq M$. *Then there is an* $\mathcal{M}_0 = \langle M_0, \ldots\rangle \prec \mathcal{M}$ *with cardinality at most* $|X| + |\mathcal{L}|$ *such that* $X \subseteq M_0$. ⊣

Beyond this result, a complete set of Skolem functions provides a *uniform* way of generating many elementary substructures of a given structure.

Ultraproducts

The basic ultraproduct construction is reviewed here partly in order to establish the notation (see Chang-Keisler [90: §4.1] for details). Let U be an ultrafilter over a set S, and for each $i \in S$ let $\mathcal{M}_i = \langle M_i, \ldots \rangle$ be a structure for some fixed language $\mathcal{L}$. $\prod_S M_i$ denotes the Cartesian product of the M_i's, i.e. the collection of all functions f with domain S satisfying $f(i) \in M_i$ for $i \in S$. For $f, g \in \prod_S M_i$ define

$$f =_U g \quad \textit{iff} \quad \{i \in S \mid f(i) = g(i)\} \in U \; .$$

Then $=_U$ is an equivalence relation on $\prod_S M_i$, so let $(f)_U$ denote the corresponding equivalence class of f and set $\prod_S M_i / U = \{(f)_U \mid f \in \prod_S M_i\}$. Finally, define the *ultraproduct* of the $\mathcal{M}_i$'s by U, a structure for $\mathcal{L}$ denoted by $\prod_S \mathcal{M}_i / U$, as follows: (i) its domain is $\prod_S M_i / U$, and (ii) for any n-ary predicate symbol in $\mathcal{L}$ interpreted in $\mathcal{M}_i$ by the n-ary relation $R_i \subseteq {}^n M_i$, the interpretation R_U in the ultraproduct is defined by:

$$\langle (f_1)_U, \ldots, (f_n)_U \rangle \in R_U \quad \textit{iff} \quad \{i \in S \mid \langle f_1(i), \ldots, f_n(i) \rangle \in R_i\} \in U \; .$$

The interpretations of the function and constant symbols are defined analogously. Note that if each R_i is the real membership relation restricted to M_i, then the corresponding interpretation E_U in the ultraproduct is given by:

$$(f)_U \; E_U \; (g)_U \quad \textit{iff} \quad \{i \in S \mid f(i) \in g(i)\} \in U \; .$$

The following is the basic *Łoś's Theorem*:

0.6 Theorem (Łoś [55]). *For a formula* $\varphi(v_1, \ldots, v_n)$ *and* $f_1, \ldots, f_n \in \prod_S M_i$,

$$\textstyle\prod_S \mathcal{M}_i / U \models \varphi[(f_1)_U, \ldots, (f_n)_U] \quad \textit{iff} \quad \{i \in S \mid \mathcal{M}_i \models \varphi[f_1(i), \ldots, f_n(i)]\} \in U \; . \qquad \dashv$$

The proof is by induction on the complexity of φ; the Axiom of Choice is needed at the existential quantifier step.

When there is a fixed $\mathcal{M} = \langle M, \ldots \rangle$ such that each $\mathcal{M}_i$ is $\mathcal{M}$, the ultraproduct is the *ultrapower* of $\mathcal{M}$ by U, denoted ${}^S\mathcal{M}/U$. In this case there is by 0.6 an elementary embedding $j\colon \mathcal{M} \to {}^S\mathcal{M}/U$ given by: $j(x) = (f_x)_U$ for $x \in M$, where f_x is the constant function: $S \to \{x\}$. Ultrapowers of V itself are formulated in §5, and variations considered in subsequent sections.

Direct Limits

The direct limit construction is formulated here for elementary embeddings of structures; tailored versions with obvious modifications are used several times in different contexts. A *directed set* is a partially ordered set $\langle S, \leq \rangle$ such that for any $i, j \in S$ there is a $k \in S$ such that $i \leq k$ and $j \leq k$. A *directed system* is a pair $\langle \langle \mathcal{M}_i \mid i \in S \rangle, \langle f_{ij} \mid i \leq j \rangle \rangle$ where $\langle S, \leq \rangle$ is a directed set, each $\mathcal{M}_i$ is a structure for a fixed language $\mathcal{L}$, and each $f_{ij}\colon \mathcal{M}_i \prec \mathcal{M}_j$ is an elementary embedding such that $f_{ik} = f_{jk} \circ f_{ij}$ for $i \le j \le k$ (so that each f_{ii} is the identity on $\mathcal{M}_i$).

A *direct limit* of such a system is a structure $\mathcal{M}$ for $\mathcal{L}$ for which there are elementary embeddings $f_i\colon \mathcal{M}_i \prec \mathcal{M}$ for $i \in S$ with $f_i = f_j \circ f_{ij}$ for $i \le j$, such that: for each x in the domain of $\mathcal{M}$, $x \in \operatorname{ran}(f_i)$ for some $i \in S$. The following proposition gives the essence of this concept:

0.7 Proposition. *Suppose that $\langle\langle \mathcal{M}_i \mid i \in S\rangle, \langle f_{ij} \mid i \le j\rangle\rangle$ is a directed system and $\mathcal{M}$ is a direct limit with corresponding embeddings $f_i\colon \mathcal{M}_i \prec \mathcal{M}$. Assume that $\mathcal{N}$ is a structure such that there are elementary embeddings $g_i\colon \mathcal{M}_i \prec \mathcal{N}$ satisfying $g_i = g_j \circ f_{ij}$ for $i \le j$. Then there is an elementary embedding $g\colon \mathcal{M} \prec \mathcal{N}$ such that $g_i = g \circ f_i$.* ⊣

To define g, for x in the domain of $\mathcal{M}$, say $x = f_i(\overline{x})$ for some $i \in S$ and $\overline{x}$ in the domain of $\mathcal{M}_i$, set $g(x) = g_i(\overline{x})$. "Une chasse sur les diagrammes" confirms that the g is a well-defined elementary embedding.

0.7 implies that any two direct limits of a directed system are isomorphic, and so one can speak of *the* direct limit – once it is established that there is one at all, which can be done as follows:

Suppose that $\langle\langle \mathcal{M}_i \mid i \in S\rangle, \langle f_{ij} \mid i \le j\rangle\rangle$ is a directed system with M_i the domain of $\mathcal{M}_i$ for $i \in S$. Set $B = \bigcup_{i \in S}\{i\} \times M_i$, a union of disjoint copies of the M_i's. Define a binary relation $\sim$ on B by:

$$\langle i, x\rangle \sim \langle j, y\rangle \ \text{ iff } \ \exists k \in S(i \le k \ \wedge \ j \le k \ \wedge \ f_{ik}(x) = f_{jk}(y)) \,.$$

It is simple to check that $\sim$ is an equivalence relation, so letting $[\langle i, x\rangle]$ be the corresponding equivalence class of $\langle i, x\rangle$, set

$$M = B/{\sim} = \{[\langle i, x\rangle] \mid \langle i, x\rangle \in B\} \,.$$

To expand M into a structure for $\mathcal{L}$, suppose for instance that $\mathcal{L}$ has an n-ary predicate symbol interpreted in $\mathcal{M}_i$ by $R_i \subseteq {}^nM_i$ for $i \in S$. Then define a corresponding $R \subseteq {}^nM$ as follows: Given $[\langle i_1, x_1\rangle], \ldots, [\langle i_n, x_n\rangle] \in M$, first find a $k \in S$ such that $i_s \le k$ for $1 \le s \le n$. Then stipulate that

$$\langle[\langle i_1, x_1\rangle], \ldots, [\langle i_n, x_n\rangle]\rangle \in R \ \text{ iff } \ \langle f_{i_1k}(x_1), \ldots, f_{i_nk}(x_n)\rangle \in R_k \,.$$

It is simple to check that R is well-defined. Interpretations of function and constant symbols are defined analogously. Let $\mathcal{M} = \langle M, R, \ldots\rangle$ be the resulting structure for $\mathcal{L}$. For each $i \in S$ define $f_i\colon \mathcal{M}_i \to \mathcal{M}$ by: $f_i(x) = [\langle i, x\rangle]$. Then straightforward arguments show that $\mathcal{M}$ is a direct limit with f_i's the verifying embeddings.

Measure and Category

These preliminaries are concluded by reviewing some basic concepts used in the study of the continuum. $\mathbb{R}$ denotes the set of reals, formalized as the Dedekind completion of the rationals. However, members of ${}^{\omega}\omega$ or $\mathcal{P}(\omega)$ are also called "reals" following set-theoretic practice. For discussing the structural properties of $\mathbb{R}$ in set theory it is convenient to work with ${}^{\omega}\omega$ instead, topologized by taking as the basic open sets

$$O(s) = \{f \in {}^{\omega}\omega \mid s \subseteq f\}$$

for $s \in {}^{<\omega}\omega$. This is known as *Baire space*, again denoted by ${}^{\omega}\omega$. This space is homeomorphic to the irrationals, and the essential features of the structural properties that will be considered are preserved in this association with $\mathbb{R}$ (see e.g. Levy [79: VII§3]). Once and for all,

we fix an enumeration $\langle \mathbf{s}_i \mid i \in \omega \rangle$ of ${}^{<\omega}\omega$ such that $|\mathbf{s}_i| \leq i$

and sequences appear after their proper initial segments, given in some effective manner. Some simple observations are made about the topology:

0.8 Exercise (ZF). *Suppose that $s, t \in {}^{<\omega}\omega$. Then:*
(a) $O(s) \cap O(t)$ is either $\emptyset$, $O(s)$, or $O(t)$.
(b) $O(s) - O(t)$ is a disjoint union of basic open sets.
(c) $O(t)$ is clopen, i.e. closed as well as open.
(d) Every open set is a disjoint union of basic open sets.

Hint. For (b) note that

$$O(s) - O(t) = \bigcup\{O(s) \cap O(u) \mid |u| = |t| \ \wedge \ u \neq t\}$$

and apply (a). For (c) take $s = \emptyset$ in (b). Finally, for (d) suppose that O is an open set, say $O = \bigcup_{j\in\omega} O(\mathbf{s}_{i_j})$ in terms of our fixed enumeration. Set $X_j = O(\mathbf{s}_{i_j}) - \bigcup_{k<j} O(\mathbf{s}_{i_k})$. Then $O = \bigcup_j X_j$ is a disjoint union, and by repeated application of (b) each X_j is in turn a disjoint union of basic open sets. ⊣

These properties are one of several advantages of working with ${}^{\omega}\omega$ and imply that, chameleon-like, ${}^{\omega}\omega$ is "zero-dimensional": For $0 < k < \omega$ let ${}^{k}({}^{\omega}\omega)$, the k-fold Cartesian product of ${}^{\omega}\omega$, be topologized with the product topology, i.e. using basic open sets of the form

$$O(\mathbf{s}_{i_1}) \times \cdots \times O(\mathbf{s}_{i_k}) \,.$$

Then unlike for $\mathbb{R}$, there is a homeomorphism: ${}^{\omega}\omega \to {}^{k}({}^{\omega}\omega)$. In fact, any bijection of ω with $\omega \cdot k$ induces such a homeomorphism. In what follows, *concepts are described for ${}^{\omega}\omega$ for notational simplicity, but their extensions to ${}^{k}({}^{\omega}\omega)$ are assumed* through these homeomorphisms or directly in terms of the corresponding product notions.

Turning first to the concept of measure, we first review the axiomatic approach of Emile Borel: For any set S and $F \subseteq \mathcal{P}(S)$,

F is a *σ-algebra on S iff* $\emptyset \in F$ and F is closed under the taking of complements and countable unions .

For such F and any $m\colon F \to [0,1]$, where $[0,1] \subseteq \mathbb{R}$ is the unit interval of reals,

m is a *(probability) measure on* F *iff*
(i) $m(\emptyset) = 0$ and $m(S) = 1$, and
(ii) if $\{X_n \mid n \in \omega\} \subseteq F$ is a pairwise disjoint collection, then $m(\bigcup_n X_n) = \sum_n m(X_n)$.

With ${}^\omega\omega$ in place of $\mathbb{R}$,

the *Borel sets* are the members of the σ-algebra $\mathcal{B}$ on ${}^\omega\omega$ generated by the $O(s)$'s .

The Borel sets form a natural hierarchy (see §12), and focal are members of the second level, the F_σ and G_δ sets in the classical but persistent terminology of Hausdorff [14]:

A is F_σ *iff* A is a union of countably many closed sets .
A is G_δ *iff* A is an intersection of countably many open sets .

To endow $\mathcal{B}$ with a measure, let m_ω be that measure on $\mathcal{P}(\omega)$ satisfying $m_\omega(\{i\}) = 2^{-(i+1)}$, so that $m_\omega(a) = \sum_{i\in a} 2^{-(i+1)}$ for any $a \subseteq \omega$. Applying a standard measure-theoretic construction (see e.g. Halmos [50: 157]), stipulate that

$m_{\mathcal{B}}$ is the product measure on $\mathcal{B}$ induced by m_ω .

$m_{\mathcal{B}}(O(s)) = \prod_{i<|s|} 2^{-(s(i)+1)}$ for each $s \in {}^{<\omega}\omega - \{\emptyset\}$, and $m_{\mathcal{B}}$ is the unique measure on $\mathcal{B}$ with this property. With ${}^\omega\omega$ in place of $\mathbb{R}$, $m_{\mathcal{B}}$ is Borel's measure from his [98:46-47], and it was extended by Lebesgue's measure from his [02], now of course a basic feature of mathematical analysis. A simple way to formulate his extension is to stipulate first for $N \subseteq {}^\omega\omega$ that

N is *null* *iff* there is a Borel set X such that $m_{\mathcal{B}}(X) = 0$ and $N \subseteq X$,

and then for $A \subseteq {}^\omega\omega$ that

A is *Lebesgue measurable* *iff* $A \triangle B$ is null for some Borel set B .

in which case the Lebesgue measure of A is

$$m_L(A) = m_{\mathcal{B}}(B) .$$

Clearly,

the Lebesgue measurable sets form a σ-algebra $\mathcal{M}_L$ extending $\mathcal{B}$,

and

m_L is a measure on $\mathcal{M}_L$ extending $m_{\mathcal{B}}$ on $\mathcal{B}$.

See e.g. Halmos [50] for the further development of the theory. Two well-known properties are stated here for later reference. The first is immediate from the standard characterization of m_L in terms of the inner and outer measures derived from $m_{\mathcal{B}}$.

0.9 Lemma.

(a) For any Lebesgue measurable set A and $\epsilon \in \mathbb{R}$ *with* $\epsilon > 0$, *there is a closed set C and an open set O such that* $C \subseteq A \subseteq O$ *and* $m_L(O - C) < \epsilon$.

(b) For any Lebesgue measurable set A, there is an F_σ *set X and a* G_δ *set Y such that* $X \subseteq A \subseteq Y$ *and* $m_L(X) = m_L(A) = m_L(Y)$. ⊣

(b) follows directly from (a) and highlights the significance of F_σ and G_δ sets.

The next theorem is the set-theoretic formulation of the familiar Fubini Theorem in the case of null subsets of the plane. In the present context, Lebesgue measure for ${}^k({}^\omega\omega)$ is just the k-fold product measure of m_L; for $A \subseteq {}^2({}^\omega\omega)$ and $y \in {}^\omega\omega$, temporarily set $A_y = \{x \mid \langle x, y\rangle \in A\}$.

0.10 Theorem. *Suppose that* $A \subseteq {}^2({}^\omega\omega)$ *is Lebesgue measurable. Then A is a null subset of* ${}^2({}^\omega\omega)$ *iff* $\{y \mid A_y$ is not a null subset of ${}^\omega\omega\}$ *is a null subset of* ${}^\omega\omega$. ⊣

The Baire property evolved from the topological classifications of René Baire. For $A \subseteq {}^\omega\omega$,

$$\mathrm{int}(A) = \bigcup\{O \mid O \text{ is open } \wedge\ O \subseteq A\}\ , \text{ and}$$
$$\mathrm{cl}(A) = \bigcap\{C \mid C \text{ is closed } \wedge\ C \supseteq A\}\ .$$

$\mathrm{int}(A)$ is the *interior* of A, the largest open set contained in A, and $\mathrm{cl}(A)$ is the *closure* of A, the smallest closed set containing A.

A is *nowhere dense* *iff* $\mathrm{int}(\mathrm{cl}(A)) = \emptyset$.
A is *meager* (or *first category*) *iff* A is a union of countably many nowhere dense sets .
A has the *Baire property* *iff* $A \triangle O$ is meager for some open set O .

It is simple to see that *if* O *is open, then the closed set* $\mathrm{cl}(O) - O$ *is nowhere dense.* Also, *open and meager sets have the Baire property* (the latter as $\emptyset$ is open). The Baire property arose from considerations related to the *Baire Category Theorem*:

0.11 Theorem (Baire [99]). *No nonempty open set is meager.*

Proof. Assume to the contrary that O is a nonempty open set such that $O = \bigcup_{j\in\omega} X_j$ where each X_j is nowhere dense. Define a sequence $\langle i_j \mid j \in \omega\rangle \in {}^\omega\omega$ such that $j < k$ implies that $O \supseteq O(\mathbf{s}_{i_j}) \supseteq O(\mathbf{s}_{i_k})$ and $|\mathbf{s}_{i_j}| < |\mathbf{s}_{i_k}|$ as follows:

Let i_0 be least such that $O(\mathbf{s}_{i_0}) \subseteq O$. Given i_j, there must be a $t \supseteq \mathbf{s}_{i_j}$ such that $O(t) \cap X_j = \emptyset$, else $\mathrm{cl}(X_j) \supseteq O(\mathbf{s}_{i_j})$ and X_j would not be nowhere dense. Let i_{j+1} be the least i such that: $O(\mathbf{s}_i) \cap X_j = \emptyset$, $\mathbf{s}_i \supseteq \mathbf{s}_{i_j}$, and $|\mathbf{s}_i| > |\mathbf{s}_{i_j}|$.

Clearly, if $x = \bigcup_j \mathbf{s}_{i_j}$, then $x \in O - \bigcup_j X_j$, which is a contradiction. ⊣

This theorem can be seen as an extension of Cantor's result that the reals are uncountable, and the proof a generalization of his original 1873 proof. The Baire

Property draws its life from the theorem: if it failed, then by a homeomorphism argument ${}^{\omega}\omega$ would be meager, and hence any $A \subseteq {}^{\omega}\omega$ would have the Baire Property.

The next lemma gives some simple consequences of the definitions:

0.12 Lemma. *Suppose that $A \subseteq {}^{\omega}\omega$ has the Baire property. Then:*

(a) ${}^{\omega}\omega - A$ has the Baire property.

(b) There is a G_δ set X and an F_σ set Y such that $X \subseteq A \subseteq Y$ and $Y - X$ is meager.

Proof. Let O be open such that $A \triangle O$ is meager. For (a), note that

$$({}^{\omega}\omega - A) \triangle ({}^{\omega}\omega - O) = A \triangle O\ ,$$

that $\mathrm{cl}(O) - O$ is nowhere dense, and hence that $({}^{\omega}\omega - A) \triangle ({}^{\omega}\omega - \mathrm{cl}(O))$ is meager.

For (b), first note that since the closure of a nowhere dense set is also nowhere dense, there is a meager F_σ set F such that $A \triangle O \subseteq F$. Then $X = O - F$ is G_δ, $X \subseteq A$, and $A - X$ is meager. By (a) there is similarly a G_δ set G such that $G \subseteq ({}^{\omega}\omega - A)$ and $({}^{\omega}\omega - A) - G$ is meager. Taking $Y = {}^{\omega}\omega - G$, the proof is complete. ⊣

Note that (b) is analogous to 0.9(b) but with the roles of the F_σ and G_δ sets reversed. By (a) and simple consequences of the definitions,

the sets having the Baire property form a σ-algebra extending $\mathcal{B}$.

Further important analogies as well as crucial differences exist between Lebesgue measurability and the Baire property (see Oxtoby [71] and Kunen [84]). The following theorem is the analogue of 0.10; like Lebesgue measure, the topological concepts extend naturally to ${}^{k}({}^{\omega}\omega)$.

0.13 Theorem (Kuratowski-Ulam [32]). *Suppose that $A \subseteq {}^{2}({}^{\omega}\omega)$ has the Baire property. Then A is a meager subset of ${}^{2}({}^{\omega}\omega)$ iff $\{y \mid A_y$ is not a meager subset of ${}^{\omega}\omega\}$ is a meager subset of ${}^{\omega}\omega$.* ⊣

Chapter 1

Beginnings

The beginning threads of the subject are picked up in its early history. §1 discusses weak inaccessibility and Mahloness, concepts that arose in the study of cardinal limit processes, and their strong versions, which led to early speculations about completeness and consistency. §2 describes Ulam's formulation of measurability, the most prominent of all large cardinal hypotheses, out of a measure problem for sets of reals. In §3 Gödel's work on L, the beginning of axiomatic set theory as a distinctive field of mathematics, is reviewed since in both reaction and generalization it shaped much of the subsequent work in large cardinals. §4 discusses weak and strong compactness, concepts that emerged from Tarski's study of infinitary languages, and establishes Hanf's result, that in a strong sense there are many inaccessibles below a measurable cardinal. The focus of §5 is on elementary embeddings and the ultrapower construction: Scott's pivotal result that if there is a measurable cardinal, then $V \neq L$; the characterization of measurability in terms of ultrapowers and elementary embeddings; and the related notion of normality. And finally §6 discusses indescribability, a natural formalization of reflection phenomena in terms of higher-order languages that provided a schematic approach to comparing large cardinals by size.

1. Inaccessibility

That volume of *Mathematische Annalen* containing Zermelo's first axiomatization [08a] of set theory also contained Hausdorff's wide-ranging paper [08] on transfinite ordertypes. While Cantor had concentrated his efforts on the rational and real ordertypes, the second-number class, and of course, the Continuum Hypothesis, Hausdorff extended mathematical investigations into the higher transfinite. Deploring all the fuss made over foundations by his contemporaries (p. 436) he ventured forth with vigor, pursuing structure for its own sake. His paper contains the first statement of the Generalized Continuum Hypothesis, the construction of the η_α sets – prototypes for saturated model theory – and for the first time, the following concept (p. 443) formulated for $\kappa > \omega$:

$$\kappa \text{ is } \textit{weakly inaccessible} \;\; \textit{iff} \;\; \kappa \text{ is a regular limit cardinal}\,.$$

(The term "inaccessible" is attributed to Kuratowski in Sierpiński-Tarski [30]; "weakly" is appended for a later distinction.)

Hausdorff observed that such a κ must satisfy $\omega_\kappa = \kappa$ and is a natural closure point for cardinal limit processes. In modern terms, a simple argument shows that if C is closed unbounded in κ, then so is $C' = \{\alpha \in C \mid |C \cap \alpha| = \alpha\}$; taking $C = \kappa$ we have $C' = \{\alpha < \kappa \mid |\alpha| = \alpha\}$, $C'' = \{\alpha < \kappa \mid \omega_\alpha = \alpha\}$, and so forth. However, Hausdorff was to write in his classic text [14: 131] that if weakly inaccessible cardinals did exist, "the least among them has such an exorbitant magnitude [exorbitanten Grösse] that it will hardly ever come into consideration for the usual [üblich] purposes of set theory". It is now well-known that the existence of weakly inaccessible cardinals cannot be established in ZFC (since $L_\kappa \models \text{ZFC}$ for such κ – see §3).

Inspired by Hausdorff's work Mahlo [11, 12, 13] ventured much further and investigated hierarchies of regular cardinals formulated in terms of higher fixed point phenomena. Given its early appearance this work is remarkable for its boldness and sophistication. (Mahlo was a student of Felix Bernstein at Halle and attended seminars of Cantor there, completing his dissertation (on geometry) in 1908. He was first to publish [13a] a construction under CH of what is now known as a Luzin set, an uncountable set of reals that has countable intersection with every meager set. Gottwald-Kreiser [84] discusses Mahlo's life and work.) Recasting Mahlo's π_α numbers,

$$\begin{aligned}
\kappa \text{ is } 0\textit{-weakly inaccessible} \;\; &\textit{iff} \;\; \kappa \text{ is regular}\,;\\
\kappa \text{ is } (\alpha+1)\textit{-weakly inaccessible} \;\; &\textit{iff} \;\; \kappa \text{ is a regular limit of } \alpha\text{-weakly}\\
&\qquad \text{inaccessible cardinals}\,; \text{ and}\\
\kappa \text{ is } \delta\textit{-weakly inaccessible} \;\; &\textit{iff} \;\; \kappa \text{ is } \alpha\text{-weakly inaccessible}\\
&\qquad \text{for every } \alpha < \delta
\end{aligned}$$

for limit ordinals $\delta > 0$. This hierarchy can be extended through diagonalization, considering next the regular limits of those ξ that are ξ-weakly inaccessible. By a simple induction argument, *if κ is β-weakly inaccessible and $\alpha < \beta$, then κ is*

α-weakly inaccessible. With Reg the class of regular cardinals and Λ the operation defined on $X \subseteq \mathrm{On}$ by

$$\Lambda(X) = \{\alpha \in X \mid |X \cap \alpha| = \alpha\} ,$$

the α-weakly inaccessible cardinals are just the members $\Lambda^{\alpha}(\mathrm{Reg})$, where the superscript indicates the number of iterative applications of Λ with intersections taken at limit stages. The hierarchy of α-weakly inaccessible cardinals can thus be seen as the consequence of imposing regularity on the process of taking cardinal limits.

The Λ operation leads to larger and larger cardinals, but typically, a new principle is needed to achieve a qualitative transcendence. Mahlo was able to formulate such a principle, using for the first time the concept of a stationary set. (The term itself comes from Bloch [53].) His ρ_0 numbers are now known through the following definition for $\kappa > \omega$:

κ is *weakly Mahlo* *iff* $\{\rho < \kappa \mid \rho \text{ is regular}\}$ is stationary in κ .

Such a κ is regular (if there were an unbounded $X \subseteq \kappa$ such that $|X| < \kappa$, then the limit points of $X - |X|$ other than κ would form a closed unbounded subset not containing any regular cardinals), and so *weakly Mahlo cardinals are weakly inaccessible.* The following illustrates what led Mahlo to these cardinals:

1.1 Proposition (Mahlo [11]). *If κ is weakly Mahlo, then κ is κ-weakly inaccessible.*

Proof. Setting $R = \{\rho < \kappa \mid \rho \text{ is regular}\}$, define sets C_α closed unbounded in κ for $\alpha < \kappa$ by recursion as follows: Set $C_0 = \kappa$. Given C_α, let $C_{\alpha+1}$ consist of the limit points of $C_\alpha \cap R$ other than κ. As κ is weakly Mahlo, $C_\alpha \cap R$ is stationary, and so $C_{\alpha+1}$ is closed unbounded. Finally, for limit $\delta > 0$ set $C_\delta = \bigcap_{\alpha<\delta} C_\alpha$. It is readily seen by induction that for $\alpha < \kappa$, $C_\alpha \cap R$ consists of the α-weakly inaccessible cardinals below κ, and hence the proof is complete. ⊣

This result is not optimal since we could take the diagonal intersection $C = \triangle_{\alpha<\kappa} C_\alpha$, infer that $C \cap R = \{\xi < \kappa \mid \xi \text{ is } \xi\text{-weakly inaccessible}\}$ is stationary in κ, and continue. Interestingly enough, Gaifman [67] showed that in a concrete sense a weakly Mahlo cardinal is the least upper bound of diagonalizing limit processes from below. Such an upward approach is not possible for the larger cardinals that will be encountered.

Mahlo formulated his ρ_α numbers by iterating the new process:

κ is *0-weakly Mahlo* *iff* κ is regular ;
κ is *$(\alpha+1)$-weakly Mahlo* *iff* $\{\xi < \kappa \mid \xi \text{ is } \alpha\text{-weakly Mahlo}\}$ is stationary in κ ; and
κ is *δ-weakly Mahlo* *iff* κ is α-weakly Mahlo for every $\alpha < \delta$

for limit ordinals $\delta > 0$. Again, the hierarchy can be extended through diagonalization, considering next those κ such that $\{\xi < \kappa \mid \xi \text{ is } \xi\text{-weakly Mahlo}\}$

is stationary in κ. By a simple induction argument, *if κ is β-weakly Mahlo and $\alpha < \beta$, then κ is α-weakly Mahlo.* The following operation defined for $X \subseteq \mathrm{On}$ is now known as *Mahlo's operation:*

$$M(X) = \{\alpha \in X \mid X \cap \alpha \text{ is stationary in } \alpha\} .$$

As for Λ and the α-weakly inaccessible cardinals, the α-weakly Mahlo cardinals are just the members of $M^\alpha(\mathrm{Reg})$; 1.1 illustrates the transcendence of M over Λ. Despite its modest debut Mahlo's operation is now a standard part of the modern theory.

Almost two decades were to pass before the initial preoccupation with the *extent* of the ordinals was enhanced by considerations involving the *width* of the set-theoretic universe as advanced by the power set operation. Sierpiński-Tarski [30] and Zermelo [30] formulated the following concept for $\kappa > \omega$:

κ is *(strongly) inaccessible iff* κ is regular and a strong limit: if $\lambda < \kappa$, then $2^\lambda < \kappa$.

The adverb "strongly" is suppressed for this preferred notion. *Assuming* GCH, *κ is inaccessible iff κ is weakly inaccessible.*

The cumulative hierarchy view $V = \bigcup_\alpha V_\alpha$ of the set-theoretic universe was emerging at this time. In its terms, the following proposition implies that the existence of inaccessible cardinals cannot be established in set theory:

1.2 Proposition. *Suppose that κ is inaccessible. Then:*

(a) If $x \subseteq V_\kappa$, then $x \in V_\kappa$ iff $|x| < \kappa$.

(b) $\langle V_\kappa, \in\rangle \models$ ZFC.

Proof. (a) In the forward direction, it suffices to show that $|V_\alpha| < \kappa$ for $\alpha < \kappa$. But this follows readily by induction on α. Conversely, suppose that $x \subseteq V_\kappa$ with $|x| < \kappa$. By the regularity of κ, $\{\mathrm{rank}(y) \mid y \in x\} \subseteq \alpha$ for some $\alpha < \kappa$, and so $x \in V_{\alpha+1} \subseteq V_\kappa$.

(b) All the axioms of ZFC except Replacement are readily seen to hold in V_α for any limit ordinal $\alpha > \omega$. To verify Replacement for V_κ, suppose that $x \in V_\kappa$ and F is any function: $x \to V_\kappa$. Then $|F``x| \leq |x| < \kappa$ and so by (a) $F``x \in V_\kappa$. ⊣

Temporarily let IC be the hypothesis: There is an inaccessible cardinal. If IC and κ is the least inaccessible cardinal, then it is readily seen that $V_\kappa \models \mathrm{ZFC} + \neg\mathrm{IC}$. Hence, IC *is not provable in* ZFC. This first independence result over ZFC was essentially observed in Zermelo [30] and was also asserted in Sierpiński-Tarski [30]. (Kuratowski [24] reported that he had considered a theory of sets – essentially ZFC *sans* Foundation – and remarked that even in the presence of Replacement "one cannot establish [on ne saurait établir]" the existence of *weakly* inaccessible cardinals. However, despite the reading of Mostowski [49: 162] this could not have been shown at the time without assuming some form of GCH.)

That the existential postulation IC immediately leads to its own independence is analogous to the situation with the Axiom of Infinity: $V_\omega \models \ulcorner$Every set is finite$\urcorner$. Similarly, if κ is the least Mahlo, then $V_\kappa \models \ulcorner$There are no Mahlo cardinals$\urcorner$. This is a typical feature of large cardinals.

The independence of IC can also be established by appealing to the Second Incompleteness Theorem of Gödel [31] since ZFC + IC $\vdash$ Con(ZFC) by 1.2(b). This further implies that Con(ZFC + IC) *is not provable from* Con(ZFC), again by Gödel's theorem. This finer result points out a qualitative difference: the very possibility of discussing formal consistency in Gödel's careful analysis far transcends the simple proof of 1.2(b). Tarski soon became aware of this distinction ([38: 87]).

With ZFC having been cast as a first-order theory 1.2(b) does not characterize inaccessibility, by the Löwenheim-Skolem Theorem (0.5). Significantly, this fact was observed only latterly (Mostowski [49], Montague-Vaught [59]) in the wake of new model-theoretic initiatives. On the other hand, a characterization can be achieved by venturing into higher-order logic. Let ZFC^2 denote the second-order version of ZFC where the schema of Replacement is replaced by a single axiom with a second-order universal quantifier. The following is established with the intended interpretation of second-order variables as ranging over arbitrary subsets of the domain.

1.3 Theorem (Zermelo [30], Shepherdson [52]). *κ is inaccessible iff* $V_\kappa \models$ ZFC^2.

Proof. The forward direction is as for 1.2(b).

For the converse, first note that κ is regular: If not, there would be an $\alpha < \kappa$ and a function $G: \alpha \to \kappa$ with a range unbounded in κ. Then $G \subseteq V_\kappa$, and so by second-order Replacement $G``\alpha \in V_\kappa$. Hence, $\sup(G``\alpha) = \kappa \in V_\kappa$, a contradiction.

It remains to establish that κ is a strong limit cardinal: If not, there would be a $\lambda < \kappa$ such that $2^\lambda \geq \kappa$. Then $\mathcal{P}(\lambda) \in V_\kappa$ by the Power Set Axiom, yet there would be a surjection $H: \mathcal{P}(\lambda) \to \kappa$. So again by second-order Replacement, $H``\mathcal{P}(\lambda) = \kappa \in V_\kappa$, a contradiction. $\dashv$

Strictly speaking, the intrusion of the second-order satisfaction relation is not essential here since ZFC^2 can be cast as a single axiom: $V_\kappa \models$ ZFC^2 is equivalent to the relativization $\varphi^{V_{\kappa+1}}$ for some first-order formula φ. John Shepherdson [52] was first to give a formal proof of 1.3. However, it is a remarkable historical happenstance that Zermelo [30] had already established an informal version:

In [30] Zermelo presented his final axiomatization of set theory, incorporating Replacement and Foundation, and moreover offered a striking, synthetic vision of sets as partaking in a succession of natural models. From his [29] clarification of the crucial notion of *definit* property used in his Separation Axiom it is evident that Zermelo couched his approach in higher-order logic. Strictly speaking, this would be the formal interpretation, and Shepherdson's criticism

([52: 227]) of Zermelo's proof of 1.3 as "insufficiently rigorous" is justified from this point of view. But Zermelo had adopted a definite anti-formalist viewpoint by then, and downgraded the significance of Skolem's remarks [23] on the relativism of first-order formalizations. For Zermelo concepts like power set and cardinal number simply had definite extensions and assumed an absolute significance.

It was in Zermelo [30] that initial segments of the cumulative hierarchy $\bigcup_\alpha V_\alpha$ were first proposed as models for the axioms of set theory. Although the hierarchy was adumbrated by Mirimanov [17] and by von Neumann [25], Zermelo was the first to actually adopt the Foundation Axiom. Moreover, he started with collections of urelements (his term) as bases for his hierarchical models. Although this device has turned out to be unnecessary for the enrichment of set theory itself, it is still important in restricted contexts (like admissible sets – see Barwise [75]).

Zermelo's main achievement was to establish a second-order categoricity of sorts for his axioms: he showed that his models are characterized up to isomorphism by two cardinals, the number of its urelements and the height of its ordinals. Moreover, there is a unique end-extension relation between any two models, based on these two invariants, so that one model is just a set in a higher domain. In order to establish these results Zermelo used the cumulative hierarchy analysis to correlate models rank by rank. Grappling with Replacement he characterized these ordinal heights of models ("Grenzzahlen") as regular fixed points of the Beth function, and hence arrived at inaccessible cardinals essentially through 1.3.

Zermelo went on to propound a dynamic view of sets that posits an endless succession of models of set theory ([30: 47]):

> The 'ultrafinite antinomies of set theory' that scientific reactionaries and anti-mathematicians refer to so assiduously and lovingly in their campaign against set theory, these seeming 'contradictions', are only due to a confusion of set theory itself, which is non-categorically determined by its axioms, with particular representing models: What appears in one model as an 'ultrafinite non-or metaset' is in the next higher one already a fully valid 'set' with cardinal number and ordinal type, and is itself the foundation stone for the construction of the new domain. The unlimited series of Cantor's ordinal numbers is matched by just as infinite a double series of essentially different set-theoretic models, the whole classical theory being manifested in each of them. The two diametrically opposite tendencies of the thinking spirit, the idea of creative progress and of comprehensive completion, which also lie at the root of the Kantian 'antinomies', find their symbolic representation and symbolic reconciliation in the transfinite series of numbers based on the concept of well-ordering. This series in its boundless progression does not have a true conclusion, only relative stopping points, namely those 'limit numbers' [i.e. inaccessible cardinals] which separate the higher from the lower model types. And thus also, the set-theoretic 'antinomies' lead, if properly understood, not to a restriction or mutilation but rather to a presently unsurveyable unfolding and enrichment, of mathematical science.

These words run counter to Cantor's (and later, Gödel's) view of a fixed set-theoretic universe, yet conform in a way to his realist conception of mathematics as consisting of all consistently conceivable objects. In any event, it is noteworthy that inaccessible cardinals played such a prominent role in an early structural analysis of the universe of sets. Tarski [38, 39a] later formalized the existence of arbitrarily many inaccessible cardinals as an axiom, phrased in such a way that most of the axioms of set theory including the Axiom of Choice are derivable. Nevertheless, Zermelo's infinitistic conception was soon to be overtaken by a growing preoccupation with the formalism of first-order logic, following the incisive analyses of Skolem and Gödel.

In analogy with the weakly Mahlo cardinals,

κ is *(strongly) Mahlo* *iff* $\{\alpha < \kappa \mid \alpha \text{ is inaccessible}\}$ is stationary in κ .

As with inaccessibility the adverb "strongly" is suppressed for this preferred notion. The hierarchies of the

α*-inaccessible* and α*-Mahlo* cardinals

are analogously defined, and the analogue of 1.1 obtains.

Almost half a century after their introduction Levy revitalized the study of Mahlo cardinals with his investigation of reflection phenomena, shifting the focus from the "weak" to the "strong" versions because of the interplay with the cumulative hierarchy (§6).

2. Measurability

This section picks up another thread, one that led to measurability, the most prominent of all large cardinal hypotheses. This development occurred in Poland, which featured a school of mathematics crucial to the foundations of mathematical logic, topology, and analysis. After the reunification of the country in 1918 Zygmunt Janiszewski at the newly reopened University of Warsaw encouraged the focusing of Polish mathematics on set theory and related areas to establish a national tradition, and the publication of a new journal to promote international research in these directions (see Kuratowski [80] and Kuzawa [68]). This was the origin of *Fundamenta Mathematicae* (genitive singular), the first specialized journal devoted to foundational issues and the main conduit of scholarship in this general area during the 1920's and 1930's (see Kuzawa [70]). The first volume, incidentally, contained the well-known problem of Suslin [20] that led to the investigation of Suslin trees in modern set theory. At Warsaw Sierpiński together with Kuratowski and Tarski were soon making fundamental contributions to set theory and the understanding of its role in mathematics. For present concerns work of Stefan Banach and the young Stanisław Ulam at Lwów (now Lviv) on an abstract measure problem turned out to be fundamental.

Modern measure theory dates back to Lebesgue's thesis [02], where he posed the *Measure Problem*. For the real line this asks: Is there a function m that associates to every bounded set of reals X a non-negative real number $m(X)$ such that:

(a) m is not identically zero .
(b) m is translation-invariant, i.e. $m(X) = m(Y)$ whenever there is a real r such that $Y = \{x + r \mid x \in X\}$, and
(c) m is countably additive, i.e. if $\{X_n \mid n \in \omega\}$ is a pairwise disjoint collection whose union is a bounded set of reals, then $m(\bigcup_n X_n) = \sum_n m(X_n)$.

Lebesgue developed his measure towards a solution to this problem, and of course, it is now integral to mathematical analysis. (For the historical development of Lebesgue measure and integration, see Hawkins [75].) Part of the problem was to decide what passed for a set of reals. Giuseppe Vitali [05] constructed a non-Lebesgue measurable set of reals from a well-ordering of the reals, thereby showing that the Measure Problem has no solution under the Axiom of Choice. This was the first explicit use of AC to construct a specific set of reals after Zermelo's formulation of the axiom, a manipulative use of a well-ordering very different from Cantor's associations of sets with well-orderings. For Lebesgue [07], Vitali's construction raised doubts not so much about the possibilities for a measure but about AC.

Vitali's counterexample used all the conditions (a)–(c). Banach proposed a generalization of the Measure Problem where (b) is replaced by a minimal necessary condition to avoid trivial solutions: $m(\{x\}) = 0$ for every x. With a proof that has a contemporary relevance Banach-Kuratowski [29] established that under CH this version of the problem also has no solution. Note that by (b) and (c) a solution m to the Measure Problem is determined by its values on $\mathcal{P}([0,1])$, the subsets of the unit interval $[0,1]$. Banach realized that his condition removed geometric considerations from the problem, so that $[0,1]$ can be replaced by an arbitrary set S. In this case, if m is not identically 0 on $\mathcal{P}(S)$, then surely $m(S) > 0$, and so we can normalize and assume that $m(S) = 1$. Thus, the measure problem of Banach [30] might as well be posed as follows: Is there a nonempty set S and a function $m\colon \mathcal{P}(S) \to [0,1]$ such that:

(i) $m(S) = 1$,
(ii) $m(\{x\}) = 0$ for every $x \in S$, and
(iii) for pairwise disjoint $\{X_n \mid n \in \omega\} \subseteq \mathcal{P}(S)$,
$m(\bigcup_n X_n) = \sum_n m(X_n)$.

Such a function will simply be called a *measure over* S (it is a measure *on* $\mathcal{P}(S)$ in the sense of §0). The following exercise points to a salient feature of measures:

2.1 Exercise. *If $T \subseteq \{X \subseteq S \mid m(X) > 0\}$ is uncountable, then there are distinct $Y, Z \in T$ such that $m(Y \cap Z) > 0$.*

Hint. Note that for some $n \in \omega$, $\{X \in T \mid m(X) > \frac{1}{n}\}$ is uncountable. ⊣

For a measure m over a set S,

m is λ-*additive* *iff* for any $\gamma < \lambda$ and pairwise disjoint
$\{X_\alpha \mid \alpha < \gamma\} \subseteq \mathcal{P}(S)$, $m(\bigcup_\alpha X_\alpha) = \sum_\alpha m(X_\alpha)$.

(Here, a transfinite sum of reals is the supremum of the sums of finite subcollections.) Banach saw that only the cardinality of the set S matters in his problem, and that in lieu of property (iii) one might as well require a stronger property.

2.2 Exercise. *Suppose that κ is the least cardinal such that there is a measure over κ. Then every measure over κ is κ-additive.*

Hint. If not, let m be a measure over κ, $\gamma < \kappa$, and $\{X_\alpha \mid \alpha < \gamma\} \subseteq \mathcal{P}(\kappa)$ pairwise disjoint such that $m(\bigcup_\alpha X_\alpha) \neq \sum_\alpha m(X_\alpha)$. Then $\gamma > \omega$ and there are only countably many α's with $m(X_\alpha) > 0$ by 2.1. Removing these through the additivity property (iii), assume without loss of generality that each $m(X_\alpha) = 0$, yet $m(\bigcup_\alpha X_\alpha) = r > 0$. Now check that $\overline{m}\colon \mathcal{P}(\gamma) \to [0,1]$ defined by

$$\overline{m}(Y) = \frac{m(\bigcup_{\alpha \in Y} X_\alpha)}{r}$$

is a measure over γ, contradicting the minimality of κ. ⊣

For $\kappa > \omega$,

κ is *real-valued measurable* *iff* there is a κ-additive measure over κ.

If m is such a measure, then clearly $m(X) = 0$ whenever $|X| < \kappa$. Hence, it is easy to see that *a real-valued measurable cardinal is regular*. Banach [30: 101] established under GCH that every real-valued measurable cardinal is weakly inaccessible.

As a student at Lwów Ulam [29] had already provided, in measure-theoretic terms, the first construction of an ultrafilter over ω using a well-ordering of $\mathcal{P}(\omega)$. (Tarski [29] announced the general result that any filter over a set can be extended to an ultrafilter over that set.) In his doctoral dissertation Ulam then established fundamental results concerning Banach's measure problem; as we shall see, this work involved a direct *generalization* of an ultrafilter over ω. Ulam first removed GCH from Banach's result above; for this purpose, he devised a useful combinatorial device now known as an *Ulam matrix:*

2.3 Proposition (Ulam [30]). *For any λ, there is a collection of sets $\{A^\xi_\alpha \mid \alpha < \lambda^+ \wedge \xi < \lambda\} \subseteq \mathcal{P}(\lambda^+)$ satisfying*

(a) $A^\xi_\alpha \cap A^\xi_\beta = \emptyset$ whenever $\alpha < \beta < \lambda^+$ and $\xi < \lambda$; and

(b) $|\lambda^+ - \bigcup_{\xi<\lambda} A^\xi_\alpha| \leq \lambda$ for each $\alpha < \lambda^+$.

Proof. For each $\eta < \lambda^+$ let $f_\eta \colon \lambda \to \eta + 1$ be surjective, and for $\alpha < \lambda^+$ and $\xi < \lambda$ set $A^\xi_\alpha = \{\eta < \lambda^+ \mid f_\eta(\xi) = \alpha\}$. Then (a) is immediate and for (b) note that $(\lambda^+ - \bigcup_{\xi<\lambda} A^\xi_\alpha) \subseteq \alpha$. ⊣

2.4 Corollary. *If κ is real-valued measurable, then κ is weakly inaccessible.*

Proof. Since κ must be regular, it remains to establish that κ is a limit cardinal. Assume to the contrary that $\kappa = \lambda^+$, and consider an Ulam matrix as in 2.3. Let m be a κ-additive measure over κ. Then for each $\alpha < \lambda^+$, there is a $\xi_\alpha < \lambda$ such that $m(A^{\xi_\alpha}_\alpha) > 0$ by 2.3(b). But then, there must be a fixed $\xi < \lambda$ such that $\xi_\alpha = \xi$ for λ^+ many α's. By 2.3(a), this contradicts 2.1. ⊣

Ulam then pointed out a major dichotomy. For a measure m over κ,

$A \subseteq \kappa$ is an *atom for m* *iff* $m(A) > 0$ yet for any $B \subseteq A$, $m(B) = m(A)$ or $m(B) = 0$, and

m is *atomless* *iff* there are no atoms for m.

Ulam drew important conclusions both from the existence of an atomless measure and from the existence of a measure with an atom. (b) of the following does not need much beyond Ulam's proof of (a).

2.5 Theorem (Ulam [30]). *Suppose that there is an atomless κ-additive measure m over κ. Then:*

(a) $\kappa \leq 2^{\aleph_0}$.

(b) There is a measure over the reals extending Lebesgue measure.

2.6 Lemma.

(i) For any $\epsilon \in \mathbb{R}$ with $\epsilon > 0$ and $X \subseteq \kappa$ with $m(X) > 0$, there is a $Y \subseteq X$ satisfying $0 < m(Y) < \epsilon$.

(ii) For any $X \subseteq \kappa$ there is a $Y \subseteq X$ satisfying $m(Y) = \frac{1}{2} \cdot m(X)$.

Proof. (i) It suffices to recursively define a $\subseteq$-descending sequence of sets X_i satisfying $0 < m(X_{i+1}) \leq \frac{1}{2} \cdot m(X_i)$ for each $i \in \omega$. But given X_i, since m is atomless, there is a partition $A \cup B = X_i$ such that $0 < m(A) \leq m(B)$; set $X_{i+1} = A$.

(ii) Recursively define a $\subseteq$-descending sequence of sets X_α such that $m(X_\alpha) \geq \frac{1}{2} \cdot m(X)$ for as long as possible, as follows: Set $X_0 = X$. If X_α has been defined, define $X_{\alpha+1}$ exactly when $m(X_\alpha) > \frac{1}{2} \cdot m(X)$, in which case it is to satisfy $X_{\alpha+1} \subseteq X_\alpha$ and $m(X_\alpha) > m(X_{\alpha+1}) \geq \frac{1}{2} \cdot m(X)$. This is possible by (i). Finally, for δ a limit ordinal define $X_\delta = \bigcap_{\alpha<\delta} X_\alpha$ exactly when X_α has been defined for each $\alpha < \delta$. There must now be an $\alpha < \omega_1$ such that $m(X_\alpha) = \frac{1}{2} \cdot m(X)$, else the collection $\{X_\alpha - X_{\alpha+1} \mid \alpha < \omega_1\}$ would contradict 2.1. ⊣

Proof of 2.5. For each $s \in {}^{<\omega}\omega$ define sets $X_s \subseteq \kappa$ as follows: Set $X_\emptyset = \kappa$. Given X_s, apply 2.6(ii) recursively to get sets $X_{s^\frown\langle i\rangle}$ for $i \in \omega$ such that

$$X_s = \bigcup_{i\in\omega} X_{s^\frown\langle i\rangle} \text{ is a disjoint union },$$

and

$$m(X_{s^\frown\langle i\rangle}) = 2^{-(i+1)} \cdot m(X_s) .$$

For each $f \in {}^\omega\omega$ set $Y_f = \bigcap_n X_{f|n}$, so that $m(Y_f) = 0$. $\kappa = \bigcup\{Y_f \mid f \in {}^\omega\omega\}$, and so m cannot be $(2^{\aleph_0})^+$-additive, and hence (a) follows.

For (b) define μ: $\mathcal{P}({}^\omega\omega) \to [0, 1]$ by

$$\mu(A) = m(\bigcup\{Y_f \mid f \in A\}) .$$

It is simple to check that μ is a (in fact κ-additive) measure over ${}^\omega\omega$. It remains to use properties of the Lebesgue measure m_L as formulated in §0 to show that μ extends m_L: By the construction of X_s for $s \in {}^{<\omega}\omega$,

$$\mu(O(s)) = m(X_s) = \prod_{i<|s|} 2^{-(s(i)+1)} = m_L(O(s)) .$$

Thus, by definition of m_L, μ and m_L agree on the Borel sets. But then, $\mu(N) = 0$ for any null set N, and consequently μ and m_L agree on the Lebesgue measurable sets. ⊣

This result provided for the first time a well-motivated example of a weakly inaccessible cardinal $\leq 2^{\aleph_0}$, and hence a plausible hypothesis implying a drastic failure of Cantor's Continuum Hypothesis. Much stronger results are now known along these lines, but almost forty years were to pass before the primary results on real-valued measurability $\leq 2^{\aleph_0}$ were established by Solovay (§§16, 17). For a portmanteau compendium of recent results on real-valued measurability, see Fremlin [93].

The other road taken by Ulam was to be even more consequential. Suppose that a κ-additive measure m over κ does have an atom $A \subseteq \kappa$. If μ is defined on $\mathcal{P}(\kappa)$ by

$$\mu(X) = \frac{m(X \cap A)}{m(A)} ,$$

then μ *is a* κ*-additive measure over* κ *with range* $\{0, 1\}$. This property can be conveniently expressed in terms of ultrafilters: For such a two-valued measure μ, set:

$$U_\mu = \{X \subseteq \kappa \mid \mu(X) = 1\} .$$

Then U_μ is a (non-principal) ultrafilter over κ. For instance, if $X, Y \in U_\mu$, then $X \cap Y \in U_\mu$: otherwise $X - Y \in U_\mu$ and $Y - X \in U_\mu$, leading to the contradiction $\mu(X \cup Y) \geq 2$. Furthermore, the κ-additivity of μ translates to a strong intersection property: Recall that a filter F is λ*-complete iff* for any $\gamma < \lambda$ and $\{X_\alpha \mid \alpha < \gamma\} \subseteq F$, $\bigcap_{\alpha<\gamma} X_\alpha \in F$. There is a simple dual characterization for ultrafilters:

2.7 Exercise. *An ultrafilter* U *is* λ*-complete iff for any* $\gamma < \lambda$ *and* $\bigcup\{Y_\alpha \mid \alpha < \gamma\} \in U$*, there is an* $\alpha < \gamma$ *such that* $Y_\alpha \in U$. ⊣

Hence, our U_μ is κ-complete. This leads to the most important concept of all large cardinal theory: For $\kappa > \omega$,

κ is *measurable iff* there is a κ-complete ultrafilter over κ .

Measurability is a direct generalization of the existence of ultrafilters over ω (which of course are ω-complete). The finite intersection property is automatically preserved when taking the union of a chain of filters, so that just a maximal principle is needed to get ultrafilters over ω. But κ-completeness for $\kappa > \omega$ is not similarly preserved, so that the existence of a κ-complete ultrafilter over κ must be explicitly postulated. Tarski [39,45] investigated the concept of measurability in the context of Boolean algebras and κ-complete prime ideals – just the dual notion.

The following result stands in contrast to 2.5(a):

2.8 Theorem (Ulam; Tarski – Ulam [30: 146]). *If* κ *is measurable, then* κ *is inaccessible.*

Proof. What remains beyond 2.4 is to establish that κ is a strong limit. Suppose that U is a κ-complete ultrafilter over κ, and assume to the contrary that $\lambda < \kappa$, yet there is an injective function $f : \kappa \to {}^\lambda 2$. For each $\alpha < \lambda$, there is an $i_\alpha < 2$ such that $X_\alpha = \{\xi < \kappa \mid f(\xi)(\alpha) = i_\alpha\} \in U$. Hence, $X = \bigcap_{\alpha<\lambda} X_\alpha \in U$, and for $\xi \in X$, $f(\xi)(\alpha) = i_\alpha$ for every $\alpha < \lambda$. But then, X can have at most one member, which is a contradiction. ⊣

Whether the least measurable cardinal is strictly larger than the least inaccessible cardinal became a focal question, and was settled only thirty years later. For a result that by present-day standards is rather straightforward this may seem like a remarkably long time – even taking into account the convulsive events that took place in Europe. However, this is a typical case of a long-standing mathematical problem suddenly solved in the wake of new techniques – and perhaps surprisingly, sufficiently strong methods were first to emerge in the semantics of infinitary languages (§4).

Meanwhile, the center stage in set theory was taken by Gödel's formulation of the constructible hierarchy and the consistency of the Axiom of Choice and the Generalized Continuum Hypothesis (§3). Nonetheless, in the growing abstraction of modern mathematics measurability occurred in various fields as a limitative concept: Mackey [44], vector lattices; Hewitt [48], rings of continuous functions; Ehrenfeucht-Łoś [54], infinite cyclic groups; and further references in Keisler-Tarski [64: 272]. It remained for Scott to establish the modern and central relevance of measurable cardinals in connection with the ultrapower construction (§5).

3. Constructibility

This section does not discuss a new large cardinal concept, but rather the major contribution to set theory of the foremost mathematical logician of our time and his speculations about large cardinals. That contribution not only launched axiomatic set theory as a distinctive field of mathematics, but stimulated subsequent work in large cardinals first in complementary reaction and then in structural generalization.

In October of 1935 Gödel informed von Neumann at the Institute for Advanced Study in Princeton that he had established the relative consistency of the Axiom of Choice. This he did, of course, by devising his constructible hierarchy L (for "law") and verifying the axiom there. Gödel conjectured that the Continuum Hypothesis would also hold in L, but he soon fell ill and only gave a proof of this and GCH two years later (the crucial idea apparently came to him during the night of June 14-15, 1937 – Gödel [86: 40]). In addition to these fundamental results he observed that in L there are delimitative counterexamples for descriptive set theory (§§12, 13).

Gödel's article [38] in the *Proceedings of the National Academy of Sciences U.S.A.* was the first announcement of these results, and the succeeding [39] provided more details in the context of ZF. To review, a set y is *definable over* a structure $\mathcal{M}$ *iff* there is a first-order formula $\varphi(v_0, \ldots, v_n)$ in the language of $\mathcal{M}$ and parameters $a_1, \ldots, a_n$ in the domain of $\mathcal{M}$ such that: $z \in y$ *iff* $\mathcal{M} \models \varphi[z, a_1, \ldots, a_n]$. For any set x,

$$\mathrm{def}(x) = \{y \subseteq x \mid y \text{ is definable over } \langle x, \in\rangle\} .$$

This is a set, being itself a subset of $\mathcal{P}(x)$ definable via the formalized satisfaction relation for $\langle x, \in\rangle$. Now define by transfinite recursion:

$$L_0 = \emptyset \ ; \ L_{\alpha+1} = \mathrm{def}(L_\alpha) \ ; \ L_\delta = \bigcup_{\alpha<\delta} L_\alpha \text{ for limit } \delta > 0 \ ;$$

and

$$L = \bigcup_\alpha L_\alpha \ .$$

Thus, L is a (definable) class, the class of *constructible* sets, and the assertion $V = L$, i.e. $\forall x(x \in L)$, is the *Axiom of Constructibility*. It is already apparent why the Axiom of Choice would hold in L: By transfinite recursion within L one can well-order L level-by-level, well ordering $L_{\alpha+1} - L_\alpha$ according to definitions and the previous well-ordering of the parameters from L_α. The verification of GCH was less obvious, evidently inspired by the analysis of Skolem [23], the source of the Löwenheim-Skolem Theorem. Devlin [84], Kunen [80], Moschovakis [80] and Drake [74] provide expositions of Gödel's results.

Gödel believed that careful epistemological analysis can lead to significant mathematical advances, and his remarkable successes amply bear out this contention. His results with L actually represent a steady intellectual development from his celebrated Incompleteness Theorem [31] which was to extend later to speculations on large cardinals. Already in footnote 48a to [31], made much of by Kreisel [80], Gödel wrote:

> ... the true reason for the incompleteness inherent in all formal systems of mathematics is that the formation of ever higher types can be continued into the transfinite ... while in any formal system at most denumerably many of them are available. For it can be shown that the undecidable propositions constructed here become decidable whenever appropriate higher types are added (for example, the type ω to the system P [Peano Arithmetic]). An analogous situation prevails for the axiom system of set theory.

Gödel's first description of L in [38] shows how he had built on this insight:

> This model, roughly speaking, consists of all "mathematically constructible" sets, where the term "constructible" is to be understood in the semiintuitionistic sense which excludes impredicative procedures. This means "constructible" sets are defined to be those sets which can be obtained by Russell's ramified hierarchy of types, if extended to include the transfinite orders. The extension to transfinite orders has the consequence that the model satisfies the impredicative axioms of set theory, because an axiom of reducibility can be proved for sufficiently high orders.

Gödel thus regarded his hierarchy as a transfinite extension of Russell's, which can be construed as a formal theory of $L_{\omega+\omega}$. Although this is a hierarchy of predicative definitions (i.e. quantifying over domains of previously formed objects), Gödel's realization was that extending the indexing of higher types through all the ordinals would lead to a completion of the axioms of set theory, as adumbrated in that footnote to [31]. Gödel stressed the historical continuity: the allusion to Russell's ill-fated Axiom of Reducibility is a clear reference to the rectification: if $x \in L_\lambda$ and λ is a cardinal in L, then for any $y \subseteq x$ in L there is a $\gamma < \lambda$ such that $y \in L_\gamma$. Thus, the impredicative power set operation is tamed in L, leading to the consistency of GCH. In later commentary on Russell's mathematical logic Gödel [44: 147] argued that with his "transfinite theorem of reducibility", "all impredicativities are reduced to one special kind, namely the existence of certain large ordinal numbers (or well-ordered sets) and the validity of recursive reasoning for them."

Interestingly enough, Gödel viewed L as an outright construction using transfinite reasoning in metamathematics. See Wang [74: 8ff]; in a letter quoted there Gödel wrote about his CH result: "... there was a special obstacle which *really* made it *practically impossible* for constructivists to discover my consistency proof. It is the fact that the ramified hierarchy, which had been invented *expressly for constructive purposes,* had to be used in an *entirely nonconstructive way.*" Gödel was aware of Zermelo's [30], and there is an affinity of sorts in the direct use of infinitary methods and the positing of successive domains. To be sure, Zermelo did not formalize his logic, while Gödel was led to transfinite types by his investigation of formal systems. In any formalization of L the extent of the ordinals as sustained by Replacement has to be accommodated, and significantly, the main statement of formal consistency about ZF in Gödel [39] appealed to what Zermelo had called "Grenzzahlen": If κ is inaccessible, then $L_\kappa \models \text{ZFC} + \text{GCH}$. κ need only be weakly inaccessible here because of the proof of GCH (Firestone-Rosser [49]).

In his monograph [40], based on lectures given at the Institute for Advanced Study during the winter of 1938-39, Gödel gave another presentation of L. This time he generated L set by set with a transfinite recursion in terms of eight elementary set generators, a sort of Gödel numbering into the transfinite. These generators were based on a finite second-order axiomatization of the Separation Schema by Paul Bernays [37], which in turn had an antecedent in a functional formulation in von Neumann [25]. Providing a rigorous formalization of his metamathematical construction Gödel now emphasized how it yields an "inner model" and finitary *relative* consistency proofs. In particular, an external appeal to an inaccessible cardinal was no longer necessary. Moreover, the new presentation highlighted the stark contrast between the elementary set formation processes and the extent of the ordinals.

Ironically, the development of set theory in these formative years may have been ill-served by [40]. Bearing the burden of authority it overshadowed [39] and obscured its model-theoretic approach, especially in the Skolem hull argument for the consistency of GCH. Only decades later was the sort of fine analysis of [40] to become an appropriate framework: Gödel himself considered L only as a contrivance for establishing consistency results, but in the mid-1960's Ronald Jensen [72] developed a "fine structure" theory for L of intrinsic interest, and the study of constructibility and its generalizations has become one of the mainstreams of modern set theory.

Expanding on his vision of completability of formal systems using higher types Gödel speculated in the 1940's about the possibility of deciding propositions with large cardinal hypotheses, particularly with respect to the Continuum Problem. In a lecture at Princeton University in December of 1946, Gödel remarked [90: 151]):

> In set theory, e.g. the successive extensions can most conveniently be represented by stronger and stronger axioms of infinity. It is certainly impossible to give a combinatorial and decidable characterization of what an axiom of infinity is but there might exist, e.g. a characterization of the following sort: An axiom of infinity is a proposition which has a certain (decidable) formal structure and which in addition is true. Such a concept of demonstrability might have the required closure property, i.e. the following could be true: Any proof for a set-theoretic theorem in the next higher system above set theory ... is replaceable by a proof from such an axiom of infinity. It is not impossible that for such a concept of demonstrability some completeness theorem would hold which would say that every proposition expressible in set theory is decidable from the present axioms plus some true assertion about the largeness of the universe of all sets.

This is a remarkably optimistic statement about the possibility of discovering new "true" axioms that will decide *every* set-theoretic proposition. Note that the following *reflection* argument for motivating large cardinals is almost explicit: (a) any set-theoretic proposition can be established in the "next higher system above set theory", say if the satisfaction relation for V were available; and (b) this use of the relation may in particular cases be replaceable by a strong axiom of infinity, say with a corresponding large cardinal playing the role of On. Gödel went on to apply his reflection argument to formulate the concept of ordinal definability (see Gödel [90: 146]). The Reflection Principle (see after 6.2) later schematized a basic argument that can be carried out in ZF.

In an accentuated form reflection came to be the main heuristic advanced for motivating various large cardinals: Largeness properties ascribable to On confront the antithetical contention that it is essentially incomprehendable, recalling Cantor's Absolute. This inability to characterize On in its open-endedness then fosters the synthetic move to a large cardinal at which such properties obtain.

The Continuum Problem was the focus of an expository article [47] by Gödel that included rare statements concerning his realist conception of mathematics. He assumed that the Continuum Hypothesis would be shown independent from ZF, and argued moreover that it must be false according to certain *a priori* intuitions. Regarding formal axioms that might resolve such questions he speculated about large cardinals in more concrete terms than in his 1946 remarks. Citing Zermelo [30] and echoing its theme Gödel argued on the basis of the cumulative hierarchy for new axioms "which assert the existence of still further iterations of the operation 'set of' ", giving inaccessible and Mahlo cardinals as examples. However, he mentioned that cardinals like these relativize to L (3.1) and hence cannot imply the failure of CH.

Gödel went on to speculate ([47: 521]):

> ... even disregarding the intrinsic necessity of some new axiom, and even in case it had no intrinsic necessity at all, a decision about its truth is possible also in another way, namely, inductively by studying its "success", that is, its fruitfulness in consequences and in particular in "verifiable" consequences, i.e. consequences demonstrable without the new axiom, whose proofs by means of the new axiom, however, are considerably simpler and easier to discover, and make it possible to condense into one proof many different proofs ... There might exist axioms so abundant in their verifiable consequences, shedding so much light upon a whole discipline, and furnishing such powerful methods for solving given problems (and even solving them, as far as that is possible, in a constructivistic way) that quite irrespective of their intrinsic necessity they would have to be assumed at least in the same sense as any well established physical theory.

This interestingly undercuts an avowedly realist position with a pragmatism that dilutes the force of "truth", but is resonant with subsequent investigations, particularly of the Axiom of Determinacy (Chapter 6). See Wang [74: 200ff] for more on Gödel's views on intrinsic necessity and pragmatic success for accepting new axioms. Gödel concluded his article with some controversial remarks on mathematical evidence against CH (cf. Martin-Solovay [70: 176], Martin [76]).

Whatever can be said about Gödel's proposals and despite latter-day references to "Gödel's program", it is unclear how much actual influence they had on subsequent developments. True, Tarski who was to make an important contribution in 1960 (§4) heard Gödel's 1946 Princeton lecture, but a text did not become generally accessible until 1965 in Gödel [65]. The article [47] was widely read, but it was addressed to non-specialists. And nowhere have Gödel's remarks been acknowledged as having been an inspiration. In the 1960's the theory of large cardinals quickly developed a self-fueling momentum, and blossomed into a sophisticated branch of set theory far overshadowing Gödel's epistemological concerns.

In an unpublished footnote 20 toward a 1966 revision of his [47] Gödel was to acknowledge ([90: 260ff]) the new developments, matters taken up in §§4, 5:

> In recent years great progress has been made in the area of axioms of infinity. In particular, some propositions have been formulated which, if consistent, are extremely strong axioms of infinity of an entirely new kind ... Dana Scott ... has proved that one of them implies the existence of non-constructible sets. That these axioms are implied by the general concept of set in the same sense as Mahlo's has not been made clear yet ... However, they are supported by strong arguments from analogy, e.g., by the fact that they follow from the existence of generalizations of Stone's representation theorem to Boolean algebras with operations on infinitely many elements.

This last presumably refers to strong compactness (see 4.1). The heuristic of *generalization* from $\aleph_0$, like reflection, also came to be used to motivate various large cardinals. Recalling Cantor's unitary view of the finite and the transfinite, large cardinal properties satisfied by $\aleph_0$ would be too accidental if they were not also ascribable to higher cardinals in an eternal recurrence.

It is now known that in 1942 Gödel had developed partial results toward the independence of the Axiom of Choice, and that he soon abandoned this work (see Moore [88: 149-151]). According to Kreisel [80: 201], "With present experience it is not too difficult to complete the proof. But something essential – in Gödel's words (in conversation): a *method* – had been missing ..." According to John Addison (Moore [88: 150]), Gödel feared that his proof would lead set-theoretic research in the wrong direction. Rather than developing relative consistency results, the concept of set should be analyzed more carefully and new axioms sought that would simply settle issues like the Continuum Hypothesis. Nonetheless, in a revised postscript toward a 1966 revision of his [47] Gödel was to declare (Gödel [90: 270]) that Paul Cohen's work on the independence of CH "... is the greatest advance in the foundations of set theory since its axiomatization", noting that it showed in particular that large cardinals have no direct bearing on CH (10.12ff).

Whatever the subsequent developments, Gödel's construction of L had established the minimum possibility for the set-theoretic universe, and large cardinals were to provide the counterweight first in reaction and then in generalization. Scott's result that measurable cardinals contradict $V = L$ (§5) inspired research that was to establish the intrinsic necessity of large cardinals for transcending such hypotheses. The generalizations of constructibility accommodating measurability (§§20, 21) led to a full-blown theory of minimal models for large cardinal hypotheses. And throughout, a crescendo of results was to amply demonstrate the pragmatic success of large cardinals in settling a large variety of questions, many about definable sets of reals. In these more subtle ways Gödel's hopes about large cardinals have been vindicated.

Inner Models

The structural generalizations of Gödel's L began with abstractions, relativizations, and some conditional independence results. With largely proof-theoretic aims in mind Shepherdson [51, 52, 53] carefully formalized and studied a general notion of "inner model". The term will be reserved for a special case: For a proper class M,

M is an *inner model iff* M is a transitive $\in$-model of ZF with On $\subseteq M$.

Such M modeling ZFC are specified by M is an *inner model of* ZFC, and so forth. These notions can be formalized for arbitrary classes M in a theory like Morse-Kelley where the class satisfaction relation is definable, but some clarification is necessary for formalization in ZF:

(i) By "class" in the ZFC context is meant definable class, i.e. there is a formula $\varphi(v_0, \ldots, v_n)$ and sets $a_1, \ldots, a_n$ such that $x \in M$ is merely *une façon de parler* for $\varphi[x, a_1, \ldots, a_n]$.

(ii) Let $\mathrm{Inn}(v)$ be the formula

$$\bigcup v \subseteq v \ \wedge \ \forall\alpha(\mathrm{def}(v \cap V_\alpha) \subseteq v) .$$

$\mathrm{Inn}(v)$ easily implies that On $\subseteq v$ by induction on ordinals and that $\{v \cap V_\alpha \mid \alpha \in \mathrm{On}\} \subseteq v$. Using this relativized rank hierarchy, it follows as for L that for any axiom σ of ZF, with σ^M the relativization of σ to M

$$(*) \qquad \vdash_{\mathrm{ZF}} \ \mathrm{Inn}(M) \rightarrow \sigma^M$$

(cf. Devlin [84: 60-63]). Moreover, for any transitive class N with On $\subseteq N$ satisfying $\vdash_{\mathrm{ZF}} \sigma^N$ for every axiom σ of ZF, it can be shown that $\vdash_{\mathrm{ZF}}$ $\mathrm{Inn}(N)$.

Since all instances of ZF axiom schema are needed in the proofs of the theorems $(*)$, (i) and (ii) constitute a formalization in a weak sense. Nonetheless, $\mathrm{Inn}(M)$ *will be what is meant by the assertion that* M *is an inner model, and* $M \models \varphi$ *will often be written for the more proper* φ^M.

The archetypical inner model is L, and $L \subseteq M$ *for any inner model* M *since* $L^M = L$. Because of this Shepherdson [53] observed that the relative consistency of theories like ZFC + $\neg$GCH cannot be established by relativization to an inner model. Furthermore, as noted by Gödel [38: 557] (and more clearly in [51: 69]), some large cardinals relativize to inner models:

3.1 Exercise. *If* M *is an inner model of* ZFC *and* κ *is an* α*-inaccessible cardinal, then* (κ is α-inaccessible)^M. *The analogous assertion holds for* α*-Mahlo cardinals.*

Hint. Show that: (a) if κ is a strong limit, then κ is a strong limit in the sense of M, and (b) if κ is regular, then κ is regular in the sense of M. Now use induction on α. $\dashv$

Despite these inherent limitations András Hajnal [56, 61] and Levy [57, 60] in their doctoral dissertations in Hungary and Israel respectively developed basic generalizations of L which will be central to our later concerns about the effect of large cardinals on the inner structures of set theory. Moschovakis [80: 489ff, 531ff] and Drake [74: 149ff] contain careful expositions on the following models, of which only the salient features are noted:

Hajnal essentially provided for a given set A the *constructible closure* $L(A)$, i.e. the smallest inner model M such that $A \in M$. If there is no concern about how the elements of A are to be incorporated into $L(A)$, the Zermelo [30] urelement basis idea can be used with the transitive closure $\mathrm{tc}(\{A\})$ to ensure that the resulting class is transitive:

$$L_0(A) = \mathrm{tc}(\{A\})\ ;\ \ L_{\alpha+1}(A) = \mathrm{def}(L_\alpha(A))\ ;$$
$$L_\delta(A) = \bigcup_{\alpha<\delta} L_\alpha(A) \text{ for limit } \delta > 0\ ;$$

and

$$L(A) = \bigcup_\alpha L_\alpha(A)\ .$$

Although $L(A)$ is indeed an inner model, *unless* $\mathrm{tc}(\{A\})$ *has a well-ordering in* $L(A)$, $L(A)$ *does not satisfy the Axiom of Choice.* $|L_\alpha(A)| = |\mathrm{tc}(\{A\})| \cdot |\alpha|$ for $\alpha \geq \omega$, a result established by induction on α.

Levy on the other hand developed for a given set A the inner model $L[A]$ of sets *constructible relative to* A, i.e. the smallest inner model M such that for every $x \in M$, $A \cap x \in M$. The idea is to define a relativized hierarchy where assertions about membership in A can be made of sets defined thus far. Let

$$\mathrm{def}^A(x) = \{y \subseteq x \mid y \text{ is definable over } \langle x, \in, A \cap x\rangle\}\ ,$$

making $A \cap x$ available as a unary relation for definitions. In analogy with L,

$$L_0[A] = \emptyset\ ;\ \ L_{\alpha+1}[A] = \mathrm{def}^A(L_\alpha[A])\ ;\ \ L_\delta[A] = \bigcup_{\alpha<\delta} L_\alpha[A] \text{ for limit } \delta > 0\ ;$$

and

$$L[A] = \bigcup_\alpha L_\alpha[A]\ .$$

Unlike for $L(A)$ what remains of A is only $A \cap L[A] \in L[A]$, so that for example $L[\mathbb{R}] = L$ for the reals $\mathbb{R}$. However, $L[A]$ is more constructive since knowledge of A is incorporated through the hierarchy of definitions, and like L, $L[A]$ *satisfies the Axiom of Choice.* $|L_\alpha[A]| = |\alpha|$ for $\alpha \geq \omega$, a result established by induction on α. The theory of $L[A]$ is best developed with structures $\langle L_\alpha[A], \in, A \cap L_\alpha[A]\rangle$ for $\mathcal{L}_\in(\dot{A})$, the language of set theory augmented by one unary predicate symbol $\dot{A}$.

As with L itself a prominent feature of the theory is the *absoluteness* of the definitions involved, which leads to the following (see Moschovakis [80: 515, 532]):

3.2 Proposition.

(a) If M is an inner model with $A \in M$, then $L(A)^M = L(A)$.

(b) If M is an inner model with $A \cap M \in M$, then $L[A \cap M]^M = L[A]$.

(c) If $A \cap L[A] = B \cap L[A]$, then $L[A] = L[B]$.

(d) $L[A] = L[A \cap L[A]] = L(A \cap L[A])$.

(e) If $A \subseteq L$, then $L[A] = L(A)$. ⊣

For future reference, some results about L and $L[A]$ are stated that begin the finer analysis of constructibility (see Devlin [84: 71, 75] or Moschovakis [80: 496, 516, 533]).

3.3 Theorem.

(a) There is a sentence σ_0 of $\mathcal{L}_\in$ such that for any transitive class N,

$$\langle N, \in\rangle \models \sigma_0 \text{ iff } N = L \vee N = L_\delta \text{ for some limit } \delta > \omega .$$

Also, there is a formula $\varphi_0(v_0, v_1)$ of $\mathcal{L}_\in$ that defines in L a well-ordering $<_L$ of L such that for any limit $\delta > \omega$, any $y \in L_\delta$, and any x,

$$x <_L y \text{ iff } x \in L_\delta \wedge \langle L_\delta, \in\rangle \models \varphi_0[x, y] .$$

(b) There is a sentence σ_1 of $\mathcal{L}_\in(\dot{A})$ where $\dot{A}$ is unary such that for any A and any transitive class N,

$$\langle N, \in, A \cap N\rangle \models \sigma_1 \text{ iff } N = L[A] \vee N = L_\delta[A] \text{ for some limit } \delta > \omega .$$

Also, there is a formula $\varphi_1(v_0, v_1)$ of $\mathcal{L}_\in(\dot{A})$ that in any $\langle L[A], \in, A \cap L[A]\rangle$ defines a well-ordering $<_{L[A]}$ of $L[A]$ such that for any limit $\delta > \omega$, any $y \in L_\delta[A]$, and any x,

$$x <_{L[A]} y \text{ iff } x \in L_\delta[A] \wedge \langle L_\delta[A], \in, A \cap L_\delta[A]\rangle \models \varphi_1[x, y] . \quad ⊣$$

The sentence σ_0 leads directly to Gödel's Condensation Lemma, the crux of his proof of GCH in L: If $\delta > \omega$ is a limit ordinal and $\langle H, \in\rangle$ is an elementary substructure of $\langle L_\delta, \in\rangle$, then $\langle H, \in\rangle$ has a transitive collapse (by 0.4) that must be of form $\langle L_\alpha, \in\rangle$ for some α (because of σ_0).

Although differing in their formal presentations, both Hajnal and Levy used a set of ordinals A so that $L[A] = L(A)$ by 3.2(e), and the distinctions were to surface only later. Hajnal and Levy (as well as Shoenfield [59] who formulated a special version of Levy's construction) used these models to establish conditional independence results of the sort: If $\neg$CH is consistent, then so is $\neg$CH together with $2^\lambda = \lambda^+$ for sufficiently large λ. Hajnal's finer analysis led to a useful fact: If $V = L[A]$ and $A \subseteq \kappa^+$, then $2^\kappa = \kappa^+$. In particular, if $\neg$CH and $A \subseteq \omega_2$ codes ω_2 distinct subsets of ω as well as injections: $\alpha \to \omega_1$ for every $\alpha < \omega_2$, then $\omega_2^{L[A]} = \omega_2$ and so $(2^{\aleph_0} = 2^{\aleph_1} = \omega_2)^{L[A]}$. More pointedly, if $2^{\aleph_0} \neq \omega_2$ is provable in ZFC, then so is CH. All this anticipated the expected independence of CH, and providing at least a semblance of continuity Cohen duly established this with his celebrated method of forcing.

4. Compactness

For almost three decades after 1930 no significant advance was made in the investigation of large cardinals, but Alfred Tarski maintained a steady interest in the subject. He visited the United States from Poland in 1939, but the outbreak of war precluded his return, and by 1942 he was established at the University of California at Berkeley. Through his initiatives he was to play a pivotal role in the flowering of mathematical logic in California, and Berkeley became the leading center for set theory in the 1960's. In particular, he and his co-workers were to make basic contributions to the theory of large cardinals through the infusion of model-theoretic methods. This section describes the early stages of these developments and brings into full play the model-theoretic preliminaries of §0.

Combinatorial elaborations had already been suggested in the early paper Erdős-Tarski [43] which at the end described various properties of cardinals implying inaccessibility (see §7). The details of implications asserted there were presented in an influential seminar conducted by Tarski and Andrzej Mostowski at Berkeley in 1958-9, and soon appeared in Erdős-Tarski [61]. It was against this backdrop that Tarski's initiatives in another direction were to lead to a real breakthrough.

Tarski [58] considered the semantics of the infinitary predicate languages $L_{\lambda\mu}$ and later raised the issue of their possible compactness. In brief, an $L_{\lambda\mu}$ language is formulated as follows: Proceeding as for the usual first-order logic, first specify a supply of *non-logical* symbols: (finitary) predicate, function, and constant symbols. These together with an allowed supply of $\max(\{\lambda, \mu\})$ many variables lead to the terms and atomic formulas. Then the usual formula generating rules are expanded to allow conjunctions $\bigwedge_{\xi<\alpha}$ and disjunctions $\bigvee_{\xi<\alpha}$ of α formulas for any $\alpha < \lambda$, and universal quantifications $\forall_{\xi<\beta}$ and existential quantifications $\exists_{\xi<\beta}$ of β variables for any $\beta < \mu$. Finally, a formula is an expression so generated with less than μ free variables, this to allow the possibility of quantificational closure. Structures for interpreting the language are as for first-order logic, and the satisfaction relation is extended to incorporate the new infinitary connectives and quantifiers in the expected way. (For book studies of these languages see Karp [64] for the formal syntax, Keisler [71] for the model theory of $L_{\omega_1\omega}$, and Dickmann [75] for the general model theory; Dickmann [85] provides an overview.) In what follows, properties of $L_{\lambda\mu}$ logic that are straightforward generalizations of those for the usual first-order, i.e. $L_{\omega\omega}$, logic are taken for granted. In subsequent sections the concepts introduced will be applied in terms of combinatorial characterizations and the results established considerably improved.

Tarski [62] formulated two natural generalizations of the well-known compactness property of $L_{\omega\omega}$: A collection of $L_{\lambda\mu}$ sentences is *satisfiable iff* it has a model under the expected interpretation of infinitary conjunction, disjunction and quantification; and is ν*-satisfiable iff* every subcollection of cardinality less than ν is satisfiable. For $\kappa > \omega$,

κ is *strongly compact iff* any collection of $L_{\kappa\kappa}$ sentences, if κ-satisfiable, is satisfiable .

κ is *weakly compact iff* any collection of $L_{\kappa\kappa}$ sentences using at most κ non-logical symbols, if κ-satisfiable, is satisfiable .

Tarski's original formulation of weak compactness had the more stringent condition $|\Sigma| = \kappa$ and does not imply the inaccessibility of κ (Boos [76]), while the one that is adopted here does (4.4) which is the modern preference.

By the early 1960's the development of the basic ultraproduct construction led to a surge of new results in model theory and an enduring interest in ultrafilters. The general construction was introduced by Jerzy Łoś [55], where the basic theorem 0.6 is implicit. Then Frayne-Morel-Scott [62] and also Kochen [61] propagated the basic theory as developed by the authors and Tarski. The former paper provided an ultraproduct proof of the Compactness Theorem, a direct generalization of which is used in the following characterization.

4.1 Proposition (Keisler-Tarski [64]). *κ is strongly compact iff for any set S, every κ-complete filter over S can be extended to a κ-complete ultrafilter over S.*

Proof. Suppose first that κ is strongly compact and F is a κ-complete filter over a set S. Using constants $\dot{X}$ for every $X \subseteq S$, let Σ be the $L_{\kappa\kappa}$ theory of $\langle S \cup \mathcal{P}(S), \in, X\rangle_{X\subseteq S}$ together with the sentences $c \in \dot{X}$ for every $X \in F$, where c is a new constant. Σ is κ-satisfiable as F is κ-complete, so let $\mathcal{M}$ model Σ by strong compactness. Now define U by:

$$X \in U \quad \textit{iff} \quad X \subseteq S \ \wedge \ \mathcal{M} \models c \in \dot{X} \ .$$

It is simple to check that U is an ultrafilter over S extending F, and Σ has the $L_{\kappa\kappa}$ sentences which ensure that U is κ-complete.

For the converse, note first that κ must be regular: If to the contrary κ were singular, then κ-completeness would readily imply κ^+-completeness for filters. But then, if U is any κ-complete ultrafilter over κ^+ extending the κ-complete (even κ^+-complete) filter $\{X \subseteq \kappa^+ \mid |\kappa^+ - X| < \kappa^+\}$, then U would also be κ^+-complete and so κ^+ would be measurable, contradicting Ulam's 2.4.

Suppose now that $\Sigma = \{\sigma_\alpha \mid \alpha < \lambda\}$ is a κ-satisfiable collection of $L_{\kappa\kappa}$ sentences. Recall that $\mathcal{P}_\kappa\lambda = \{x \subseteq \lambda \mid |x| < \kappa\}$. For any $x \in \mathcal{P}_\kappa\lambda$, let $\mathcal{M}_x$ be a structure for the language of Σ so that $\mathcal{M}_x \models \bigwedge_{\alpha\in x} \sigma_\alpha$. With the availability of $\mathcal{M}_\lambda$ we can assume that $\lambda \geq \kappa$. As

$$\{\{x \in \mathcal{P}_\kappa\lambda \mid y \subseteq x\} \mid y \in \mathcal{P}_\kappa\lambda\}$$

generates a κ-complete filter over $\mathcal{P}_\kappa\lambda$ by the regularity of κ, let U be a κ-complete ultrafilter over $\mathcal{P}_\kappa\lambda$ extending this filter. Consider the ultraproduct $\mathcal{M} = \prod_{\mathcal{P}_\kappa\lambda} \mathcal{M}_x / U$. It is straightforward to check that, essentially by the same proof as for $L_{\omega\omega}$, Łoś's Theorem 0.6 holds for $L_{\kappa\kappa}$ and ultraproducts by κ-complete ultrafilters. Since for any $\alpha < \lambda$,

$$\{x \in \mathcal{P}_\kappa\lambda \mid \mathcal{M}_x \models \sigma_\alpha\} \supseteq \{x \in \mathcal{P}_\kappa\lambda \mid \alpha \in x\} \in U \ ,$$

it follows that $\mathcal{M} \models \sigma_\alpha$. ⊣

An analogous characterization exists for weak compactness in terms of an appropriately curtailed filter extension property (Keisler-Tarski [64: 288]).

The next two results were first observed in equivalent formulations.

4.2 Corollary (Erdős-Tarski [43: 328]). *If κ is strongly compact, then κ is measurable.*

Proof. κ is regular by the proof of 4.1. The filter $\{X \subseteq \kappa \mid |\kappa - X| < \kappa\}$ is consequently κ-complete, and any κ-complete ultrafilter over κ extending it verifies the measurability of κ. ⊣

4.3 Proposition (Erdős-Tarski [43: 328]). *If κ is measurable, then κ is weakly compact.*

Proof. Suppose that Σ be a κ-satisfiable collection of $L_{\kappa\kappa}$ sentences using at most κ non-logical symbols. Then it is simple to see that $|\Sigma| \leq \kappa^{<\kappa}$. But measurable cardinals are inaccessible, so $|\Sigma| \leq \kappa$. With $\langle \sigma_\alpha \mid \alpha < \kappa \rangle$ enumerating Σ we can now proceed as for 4.1:

For each $\beta < \kappa$ let $\mathcal{M}_\beta \models \bigwedge_{\alpha<\beta} \sigma_\alpha$. Let U be any κ-complete ultrafilter over κ, and set $\mathcal{M} = \prod_\kappa \mathcal{M}_\beta / U$. Then for any $\alpha < \kappa$,

$$\{\beta < \kappa \mid \mathcal{M}_\beta \models \sigma_\alpha\} \supseteq \{\beta < \kappa \mid \beta > \alpha\} \in U \ ,$$

and so $\mathcal{M} \models \sigma_\alpha$. ⊣

Finally, weak compactness entails inaccessibility:

4.4 Proposition. *If κ is weakly compact, then κ is inaccessible.*

Proof. To show that κ is regular, assume to the contrary that $X \subseteq \kappa$ is unbounded yet $|X| < \kappa$. Then for distinct constants c and c_α for $\alpha < \kappa$,

$$\{c \neq c_\alpha \mid \alpha < \kappa\} \cup \{\bigvee_{\beta \in X} \bigvee_{\alpha<\beta} c = c_\alpha\}$$

is κ-satisfiable yet not satisfiable – a contradiction.

To establish that κ is a strong limit, assume to the contrary that there is a $\lambda < \kappa$ such that $2^\lambda \geq \kappa$. Then for distinct constants c_α and d^i_α for $\alpha < \lambda$ and $i < 2$,

$$\{\bigwedge_{\alpha<\lambda}[(c_\alpha = d^0_\alpha \vee c_\alpha = d^1_\alpha) \wedge d^0_\alpha \neq d^1_\alpha]\} \ \cup \ \{\bigvee_{\alpha<\lambda}(c_\alpha \neq d^{f(\alpha)}_\alpha) \mid f \in {}^\lambda 2\} \ .$$

is not satisfiable, else any interpretation of $\langle c_\alpha \mid \alpha < \lambda \rangle$ would correspond to a function: $\lambda \to 2$ different from every member of ${}^\lambda 2$, yet by similar reasoning any proper subset is satisfiable – a contradiction. ⊣

Presumably with the old question of whether the least measurable cardinal is strictly larger than the least inaccessible cardinal in mind, Tarski suggested to his student William Hanf at Berkeley that he investigate the possible weak compactness of inaccessible cardinals. Using appropriate $L_{\kappa\kappa}$ sentences Hanf [64] in 1960 was able to establish that there are in fact many inaccessible cardinals below a weakly compact cardinal; notably, the early proofs were directly analogous to Tarski's construction of a consistent yet ω-inconsistent theory. Thus, the least measurable cardinal *is, a fortiori,* strictly larger than the least inaccessible cardinal (Tarski [62]).

This Hanf-Tarski breakthrough was the first result about the size of measurable cardinals since Ulam's original paper. It was greeted by Abraham Robinson [62: 78] as "a spectacular success" for metamathematical methods. Tarski himself opined ([62: 125]) that it was "contrary to expectations"; presumably, inaccessibility should have sufficed to generalize the compactness property of $L_{\omega\omega}$. Several times in the development of set theory, such key results have reoriented the collective set-theoretic intuition and spurred a spate of new research. Hanf's work radically altered size intuitions about problems that were coming to be understood in terms of large cardinals.

Several methods are now known to establish Hanf's result, two quickly discovered in its wake: Keisler [62] via ultrapowers and Hanf-Scott [61] via indescribability (§6). We shall proceed without much ado from an equivalent formulation of weak compactness. Like the recursive functions weak compactness has many diverse characterizations, which speaks to the naturalness and centrality of the concept. More will be encountered in coming sections as different themes are pursued. Most of the equivalences were established in the early 1960's, and later summarized in Silver [71: 62] and Devlin [75] – see also Comfort-Negrepontis [74: 185] for topological equivalences and Dickmann [75: 184] for model-theoretic equivalences.

H. Jerome Keisler established the following result, generalizing the use of the Compactness Theorem to get proper extensions of models. A student of Tarski at Berkeley and later a prominent model theorist, Keisler made important contributions at this formative stage by providing useful model-theoretic characterizations as well as ultraproduct proofs (cf. 5.6).

4.5 Theorem (Keisler [62, 62a]). *κ is weakly compact iff κ has the Extension Property: for any $R \subseteq V_\kappa$ there is a transitive set $X \neq V_\kappa$ and an $S \subseteq X$ such that $\langle V_\kappa, \in, R\rangle \prec \langle X, \in, S\rangle$.*

Remarks. $\prec$ denotes the assumed concept of elementary substructure; it is reviewed in a larger setting at the beginning of §5. Note that since $X \neq V_\kappa$ is to be transitive, $\kappa \in X$.

Proof. Suppose first that κ is weakly compact and $R \subseteq V_\kappa$. κ is inaccessible by 4.4, so that $|V_\kappa| = \kappa$. Using distinct constants c and $\dot{x}$ for each $x \in V_\kappa$, let Σ be the $L_{\kappa\kappa}$ theory of

$$\langle V_\kappa, \in, R, x \rangle_{x \in V_\kappa}$$

together with the sentences $\ulcorner c$ is an ordinal$\urcorner$ and $\ulcorner c \neq \dot{\alpha} \urcorner$ for each $\alpha < \kappa$. Then Σ is clearly κ-satisfiable, and so by weak compactness it is satisfiable. Of course, the Axiom of Extensionality is in Σ. Also, well-foundedness is readily seen to be expressible already in $L_{\omega_1\omega_1}$, and so Σ has a member saying $\in$ is well-founded. Hence, by the Collapsing Lemma 0.4 Σ has a transitive model

$$\langle X, \in, S, \overline{x}, \gamma \rangle_{x \in V_\kappa}$$

where γ interprets c. For any $x \in V_\kappa$, Σ contains the sentence

$$\forall v(v \in \dot{x} \leftrightarrow \bigvee\nolimits_{y \in x} v = \dot{y}) ,$$

so by induction on rank, $x \in V_\kappa$ implies that $\overline{x} = x$. Clearly γ is an ordinal $\geq \kappa$ so $X \neq V_\kappa$, and hence the reduct $\langle X, \in, S \rangle$ satisfies all the requirements.

For the converse, first observe that the Extension Property for κ implies its inaccessibility: Assume first that there is a $\mu < \kappa$, a cofinal map $F\colon \mu \to \kappa$, and a proper extension $\langle V_\kappa, \in, F \rangle \prec \langle X, \in, S \rangle$ where X is transitive. Then $\exists x (F``\mu \subseteq x)$ fails in the first structure yet $\exists x (S``\mu \subseteq x)$ holds in the second, which is a contradiction. Assume next that there is a $\nu < \kappa$ such that $\kappa \leq 2^\nu$. Then $2^\nu \leq |V_{\nu+1}|$ and $V_{\nu+1} \in V_\kappa$, and so a surjection $G\colon V_{\nu+1} \to \kappa$ can be used to again derive a contradiction.

Next, note that for any inaccessible λ the following two Löwenheim-Skolem theorems for $L_{\lambda\lambda}$ are verifiable by essentially the same arguments as for $L_{\omega\omega}$ (0.5):

(i) If σ is a satisfiable $L_{\lambda\lambda}$ sentence, then it has a model of cardinality less than λ.

(ii) If Σ is a satisfiable collection of $L_{\lambda\lambda}$ sentences of cardinality at most λ, then it has a model of cardinality at most λ.

Suppose now that Σ is a κ-satisfiable collection of $L_{\kappa\kappa}$ sentences using at most κ non-logical symbols. The corresponding $L_{\kappa\kappa}$ language has κ formulas by the inaccessibility of κ, and so we can carry out an arithmetization of the language that codes formulas with members of V_κ and formulate the satisfaction relation for set structures in V_κ as a class of V_κ. Let $R_1 \subseteq V_\kappa$ code this satisfaction relation, and construing Σ to be a subset of V_κ through the arithmetization let $R_2\colon \kappa \to \Sigma$ be surjective. Then by (i) above

$$\langle V_\kappa, \in, R_1, R_2 \rangle \models \forall \alpha (R_2``\alpha \text{ has a model}) .$$

By the Extension Property (with $\langle R_1, R_2 \rangle$ coded by a single subset of V_κ) let $\langle V_\kappa, \in, R_1, R_2 \rangle \prec \langle X, \in, S_1, S_2 \rangle$ be a proper extension where X is transitive. By elementarity $S_2``\kappa = R_2``\kappa = \Sigma$ and $\langle X, \in, S_1, S_2 \rangle \models \Sigma$ has a model. But by the inaccessibility of κ, $\langle V_\kappa, \in \rangle$ and hence $\langle X, \in \rangle$ models ZFC, and since X is transitive and $\kappa \in X$, $\langle X, \in \rangle \models \kappa$ is inaccessible (cf. 3.1). It follows by (ii) above that

$$\langle X, \in, S_1, S_2\rangle \models \Sigma \text{ has a model } \mathcal{M} \text{ with domain } \subseteq \kappa\ .$$

Finally, it can be assumed that the $L_{\kappa\kappa}$ satisfaction relation S_1 for $\mathcal{M}$ is upward persistent (in fact absolute) between X and V as ${}^{<\kappa}\kappa \subseteq V_\kappa \subseteq X$: as in the $L_{\omega\omega}$ case the relation for $\mathcal{M}$ is Σ_1 (and also Π_1). Hence, $\mathcal{M}$ really models Σ in V.⊣

Although $\langle X, \in\rangle \models \kappa$ is inaccessible, it cannot in general be asserted that $\langle X, \in\rangle \models \kappa$ is weakly compact: X may not contain all the needed models of κ-satisfiable sets of sentences. Otherwise, $\langle V_\kappa, \in\rangle \models \exists\alpha(\alpha$ is weakly compact) by elementarity, and hence by a simple argument there really would be a weakly compact cardinal less than κ. This cannot happen, of course, if κ is the least weakly compact cardinal. Large cardinal theory is replete with this sort of near miss.

Toward Hanf's result, a result about reflecting stationary sets is derived:

4.6 Proposition. *Suppose that κ is weakly compact. Then:*

(a) If A is stationary in κ, then there is an inaccessible $\lambda < \kappa$ such that $A \cap \lambda$ is stationary in λ.

(b) If A_α is stationary in κ for each $\alpha < \kappa$, then there is an inaccessible $\lambda < \kappa$ such that $A_\alpha \cap \lambda$ is stationary in λ for each $\alpha < \lambda$.

Proof. (a) Assume to the contrary that for each $\xi < \kappa$, either ξ is not inaccessible or else there is a C_ξ closed unbounded in ξ such that $C_\xi \cap A = \emptyset$. Let $R\colon \kappa \to V_\kappa$ be defined by: $R(\xi) = C_\xi$ if ξ is inaccessible and $= \emptyset$ otherwise. By 4.5 (with $\langle A, R\rangle$ coded by a single subset of V_κ) let $\langle V_\kappa, \in, A, R\rangle \prec \langle X, \in, B, S\rangle$ be a proper extension with X transitive. Then

$$\begin{aligned}\langle X, \in, B, S\rangle \models\ & \kappa \text{ is inaccessible } \wedge \\ & S(\kappa) \text{ is closed unbounded in } \kappa\ \wedge\ S(\kappa) \cap B = \emptyset\ .\end{aligned}$$

But then, $S(\kappa)$ really is closed unbounded, and since $B \cap \kappa = A$, $S(\kappa) \cap A = \emptyset$ contradicting the assumption that A is stationary.

(b) The argument is similar, using the assumption $C_\xi \cap A_\alpha = \emptyset$ for some $\alpha < \xi$ and $A = \{\langle\alpha, \beta\rangle \mid \beta \in A_\alpha\}$ instead. ⊣

4.7 Corollary (Hanf [64]). *If κ is weakly compact, then κ is κ-Mahlo.*

Proof. We show that κ is α-Mahlo for $\alpha \leq \kappa$ by induction on α. The basis and limit cases are immediate, so it remains to argue from α to $\alpha + 1$:

Suppose that κ is α-Mahlo. Then for $\beta < \alpha$, $A_\beta = \{\xi < \kappa \mid \xi \text{ is } \beta\text{-Mahlo}\}$ is stationary in κ. Let C be closed unbounded in κ. Then by 4.6(b) there is an inaccessible $\lambda < \kappa$ such that $C \cap \lambda$ and $A_\beta \cap \lambda$ for $\beta < \alpha$ are all stationary in λ. Hence, $\lambda \in C$ and λ is α-Mahlo. Since this obtains for any such C, $\{\xi < \kappa \mid \xi \text{ is } \alpha\text{-Mahlo}\}$ is stationary in κ, i.e. κ is $(\alpha + 1)$-Mahlo. ⊣

The full force of 4.6(b) leads to the conclusion that $\{\xi < \kappa \mid \xi \text{ is } \xi\text{-Mahlo}\}$ is stationary in κ. Indeed, Hanf [64] extended the Mahlo hierarchy with a diagonalization operator related to 4.6(b) and established that weakly compact cardinals are high in this hierarchy. He wrote ([64: 313]): "However, it turns out that for each class defined in such a constructive way, it appears almost certain on the basis of "naïve" set theory (or what is sometimes called "Cantor's absolute") that not all cardinals belong to the class." That is, his arguments in no way disallow weakly compact cardinals.

Tarski [62: 134] did make a clear distinction between inaccessible and weakly compact ("not strongly incompact") cardinals:

> For one thing, the belief in the existence of inaccessible cardinals $> \omega$ (and even of arbitrarily large cardinals of this kind) seems to be a natural consequence of basic intuitions underlying the 'naive' set theory and referring to what can be called 'Cantor's absolute'. On the contrary, we see at this moment no cogent intuitive reasons which could induce us to believe in the existence of cardinals $> \omega$ that are not strongly incompact, or which at least would make it very plausible that the hypothesis stating the existence of such cardinals is consistent with familiar axiom systems of set theory.

However, in a contrasting passage he wrote later on the page:

> We would of course fully dispose of all the problems involved if we decided to enrich the axiom system of set theory by including (so to speak, on a permanent basis) a statement which precludes the existence of 'very large' cardinals, e.g. by a statement to the effect that every cardinal $> \omega$ is strongly incompact [not weakly compact]. Such a decision, however, would be contrary to what is regarded by many as one of the main aims of research in the foundations of set theory, namely the axiomatization of increasingly large segments of 'Cantor's absolute'. Those who share this attitude are always ready to accept new 'construction principles', new axioms securing the existence of new classes of 'large' cardinals (provided they appear to be consistent with old axioms), but are not prepared to accept any axioms precluding the existence of such cardinals – unless this is done on a strictly temporary basis, for the restricted purpose of facilitating the metamathematical discussion of some axiom systems of set theory.

Although Tarski's words are somewhat guarded, the influence of Cantor's realist conception, perhaps refined by Gödel's remarks, is evident.

Keisler and Tarski detailed the results and interconnections of the theory to date in their comprehensive paper [64]. Although they believed that the original metamathematical methods "provide the intuitive and deductively simplest approach" to the results, they nevertheless labored with ultrapower proofs to provide a "purely mathematical treatment". They continued ([64: 226]):

> We have been motivated by the realization of the practical fact that the knowledge of metamathematics is not sufficiently widespread and may be defective among mathematicians who would otherwise be intensely interested in the topic discussed, and to a certain extent also by some (irrational) inclination toward puritanism in methods. As will be seen from the remarks below, we do not feel that we have been entirely successful in our undertaking.

Indeed, they first established that measurable cardinals are high in the Mahlo hierarchy using the ultrapower approach evolving from Keisler [62] (see 5.14), but only sketched the sharper results for weakly compact cardinals because of the expository difficulties involved. It is never simple to sublimate the satisfaction relation!

As large cardinal theory developed, metamathematical methods became basic to the subject in the interplay of inner models, forcing, and elementary embeddings. Symptomatic of this is how the details of structures and satisfaction so painstakingly set out in the early papers were soon replaced by casual expositions in informal rigor. With this development the field became an advanced and sophisticated part of the mathematical enterprise with less direct concern for foundational issues.

The discussion in Keisler-Tarski [64] is framed around the classes C_0, C_1 and C_2 of cardinals that are, respectively, not weakly compact, not measurable, and not strongly compact. To the set theorist of today, flipping through its pages is a curious experience, as he repeatedly encounters theorems which he himself would state in dual or contrapositive form. The several theorems with hypotheses of form "if C_i contains all the cardinals", while not necessarily espousing these hypotheses according to the previous Tarski quotation, are at least suggestive of a uniform, orderly cosmos delimited by pathologies beyond. In the next wave of research by Gaifman, Rowbottom, Silver, Solovay, Kunen, and others, such universal constraints were replaced by existential postulations and positive implications about the possible new richness of the set-theoretic universe. Perhaps it is a coincidence, but it was by this time that the terminological move from the half-empty notion of prime ideal to the half-full dual notion of ultrafilter was completed in model theory.

5. Elementary Embeddings

In the early 1960's set theory was veritably transformed by structural initiatives based on new possibilities for constructing well-founded models and establishing relative consistency results. This of course was due largely to the creation of forcing by Cohen, who came upon a remarkably fertile method for producing extensions of models of set theory. Fine mathematicians like Solovay quickly perceived the possibilities abounding, and within a few years the method of forcing was systematized, and a cornucopia of relative consistency results was being produced (§10).

A seminal result of Scott provided the impetus in another direction, the investigation of elementary embeddings. Scott had begun to work in the 1950's, initially with Tarski at Berkeley, on a variety of foundational issues. It is the nuts and bolts of forcing to establish the relative consistency with ZF of $V \neq L$, but Scott [61] was the first to realize the possibility by showing: *If there is a measurable cardinal, then* $V \neq L$. The difference is crucial: instead of a relative consistency result he had established a direct implication from a prior principle. It was in such ways that Gödel's hopes for large cardinals were to be vindicated; his axiom $V = L$ had imposed a consistent delimitation on ZFC, and developments from Scott's result were to establish the necessity of large cardinals for securing a real transcendence over L (§9).

Scott's work began the liberal use of manipulative proper class constructions in set theory and led to a structural characterization of measurable cardinals via ultrapowers and elementary embeddings. Setting the stage Keisler [62] (and in full exposition Keisler-Tarski [64]) had used ultrapowers within set theory to provide an alternate proof of Hanf's result 4.6, and Scott then exploited the global approach of taking an ultrapower of V itself. As with inner models the formalization of such class concepts requires care:

We first develop the model-theoretic concept of elementary embedding for structures with possibly proper class domains, remembering that such structures can be recast as single classes through definability. To review, for structures $\mathcal{M}_0 = \langle M_0, \ldots \rangle$ and $\mathcal{M}_1 = \langle M_1, \ldots \rangle$ for a language $\mathcal{L}$ an injective function $j\colon M_0 \to M_1$ is an *elementary embedding* of $\mathcal{M}_0$ into $\mathcal{M}_1$, denoted $j\colon \mathcal{M}_0 \prec \mathcal{M}_1$, *iff* it satisfies the *elementarity schema*: for any formula $\varphi(v_1, \ldots, v_n)$ of $\mathcal{L}$ and $x_1, \ldots, x_n \in M_0$,

$$(*) \qquad \mathcal{M}_0 \models \varphi[x_1, \ldots, x_n] \textit{ iff } \mathcal{M}_1 \models \varphi[j(x_1), \ldots, j(x_n)]\,.$$

If j is also the identity map on M_0, then $\mathcal{M}_0$ is an *elementary substructure* of $\mathcal{M}_1$, denoted $\mathcal{M}_0 \prec \mathcal{M}_1$. Most structures to be considered will be for some extension of $\mathcal{L}_\in$, the language of set theory. A further specialization is to $\in$-models, structures of form $\mathcal{M} = \langle M, \in \rangle$, usually denoted by just M. If $\mathcal{M}_0$ and $\mathcal{M}_1$ above are both of this form, $(*)$ can be expressed in terms of relativized formulas of $\mathcal{L}_\in$:

$$(**) \qquad \varphi^{M_0}[x_1, \ldots, x_n] \text{ iff } \varphi^{M_1}[j(x_1), \ldots, j(x_n)] .$$

Similarly, if $t(v_1, \ldots, v_n)$ is a definable term of $\mathcal{L}_\in$, $x_1, \ldots, x_n \in M_0$, and $t^{M_0}[x_1, \ldots, x_n] \in M_0$, then

$$j(t^{M_0}[x_1, \ldots, x_n]) = t^{M_1}[j(x_1), \ldots, j(x_n)] .$$

All this is formalizable for classes in terms of the satisfaction relation for classes, definable in a theory like Morse-Kelley, but not in ZFC because of the Gödel-Tarski undefinability of truth. However, we will be dealing primarily with set structures for which the formalization can be carried out, or inner models, for which a satisfactory approach to formalization is available:

Let j be as specified before $(*)$ with $\mathcal{L}$ an extension of $\mathcal{L}_\in$. Referring to the Levy hierarchy of formulas (§0),

j is a Σ_n*-elementary embedding*, denoted $j\colon \mathcal{M}_0 \prec_n \mathcal{M}_1$, *iff* $(*)$ holds for the Σ_n formulas .

If j is also the identity map on M_0, then $\mathcal{M}_0$ is a Σ_n*-elementary substructure* of $\mathcal{M}_1$, denoted $\mathcal{M}_0 \prec_n \mathcal{M}_1$. In particular, for any class M, that $M \prec_n V$ (understood as $\langle M, \in\rangle \prec_n \langle V, \in\rangle$) is just the assertion that the Σ_n formulas are absolute for M. Note that $j\colon \mathcal{M}_0 \prec_n \mathcal{M}_1$ implies that $(*)$ also holds for the Π_n formulas, and in fact for all Boolean combinations of Σ_n formulas. Also, if $\mathcal{M}_0 = \langle M_0, \in\rangle$, $\mathcal{M}_1 = \langle M_1, \in\rangle$, $M_0 \subseteq M_1$, and both are transitive, then $\mathcal{M}_0 \prec_0 \mathcal{M}_1$ since Σ_0 formulas are absolute for transitive classes.

For any particular integer n, the satisfaction relation for Σ_n formulas for a transitive proper class structure is formalizable in ZF (§0). Thus, the concept of Σ_n-elementary embedding *is* formalizable in ZF even for such classes. Focusing on inner models, some simple observations can be made using 0.2, according to which $\ulcorner\text{rank}(v_0) = v_1\urcorner$ is Δ_1^{ZF} and $\ulcorner V_{v_0} = v_1\urcorner$ is Π_1^{ZF}.

5.1 Proposition. *Suppose that M_0 and M_1 are inner models and $j\colon M_0 \prec_1 M_1$. Then:*

(a) For any ordinal α, $j(\alpha)$ is an ordinal $\geq \alpha$.

(b) Suppose that j is not the identity, and either $M_1 \subseteq M_0$ or $M_0 \models$ AC. Then $j(\delta) > \delta$ for some ordinal δ.

(c) For any particular integer n, $j\colon M_0 \prec_n M_1$.

Proof. (a) Being an ordinal is Σ_0^{ZF}: transitive and linearly ordered by $\in$. That $j(\alpha) \geq \alpha$ follows by induction.

(b) Suppose first that $M_1 \subseteq M_0$. Let x be of least rank such that $j(x) \neq x$, and set $\delta = \text{rank}(x)$. Note that $x \subseteq j(x)$, since $y \in x$ implies that $y = j(y) \in j(x)$. Hence, there must be a $z \in j(x) - x$. If $\text{rank}(j(x)) = \delta$, then since $z \in M_1 \subseteq M_0$, j can be applied to get $j(z) = z \in j(x)$, leading to the contradictory $z \in x$. Hence, $\text{rank}(j(x)) > \delta$, but also $\text{rank}(j(x)) = j(\delta)$ by 0.2.

Suppose next that $M_0 \models$ AC. Assume to the contrary that $j(\delta) = \delta$ for every ordinal δ. Note first that for any set $x \in M_0$ of *ordinals*, $j(x)$ is also a set of ordinals, and so the argument of the previous paragraph shows that $j(x) = x$. How transitive $\in$-models of ZFC are determined by their sets of ordinals is now exploited:

Suppose that $a \in M_0$. Let $b = \mathrm{tc}(\{a\})$, the transitive closure of $\{a\}$ (in V and hence also in M_0). Since $M_0 \models \mathrm{AC}$, there is a bijection $e \in M_0$ between some ordinal γ and b. Let $E \in M_0$ be the binary relation on γ defined by: $\langle \alpha, \beta \rangle \in E$ *iff* $e(\alpha) \in e(\beta)$. Identifying E with a set of ordinals through a pairing function, it follows from the previous paragraph that $j(E) = E$. Now the function e verifies that E is a well-founded relation in V as well as in M_0. Since every non-empty subset of γ has an E-minimal element in V, this also holds in M_1, i.e. E is well-founded in M_1 as well. The Collapsing Lemma 0.4 can now be applied in M_1 as $\langle \gamma, E \rangle$ is also extensional there, and by uniqueness the transitive collapse is $\langle b, \in \rangle \in M_1$. Since j is Σ_1-elementary, $j(E) = E$, and $\langle b, \in \rangle$ is the unique transitive collapse of $\langle \gamma, E \rangle$ in both M_0 and M_1, it follows that $j(b) = b$. Using 0.2 it then follows that $j(a) = a$, since a is definable as that member of b of largest rank. But $a \in M_0$ was arbitrary, and so j is the identity – a contradiction.

(c) The proof proceeds by informal induction up to any particular integer: Assume that $j\colon M_0 \prec_n M_1$ with $n \geq 1$, and suppose that φ is Σ_{n+1}, say $\exists v_0 \psi(v_0, v_1)$ where ψ is Π_n for simplicity, and that $x \in M_0$. If $M_0 \models \varphi[x]$, then $M_0 \models \psi[y, x]$ for some $y \in M_0$. Hence $M_1 \models \psi[j(y), j(x)]$ and so $M_1 \models \varphi[j(x)]$.

For the converse, suppose that $M_1 \models \varphi[j(x)]$, say $M_1 \models \psi[z, j(x)]$. By (a) there must be an α such that $z \in V^{M_1}_{j(\alpha)}$, and by 0.2, $V^{M_1}_{j(\alpha)} = j(V^{M_0}_\alpha)$. Hence,

$$M_1 \models \exists v_0 \in v_2 \psi[j(x), j(V^{M_0}_\alpha)] .$$

($\exists v_0 \in v_2 \psi$ is a formula in two free variables v_1 and v_2, interpreted respectively by $j(x)$ and $j(V^{M_0}_\alpha)$.) But as observed in §0, $\exists v_0 \in v_2 \psi$ is still Π^{ZF}_n by applications of Replacement. Hence,

$$M_0 \models \exists v_0 \in v_2 \psi[x, V^{M_0}_\alpha] ,$$

and so $M_0 \models \varphi[x]$. ⊣

(c) shows that for inner models Σ_1-elementary embeddings provide an adequate formalization in ZF of the informal concept of elementary embedding, since in any particular argument only finitely many instances of the elementarity schema are needed. The argument for (c) works under rather general conditions on arbitrary models of ZF as long as j“On^{M_0} is cofinal in On^{M_1}; see Gaifman [74], which provides another approach to formalization in many cases.

Henceforth, with (c) in mind but anticipating more general situations,

> by *elementary embedding* between transitive proper class $\in$-structures is meant Σ_1-elementary embedding .

Also, to avoid trivialities it will be implicit that

> elementary embeddings are not the identity on their domains .

The least ordinal as in 5.1(b) will thus play a critical role: For any $j\colon N \to M$,

$\delta = \text{crit}(j)$, the *critical point* of j, *iff* δ is the least ordinal moved by j .

Generally, elementary embeddings will be handled informally, with remarks on formalizability in ZFC made only when it may be problematic. Such a circumstance involves quantification over proper class elementary embeddings. However, in each particular case the putative quantification will be one that can be recast as one over sets. Formalizability is usually a by-product of the developing theory, and elementary embedding formulations, particularly of strong hypotheses, are often more perspicuous.

Taking the ultrapower of the universe V by an ultrafilter U over a set S is our next concern. The general ultraproduct construction was reviewed in §0 and applied in §4. The first obstacle in formally defining an ultrapower of V is that for any $f\colon S \to V$, its U equivalence class

$$(f)_U = \{g \mid g\colon S \to V \ \wedge \ \{i \in S \mid f(i) = g(i)\} \in U\}$$

is a proper class. However, using a device from Scott [55] let

$$(f)_U^0 = \{g \mid g \in (f)_U \wedge \forall h(h \in (f)_U \to \ \text{rank}(g) \leq \text{rank}(h))\} ,$$

i.e. those members of $(f)_U$ of minimal rank. Then $(f)_U^0$ is a set, and so the domain of the ultrapower can be formulated as the class

$${}^S V/U = \{(f)_U^0 \mid f\colon S \to V\}$$

definable from U, with its membership relation E_U defined as expected:

$$(f)_U^0 \ E_U \ (g)_U^0 \ \textit{ iff } \ \{i \in S \mid f(i) \in g(i)\} \in U \ .$$

The ultrapower can then be defined by:

$$\text{Ult}(V, U) = \langle {}^S V/U, E_U \rangle \ .$$

More generally, if N is an inner model of ZFC and $\langle N, \in, U \rangle \models U$ is an ultrafilter over a set S, an analogous procedure leads to the ultrapower of N by U using functions in ${}^S N \cap N$:

$$\text{Ult}(N, U) = \langle {}^S N/U, E_U \rangle \ .$$

It is not necessary to assume that $U \in N$.

Continuing in terms of V, the needed version of Łoś's fundamental result can be established by the usual induction on formula complexity:

5.2 Theorem. *For any formula $\varphi(v_1, \ldots, v_n)$ of $\mathcal{L}_\in$ and $f_1, \ldots, f_n$ functions: $S \to V$,*

$$\text{Ult}(V, U) \models \varphi[(f_1)_U^0, \ldots, (f_n)_U^0] \ \textit{ iff } \ \{i \in S \mid \varphi[f_1(i), \ldots, f_n(i)]\} \in U \ . \ \dashv$$

As this depends on the satisfaction relation for proper classes, it cannot be formulated as a single theorem of ZFC but only as a schema of theorems, one for each particular φ.

Finally, a strong hypothesis on U leads to inner models corresponding to ultrapowers:

5.3 Proposition. *U is ω_1-complete iff E_U is well-founded.*

Proof. In the forward direction, if there were a sequence $\langle (f_n)^0_U \mid n \in \omega \rangle$ such that $(f_{n+1})^0_U \; E_U \; (f_n)^0_U$ for each $n \in \omega$, then $\bigcap_n \{i \in S \mid f_{n+1}(i) \in f_n(i)\} \neq \emptyset$ would lead to an infinite descending $\in$-sequence of sets.

Conversely, if there were sets $\{X_n \mid n \in \omega\} \subseteq U$ yet $\bigcap_{n \in \omega} X_n \notin U$, then define $g_k \colon S \to V$ for each $k \in \omega$ by:

$$g_k(i) = \begin{cases} n - k & \text{if } i \in (\bigcap_{m<n} X_m) - X_n \text{ and } n \geq k \,, \\ 0 & \text{otherwise} \,. \end{cases}$$

Then $\{i \in S \mid g_{k+1}(i) \in g_k(i)\} \supseteq \bigcap_{m \leq k} X_m - \bigcap_{n \in \omega} X_n \in U$ for $k \in \omega$, and so $\langle (g_n)^0_U \mid n \in \omega \rangle$ confirms that E_U is ill-founded. $\dashv$

In connection with the Collapsing Lemma 0.4 note that E_U is set-like: Suppose that $(g)^0_U \; E_U \; (f)^0_U$, and let $g_0 \in (g)^0_U$. Define $g_1 \colon S \to V$ by

$$g_1(i) = \begin{cases} g_0(i) & \text{if } g_0(i) \in f(i) \,, \\ 0 & \text{otherwise} \,. \end{cases}$$

Then $g_1 \in (g)^0_U$ and $\text{rank}(g_1) \leq \text{rank}(f)$. Hence, $\text{rank}((g)^0_U) \leq \text{rank}(f) + 1$, and so $\{(g)^0_U \mid (g)^0_U \; E_U \; (f)^0_U\} \subseteq V_{\text{rank}(f)+2}$ and is therefore a set.

If U is ω_1-complete, it thus follows from 5.3 and 0.4 that there is a transitive class M_U and an isomorphism

$$\pi_U \colon \text{Ult}(V, U) \to \langle M_U, \in \rangle \,.$$

It is immediate from Łoś's theorem 5.2 that *M_U is an inner model of* ZFC. Henceforth to be used is the preferred notation

$$[f]_U = \pi_U((f)^0_U) \quad \text{for } f \colon S \to V \,,$$

emphasizing that f represents the set $[f]_U$ in the inner model M_U. That f need only be defined on a set $X \subseteq S$ in U will often be used without further mention. Finally, for any set x let f_x be the constant function: $S \to \{x\}$. Then setting

$$j_U(x) = [f_x]_U \quad \text{for } x \in V \,,$$

$j_U \colon V \prec M_U$ by 5.2, the usual canonical embedding into the ultrapower modulated by its transitive collapse. The situation will usually be summed up by

$$j_U \colon V \prec M_U \cong \text{Ult}(V, U) \,.$$

More generally, if N is an inner model of ZFC, $\langle N, \in, U\rangle \models U$ is an ultrafilter over a set S, and U is ω_1-complete (in V), then an analogous procedure using functions in ${}^S N \cap N$ leads to

$$j_U\colon N \prec M_U \cong \mathrm{Ult}(N, U)\ .$$

Again, it is not necessary to assume that $U \in N$. Subscripts will usually be suppressed when clear from the context.

With these preliminaries, we can turn at last to measurable cardinals:

5.4 Proposition. *Suppose that U is a κ-complete ultrafilter over a measurable cardinal κ with corresponding embedding $j\colon V \prec M \cong \mathrm{Ult}(V, U)$. Then* $\mathrm{crit}(j) = \kappa$.

Proof. First observe that $j(\alpha) = \alpha$ for every $\alpha < \kappa$: If not, let $\alpha < \kappa$ be the least such that $j(\alpha) > \alpha$. If $[f] = \alpha$, then $\{\xi < \kappa \mid f(\xi) < \alpha\} \in U$, and so by κ-completeness there is a $\beta < \alpha$ such that $\{\xi < \kappa \mid f(\xi) = \beta\} \in U$. But then, $[f] = j(\beta) = \beta$ is a contradiction.

Next, since U is κ-complete, it contains no bounded subsets of κ. So, with id: $\kappa \to \kappa$ the identity map on κ, for any $\alpha < \kappa$, $\{\xi \mid \alpha < \xi < \kappa\} \in U$ implies that $\alpha = j(\alpha) < [\mathrm{id}] < j(\kappa)$. Hence, $\kappa \le [\mathrm{id}] < j(\kappa)$, and $\kappa = \mathrm{crit}(j)$. ⊣

Here is Scott's original argument, which builds on the Keisler [62] ultraproduct proof of the Hanf-Tarski result.

5.5 Corollary (Scott [61]). *If there is a measurable cardinal, then $V \neq L$.*

Proof. Building on 5.4 suppose that κ is the least measurable cardinal, and assume to the contrary that $V = L$. Then $M \subseteq L$, so $M = L$. But also by elementarity, $(j(\kappa)$ is the least measurable cardinal$)^M$, contradicting $j(\kappa) > \kappa$. ⊣

The following result is a converse to 5.4, and was first noticed in a local form.

5.6 Theorem (Keisler [62a]). *If there is an elementary embedding $j\colon V \prec M$ for some inner model M, then its critical point is a measurable cardinal.*

Proof. Set $\delta = \mathrm{crit}(j)$. Then $\delta > \omega$ since every ordinal $\le \omega$ is definable and j is elementary. Now define U by:

$$X \in U \quad \textit{iff} \quad X \subseteq \delta \ \wedge\ \delta \in j(X)\ .$$

The proof is completed by showing that U is a (non-principal) δ-complete ultrafilter over δ. (This in particular implies that δ is regular and hence a cardinal.) Note first that $\delta \in U$, yet $\alpha < \delta$ implies that $\{\alpha\} \notin U$ since $j(\{\alpha\}) = \{\alpha\}$. To check that U is an ultrafilter is straightforward. Finally, to verify δ-completeness, suppose that $\gamma < \delta$ and $\mathcal{X} \in {}^{\gamma}U$. Then $\delta \in \bigcap\{j(\mathcal{X}(\alpha)) \mid \alpha < \gamma\}$. But since $j(\alpha) = \alpha$ for $\alpha \leq \gamma$, $j(\mathcal{X})$ is a function with domain γ such that $j(\mathcal{X})(\alpha) = j(\mathcal{X}(\alpha))$ for every $\alpha < \gamma$. Hence,

$$j(\bigcap\{\mathcal{X}(\alpha) \mid \alpha < \gamma\}) = \bigcap\{j(\mathcal{X})(\alpha) \mid \alpha < \gamma\} = \bigcap\{j(\mathcal{X}(\alpha)) \mid \alpha < \gamma\},$$

and so $\bigcap\{\mathcal{X}(\alpha) \mid \alpha < \gamma\} \in U$. ⊣

Since the quantifier $\exists j$ over classes j cannot be formalized in ZFC, 5.6 can only be regarded as a schema of theorems, one for each j, or else be formalized with V replaced by a V_α with $\alpha > \mathrm{crit}(j)$ as indeed was Keisler's approach. In any case, measurable cardinals exist exactly when there are non-trivial elementary embeddings of (initial segments of) the universe. Consequently, if the investigation of well-founded models of set theory is to move beyond restrictions via inner models and extensions via forcing and to comprehend elementary embeddings, then measurable cardinals become intrinsically necessary.

The proof of 5.6 shows how critical points of elementary embeddings can be used like principal generators for ultrafilters, which in turn lead to ultrapowers and elementary embeddings. Not every elementary embedding is an ultrapower embedding (as will become clear in subsequent sections), and the switch to ultrapowers has definite advantages since the concrete structure yields more information:

5.7 Proposition. *Suppose that U is a κ-complete ultrafilter over $\kappa > \omega$ and $j\colon V \prec M \cong \mathrm{Ult}(V, U)$ the corresponding embedding. Then:*

(a) $j(x) = x$ for every $x \in V_\kappa$, and so $V_\kappa^M = V_\kappa$; $j(X) \cap V_\kappa = X$ for every $X \subseteq V_\kappa$, and so $V_{\kappa+1}^M = V_{\kappa+1}$; and $\kappa^{+M} = \kappa^+$.

(b) $2^\kappa \leq (2^\kappa)^M < j(\kappa) < (2^\kappa)^+$.

(c) If θ is a strong limit cardinal of cofinality $\neq \kappa$, then $j(\theta) = \theta$.

(d) ${}^\kappa M \subseteq M$ yet ${}^{\kappa^+}M \nsubseteq M$, i.e. M is closed under the taking of arbitrary κ-sequences, but not of arbitrary κ^+-sequences.

(e) $U \notin M$.

Proof. (a) The first assertion follows from 5.4 and the rank argument for 5.1(b); the rest follow in sequence, with $\kappa^{+M} = \kappa^+$ a consequence of M containing every well-ordering of κ.

(b) $2^\kappa \leq (2^\kappa)^M$ since $\mathcal{P}(\kappa)^M = \mathcal{P}(\kappa)$ by (a) and $M \subseteq V$. $(2^\kappa)^M < j(\kappa)$ since $j(\kappa)$ is inaccessible in M. Finally, $j(\kappa) = \{[f] \mid f \in {}^\kappa\kappa\}$ so that $j(\kappa) < (2^\kappa)^+$.

(c) It suffices to take $\theta > \kappa$, suppose that $[f] < j(\theta)$, and then show that $[f] < \theta$: We can assume that $f(\xi) < \theta$ for every $\xi < \kappa$, and so since $\mathrm{cf}(\theta) \neq \kappa$, there is an $\alpha < \theta$ such that $\{\xi < \kappa \mid f(\xi) \leq \alpha\} \in U$. (If $\mathrm{cf}(\theta) < \kappa$, this follows from κ-completeness, and if $\mathrm{cf}(\theta) > \kappa$, we could take $\alpha = \sup(\mathrm{ran}(f))$.) Thus,

$$[f] \leq j(\alpha) = \{[g] \mid g \in {}^\kappa\alpha\} < \theta$$

as θ is a strong limit.

(d) Suppose that $\{[f_\alpha] \mid \alpha < \kappa\} \subseteq M$; a $g: \kappa \to V$ must be found so that $[g] = \langle [f_\alpha] \mid \alpha < \kappa \rangle$. Let $h: \kappa \to \kappa$ be so that $[h] = \kappa$. For each $\xi < \kappa$, let $g(\xi)$ be that function with domain $h(\xi)$ satisfying $(g(\xi))(\alpha) = f_\alpha(\xi)$. By the Łoś theorem 5.2, $[g]$ is a function with domain $[h] = \kappa$, and for each $\alpha < \kappa$, $[g](\alpha) = [f_\alpha]$.

The second assertion will be established by showing that $j``\kappa^+ \notin M$: $j``\kappa^+$ is cofinal in $j(\kappa^+)$, for if $[f] < j(\kappa^+)$, we can assume that $f(\xi) < \kappa^+$ for every $\xi < \kappa$, take $\alpha = \sup(\operatorname{ran}(f)) < \kappa^+$, and see that $[f] < j(\alpha)$. But $j``\kappa^+$ has ordertype $\kappa^+$, which by (b) is less than $j(\kappa^+)$, and so $j``\kappa^+ \in M$ would contradict $M \models j(\kappa^+)$ is regular.

(e) ${}^\kappa\kappa = ({}^\kappa\kappa)^M \in M$ by (a). If $U \in M$, then the map sending $f \in {}^\kappa\kappa$ to $[f]$ would also be in M. But then, $j(\kappa) < (2^\kappa)^{+M}$ by the argument for (b), contradicting the inaccessibility of $j(\kappa)$ in M. ⊣

The point of (c) is that there is a definable, proper class of ordinals fixed by j. (b) or (e) imply that $M \neq V$, so that there can be no ultrapower embedding: $V \prec V$; In fact, there can be no embedding: $V \prec V$ of any sort (23.12), and this was to be a watershed for the overall theory. That $V^M_{\kappa+1} = V_{\kappa+1}$ leads to a striking observation about how the measurability of κ controls the size of 2^κ:

5.8 Corollary (Scott). *If κ is measurable and $2^\alpha = \alpha^+$ for every $\alpha < \kappa$, then $2^\kappa = \kappa^+$.*

Proof. With $j: V \prec M \cong \mathrm{Ult}(V, U)$ as before, $\kappa < j(\kappa)$ and elementarity implies that $(2^\kappa)^M = \kappa^{+M}$. But then, $2^\kappa \leq (2^\kappa)^M = \kappa^{+M} = \kappa^+$. ⊣

Similarly, *if κ is measurable and $2^\alpha \leq \alpha^{++}$ for every $\alpha < \kappa$, then $2^\kappa \leq \kappa^{++}$*, and so forth. Actually getting a measurable cardinal κ satisfying $2^\kappa > \kappa^+$ turned out to require strong hypotheses, and the investigation of this possibility was to lead to the development of important new forcing techniques (see volume II).

The Czech mathematician Petr Vopěnka [62] independently derived Scott's result 5.5 from a difference in cardinal arithmetic between V and M as for 5.7(b), since $j(\kappa)$ is inaccessible in M. Vopěnka and his student Karel Hrbáček then established a global generalization of Scott's result for the inner models $L(A)$ defined in §3:

5.9 Theorem (Vopěnka-Hrbáček [66]). *If there is a strongly compact cardinal, then $V \neq L(A)$ for any set A.*

Proof. Suppose that κ is strongly compact, and assume to the contrary that $V = L(A)$ for some set A, which we can take to be transitive. Set $\lambda = \max(\{\kappa, |A|\})^+$. By 4.1, the κ-complete filter generated by $\{\{\xi \mid \alpha \leq \xi < \lambda\} \mid \alpha < \lambda\}$ can be extended to a κ-complete ultrafilter U over λ. Let

$$j: V \prec M \cong \mathrm{Ult}(V, U) \quad \text{with } M = \{[f] \mid f: \lambda \to V\}$$

as usual.

Now let $\mathrm{Ult}^-(V, U)$ be that substructure of $\mathrm{Ult}(V, U)$ with domain consisting only of those equivalence classes containing functions $f\colon \lambda \to V$ such that $|\mathrm{ran}(f)| < \lambda$. Łoś's Theorem also holds for $\mathrm{Ult}^-(V, U)$: For the induction step to the existential quantifier, note that if

$$\{\alpha < \lambda \mid \exists v_{n+1}\varphi[f_1(\alpha), \ldots, f_n(\alpha)]\} \in U ,$$

where each $|\mathrm{ran}(f_i)| < \lambda$, then there are less than λ n-tuples $\langle f_1(\alpha), \ldots, f_n(\alpha)\rangle$ involved, and so there is a $g\colon \lambda \to V$ with $|\mathrm{ran}(g)| < \lambda$ such that

$$\{\alpha < \lambda \mid \varphi[f_1(\alpha), \ldots, f_n(\alpha), g(\alpha)]\} \in U .$$

$\mathrm{Ult}^-(V, U)$ is well-founded, so let N be its transitive collapse and

$$k\colon V \prec N \cong \mathrm{Ult}^-(V, U) .$$

For $f\colon \lambda \to V$ with $|\mathrm{ran}(f)| < \lambda$ let $[f]^-$ denote that element of N corresponding to the equivalence class of f in $\mathrm{Ult}^-(V, U)$. Next, let $i\colon N \to M$ be defined by $i([f]^-) = [f]$; it is readily seen that this definition does not depend on the choice of f representing $[f]^-$. Then:

(i) i is elementary, and $j = i \circ k$.
(ii) $i(\alpha) = \alpha$ for every $\alpha < k(\lambda)$, and $i(k(A)) = k(A)$.
(iii) $k(\lambda) = \sup(\{k(\alpha) \mid \alpha < \lambda\}) \le [\mathrm{id}] < j(\lambda)$, where id: $\lambda \to \lambda$ is the identity map on λ .

For (ii) it can in fact be proved by induction on $\in$ that $[f]^- = [f]$ for any f such that the transitive closure of $\mathrm{ran}(f)$ has cardinality less than λ. For (iii) the equality follows from the observation that if $\{\xi < \lambda \mid g(\xi) < \lambda\} \in U$ and $|\mathrm{ran}(g)| < \lambda$, then $\{\xi < \lambda \mid g(\xi) < \alpha\} \in U$ for some $\alpha < \lambda$ by the regularity of λ; the rest follows as for 5.4.

A contradiction can now be derived as follows: Since V is the class $L(A)$ definable from A and k is elementary, N is the class $L(k(A))^N$ definable from $k(A)$. It follows from 3.2(a) that $N = L(k(A))^N = L(k(A))$. Similarly, since i is elementary and $i(k(A)) = k(A)$, $M = L(k(A)) = N$. Now λ is a successor cardinal, say of λ_0. So in M, $j(\lambda)$ is the successor of $j(\lambda_0)$ and in N, $k(\lambda)$ is the successor of $k(\lambda_0)$. But $k(\lambda_0) = i(k(\lambda_0)) = j(\lambda_0)$ by (ii), so $M = N$ implies that $k(\lambda) = j(\lambda)$. This contradicts (iii). ⊣

Normality

Scott devised a further means of extracting information from the ultrapower construction, particularly about reflection phenomena at measurable cardinals. The key concept that he isolated has a combinatorial formulation in terms of filters. For a filter F over λ,

F is *normal* *iff* for any $\langle X_\alpha \mid \alpha < \lambda\rangle \in {}^\lambda F$ its diagonal intersection $\Delta_{\alpha<\lambda} X_\alpha = \{\xi < \lambda \mid \xi \in \bigcap_{\alpha<\xi} X_\alpha\} \in F$.

A filter over a cardinal λ contains the final segments $\{\xi \mid \alpha \leq \xi < \lambda\}$ for every $\alpha < \lambda$ by a §0 convention. By setting $X_\alpha = \lambda$ for α sufficiently large it follows that *normality subsumes λ-completeness.* Also, the diagonal intersection of $\langle\{m \in \omega \mid n+1 < m\} \mid n \in \omega\rangle$ is empty. Hence, *if there is a normal filter over λ, then λ is regular and uncountable.* For such λ, the closed unbounded filter $\mathcal{C}_\lambda$ is normal (0.1(b)). Oswald Veblen [08] first took the diagonal intersection of closed unbounded sets, although he did so only in the special case of $\langle C_\alpha \mid \alpha < \lambda\rangle$ where for each $0 < \alpha < \lambda$, $\xi \in C_\alpha$ *iff* ξ is a limit point of C_β for every $\beta < \alpha$. That Fodor's regressive function lemma 0.1(c) is just a dual formulation was soon realized as a general and useful fact about normal filters. For a filter F over a set S and $X \subseteq S$, X is *F-stationary iff* $X \cap Z \neq \emptyset$ for every $Z \in F$.

5.10 Exercise. *A filter F over λ is normal iff for any F-stationary X and regressive $f\colon X \to \lambda$ there is an $\alpha < \lambda$ such that $f^{-1}(\{\alpha\})$ is F-stationary in λ.*

Hint. In the forward direction, if for every $\alpha < \lambda$, $f^{-1}(\{\alpha\})$ is not F-stationary so that $\lambda - f^{-1}(\{\alpha\}) \in F$, then $X \cap \Delta_{\alpha<\lambda}(\lambda - f^{-1}(\{\alpha\})) = \emptyset$.

Conversely, if $\langle X_\alpha \mid \alpha < \lambda\rangle \in {}^\lambda F$ yet $\Delta_{\alpha<\lambda} X_\alpha \notin F$, then $\lambda - \Delta_{\alpha<\lambda} X_\alpha$ is F-stationary. Define f on this set by: $f(\xi) =$ least α such that $\xi \notin X_\alpha$. Then f is regressive, yet $f^{-1}(\{\alpha\}) \cap X_\alpha = \emptyset$ for every $\alpha < \lambda$. ⊣

It follows that *any normal filter over λ extends $\mathcal{C}_\lambda$:* If C is closed unbounded in λ, then $f(\xi) = \sup(C \cap \xi) < \xi$ whenever $\xi \notin C \cup \{0\}$, yet for each $\alpha < \lambda$, $\{\xi \mid f(\xi) = \alpha\}$ is a bounded subset of λ.

In the first result 5.4 about measurable cardinals and ultrapowers the identity map id: $\kappa \to \kappa$ had been used as a convenience. Scott saw that significant results can be derived by making it a focus of attention:

5.11 Exercise (Scott – Keisler-Tarski [64: 244]). *For a κ-complete ultrafilter U over $\kappa > \omega$ the following are equivalent:*

(a) U is normal.

(b) Whenever $f \in {}^\kappa\kappa$ and $\{\xi < \kappa \mid f(\xi) < \xi\} \in U$, there is an $\alpha < \kappa$ such that $\{\xi < \kappa \mid f(\xi) = \alpha\} \in U$.

(c) $[\mathrm{id}]_U = \kappa$, where id: $\kappa \to \kappa$ *is the identity map on κ.*

Hint. The equivalence of (a) and (b) follows from 5.10. ⊣

5.12 Exercise (Scott – Keisler-Tarski [64: 241]). *If κ is measurable, then there is a normal ultrafilter over κ.*

Hint. Suppose that U is any κ-complete ultrafilter over κ, and $f \in {}^\kappa\kappa$ satisfies $[f]_U = \kappa$. Then

$$\{X \subseteq \kappa \mid f^{-1}(X) \in U\}$$

is a normal ultrafilter over κ. ⊣

Thus, any κ-complete ultrafilter can be "normalized" by a simple projection process. The proof of 5.6 provided a canonical way of defining an ultrafilter from an elementary embedding; it is simple to check that *the ultrafilter U defined in the proof of 5.6 is normal.* Normality leads to simple but informative properties about the corresponding ultrapower that foreshadow later work:

5.13 Proposition. *Suppose that U is a normal ultrafilter over κ. Then:*

(a) $M_U = \{j_U(f)(\kappa) \mid f\colon \kappa \to V\}$.

(b) If U had been defined as for 5.6 from a $j\colon V \prec M$ with critical point κ, then there is an elementary embedding $k\colon M_U \prec M$ such that $k \circ j_U = j$. Moreover, if $j = j_N$ for some normal ultrafilter N over κ, then $U = N$ and so k is the identity.

Proof. (a) If $x \in M_U$, then $x = [f]_U$ for some $f\colon \kappa \to V$. But then, by 5.11(c), $j_U(f)(\kappa) = j_U(f)([\mathrm{id}]_U) = [f]_U$, since the last equality is just a schizophrenic translation of $\{\xi < \kappa \mid f(\xi) = f(\xi)\} \in U$.

(b) Define $k\colon M_U \to M$ by $k([f]_U) = j(f)(\kappa)$. Then for any formula $\varphi(v_1, \ldots, v_n)$,

$$M_U \models \varphi[[f_1]_U, \ldots, [f_n]_U] \;\; \textit{iff} \;\; \{\xi < \kappa \mid \varphi[f_1(\xi), \ldots, f_n(\xi)]\} \in U$$
$$\textit{iff} \;\; M \models \varphi[j(f_1)(\kappa), \ldots, j(f_n)(\kappa)]$$

by the definition of U. Hence, k is elementary, and that $k \circ j_U = j$ is clear.

If $j = j_N$ where N is a normal ultrafilter over κ, then for any $X \subseteq \kappa$,

$$X \in U \;\; \textit{iff} \;\; \kappa \in j_N(X) \;\; \textit{iff} \;\; \{\xi < \kappa \mid \xi \in X\} = X \in N$$

by the normality of N. ⊣

(a) asserts that M_U is a simple closure of the minimal information $\{j_U(f) \mid f\colon \kappa \to V\} \cup \{\kappa\}$, and (b), that M_U is minimal in a natural sense among inner models provided by elementary embeddings. From another vantage point, that $[\mathrm{id}]_U = \kappa$ leads to rather pathological consequences about U that belie the term normal. For example, since $V_{\kappa+1} \subseteq M_U$ by 5.7(a), $M_U \models \kappa = [\mathrm{id}]_U$ is inaccessible. Hence, the sparse set $\{\alpha < \kappa \mid \alpha \text{ is inaccessible}\}$ is in U even though ultrafilters intuitively consist of large sets. Moreover, as a model for arguments with weakly compact cardinals Keisler-Tarski [64] showed that normal ultrafilters are closed under Mahlo's operation $M(X) = \{\alpha \in X \mid X \cap \alpha \text{ is stationary in } \alpha\}$. A short model-theoretic proof is given here; the original proof was later adapted to a more general setting by Solovay (16.8(c)).

5.14 Proposition (Keisler-Tarski [64: 248]). *Suppose that U is a normal ultrafilter over κ. If $S \subseteq \kappa$ is stationary in κ, then*

$$\{\alpha < \kappa \mid S \cap \alpha \text{ is stationary in } \alpha\} \in U\,.$$

Hence, if $X \in U$, then $M(X) \in U$.

Proof. By 5.7(a), $j_U(S) \cap \kappa = S$ and so $j_U(S) \cap \kappa$ is stationary in κ in M_U. Since $[\mathrm{id}]_U = \kappa$, it follows that $\{\alpha < \kappa \mid S \cap \alpha \text{ is stationary in } \alpha\} \in U$. The second assertion follows since normal filters contain every closed unbounded set, and so any $X \in U$ is stationary. ⊣

More can be said outright using a further reflection argument:

5.15 Proposition. *If U is a normal ultrafilter over κ, then*

$$\{\alpha < \kappa \mid \alpha \text{ is weakly compact}\} \in U .$$

Proof. It suffices to show that κ is weakly compact in M_U. Since weak compactness is characterized by the Extension Property of 4.5, κ has this property, and it suffices to check that κ has this property in M_U. But if $R \subseteq V_\kappa$, there is a transitive $X \neq V_\kappa$ and an $S \subseteq X$ such that $\langle V_\kappa, \in, R\rangle \prec \langle X, \in, S\rangle$ exactly when there is such an X satisfying $|X| = \kappa$, by the Löwenheim-Skolem Theorem 0.5. Hence, there is such a structure $\langle X, \in, S\rangle$ coded as a subset of V_κ. Also, the appropriate satisfaction relation can be coded as a subset of V_κ. But $V_{\kappa+1} \subseteq M_U$, and so the proof is complete. ⊣

Hence, the weak compactness of a measurable cardinal κ reflects, and κ is the κth weakly compact cardinal. This was first established by Hanf and Scott (6.5). As expected, there is a limit to this sort of reflection:

5.16 Proposition (Solovay). *For any measurable cardinal κ, there is a normal ultrafilter U over κ such that*

$$\{\alpha < \kappa \mid \alpha \text{ is not measurable}\} \in U .$$

Proof. Proceeding by induction let κ be measurable, U a normal ultrafilter over κ, and set $A = \{\alpha < \kappa \mid \alpha \text{ is measurable}\}$. If $A \notin U$, then we are done. Otherwise, for each $\alpha \in A$ inductively let U_α be a normal ultrafilter over α such that $A \cap \alpha \notin U_\alpha$. Define W over κ by:

$$X \in W \textit{ iff } X \subseteq \kappa \wedge \{\alpha \in A \mid X \cap \alpha \in U_\alpha\} \in U .$$

It is simple to check that W is a normal ultrafilter over κ such that $A \notin W$. ⊣

Finally, Scott's original proof of 5.8, implicit in Hanf-Scott [61], is given for which a weaker hypothesis suffices:

5.17 Proposition (Scott). *If there is a normal ultrafilter U over κ such that $\{\alpha < \kappa \mid 2^\alpha = \alpha^+\} \in U$, then $2^\kappa = \kappa^+$.*

Proof. Since $[\mathrm{id}]_U = \kappa$, $(2^\kappa)^{M_U} = \kappa^{+M_U}$, and the argument can be completed as for 5.8. ⊣

This argument was first published in Vopěnka [65], which also noted that if $\{\alpha < \kappa \mid 2^\alpha \leq \alpha^{++}\} \in U$, then $2^\kappa \leq \kappa^{++}$, and so forth.

Combinatorial features of the use of normality eventually emerged in generalizations, and besides having combinatorial ramifications normality was to play a crucial role in the structural investigation of inner models of large cardinals (§§19, 20). All in all, it is remarkable that a combinatorial contingency noticed by Ulam in his salad days in Lwów should emerge so transfigured by the infusion of ultrapowers and normality.

6. Indescribability

The formalization of reflection properties was one of the early developments in the increasing reliance on model-theoretic approaches in set theory. After Richard Montague [55, 61] studied reflection phenomena and showed that ZFC is not finitely axiomatizable in a strong sense, Levy [60a] established their broader significance and the close involvement of the Mahlo hierarchy. Then Hanf-Scott [61] postulated reflection properties directly for structures $\langle V_\kappa, \in, R\rangle$ where $R \subseteq V_\kappa$ in terms of higher-order languages. They showed that this generalization in the direction of higher types rather than infinitely long formulas results in a schematic approach to comparing large cardinals by size, and moreover provides a nice characterization of weak compactness. The theory was further elaborated by Levy [71], and led in the 1970's to important analogues in the theory of admissible sets and inductive definitions.

Levy's initial observations lay the groundwork. The following lemma illustrates the contemporary importance of closed unbounded sets owing to the infusion of model-theoretic techniques:

6.1 Lemma. *Suppose that κ is inaccessible and $R \subseteq V_\kappa$. Then*

$$\{\alpha < \kappa \mid \langle V_\alpha, \in, R \cap V_\alpha\rangle \prec \langle V_\kappa, \in, R\rangle\}$$

is closed unbounded in κ.

Proof. That this set is closed is immediate. To show that it is unbounded, let $\alpha < \kappa$ be arbitrary. Define $\alpha_n < \kappa$ for $n \in \omega$ by recursion as follows: Set $\alpha_0 = \alpha$. Given $\alpha_n < \kappa$ define α_{n+1} to be the least $\beta \geq \alpha_n$ such that whenever $y_1, \ldots, y_k \in V_{\alpha_n}$ and $\langle V_\kappa, \in, R\rangle \models \exists v_0\varphi[y_1, \ldots, y_k]$ for some formula φ, there is an $x \in V_\beta$ such that $\langle V_\kappa, \in, R\rangle \models \varphi[x, y_1, \ldots, y_k]$. Since κ is inaccessible, $|V_{\alpha_n}| < \kappa$ and so $\alpha_{n+1} < \kappa$. Finally, set $\alpha = \sup(\{\alpha_n \mid n \in \omega\})$. Then $\langle V_\alpha, \in, R \cap V_\alpha\rangle \prec \langle V_\kappa, \in, R\rangle$ by the usual (Tarski) criterion for elementary substructure. ⊣

The first part of the next proposition recalls 1.3, and is really a synoptic version.

6.2 Proposition (Levy [60a]).

(a) κ is inaccessible iff for any $R \subseteq V_\kappa$ there is an $\alpha < \kappa$ such that $\langle V_\alpha, \in, R \cap V_\alpha\rangle \prec \langle V_\kappa, \in, R\rangle$.

(b) κ is Mahlo iff for any $R \subseteq V_\kappa$ there is an inaccessible cardinal $\alpha < \kappa$ such that $\langle V_\alpha, \in, R \cap V_\alpha\rangle \prec \langle V_\kappa, \in, R\rangle$.

Proof. The forward directions are immediate by 6.1. For the converses, it is simple to see that $\kappa > \omega$. Note that κ is regular: If not, there would be a $\beta < \kappa$ and a function $F\colon \beta \to \kappa$ with range unbounded in κ. Set $R = \{\beta\} \cup F$. By hypothesis there is an $\alpha < \kappa$ such that $\langle V_\alpha, \in, R \cap V_\alpha\rangle \prec \langle V_\kappa, \in, R\rangle$. Since β is the single ordinal in R, $\beta \in V_\alpha$ by elementarity. But this is a contradiction since the domain of $F \cap V_\alpha$ cannot be all of β.

Next, κ is a strong limit: If not, there would be a $\lambda < \kappa$ such that $2^\lambda \geq \kappa$. Let $G: \mathcal{P}(\lambda) \to \kappa$ be surjective and set $R = \{\lambda + 1\} \cup G$. By hypothesis, there is an $\alpha < \kappa$ such that $\langle V_\alpha, \in, R \cap V_\alpha \rangle \prec \langle V_\kappa, \in, R \rangle$. $\lambda + 1 \in V_\alpha$ and so $\mathcal{P}(\lambda) \in V_\alpha$, but this is a contradiction as before.

Finally, for (b) κ is Mahlo: If not, there would be a C closed unbounded in κ containing no inaccessible cardinals. By the hypothesis from (b) there is an inaccessible $\alpha < \kappa$ such that $\langle V_\alpha, \in, C \cap V_\alpha \rangle \prec \langle V_\kappa, \in, C \rangle$. By elementarity $C \cap \alpha$ is unbounded in α. But then, $\alpha \in C$, which is a contradiction. ⊣

The *Reflection Principle* for ZF of Montague [61] and Levy [60a] asserts that for any formula $\varphi(v_1, \ldots, v_n)$ and any β, there is a limit ordinal $\alpha > \beta$ such that for any $x_1, \ldots, x_n \in V_\alpha$,

$$\varphi[x_1, \ldots, x_n] \ \textit{iff} \ \varphi^{V_\alpha}[x_1, \ldots, x_n] .$$

This principle is implicit in remarks of Gödel made in 1946 (see Gödel [90: 146]). Classes R can be incorporated through their definitions to achieve a certain resemblance to the latter part of 6.2(a) with V replacing V_κ, but there are crucial differences: Only one formula can be reflected, not all formulas as with elementary substructures; indeed, the arithmetization of syntax cannot be completed in ZF to find an α such that $V_\alpha \prec V$ because of the undefinability of truth. Also, *arbitrary* $R \subseteq V_\kappa$ are allowed in 6.2(a); this incorporates a bit of second-order logic to attain the requisite strength (cf. 1.3 and the discussion before).

Levy [60a: 234] observed that the Reflection Principle is equivalent to the Axiom of Infinity and the Replacement Schema in the presence of the other ZF axioms. Asserting this principle within some V_κ with arbitrary $R \subseteq V_\kappa$ necessitates the inaccessibility of κ by the proof of 6.2(a). Next consider the reflection of V to a V_α for α inaccessible; asserting this new principle within some V_κ with arbitrary $R \subseteq V_\kappa$ necessitates the Mahloness of κ by the proof of 6.2(b), and so forth. This is reflecting reflection: a reflection scheme is first formulated and then is itself reflected! In this way Levy developed a hierarchy of set theories that correspond to the α-Mahlo cardinals and showed how the iterative formalization of reflection illuminates Mahlo's original scheme, formulated half a century before.

The formulations of Hanf and Scott using higher-order languages are the main topics of this section. The Π^m_n and Σ^m_n formulas for $m, n \in \omega$ were described in §0; for Q either Π^m_n or Σ^m_n,

κ *is Q-indescribable iff* for any $R \subseteq V_\kappa$ and Q sentence φ such that $\langle V_\kappa, \in, R \rangle \models \varphi$, there is an $\alpha < \kappa$ such that $\langle V_\alpha, \in, R \cap V_\alpha \rangle \models \varphi$.

Over structures of form $\langle V_\alpha, \in, R \rangle$ with α a limit and $R \subseteq V_\alpha$, every higher-order formula is equivalent by standard means to a Π^m_n or Σ^m_n formula for some m and

n with blocks of like quantifiers contracted into one. Including arbitrary $R \subseteq V_\kappa$ bolsters the theory with nice consequences, although the subject can be pursued without the R's (cf. 6.9). As V_κ is closed under pairing, the definition is equivalent to one where R is replaced by any finite number of finitary relations; this will often be invoked in what follows.

Extensions of the definition can be pursued using transfinite types, as was done by Jensen (see Drake [74: 284ff]). Although there may be a traditional bias for finite types in language, several indescribability results have sweeping proofs that accommodate stronger modes of expressibility. However, as for the Mahlo hierarchy transfinite types may lead to larger cardinals, but not to a qualitative transcendence. So, the following weak overall notion is adopted, keeping in mind that stronger indescribability properties are often derivable:

κ is *totally indescribable iff* κ is Π^m_n-indescribable for every $m, n \in \omega$.

Following upon Levy [60a], Bernays [61] postulated the full second-order reflection principle for the universe of sets, which amounts to ascribing Π^1_n-indescribability to On for every $n \in \omega$. Pursuing his second-order approach to set theory he showed that his principle implies the Global Axiom of Choice: There is a class function F such that for any non-empty set x, $F(x) \in x$. These investigations have been continued by Rolando Chuaqui [78, 81].

Turning to the subject at hand, some observations are first made for the case $m = 1$:

6.3 Proposition.
(a) For any n, κ is Σ^1_{n+1}-indescribable iff κ is Π^1_n-indescribable.
(b) κ is Σ^1_1-indescribable iff κ is inaccessible.

Proof. (a) In the substantive direction, suppose that $R \subseteq V_\kappa$ and $\langle V_\kappa, \in, R\rangle \models \exists X\varphi(X)$ say, where X is type 2 and φ is Π^1_n. So, there is an $S \subseteq V_\kappa$ such that $\langle V_\kappa, \in, R\rangle \models \varphi[S]$. But then, X can be replaced in φ by a new predicate symbol and Π^1_n-indescribability applied to $\langle V_\kappa, \in, R, S\rangle$.

(b) This follows by reducing to first-order formulas by (a) and using the arguments for 6.2(a). ⊣

What about Π^1_1-indescribability? Hanf and Scott found that it provides a nice characterization:

6.4 Theorem (Hanf-Scott [61]). *κ is Π^1_1-indescribable iff κ is weakly compact.*

Proof. We can argue in terms of the 4.5 characterization of weak compactness in terms of the Extension Property. For this purpose, suppose first that κ is inaccessible and $R \subseteq V_\kappa$. If there is a transitive $X \neq V_\kappa$ and an $S \subseteq X$ such that $\langle V_\kappa, \in, R\rangle \prec \langle X, \in, S\rangle$, then there is such an X satisfying $|X| = |V_\kappa|$ by the Löwenheim-Skolem Theorem 0.5. Using an injection of X into V_κ, this in turn amounts to asserting that there is an $A \subseteq V_\kappa$ coding a structure $\langle Y, E, T\rangle$ with E

a binary relation satisfying the hypotheses of the Collapsing Lemma 0.4, and a proper elementary embedding $J\colon \langle V_\kappa, \in, R\rangle \to \langle Y, E, T\rangle$. That E is well-founded is a first-order assertion in $\langle V_\kappa, \in, E\rangle$ saying that there are no infinite E-descending sequences, since ${}^\omega V_\kappa \subseteq V_\kappa$ by the inaccessibility of κ. That J is elementary is expressible in terms of the satisfaction relation, formalizable as usual through the existence of a satisfaction sequence K defined by recursion on formula complexity. Thus, when properly formalized the Extension Property holds for $R \subseteq V_\kappa$ exactly when $\langle V_\kappa, \in, R\rangle$ satisfies a Σ^1_1 sentence $\exists A \exists J \exists K \rho$ where ρ is first-order.

Suppose now that κ is Π^1_1-indescribable. Then κ is inaccessible by 6.3. Assume to the contrary that the Extension Property fails for some $R \subseteq V_\kappa$. Then $\langle V_\kappa, \in, R\rangle \models \sigma$ for a Π^1_1 sentence σ equivalent to the negation of the Σ^1_1 sentence above. It is simple to devise a Π^1_1 sentence τ such that $\langle V_\delta, \in\rangle \models \tau$ *iff* δ is inaccessible. Finally, by 6.1,

$$C = \{\beta < \kappa \mid \langle V_\beta, \in, R \cap V_\beta\rangle \prec \langle V_\kappa, \in, R\rangle\}$$

is closed unbounded in κ, and so $\langle V_\kappa, \in, R, C\rangle \models \psi$, where ψ is a first-order sentence asserting that C is closed unbounded.

By Π^1_1-indescribability there is an $\alpha < \kappa$ satisfying

$$\langle V_\alpha, \in, R \cap V_\alpha, C \cap V_\alpha\rangle \models \sigma \wedge \tau \wedge \psi$$

so that in particular α is inaccessible and $\alpha \in C$. By an iterated Skolem hull argument, there is an $\langle X_\alpha, \in, S_\alpha\rangle \prec \langle V_\kappa, \in, R\rangle$ such that: X_α is transitive, V_α is a proper subset of X_α, and $|X_\alpha| = |V_\alpha|$. Having arranged $\alpha \in C$, $\langle V_\alpha, \in, R \cap V_\alpha\rangle \prec \langle X_\alpha, \in, S_\alpha\rangle$. But this contradicts $\langle V_\alpha, \in, R \cap V_\alpha\rangle \models \sigma$.

For the converse, suppose that κ is weakly compact and $\langle V_\kappa, \in, R\rangle \models \varphi$ where φ is Π^1_1. By the Extension Property there is a transitive $X \neq V_\kappa$ and an $S \subseteq X$ such that $\langle V_\kappa, \in, R\rangle \prec \langle X, \in, S\rangle$. In particular, $\langle X, \in\rangle$ is a transitive model of ZFC with $\kappa \in X$, and so $V^X_\kappa = V_\kappa \in X$. Since the only type 2 quantifiers of φ are universal and persist downward when there are possibly fewer subsets of the domain,

$$\langle X, \in, S\rangle \models \exists\alpha(\langle V_\alpha, \in, S \cap V_\alpha\rangle \models \varphi)\ ,$$

since κ is such an α and $S \cap V_\kappa = R$. This sentence is first-order, so also holds in $\langle V_\kappa, \in, R\rangle$ by elementarity. Since κ is inaccessible, this means that there really is an $\alpha < \kappa$ such that $\langle V_\alpha, \in, R \cap V_\alpha\rangle \models \varphi$, and the proof is complete. ⊣

This is a subtle phenomenon: the equivalence of an existential *extension* property and a universal *reflection* property. Results like 4.7 on the size of weakly compact cardinals in terms of the Mahlo hierarchy can also be derived via Π^1_1-indescribability.

Finally, Hanf and Scott made a general statement about reflection at a measurable cardinal that subsumes 5.14 and 5.15:

6.5 Proposition (Hanf-Scott [61]). *Measurable cardinals are Π^2_1-indescribable. Moreover, if U is a normal ultrafilter over κ, $R \subseteq V_\kappa$, and $\langle V_\kappa, \in, R\rangle \models \varphi$ where φ is Π^2_1, then*

$$\{\alpha < \kappa \mid \langle V_\alpha, \in, R \cap V_\alpha\rangle \models \varphi\} \in U \ .$$

Proof. Let $j\colon V \prec M \cong \mathrm{Ult}(V, U)$. $V_{\kappa+1} \subseteq M$ by 5.7(a), so that second-order sentences about $\langle V_\kappa, \in\rangle$ are absolute between V and M. But since the only type 3 quantifiers of φ are universal and persist downward when there are possibly fewer subsets of $V_{\kappa+1}$, $(\langle V_\kappa, \in, R\rangle \models \varphi)^M$. The result now follows by normality since $j(R) \cap V_\kappa = R$. ⊣

Kunen (see Solovay [71] or Fremlin [93]) established a kind of Π^2_1-indescribability for real-valued measurable cardinals, which for instance has the following application: *If $\kappa \leq 2^{\aleph_0}$ is real-valued measurable, then there is a non-Lebesgue measurable set of reals of cardinality less than κ.* Also, Kunen [71] considered a weak version of indescribability that does not entail inaccessibility, and obtained results below $2^{\aleph_0}$. (The formulation of Hanf-Scott [61] had actually been in terms of the weak version, with the inaccessibility of κ an ambient hypothesis.)

There is a Σ^2_1 description over $\langle V_\kappa, \in\rangle$ of the measurability of κ, asserting $\exists U \in \mathcal{P}^2(\kappa)$ satisfying a second-order formula. 6.5 is hence the best possible in the sense that the least measurable cardinal is not Σ^2_1-indescribable. However:

6.6 Proposition (Vaught [63a]). *If κ is measurable and U is a normal ultrafilter over κ, then*

$$\{\alpha < \kappa \mid \alpha \text{ is totally indescribable}\} \in U \ .$$

Proof. Let $j\colon V \prec M \cong \mathrm{Ult}(V, U)$. By normality it suffices to show that $M \models$ κ is totally indescribable. The following argument imposes little restriction on the complexity of φ: Suppose that

$$M \models (R \subseteq V_\kappa \ \wedge \langle V_\kappa, \in, R\rangle \models \varphi) \ .$$

Then

$$M \models \exists\alpha < j(\kappa)(\langle V_\alpha, \in, j(R) \cap V_\alpha\rangle \models \varphi) \ ,$$

since κ is such an α and $j(R) \cap V_\kappa = R$. By elementarity,

$$\exists\alpha < \kappa(\langle V_\alpha, \in, R \cap V_\alpha\rangle \models \varphi) \ ,$$

and this then holds in M as well, since $V^M_\kappa = V_\kappa$. ⊣

Hence, there are many cardinals below the least measurable cardinal κ that may have stronger reflection properties than κ itself. On the other hand, unlike measurability Π^m_n-indescribability is compatible with $V = L$. The argument appeals to specific properties of L, unlike the relativization of α-inaccessible and α-Mahlo cardinals to inner models (3.1):

6.7 Theorem. *Suppose that Q is either Π^m_n or Σ^m_n with either $m > 1$ or else $m = 1$ and $n > 0$, and κ is Q-indescribable. Then* (κ is Q-indescribable)^L. *In particular, if κ is weakly compact, then* (κ is weakly compact)^L.

Proof. For only this proof let δ^+ denote the first cardinal greater than δ and $\delta^{+(i)}$ the ith cardinal greater than δ *in the sense of L*. It is first shown that for $0 < i < \omega$ and inaccessible λ, to each Π^r_s (resp. Σ^r_s) formula $\varphi(X_1, \ldots, X_t)$ for $s > 0$ a Π^{r+i}_s (resp. Σ^{r+i}_s) formula $\tilde{\varphi}(Y_1, \ldots, Y_t)$ can be associated where $\text{type}(Y_k) = \text{type}(X_k) + i$ for $1 \leq k \leq t$, and to each $A \in \mathcal{P}^j(L_{\lambda^{+(i)}})$ for $j \in \omega$ (allowing $\mathcal{P}^0(L_{\lambda^{+(i)}}) = L_{\lambda^{+(i)}}$) a set $\tilde{A} \in \mathcal{P}^{j+i}(V_\lambda) = V_{\lambda+j+i}$ can be associated so that

$$(*) \qquad L_{\lambda^{+(i)}} \models \varphi[A_1, \ldots, A_t] \;\textit{ iff }\; V_\lambda \models \tilde{\varphi}[\tilde{A}_1, \ldots, \tilde{A}_t]\,.$$

Although this will only be needed for $r = 0$ and φ a sentence, the complexity must be maintained for the inductive argument.

For the case $i = 1$, let σ_0 be the sentence of 3.3(a) so that for any transitive set model N of σ_0, $N = L_\alpha$ for some α. By the proof of GCH in L, if $A \in L_{\lambda^+}$, $A \in L_\gamma$ for some $\gamma < \lambda^+$. Since $|L_\gamma| \leq \lambda$, this amounts to the assertion: There is a $B \subseteq V_\lambda$ and E a binary relation on B satisfying the hypotheses of the Collapsing Lemma 0.4 such that $\langle B, E\rangle \models \sigma_0$, and for that map H inductively defined by $H(x) = \{H(y) \mid y \; E \; x\}$ as in that lemma's proof there is a $z \in B$ such that $H(z) = A$. Coding $\langle B, E, z\rangle$ as a subset $\tilde{A}$ of V_λ, this assertion is formalizable as a Σ^1_1 formula $\psi(Y)$ over V_λ with $\tilde{A}$ interpreting type 2 variable Y by the argument of the first paragraph of the proof of 6.4.

Proceed now by recursion to define $\tilde{\varphi}$ for first-order φ: For atomic formulas, if $A_1, A_2 \in L_{\lambda^+}$, both $A_1 \in A_2$ and $A_1 = A_2$ can be rendered as Σ^1_1 assertions over V_λ about $\tilde{A}_1$ and $\tilde{A}_2$ using the ψ above. If φ is $\exists X \varphi_0(X, X_1, \ldots, X_t)$ and $\tilde{\varphi}_0(Y, Y_1, \ldots, Y_t)$ corresponds to φ_0, then

$$\begin{aligned} L_{\lambda^+} \models \varphi[A_1, \ldots, A_t] \;&\textit{iff}\; L_{\lambda^+} \models \varphi_0[A, A_1, \ldots, A_t] \text{ for some } A \in L_{\lambda^+} \\ &\textit{iff}\; V_\lambda \models \tilde{\varphi}_0[\tilde{A}, \tilde{A}_1, \ldots, \tilde{A}_t] \text{ by induction} \\ &\textit{iff}\; V_\lambda \models \exists Y(\psi \wedge \tilde{\varphi}_0)[\tilde{A}_1, \ldots, \tilde{A}_t]\,. \end{aligned}$$

Hence, $\tilde{\varphi}$ can be taken to be $\exists Y(\psi \wedge \tilde{\varphi}_0)$, observing that if φ is Σ^0_s, then $\tilde{\varphi}$ is Σ^1_s by induction. The sentential connectives are immediate, and so the argument is complete for first-order φ. For higher-order φ, corresponding sets $\tilde{A}$ can be recursively associated with sets A in a straightforward manner. This completes the case $i = 1$.

For $i > 1$, proceed by induction: Using the $i = 1$ argument with $L_{\lambda^{+(i+1)}}$ replacing L_{λ^+} and $L_{\lambda^{+(i)}}$ replacing V_λ, first translate down from $L_{\lambda^{+(i+1)}}$ to $L_{\lambda^{+(i)}}$; then use the induction hypothesis to translate down to V_λ. This is where the general r is needed. The translation down from $L_{\lambda^{+(i+1)}}$ to $L_{\lambda^{+(i)}}$ involves one new difficulty: For the V_λ case, the first-order expressibility of the well-foundedness of E depended on ${}^\omega V_\lambda \subseteq V_\lambda$ as in the cited proof of 6.4. In the current situation, only ${}^{\omega}L_{\lambda^{+(i)}} \cap L \subseteq L_{\lambda^{+(i)}}$ is known from the proof of GCH in L. However, this suffices by a basic fact about inner models and well-foundedness: First, for $A \in L_{\lambda^{+(i+1)}}$, $A \in L_\gamma$ for some γ satisfying $|L_\gamma|^L \le \lambda^{+(i)}$, so that we can assume that the corresponding $E \in L_{\lambda^{+(i)}}$. Now if E is well-founded in L, then E is well-founded in V by absoluteness (0.3). Hence, it is enough to assert that there are no infinite E-descending sequences in L, and for this ${}^{\omega}L_{\lambda^{+(i)}} \cap L \subseteq L_{\lambda^{+(i)}}$ suffices.

Proceeding toward the main argument for the theorem, note that if λ is inaccessible, then it is inaccessible in L and so $(V_\lambda)^L = L_\lambda$ by 1.2(a) and induction on rank. For reflecting down to such a situation, note (as for 6.4) that there is a Π^1_1 sentence τ such that $V_\delta \models \tau$ *iff* δ is inaccessible.

Finally, suppose that κ is Π^m_n-indescribable with m, n as hypothesized. (The Σ^m_n case is analogous, with the special Σ^1_1 case following separately from 6.3(b).) To verify Π^m_n-indescribability in L, note first that by the inaccessibility of κ, $(V_\kappa)^L = L_\kappa$. So assume that

$$R \in \mathcal{P}(L_\kappa) \cap L \text{ and } (\langle L_\kappa, \in, R\rangle \models \varphi_0)^L$$

where φ_0 is Π^m_n. By the proof of GCH in L, $\mathcal{P}(L_\kappa) \cap L \subseteq L_{\kappa^+}$ and inductively $\mathcal{P}^i(L_\kappa) \cap L \subseteq L_{\kappa^{+(i)}}$ for each $i \in \omega$. Hence, our assertion can be rendered as

$$\langle L_{\kappa^{+(m)}}, \in, R\rangle \models \varphi_1 ,$$

where φ_1 is a Π^0_n sentence corresponding to φ_0. By previous remarks, this in turn translates to

$$\langle V_\kappa, \in, \tilde{R}\rangle \models \tilde{\varphi}_1$$

where $\tilde{\varphi}_1$ is Π^m_n. The result now follows by reflecting $\tilde{\varphi}_1 \wedge \tau$ down to an $\alpha < \kappa$ and translating backwards to get

$$(\langle L_\alpha, \in, R \cap L_\alpha\rangle \models \varphi_0)^L . \qquad \dashv$$

The following result on universal formulas is a routine application of the satisfaction relation. (See Levy [71:208], Drake [74:272], or Devlin [75:96ff] for details.)

6.8 Proposition. *For any $m, n > 0$ there is a Π^m_n formula $\psi_{mn}(X, Y)$ with X type 2 and Y type 1 such that: for any Π^m_n formula $\varphi(X)$ there is a $k \in \omega$ such that for any limit ordinal α and $R \subseteq V_\alpha$,*

$$\langle V_\alpha, \in\rangle \models \varphi[R] \text{ iff } \langle V_\alpha, \in\rangle \models \psi_{mn}[R, k] .$$

There is an Σ^m_n formula $\psi'_{mn}(X, Y)$ with the analogous property for Σ^m_n formulas $\varphi(X)$. $\dashv$

A shift from a unary predicate to a type 2 variable was made to get the following corollary:

6.9 Corollary.
(a) For any n there is a Π^1_{n+1} *sentence* χ_{1n} *such that for any* κ,

$$\langle V_\kappa, \in\rangle \models \chi_{1n} \text{ iff } \kappa \text{ is } \Pi^1_n\text{-indescribable} .$$

(By 6.3 this subsumes the Σ^1_n *cases.)*
(b) If $m > 1$, *for any* $n > 0$ *there is a* Π^m_n *sentence* χ_{mn} *such that for any* κ,

$$\langle V_\kappa, \in\rangle \models \chi_{mn} \text{ iff } \kappa \text{ is } \Sigma^m_n\text{-indescribable} .$$

There is a Σ^m_n *sentence that similarly characterizes* Π^m_n*-indescribability.*

Proof. (a) For $n = 0$, let χ_{10} be any Π^1_1 description of inaccessibility. For $n > 0$, in terms of ψ_{1n} from 6.8 let χ_{1n} be

$$\forall X \forall Y(\psi_{1n}(X, Y) \to \exists\alpha > 0(\alpha \text{ is a limit } \wedge \langle V_\alpha, \in\rangle \models \psi_{1n}(X \cap V_\alpha, Y))) ,$$

appropriately formalized with the satisfaction relation for sets. Because of the $\forall X$ and the occurrence of $\psi_{1n}(X, Y)$ to the left of $\to$ this is Π^1_{n+1}.

(b) This is like (a), except that for $m > 1$ the $\forall X$ can be subsumed in the prenexing procedure for classifying the resulting formula. ⊣

Set

$$\pi^m_n = \text{least } \Pi^m_n\text{-indescribable cardinal , and}$$
$$\sigma^m_n = \text{least } \Sigma^m_n\text{-indescribable cardinal ,}$$

with the assumption implicit in the use of this notation that such cardinals exist. The following is a consequence 6.3 and 6.9.

6.10 Proposition (Hanf-Scott [61]).
(a) σ^1_1 *is the least inaccessible cardinal.*
(b) $\pi^1_1 = \sigma^1_2 < \pi^1_2 = \sigma^1_3 < \ldots$
(c) For $m > 1$ *and* $n > 0$, $\sigma^m_n \neq \pi^m_n$, *and* $\pi^m_n < \sigma^m_{n+1}, \pi^m_{n+1}$. ⊣

The order relationship between σ^m_n and π^m_n for $m > 1$ and $n > 0$ is discussed at the end of the section.

Levy [71] carried out a systematic study of the sizes of indescribable cardinals, extending aspects of Keisler-Tarski [64]. Most notably, he vitalized the idea implicit in that paper of investigating definable filters, an idea to be used in later contexts, and considerably extended 6.10. The starting point of his approach is that various large cardinal properties are not only attributable to cardinals, but also to their subsets. For $X \subseteq \kappa$ and Q either Π^m_n or Σ^m_n,

X is Q-indescribable in κ *iff* for any $R \subseteq V_\kappa$ and Q sentence φ such that $\langle V_\kappa, \in, R\rangle \models \varphi$, there is an $\alpha \in X$ such that $\langle V_\alpha, \in, R \cap V_\alpha\rangle \models \varphi$.

This leads in turn to the consideration of the collections

$$\begin{aligned} I =&\{X \subseteq \kappa \mid X \text{ is not } Q\text{-indescribable in } \kappa\}\ , \\ &\{X \subseteq \kappa \mid X \text{ is } Q\text{-indescribable in } \kappa\}\ , \text{ and} \\ F =&\{X \subseteq \kappa \mid \kappa - X \text{ is not } Q\text{-indescribable in } \kappa\} \end{aligned}$$

of the negligible, non-negligible, and all but negligible subsets of κ with respect to Q-indescribability. It is simple to check that κ is Q-indescribable *iff* F is a (proper) filter, the *Q-indescribable filter over κ*. (F conforms to our §0 convention about containing all the final segments $\{\xi \mid \alpha < \xi < \kappa\}$, since the assertion in $\langle V_\kappa, \in, \{\alpha\}\rangle$ that $\{\alpha\}$ is not empty cannot be reflected down to any V_ξ with $\xi \leq \alpha$. Levy styled the members of F *weakly Q-enforceable at κ*, and did not refer explicitly to the filter.) I is then the ideal dual to F and the F-stationary sets are just the Q-indescribable in κ sets. The Π^1_1-indescribable filter is also known as the *weakly compact filter* because of 6.4. In the use of this terminology it is assumed that the filters are indeed proper, i.e. that the ambient κ has the requisite strength.

These definable filters have a crucial property:

6.11 Proposition (Levy [71]). *For $m, n > 0$ the Π^m_n-indescribable and Σ^m_n-indescribable filters over κ are normal.*

Proof. Let F be the Π^m_n-indescribable filter over κ (the Σ^m_n case is analogous). Suppose that $X \subseteq \kappa$ and $f\colon X \to \kappa$ is regressive. Assuming that $f^{-1}(\{\alpha\})$ is not Π^m_n-indescribable in κ for any $\alpha < \kappa$, it suffices to establish that X is not Π^m_n-indescribable in κ:

Invoking the universal formula of 6.8, it can be assumed that for each $\alpha < \kappa$ there is an $R_\alpha \subseteq V_\kappa$ and a $k_\alpha \in \omega$ such that $\langle V_\kappa, \in\rangle \models \psi_{mn}[R_\alpha, k_\alpha]$ yet $\langle V_\xi, \in\rangle \models \neg\psi_{mn}[R_\alpha \cap V_\xi, k_\alpha]$ for any $\xi \in X$ with $f(\xi) = \alpha$. Set

$$\begin{aligned} R &= \{\langle \alpha, \beta\rangle \mid \alpha < \kappa \ \wedge\ \beta \in R_\alpha\}\ , \text{ and} \\ T &= \{\langle \alpha, k_\alpha\rangle \mid \alpha < \kappa\}\ . \end{aligned}$$

Let τ be any first-order sentence such that if $\langle V_\delta, \in\rangle \models \tau$, then δ is a non-zero limit ordinal. Then

$$\langle V_\kappa, \in, R, T\rangle \models \tau \ \wedge\ \forall\alpha\forall U\forall v(U = R``\{\alpha\} \ \wedge\ v = T(\alpha) \to \psi_{mn}(U, v))\ .$$

Properly formalized this sentence is Π^m_n, and because of τ any ξ that reflects it satisfies

$$\begin{aligned} R \cap V_\xi &= \{\langle \alpha, \beta\rangle \mid \alpha < \xi \ \wedge\ \beta \in R_\alpha \cap V_\xi\}\ , \text{ and} \\ T \cap V_\xi &= T | \xi\ , \end{aligned}$$

and so $\xi \notin X$ as f is regressive. Hence, X is not Π^m_n-indescribable in κ. ⊣

Note that for any $R \subseteq V_\kappa$ and Π^m_n sentence φ such that $\langle V_\kappa, \in, R\rangle \models \varphi$,

$$\{\alpha < \kappa \mid \langle V_\alpha, \in, R \cap V_\alpha\rangle \models \varphi\}$$

is in the Π^m_n-indescribable filter, and similarly for Σ^m_n. This leads to the following:

6.12 Exercise (Levy [71]).

(a) For any n, $\{\alpha < \kappa \mid \alpha \text{ is } \Pi^1_n\text{-indescribable in } \alpha\}$ *is in the* Π^1_{n+1}-*indescribable filter over* κ.

(b) For $m > 1$ *and* $n > 0$, $\{\alpha < \kappa \mid \alpha \text{ is } \Sigma^m_n\text{-indescribable in } \alpha\}$ *is in the* Π^m_n-*indescribable filter over* κ, *and* $\{\alpha < \kappa \mid \alpha \text{ is } \Pi^m_n\text{-indescribable in } \alpha\}$ *is in the* Σ^m_n-*indescribable filter over* κ.

(c) If S *is stationary in* κ, *then* $\{\alpha < \kappa \mid S \cap \alpha \text{ is stationary in } \alpha\}$ *is in the* Π^1_1-*indescribable filter over* κ.

Hint. (a) and (b) follow from the sentences specified in 6.9, and (c) follows from devising a Π^1_1 sentence σ such that for any δ and $A \subseteq \delta$, $\langle V_\delta, \in, A\rangle \models \sigma$ *iff* A is stationary in δ. ⊣

Any normal filter extends the closed unbounded filter (5.10ff), so the sets exhibited in (a) and (b) are stationary in κ. Moreover, (c) implies that *the indescribable filters are closed under Mahlo's operation* and strengthens 4.6. 6.12 typifies how in hierarchies of large cardinals, a cardinal at one level defines for itself its transcendent size relative to cardinals at lower levels.

A fruitful offshoot of the study of large cardinals has been the investigation of their various analogues in restricted contexts. After all, natural ideas first germinated in the maximal setting of set theory ought to thrive in more focused situations. The first substantive move in this direction was made in the early 1970's in the theory of inductive definitions allowing non-monotone operators. With the admissible ordinals playing the role of regular cardinals, natural analogues of Mahlo and indescribable cardinals were developed in this context. Wayne Richter and Peter Aczel in their [74] and Aczel [77: 772ff] provide the details. A second wave of activity began in 1980 with study of Σ_n admissibility in terms of reflection and partition properties by Evangelos Kranakis [82, 82a, 83]. This was extended by, e.g. Kranakis and Iain Phillips in their [84], and expanded to the study of analogues of measurable cardinals by Matthew Kaufman and Kranakis in their [84] and by Josephus Baeten [86].

Penetrating work by Stål Aanderaa [74] in the earlier context of inductive definitions motivated Yiannis Moschovakis to establish the following:

6.13 Theorem (Moschovakis [76]). *If* $V = L$, $m > 1$, *and* $n > 0$, *then* $\sigma^m_n < \pi^m_n$. ⊣

Devlin [75: 100] sketches a proof. This is a curious turn of events! The use of effective analogues for large cardinals provided new characterizations and insights, which in turn reverberated back to the original context.

Whether some restrictive hypothesis like $V = L$ is necessary in 6.13 was left unresolved for some time, until Kai Hauser established a contrasting result with sophisticated forcing techniques:

6.14 Theorem (Hauser [91,92]). *If $m > 1$, $n > 0$, and there is a Σ^m_n-indescribable cardinal and a Π^m_n-indescribable cardinal below it, then there is a generic extension with such cardinals in which $\sigma^m_n > \pi^m_n$.* ⊣

This was somewhat unexpected, since the straightforward definitions and the $m = 1$ precedent suggested an absolute result in the Moschovakis direction. (The ordering of the cardinals in the hypothesis of the theorem is necessary, since from $\sigma^m_n > \pi^m_n$ one can establish the *consistency* of: there are both Σ^m_n-indescribable cardinals and Π^m_n-indescribable cardinals.)

The study of indescribability set the stage for the general investigation of large cardinals in terms of reflection phenomena, and also introduced an important technique, the use of definable normal filters. 6.14 may have resolved what had been the major open problem about indescribability, but in any case the concept provides a useful framework for the analysis of the relative size of various large cardinals.

Chapter 2

Partition Properties

This chapter describes the progression of ideas and results emerging from partition properties first considered by Erdős and culminating in Silver's results about the existence of the set of integers $0^\#$, a principle of transcendence over L. This development incorporated both the refined analysis of combinatorics as well as the full play of model-theoretic techniques, and provided formulations that have come to be regarded as basic. §7 explores partitions of n-tuples, developing the related tree property and further characterizations of weak compactness, and introduces partitions of all finite subsets. §8 gives Rowbottom's model-theoretic characterizations and results about L, and explores the related concepts of Rowbottom and Jónsson cardinals. Finally, §9 presents Silver's definitive work on sets of indiscernibles and the implications for L of the existence of $0^\#$.

7. Partitions and Trees

Although an itinerant mathematician for most of his life, Paul Erdős has been the prominent figure of a strong Hungarian tradition in the concrete mathematics of combinatorics, and through some initial results he introduced major initiatives into the detailed combinatorial study of the transfinite. Erdős and his collaborators simply viewed the transfinite numbers as a combinatorially rich source of intrinsically interesting problems. The lines of inquiry that they pursued soon developed a momentum of their own, with the concrete questions about graphs and mappings having a natural appeal through their immediacy. Some of the initial speculations led to important correlations with weak compactness, and these ramifications are discussed in this section. Enough was known by the late 1950's so that the result that the least inaccessible cardinal is not measurable could have been deduced *before* the Hanf-Tarski work. That this combinatorial research should have some bearing on large cardinals was not surprising, but that it was to play a central role in their structural elaboration in the 1960's (§§8, 9) was rather unexpected.

Despite later developments the story begins with a problem in formal logic. A couple of years before Gödel established the Incompleteness Theorem [31], Ramsey [30] demonstrated the decidability of the class of $\exists\forall$ formulas with identity (see Dreben-Goldfarb [79] for a general framework). It was for this purpose that he proved his well-known finite combinatorial theorem. Up at Cambridge and brother to a later Archbishop of Canterbury, Ramsey through his association with Bertrand Russell and Ludwig Wittgenstein is a pivotal figure in the philosophy of mathematics. With the latter at his side, Ramsey died tragically at the age of 26. Skolem [33] sharpened Ramsey's work, but his theorem did not become widely known until Erdős-Szekeres [35] rediscovered it and applied it to a problem of combinatorial geometry. Today, Ramsey Theory is a thriving field of combinatorics (see Graham-Rothschild-Spencer [90]).

Ramsey also established an infinite version of his combinatorial theorem which is just as well-known, and it was the investigation of its analogues in the transfinite that led to large cardinal properties. In the joint paper Erdős-Tarski [43] weakly inaccessible cardinals were incorporated into a discussion about maximal mutually disjoint families. But enticingly, that paper ended with an intriguing list of six combinatorial problems (and announced some interconnections between them in a footnote) whose positive solutions amounted to the existence of either a strongly compact, measurable, or weakly compact cardinal. It is evident that here the authors were motivated by strong properties of ω to formulate *direct combinatorial generalizations*. They took a distinctly empirical approach to foundational issues, considering their problems as lines of inquiry towards possible new axioms. They speculated ([43: 328ff]):

> The difficulties which we meet in attempting to solve the problems under consideration do not seem to depend essentially on the nature of inaccessible numbers. In most cases the difficulties seem to arise from lack of devices which enable us to construct maximal sets which are closed under certain infinite operations. It is quite possible that a complete solution of these problems would require new axioms which would differ considerably in

their character not only from the usual axioms of set theory, but also from those hypotheses whose inclusion among the axioms has previously been discussed in the literature and mentioned previously in this paper (e.g., the existential axioms which secure the existence of inaccessible numbers, or from hypotheses like that of Cantor which establish arithmetical relations between the cardinal numbers).

At first, it might seem as if the possibility of such prior principles has not been realized, since the positive solutions to their problems have simply been adopted as large cardinal hypotheses. But, reflection principles studied in the 1960's (§6) have justified "maximal sets which are closed under certain infinite operations". In particular, the characterization of weak compactness via Π^1_1-indescribability has provided a local explanatory principle leading to the positive solutions of the combinatorial problems, and the second-order indescribability principle of Bernays [61], essentially the Π^1_n-indescribability of On for every $n \in \omega$, a global one.

The program initiated by Erdős is now described, so that the related problems of Erdős-Tarski [43] can be considered in context. Richard Rado was Erdős' main collaborator in this direction in the 1950's, and Hajnal, since then. It is interesting to note that like his earlier compatriot von Neumann, Hajnal's initial work was in the axiomatics of set theory (see before 3.2), but his concerns were quite different afterwards. A general framework called a *partition calculus* was developed by Erdős-Rado [56], and the starting point is a special case of their *ordinary partition symbol*. Recall that for $x \subseteq \mathrm{On}$, $[x]^\gamma = \{y \subseteq x \mid y \text{ has ordertype } \gamma\}$. The *ordinary partition relation*

$$\beta \longrightarrow (\alpha)^\gamma_\delta$$

asserts that for any $f\colon [\beta]^\gamma \to \delta$, there is an $H \in [\beta]^\alpha$ *homogeneous* for f: $|f``[H]^\gamma| \leq 1$. In other words, for any partition of the ordertype γ subsets of β into δ cells there is an $H \subseteq \beta$ of ordertype α all of whose ordertype γ subsets lie in the same cell. The negation of this and like relations is indicated with a $\not\longrightarrow$ replacing the $\longrightarrow$. The idea behind this "arrow" notation is that the relation is obviously preserved upon making the β on the left larger, or making any of the α, γ, δ on the right smaller, as long as the order relationship between γ and α is preserved, e.g. the trivially true case $\alpha \leq \gamma$ can become false when $\gamma < \alpha$. Of course, δ can be taken to be a cardinal, and if α is a cardinal, then the least β satisfying the relation must also be a cardinal. Ramsey's infinitary theorem is the assertion $\omega \longrightarrow (\omega)^m_n$ for every $m, n \in \omega$ and is established by 7.7.

An early comment was that γ must be finite in the presence of the Axiom of Choice:

7.1 Proposition (Erdős-Rado [52: 434]). *For any κ, $\kappa \not\longrightarrow (\omega)^\omega_2$.*

Proof. Let $\prec$ well-order $[\kappa]^\omega$, and define $f\colon [\kappa]^\omega \to 2$ by $f(s) = 0$ if every $t \in [s]^\omega - \{s\}$ satisfies $s \prec t$, and $f(s) = 1$ otherwise. Then no $x \in [\kappa]^\omega$ can be homogeneous for f: If y is the $\prec$-least member of $[x]^\omega$, then $f(y) = 0$. But taking any infinite increasing $\subset$-chain $x_0 \subset x_1 \subset x_2 \ldots \subset x$ of infinite sets, $f(x_n) = 0$ for every n would imply that $\ldots x_2 \prec x_1 \prec x_0$, contrary to $\prec$ being a well-ordering. ⊣

Erdős-Rado [56] provided some basic results for the general ordinary partition relation. Working independently Djuro Kurepa [59] also obtained similar results. Later, an almost complete theory for cardinals was given in Erdős-Hajnal-Rado [65] assuming GCH. Incorporating the further work of Shelah [75] the book Erdős-Hajnal-Máté-Rado [84] then extended the discussion with Byzantine detail to the general situation without GCH. What lies at the heart of these matters is a basic argument for producing large homogeneous sets using trees.

A *tree* is a partially ordered set $\langle T, <_T \rangle$ such that for any $t \in T$ the set $\{u \in T \mid u <_T t\}$ of $<_T$-predecessors of t is well-ordered by $<_T$. The *αth level* of T consists of those $t \in T$ whose set of $<_T$-predecessors has ordertype α under $<_T$. The *height* of T is the least α such that the αth level of T is empty. A *chain* of T is a linearly ordered subset, and an *antichain* of T is a subset consisting of pairwise $<_T$-incomparable elements. For example, any level is an antichain. A *branch* of T is a maximal chain of T, a *cofinal branch* of T is a branch with members at every non-empty level of T, and an *α-branch* of T is a branch whose ordertype under $<_T$ is α. A *subtree* of T is a subset $\overline{T} \subseteq T$ together with the induced ordering on $\overline{T}$. Although not always done in set theory, it is assumed that the 0th level of a tree consists of a single element, called the *root*.

Trees abound in contemporary set theory as basic combinatorial objects. The first systematic study of trees was carried out in Kurepa's Paris thesis [35] with Fréchet, where several tree and linear order equivalences are derived, e.g. for Suslin's Problem [35: 127ff]. See Levy [79: IX§2] for the basic combinatorics, Devlin [84] for trees in L, and Todorčević [84] for a magisterial discussion. The basic example is $\langle {}^{<\gamma}x, \subset \rangle$, the set ${}^{<\gamma}x = \bigcup_{\alpha<\gamma} {}^{\alpha}x$ ordered by proper inclusion, which for $x \neq \emptyset$ is a tree of height γ with αth level ${}^{\alpha}x$.

The following lemma provides a basic analysis of partitions in terms of trees and is known in general form as the Ramification Lemma (cf. Erdős-Hajnal-Rado [65: §6]). It is used to establish a close relationship between large homogeneous sets for partitions and long chains in trees, and as such provides the basis for the quintessential *tree argument* in infinitary combinatorics: Instead of building large homogeneous sets directly by recursion, one argues that because a tree must have long chains, *some* sufficiently large homogeneous set must exist. The first arguments of this sort were those appealing to the well-known tree lemma of Dénes König [27], a generalization of which will be considered shortly. Recall our §0 convention that $f(\alpha_1, \ldots, \alpha_n)$ is written for $f(\{\alpha_1, \ldots, \alpha_n\})$ with the understanding that $\alpha_1 < \ldots < \alpha_n$.

7.2 Lemma. *Suppose that* $2 \leq n < \omega$, σ *is a cardinal, and* $f\colon [\kappa]^n \to \sigma$. *Then there is a tree* $\langle \kappa, <_f \rangle$ *such that:*

(a) If $\xi <_f \eta$, *then* $\xi < \eta$.

(b) If $\xi_1 <_f \ldots <_f \xi_{n-1} <_f \delta <_f \eta$, *then*

$$f(\xi_1, \ldots, \xi_{n-1}, \delta) = f(\xi_1, \ldots, \xi_{n-1}, \eta)\ .$$

(c) For each $\alpha < \kappa$, the αth level of the tree has cardinality at most $\sigma^{|\omega+\alpha|}$, and is finite if both σ and α are finite.

Proof. The tree is defined by recursion, initially letting the first n levels be $0 <_f 1 <_f \ldots <_f n-1$. Suppose that $n \leq \eta < \kappa$, and every $\xi < \eta$ has already been put on the tree, i.e. $<_f |(\eta \times \eta)$ has been defined. Then choose a downward closed chain b of the tree thus far constructed maximal with respect to having the property:

$$\text{for any } \xi_1 <_f \ldots <_f \xi_n \text{ all in } b,\ f(\xi_1, \ldots, \xi_n) = f(\xi_1, \ldots, \xi_{n-1}, \eta)\,,$$

and stipulate that $\xi <_f \eta$ for every $\xi \in b$.

(Actually, this b is unique: If $b_1 \neq b_2$ were chains that met the conditions for b, let $c = b_1 \cap b_2$, γ_1 the $<_f$-least element of $b_1 - c$, and γ_2 the $<_f$-least element of $b_2 - c$, say with $\gamma_1 < \gamma_2$. Then if $\xi_1 <_f \ldots <_f \xi_{n-1}$ are all in c,

$$f(\xi_1, \ldots, \xi_{n-1}, \gamma_1) = f(\xi_1, \ldots, \xi_{n-1}, \eta) = f(\xi_1, \ldots, \xi_{n-1}, \gamma_2)\,.$$

But this is contradictory, since γ_2 had been put atop c, yet c did not meet the maximality condition at the time because of $c \cup \{\gamma_1\}$.)

$\langle \kappa, <_f \rangle$ is a tree satisfying (a) and (b). To complete the proof, (c) is established by induction: At limit ordinals $\alpha > 0$, first note that if b is a maximal chain through that part of the tree below the αth level, then b has at most one successor at the αth level: If $\eta_1 < \eta_2$ were two successors, for any $\xi_1 <_f \ldots <_f \xi_{n-1}$ all in b, taking a $\xi \in b$ such that $\xi_{n-1} <_f \xi$ as α is a limit,

$$f(\xi_1, \ldots, \xi_{n-1}, \eta_1) = f(\xi_1, \ldots, \xi_{n-1}, \xi) = f(\xi_1, \ldots, \xi_{n-1}, \eta_2)\,.$$

But this is contradictory, since η_2 had been put atop b, yet b was not maximal at the time because of $b \cup \{\eta_1\}$. Thus, the cardinality of the αth level is at most the cardinality of the set of maximal chains through that part of the tree below the αth level. By induction this is at most $\prod_{\beta<\alpha} \sigma^{|\omega+\beta|} = \sigma^{|\alpha|}$ as required.

For the argument at successor ordinals, first note that if ξ is at the αth level, then ξ has at most $\sigma^{|\alpha|^{n-2}}$ immediate successors at the $(\alpha+1)$st level: If $\eta_1 < \eta_2$ are both immediate successors of ξ, then there must be some $\xi_1 <_f \ldots <_f \xi_{n-1} \leq_f \xi$ such that

$$(*) \qquad f(\xi_1, \ldots, \xi_{n-1}, \eta_1) \neq f(\xi_1, \ldots, \xi_{n-1}, \eta_2)$$

else η_2 would not have been an immediate successor, as before. In fact, ξ_{n-1} must equal ξ, else $(*)$ would be an equality just as in the previous paragraph. There are at most $|\alpha|^{n-2}$ $(n-2)$-tuples $\langle \xi_1, \ldots, \xi_{n-2} \rangle$ as ξ is at the αth level, and each immediate successor η of ξ determines for each such $(n-2)$-tuple a value $f(\xi_1, \ldots, \xi_{n-2}, \xi, \eta) < \sigma$. Since these values must be different as in $(*)$ for distinct immediate successors, there must be at most $\sigma^{|\alpha|^{n-2}}$ immediate successors.

It is now simple to complete the argument: By induction, the αth level has cardinality at most $\sigma^{|\omega+\alpha|} \cdot \sigma^{|\alpha|^{n-2}} \leq \sigma^{|\omega+\alpha|}$. Also, if σ and α are both finite and by induction the αth level is finite, then so is the $(\alpha+1)$st level since $\sigma^{|\alpha|^{n-2}}$ is finite. ⊣

To illustrate how this lemma is used, with it we establish what is called *the* Erdős-Rado Theorem, which shows in particular that for any α, δ, and $n \in \omega$ there is a κ such that $\kappa \longrightarrow (\alpha)^{n+1}_{\delta}$. The case $n = 1$ had occurred in Erdős [42]; the full theorem was established independently by Kurepa [59]. For the statement of the theorem, define the first few Beth numbers based at a cardinal κ by: $\text{beth}_0(\kappa) = \kappa$ and $\text{beth}_{n+1}(\kappa) = 2^{\text{beth}_n(\kappa)}$ for $n \in \omega$.

7.3 Theorem (Erdős-Rado [56]). *For any κ and $n \in \omega$,*

$$\text{beth}_n(\kappa)^+ \longrightarrow (\kappa^+)^{n+1}_{\kappa} .$$

Proof. Proceed by induction on n. The case $n = 0$ follows from the regularity of κ^+. Assume now that the assertion is true for n, and suppose that $f\colon [\text{beth}_{n+1}(\kappa)^+]^{n+2} \to \kappa$. Let $<_f$ be the corresponding tree ordering on $\text{beth}_{n+1}(\kappa)^+$ as in the lemma. Then the $\text{beth}_n(\kappa)^+$th level of the tree is not empty: Otherwise, (c) of the lemma would lead to the contradiction

$$\text{beth}_{n+1}(\kappa)^+ \leq \bigcup\{\kappa^{|\alpha|} \mid \alpha < (\text{beth}_n(\kappa))^+\} \leq 2^{\text{beth}_n(\kappa)} = \text{beth}_{n+1}(\kappa) .$$

Let η be at the $\text{beth}_n(\kappa)^+$th level and consider the chain $C = \{\xi \mid \xi <_f \eta\}$. By (a) and (b) of the lemma the function $g\colon [C]^{n+1} \to \kappa$ given by

$$g(\xi_1, \ldots, \xi_{n+1}) = f(\xi_1, \ldots, \xi_{n+1}, \delta) \text{ for some (any) } \delta \in C - (\xi_{n+1} + 1)$$

is well-defined. But then, the inductive hypothesis can be applied to g to extract a subset of C of cardinality κ^+, homogeneous for the original f. ⊣

7.3 is known to be best possible in the sense that $\text{beth}_n(\kappa)^+$ cannot be replaced by $\text{beth}_n(\kappa)$ (Erdős-Hajnal-Rado [65]). This follows from their "stepping-up" lemma ([65: 118ff]), starting from two basic counterexamples for $n = 1$ that had been noticed much earlier:

7.4 Exercise (Gödel – Erdős-Kakutani [43: 459]). *For any κ,*

$$2^{\kappa} \not\longrightarrow (3)^2_{\kappa} .$$

Hint. With $\langle f_\alpha \mid \alpha < 2^\kappa \rangle$ enumerating ${}^{\kappa}2$ consider $F\colon [2^\kappa]^2 \to \kappa$ defined by: $F(\alpha, \beta) =$ least ξ such that $f_\alpha(\xi) \neq f_\beta(\xi)$. ⊣

7.5 Proposition (Sierpiński [33]; Kurepa [41: 487]). *For any κ,*

$$2^{\kappa} \not\longrightarrow (\kappa^+)^2_2 .$$

Proof. Let $<$ be the lexicographic ordering of ${}^{\kappa}2$, i.e. for $f \neq g \in {}^{\kappa}2$, $f < g$ *iff* $f(\alpha) < g(\alpha)$ for the least $\alpha < \kappa$ such that $f(\alpha) \neq g(\alpha)$. With $\langle f_\alpha \mid \alpha < 2^\kappa \rangle$

enumerating ${}^{\kappa}2$ define $F\colon [2^\kappa]^2 \to 2$ by: $F(\alpha, \beta) = 0$ if $f_\alpha < f_\beta$, and $F(\alpha, \beta) = 1$ otherwise.

Assume to the contrary that there is an $H \in [2^\kappa]^{\kappa^+}$ homogeneous for F. Take $F``[H]^2 = \{0\}$; the other case is entirely analogous. For each $\alpha \in H$ let $d(\alpha)$ be the least ordinal $\xi < \kappa$ such that for some $\gamma \in H$, $f_\alpha|\xi = f_\gamma|\xi$ yet $f_\alpha(\xi) < f_\gamma(\xi)$. For $\xi < \kappa$, set

$$H_\xi = \{\alpha \in H \mid d(\alpha) = \xi\}\ .$$

Note that for any $\xi < \kappa$, if $\alpha, \beta \in H_\xi$, then $f_\alpha|\xi = f_\beta|\xi$. Thus, if $\alpha \in H_\xi$ and $\gamma \in H$ is such that $f_\alpha|\xi = f_\gamma|\xi$ yet $f_\alpha(\xi) < f_\gamma(\xi)$, then any other member of H_ξ has this property with respect to the *same* γ. Hence, $H_\xi \subseteq \{\alpha \in H \mid f_\alpha < f_\gamma\}$, and so $|H_\xi| \le \kappa$, since by homogeneity H is well-ordered by $<$ in ordertype κ^+. However, this contradicts the regularity of κ^+ since $H = \bigcup_{\xi<\kappa} H_\xi$. ⊣

For any λ, using 7.3 there is a sequence $\langle \lambda_k \mid k \in \omega\rangle$ such that $\lambda_0 = \lambda$ and for $k \in \omega$, $\lambda_{k+1} \longrightarrow (\alpha)^n_\delta$ for any $\alpha, \delta < \lambda_k$ and $n \in \omega$. It follows that if $\kappa = \sup(\{\lambda_k \mid k \in \omega\})$, then $\kappa > \lambda$ and $\kappa \longrightarrow (\alpha)^n_\delta$ for any $\alpha, \delta < \kappa$ and $n \in \omega$. The following result is about the next interesting possibility, directly generalizing Ramsey's $\omega \longrightarrow (\omega)^2_2$:

7.6 Corollary (Erdős). *If $\kappa > \omega$ and $\kappa \longrightarrow (\kappa)^2_2$, then κ is inaccessible.*

Proof. First, κ must be regular: Otherwise, κ would be a disjoint union $\kappa = \bigcup_{\xi<\gamma} X_\xi$, where $\gamma < \kappa$ and $|X_\xi| < \kappa$ for each $\xi < \gamma$. Then the function $f\colon [\kappa]^2 \to 2$ defined by $f(\alpha, \beta) = 0$ if α and β are in the same X_ξ, and $f(\alpha, \beta) = 1$ otherwise, cannot have a homogeneous set of cardinality κ.

Next, κ must be a strong limit: Otherwise, there would be a $\lambda < \kappa$ such that $\kappa \le 2^\lambda$. But then, $\kappa \longrightarrow (\kappa)^2_2$ would contradict 7.5 for λ. ⊣

The weaker conclusion that κ is weakly inaccessible is almost explicit in Erdős [42] (cf. Erdős-Tarski [43: 328]).

The Tree Property

The proof of 7.3 shows how the existence of large homogeneous sets can be related to the existence of long chains in trees. For $\kappa \longrightarrow (\kappa)^2_2$, the key feature is a strong property considered by Kurepa [35] that replaces the cardinality argument of 7.3: A *κ-tree* is a tree of height κ each of whose levels has cardinality less than κ.

κ has *the tree property iff* every κ-tree has a cofinal branch .

This generalizes the well-known tree lemma of König [27], which is just the assertion that *ω has the tree property*. The following result will shortly be seen to provide another characterization of weak compactness.

7.7 Exercise (Ramsey [30] for $\kappa = \omega$; Erdős-Tarski [43: 328] for $n = \lambda = 2$). *If $\kappa = \omega$, or is inaccessible and has the tree property, then for any $n \in \omega$ and $\lambda < \kappa$, $\kappa \longrightarrow (\kappa)^n_\lambda$.*

Hint. Use the inductive argument for 7.3; for the case $\kappa = \omega$, apply the last clause of 7.2(c). $\dashv$

Ramsey's original proof was essentially this, although he did not formulate König's lemma explicitly.

As mentioned in §4 the details of implications announced at the end of Erdős-Tarski [43] were provided in a seminar at Berkeley in 1958-9 by Mostowski and Tarski, and appeared in Erdős-Tarski [61]. In addition to the crucial Hanf (§4) and Scott (§5) results that followed, the tree and partition equivalences for weak compactness were soon worked out:

7.8 Theorem (Erdős-Tarski [43, 61], Hanf [64a], and Monk-Scott [64]). *The following are equivalent for $\kappa > \omega$:*

(a) κ is weakly compact.

(b) κ is inaccessible and has the tree property.

(c) $\kappa \longrightarrow (\kappa)^n_\lambda$ for every $n < \omega$ and $\lambda < \kappa$.

(d) $\kappa \longrightarrow (\kappa)^2_2$.

Proof. (a) $\to$ (b). Inaccessibility is a consequence of weak compactness. To establish the tree property a typical compactness argument can be used: Let $\langle T, <_T \rangle$ be a κ-tree. To each $t \in T$ associate a propositional (0-ary predicate) symbol P_t, and consider the collection of $L_{\kappa\omega}$ sentences consisting of: disjunctions $\bigvee\{P_t \mid t \text{ is at the } \alpha\text{th level of } T\}$ for $\alpha < \kappa$, and $\neg(P_t \wedge P_{t'})$ for $<_T$-incomparable $t, t' \in T$. Since T has height κ, this collection of κ sentences is κ-satisfiable. Hence by weak compactness, it is satisfiable, say by a model $\mathcal{M}$. $\{t \in T \mid \mathcal{M} \models P_t\}$ is then a cofinal branch through T.

(b) $\to$ (c) follows from 7.7, and (c) $\to$ (d) is immediate. To complete the proof, it will be convenient to establish (d) $\to$ (b) $\to$ (a).

(d) $\to$ (b). Inaccessibility follows from 7.6. To establish the tree property, let $\langle T, <_T \rangle$ be a κ-tree. Since $|T| = \kappa$, it can be assumed that $T = \kappa$. If $\xi < \kappa$ is at a level $\geq \alpha$, let $\pi_\alpha(\xi)$ be its predecessor at level α (allowing $\pi_\alpha(\xi) = \xi$ if ξ is *at* the αth level). $<_T$ can be extended to a linear ordering $\prec$ of κ by specifying how it orders two $<_T$-incomparable elements ξ, η as follows: Let α be the least such that $\pi_\alpha(\xi) \neq \pi_\alpha(\eta)$, and define $\xi \prec \eta$ *iff* $\pi_\alpha(\xi) < \pi_\alpha(\eta)$.

Now define $f\colon [\kappa]^2 \to 2$ by $f(\xi, \eta) = 0$ if $\xi \prec \eta$, and $f(\xi, \eta) = 1$ otherwise. By $\kappa \longrightarrow (\kappa)^2_2$ let $H \in [\kappa]^\kappa$ be homogeneous for f. Since every level of the tree has cardinality less than κ, there is a $\rho_\alpha < \kappa$ such that if $\rho_\alpha < \xi$ and $\xi \in H$ then ξ is at a level $\geq \alpha$. By definition of $\prec$ if two ordinals $\gamma \prec \delta$ are both at levels $\geq \alpha$, then $\pi_\alpha(\gamma) \prec \pi_\alpha(\delta)$ or $\pi_\alpha(\gamma) = \pi_\alpha(\delta)$. Hence, if $f``[H]^2 = \{0\}$, then $\langle \pi_\alpha(\xi) \mid \rho_\alpha < \xi \wedge \xi \in H \rangle$ is non-$\prec$-decreasing in $\xi$, and if $f``[H]^2 = \{1\}$, then it is non-$\prec$-increasing. In either case there is a $\sigma_\alpha < \kappa$ and a b_α such that if $\sigma_\alpha < \xi$ and $\xi \in H$, then $\pi_\alpha(\xi) = b_\alpha$. Clearly, $\langle b_\alpha \mid \alpha < \kappa \rangle$ is a cofinal branch since any two b_α's have a common $<_T$-successor.

(b) $\to$ (a). The Extension Property (4.5) for weak compactness will be verified. So, suppose that $R \subseteq V_\kappa$; a transitive set $X \neq V_\kappa$ and an $S \subseteq X$ must be found so that $\langle V_\kappa, \in, R\rangle \prec \langle X, \in, S\rangle$. By the inaccessibility of κ and 6.1,

$$\{\alpha < \kappa \mid \langle V_\alpha, \in, R \cap V_\alpha\rangle \prec \langle V_\kappa, \in, R\rangle\}$$

is closed unbounded in κ; let $\langle \alpha_\xi \mid \xi < \kappa\rangle$ be its increasing enumeration.

A tree $\langle T, <_T\rangle$ is now defined. Fix a complete set of Skolem functions (§0) for $\langle V_\kappa, \in, R\rangle$, and define Skolem hulls for any $X \subseteq V_\kappa$ in what follows as being with respect to this particular set. For $\alpha_\xi < \beta < \kappa$ let

$$H(\xi, \beta) = \text{the Skolem hull of } V_{\alpha_\xi} \cup \{\beta\} \text{ in } \langle V_\kappa, \in, R\rangle$$

so that

$$\langle V_{\alpha_\xi}, \in, R \cap V_{\alpha_\xi}\rangle \prec H(\xi, \beta)\ .$$

Define

$$H(\xi, \beta) \approx H(\overline{\xi}, \overline{\beta}) \text{ iff } \xi = \overline{\xi} \text{ and there is an isomorphism between the two structures fixing } V_{\alpha_\xi} \text{ and sending } \beta \text{ to } \overline{\beta}\ .$$

Clearly, $\approx$ is an equivalence relation, and so let $[H(\xi, \beta)]$ denote the equivalence class of $H(\xi, \beta)$. The elements of our tree T are to be the $[H(\xi, \beta)]$'s. Finally, set

$$[H(\xi, \beta)] <_T [H(\overline{\xi}, \overline{\beta})] \text{ iff } \xi < \overline{\xi},\ \beta \leq \overline{\beta}, \text{ and } H(\xi, \beta) \text{ is isomorphic to the Skolem hull of } V_{\alpha_\xi} \cup \{\overline{\beta}\} \text{ in } H(\overline{\xi}, \overline{\beta})\ .$$

That $\langle T, <_T\rangle$ is indeed a tree is not difficult to see; note that the ξth level of T is $\{[H(\xi, \beta)] \mid \alpha_\xi < \beta < \kappa\}$. That $\langle T, <_T\rangle$ is a κ-tree follows from the inaccessibility of κ, there being at most $2^{|V_{\alpha_\xi}|} < \kappa$ Skolem hulls up to isomorphism generated by $V_{\alpha_\xi} \cup \{x\}$ for sets x.

Hence, by the tree property there is a κ-branch $\langle [H(\xi, \beta_\xi)] \mid \xi < \kappa\rangle$ through T. By definition of $<_T$, whenever $\xi \leq \eta < \kappa$ there is an elementary embedding $i_{\xi\eta}$: $H(\xi, \beta_\xi) \prec H(\eta, \beta_\eta)$ that fixes V_{α_ξ} so that $i_{\xi\eta}(\beta_\xi) = \beta_\eta$. From the construction it is seen that $\xi \leq \eta \leq \rho < \kappa$ implies that $i_{\xi\rho} = i_{\eta\rho} \circ i_{\xi\eta}$. Thus, the direct limit (§0) can be formed, and it is well-founded as $\mathrm{cf}(\kappa) > \omega$. The transitive collapse (0.4) is then an elementary extension $\langle X, \in, S\rangle$ of $\langle V_\kappa, \in, R\rangle$. Since the β_ξ's get identified together to correspond to an ordinal $\beta \in X$ such that $\beta \geq \kappa$, $\langle X, \in, S\rangle$ is as required. ⊣

The proof of (a) $\to$ (b) shows that weak compactness is equivalent to its defining property restricted to propositional logic for $L_{\kappa\omega}$. Be that as it may, nowadays the roots of the concept in infinitary languages are usually passed over in favor of its more immediate formulations, particularly (b). Just as König's lemma is the mathematical essence of the compactness property of $L_{\omega\omega}$, so the tree property provides the gist of the generalization. Further tree-related equivalences for weak compactness have been provided by Baumgartner [75: 123], Shelah [79], Erdős-Hajnal-Máté-Rado [84: §31] (also due to Baumgartner), and

Todorčević [87]. Harkening back to the efforts of Keisler-Tarski [64] (§4), much of the interest has been in devising avowedly combinatorial proofs of the results on the size of weakly compact cardinals.

The tree property, decoupled from inaccessibility and hence from the partition property $\kappa \longrightarrow (\kappa)^2_2$, has come to be regarded as intrinsically interesting in its own right, especially after the advent of forcing which led to relative consistency results about accessible cardinals. That ω_1 does not have the tree property is a 1934 result of Nathan Aronszajn, presented in Kurepa [35: 96] and acknowledged (inaccurately) in Erdős-Tarski [43: 328]. A fellow student with Kurepa in Paris at the time, Aronszajn may seem a shadowy figure in the annals of logic, but he became well-known as a functional analyst (e.g. [52]). With but one paper [52a] in set theory and on a topic unrelated to trees, it is a happenstance of mathematics that his name has become commonplace in infinite combinatorics:

A κ-*Aronszajn tree* is a κ-tree with no cofinal branch .

Hence, a κ-Aronszajn tree is simply a counterexample to the tree property for κ. Note that *if κ is singular, then there is a κ-Aronszajn tree*: Let $\kappa - \{0\} = \bigcup_{\alpha<\gamma} X_\alpha$ be a disjoint union with $\gamma < \kappa$ and $|X_\alpha| < \kappa$ for each $\alpha < \gamma$, and consider $\langle \kappa, <_T \rangle$ where $\xi <_T \zeta$ *iff* $\xi < \zeta$ and $\xi, \zeta \in \{0\} \cup X_\alpha$ for some $\alpha < \gamma$. For regular κ, uniformly thin κ-trees do have cofinal branches:

7.9 Proposition (Kurepa [35: 80]). *Suppose that κ is regular, $\lambda < \kappa$, and $\langle T, <_T \rangle$ is a κ-tree each of whose levels has cardinality less than λ. Then $\langle T, <_T \rangle$ has a cofinal branch.*

Proof. It can be assumed, by trimming the tree if necessary, that distinct members at a limit level have a distinct $<_T$-predecessors at some earlier level. It can further be assumed that $T = \kappa$. Let T_α denote the αth level of the tree, so that $|T_\alpha| < \lambda$. The result is first established assuming that λ is regular:

For each $\alpha < \kappa$ with $\mathrm{cf}(\alpha) = \lambda$, choose a $\xi_\alpha \in T_\alpha$. Then for such α there is an $f(\alpha) < \alpha$ such that for any $\xi \in T_\alpha - \{\xi_\alpha\}$, ξ and ξ_α have no common $<_T$-predecessor in $T_{f(\alpha)}$. Since $\{\alpha < \kappa \mid \mathrm{cf}(\alpha) = \lambda\}$ is stationary in κ, it has (0.1(c)) a subset X unbounded in κ on which f is constant, say with value γ. It follows from $|T_\gamma| < \kappa$ that there is a $Y \subseteq X$ unbounded in κ such that one fixed member of T_γ is a $<_T$-predecessor to every member of $\{\xi_\alpha \mid \alpha \in Y\}$. But then, $\{\xi_\alpha \mid \alpha \in Y\}$ is readily seen to be a chain in $\langle \kappa, <_T \rangle$, and hence it determines a cofinal branch.

For singular $\lambda < \kappa$, there is a regular $\nu < \lambda$ so that $Y = \{\alpha < \kappa \mid |T_\alpha| < \nu\}$ is unbounded in κ. By the previous argument, the subtree determined by $\bigcup_{\alpha \in Y} T_\alpha$ has a cofinal branch. But any such branch determines a cofinal branch through the original tree. ⊣

Specker generalized Aronszajn's original result under a cardinality assumption. A proof based on the elegant "minimal walk" construction of Stevo Todorčević is provided:

7.10 Theorem (Aronszajn for $\kappa = \omega$ – Kurepa [35: 96]; Specker [49]). *If κ is regular and $2^{<\kappa} = \kappa$, then there is a κ^+-Aronszajn tree.*

Proof (Todorčević [87]). For each $\alpha < \kappa^+$ let C_α be closed unbounded in α of ordertype $\leq \kappa$; successor ordinals can be accommodated by just requiring $\xi \in C_{\xi+1}$. Define $\rho_\beta \colon \beta \to {}^{<\omega}[\mathcal{P}(\beta)]$ for each $\beta < \kappa^+$ as follows: For any $\alpha < \beta$, let $\beta_0^\alpha = \beta$ and $\beta_{i+1}^\alpha = \min(C_{\beta_i^\alpha} - \alpha) < \beta_i^\alpha$ until $\beta_n^\alpha = \alpha$; then set

$$\rho_\beta(\alpha) = \langle C_{\beta_i^\alpha} \cap \alpha \mid i < n \rangle \, .$$

Several properties are noted:

(i) If $\xi < \alpha < \beta < \kappa^+$, then there is a unique $j \in \omega$ such that $\beta_i^\xi = \beta_i^\alpha$ for $i \leq j$ and $\xi \leq \beta_{j+1}^\xi < \alpha$.

(ii) If $\xi < \alpha < \beta < \kappa^+$ and $\rho_\beta(\xi)$ and $\rho_\beta(\alpha)$ have the same length n, then for some $j < n$, $C_{\beta_j^\xi} \cap \xi$ is a proper initial segment of $C_{\beta_j^\alpha} \cap \alpha$.

(iii) If $\alpha < \beta < \gamma$ and $\rho_\beta(\alpha) = \rho_\gamma(\alpha)$, then $\rho_\beta | \alpha = \rho_\gamma | \alpha$.

For (ii), take j as in (i); then $\beta_j^\xi = \beta_j^\alpha$ and $C_{\beta_j^\xi} \cap \xi$ is a proper initial segment of $C_{\beta_j^\alpha} \cap \alpha$ as β_{j+1}^ξ belongs to the latter set. For (iii), suppose that $\xi < \alpha$. Let j be maximal such that $\beta_j^\xi = \beta_j^\alpha$ and $\gamma_j^\xi = \gamma_j^\alpha$. Then for $i \leq j$, the consequence $C_{\beta_i^\alpha} \cap \alpha = C_{\gamma_i^\alpha} \cap \alpha$ of $\rho_\beta(\alpha) = \rho_\gamma(\alpha)$ implies that $C_{\beta_i^\xi} \cap \xi = C_{\gamma_i^\xi} \cap \xi$. Moreover, $\beta_{j+1}^\xi < \alpha$ or $\gamma_{j+1}^\xi < \alpha$ by (i). In either case, the consequence $C_{\beta_j^\alpha} \cap \alpha = C_{\gamma_j^\alpha} \cap \alpha$ of $\rho_\beta(\alpha) = \rho_\gamma(\alpha)$ implies that $\beta_{j+1}^\xi = \gamma_{j+1}^\xi$, and thus that $\beta_i^\xi = \gamma_i^\xi$ for $i > j$. Hence, $\rho_\beta(\xi) = \rho_\gamma(\xi)$.

Set

$$T = \{\rho_\beta | \alpha \mid \alpha \leq \beta < \kappa^+\} \, .$$

Then $\langle T, \subset \rangle$ is a tree of height κ^+ which is κ^+-Aronszajn: First, if b were a cofinal branch through T, then $f = \bigcup b$ would be a function with domain κ^+. Since the ordertypes of the C_α's are at most κ, there would be an $X \subseteq \kappa^+$ with $|X| = \kappa^+$ and fixed $n \in \omega$ and ordinals $\zeta_0, \ldots, \zeta_{n-1}$ such that for any $\alpha \in X$, $f(\alpha)$ has length n and its ith element has ordertype ζ_i. But then b could not have been a branch by (ii).

To complete the proof, note that for any $\alpha < \kappa^+$ the definition of the ρ_β's and (iii) imply that the αth level of T has cardinality at most

$$|\{C_\beta \cap \alpha \mid \alpha \leq \beta < \kappa^+\}| \leq \kappa^{<\kappa} = \kappa$$

since $2^{<\kappa} = \kappa$ and κ is regular. ⊣

The original constructions of κ^+-Aronszajn trees ensured a further property that has become pivotal in subsequent developments:

A κ-tree is *special iff* it is the union of fewer than κ antichains .

Clearly, *a special κ-tree is a κ-Aronszajn tree.* The Todorčević construction also provides special κ^+-trees:

7.11 Exercise. *With the hypotheses of 7.10, there is a special κ^+-tree.*

Hint. Modify the previous construction by taking instead $\overline{\rho}_\beta \colon \beta \to {}^{<\omega}\kappa$ given by

$$\overline{\rho}_\beta(\alpha) = \langle \mathrm{ot}(C_{\beta_i^\alpha} \cap \alpha) \mid i < n \rangle \ ,$$

and setting

$$\overline{T} = \{\overline{\rho}_\beta | (\alpha + 1) \mid \alpha < \beta < \kappa^+\} \ .$$

By (ii) above each $\overline{\rho}_\beta$ is an injective function, so that for $s \in {}^{<\omega}\kappa$,

$$A_s = \{f \in \overline{T} \mid \exists \alpha (\mathrm{dom}(f) = \alpha + 1 \ \wedge \ f(\alpha) = s)\}$$

is an antichain of $\langle \overline{T}, \subset \rangle$, and $\overline{T} = \bigcup \{A_s \mid s \in {}^{<\omega}\kappa\}$. ⊣

In contrast, the first inkling that cardinals $\leq 2^{\aleph_0}$ can have the tree property was provided by the following result.

7.12 Proposition (Silver [66, 71]). *If κ is real-valued measurable, then κ has the tree property.*

This is a simple consequence of 7.9, as is later pointed out in context (16.4(c)). Few other direct implications in ZFC about the tree property were established until the late 1980's. On the other hand, consistency results in the early 1970's involving large cardinals have considerably clarified the possibilities at accessible cardinals like ω_2 (see The Tree Property in volume II).

Partitions of All Finite Subsets

Erdős, Hajnal, and Rado formulated a further partition relation that was to invite the infusion of model-theoretic techniques more directly into the study of large cardinals. Recall that for $x \subseteq \mathrm{On}$, $[x]^{<\omega} = \bigcup_{n \in \omega} [x]^n$. The following partition relation first occurred in Erdős-Hajnal [58: 113]:

$$\beta \longrightarrow (\alpha)^{<\omega}_\delta$$

asserts that for any $f \colon [\beta]^{<\omega} \to \delta$ there is an $H \in [\beta]^\alpha$ *homogeneous* for f: $|f``[H]^n| \leq 1$ for every $n \in \omega$. In other words, for every n, H is homogeneous for $f|[\beta]^n$ in the former sense. For $\alpha \geq \omega$,

the *Erdős cardinal* $\kappa(\alpha)$ is the least λ such that $\lambda \longrightarrow (\alpha)^{<\omega}_2$.

Implicit in the use of this notation will be the assumption that there *is* some λ such that $\lambda \longrightarrow (\alpha)^{<\omega}_2$; unlike for the ordinary partition relation, this cannot be established in ZFC (7.15(b)). Next, the fixed points of the sequence of Erdős cardinals are specified:

$$\kappa \text{ is } \textit{Ramsey iff } \kappa \longrightarrow (\kappa)_2^{<\omega} .$$

Thus, *Ramsey cardinals are weakly compact*. Historically, the first comment on these partition relations was the following, which showed that the term "Ramsey" is not altogether appropriate.

7.13 Exercise (Erdős-Rado [52: 435]). *ω is not Ramsey.*

Hint. f: $[\omega]^{<\omega} \to 2$ given by $f(k_1, \ldots, k_n) = 0$ if $k_1 \leq n$, and $= 1$ otherwise, has no infinite homogeneous set. ⊣

The following proofs are typical of the early combinatorics:

7.14 Proposition. *Suppose that $\alpha \geq \omega$. Then:*

(a) If $\alpha < \beta$, then $\kappa(\alpha) < \kappa(\beta)$.

(b) $\kappa(\alpha)$ is regular.

(c) κ is Ramsey iff for any $\gamma < \kappa$, $\kappa \longrightarrow (\kappa)_\gamma^{<\omega}$.

Proof. (a) For each $\gamma < \kappa(\alpha)$ let f_γ: $[\gamma]^{<\omega} \to 2$ have no homogeneous set of ordertype α. Define g: $[\kappa(\alpha)]^{<\omega} \to 2$ by $g(\xi_1, \ldots, \xi_n) = f_{\xi_n}(\xi_1, \ldots, \xi_{n-1})$ (and $=$ 0 for $n = 1$). If H is homogeneous for g, then for $\gamma < \kappa(\alpha)$ with $\gamma \in H$, $H \cap \gamma$ is homogeneous for f_γ and so has ordertype $< \alpha$. Hence, H has ordertype $\leq \alpha$, and so $\kappa(\alpha) < \kappa(\beta)$.

(b) Assume to the contrary that $\kappa(\alpha)$ is singular. Then there is a non-decreasing function h: $\kappa(\alpha) \to \delta$ where $\delta < \kappa(\alpha)$ and $|h^{-1}(\{\gamma\})| < \kappa(\alpha)$ for each $\gamma < \delta$. Let f: $[\delta]^{<\omega} \to 2$ and f_γ: $[h^{-1}(\{\gamma\})]^{<\omega} \to 2$ for $\gamma < \delta$ each have no homogeneous set of ordertype α. Define g: $[\kappa(\alpha)]^{<\omega} \to 2$ by

$$g(\xi_1, \ldots, \xi_n) = \begin{cases} 0 & \text{if } n = 2 \text{ and } h(\xi_1) = h(\xi_2) , \\ 1 & \text{if } n = 2 \text{ and } h(\xi_1) < h(\xi_2) , \\ f_\gamma(\xi_3, \ldots, \xi_n) & \text{if } n > 2 \text{ and } h(\xi_1) = \ldots = h(\xi_n) = \gamma , \\ f(h(\xi_3), \ldots, h(\xi_n)) & \text{if } n > 2 \text{ and } h(\xi_1) < \ldots < h(\xi_n) , \\ 0 & \text{otherwise} . \end{cases}$$

Let $\overline{H} \in [\kappa(\alpha)]^\alpha$ be homogeneous for g, and ρ_1 and ρ_2 the first two elements of $\overline{H}$. Since $\alpha \geq \omega$, $H = \overline{H} - \{\rho_1, \rho_2\}$ also has ordertype α. If $g``[\overline{H}]^2 = \{0\}$, then $h``\overline{H} = \{\gamma\}$ for some $\gamma < \delta$, and using ρ_1 and ρ_2 as ξ_1 and ξ_2 in the third clause of g it follows that H is homogeneous for f_γ, a contradiction. If $g``[\overline{H}]^2 = \{1\}$, then $h``\overline{H}$ has ordertype α, and using ρ_1 and ρ_2 as ξ_1 and ξ_2 in the fourth clause of g, it follows that $h``H$ is homogeneous for f, also a contradiction. $\kappa(\alpha)$ is hence regular.

(c) Suppose that $f\colon [\kappa]^{<\omega} \to \gamma$ with $\gamma < \kappa$. Define $g\colon [\kappa]^{<\omega} \to 2$ by: $g(\xi_1, \ldots, \xi_n) = 0$ if $n = 2m$ and $f(\xi_1, \ldots, \xi_m) = f(\xi_{m+1}, \ldots, \xi_n)$, and $g(\xi_1, \ldots, \xi_n) = 1$ otherwise. Let $H \in [\kappa]^\kappa$ be homogeneous for g, and $n = 2m$. Since $\gamma < \kappa$, there must be $s, t \in [H]^m$ with $\max(s) < \min(t)$ such that $f(s) = f(t)$. Thus, $g(s \cup t) = 0$ and so by homogeneity $g``[H]^n = \{0\}$. For any $u, v \in [H]^m$, if $w \in [H]^m$ with $\max(u), \max(v) < \min(w)$, then $f(u) = f(w) = f(v)$. Hence H is homogeneous for f as well. ⊣

For cardinals α Erdős-Hajnal-Rado [65: §17] labored with combinatorial methods to show that $\kappa(\alpha)$ must be large, but far better results were soon to be achieved using model-theoretic techniques. The following basic result generalizing 7.14(c) was first established by these means; the proof given here is simpler and based on an idea from Baumgartner-Galvin [78]:

7.15 Proposition (Silver [66]). *Suppose that $\alpha \geq \omega$ is a limit ordinal. Then:*

(a) For any $\gamma < \kappa(\alpha)$, $\kappa(\alpha) \longrightarrow (\alpha)^{<\omega}_\gamma$.

(b) $\kappa(\alpha)$ is inaccessible.

Proof. (a) Set $\kappa = \kappa(\alpha)$, and note first that by the proof of 7.14(c), $\kappa \longrightarrow (\alpha)^{<\omega}_t$ for any $t \in \omega$.

Suppose now that $f\colon [\kappa]^{<\omega} \to \gamma$ with $\gamma < \kappa$, and let $g\colon [\gamma]^{<\omega} \to 2$ have no homogeneous set of ordertype α. Define $h\colon [\kappa]^{<\omega} \to 4$ by setting $h(\xi_1, \ldots, \xi_n) = 0$ unless $n = 2^i 3^j$ for some $i, j > 0$, in which case:

$$h(\xi_1, \ldots, \xi_n) = \begin{cases} 0 & \text{if } f(\xi_1, \ldots, \xi_i) = f(\xi_{i+1}, \ldots, \xi_{2i})\,, \\ 1 & \text{if } f(\xi_1, \ldots, \xi_i) > f(\xi_{i+1}, \ldots, \xi_{2i})\,, \\ 2 & \text{if } \langle f(\xi_{ik+1}, \ldots, \xi_{i(k+1)}) \mid k < j \rangle\,, \\ & \quad \text{is an increasing enumeration of} \\ & \quad j \text{ ordinals homogeneous for } g\,, \text{ and} \\ 3 & \text{otherwise}\,. \end{cases}$$

By the initial remark, there is an $H \in [\kappa]^\alpha$ homogeneous for h. If $h``[H]^{<\omega} = \{0\}$, then $H$ is homogeneous for $f$, by the argument for 7.14(c). So, assume to the contrary that for some $\overline{n} = 2^{\overline{i}} 3^{\overline{j}}$, $h``[H]^{\overline{n}} \neq \{0\}$ to derive a contradiction:

Note first that $h``[H]^{\overline{n}} \neq \{1\}$ also, else there would be an infinite descending sequence of ordinals. If $\langle \zeta_\beta \mid \beta < \alpha \rangle$ is the increasing enumeration of H, set $\eta_\beta = f(\zeta_{\overline{i}\beta+1}, \ldots, \zeta_{\overline{i}(\beta+1)})$ for every $\beta < \alpha$, possible since α is a limit ordinal. $\langle \eta_\beta \mid \beta < \alpha \rangle$ must be a strictly increasing sequence, since $h``[H]^{\overline{n}} \neq \{0\}, \{1\}$. In particular, for any natural number of form $n = 2^{\overline{i}} 3^j$ for arbitrary $j > 0$, it must also be the case that $h``[H]^n \neq \{0\}, \{1\}$. It will now be established that $h``[H]^n = \{2\}$ for every such n. This would imply that $\{\eta_\beta \mid \beta < \alpha\}$ is homogeneous for g, since every finite subset of it is, contradicting the choice of g and thereby completing the proof.

To do this for a given $n = 2^{\bar{i}}3^j$ with $j > 0$, apply Ramsey's Theorem 7.7 j times to get an infinite $W \subseteq \{\eta_\beta \mid \beta < \omega\}$ homogeneous for every $g|[\gamma]^k$ with $k < j$. Let $\eta_{\beta_1} < \ldots < \eta_{\beta_j}$ be the first j elements of W. Then h on any n-tuple starting with

$$\zeta_{\bar{i}\beta_1+1}, \ldots, \zeta_{\bar{i}(\beta_1+1)}, \zeta_{\bar{i}\beta_2+1}, \ldots, \zeta_{\bar{i}(\beta_2+1)}, \ldots, \zeta_{\bar{i}\beta_j+1}, \ldots, \zeta_{\bar{i}(\beta_j+1)}$$

has value 2. Hence, $h``[H]^n = \{2\}$ by homogeneity.

(b) Since $\kappa(\alpha)$ is regular by 7.14(b), it remains to show that it is a strong limit. But if $\lambda < \kappa$ yet $2^\lambda \geq \kappa$, then 7.4 for λ would imply that $\kappa \not\rightarrow (3)^2_\lambda$, contradicting (a). ⊣

In contrast to 7.15(b), $\kappa(\alpha + n + 1)$ for $\alpha \geq \omega$ and $n < \omega$ is a cardinal accessible from $\kappa(\alpha)$; in the notation of 7.3, $\kappa(\alpha + n + 1) = \text{beth}_n(\kappa(\alpha))^+$ (S. Thompson – see Drake [74: 221]).

The first inkling that inaccessibility was involved at all was a special case of 7.15(b) attributed to Géza Fodor in Erdős-Hajnal [58: 125]. Erdős and Hajnal also established that *measurable cardinals are Ramsey*. In their [62], they essentially observed the simple 7.14(a), and pointed out how the result that the least inaccessible cardinal is not measurable could thus have been deduced by these straightforward means, a couple of years before the Hanf-Tarski result. This would have been a considerable coup for the Hungarians and their partition calculus, and one is left to speculate on how it was missed. Erdős was surely aware of the problem from Erdős-Tarski [43], but the thrust of Erdős-Hajnal [58] was in a different direction, and the Fodor result may have been regarded as anomalous and peripheral. To be sure, this combinatorial approach falls far short of the Hanf breakthrough, and does not even suffice to establish that measurable cardinals are Mahlo.

Further showing the model-theoretic approach to advantage Reinhardt and Silver observed that even $\kappa(\omega)$ is larger than the least weakly compact cardinal:

7.16 Proposition (Reinhardt-Silver [65]). *There is a totally indescribable cardinal below $\kappa(\omega)$.*

A proof of this result is sketched in the appropriate setting (9.18). Erdős cardinals have since been generalized and shown to be large in natural hierarchical terms (see Subtle Properties in volume II).

Using normality the Ramsey property of measurable cardinals was established in a strong sense by Rowbottom:

7.17 Theorem (Rowbottom [64, 71]). *Suppose that κ is measurable and U is a normal ultrafilter over κ. Then if $f\colon [\kappa]^{<\omega} \to \gamma$ where $\gamma < \kappa$, there is a set in U homogeneous for f.*

Proof. If for each $n \in \omega$ there were sets $X_n \in U$ homogeneous for $f|[\kappa]^n$, then $\bigcap_{n\in\omega} X_n \in U$ would be as required. Thus, it suffices to establish the following for every $n \in \omega$: for any $g\colon [\kappa]^n \to \gamma$ with $\gamma < \kappa$, there is a set in U homogeneous for g.

Proceeding by induction, the $n = 1$ case is clear from the κ-completeness of U. So, assume that the assertion holds for $n \geq 1$, and suppose that $g\colon [\kappa]^{n+1} \to \gamma$ where $\gamma < \kappa$. For each $s \in [\kappa]^n$ define $g_s\colon \kappa \to \gamma$ by:

$$g_s(\beta) = \begin{cases} g(s \cup \{\beta\}) & \text{if } \max(s) < \beta\ , \\ 0 & \text{otherwise}\ . \end{cases}$$

By κ-completeness, for each $s \in [\kappa]^n$ there is a $\delta_s < \gamma$ and a $Y_s \in U$ such that $g_s``Y_s = \{\delta_s\}$. By induction hypothesis, there is a fixed $\delta < \gamma$ and a $Z \in U$ such that $s \in [Z]^n$ implies that $\delta_s = \delta$. Finally, by κ-completeness $Z_\alpha = \bigcap\{Y_s \mid \max(s) \leq \alpha\} \in U$ for each $\alpha < \kappa$, so that by normality $H = Z \cap \Delta_{\alpha<\kappa} Z_\alpha \in U$. The proof is completed by checking that $g``[H]^{n+1} = \{\delta\}$: Suppose that $t \in [H]^{n+1}$, written $t = s \cup \{\beta\}$ where $\max(s) < \beta$. Then $g(t) = g_s(\beta) = \delta_s = \delta$ since $\beta \in Z_{\max(s)} \subseteq Y_s$ and $s \in [Z]^n$. ⊣

7.18 Corollary (Erdős-Hajnal [58: 125]). *Measurable cardinals are Ramsey.* ⊣

Rowbottom also observed that the least Ramsey cardinal is not measurable; this follows quickly from a typical indescribability argument:

7.19 Exercise. *If κ is measurable and U is a normal ultrafilter over κ, then*

$$\{\alpha < \kappa \mid \alpha \text{ is Ramsey}\} \in U\ .$$

Hint. There is a Π^1_2 sentence φ such that κ is Ramsey *iff* $\langle V_\kappa, \in\rangle \models \varphi$, so the result follows from 6.5. ⊣

In the next section partitions of all finite subsets are correlated with model-theoretic transfer principles, and drastic consequences for L are deduced.

8. Partitions and Structures

Frederick Rowbottom, a student of Keisler at the University of Wisconsin, investigated partition properties of measurable cardinals in his doctoral dissertation [64]. An important result of his appeared at the end of the previous section, and here we pursue his main line of argument, based on a basic characterization of model-theoretic transfer principles. With this conceptual breakthrough he was able to derive substantial statements about the distance between L and V.

A structure $\mathcal{A} = \langle A, R, \ldots \rangle$ for a *countable* first-order language with a distinguished unary predicate interpreted by $R \subseteq A$ is of *type* $\langle \kappa, \lambda \rangle$ *iff* $|A| = \kappa$ and $|R| = \lambda$.

$\langle \kappa, \lambda \rangle \twoheadrightarrow \langle \mu, \nu \rangle$ *iff* whenever $\mathcal{A}$ is of type $\langle \kappa, \lambda \rangle$, there is a $\mathcal{B} \prec \mathcal{A}$ of type $\langle \mu, \nu \rangle$.

$\langle \kappa, \lambda \rangle \twoheadrightarrow \langle \mu, <\nu \rangle$ *iff* whenever $\mathcal{A}$ is of type $\langle \kappa, \lambda \rangle$, there is a $\mathcal{B} \prec \mathcal{A}$ of type $\langle \mu, \rho \rangle$ for some $\rho < \nu$.

These are evidently strong, two-cardinal versions of the Löwenheim-Skolem Theorem. The transfer principle $\langle \kappa, \lambda \rangle \twoheadrightarrow \langle \mu, \nu \rangle$ is known as *Chang's Conjecture for the pairs* (κ, λ), (μ, ν), after the attribution to Chen-Chung Chang in Vaught [63: 309]. A test case was soon singled out:

$$\textit{Chang's Conjecture} \text{ is } \langle \omega_2, \omega_1 \rangle \twoheadrightarrow \langle \omega_1, \omega \rangle .$$

Of course, this is equivalent to $\langle \omega_2, \omega_1 \rangle \twoheadrightarrow \langle \omega_1, <\omega_1 \rangle$ since an infinite unary predicate cannot have a finite counterpart in any elementary substructure. Generally speaking, the less stringent $\langle \mu, <\nu \rangle$ formulation is more immediate for large cardinal postulations, but further analysis can produce a range of $\langle \mu, \nu \rangle$ conclusions (cf. 8.5(b)).

Rowbottom established an equivalence between the model-theoretic $\twoheadrightarrow$ concept and partition relations somewhat different from the types that have been discussed. In a nice reversal, this major application preceded the introduction of the partition symbol:

First, the *square brackets partition relation*

$$\beta \longrightarrow [\alpha]^\gamma_\delta$$

of Erdős-Hajnal-Rado [65: 144] asserts that for any $f\colon [\beta]^\gamma \to \delta$ there is an $H \in [\beta]^\alpha$ such that $f``[H]^\gamma \neq \delta$. That is, f on $[H]^\gamma$ omits at least one value, a far weaker conclusion than for the ordinary partition relation. As with that relation the focus is mostly on the case $\gamma < \omega$, although Erdős-Hajnal [66] was able to establish with a clever use of the Axiom of Choice that $\kappa \not\longrightarrow [\kappa]^\omega_\kappa$ for any κ (23.13), a result that was to impose an ultimate limitation on large cardinal hypotheses. The main incentives in the study of square bracket relations lie in the investigation of their possible *negations* as strong combinatorial propositions (see Erdős-Hajnal-Máté-Rado [84] and Todorčević [87]).

The version

$$\beta \longrightarrow [\alpha]^{\gamma}_{\delta,<\eta}$$

asserts that for any $f\colon [\beta]^{\gamma} \to \delta$ there is an $H \in [\beta]^{\alpha}$ such that $|f``[H]^{\gamma}| < \eta$.

Finally, as with the ordinary partition relation Erdős-Hajnal-Rado [65: 156] considered stronger forms:

$$\beta \longrightarrow [\alpha]^{<\omega}_{\delta}$$

asserts that for any $f\colon [\beta]^{<\omega} \to \delta$ there is an $H \in [\beta]^{\alpha}$ such that $f``[H]^{<\omega} \neq \delta$, and

$$\beta \longrightarrow [\alpha]^{<\omega}_{\delta,<\eta}$$

has the analogous meaning.

The basic Rowbottom discovery can now be stated:

8.1 Theorem (Rowbottom [64, 71]). *Suppose that $\kappa \geq \lambda$ and $\kappa \geq \mu \geq \nu > \omega$. Then the following are equivalent:*

(a) $\langle \kappa, \lambda \rangle \twoheadrightarrow \langle \mu, < \nu \rangle$.

(b) $\kappa \to [\mu]^{<\omega}_{\lambda,<\nu}$.

Proof. In the forward direction, suppose that $f\colon [\kappa]^{<\omega} \to \lambda$. Consider the structure

$$\mathcal{A} = \langle \kappa, \lambda, \in, f|[\kappa]^n \rangle_{n\in\omega} .$$

By hypothesis, there is an $H \in [\kappa]^{\mu}$ with $|\lambda \cap H| < \nu$ such that

$$\langle H, \lambda \cap H, \in, f|[H]^n \rangle_{n\in\omega} \prec \mathcal{A} .$$

Clearly, this H is as desired.

For the converse, we Skolemize of course. Suppose that $\mathcal{A} = \langle A, R, \ldots \rangle$ is a structure of type $\langle \kappa, \lambda \rangle$; it can be assumed by relabeling that $A = \kappa$ and $R = \lambda$. Let $\{h_n \mid n \in \omega\}$ be a complete set of Skolem functions for $\mathcal{A}$ (§0), say with h_n $k(n)$-ary where $k(n) \leq n$. Define $f\colon [\kappa]^{<\omega} \to \lambda$ by

$$f(\xi_1, \ldots, \xi_n) = \begin{cases} h_n(\xi_1, \ldots, \xi_{k(n)}) & \text{if this is less than } \lambda \text{ , and} \\ 0 & \text{otherwise .} \end{cases}$$

By (b), let $H \in [\kappa]^{\mu}$ be such that $|f``[H]^{<\omega}| < \nu$, and set $B = \bigcup_n h_n``[H]^{k(n)}$. Then $|B| = \mu$ and B is the domain of a structure $\langle B, \lambda \cap B, \ldots \rangle \prec \mathcal{A}$. Also, $|\lambda \cap B| < \nu$ as $\lambda \cap B \subseteq f``[H]^{<\omega}$, and the proof is complete. ⊣

It is a measure of modern sophistication that this characterization is nowadays regarded as entirely straightforward, with the two statements considered synoptic; of course, we are standing on the shoulders of Skolem [23].

Some forms of these transfer principles have simpler combinatorial equivalences. In fact, before Erdős, Hajnal, and Rado developed their notations, they had pondered ([65: 154]) the relation $\omega_2 \longrightarrow [\omega_1]^2_{\omega_1,<\omega_1}$ in connection with a problem of Ulam, a relation that turns out to be equivalent to Chang's Conjecture. As an aside, a generalization of this is established:

8.2 Proposition (Erdős-Hajnal [74: 275] for $n = 2$). *For* $1 < n < \omega$,

$$\omega_n \longrightarrow [\omega_1]^{<\omega}_{\omega_1, <\omega_1} \quad \textit{iff} \quad \omega_n \longrightarrow [\omega_1]^n_{\omega_1, <\omega_1} .$$

In particular, Chang's Conjecture holds iff $\omega_2 \longrightarrow [\omega_1]^2_{\omega_1, <\omega_1}$.

Proof. The following will be established for $n \in \omega$ by induction:

($*$) For any $f\colon [\omega_n]^{<\omega} \to \omega_1$, there is a $g\colon [\omega_n]^n \to \omega_1$ such that: for any $s \in [\omega_n]^{<\omega}$ with $|s| \geq n$, there is a $t \in [s]^n$ satisfying $\max(f``[s]^{<\omega}) \leq g(t)$.

The proof would then be complete, since given such f with corresponding g, if $H \subseteq \omega_n$ and $g``[H]^n$ is countable, then $\sup(f``[H]^{<\omega}) \leq \sup(g``[H]^n) < \omega_1$ and so $f``[H]^{<\omega}$ is also countable.

($*$) for $n = 0$ is immediate, as $[\omega_0]^0 = \{0\}$ and so $g(0) = \sup(\mathrm{ran}(f))$ works.

Suppose next that ($*$) holds for n, and $f\colon [\omega_{n+1}]^{<\omega} \to \omega_1$. For $\alpha < \omega_{n+1}$ let $p_\alpha\colon \alpha \to \omega_n$ be injective, and define $f_\alpha\colon [\omega_n]^{<\omega} \to \omega_1$ by

$$f_\alpha(x) = \max(f``[p_\alpha^{-1}(x) \cup \{\alpha\}]^{<\omega}) .$$

By the induction hypothesis, let $g_\alpha\colon [\omega_n]^n \to \omega_1$ verify ($*$) for f_α. Finally, define $g\colon [\omega_{n+1}]^{n+1} \to \omega_1$ by

$$g(t) = g_\alpha(p_\alpha``(t - \{\alpha\})) \text{ where } \alpha = \max(t) .$$

If $|s| \geq n + 1$ where $\alpha = \max(s)$, then

$$\begin{aligned} \max(f``[s]^{<\omega}) = f_\alpha(p_\alpha``(s - \{\alpha\})) &\leq g_\alpha(y) \text{ for some } y \in [p_\alpha``(s - \{\alpha\})]^n \\ &= g(t) \text{ where } t = p_\alpha^{-1}(y) \cup \{\alpha\} , \end{aligned}$$

and so the proof is complete. ⊣

The exponent here is best possible in the sense that with GCH, $\omega_n \longrightarrow (\omega_2)^{n-1}_{\omega_1}$ by the Erdős-Rado Theorem 7.3. Todorčević [94] established in 1991 that Chang's Conjecture is in fact equivalent to $\omega_2 \longrightarrow [\omega_1]^3_{\omega_1}$; the exponent here is best possible in the sense that with CH, 7.3 implies that $\omega_2 \longrightarrow (\omega_1)^2_\omega$, and so $\omega_2 \longrightarrow [\omega_1]^2_{\omega_1}$. Silver established in 1967 that if there is a κ satisfying $\kappa \longrightarrow (\omega_1)^{<\omega}_2$, then there is a forcing extension in which Chang's Conjecture holds (see Kurepa's Hypothesis and Chang's Conjecture in volume II).

Rowbottom Cardinals

The conspicuous feature of measurable cardinals that concerned Rowbottom is the following. For $\omega < \nu < \kappa$,

$$\kappa \text{ is } \nu\text{-}\textit{Rowbottom} \quad \textit{iff} \quad \kappa \longrightarrow [\kappa]^{<\omega}_{\lambda,<\nu} \text{ for any } \lambda < \kappa \text{ , and}$$
$$\kappa \text{ is } \textit{Rowbottom} \quad \textit{iff} \quad \kappa \text{ is } \omega_1\text{-Rowbottom .}$$

By 7.14(c), *a Ramsey cardinal is Rowbottom.*

Gaifman (see 9.1) and Rowbottom independently and almost concurrently established the first informative structural results about the incompatibility of large cardinals with $V = L$. Whereas both Scott (5.5) and Gaifman made global use of ultrapowers, Rowbottom showed how strong local conclusions can be derived from partition properties alone. The following theorem typifies his results; although they were soon subsumed by the definitive work of Silver (§9), Gaifman and Rowbottom were first to make such remarkable statements about L and the low levels of the cumulative hierarchy. The proof is notably short, benefiting somewhat from hindsight:

8.3 Theorem (Rowbottom [64, 71]). *Suppose that there are $\kappa > \lambda > \omega$ and $\kappa \geq \mu > \omega$ such that $\langle \kappa, \lambda \rangle \twoheadrightarrow \langle \mu, < \omega_1 \rangle$. (For instance, suppose that there is a Rowbottom cardinal, or there is a κ satisfying $\kappa \longrightarrow (\omega_1)^{<\omega}_2$, or Chang's Conjecture $\langle \omega_2, \omega_1 \rangle \twoheadrightarrow \langle \omega_1, \omega \rangle$ holds.) Then ω_1 is inaccessible in L, and so in particular $\mathcal{P}(\omega)^L$ is countable.*

Proof. It suffices to show for any $\alpha < \omega_1$ that $\mathcal{P}(\alpha)^L$ is countable, for then ω_1 being regular must be inaccessible in L.

Given such an α, let

$$\mathcal{A} = \langle L_\kappa, L_\lambda, \in, \beta \rangle_{\beta \leq \alpha} \ ,$$

a structure in a countable language. By hypothesis, there is a

$$\mathcal{B} = \langle X, R, \in, \beta \rangle_{\beta \leq \alpha} \prec \mathcal{A}$$

with $|X| = \mu$ and $|R| < \omega_1$. By 0.4, $\mathcal{B}$ has a transitive collapse $\mathcal{T}$, and by 3.3(a), $\mathcal{T}$ is of form

$$\mathcal{T} = \langle L_\gamma, S, \in, \beta \rangle_{\beta \leq \alpha}$$

for some $\gamma \geq \mu$. Let i be the inverse of the collapsing isomorphism, so that $i\colon \mathcal{T} \prec \mathcal{A}$ with $i(\beta) = \beta$ for $\beta \leq \alpha$. If $x \in \mathcal{P}(\alpha) \cap L_\gamma$, this implies that $i(x) = x$, and since $\mathcal{P}(\alpha)^L \subseteq L_{\omega_1} \subseteq L_\lambda$, it follows that $x \in S$ by the elementarity of i. But also $\mathcal{P}(\alpha) \cap L_\gamma = \mathcal{P}(\alpha)^L$ since $\gamma \geq \mu > \omega$, and so $\mathcal{P}(\alpha)^L \subseteq S$ and is therefore countable. ⊣

Historically, Rowbottom's first insight was the conclusion that any ordinal definable in L (like $|\mathcal{P}(\omega)|^L$, $|\mathcal{P}(\mathcal{P}(\omega))|^L$, and so forth) is countable.

As one might have guessed, the restriction to countable languages in 8.1 and 8.3 is not essential. Rowbottom made a general statement along the lines of 8.1 about getting homogeneous sets simultaneously for *collections* of functions corresponding to many Skolem functions, and then observed that Ramsey cardinals provide the requisite strength. The following are useful variants with single functions:

8.4 Theorem (Rowbottom [64, 71]). *Suppose that $\langle A, R, \ldots\rangle$, where $R \subseteq A$, is a structure for a (first-order) language with less than ν non-logical symbols, and $|A| \geq \kappa$ and $|R| = \lambda < \kappa$. Assume that either*

(a) $\lambda^{<\nu} < \kappa$ and κ is ν-Rowbottom, or

(b) $\lambda = \nu$ is regular and $\kappa \longrightarrow [\kappa]^{<\omega}_{\nu,<\nu}$.

Then there is a $\langle B, R \cap B, \ldots\rangle \prec \langle A, R, \ldots\rangle$ such that $|B| = \kappa$ and $|R \cap B| < \nu$.

Proof. Assuming that $A \supseteq \kappa$ and $R = \lambda$ by relabeling and following the proof of 8.1, let $\{h_\alpha \mid \alpha < \rho\}$ be a complete set of Skolem functions for the language, where $\rho < \nu$ and h_α is $k(\alpha)$-ary.

First assume (a). Let f be the function on $[\kappa]^{<\omega}$ defined by

$$f(\xi_1, \ldots, \xi_n) = \{h_\alpha(\xi_1, \ldots, \xi_n) \mid \alpha < \rho \ \wedge \ n = k(\alpha) \ \wedge \ h_\alpha(\xi_1, \ldots, \xi_n) \in \lambda\}\,.$$

Thus $|\mathrm{ran}(f)| \leq \lambda^\rho$, and so there is an $H \in [\kappa]^\kappa$ such that $|f``[H]^{<\omega}| < \nu$. Set $B = \bigcup_{\alpha<\rho} h_\alpha``[H]^{k(\alpha)}$. Then $\langle B, \lambda \cap B, \ldots\rangle \prec \langle A, \lambda, \ldots\rangle$, and

$$|\lambda \cap B| \leq |\textstyle\bigcup f``[H]^{<\omega}| < \rho \cdot |f``[H]^{<\omega}| < \nu\,.$$

Assuming (b) instead, define $\overline{f}$ on $[\kappa]^{<\omega}$ from the f above by

$$\overline{f}(\xi_1, \ldots, \xi_n) = \sup(f(\xi_1, \ldots, \xi_n))\,.$$

Then $\mathrm{ran}(\overline{f}) \subseteq \nu$ by the regularity of $\lambda = \nu$. The argument can now be completed as before, invoking the regularity of ν again to insure for the corresponding B that $|\lambda \cap B| < \nu$. ⊣

8.5 Corollary. *Suppose that κ is a Ramsey cardinal. Then:*

(a) For any infinite $\alpha < \kappa$, $|\mathcal{P}(\alpha)^L| = |\alpha|$, and so every uncountable regular $\lambda < \kappa$ is inaccessible in L.

(b) For any $\nu \leq \lambda < \kappa$, $\langle\kappa, \lambda\rangle \twoheadrightarrow \langle\kappa, \nu\rangle$.

Proof. (a) This follows from the proof of 8.3 using uncountable languages.

(b) Given a structure $\langle A, R, \ldots\rangle$ of type $\langle\kappa, \lambda\rangle$ (in a countable language), let $S \in [R]^\nu$ and consider the expansion $\langle A, R, \ldots, x\rangle_{x\in S}$. Applying 8.4 with its ν replaced by ν^+, this expansion has an elementary substructure

$$\langle B, R \cap B, \ldots, b_x\rangle_{x\in S}$$

such that $|B| = \kappa$ and $|R \cap B| \leq \nu$. But then, $|R \cap B| = \nu$ because of the b_x's, and the reduct $\langle B, R \cap B, \ldots\ \rangle$ to the original language is of type $\langle\kappa, \nu\rangle$. ⊣

The postulation of Rowbottom cardinals marked a historic reversal, in that it was designed for strong implications like 8.3 but does not appear to have substantial size consequences. The only evident constraint is the following:

8.6 Exercise. *For any* λ,

(a) $\lambda^+ \not\rightarrow [\lambda+1]^2_{\lambda,<\lambda}$.

(b) $\lambda \not\rightarrow [\lambda]^1_{\mathrm{cf}(\lambda),<\mathrm{cf}(\lambda)}$.

Hence, a ν*-Rowbottom cardinal* κ *is either weakly inaccessible or has cofinality less than* ν.

Hint. (a) For each $\beta < \lambda^+$ let $f_\beta : \beta \to \lambda$ be injective, and consider $g : [\lambda^+]^2 \to \lambda$ defined by: $g(\alpha, \beta) = f_\beta(\alpha)$.

(b) Let $\langle \gamma_\alpha \mid \alpha < \mathrm{cf}(\kappa) \rangle$ be cofinal in κ, and consider $h : \lambda \to \mathrm{cf}(\lambda)$ defined by: $h(\xi) =$ the least α such that $\xi < \gamma_\alpha$. ⊣

Karel Prikry showed that singular cardinals can exhibit strong partition properties:

8.7 Theorem (Prikry [70]). *Suppose that* κ *is a singular limit of measurable cardinals. Then* κ *is* $\mathrm{cf}(\kappa)^+$*-Rowbottom.*

Proof. Set $\nu = \mathrm{cf}(\kappa)$, and let $\langle \kappa_\alpha \mid \alpha < \nu \rangle$ be the increasing enumeration of a set closed unbounded in κ with $\kappa_0 = \nu$ and $\kappa_{\alpha+1}$ measurable for each $\alpha < \nu$. Let $U_{\alpha+1}$ be a normal ultrafilter over $\kappa_{\alpha+1}$ for $\alpha < \nu$, and define D by:

$$X \in D \text{ iff } X \subseteq \kappa \ \wedge\ \exists \beta < \nu \forall \alpha (\beta \le \alpha < \nu \to X \cap \kappa_{\alpha+1} \in U_{\alpha+1}) ,$$

so that D is a uniform filter over κ. For $s \in [\kappa]^{<\omega}$ set

$$\mathrm{type}(s) = \{ \langle \alpha, k \rangle \in \nu \times \omega \mid |s \cap (\kappa_{\alpha+1} - \kappa_\alpha)| = k > 0 \} .$$

The following will be established:

(∗) For any $f : [\kappa]^{<\omega} \to \gamma$ with $\gamma < \kappa$ there is an $H \in D$ such that: for any $s_0, s_1 \in [H]^{<\omega}$ with $\mathrm{type}(s_0) = \mathrm{type}(s_1)$, $f(s_0) = f(s_1)$.

With at most ν possibilities for type(s) it will follow that $|f``[H]^{<\omega}| \le \nu$, and hence that κ is ν^+-Rowbottom.

Consider for $n \in \omega$ the assertion:

$(*)_n$ For any $f : [\kappa]^{<\omega} \to \gamma$ with $\gamma < \kappa$ there is an $H \subseteq \kappa$ such that:
  (i) for every $\kappa_{\alpha+1} > \gamma$, $H \cap \kappa_{\alpha+1} \in U_{\alpha+1}$, and
  (ii) for any $s_0, s_1 \in [H]^{<\omega}$ with $\mathrm{type}(s_0) = \mathrm{type}(s_1)$ and $|\mathrm{type}(s_0)| = n$, $f(s_0) = f(s_1)$.

If $f : [\kappa]^{<\omega} \to \gamma$ with $\gamma < \kappa$ and for each $n \in \omega$ there is an H_n satisfying $(*)_n$ for f, then by (i), $\bigcap_n H_n \in D$, and by (ii), (∗) is satisfied with $H = \bigcap_n H_n$. Thus, it suffices to establish $(*)_n$ for every $n \in \omega$ by induction:

$(*)_0$ holds vacuously, so proceeding to the inductive step, suppose that $(*)_n$ holds, and let $f : [\kappa]^{<\omega} \to \gamma$ with $\gamma < \kappa$. For $\alpha < \nu$ with $\kappa_{\alpha+1} > \gamma$ and $t \in [\kappa_\alpha]^{<\omega}$, define $f_t : [\kappa_{\alpha+1}]^{<\omega} \to \gamma$ by: $f_t(u) = f(t \cup u)$. By 7.17 there is an $X^\alpha_t \in U_{\alpha+1}$ homogeneous for f_t. For $\alpha < \nu$ with $\kappa_{\alpha+1} > \gamma$, set

$$X_\alpha = \bigcap\{X_t^\alpha \mid t \in [\kappa_\alpha]^{<\omega}\} - \kappa_\alpha$$

so that $X_\alpha \in U_{\alpha+1}$ by $\kappa_{\alpha+1}$-completeness.

Next, for any $\langle \alpha, k\rangle \in \nu \times \omega$ define $f_{\alpha,k}$: $[\kappa]^{<\omega} \to \gamma$ by

$$f_{\alpha,k}(t) = \begin{cases} f(t \cup u) & \text{if } t \in [\kappa_\alpha]^{<\omega}, \text{ where } u \text{ is some (any) member of } [X_\alpha]^k \text{ , and} \\ 0 & \text{otherwise .} \end{cases}$$

$f_{\alpha,k}$ is well-defined by definition of X_α. Let $H_{\alpha,k} \subseteq \kappa$ be for $f_{\alpha,k}$ as provided by $(*)_n$.

Finally, set

$$H = \bigcup\{X_\alpha \mid \alpha < \nu \ \wedge \ \gamma < \kappa_{\alpha+1}\} \cap \bigcap\{H_{\alpha,k} \mid \langle \alpha, k\rangle \in \nu \times \omega\} .$$

Then for $\kappa_{\alpha+1} > \gamma$, $H \cap \kappa_{\alpha+1} \in U_{\alpha+1}$ by ν^+-completeness. Suppose now that $s_0, s_1 \in [H]^{<\omega}$ with $\text{type}(s_0) = \text{type}(s_1)$ and $|\text{type}(s_0)| = n+1$. Let α be maximal such that $\langle \alpha, k\rangle \in \text{type}(s_0)$ for some k, and set $s_0 = t_0 \cup u_0$ and $s_1 = t_1 \cup u_1$ where $\text{type}(t_0) = \text{type}(t_1)$ with $|\text{type}(t_0)| = n$ and $\text{type}(u_0) = \text{type}(u_1) = \{\langle \alpha, k\rangle\}$. Then $f(s_0) = f_{\alpha,k}(t_0) = f_{\alpha,k}(t_1) = f(s_1)$, confirming $(*)_{n+1}$ for f. ⊣

Prikry also showed that a measurable cardinal can be maintained a Rowbottom cardinal while changing its cofinality to ω by forcing (18.6). However, the following is a major question that has remained impenetrable.

8.8 Question. *Can ω_ω be Rowbottom?*

For related results, see 8.16 and remarks surrounding.

Several observations about how ν-Rowbottom cardinals constrain power set cardinalities, although not difficult, were made considerably later.

8.9 Proposition (Prikry [75]). *Suppose that $2^{<\nu} = \kappa$ and κ is a regular ν-Rowbottom cardinal. Then for $\nu \leq \lambda < \kappa$, $2^\lambda = \kappa$.*

Proof. As $\nu < \kappa = \text{cf}(\kappa)$, there is a $\mu < \nu$ such that $2^\mu = \kappa$. Proceeding by induction, suppose then that $\lambda < \kappa$ and for $\mu \leq \lambda' < \lambda$, $2^{\lambda'} = \kappa$. If λ is singular, a well-known argument establishes that $2^\lambda = \kappa$: Let $\langle \gamma_\alpha \mid \alpha < \text{cf}(\lambda)\rangle$ be cofinal in λ, and for $X \subseteq \lambda$ let f_X on $\text{cf}(\lambda)$ be defined by: $f_X(\alpha) = X \cap \gamma_\alpha$. Then $X \neq Y$ implies that $f_X \neq f_Y$, and so $2^\lambda \leq (2^{<\lambda})^{\text{cf}(\lambda)} = 2^{<\lambda} = \kappa$.

Suppose next that λ is regular. Similarly considering f_X on λ for $X \subseteq \lambda$ defined by $f_X(\alpha) = X \cap \alpha$, $2^{<\lambda} = \kappa$ implies that there is a family $\mathcal{F}$ of 2^λ functions: $\lambda \to \kappa$ that are *almost disjoint*, i.e. for $f \neq g$ both in $\mathcal{F}$, $f(\alpha) \neq g(\alpha)$ for sufficiently large $\alpha < \lambda$.

Assume now to the contrary that $2^\lambda > \kappa$. Since κ is regular, for each $f \in \mathcal{F}$ there is a $\delta_f < \kappa$ such that $\text{ran}(f) \subseteq \delta_f$. Consequently, there is a $\delta < \kappa$ and a $\mathcal{G} \subseteq \mathcal{F}$ with $|\mathcal{G}| = \kappa$ such that $f \in \mathcal{G}$ implies that $\delta_f = \delta$. Enumerating

$\mathcal{G}$ as $\langle f_\xi \mid \xi < \kappa \rangle$, define $F\colon [\kappa]^2 \to \lambda$ by: $F(\xi, \zeta) =$ the least β such that $f_\xi(\alpha) \neq f_\zeta(\alpha)$ for $\beta \leq \alpha < \lambda$. Since κ is ν-Rowbottom, there is an $H \in [\kappa]^\kappa$ such that $|F``[H]^2| < \nu$. Now $\lambda$ is regular, and the assumption $2^\lambda > \kappa$ together with the hypothesis $2^{<\nu} = \kappa$ implies that $\nu \leq \lambda$, and so $\gamma = \sup(F``[H]^2) < \lambda$. But then, $\{f_\xi(\gamma) \mid \xi \in H\}$ consists of κ ordinals all less than λ, which is a contradiction. ⊣

It follows that *if* 2^ν *is* ν^+*-Rowbottom, then for* $\nu \leq \lambda < 2^\nu$, $2^\lambda = 2^\nu$. The following result complements 8.9.

8.10 Proposition (Tryba [86]). *Suppose that* $2^{<\nu} < \kappa$, *and* κ *is* ν*-Rowbottom. Then* κ *is a strong limit cardinal.*

Proof. Assume to the contrary that $\lambda < \kappa$ yet $2^\lambda \geq \kappa$. Arguing as for 8.1, let $\langle M, \in \rangle \prec \langle V_{\kappa+\omega}, \in \rangle$ be such that $|M| = \kappa = |M \cap \kappa|$ and $|M \cap \lambda| < \nu$. Let N be the transitive collapse of M, and $j\colon \langle N, \in \rangle \prec \langle V_{\kappa+\omega}, \in \rangle$ the inverse of the collapsing isomorphism, so that $j(\kappa) = \kappa$ and $j(\delta) = \lambda$ for some $\delta < \nu$. By assumption there is an injection: $\kappa \to \mathcal{P}(\lambda)$ in $V_{\kappa+\omega}$, and so there is an injection: $\kappa \to \mathcal{P}(\delta)$ in N. But this contradicts the hypothesis $2^{<\nu} < \kappa$. ⊣

In particular, *if* ω_ω *is Rowbottom and* $2^{\aleph_0} < \omega_\omega$, *then* ω_ω *is a strong limit cardinal.* Further results along these lines appear in Tryba [86]. A similar argument establishes the following:

8.11 Exercise (Tryba [81]). *If* κ *is* ν*-Rowbottom and there is a limit cardinal* μ *such that* $\nu \leq \mu < \kappa$, *then* κ *is a limit of limit cardinals.*

Hint. Assume to the contrary that λ is the largest limit cardinal less than κ. Let $j\colon \langle N, \in \rangle \prec \langle V_{\kappa+\omega}, \in \rangle$ be as in the proof of 8.10 for this λ. Now note that μ is a limit cardinal in M to derive a contradiction. ⊣

It follows that $\omega_{\omega+\omega}$ *is not Rowbottom, and in fact the least Rowbottom cardinal* $> \omega_\omega$ *is* $\geq \omega_{\omega\cdot\omega}$. Hypothesizing that μ is an inaccessible cardinal, a Mahlo cardinal, etc., lead to analogous results about the size of κ. Also, the same means establish that *if* κ *is* ν*-Rowbottom and* $\omega_\nu < \kappa$, *then* $\kappa = \omega_\kappa$.

Little else is known about ν-Rowbottom cardinals combinatorially or consistency-wise which is not already a consequence of a related concept, to which we now turn.

Jónsson Cardinals

An initiative from a different quarter completed the transition from the study of ordinary partition relations to the consideration of an intrinsically interesting algebraic problem. Bjarni Jónsson, a student of Tarski and a prime exponent of transfinite universal algebra, posed in 1962 the following problem (see Jónsson [72: 3.9] for the context): For a cardinal κ, does every algebra of cardinality κ

have a proper subalgebra of the same cardinality? Here, an *algebra* is a structure $\mathcal{A} = \langle A, f_n \rangle_{n \in \omega}$ where for each n, f_n: $[A]^{k(n)} \to A$ for some $k(n) \in \omega$; and a *subalgebra* of it is a structure of form $\mathcal{A}_0 = \langle A_0, f_n | [A_0]^{k(n)} \rangle_{n \in \omega}$ where $A_0 \subseteq A$ and f_n"$[A_0]^{k(n)} \subseteq A_0$. Thus, these are standard model-theoretic notions for a countable language with only function symbols. Jónsson queried specifically about algebras with only finitely many operations, but the difference is inessential because of coding. A *Jónsson algebra* is an algebra without a proper subalgebra of the same cardinality. A *Jónsson cardinal* is one that has *no* Jónsson algebras of that cardinality, or stated affirmatively,

κ is *Jónsson* *iff* every algebra of cardinality κ has a proper subalgebra of the same cardinality .

Jónsson's problem is thus: Are there any Jónsson cardinals?

As with Rowbottom cardinals, Jónsson cardinals were characterized combinatorially and model-theoretically:

8.12 Exercise (Erdős-Hajnal [66] for (b), Keisler-Rowbottom [65] for (c)). *The following are equivalent:*

(a) κ *is Jónsson.*

(b) $\kappa \longrightarrow [\kappa]^{<\omega}_{\kappa}$.

(c) Any structure for a countable first-order language with domain of cardinality κ *has a proper elementary substructure with domain of the same cardinality.*

Hint. For (a) $\to$ (b), if f: $[\kappa]^{<\omega} \to \kappa$, consider the algebra $\mathcal{A} = \langle \kappa, f | [\kappa]^n \rangle_{n \in \omega}$. (b) $\to$ (c) is as for 8.1. ⊣

Keisler and Rowbottom thus saw that *if* κ *is* ν*-Rowbottom for some* $\nu < \kappa$*, then* κ *is Jónsson*. By Skolemizing, they were able to incorporate predicates and *elementary* substructures into Jónsson's problem, and establish an extension of Scott's original result on L:

8.13 Proposition (Keisler-Rowbottom [65]). *If there is a Jónsson cardinal, then* $V \neq L$.

Proof. Let κ be Jónsson, and by 8.12(c) let $\langle B, \in \rangle \prec \langle L_\kappa, \in \rangle$ with $|B| = \kappa$ and $B \neq L_\kappa$. By 0.4, $\langle B, \in \rangle$ has a transitive collapse, and by 3.3(a), it must be $\langle L_\kappa, \in \rangle$. Let i: $L_\kappa \prec L_\kappa$ then be the inverse of the collapsing isomorphism. Since $B \neq L_\kappa$, i cannot be the identity and so has a critical point δ by the argument for 5.1(b). Now define U by:

$$X \in U \quad \textit{iff} \quad X \in \mathcal{P}(\delta)^L \ \wedge \ \delta \in i(X) .$$

Since $\mathcal{P}(\delta)^L \subseteq L_{\delta^+} \subseteq L_\kappa$, if $V = L$, then (cf. 5.6) U is a δ-complete ultrafilter over δ and hence that δ is a measurable cardinal. But this contradicts Scott's result 5.5. ⊣

The conclusion here is weaker than for 8.3, but in any case these results were soon to be subsumed by the incisive work of Kunen (21.4) who showed that the strongest conclusions in this direction can be derived from these hypotheses. He also showed (20.23) that it is consistent with the existence of a measurable cardinal that a cardinal is Jónsson *iff* it is Ramsey. However, even less is known in ZFC about the size of Jónsson cardinals than of Rowbottom cardinals, and this has become a major area of investigation; the initial observations were the following:

8.14 Proposition.

(a) ω is not Jónsson.

(b) (Chang, Rowbottom, Erdős-Hajnal [66]) *If κ is not Jónsson, neither is κ^+.*

(c) (Erdős-Hajnal-Rado [65: 145]) *If $2^\kappa = \kappa^+$, then $\kappa^+ \not\to [\kappa^+]^2_{\kappa^+}$ and hence κ^+ is not Jónsson.*

(d) (Rowbottom – Devlin [73: 311]) *The least Jónsson cardinal is either weakly inaccessible or singular of cofinality ω.*

Proof. (a) If $f \in {}^\omega\omega$ is defined by $f(0) = 0$ and $f(n) = n - 1$ otherwise, then $\langle \omega, f \rangle$ is a Jónsson algebra.

(b) For $\kappa \leq \alpha < \kappa^+$, using a bijection let f_α witness $\alpha \not\to [\kappa]^{<\omega}_\alpha$. Define $g\colon [\kappa^+]^{<\omega} \to \kappa^+$ by: $g(s) = f_\alpha(s - \{\alpha\})$ if $\alpha = \max(s) \geq \kappa$, and $g(s) = 0$ otherwise. Then g witnesses $\kappa^+ \not\to [\kappa^+]^{<\omega}_{\kappa^+}$: If $X \in [\kappa^+]^{\kappa^+}$ and $\beta < \kappa^+$ is arbitrary, let $\alpha \in X$ such that $\beta < \alpha$ and $|X \cap \alpha| = \kappa$. Then for some $t \in [X \cap \alpha]^{<\omega}$, $\beta = f_\alpha(t) = g(t^\frown\{\alpha\})$.

(c) Using $2^\kappa = \kappa^+$ let $\{X_\alpha \mid \kappa \leq \alpha < \kappa^+\} = [\kappa^+]^\kappa$ with $X_\alpha \subseteq \alpha$. Then there is an $f\colon [\kappa^+]^2 \to \kappa^+$ so that for any $\kappa \leq \beta < \kappa^+$,

$$(*) \qquad \text{if } \kappa \leq \alpha < \beta \text{ and } \eta < \beta, \text{ there is a } \xi \in X_\alpha \text{ such that } f(\xi, \beta) = \eta\,.$$

To show this, for such β let $\{\langle \alpha_i, \eta_i \rangle \mid i < \kappa\} = \{\langle \alpha, \eta \rangle \mid \kappa \leq \alpha < \beta \wedge \eta < \beta\}$, recursively define $\{\xi_i \mid i < \kappa\}$ so that $\xi_i \in X_{\alpha_i} - \{\xi_j \mid j < i\}$, and then stipulate that f satisfy $f(\xi_i, \beta) = \eta_i$ for each $i < \kappa$.

To verify that f witnesses $\kappa^+ \not\to [\kappa^+]^2_{\kappa^+}$, let $S \in [\kappa^+]^{\kappa^+}$ and $\eta < \kappa^+$ be arbitrary. Take any $\alpha < \kappa^+$ so that $X_\alpha \subseteq S$, and let $\beta \in S - (\max(\{\eta, \alpha\}) + 1)$. Then $(*)$ implies that there is a $\xi \in X_\alpha$ such that $f(\xi, \beta) = \eta$.

(d) Suppose that κ is the least Jónsson cardinal. Then κ is an uncountable limit cardinal by (a) and (b), so assume to the contrary that $\omega < \mathrm{cf}(\kappa) = \lambda < \kappa$. Let $\langle \mu_\alpha \mid \alpha < \lambda \rangle$ be an increasing sequence of cardinals closed and unbounded in κ with $\lambda < \mu_0$. For each $\alpha < \lambda$ let f_α witness $\mu_\alpha \not\to [\mu_\alpha]^{<\omega}_{\mu_\alpha}$, and define $f\colon [\kappa]^{<\omega} \to \kappa$ by:

$$f(s) = \begin{cases} f_\alpha(s - \{\alpha\}) & \text{if } \alpha = \min(s) < \lambda \text{ and } \max(s) < \mu_\alpha\,, \text{ and} \\ 0 & \text{otherwise}\,. \end{cases}$$

Also, let $g\colon [\kappa]^{<\omega} \to \kappa$ be any extension of a function witnessing $\lambda \not\to [\lambda]^{<\omega}_\lambda$, and let $h\colon \kappa \to \lambda$ be defined by $h(\xi) =$ that α such that $\mu_\alpha \leq \xi < \mu_{\alpha+1}$.

We shall show that

$$\langle \kappa, f|[\kappa]^n, g|[\kappa]^n, h\rangle_{n\in\omega}$$

is a Jónsson algebra, arriving at a contradiction. So, suppose that $X \in [\kappa]^\kappa$ is the domain of a subalgebra. $|X \cap \lambda| = \lambda$ because of h, and so $X \cap \lambda = \lambda$ because of g. Assume now that $\xi < \kappa$ is arbitrary. Let $\alpha_0 < \lambda$ such that $\xi < \mu_{\alpha_0}$, and by recursion choose $\alpha_n \leq \alpha_{n+1} < \lambda$ such that $|X \cap \mu_{\alpha_{n+1}}| \geq \mu_{\alpha_n}$. Set $\beta = \sup(\{\alpha_n \mid n \in \omega\})$ so that $\beta < \lambda$ as $\lambda = \mathrm{cf}(\kappa) > \omega$. Then $|X \cap \mu_\beta| = \mu_\beta$ since the μ_α's form a closed set of ordinals. Hence, f_β"$[(X \cap \mu_\beta) - \lambda]^{<\omega} = \mu_\beta$, and so by definition of f and the fact that $\beta \in \lambda = X \cap \lambda$, $\xi \in X$. Hence $X = \kappa$, and with this contradiction the proof is complete. ⊣

The proof of (b) was considerably elaborated by Shelah [80] to show that there is a Jónsson algebra of cardinality $\aleph_1$ which is a group, answering an old question of Kurosh.

Eugene Kleinberg observed that Jónsson cardinals have Rowbottom-like properties:

8.15 Proposition (Kleinberg [73]).

(a) If κ is Jónsson, then there is an $\alpha < \kappa$ such that $\kappa \longrightarrow [\kappa]^{<\omega}_\alpha$. If δ is the least such α, then δ is regular and $\kappa \longrightarrow [\kappa]^{<\omega}_{\delta,<\delta}$.

(b) If $\kappa \longrightarrow [\kappa]^{<\omega}_\lambda$ and λ is not Jónsson, then $\kappa \longrightarrow [\kappa]^{<\omega}_{\lambda,<\lambda}$.

(c) The least Jónsson cardinal κ is ν-Rowbottom for some $\nu < \kappa$; in fact, ν can be taken to be the δ of (a) for the Jónsson cardinal.

Proof. (a) Assume that for each $\alpha < \kappa$, there is an $f_\alpha\colon [\kappa]^{<\omega} \to \alpha$ witnessing $\kappa \not\to [\kappa]^{<\omega}_\alpha$. Then it is simple to check that $f\colon [\kappa]^{<\omega} \to \kappa$ defined by $f(s) = f_\alpha(s - \{\alpha\})$ where $\alpha = \min(s)$ (and $f(\emptyset) = 0$) witnesses $\kappa \not\to [\kappa]^{<\omega}_\kappa$, which is a contradiction.

To verify the assertion about δ, it suffices to show that $\kappa \longrightarrow [\kappa]^{<\omega}_{\mathrm{cf}(\delta),<\mathrm{cf}(\delta)}$. So, assume to the contrary that this has a counterexample g. Let $e\colon \mathrm{cf}(\delta) \to \delta$ be cofinal, and f_α for $\alpha < \delta$ as above. Define $h\colon [\kappa]^{<\omega} \to \delta$ by setting $h(\xi_1, \ldots, \xi_n) = 0$ unless $n = 2^i 3^j$ for some $i, j > 0$, in which case

$$h(\xi_1, \ldots, \xi_n) = f_{e(g(\xi_1,\ldots,\xi_i))}(\xi_{i+1}, \ldots, \xi_{i+j}) .$$

Then it is straightforward to check that h witnesses $\kappa \not\to [\kappa]^{<\omega}_\delta$, which is a contradiction.

(b) We shall verify that $\langle\kappa, \lambda\rangle \twoheadrightarrow \langle\kappa, <\lambda\rangle$ (cf. 8.1). So suppose that $\mathcal{S} = \langle A, R, \ldots\rangle$ is a structure of type $\langle\kappa, \lambda\rangle$ where it can be assumed that $A = \kappa$ and $R = \lambda$, and let g be any function: $[\kappa]^{<\omega} \to \kappa$ extending a function witnessing $\lambda \not\to [\lambda]^{<\omega}_\lambda$. Expand $\mathcal{S}$ to a structure $\mathcal{A} = \langle\kappa, \lambda, \ldots, g|[\kappa]^n\rangle_{n\in\omega}$, and define a function $f\colon [\kappa]^{<\omega} \to \lambda$ just as for 8.1 from a complete set of Skolem functions

$\{h_n \mid n \in \omega\}$ for $\mathcal{A}$. By hypothesis there is an $H \in [\kappa]^\kappa$ such that $f``[H]^{<\omega} \neq \lambda$. Set $A_0 = \bigcup_n h_n``[H]^{k(n)}$, so that

$$\mathcal{A}_0 = \langle A_0, \lambda \cap A_0, \ldots, g|[A_0]^n \rangle_{n \in \omega} \prec \mathcal{A} .$$

If $|\lambda \cap A_0| = \lambda$, then $\lambda \cap A_0 = \lambda$ because of g, contradicting $f``[H]^{<\omega} \neq \lambda$. Hence, the reduct of $\mathcal{A}_0$ to the original language without g is an elementary substructure of $\mathcal{S}$ of type $\langle \kappa, \rho \rangle$ for some $\rho < \lambda$.

(c) This follows in a straightforward manner from (a) and (b). ⊣

By 8.14(a)(b), ω_ω is ostensibly the least possibility for a Jónsson cardinal. Whether this can be realized is unresolved, and is essentially the same question as 8.8 by 8.15(c) and the proof of the next result, which will be deferred until our forcing context has been established (10.18).

8.16 Theorem (Kleinberg [72, 79]). *The following theories are equiconsistent:*

(a) ZFC + $\exists\kappa(\kappa$ is Rowbottom).

(b) ZFC + $\exists\kappa(\kappa$ is Jónsson).

Devlin [73] also observed with simple forcing arguments that: (i) Con (ZFC + $\exists\kappa(\kappa$ is Jónsson)) implies Con(ZFC + $\exists\kappa(\kappa$ is Jónsson $\wedge\ \kappa \leq 2^{\aleph_0}$)), and (ii) Con(ZFC + $\exists\kappa(\kappa$ is Ramsey)) implies Con(ZFC + $\exists\kappa(\kappa$ is Jónsson) + ⌜the least Jónsson cardinal is not Rowbottom⌝). Then Silver in 1974 established with an elegant argument that if ω_ω is Jónsson and $2^{\aleph_0} < \omega_\omega$, then ω_ω is measurable in an inner model (see Kanamori-Magidor [78: 129ff] for a proof).

After this work, save for a paper Shelah [78] there was a hiatus in the study of Jónsson cardinals *per se* until a resurgence of interest in the 1980's. Then developments in inner model theory established that the existence of accessible Jónsson cardinals has substantial large cardinal consequences. Extending results of Donder-Koepke [83], Koepke [88] showed for example that *if there is a Jónsson cardinal κ such that $\kappa = \omega_\xi$ for some $\xi < \kappa$ or κ is regular but not κ-Mahlo, then for any λ there is an inner model with λ measurable cardinals.* He also showed that Prikry's 8.7 is sharp in consistency strength for uncountable cofinalities: *if there is a Jónsson cardinal κ such that $\omega <$ cf$(\kappa) < \kappa$, then there is an inner model with* cf(κ) *measurable cardinals.* Then Koepke [89] established that the existence of a successor Jónsson cardinal, the most acute possibility, has far stronger consequences than having many measurable cardinals. This complemented important developments concerning such Jónsson cardinals:

First came the following observation:

8.17 Proposition (Tryba [84], Woodin). *Suppose that λ is regular and there is an S stationary in λ such that $S \cap \alpha$ is not stationary in α for any limit ordinal $\alpha < \lambda$. Then λ is not Jónsson. In particular, if κ is regular, then κ^+ is not Jónsson.*

Proof. By a well-known result of Solovay that will be later established in context (16.9) there is a partition $S = \bigcup_{\eta<\lambda} S_\eta$ of S into disjoint stationary sets S_η.

Let $f\colon S \to \lambda$ be defined by: $f(\xi) =$ that η such that $\xi \in S_\eta$. Next, let $\langle A, \in\rangle \prec \langle V_{\lambda+\omega}, \in\rangle$ with $\lambda \cup \{S, f\} \subseteq A$ and $|A| = \lambda$, and $g\colon \lambda \to A$ a bijection. Now consider the structure

$$\mathcal{A} = \langle A, \in, \{S\}, f, g\rangle .$$

Because of g, whenever $\mathcal{B} = \langle B, \in, \{S\}, f|B, g|B\rangle \prec \mathcal{A}$, $|B| = \lambda$ *iff* $|B \cap \lambda| = \lambda$, and $B = A$ *iff* $\lambda \subseteq B$. So, to establish that λ is not Jónsson it suffices to show by 8.12(c) that for such $\mathcal{B}$, $|B \cap \lambda| = \lambda$ implies that $\lambda \subseteq B$:

Let $C = \{\xi < \lambda \mid \sup(B \cap \xi) = \xi\}$, a set closed unbounded in λ since $|B \cap \lambda| = \lambda$. Then $C \cap S \subseteq B$: Assume to the contrary that $\xi \in (C \cap S) - B$. Let α be the least member of $B \cap \lambda$ above ξ. Then α must be a limit ordinal as $\langle B, \in\rangle \prec \langle V_{\lambda+\omega}, \in\rangle$ implies that B is closed under ordinal predecessors. So $S \in B$, $\langle B, \in\rangle \prec \langle V_{\lambda+\omega}, \in\rangle$, and the hypothesis on S imply that there is a $D \in B$ such that D is closed unbounded in α in the sense of $\mathcal{B}$ and $D \cap S = \emptyset$. But then, $\xi \in C$ and $\sup(B \cap \alpha) = \xi$ implies that $D \cap B \cap \xi$ is unbounded in ξ, reaching the contradiction $\xi \in D \cap S$.

For any $\eta < \lambda$, it follows that $C \cap S_\eta \subseteq B$, and since $C \cap S_\eta \neq \emptyset$, there is a $\xi \in S_\eta \cap B$. But then, $g(\xi) = \eta \in B$. Hence, $\lambda \subseteq B$.

The last assertion follows by taking $S = \{\alpha < \kappa^+ \mid \mathrm{cf}(\alpha) = \kappa\}$: For any limit $\beta < \kappa^+$ there is a set closed unbounded in β consisting of ordinals of cofinality less than κ, so that $S \cap \beta$ is not stationary in β. ⊣

With some elegant combinatorics, Todorčević improved this to the denial of a partition relation for pairs:

8.18 Theorem (Todorčević [87: 285]). *Suppose that λ is regular and there is an S stationary in λ such that $S \cap \alpha$ is not stationary in α for any limit ordinal $\alpha < \lambda$. Then $\lambda \not\longrightarrow [\lambda]^2_\lambda$. In particular, if κ is regular, then $\kappa^+ \not\longrightarrow [\kappa^+]^2_{\kappa^+}$.*

Note that this considerably strengthens the implications 4.6(a) and 7.8(d) from weak compactness. It also shows that the assumption $2^\kappa = \kappa^+$ is unnecessary in 8.14(c) when κ is regular. Whether or not $\omega_1 \not\longrightarrow [\omega_1]^2_{\omega_1}$ had been a prominent problem of infinitary combinatorics for two decades (see Problem 15, Erdős-Hajnal [71]).

The question of whether κ^+ can be Jónsson when κ is singular (without assuming $2^\kappa = \kappa^+$) remained, and seemed quite difficult. Although Shelah [78] and Tryba [86, 87] imposed several constraints, the focal case $\kappa = \omega_\omega$ seemed destined to be resolved eventually by a positive consistency result relative to some strong hypothesis. So, it was somewhat unexpected when Shelah established in 1988 that unless a singular cardinal κ satisfies some stringent conditions, κ^+ is not Jónsson. This was one of the many fruits of Shelah's rich *pcf* theory for the study of powers of singular cardinals; see Burke-Magidor [90] or Jech [92] for an exposition. The following theorem is illustrative; the hypothesis can be weakened further.

8.19 Theorem (Shelah). *Suppose that a singular cardinal* κ *is not the limit of regular Jónsson cardinals. Then* κ^+ *is not Jónsson. In particular,* $\omega_{\omega+1}$ *is not Jónsson.*

Todorčević also refined this to the partition relation for pairs; see Burke-Magidor [90: 224].

8.20 Theorem (Todorčević). *Suppose that a singular cardinal* κ *is not the limit of regular cardinals* λ *satisfying* $\lambda \longrightarrow [\lambda]^2_\lambda$. *Then* $\kappa^+ \not\longrightarrow [\kappa^+]^2_{\kappa^+}$. *In particular,* $\omega_{\omega+1} \not\longrightarrow [\omega_{\omega+1}]^2_{\omega_{\omega+1}}$.

The overall question still remains open:

8.21 Question. *If* κ *is singular, does* $\kappa^+ \not\longrightarrow [\kappa^+]^2_{\kappa^+}$*?*

As was elaborated in some detail, interest in combinatorial ramifications of Rowbottom's work has persisted to the present day, primarily because of the enticing possibility that strong partition properties can obtain in the low levels of the cumulative hierarchy. We now return to the mid-1960's, and Silver's dramatic amplification concerning L based on a concept that was just coming into prominence: *set of indiscernibles for a structure.*

9. Indiscernibles and $0^\#$

Jack Silver was first a student at Berkeley and then joined its faculty in the 1960's, a time of great activity there in set theory. His remarkable results on $0^\#$ ("zero sharp") appeared in his 1966 dissertation [66], rivaled only by Kunen's (see §19) of a couple of years later for its impact on the development of set theory. This section is devoted to Silver's results. §14 discusses the definability of $0^\#$ and related issues about the forcing method, a line of inquiry concurrently initiated by Solovay. He and Silver independently isolated $0^\#$ in the context of Silver's analysis of L.

Silver was motivated by and extended the groundbreaking work of Gaifman. A student of Tarski, Gaifman wrote his 1962 Berkeley dissertation on Boolean algebras. Soon afterwards, he had turned to an altogether different subject, and established a strong extension of Scott's result on the incompatibility of measurable cardinals and $V = L$:

9.1 Theorem (Gaifman [64]). *Suppose that there is a measurable cardinal. Then:*

(a) There is a closed unbounded class $C \subseteq \{\alpha \mid L_\alpha \prec L\}$ such that for any infinite β, there is an $\alpha > \beta$ with $|\alpha| = |\beta|$ and $\alpha \in C$. This implies that $|\mathcal{P}(x)^L| = |x|$ for any infinite $x \in L$.

(b) There is an $a \subseteq \omega$ such that in $L[a]$, C is a definable class and (a) holds. This implies that arbitrarily large cardinals in L are no longer cardinals in $L[a]$. ⊣

Gaifman achieved these results using his method of iterated ultrapowers, a method to be further exploited by Kunen (§19). With (a) Gaifman had independently derived Rowbottom's local conclusion 8.5(a), but moreover had shown for the first time that drastic *global* consequences hold for L in the presence of a measurable cardinal. Gaifman was not far away from some of Silver's results on $0^\#$: $0^\#$ is a subset of ω satisfying (b), and it uniquely determines a closed unbounded class satisfying (a); beyond that, it incorporates a specific generating scheme for L. The existence of $0^\#$ was a genuinely new structural principle, and its isolation established the *intrinsic necessity* of large cardinals for transcendence over L.

The story of $0^\#$ begins in model theory. Investigating the problem of getting models of theories with a large number of automorphisms, Andrzej Ehrenfeucht and Mostowski developed the concept of indiscernibility, and brought Ramsey's theorem 7.7 into model-theoretic prominence. For $\mathcal{M}$ a structure and X a subset of the domain of $\mathcal{M}$ linearly ordered by $<$ (not necessarily a relation of $\mathcal{M}$), $\langle X, < \rangle$ is a *set of indiscernibles for $\mathcal{M}$ iff* for every formula $\varphi(v_1, \ldots, v_n)$ in the language of $\mathcal{M}$ and $x_1 < \ldots < x_n$ and $y_1 < \ldots < y_n$ all in X,

$$\mathcal{M} \models \varphi[x_1, \ldots, x_n] \;\; \textit{iff} \;\; \mathcal{M} \models \varphi[y_1, \ldots, y_n] \,.$$

That is, for each $n \in \omega$ all increasing n-tuples from X have the same first-order properties in $\mathcal{M}$. Variants on this terminology, e.g. with $\langle X, < \rangle$ replaced by X

when the $<$ is clear, should be unambiguous. The following became a basic ingredient of several important results:

9.2 Theorem (Ehrenfeucht-Mostowski [56]). *Suppose that T is a theory with infinite models and $\langle X, <\rangle$ is a linearly ordered set. Then there is a model $\mathcal{M}$ of T such that X is included in its domain and is a set of indiscernibles for $\mathcal{M}$.*

Proof. Expand the language of T by introducing new constants c_x for each $x \in X$ and consider the theory $\overline{T} = T \cup \{c_x \neq c_y \mid x \neq y \text{ both in } X\} \cup \{\varphi(c_{x_1}, \ldots, c_{x_n}) \leftrightarrow \varphi(c_{y_1}, \ldots, c_{y_n}) \mid \varphi(v_1, \ldots, v_n)$ is a formula in the language of T and $x_1 < \ldots < x_n$ and $y_1 < \ldots < y_n$ all in $X\}$. It suffices to show that $\overline{T}$ is consistent, which by the Compactness Theorem amounts to showing that every finite subset of $\overline{T}$ is satisfiable.

Assume then that $S \subseteq \overline{T}$ is finite. Let $\mathcal{A}$ be an infinite model of T, and $\{a_i \mid i \in \omega\}$ distinct members of the domain of $\mathcal{A}$. Let m be the number of new constants appearing among the members of S, and for $k \leq m$ define f_k on $[\omega]^k$ by:

$$\begin{aligned} f_k(i_1, \ldots, i_k) = \ & \{\varphi(v_1, \ldots, v_k) \mid \\ & \varphi(c_{x_1}, \ldots, c_{x_k}) \leftrightarrow \varphi(c_{y_1}, \ldots, c_{y_k}) \in S \ \wedge \ \mathcal{A} \models \varphi[a_{i_1}, \ldots, a_{i_k}]\} \, . \end{aligned}$$

Since S is finite, the range of f_k is finite, so applying Ramsey's Theorem 7.7 m times there is an $H \in [\omega]^\omega$ homogeneous for each f_k. Hence, $\mathcal{A}$ satisfies S with any m elements of $\{a_i \mid i \in H\}$ assigned to the new constants appearing in S in corresponding increasing order. ⊣

Actually, Ramsey's original application [30] of the finite version of his theorem was essentially to get indiscernibles in a finite model context. Michael Morley [65, 65a] drew attention to 9.2 by applying it to derive important results in model theory. Since its proof proceeds by a compactness argument, it gives little information about the resulting model. Silver pointed out that strong partition relations directly provide indiscernibles for structures:

9.3 Theorem (Silver [66, 71]). *For infinite limit ordinals α, $\kappa \longrightarrow (\alpha)_2^{<\omega}$ iff for any structure $\mathcal{M}$ for a countable language with κ a subset of its domain, there is a set of indiscernibles $X \in [\kappa]^\alpha$ for $\mathcal{M}$.*

Proof. Let $\{\varphi_n \mid n \in \omega\}$ enumerate the formulas of the language so that φ_n has at most the variables $v_1, \ldots, v_{k(n)}$ free where $k(n) \leq n$. Define $f \colon [\kappa]^{<\omega} \to 2$ by: $f(\xi_1, \ldots, \xi_n) = 0$ if $\mathcal{M} \models \varphi_n[\xi_1, \ldots, \xi_{k(n)}]$, and $f(\xi_1, \ldots, \xi_n) = 1$ otherwise. Then any set homogeneous for f with ordertype a limit ordinal is a set of indiscernibles for $\mathcal{M}$.

Conversely, suppose that $f \colon [\kappa]^{<\omega} \to 2$ and X is a set of indiscernibles for the structure $\langle \kappa, \in, f \restriction [\kappa]^n \rangle_{n \in \omega}$. Then X is homogeneous for f. ⊣

Silver realized that because of the uniformity of the constructible hierarchy L, if a sufficiently rich structure had enough ordinal indiscernibles, the theory of the structure fueled by the class of ordinals in the role of indiscernibles can be used to generate L. This harkens back to Gödel's original impredicative use of the class of ordinals to construct L by extending type theory, but the indiscernible generation shifts the weight of the construction squarely on the theory and leads to striking consequences about the distance between V and L.

Silver's work is now cast into a series of lemmata that highlight the consequences of the various conditions leading to $0^{\#}$. An observation of later significance is that these results do not depend on the Axiom of Choice, and how each apparent use can be avoided is described at its juncture.

By 3.3(a) there is a formula $\varphi_0(v_0, v_1)$ that defines in L a well-ordering $<_L$ of L such that: for any limit $\delta > \omega$ and $x, y \in L_\delta$, $x <_L y$ *iff* $L_\delta \models \varphi_0[x, y]$. For each formula $\varphi(v_0, \ldots, v_m)$ of $\mathcal{L}_\in$, define the *canonical Skolem term* t_φ for φ using φ_0 as follows:

$$t_\varphi(v_1, \ldots, v_m) = v_0 \ \textit{ iff } \ (\forall v_{m+2} \neg\varphi(v_{m+2}, v_1, \ldots, v_m) \ \wedge \ v_0 = \emptyset) \ \vee$$
$$(\varphi(v_0, v_1, \ldots, v_m) \ \wedge \ \forall v_{m+1}(\varphi_0(v_{m+1}, v_0) \to \neg\varphi(v_{m+1}, v_1, \ldots, v_m))) \, .$$

For any $\mathcal{M} = \langle M, E\rangle$ satisfying the requisite well-ordering properties with φ_0 (e.g. $\mathcal{M}$ is elementarily equivalent to some $\langle L_\delta, \in\rangle$ with δ a limit ordinal $> \omega$), the corresponding expansion

$$\langle M, E, t_\varphi^{\mathcal{M}}\rangle_\varphi$$

can be considered where the well-defined interpretation $t_\varphi^{\mathcal{M}}$ is a Skolem function for φ such that $t_\varphi^{\mathcal{M}}(x_1, \ldots, x_m)$ is the least y according to φ_0 satisfying $\mathcal{M} \models \varphi[y, x_1, \ldots, x_m]$ when one exists. Note that $\{t_\varphi^{\mathcal{M}} \mid \varphi$ is a formula of $\mathcal{L}_\in\}$ is already closed under functional composition by definability, and hence is a complete set of Skolem functions for $\mathcal{M}$. Consequently, for $X \subseteq M$ *the* Skolem hull of X in $\mathcal{M}$ can be taken to be well-defined, with domain

$$\{t_\varphi^{\mathcal{M}}(x_1, \ldots, x_m) \mid \varphi \text{ is a formula of } \mathcal{L}_\in \text{ and } x_1, \ldots, x_m \in X\} \, .$$

This set is canonical: It coincides with the collection of those $x \in M$ such that $\{x\}$ is definable over $\mathcal{M}$ using parameters from X, since that collection is included in any elementary substructure of $\mathcal{M}$ that includes X.

By *Skolem term* is meant one of the t_φ's in this section

although their specific form will not matter until they are taken up again in subsequent sections.

Theories with indiscernibles are considered next. Let $\mathcal{L}_\in^*$ be $\mathcal{L}_\in$ augmented by constants $\{c_k \mid k \in \omega\}$. By an EM *blueprint* (for Ehrenfeucht-Mostowski) is meant the theory in $\mathcal{L}_\in^*$ of some structure

$$\langle L_\delta, \in, x_k\rangle_{k\in\omega}$$

where δ is a limit ordinal $> \omega$ and $\{x_k \mid k \in \omega\}$ is a set of ordinal indiscernibles for $\langle L_\delta, \in \rangle$ indexed in increasing order. A basic observation is that for any limit ordinal $\delta > \omega$, any infinite set of ordinal indiscernibles for $\langle L_\delta, \in \rangle$ uniquely determines an EM blueprint: just take the theory of $\langle L_\delta, \in, x_k \rangle_{k \in \omega}$ for any increasing subsequence $\langle x_k \mid k \in \omega \rangle$ of the indiscernibles.

For a theory T in $\mathcal{L}^*_\in$, let T^- denote its restriction to $\mathcal{L}_\in$, i.e. those sentences of T with no occurrence of any c_k.

9.4 Lemma. *Suppose that T is an* EM *blueprint. Then for any α there is a model $\mathcal{M} = \mathcal{M}(T, \alpha)$ of T^- unique up to isomorphism such that:*

(a) There is a set X of ordinals in the sense of $\mathcal{M}$ of ordertype α (under $\mathcal{M}$'s ordinal ordering) that constitutes a set of indiscernibles for $\mathcal{M}$. Moreover, for any formula $\varphi(v_1, \ldots, v_n)$ of $\mathcal{L}_\in$ an increasing n-tuple from X satisfies φ in $\mathcal{M}$ exactly when $\varphi(c_0, \ldots, c_{n-1}) \in T$.

(b) The Skolem hull of X in $\mathcal{M}$ is again $\mathcal{M}$.

Proof. By definition of EM blueprint, T has at least one infinite model. So, the proof of 9.2 shows that there is a model satisfying (a). Taking the Skolem hull of the indiscernibles in this model results in a model $\mathcal{M}$ that also satisfies (b). (Suppose that the usual Henkin construction for the Completeness Theorem had been used for getting the model satisfying (a), using the language $\mathcal{L}^\alpha_\in$ with a constant for each member of α. Then the resulting "term" model already satisfies (b), and no appeal to the Axiom of Choice is necessary as the terms of $\mathcal{L}^\alpha_\in$ inherit a well-ordering from α.)

Suppose next that there were two such models, $\mathcal{M}$ and $\overline{\mathcal{M}}$, with corresponding sets of indiscernibles X and $\overline{X}$. Since X and $\overline{X}$ have ordertype α, let $h\colon X \to \overline{X}$ be the order isomorphism. Then h extends to an isomorphism between $\mathcal{M}$ and $\overline{\mathcal{M}}$: By (b), any element of $\mathcal{M}$ is of form $t^{\mathcal{M}}(x_1, \ldots, x_n)$ for some Skolem term t and $x_1, \ldots, x_n$ in X, and similarly for $\overline{\mathcal{M}}$ and $\overline{X}$. Hence, it suffices to show that the map $\tilde{h}$ given by

$$\tilde{h}(t^{\mathcal{M}}(x_1, \ldots, x_n)) = t^{\overline{\mathcal{M}}}(h(x_1), \ldots, h(x_n))$$

is an isomorphism: For any formula $\varphi(v_1, \ldots, v_n)$ and increasing n-tuple $\langle x_1, \ldots, x_n \rangle$ drawn from X,

$$\mathcal{M} \models \varphi[x_1, \ldots, x_n] \ \textit{iff} \ \varphi(c_0, \ldots, c_{n-1}) \in T \ \textit{iff} \ \overline{\mathcal{M}} \models \varphi[h(x_1), \ldots, h(x_n)] .$$

Skolem terms are *definable*, so using particular formulas asserting the equality of Skolem terms it can be checked that $\tilde{h}$ is well-defined, i.e. does not depend on the term descriptions, and that it is injective. Similarly, it preserves membership, and clearly it is surjective. ⊣

Particularly portentous is when $\mathcal{M}(T, \alpha)$ is well-founded; it then has a transitive collapse, which must be of the form $\langle L_\delta, \in \rangle$ by 3.3(a). In that case,

$$\mathcal{M}(T, \alpha) \text{ is identified with } \langle L_\delta, \in \rangle .$$

By a standing convention $\langle L_\delta, \in \rangle$ is usually denoted by L_δ.

9.5 Lemma. *Suppose that T is an* EM *blueprint. Then $\mathcal{M}(T, \alpha)$ is well-founded for every α iff*

(I) $\quad \mathcal{M}(T, \alpha)$ is well-founded for every $\alpha < \omega_1$.

Proof. Assume that for some α, $\mathcal{M}(T, \alpha) = \langle M, E \rangle$ is ill-founded. Let

$$\langle a_i \mid i \in \omega \rangle \in {}^{\omega}M \text{ with } a_{i+1} \; E \; a_i \text{ for every } i \in \omega .$$

(This does not require the Axiom of Choice as $\langle M, E \rangle$ inherits a well-ordering from α, and so such a sequence is definable by recursion.) Each a_i is of form $t^{\langle M, E \rangle}(x_1, \ldots, x_j)$ for some Skolem term t and indiscernibles $x_1, \ldots, x_j$, so let Y be the countable set consisting of the indiscernibles involved in these terms. If $\mathcal{N}$ is the Skolem hull of Y in $\mathcal{M}(T, \alpha)$, then $\mathcal{N}$ is ill-founded, yet $\mathcal{N} \cong \mathcal{M}(T, \beta)$ where $\beta < \omega_1$ is the ordertype of Y. ⊣

This result motivates the use of a strong partition property to produce EM blueprints with arbitrarily large well-founded models:

9.6 Lemma. *Suppose that there is a κ satisfying $\kappa \longrightarrow (\omega_1)_2^{<\omega}$. Then there is an* EM *blueprint satisfying* (I) *of 9.5.*

Proof. By 9.3, L_κ has an uncountable set of ordinal indiscernibles. Let T be the corresponding EM blueprint. Then for any $\alpha < \omega_1$, $\mathcal{M}(T, \alpha)$ is well-founded since it is isomorphic to the Skolem hull in L_κ of the first α indiscernibles. ⊣

(I) already yields Rowbottom's conclusion 8.3:

9.7 Lemma. *If there is an* EM *blueprint satisfying* (I)*, then $\mathcal{P}(\omega)^L$ is countable.*

Proof. $\mathcal{M}(T, \omega_1) = L_\delta$ for some $\delta \geq \omega_1$, so that $\mathcal{P}(\omega)^L \subseteq L_\delta$. In particular, if $a \in \mathcal{P}(\omega)^L$, then $a = t^{L_\delta}(x_0, \ldots, x_n)$ for some Skolem term t and indiscernibles $x_0 < \ldots < x_n$. Let $\langle z_i \mid i \in \omega \rangle$ be the increasing enumeration of the first ω indiscernibles. Since each $k \in \omega$ is definable,

$$k \in t^{L_\delta}(x_0, \ldots, x_n) \;\; \textit{iff} \;\; k \in t^{L_\delta}(z_0, \ldots, z_n)$$

and hence $a = t^{L_\delta}(z_0, \ldots, z_n)$. Since there are only countably many such forms, $\mathcal{P}(\omega)^L$ is countable. ⊣

The hypothesis of 9.6 entails the existence of an EM blueprint satisfying two further conditions, from which some remarkable conclusions can be drawn. On the basis of its proof, specify that

(i) ρ is the least limit ordinal such that L_ρ has a set of ordinal indiscernibles of ordertype ω_1;
(ii) H is such a set of indiscernibles with the least possible ωth element; and
(iii) T_0 is the corresponding EM blueprint.

9.8 Lemma. *The following condition holds for* $T = T_0$*:*

(II) For any n-ary Skolem term t, T contains the sentence: $t(c_0, \ldots, c_{n-1}) \in \mathrm{On} \rightarrow t(c_0, \ldots, c_{n-1}) < c_n$.

Proof. Assume to the contrary that

$$t(c_0, \ldots, c_{n-1}) \in \mathrm{On} \ \wedge \ c_n \leq t(c_0, \ldots, c_{n-1})$$

is in T_0 for some t. Let $z_0 < \ldots < z_{n-1}$ be the first n members of our fixed set H of indiscernibles for L_ρ, and set $\overline{H} = H - \{z_0, \ldots, z_{n-1}\}$ and $\delta = t^{L_\rho}(z_0, \ldots, z_{n-1}) < \rho$. By our assumption and indiscernibility, $\overline{H} \subseteq \delta$. Moreover, δ can be taken to be a limit ordinal. (If $\delta = \overline{\delta} + k$ where $k \in \omega$, there are members of $\overline{H}$ below $\overline{\delta}$ and so $c_n \leq t(c_0, \ldots, c_{n-1}) - k$ holds, and hence $\overline{H} \subseteq \overline{\delta}$.)

It will now be shown that $\overline{H}$ is a set of indiscernibles for L_δ, contradicting the minimal choice of ρ: Suppose that $x_1 < \ldots < x_m$ and $y_1 < \ldots < y_m$ are all in $\overline{H}$. Then for any formula $\varphi(v_1, \ldots, v_m)$,

$$(*) \qquad L_\delta \models \varphi[x_1, \ldots, x_m]$$

is equivalent to $L_\rho \models \varphi^{L_\delta}[x_1, \ldots, x_m]$, this relativization being possible as $L_\delta \subseteq L_\rho$ are transitive. Noting that L_δ is definable in L_ρ from δ (see Devlin [84: II§2]) and incorporating the definition of δ, $(*)$ is equivalent to

$$L_\rho \models \overline{\varphi}[x_1, \ldots, x_m, z_0, \ldots, z_{n-1}]$$

for some $\overline{\varphi}$. Analogously,

$$L_\delta \models \varphi[y_1, \ldots, y_m] \ \ \textit{iff} \ \ L_\rho \models \overline{\varphi}[y_1, \ldots, y_m, z_0, \ldots, z_{n-1}] \, .$$

Hence, the indiscernibility of H for L_ρ implies the indiscernibility of $\overline{H}$ for L_δ. ⊣

The following characterization is immediate:

9.9 Lemma. *An* EM *blueprint* T *satisfies* (II) *of 9.8 iff for any infinite limit ordinal* α*, the set of indiscernibles corresponding to* $\mathcal{M}(T, \alpha)$ *is cofinal in the ordinals of the structure.* ⊣

Rowbottom was involved in this work, particularly in connection with the following further condition (Devlin [73: 195]; Silver [71: 77]):

9.10 Lemma. *The following condition holds for* $T = T_0$*:*

(III) For any $(m+n+1)$-ary Skolem term t, T contains the sentence:
$$t(c_0, \ldots, c_{m+n}) < c_m \;\rightarrow$$
$$t(c_0, \ldots, c_{m+n}) = t(c_0, \ldots, c_{m-1}, c_{m+n+1}, \ldots, c_{m+2n+1}) \,.$$

Remark. This is known as the *remarkable* condition. By a simple indiscernibility argument, the conclusion can be replaced by the self-refinement

$$t(c_0, \ldots, c_{m+n}) = t(c_0, \ldots, c_{m-1}, c_{k_1}, \ldots, c_{k_{n+1}})$$

for any $k_1 < \ldots < k_{n+1}$ with $m \leq k_1$. Without further comment, this is how (III) will be invoked.

Proof. It can be assumed that

$$t(c_0, \ldots, c_{m+n}) < c_m$$

is in T_0. Let $H = \bigcup\{s_\xi \mid \xi < \omega_1\}$ be a disjoint partition into sets consisting of consecutive elements of H such that $|s_0| = m$, and $|s_\eta| = n+1$ and $\max(s_\xi) < \min(s_\eta)$ for $0 \leq \xi < \eta < \omega_1$. Let $t(s_0, s_\xi)$ for $0 < \xi < \omega_1$ denote $t^{L_\rho}(x_0, \ldots, x_{m-1}, y_0^\xi, \ldots, y_n^\xi)$, where $s_0 = \{x_0, \ldots, x_{m-1}\}$ and $s_\xi = \{y_0^\xi, \ldots, y_n^\xi\}$ in increasing order. It suffices by indiscernibility to derive a contradiction from the assumption that $t(s_0, s_\xi) \neq t(s_0, s_\eta)$ for some (and hence any) $0 < \xi < \eta < \omega_1$.

If $t(s_0, s_\xi) > t(s_0, s_\eta)$ for $0 < \xi < \eta < \omega_1$, there would be an infinite descending sequence of ordinals. On the other hand, if $t(s_0, s_\xi) < t(s_0, s_\eta)$ for $0 < \xi < \eta < \omega_1$ then it is simple to see that $\{t(s_0, s_\xi) \mid 0 < \xi < \omega_1\}$ would be a set of indiscernibles for L_ρ. However, the first element y_0^ω of s_ω is the ωth element of H and $t(s_0, s_\omega) < y_0^\omega$ by our initial assumption, contradicting the minimality of the ωth element of H. ⊣

We now proceed to analyze those EM blueprints T that satisfy conditions (I)–(III). For such a T and for any α, temporarily let

$$\langle \iota_\xi^{T,\alpha} \mid \xi < \alpha \rangle$$

denote the corresponding increasing sequence of indiscernibles for $\mathcal{M}(T, \alpha)$. (III) has the following primary consequence:

9.11 Lemma. *If* T *is an* EM *blueprint satisfying* (I)–(III) *and* $\omega \leq \alpha \leq \beta$ *with* α *a limit ordinal, then the Skolem hull of* $\{\iota_\xi^{T,\beta} \mid \xi < \alpha\}$ *in* $\mathcal{M}(T, \beta)$ *is* L_ι*, where* $\iota = \iota_\alpha^{T,\beta}$*. Consequently,*

$$\mathcal{M}(T, \alpha) = L_\iota \;\text{ and }\; \iota_\xi^{T,\alpha} = \iota_\xi^{T,\beta} \;\text{ for every }\; \xi < \alpha \,.$$

Proof. Let $\mathcal{N}$ be the stated Skolem hull. It suffices to show that $\mathrm{On}^{\mathcal{N}} = \iota_\alpha^{T,\beta}$, as the second sentence follows from the definition and uniqueness of $\mathcal{M}(T, \alpha)$. For convenience, the superscripts ${}^{T,\beta}$ for the indiscernibles will be suppressed in what follows.

If σ is an ordinal in $\mathcal{N}$, then for some Skolem term t and $\xi_0 < \ldots < \xi_{n-1} < \alpha$,

$$\sigma = t^{\mathcal{M}(T,\beta)}(\iota_{\xi_0}, \ldots, \iota_{\xi_{n-1}}) < \iota_{\xi_n} < \iota_\alpha$$

by (II) and as α is a limit. Conversely, if $\tau < \iota_\alpha$, then

$$\tau = u^{\mathcal{M}(T,\beta)}(\iota_{\zeta_0}, \ldots, \iota_{\zeta_{m-1}}, \iota_{\eta_0}, \ldots, \iota_{\eta_n}) \; < \; \iota_\alpha$$

for some Skolem term u and indiscernibles in increasing order with $\zeta_{m-1} < \alpha \leq \eta_0$. By (III),

$$\tau = u^{\mathcal{M}(T,\beta)}(\iota_{\zeta_0}, \ldots, \iota_{\zeta_{m-1}}, \iota_{\zeta_{m-1}+1}, \ldots, \iota_{\zeta_{m-1}+n+1}) \, ,$$

and this is in $\mathcal{N}$ as α is a limit ordinal. ⊣

Thus, for any EM blueprint T satisfying (I)–(III) and any ξ, unambiguously set

$$\iota_\xi^T = \iota_\xi^{T,\alpha} \text{ for some (any) limit ordinal } \alpha > \xi \, , \text{ and}$$
$$I^T = \{\iota_\xi^T \mid \xi \in \mathrm{On}\} \, .$$

9.11 has a host of consequences:

9.12 Lemma. *Suppose that T is an* EM *blueprint satisfying* (I)–(III). *Then:*
(a) $L_{\iota_\xi^T} \prec L_{\iota_\zeta^T}$ *when* $\xi < \zeta$.
(b) $|\iota_\xi^T| = |\xi| + \aleph_0$ *for every* ξ.
(c) I^T *is a closed unbounded class of ordinals.*
(d) For any cardinal $\lambda > \omega$, $\iota_\lambda^T = \lambda \in I^T$ *and so* $\mathcal{M}(T,\lambda) = L_\lambda$.
(e) T is the only EM *blueprint satisfying* (I)–(III).

Proof. (a) for infinite limit ordinals $\xi < \zeta$ follows directly from 9.11. Consequently, it holds for arbitrary $\xi < \zeta$ by an indiscernibility argument within some sufficiently large $L_{\iota_\beta^T}$.

If α is an infinite limit ordinal, then $L_{\iota_\alpha^T}$ is the Skolem hull in itself of $\{\iota_\xi^T \mid \xi < \alpha\}$ by 9.11. Hence, $|\iota_\alpha^T| = |\alpha|$ and so (b) follows for every ξ. (c) also follows, since $\{\iota_\xi^T \mid \xi < \alpha\}$ is cofinal in ι_α^T by 9.9.

(d) is a consequence of (b) and (c).

Finally, for (e) note that T is the theory of $\langle L_{\omega_\omega}, \in, \omega_{n+1}\rangle_{n\in\omega}$ by (d). ⊣

Assuming (a) and (d), L is the union of the elementary chain $\langle L_\lambda \mid \lambda > \omega$ is a cardinal$\rangle$, so that the satisfaction relation for L can be defined in ZFC by:

$$L \models \varphi[a_1, \ldots, a_n] \textit{ iff } L_\lambda \models \varphi[a_1, \ldots, a_n] \text{ for some (any) } \lambda > \omega \text{ such that } a_1, \ldots, a_n \in L_\lambda \, .$$

In particular, a truth definition for L can be given by

$$L \models \sigma \ \textit{iff} \ L_{\omega_1} \models \sigma$$

for sentences σ. Thus, the set of (Gödel numbers of) sentences true in L is *constructible.* It cannot be *definable* in L by the undefinability of truth, but of course (the real) ω_1 being indiscernible is not definable in L. With the satisfaction relation for L in hand, assertions like the following are directly formalizable in ZFC.

9.13 Lemma. *Suppose that T is an* EM *blueprint satisfying* (I)–(III). *Then:*

(a) $L_{\iota_\xi^T} \prec L$ for every ξ.

(b) C is a closed unbounded class of ordinal indiscernibles for L such that the Skolem hull of C in L is again L iff $C = I^T$.

Proof. It is only necessary to verify the forward direction of (b): Note that $C \cap I^T$ is infinite, so that the EM blueprint corresponding to C is again T. Let $h: C \to I^T$ be the order-preserving bijection. It follows by the uniqueness argument for 9.4 that h extends to an isomorphism $\tilde{h}: L \to L$. But then, $\tilde{h}$ must be the identity (else it would move some ordinal, removing it from the range). Hence, $C = I^T$. ⊣

With 9.12(e) in hand, stipulate that

$0^\#$ is that unique EM blueprint satisfying (I)–(III)

if there is one, acceding to the accepted solecism

$0^\#$ exists

for: There is an EM blueprint satisfying (I)–(III). Through a recursive arithmetization of $\mathcal{L}_\in^*$, $0^\#$ *is regarded as a subset of* ω. While Silver had concentrated on the theory, Solovay [67], the source of the # notation, emphasized its construal as a subset of ω. The sense of "$0^\#$ exists" will be clarified by a later absoluteness result (14.12) according to which the *set* $0^\# \notin L$, yet remarkably there is a formula that defines $0^\#$ in any model of ZF containing it. Dropping the superscript T in 9.12 and 9.13 by uniqueness, the following is a Hauptsatz summarizing Silver's results:

9.14 Theorem (Silver [66, 71]).

(a) $0^\#$ exists iff some L_δ has an uncountable set of indiscernibles. Hence, if $\kappa \longrightarrow (\omega_1)_2^{<\omega}$ for some κ (e.g. if κ is measurable), then $0^\#$ exists.

(b) $0^\#$ exists iff there is a class I of ordinals characterized by: I is a closed unbounded class of indiscernibles for L such that the Skolem hull of I in L is again L. Moreover, with $\langle \iota_\xi \mid \xi \in \mathrm{On} \rangle$ the increasing enumeration of I it has the following properties: If $\xi \leq \zeta$, then $L_{\iota_\xi} \prec L_{\iota_\zeta} \prec L$ and $|\iota_\xi| = |\xi| + \aleph_0$ so that I contains every uncountable cardinal, and for any limit ordinal $\alpha \geq \omega$ the Skolem hull of $\{\iota_\xi \mid \xi < \alpha\}$ in L_{ι_α} is again L_{ι_α}. ⊣

Thus, Gaifman's 9.1 conclusions have been extended in an ultimate way: the existence of $0^{\#}$ is an intrinsic principle that provides a complete scheme for constructing L using one theory and a generating class of ordinal indiscernibles. Silver also reduced Gaifman's hypothesis to $\kappa \longrightarrow (\omega_1)_2^{<\omega}$. With the next result he showed that this reduction is sharp in the hierarchy of Erdős cardinals:

9.15 Theorem (Silver [66, 70]). *Suppose that M is an inner model of* ZFC, *$\alpha < \omega_1^M$, and $\kappa \longrightarrow (\alpha)_\delta^{<\omega}$. Then $(\kappa \longrightarrow (\alpha)_\delta^{<\omega})^M$. Hence,*

$$\mathrm{Con}(\mathrm{ZFC} + \exists\kappa\forall\alpha < \omega_1(\kappa \longrightarrow (\alpha)_2^{<\omega}))$$

implies

$$\mathrm{Con}(\mathrm{ZFC} + \exists\kappa\forall\alpha < \omega_1(\kappa \longrightarrow (\alpha)_2^{<\omega}) + 0^{\#} \text{ does not exist}) .$$

Proof. Suppose that $f\colon [\kappa]^{<\omega} \to \delta$ with $f \in M$. A set in M of ordertype α must be found homogeneous for f. Since $\alpha < \omega_1^M$, let $g\colon \omega \to \alpha$ be a bijection with $g \in M$. Set $D = \{d \mid d$ is an order-preserving injection: $g``n \to \kappa$ for some n, whose range is homogeneous for $f\}$, and define a partial ordering $\prec$ on D by: $d \prec \bar{d}$ *iff* $d \supset \bar{d}$. Since $g \in M$ and $(\kappa^{<\omega})^M = \kappa^{<\omega}$, $\langle D, \prec\rangle \in M$. It is straightforward to see that the following is a theorem of ZFC:

$\prec$ is ill-founded *iff* there is a set of ordertype α homogeneous for f .

It thus follows that $\prec$ is ill-founded in V. But then, $\prec$ is ill-founded in M by absoluteness (0.3), and hence the above theorem used in M yields the desired result.

The last assertion follows from taking $M = L$; by 9.7 (there is no EM blueprint satisfying (I))^L. ⊣

Note that M need only model one theorem of ZFC of sufficient strength to carry out the foregoing argument, and that the superscript "$<\omega$" can be replaced by "n" for any particular $n \in \omega$.

There have been further developments in the study of combinatorial properties that imply the existence of $0^{\#}$. Kunen showed that if there is a Jónsson cardinal, then $0^{\#}$ exists (21.4). Devlin-Paris [73] showed how to get $0^{\#}$ from a combinatorial consequence of $\kappa \longrightarrow (\omega_1)_2^{<\omega}$. Baumgartner-Galvin [78] formulated a generalized version of Erdős cardinals sensitive to the possible EM blueprints produced, and provided sharp implications about $0^{\#}$ in their framework.

Finally, there is a straightforward way to get an equivalence, at the cost of incorporating inner model hypotheses. For an inner model M,

$$\eta \overset{M}{\longrightarrow} (\alpha)_2^{<\omega}$$

asserts that for every function $f\colon [\eta]^{<\omega} \to 2$ such that $f \in M$, there is an $X \in [\eta]^\alpha$ (not necessarily in M) homogeneous for f.

9.16 Exercise (Gloede [72: 153]). *The following are equivalent:*

(a) $0^\#$ *exists.*

(b) For every cardinal $\lambda > \omega$, $\lambda \xrightarrow{L} (\lambda)_2^{<\omega}$.

(c) For some η, $\eta \xrightarrow{L} (\omega_1)_2^{<\omega}$. ⊣

If $0^\#$ does exist, it is not unexpected that there are consequences of large cardinal character, especially about the indiscernibles $\{\iota_\xi \mid \xi \in \mathrm{On}\}$ themselves.

9.17 Theorem. *Assume that* $0^\#$ *exists. Then:*

(a) There is an elementary embedding: $L \prec L$.

(b) (ι_ξ is totally indescribable)^L *for every* ξ.

(c) $|\mathcal{P}(x)^L| = |x|$ *for every infinite* $x \in L$.

(d) $(\iota_\xi \to (\alpha)_2^{<\omega})^L$ *for every* ξ *and* $\alpha < \omega_1^L$.

Proof. (a) Any order-preserving injection h from $\{\iota_\xi \mid \xi \in \mathrm{On}\}$ into itself extends uniquely to an $\tilde{h}$: $L \prec L$ defined by:

$$\tilde{h}(t^L(\iota_{\xi_1}, \ldots, \iota_{\xi_n})) = t^L(h(\iota_{\xi_1}), \ldots, h(\iota_{\xi_n}))$$

(cf. the proof of 9.4). Any such h which is not the identity thus induces an elementary embedding that moves some ordinal.

(b) By indiscernibility, it suffices to show that ι_0 is totally indescribable in L. Let h be any map as above such that $h(\iota_0) > \iota_0$. Then ι_0 is the critical point of $\tilde{h}$: $L \prec L$: If $\delta = t^L(\iota_{\xi_1}, \ldots, \iota_{\xi_n}) < \iota_0$ with $\xi_1 < \ldots < \xi_n$, then

$$\tilde{h}(\delta) = t^L(h(\iota_{\xi_1}), \ldots, h(\iota_{\xi_n})) = t^L(\iota_{\xi_1}, \ldots, \iota_{\xi_n}) = \delta$$

by (III). The argument for 6.6 now shows that ι_0 must be totally indescribable in L.

(c) By (b), the (real) cardinal $|x|^+$ is inaccessible in L.

(d) It suffices by indiscernibility to establish the property assuming that ι_ξ is a cardinal $\lambda > \omega$. Suppose then that f: $[\lambda]^{<\omega} \to 2$ and $f \in L$. Then $f = t^L(\iota_{\xi_0}, \ldots, \iota_{\xi_{m-1}}, \iota_{\zeta_0}, \ldots, \iota_{\zeta_n})$ say, with the indiscernibles in increasing order and $\iota_{\xi_{m-1}} < \lambda = \iota_\lambda \leq \iota_{\zeta_0}$. By an indiscernibility argument it follows that $\{\iota_{\xi_{m-1}+1+\beta} \mid \beta \in \alpha\}$ is homogeneous for f. Hence by 9.15, there is a homogeneous set in L of ordertype α. ⊣

Digressing briefly, a proof of 7.16 is sketched as previously promised:

9.18 Exercise (Reinhardt-Silver [65]). *There is a totally indescribable cardinal below* $\kappa(\omega)$.

Hint. Let $\kappa = \kappa(\omega)$, W a well-ordering of V_κ, and I_0 a set of indiscernibles of ordertype ω for $\langle V_\kappa, \in, W\rangle$. Let $H \prec V_\kappa$ be the Skolem hull of I_0 in V_κ with respect to Skolem functions defined with W, H_0 the transitive collapse of H, and e the inverse of the collapsing isomorphism. Note that H_0, H, and V_κ are all models of ZFC since κ is inaccessible. As in the proof of 9.17(b), any order-preserving injection $h: I_0 \to I_0$ which is not the identity induces an elementary embedding $\tilde{h}: H_0 \prec H_0$ whose critical point δ satisfies $H_0 \models \delta$ is totally indescribable. But then, $e(\delta)$ is totally indescribable, since it is so in H and hence in V_κ. ⊣

Sharps

The formulation of $0^\#$ for L can be relativized in a straightforward way to inner models $L[a]$ (§3) for sets $a \subseteq \mathrm{On}$ to produce corresponding sets $a^\# \subseteq \mathrm{On}$. Hence, # can be considered as an operation on sets of ordinals, with $0^\#$ for $L = L[0]$. Suppose that a is a set of ordinals, say with $\alpha = \bigcup a$, which can be taken to be infinite to avoid trivialities. An EM *blueprint for a* is the theory of some structure

$$\langle L_\delta[a], \in, x_k, a, \xi \rangle_{k\in\omega, \xi\leq\alpha}$$

where $\delta > \alpha$ is a limit ordinal and $\{x_k \mid k \in \omega\}$ is a set of ordinal indiscernibles for $\langle L_\delta[a], \in, a, \xi \rangle_{\xi\leq\alpha}$ indexed in increasing order. To get an EM blueprint for a satisfying (the analogues of) the conditions (I)–(III), uncountably many ordinal indiscernibles are needed for a structure $\mathcal{M}$ for a language with $|\alpha|$ non-logical symbols. For this purpose, instead of the special argument of 9.3, if $\{\varphi_\zeta \mid \zeta < |\alpha|\}$ enumerates the formulas of the language with φ_ζ having at most the variables $v_1, \ldots, v_{k(\zeta)}$ free, the function

$$f(\xi_1, \ldots, \xi_n) = \{\zeta < |\alpha| \mid k(\zeta) \leq n \ \wedge\ \mathcal{M} \models \varphi_\zeta[\xi_1, \ldots, \xi_{k(\zeta)}]\}$$

can be used. Consequently, a cardinal κ satisfying the partition property

$$\kappa \longrightarrow (\omega_1)^{<\omega}_{2^{|\alpha|}}$$

suffices. All further details go through with little change to procure a unique EM blueprint $a^\#$ satisfying (I)–(III) which through coding can be regarded as a subset of $|\alpha|$, and a corresponding closed unbounded class of ordinal indiscernibles I_a for $\langle L[a], \in, a, \xi \rangle_{\xi\leq\alpha}$. $a^\#$ for $a \subseteq \omega$ hold the main interest because of the intrinsic importance of the reals, construed here as subsets of ω. (From 14.16 on, a shift is made to $a^\#$ for reals a construed as members of ${}^\omega\omega$.) No excursion into uncountable languages is then necessary:

9.19 Theorem. *If $\kappa \longrightarrow (\omega_1)^{<\omega}_2$ for some κ, then $a^\#$ exists for every $a \subseteq \omega$.* ⊣

The isolation of $0^\#$ was the culmination of a flow of ideas from early efforts in the partition calculus, through the infusion of model-theoretic techniques, to the investigation of L. After Silver's results on $0^\#$ together with Solovay's on its definability (§14) established its intrinsic importance, further results provided new structural insights. Kunen soon showed (21.1) that the converse of 9.17(a) holds: *If there is an elementary embedding $j: L \prec L$, then $0^\#$ exists*. This result provided a simple paradigm for transcendence over inner models M, namely the existence of an elementary embedding: $M \prec M$. Then in 1974, Jensen established his celebrated Covering Theorem (see volume II), easily the most important result of the 1970's in set theory: $0^{\#}$ *does not exist iff for any uncountable set* $X \subseteq \mathrm{On}$, *there is a* $Y \in L$ *such that* $Y \supseteq X$ *and* $|Y| = |X|$. Not only does this result provide a surprisingly weak condition for the existence of $0^{\#}$, but it implies in the absence of $0^{\#}$ that some of the regular behavior of cofinalities in L lifts to V. These results have buttressed the existence of $0^{\#}$ as the focal hypothesis of transcendence over L.

Chapter 3

Forcing and Sets of Reals

This chapter describes the first advances using Cohen's method of forcing that involved large cardinals and the first applications of large cardinals in descriptive set theory. Cohen's creation transformed set theory, and large cardinal hypotheses played an increasingly prominent role as a consequence. §10 discusses the development of forcing, reviews the basic theory, and then focuses on mild extensions and the Levy collapse. §11 is devoted to Solovay's inspiring result that if there is an inaccessible cardinal, then in an inner model of a forcing extension, every set of reals is Lebesgue measurable. §12 reviews the historical development of descriptive set theory and establish a working context, one in which the classical results are established in §13 through to a delimitation established by Gödel with L. This sets the stage for the further results about large cardinals and projective sets, a major direction of set-theoretic research from the mid-1960's onwards. §14 describes Solovay's germinal work on $\mathbf{\Sigma}^1_2$ sets that grew out of his Lebesgue measurability result, and his results and conjectures on the definability of sharps. Then §15 describe how Martin used sharps to extend the methods of classical descriptive set theory to the analysis of $\mathbf{\Sigma}^1_3$ sets.

10. Development of Forcing

In April of 1963 the analyst Paul Cohen of Stanford University circulated notes sketching proofs of the independence of the Axiom of Choice from ZF and of the Continuum Hypothesis from ZFC. These proofs, of course, were the inaugural examples of his technique of forcing for extending models of set theory. Cohen lectured on his results on 3 May 1963 at the Institute for Advanced Study, and then two articles [63, 64] on the CH result were quickly communicated by Gödel to the *Proceedings of the National Academy of Sciences U.S.A.*, where his own consistency results with L had appeared a quarter of a century before.

Stirred by clear indications of a general procedure in Cohen's breakthrough, logicians close at hand soon developed the theory of forcing and began to establish further results with it. By mid-1963 Solomon Feferman at Stanford had results in first- and second-order arithmetic and on the Axiom of Choice, and arriving there in the summer Levy quickly obtained further results elaborating AC. Then two results displaying a notable sophistication were established at Princeton University: William Easton's [64, 70] in late 1963 on powers of regular cardinals by class forcing, and Solovay's [65b, 70] in mid-1964 that if there is an inaccessible cardinal, then in an inner model of a forcing extension, every set of reals is Lebesgue measurable. This preparatory section discusses the development of the forcing method and sets the stage for its interplay with large cardinals, especially toward Solovay's result. See Moore [88] for more on the origins of forcing.

Forcing provided a remarkably general and flexible scheme with strong intuitive underpinnings for establishing relative consistency and independence. According to Scott (Bell [85: ix]): "Set theory could never be the same after Cohen, and there is simply no comparison whatsoever in the sophistication of our knowledge about models of set theory today as contrasted to the pre-Cohen era." In retrospect one can point to precursors of forcing in the Beth semantics for intuitionistic logic (see Dummett [77: §5]); in the study of 2-valued homomorphisms of Boolean-valued models (see Rasiowa-Sikorski [63: VIII]); in Stephen Kleene and Emil Post's argument for producing incomparable Turing degrees (see Lerman [83: II§2]); and especially in Clifford Spector's two-quantifier argument for producing minimal Turing degrees (see Lerman [83: V§2]). However, Cohen was unaware of these adumbrations and started with simple and basic intuitions. His particular achievement lies in devising a concrete procedure for extending models of set theory in a minimal fashion without altering the ordinals. Scott continued (Bell [85: ix]):

> I knew almost all the set-theoreticians of the day, and I think I can say that no one could have guessed that the proof would have gone in just this way. Model-theoretic methods had shown us how many *non-standard* models there were; but Cohen, starting from very primitive first principles, found the way to keep the models *standard* (that is, with a well-ordered collection of ordinals).

Set theory had undergone a sea-change, and beyond how the subject was enriched, it is difficult to convey the strangeness of it.

Beyond his initial articles [63, 64] Cohen provided a further account [65] and then a full exposition in the monograph [66] based on his lectures at Harvard University in the spring of 1965. Later, he expressed his own philosophical views on the foundations of set theory in a contribution [71] to the Axiomatic Set Theory conference held during the summer of 1967 at the University of California at Los Angeles (U.C.L.A.). By all accounts this was one of those rare, highly exhilarating conferences that both summarized the progress and focused the energy of a new field opening up.

Cohen first discussed and compared the realist and formalist positions in mathematics. He considered ([71: 11ff]) that as regards the existence of a measurable cardinal "there seems absolutely no intuitively convincing evidence for either rejection or acceptance." Concerning CH he reckoned that it may well come to be accepted as false ([71: 12]): "The justification might be that the continuum, which is given to us by the power set construction, is not accessible by any process which attempts to build up cardinals from below by means of a construction based on the Replacement Axiom. Thus C [i.e. $2^{\aleph_0}$] would be considered greater than $\aleph_1, \aleph_n, \aleph_\omega$, etc." Presumably then, $2^{\aleph_0}$ would have to be weakly inaccessible. Cohen's speculation here was from a realist perspective, one advocated more directly in [66: 151].

Acknowledging the influence of Robinson [65], Cohen [71: 13] then came down on the formalist side. He continued (p. 14):

> Another example of how repeated exposure dulls the critical faculties is the Axiom of Inaccessible Cardinals. The usual justification given for assuming it is a negative one, namely that it is unreasonable to think that every set is accessible. The analogy is drawn with the passage from finite to infinite sets. If one performs by induction a transfinite sequence of suitable closure operations, it is argued, we can still go further and presumably find an inaccessible cardinal beyond. However, I feel that this is a specious argument, since it tends rather to justify the existence of a standard model for set theory, which is an incomparably weaker assertion. A more honest reason for accepting inaccessible cardinals is that experience has shown that they do not lead to contradictions and we have developed some kind of intuition that no such contradiction exists.

This point of view stands in distinct opposition to those of Zermelo [30](§1) and Gödel [47](§4).

Cohen's concluding remarks notably anticipate later progress in large cardinals, but are tempered by an undercurrent of doubt and uncertainty:

> Yet mathematics may be likened to a Promethean labor, full of life, energy and great wonder, yet containing the seed of an overwhelming self-doubt. It is good that only rarely do we pause to review the situation and to set down our thoughts on these deepest questions. During the rest of our mathematical lives we watch and perhaps partake in the glorious procession. Great questions of set theory that seemed untouchable eventually yield. New axioms are investigated and larger and larger cardinals are somehow brought closer to our intuition. Through all this, number theory stands as a shining beacon. If, as I hope does not happen too often, our doubts begin to overpower us, we retreat back into the safe confines of number theory until refreshed, we venture out again into the unsafe ground of set theory. This is our fate, to live with doubts, to pursue a subject whose

> absoluteness we are not certain of, in short to realize that the only "true" science is itself of the same mortal, perhaps empirical, nature as all other human undertakings.

It is notable that in comments made two decades later, Cohen was to stress the formal solution of the Continuum Problem, without further speculating on the falsity of CH (Albers *et al.* [90: 54]):

> Certainly Gödel himself had the platonic view that the question demanded an absolute answer and that, therefore, neither his proof of the consistency of the continuum hypothesis with the axioms of set theory nor mine of its independence from them was the final answer. My personal view is that I regard the present solution of the problem as very satisfactory. I think that it is the only possible solution. It gives one a feeling for what's possible and what's impossible, and in that sense I feel that one should be very satisfied...There will be philosophical papers, but I don't think any mathematical paper will say that there is any answer other than the answer that it's undecidable.

Forcing was quickly forged into a general method, particularly by the efforts of Solovay. He above all epitomized this period of great expansion in set theory, with his mathematical sophistication and fundamental results with forcing, in large cardinals, and in descriptive set theory. Following initial graduate study in differential topology, Solovay focused his energies on set theory after attending Cohen's 3 May 1963 lecture. He first extended the independence of CH by characterizing the possibilities for the size of the continuum ([63, 65]), and then generalized forcing to arbitrary partial orders and dense sets. He next established his Lebesgue measurability result during March-July 1964 ([70: 1]). Then with Stanley Tennenbaum he worked out the iterated forcing proof of the consistency of the classical Suslin Hypothesis in June 1965 (Solovay-Tennenbaum [71: 201]).

Complications in describing when a formula holds over a range of generic extensions soon led Solovay to the idea of a *Boolean value* for the formula. It turned out that Petr Vopěnka had developed a similar concept in a reworking [64] of the independence of CH. (Notably, this paper contains the remark that any inaccessible κ can be made weakly but not strongly inaccessible by rendering $2^{\aleph_0} \geq \kappa$ and preserving cardinals.) The conception was generalized and simplified in a series of papers on the so-called ∇-models from the active Prague seminar founded by Vopěnka (see Hájek [71: 78]), culminating in the exposition Vopěnka [67]. However, the earlier papers did not have much impact, partly because of an involved formalism in which formulas were valued in a complete lattice rather than Boolean algebra. Working independently, Scott and Solovay developed the idea of recasting forcing entirely in terms of Boolean-valued models, and came up with a formulation essentially equivalent to Vopěnka's. The method is completely general, since any partial order can be "completed" in a natural way to get a complete Boolean algebra. Scott popularized this approach in his own reworking [67] of the independence of CH and in lecture notes distributed at the 1967 U.C.L.A. conference.

Boolean-valued models first showed how to avoid Cohen's ramified language connection with the L hierarchy and his dependence on a countable $\in$-model. With their elegant algebraic trappings and seemingly more complete information, they held the promise of being the right approach to independence results. J. Barkley Rosser wrote a book on the subject entitled *Simplified Independence Proofs* [69], and the initial iterated forcing paper Solovay-Tennenbaum [71] was written in terms of Boolean-valued models. But already at U.C.L.A., the lucid lectures of Joseph Shoenfield [71] had demonstrated that forcing with partial orders can accommodate the gist of the Boolean approach in a straightforward manner. Moreover, Boolean-valued models were soon found to be too abstract and unintuitive for establishing *new* consistency results, so that within a few years set theorists were generally working with partially orders. It is a testament to Cohen's concrete approach that in this return from abstraction, even the use of ramified languages has played an essential role in careful forcing arguments at the interface of recursion theory and set theory.

During this period new set-theoretic results were being established at a bewilderingly fast pace, in the first wave of applications of forcing and the coordinated elaboration of large cardinals. A timely survey by Adrian Mathias, circulated in typescript soon after the U.C.L.A. conference and eventually appearing as Mathias [79], played an influential role in communicating and systematizing these developments. In particular, it was a vital source for graduate students at Berkeley who were later to become prominent set theorists, like James Baumgartner, Richard Laver, and William Mitchell.

Forcing Preliminaries

The basics about forcing through two-step iterations as presented in the texts Kunen [80], Jech [78], or Bell [85] are assumed. By and large, the rather standard partial order approach and terminology of the first text is adopted; the Boolean-valued approach is integrated into the second and prominent in the third. Anticipating the wide-ranging use of forcing in volume II, the salient and sometimes distinctive features are described in what follows:

For forcing purposes, a *partial order* (p.o.) is a partially ordered set $\langle P, \leq \rangle$ that has a maximum element denoted by $\mathbb{1}$, and for $p, q \in P$ (the *conditions*)

$$p \leq q \quad \textit{iff} \quad p \text{ extends or refines } q \ ,$$

i.e. less (in $\leq$) is more (informative). This conforms to the Boolean-valued approach and suggests that p allows fewer models than q. For simplicity $\langle P, \leq \rangle$ is usually denoted by P, but if there is a need to be specific, $\leq_P$ is written for $\leq$ and $\mathbb{1}_P$ for $\mathbb{1}$.

It is not necessary to assume that $\leq$ is antisymmetric (see Kunen [80: 52ff]) and in iterated forcing arguments the naturally defined orders are not. However, for convenience in such situations

$$p \text{ and } q \text{ are identified whenever } p \leq q \text{ and } q \leq p$$

(so that $\mathbb{1}_P$ is *the* maximum element), effectively passing to a p.o. of equivalence classes, but avoiding awkward notation.

Set:

$$p \parallel q \quad \textit{iff} \quad p \text{ and } q \text{ are compatible, i.e. } \exists r(r \le p \ \wedge \ r \le q)\,.$$
$$p \perp q \quad \textit{iff} \quad p \text{ and } q \text{ are incompatible, i.e. } \neg(p \parallel q)\,.$$

For $D \subseteq P$,

$$\begin{aligned}
D \text{ is } \textit{dense (in P)} &\quad \textit{iff} \quad \forall p \in P \exists d \in D(d \le p)\,. \\
D \text{ is } \textit{predense (in P)} &\quad \textit{iff} \quad \forall p \in P \exists d \in D(d \parallel p)\,. \\
D \text{ is } \textit{dense open (in P)} &\quad \textit{iff} \quad D \text{ is dense} \wedge \forall d \in D \forall p \in P(p \le d \ \rightarrow \ p \in D)\,. \\
D \text{ is } \textit{an antichain (in P)} &\quad \textit{iff} \quad \forall p, q \in D(p \ne q \ \rightarrow \ p \perp q)\,. \\
D \text{ is an } \textit{antichain below p} &\quad \textit{iff} \quad D \text{ is an antichain in } \{q \mid q \le p\}\,.
\end{aligned}$$

If D is either dense or a maximal antichain, then it is predense.

Suppose that M is a transitive $\in$-model of ZFC. Then for any set x,

$$M[x] \text{ is the smallest transitive } \in\text{-model of ZFC including } M \cup \{x\},$$

assuming one exists. For a p.o. $P \in M$ and $G \subseteq P$,

$$\begin{aligned}
&G \text{ is } \textit{P-generic over M iff} \\
&\qquad \text{(i) } \forall p, q \in G \exists r \in G(r \le p \ \wedge \ r \le q)\,, \\
&\qquad \text{(ii) } \forall p \in G \forall q \in P(p \le q \ \rightarrow \ q \in G)\,, \text{ and} \\
&\qquad \text{(iii) } \forall D \subseteq P\ (D \text{ dense in } P \ \wedge \ D \in M \rightarrow \ G \cap D \ne \emptyset)\,.
\end{aligned}$$

Here, "dense" can be replaced by "predense". For such G, it is simple to see that for any maximal antichain $A \subseteq P$ with $A \in M$, $A \cap G$ has exactly one element; more generally, if $p \in G$ and A is a maximal antichain below p with $A \in M$, then $A \cap G$ has exactly one element. In forcing arguments, M is the *ground model* and $M[G]$ the *generic extension*. Trivial p.o.'s leading to $M = M[G]$ are allowed to facilitate the statements and proofs of theorems.

No particular formalization of the forcing language is adopted, but the following are specified about the names (forcing terms): *Names* (or *P-names* if there is a need to be specific) are denoted by dotted letters, like $\dot{t}$, and their interpretations in a generic extension $M[G]$ by $\dot{t}^G$ or $\dot{t}^{M[G]}$. This approach is taken with terms definable in the language of set theory, e.g.

$$\dot{\omega}_2 \text{ is a name for the second uncountable cardinal}$$

in the generic extension. A basic observation is that a set x belongs to $M[G]$ *iff* x is definable in $M[G]$ from G and finitely many elements of M, and hence corresponds to a name. If $x \in M$, then $\check{x}$ denotes the canonical name for x, so that $\check{x}^G = x$ for every P-generic G over M. As is customary, α is written instead of $\check{\alpha}$ for ordinals $\alpha \in M$, since ordinals are absolute for all generic extensions. $\dot{G}$ is the canonical name for the generic object, and $\check{V}$ the canonical name for the ground model. By this means, the ground model can be considered a (definable) class, and hence an inner model, in any generic extension.

The forcing relation for P is denoted by

$$p \Vdash_P \varphi \,,$$

and

$$\Vdash_P \varphi \text{ iff } \mathbb{1}_P \Vdash_P \varphi \,, \text{ and}$$
$$p \parallel_P \varphi \text{ iff } p \text{ decides } \varphi, \text{ i.e. } p \Vdash_P \varphi \text{ or } p \Vdash_P \neg\varphi \,.$$

The subscript P is suppressed when clear from the context, which should also differentiate from $\parallel$ used for compatibility of conditions. Specific formulas φ are rendered in mathematical English, only set apart by the Quinean quotes $\ulcorner \urcorner$ when a misreading is possible. $\Vdash$ is the standard, deductively closed relation, historically called "weak forcing" and differing from Cohen's original relation (see Kunen [80: 235ff]). Among its well-known properties is the *Maximal Principle*:

If $p \Vdash \exists x \varphi(x, \dot{t}_1, \ldots, \dot{t}_n)$, then for some name $\dot{t}$, $p \Vdash \varphi(\dot{t}, \dot{t}_1, \ldots, \dot{t}_n)$.

This requires the Axiom of Choice in the ground model and is crucial in several arguments. Finally, the canonical names $\dot{G}$ for the generic object and $\check{V}$ for the ground model satisfy the following for any $p, q \in P$ and name $\dot{t}$, in conformance with the recursive definition of the forcing relation for atomic formulas:

$$p \Vdash \check{q} \in \dot{G} \text{ iff } \forall r \le p \exists s \le r(s \le q) \,.$$
$$p \Vdash \dot{t} \in \check{V} \text{ iff } \forall r \le p \exists s \le r \exists x(s \Vdash \check{x} = \dot{t}) \,.$$

Forcing consistency results are usually stated syntactically, in terms of the existence of a p.o. P so that for the desired assertion φ, $\Vdash_P \varphi$. This succinct formulation makes the closest approach to the corresponding finitary relative consistency result, as $\Vdash_P \sigma$ for any ZFC axiom σ and $\Vdash_P$ is deductively closed. However, as is the usual practice proofs are often given semantically as foreshadowed by the discussion of transitive $\in$-models, and corresponding to the syntactic formulations V is regarded as the ground model and generic extensions $V[G]$ are "taken". With this in mind,

P-generic means P-generic over V.

This of course runs counter to the Cantor-Gödel view of the class $V = \{x \mid x = x\}$ as the universe of *all* sets, and the travesty is often compounded by talking of extensions of $V[G]$. However, it is well-known that this semantic approach can be regarded as merely *une façon de parler* for formalizable arguments about forcing relations, and it not only taps a vital source of intuitions about forcing but is often the only way to render arguments intelligible.

An important feature of the theory is that generic extensions can be identified through embeddings of the corresponding p.o.'s: For p.o.'s P and Q,

$i \colon P \to Q$ is a *complete embedding iff*

(i) $\forall p, q \in P(p \le q \rightarrow i(p) \le i(q))$,
(ii) $\forall p, q \in P(p \perp q \rightarrow i(p) \perp i(q))$, and
(iii) if A is a maximal antichain in P , then $i``A$ is a maximal antichain in Q .

$i \colon P \to Q$ is a *dense embedding iff* i satisfies (i) and (ii) above, and $i``P$ is dense in Q .

These definitions do not require i to be injective. A dense embedding is easily seen to be complete.

$$P \cong Q$$

asserts there is a a p.o. isomorphism: $P \to Q$. Kunen [80: 218ff] has an equivalent definition of "complete" and a proof of the following proposition. Toward its statement, for $X \subseteq P$ set

$$\bar{\imath}(X) = \{q \in Q \mid \exists p \in X(i(p) \leq q)\} .$$

10.1 Proposition. *Suppose that* $i \colon P \to Q$ *is complete.*

(a) If H *is* Q*-generic, then* $i^{-1}(H)$ *is* P*-generic and so* $V[i^{-1}(H)] \subseteq V[H]$. *If* i *is dense, then* $\bar{\imath}(i^{-1}(H)) = H$ *and so* $V[i^{-1}(H)] = V[H]$.

(b) If i *is dense and* G *is* P*-generic, then* $\bar{\imath}(G)$ *is* Q*-generic,* $i^{-1}(\bar{\imath}(G)) = G$, *and so* $V[\bar{\imath}(G)] = V[G]$. ⊣

The arguments are straightforward and the latter assertions follow from $i \in V$. Generic extensions are often identified using this result when P is actually isomorphic to a dense subset of Q, but the basic process of passing to a separative quotient (see Jech [03: 204ff]) uses the weaker hypotheses:

For a p.o. P,

$$P \text{ is } \textit{separative} \textit{ iff } \forall p, q \in P(p \not\leq q \;\to\; \exists r \leq p(r \perp q)) .$$

By and large, p.o.'s that occur in practice are separative, and the incentive behind the concept is to insure a certain orderliness: separative p.o.'s are isomorphic to a dense subset of their corresponding complete Boolean algebras, and are characterized by a simple but useful observation:

10.2 Lemma. *A p.o.* P *is separative iff* $\forall p, q \in P(p \leq q \;\leftrightarrow\; p \Vdash \check{q} \in \dot{G})$. ⊣

For a p.o. P and $p, q \in P$,

$$p \approx q \textit{ iff } \forall r \in P(r \parallel p \;\leftrightarrow\; r \parallel q) ,$$

and

$$[p] = \{q \mid q \approx p\} .$$

$\approx$ is an equivalence relation on P, and $[p]$ the corresponding equivalence class of p.

The *separative quotient of* P is $P/{\approx} = \{[p] \mid p \in P\}$

ordered by

$$[p] \leq [q] \textit{ iff } \forall r \in P(r \leq p \;\to\; r \parallel q) .$$

By the following simple observation separative p.o.'s suffice to get all generic extensions:

10.3 Lemma. *The map sending p to $[p]$ is a dense embedding of P onto $P/\approx$.* ⊣

Familiarity is assumed with chain, closure, and density (distributivity) conditions on p.o.'s and their salutary effects on generic extensions, but because of some variance in usage definitions are given: For a p.o. P,

P has the *κ-chain condition* (κ-c.c.) *iff* for any antichain A in P, $|A| < \kappa$.

P is *κ-closed iff* whenever $\gamma < \kappa$ and $\{p_\alpha \mid \alpha < \gamma\} \subseteq P$ with $p_\beta \leq p_\alpha$ for $\alpha < \beta < \gamma$, there is a $p \in P$ such that $p \leq p_\alpha$ for every $\alpha < \gamma$.

P is *κ-Baire iff* whenever $\gamma < \kappa$ and D_α for each $\alpha < \gamma$ is dense open, so is $\bigcap_{\alpha<\gamma} D_\alpha$.

These are all "less than κ" notions; κ^+-Baire is also called κ-distributive because of the connection with distributivity laws in Boolean algebras. Familiarity is assumed with how the κ-c.c is often deduced from a combinatorial fact known as the *Δ-system Lemma* (see Kunen [80: 49ff]):

10.4 Proposition (Šanin [46] for $\lambda = \omega$). *Suppose that $\lambda < \kappa$, κ is regular, and $\alpha < \kappa$ implies that $\alpha^{<\lambda} < \kappa$. If $|A| = \kappa$ and $x \in A$ implies that $|x| < \lambda$, then there is a $B \subseteq A$ with $|B| = \kappa$ such that B forms a Δ-system, i.e.*

there is an r such that for distinct $x, y \in B$, $x \cap y = r$.

In particular, any uncountable set consisting of finite sets has an uncountable subset that forms a Δ-system. ⊣

It is well known that chain conditions establish upper bounds on powers of cardinals:

10.5 Proposition. *Suppose that P is a p.o. with the κ-c.c., and for a given λ, set $\theta = (|P|^{<\kappa})^\lambda$. Then*

$$\Vdash_P \ulcorner 2^\lambda \leq \theta \urcorner .$$ ⊣

To complete the review, what is assumed of *iterated forcing*, first elaborated by Solovay, is summarized starting with *product forcing*: Suppose that P and Q are p.o.'s and consider the product p.o. $P \times Q$ ordered by

$$\langle p_0, q_0 \rangle \leq \langle p_1, q_1 \rangle \text{ iff } p_0 \leq_P p_1 \wedge q_0 \leq_Q q_1 .$$

Then forcing with $P \times Q$ is characterizable as an iteration (see Kunen [80: 252ff] or Jech [03: 229ff]):

10.6 Proposition (Solovay [70]). *G is $P \times Q$-generic iff $G = G_0 \times G_1$, where G_0 is P-generic, G_1 is Q-generic over $V[G_0]$, and so $V[G] = V[G_0][G_1]$.* ⊣

More generally, products of many p.o.'s can be taken. Suppose that P_α is a p.o. for every $\alpha < \delta$. For $p \in \prod_{\alpha<\delta} P_\alpha$, define

$$\mathrm{supt}(p) = \{\alpha < \delta \mid p(\alpha) \neq 1\!\!1_{P_\alpha}\} ,$$

the *support* of p. Then the *κ-product* of $\langle P_\alpha \mid \alpha < \delta\rangle$ is

$$\{p \in \textstyle\prod_{\alpha<\delta} P_\alpha \mid |\mathrm{supt}(p)| < \kappa\}$$

ordered by

$$p \leq q \;\; \textit{iff} \;\; \forall\alpha < \delta(p(\alpha) \leq_{P_\alpha} q(\alpha)) .$$

For example, what is meant by "adding λ many Cohen reals" is forcing with the $\aleph_0$-product of λ copies of the p.o. for adding a Cohen real. The *Easton product* of $\langle P_\alpha \mid \alpha < \delta\rangle$ is

$$\{p \in \textstyle\prod_{\alpha<\delta} P_\alpha \mid \forall\lambda(\lambda \text{ regular} \to |\mathrm{supt}(p) \cap \lambda| < \lambda)\}$$

ordered coordinate-wise as before. This product was devised by Easton for the first forcing argument using a product of many p.o.'s, indeed a proper class of them:

10.7 Theorem (Easton [64, 70]). *Suppose that* GCH *holds and F is a class function from the class of regular cardinals to the class of cardinals such that for regular $\kappa \leq \lambda$: $F(\kappa) \leq F(\lambda)$ and $\kappa <$ cf$(F(\kappa))$. Then there is a generic extension preserving cardinals and cofinalities in which $2^\kappa = F(\kappa)$ for every regular κ.* $\dashv$

The forcing was with the Easton product of $\langle P_\alpha \mid \alpha \in \mathrm{On}\rangle$ with P_α the usual notion for adjoining $F(\alpha)$ subsets of α if α is regular and trivial otherwise. The following lemma played a crucial role in the product analysis.

10.8 Lemma. *Suppose that P and Q are p.o.'s and λ is such that P has the λ^+-c.c. and Q is λ^+-closed. If G is $P \times Q$-generic over V with $G = G_0 \times G_1$ as in 10.6, then any $f\colon \lambda \to V$ in $V[G]$ is in $V[G_0]$. In particular, $\mathcal{P}(\lambda) \cap V[G] = \mathcal{P}(\lambda) \cap V[G_0]$.* $\dashv$

Of course, how to force with a proper class of conditions had to be worked out as well. For details on this and product forcing in general, see Jech [03: §15].

10.7 completely solved the problem of determining the possible values of $F(\kappa) = 2^\kappa$ for regular κ; the *only* restrictions are that F be non-decreasing and consistent with the familiar König result: $\kappa <$ cf$(F(\kappa))$. The *Singular Cardinals Problem* is to determine the possible values of 2^κ for singular κ, a much more difficult problem which is far from being solved. It is remarkable that large cardinals are closely intertwined with this basic problem, and all the more remarkable that their intervention turns out to be a necessary one (see volume II).

Suppose that one wants to force with a p.o. P to get $V[G_0]$ and then with a p.o. $\dot{Q}^{G_0}$, not necessarily in V, to get $V[G_0][G_1]$. Then one might as well assume $\Vdash_P \dot{Q}$ is a p.o., and consider

$$P * \dot{Q} = \{\langle p, \dot{q}\rangle \mid p \in P \ \wedge \ \Vdash_P \dot{q} \in \dot{Q}\}$$

ordered by

$$\langle p_0, \dot{q}_0\rangle \ \leq \ \langle p_1, \dot{q}_1\rangle \ \textit{ iff } \ p_0 \leq_P p_1 \ \wedge \ p_0 \Vdash_P \dot{q}_0 \leq_{\dot{Q}} \dot{q}_1 \ .$$

As it stands, $P * \dot{Q}$ may be a proper class because of the possible $\dot{q}$'s, and $\leq$ is not antisymmetric, but according to previous remarks

$\langle p_0, \dot{q}_0\rangle$ and $\langle p_1, \dot{q}_1\rangle$ are identified whenever $\langle p_0, \dot{q}_0\rangle \ \leq \ \langle p_1, \dot{q}_1\rangle \ \leq \ \langle p_0, \dot{q}_0\rangle$,

i.e. $p_0 = p_1 \ \wedge \ p_0 \Vdash_P \dot{q}_0 = \dot{q}_1$. In reasonable formalizations of names, there are "set" many equivalence classes under this identification, and natural representatives can be chosen from each to comprise what should be meant by $P * \dot{Q}$ (Kunen [80: 269]). Henceforth,

it is assumed that $P * \dot{Q}$ is a set

under such a contraction, but $\langle p, \dot{q}\rangle$ will be used for its equivalence class to avoid awkward notation. Under a natural identification, every P-name is also considered to be a $P*\dot{Q}$-name. Moreover, after justifying p.o. isomorphisms like $R \cong P*\dot{Q}$, such p.o.'s are often identified for notational convenience, even though they may not be extensionally equal. Forcing with $P * \dot{Q}$ is characterizable as forcing first with P and then with $\dot{Q}$ (see Kunen [80: 268ff] or Jech [03: 267ff]):

10.9 Proposition (Solovay-Tennenbaum [71]).

(a) Suppose that G_0 is P-generic and G_1 is $\dot{Q}^{G_0}$-generic over $V[G_0]$. Set

$$G_0 * G_1 = \{\langle p, \dot{q}\rangle \in P * \dot{Q} \mid p \in G_0 \ \wedge \ \dot{q}^{G_0} \in G_1\} \ .$$

Then $G_0 * G_1$ is $P * \dot{Q}$-generic and $V[G_0 * G_1] = V[G_0][G_1]$.

(b) Conversely, suppose that G is $P * \dot{Q}$-generic, and set

$$G_0 = \{p \in P \mid \exists \dot{q}(\langle p, \dot{q}\rangle \in G)\} \ , \text{ and}$$
$$G_1 = \{\dot{q}^{G_0} \mid \exists p(\langle p, \dot{q}\rangle \in G)\} \ .$$

Then G_0 is P-generic, G_1 is $\dot{Q}^{G_0}$-generic over $V[G_0]$, and $G = G_0 * G_1$. ⊣

10.6 is essentially the special case of this result where $\dot{Q}$ is $\check{Q}$ for some $Q \in V$, when $P \times Q$ is isomorphic to a dense subset of $P * \dot{Q}$.

The following structure theorem applies 10.9(b) to show that every ZFC model intermediate between the ground model and a generic extension is itself a generic extension. This result about the generality of forcing has a natural proof in terms of Boolean-valued models (see Jech [03: 247ff]) but may also be established with partial orders.

10.10 Proposition. *Suppose that R is a p.o., G is R-generic over V, and N is a transitive $\in$-model of* ZFC *such that $V \subseteq N \subseteq V[G]$. Then there is a p.o. P and a P-name $\dot{Q}$ such that $\Vdash_P \dot{Q}$ is a p.o. with $P * \dot{Q}$ isomorphic to R so that, thus identifying R with $P * \dot{Q}$, if G_0 is as in 10.9(b), $V[G_0] = N$.* ⊣

The initial observation here is that transitive $\in$-models of ZFC are determined by their sets of ordinals (see Jech [03: 196ff]) so that one can deduce that $N = V[x]$ for some $x \subseteq \mathrm{On}$. Serge Grigorieff [75] investigated the situation for ZF models N.

10.9 was first used in an argument that essentially established the consistency of Martin's Axiom together with ¬CH. This well-known axiom is briefly reviewed:

For P a p.o., $\mathcal{D}$ a collection of dense subsets of P, and $G \subseteq P$,

$$\begin{aligned} G \text{ is } \mathcal{D}\text{-generic} \ \ \textit{iff} \ \ &\text{(i) } \forall p, q \in G \exists r \in G (r \leq p \wedge r \leq q) , \\ &\text{(ii) } \forall p \in G \forall q \in P (p \leq q \rightarrow q \in G) , \text{ and} \\ &\text{(iii) } \forall D \in \mathcal{D} (G \cap D \neq \emptyset) . \end{aligned}$$

For any κ, *Martin's Axiom for κ* (MA(κ)) is:

> If P is a p.o. satisfying the ω_1-c.c. and $\mathcal{D}$ is a collection of at most κ dense sets, then there is a $G \subseteq P$ which is $\mathcal{D}$-generic.

Simply *Martin's Axiom* (MA) is:

$$\forall \kappa < 2^{\aleph_0} (\mathrm{MA}(\kappa)) .$$

The clear motivation here is to justify internal generic objects when genericity is restricted to small collections $\mathcal{D} \in V$ of dense sets. It is well-known that MA(ω) is a theorem of ZFC and that MA($2^{\aleph_0}$) and MA with the ω_1-c.c. restriction deleted are both inconsistent.

10.11 Theorem (Solovay-Tennenbaum [71]). *Suppose that κ is a cardinal satisfying $\kappa^{<\kappa} = \kappa > \omega$. Then there is a ω_1-c.c. p.o. P of cardinality κ such that $\Vdash_P \mathrm{MA} \wedge 2^{\aleph_0} = \kappa$.* ⊣

In particular, starting with a model satisfying GCH, MA can be arranged with $2^{\aleph_0}$ any prescribed uncountable regular cardinal. For details on MA see Jech [03: §16], or Kunen [80: II§2] where it is axiomatically presented before forcing *per se* is discussed. Fremlin [84] is a portmanteau book on the consequences of MA.

Mild Extensions

To test the waters, some straightforward observations about forcing and large cardinals are discussed which, while not noticed at the outset, have a direct bearing on Gödel's earlier speculations on the Continuum Problem (§3). Forcing provided the means to show large cardinals do *not* decide the power of the continuum. Cohen himself had noted ([64: 110]) that his argument for the independence of CH preserves inaccessible and ("It seems probable") Mahlo cardinals, and Levy and Solovay showed that measurable cardinals are also preserved. In fact, their argument is a generic one for showing that, by and large, large cardinals κ are preserved in any *mild extension*, i.e. via a forcing with a p.o. of cardinality less than κ. The following useful results pursue this theme through several cases:

10.12 Proposition. *Suppose that κ is inaccessible and P is a p.o. such that $|P| < \kappa$. Then*

$$\Vdash_P \kappa \text{ is inaccessible} .$$

Proof. This follows from the $|P|^+$-c.c.: $\Vdash \kappa$ is regular, and 10.5 implies that $\Vdash \kappa$ is a strong limit. ⊣

10.13 Proposition. *Suppose that κ is Mahlo and P is a p.o. such that $|P| < \kappa$. Then*

$$\Vdash_P \kappa \text{ is Mahlo} .$$

The proof relies on a generally useful observation:

10.14 Lemma. *Suppose that $\kappa > \omega$ is regular, P is a p.o. with the κ-c.c., $p \in P$, and $p \Vdash \dot{C}$ is closed unbounded in κ. Then there is a C_0 closed unbounded in κ such that $p \Vdash \check{C}_0 \subseteq \dot{C}$.*

Proof. For $\alpha < \kappa$, set

$$X_\alpha = \{\xi < \kappa \mid \exists q \leq p(q \Vdash \xi \text{ is the least member of } \dot{C} - (\alpha + 1))\} .$$

By the κ-c.c., $|X_\alpha| < \kappa$, so set $f(\alpha) = \sup(X_\alpha) < \kappa$. Finally, let

$$C_0 = \{\beta < \kappa \mid \forall\alpha < \beta(f(\alpha) < \beta)\} ,$$

a closed unbounded set.

To show that $p \Vdash \check{C}_0 \subseteq \dot{C}$, assume to the contrary that there is a $q \leq p$ and a $\beta \in C_0$ such that $q \Vdash \beta \notin \dot{C}$. Then for some $r \leq q$ and $\alpha < \beta$, $r \Vdash \alpha = \sup(\dot{C} \cap \beta)$. For some $s \leq r$ and $\xi \in X_\alpha$, $s \Vdash \xi$ is the least member of $\dot{C} - (\alpha + 1)$. But $\xi < \beta$ as $\beta \in C_0$, contradicting the choice of α. ⊣

Proof of 10.13. Suppose that $p \Vdash \dot{C}$ is closed unbounded in κ. By 10.12, if λ is inaccessible and $|P| < \lambda$, then $\Vdash \lambda$ is inaccessible. Since κ is Mahlo, by 10.14 there are such $\lambda < \kappa$ satisfying $p \Vdash \lambda \in \dot{C}$. ⊣

10.15 Proposition (Levy [64], Solovay [65a], Levy-Solovay [67]). *Suppose that κ is measurable and P is a p.o. such that $|P| < \kappa$. Then*

$$\Vdash_P \kappa \text{ is measurable .}$$

Proof. Let U be a κ-complete ultrafilter over κ, and

$$\Vdash \dot{U} = \{Y \subseteq \kappa \mid \exists X \in \check{U}(X \subseteq Y)\} ,$$

the filter generated by $\check{U}$. It will in fact be shown that $\Vdash \dot{U}$ is a κ-complete ultrafilter over κ. The following claim can serve a larger purpose:

(∗) If $p \Vdash \dot{\tau}\colon \kappa \to \check{V}$, then there are $q \leq p$, $Z \in U$, and $f\colon Z \to V$ such that $q \Vdash \dot{\tau}|\check{Z} = \check{f}$.

To show this, for each $r \leq p$ let $X_r = \{\xi \in \kappa \mid \exists x(r \Vdash \dot{\tau}(\xi) = \check{x})\}$. Then $\bigcup_{r \leq p} X_r = \kappa$ and $|P| < \kappa$, so by κ-completeness there is a $q \leq p$ such that $X_q \in U$. But then, for $f\colon X_q \to V$ defined by $f(\xi) =$ that x such that $q \Vdash \dot{\tau}(\xi) = \check{x}$ (unique as $\dot{\tau}$ is forced to be a function), it follows that $q \Vdash \dot{\tau}|\check{X}_q = \check{f}$.

To complete the proof, note first that $\Vdash \dot{U}$ is non-principal, so it remains to verify that $\Vdash \dot{U}$ is ultra and κ-complete. For this it suffices by simple arguments (cf. 2.7) to show that if $\gamma < \kappa$ and $\Vdash \dot{\tau}\colon \kappa \to \gamma$, then there is an α such that $\Vdash \dot{\tau}^{-1}(\{\alpha\}) \in \dot{U}$:

By the claim, there are $Z \in U$ and $f\colon Z \to \gamma$ such that $\Vdash \dot{\tau}|\check{Z} = \check{f}$. But U is κ-complete, so there is an $\alpha < \gamma$ such that $A = f^{-1}(\{\alpha\}) \in U$. Hence, $\Vdash \check{A} \cap \check{Z} \subseteq \dot{\tau}^{-1}(\{\alpha\}) \in \dot{U}$. ⊣

A further consequence of the claim (∗) is that if U is normal, then $\Vdash \dot{U}$ is normal.

10.16 Exercise (Levy-Solovay [67]). *10.15 holds with "measurable" replaced by "Ramsey" or by "weakly compact".*

Hint. For Ramsey, if $p \Vdash \dot{\tau}\colon [\kappa]^{<\omega} \to 2$, define $g\colon [\kappa]^{<\omega} \to \mathcal{P}(P \times 2)$ by $g(s) = \{\langle q, i\rangle \mid q \leq p \wedge q \Vdash \dot{\tau}(\check{s}) = i\}$. Show that if H is homogeneous for g, then $p \Vdash \check{H}$ is homogeneous for $\dot{\tau}$. ⊣

Similar arguments can be devised for a variety of large cardinals. The problem of preserving large cardinals when forcing with p.o.'s of large cardinality was to lead to important developments both in the theory of forcing as well as large cardinals.

The Levy Collapse

The first announcement after Cohen's breakthrough of results using forcing was made four months later, and notably one result involved an inaccessible cardinal. Motivated by questions of definability Levy [63, 70] formulated a gentle collapse of an inaccessible cardinal that has become basic to the theory, and experience has shown that it often suffices for relating large cardinals with combinatorial properties of ω_1 and ω_2 in equiconsistency results. For clarity and later convenience, a general setting is first pursued:

For $S \subseteq \mathrm{On}$ and λ regular,

$$\begin{aligned}\mathrm{Col}(\lambda, S) = \{p \mid p \text{ is a function } \wedge\ |p| < \lambda\ \wedge\ \mathrm{dom}(p) \subseteq S \times \lambda \\ \wedge\ \forall\langle\alpha, \xi\rangle \in \mathrm{dom}(p)(p(\alpha, \xi) = 0 \vee p(\alpha, \xi) \in \alpha)\}\end{aligned}$$

ordered by $p \leq q$ *iff* $p \supseteq q$. Thus, $\mathrm{Col}(\lambda, S)$ is a p.o. for generically adjoining surjections: $\lambda \to \alpha$ for every $\alpha \in S$. $\mathrm{Col}(\omega, \{\omega\})$ is essentially Cohen's original p.o. for adjoining a subset of ω, a Cohen real; $\mathrm{Col}(\lambda, \{\lambda\})$ its generalization for adjoining a Cohen subset of λ; and $\mathrm{Col}(\lambda, \{\kappa\})$ for $\kappa > \lambda$ the standard p.o. for collapsing $|\kappa|$ to λ. The following lemma lists some elementary properties:

10.17 Lemma.

(a) $\mathrm{Col}(\lambda, S)$ *is λ-closed.*

(b) Suppose that $S = X \cup Y$ is a disjoint union, and set $P_0 = \mathrm{Col}(\lambda, X)$ and $P_1 = \mathrm{Col}(\lambda, Y)$. Then G is $\mathrm{Col}(\lambda, S)$-generic iff $G = \{p \cup q \mid p \in G_0 \wedge q \in G_1\}$, where G_0 is P_0-generic, $\mathrm{Col}(\lambda, Y)^{V[G_0]} = P_1$, and G_1 is P_1-generic over $V[G_0]$.

(c) If κ is regular, $\kappa > \lambda$, and either κ is inaccessible or $\lambda = \omega$, then $\mathrm{Col}(\lambda, \kappa)$ has the κ-c.c.

(d) If $\mathrm{Col}(\lambda, \kappa)$ has the κ-c.c., then forcing with it preserves cardinals $\leq \lambda$ and $\geq \kappa$.

(e) Suppose that κ is regular, $\mathrm{Col}(\lambda, \kappa)$ has the κ-c.c., and G is $\mathrm{Col}(\lambda, \kappa)$-generic. Then for any $x \in V[G]$ with $x \colon \gamma \to \mathrm{On}$ where $\gamma < \kappa$, there is a $\delta < \kappa$ such that $x \in V[G \cap \mathrm{Col}(\lambda, \delta)]$.

Proof. (a) This is obvious.

(b) This follows from 10.6, since the association of $p \in \mathrm{Col}(\lambda, S)$ with $\langle p|(X \times \lambda), p|(Y \times \lambda)\rangle$ is an isomorphism between $\mathrm{Col}(\lambda, S)$ and the product p.o. $P_0 \times P_1$; that $\mathrm{Col}(\lambda, Y)^{V[G_0]} = P_1$ follows from (a) since forcing with P_0 adds no new ordinal sequences of length less than λ.

(c) This follows by a typical application of the Δ-system lemma 10.4: Given $\{p_\beta \mid \beta < \kappa\} \subseteq \mathrm{Col}(\lambda, \kappa)$, the lemma together with the hypothesis imply that there is an $X \in [\kappa]^\kappa$ and an $r \subseteq \kappa \times \lambda$ such that $\beta \neq \gamma$ both in X implies that $\mathrm{dom}(p_\beta) \cap \mathrm{dom}(p_\gamma) = r$. By definition of $\mathrm{Col}(\lambda, \kappa)$ and $|r| < \kappa$, there is a further $Y \in [X]^\kappa$ such that $\beta \neq \gamma$ both in Y implies that $p_\beta | r = p_\gamma | r$. But for such β and γ, p_β and p_γ are compatible.

(d) This uses (a).

(e) Let $\dot{x}$ be such that $\dot{x}^G = x$, and for each $\alpha < \gamma$ let $A_\alpha \subseteq \mathrm{Col}(\lambda, \kappa)$ be a maximal antichain such that for each $p \in A_\alpha$ there is a ξ such that $p \Vdash \dot{x}(\alpha) = \xi$. $|A_\alpha| < \kappa$ for each $\alpha < \gamma$ by the κ-c.c. so that $|\bigcup_{\alpha<\gamma} A_\alpha| < \kappa$, and hence by the regularity of κ there is a $\delta < \kappa$ such that $p \in \bigcup_{\alpha<\gamma} A_\alpha$ implies that

$\mathrm{dom}(p) \subseteq \delta \times \lambda$. Finally, x is definable from $G \cap \mathrm{Col}(\lambda, \delta)$ by: $x(\alpha) = \xi$ *iff* $p \Vdash \dot{x}(\alpha) = \xi$ for the unique $p \in G \cap A_\alpha$. ⊣

For $\kappa > \lambda$,

$$\mathrm{Col}(\lambda, \kappa) \text{ is a } \textit{Levy collapse} \ ,$$

and when it has the κ-c.c.,

$$\mathrm{Col}(\lambda, \kappa) \ \textit{collapses } \kappa \textit{ to } \lambda^+$$

because of 10.17(d). This is part of the parlance; forcing with $\mathrm{Col}(\lambda, \kappa)$ does *not* collapse κ as a cardinal, but gently renders it the successor cardinal of λ by collapsing all cardinals strictly between λ and κ. Levy's original p.o. was $\mathrm{Col}(\omega, \kappa)$ with κ inaccessible, and usually 10.17(c) is invoked with inaccessible κ.

Digressing briefly, a proof is next provided of 8.16, the equiconsistency of Rowbottom and Jónsson cardinals. One direction is immediate, as Rowbottom cardinals are Jónsson, so the next proposition completes the proof. It is established by a mild extension argument with two twists: 10.17(c) is invoked for $\lambda = \omega$, and the large cardinal property to be preserved does not entail (strong) inaccessibility.

10.18 Proposition (Kleinberg [72, 79]). *If κ is the least Jónsson cardinal, then there is a p.o. P such that*

$$\Vdash_P \kappa \text{ is Rowbottom} \ .$$

Proof. By 8.15 there is a regular $\nu < \kappa$ such that

$$(*) \qquad \kappa \text{ is } \nu\text{-Rowbottom and } \kappa \not\to [\kappa]^{<\omega}_\alpha \text{ for any } \alpha < \nu \ .$$

By 10.17(c), $P = \mathrm{Col}(\omega, \nu)$ has the ν-c.c., and so renders $\nu = \omega_1$ in any generic extension. Hence, it suffices to show that

$$\Vdash_P \kappa \text{ is } \nu\text{-Rowbottom} \ .$$

To this end, suppose that $p \in P$, $\gamma < \kappa$, and $p \Vdash \dot{f}\colon [\kappa]^{<\omega} \to \gamma$. For each $s \in [\kappa]^{<\omega}$ let $A_s \subseteq P$ be a maximal antichain below p such that for each $r \in A_s$ there is an $\alpha_r < \gamma$ such that $r \Vdash \dot{f}(\check{s}) = \alpha_r$. $|A_s| < \nu$ by the ν-c.c., so by $(*)$ there is a $g_s\colon [\kappa]^{<\omega} \to A_s$ such that for any $X \in [\kappa]^\kappa$, $g_s``X^{<\omega} = A_s$. Now define $h\colon [\kappa]^{<\omega} \to \gamma$ by setting $h(\xi_1, \ldots, \xi_n) = 0$ unless $n = 2^i 3^j$ for some $i, j > 0$, in which case:

$$h(\xi_1, \ldots, \xi_n) = \text{the } \alpha \text{ such that } g_{\{\xi_1,\ldots,\xi_i\}}(\xi_{i+1}, \ldots, \xi_{i+j}) \Vdash \dot{f}(\xi_1, \ldots, \xi_i) = \alpha \ .$$

By $(*)$ there is an $H \in [\kappa]^\kappa$ such that setting $E = h``[H]^{<\omega}$, $|E| < \nu$. To complete the proof, it suffices to show that $p \Vdash \dot{f}``[\check{H}]^{<\omega} \subseteq \check{E}$.

To do this, suppose that $q \leq p$, and for some $s \in [H]^{<\omega}$ and $\alpha < \gamma$, $q \Vdash \dot{f}(s) = \alpha$. By definition of A_s, there is an $r \in A_s$ compatible with q, and since

$p \Vdash \dot{f}$ is a function, $\alpha = \alpha_r$. By definition of g_s, there is a $t \in [H-(\max(s)+1)]^{<\omega}$ such that $g_s(t) = r$. Finally, if $s = \{\xi_1, \ldots, \xi_i\}$ and $t = \{\xi_{i+1}, \ldots, \xi_{i+j}\}$, set $n = 2^i 3^j$ and choose $\xi_{i+j+1} < \ldots < \xi_n$ all from $H - (\max(t) + 1)$. Then it is simple to check that $h(\xi_1, \ldots, \xi_n) = \alpha \in E$, and so the proof is complete. $\dashv$

Returning to the main development, Levy's initial result with his collapse is discussed in the context of Solovay's result in §11, and both use a fundamental property of $\mathrm{Col}(\lambda, S)$. For a p.o. P,

P is *weakly homogeneous iff* for any $p, q \in P$ there is an automorphism e of P such that $e(p) \parallel q$.

This property arose in the independence proofs for AC. The "weakly" acknowledges that a Boolean algebra B is called *homogeneous* if for $a, b \in B$ with neither a nor b equal to either the zero or the one of B, there is an automorphism e of B such that $e(a) = b$. The first part of the next proposition is the *raison d'être* for weak homogeneity.

10.19 Proposition.
(a) If a p.o. P is weakly homogeneous, then for any formula $\varphi(v_1, \ldots, v_n)$ in the forcing language and $x_1, \ldots, x_n \in V$, either $\Vdash_P \varphi(\check{x}_1, \ldots, \check{x}_n)$ or else $\Vdash_P \neg\varphi(\check{x}_1, \ldots, \check{x}_n)$.
(b) $\mathrm{Col}(\lambda, S)$ *is weakly homogeneous.*

Proof. (a) Assume to the contrary that $p \Vdash \varphi(\check{x}_1, \ldots, \check{x}_n)$ and $q \Vdash \neg\varphi(\check{x}_1, \ldots, \check{x}_n)$ for some $p, q \in P$, and let e be an automorphism of P such that $e(p) \parallel q$. e induces a bijection: P-names $\to$ P-names that fixes $\check{x}$ for every $x \in V$ and preserves the forcing relation (see Kunen [80: 222]). Thus, $e(p) \Vdash \varphi(\check{x}_1, \ldots, \check{x}_n)$, but this contradicts $e(p) \parallel q$.

(b) Given $p, q \in \mathrm{Col}(\lambda, S)$ it is easy to find a bijection $f\colon \lambda \to \lambda$ such that for $\langle \alpha, \xi \rangle \in \mathrm{dom}(p)$ and $\langle \beta, \zeta \rangle \in \mathrm{dom}(q)$, $f(\xi) \neq \zeta$. f induces an automorphism e of $\mathrm{Col}(\lambda, S)$ that to $r \in \mathrm{Col}(\lambda, S)$ associates $e(r)$ specified by

$$\langle \alpha, f(\xi) \rangle \in \mathrm{dom}(e(r)) \textit{ iff } \langle \alpha, \xi \rangle \in \mathrm{dom}(r) \wedge e(r)(\alpha, f(\xi)) = r(\alpha, \xi) .$$

Since $\mathrm{dom}(e(p)) \cap \mathrm{dom}(q) = \emptyset$, $e(p) \parallel q$. $\dashv$

Finally, Solovay established a result specific to $\mathrm{Col}(\omega, \kappa)$ for his Lebesgue measurability result (§11). The natural proof uses what amounts to a characterization of $\mathrm{Col}(\omega, \{\alpha\})$ of intrinsic interest:

10.20 Proposition. *Suppose that P is a separative p.o. such that $|P| \leq |\alpha|$ and*

$$\Vdash_P \exists f (f\colon \omega \to \alpha \text{ is surjective} \wedge f \notin \check{V}) .$$

Then there is an injective, dense embedding of a dense subset of $\mathrm{Col}(\omega, \{\alpha\})$ *into P.*

Remarks. The hypothesis is dichotomous in that $f\colon \omega \to \alpha$ being surjective is only material when α is uncountable, and $f \notin \check{V}$, when α is countable. For $\alpha = \omega$, the result densely associates any non-trivial countable p.o. with the Cohen p.o. $\mathrm{Col}(\omega, \{\omega\})$. Finally, by this result and 10.1, the generic extensions by such P coincide with the generic extensions by $\mathrm{Col}(\omega, \{\alpha\})$; the separativity of P is extraneous in this regard since the separative quotient can first be taken and 10.3 invoked.

Proof. Set $\nu = |\alpha|$. Note first that the hypotheses imply that for any $p \in P$ there is a maximal antichain below p of cardinality ν (so that in particular $|P| = \nu$): If $\nu = \omega$, this follows from $\Vdash f \notin \check{V}$ since any condition must have incompatible extensions. If $\nu > \omega$, this follows from the collapse of ν to ω and the consequent failure of the ν-c.c.

The dense subset of $\mathrm{Col}(\omega, \{\alpha\})$ in question is

$$D_\alpha = \{p \mid \exists n \in \omega (p\colon \{\alpha\} \times n \to \alpha)\} ,$$

i.e. those members of $\mathrm{Col}(\omega, \{\alpha\})$ which are not only finite but have domains of form $\{\alpha\} \times n$ for some $n \in \omega$. By the maximal principle, there is a term $\dot{g}$ such that $\Vdash \dot{g}\colon \omega \to \dot{G}$ is surjective (where $\dot{G}$ is the canonical name for the generic object). Proceed now by recursion on $|p|$ for $p \in D_\alpha$ to define an injective, dense embedding $e\colon D_\alpha \to P$:

Set $e(\emptyset) = \mathbb{1}_P$. Having defined $e(p)$ for a $p \in D_\alpha$ with $\mathrm{dom}(p) = \{\alpha\} \times n$, let $\langle a^p_\xi \mid \xi < \alpha \rangle$ enumerate a maximal antichain below $e(p)$ of cardinality ν. By refining if necessary it can be assumed that each $a^p_\xi \Vdash \dot{g}(n) = \check{r}$ for some $r \in P$. Set $e(p \cup \{\langle \langle \alpha, n \rangle, \xi \rangle\}) = a^p_\xi$ for each $\xi < \alpha$.

Clearly, this definition provides an isomorphism $e\colon D_\alpha \to e``D_\alpha$. To show that $e``D_\alpha$ is dense in P, let $r \in P$ be arbitrary. Since $r \Vdash \check{r} \in \dot{G}$ by separativity, there is an $s \leq r$ and an $n \in \omega$ such that $s \Vdash \dot{g}(n) = \check{r}$. By construction there is a $p \in D_\alpha$ with $|p| = n + 1$ such that $e(p) \parallel s$, and since $e(p)$ had also decided $\dot{g}(n)$, $e(p) \Vdash \dot{g}(n) = \check{r}$. But then, $e(p) \Vdash \check{r} \in \dot{G}$, and so by separativity and 10.2, $e(p) \leq r$. ⊣

10.21 Proposition (Solovay [70]). *Suppose that $\kappa > \omega$ is regular and G is $\mathrm{Col}(\omega, \kappa)$-generic. Then for any $x \in V[G]$ with $x\colon \omega \to \mathrm{On}$, there is an H which is $\mathrm{Col}(\omega, \kappa)$-generic over $V[x]$ such that $V[G] = V[x][H]$.*

Proof. By 10.17(e), there is a $\delta < \kappa$ such that $x \in V[G \cap \mathrm{Col}(\omega, \delta)]$. Set

$$\begin{aligned} G_0 &= G \cap \mathrm{Col}(\omega, \delta) , \\ G_1 &= G \cap \mathrm{Col}(\omega, \{\delta\}) , \text{ and} \\ G_2 &= G \cap \mathrm{Col}(\omega, \kappa - (\delta + 1)) . \end{aligned}$$

By 10.17(b), G_0 is $\mathrm{Col}(\omega, \delta)$-generic, and so by 10.3 and 10.10 there is a separative p.o. $P \in V[x]$ and a P-generic H_0 over $V[x]$ such that $V[x][H_0] = V[G_0]$. Argue in $V[x]$:

Set $Q = P \times \mathrm{Col}(\omega, \{\delta\})$. By 10.6, $V[x][H_0][G_1]$ is a generic extension over $V[x]$ using Q. Clearly, $|Q| \leq |\mathrm{Col}(\omega, \delta + 1)| = |\delta|$, and because of the second component of Q, $\Vdash_Q \exists f(f\colon \omega \to \delta \text{ is surjective } \wedge\ f \notin \check{V})$. Hence, by 10.20 there is an H_1 $\mathrm{Col}(\omega, \{\delta\})$-generic over $V[x]$ such that $V[x][H_1] = V[x][H_0][G_1]$. Also by 10.20 applied to the p.o. $\mathrm{Col}(\omega, \delta+1)$, there is an H_2 $\mathrm{Col}(\omega, \delta+1)$-generic over $V[x]$ such that $V[x][H_2] = V[x][H_1]$.

Finally, by several applications of 10.17(b),

$$V[G] = V[G_0][G_1][G_2] = V[x][H_0][G_1][G_2] = V[x][H_2][G_2]$$

and $H_2 \cup G_2$ is $\mathrm{Col}(\omega, \kappa)$-generic over $V[x]$. ⊣

That any countable sequence of ordinals can be "absorbed" into the ground model in this way plays an important role in the next section.

11. Lebesgue Measurability

This section is devoted to the following result:

11.1 Theorem (Solovay [65b, 70]). *Suppose that κ is an inaccessible cardinal and G is* $\mathrm{Col}(\omega, \kappa)$*-generic. Then $V[G]$ has an inner model satisfying:*

(a) Every set of reals is Lebesgue measurable.
(b) Every set of reals has the Baire property.
(c) Every set of reals has the perfect set property.
(d) The Principle of Dependent Choices (DC).

Solovay established this result in the Spring of 1964, just a year after Cohen's creation of forcing. Cohen had in fact mentioned the possibility of (a) in a lecture in July 1963 attended by Solovay. The result was remarkable for its early sophistication and revealed what standard of argument was possible with forcing.

Like Cohen's results, 11.1 resolved classical problems dating back to the turn of the century and before. Lebesgue measurability and the Baire property were formulated in §0. As discussed in §2 Lebesgue had intended his measure as a solution to his Measure Problem, but Vitali in an early explicit use of the Axiom of Choice constructed a non-Lebesgue measurable set of reals from a well-ordering of the reals. In fact, it can be shown that Vitali's set does not possess the Baire property either. (Another counterexample is described after 11.4.) However, the substance and import of 11.1 is ensured by its (d), the *Principle of Dependent Choices*:

$$(\mathrm{DC}) \qquad \forall X \forall R (X \neq \emptyset \ \wedge \ R \subseteq X \times X \ \wedge \ \forall x \in X \exists y \in X (\langle x, y \rangle \in R \\ \to \exists f \in {}^{\omega}X \forall n \in \omega (\langle f(n), f(n+1) \rangle \in R)) \ .$$

Formulated by Bernays [42: 86] (with a precedent in a short-lived journal, Oswald Teichmüller [39: 568]) this choice principle readily implies, and in fact is equivalent to having, the useful characterization of well-foundedness in terms of the lack of infinite descending chains. Moreover, it readily implies the *Countable Axiom of Choice*,

(AC_ω) Every countable set consisting of non-empty sets has a choice function ,

and in turn the regularity of ω_1. (Jensen [66] showed that AC_ω does not imply DC.) Being handily sufficient for the development of measure theory, DC bolsters Solovay's model as a *bona fide* one for mathematical analysis. Lebesgue's solution was thus vindicated by 11.1 to the extent that it showed that some substantial use of AC is necessary to produce a non-measurable set.

Lebesgue measurability and the Baire property are often linked together, and the proofs of 11.1(a) and (b) have parallel arguments. On the other hand, the following exercise highlights how being null and being meager are quite different indications of a set's sparsity. In it and throughout this section, the terminology developed in §0 for measure and category is put into service; in particular, by "real" is usually meant member of ${}^{\omega}\omega$; m_L denotes Lebesgue measure; and

$$\langle \mathbf{s}_i \mid i \in \omega \rangle \text{ is a fixed enumeration of } {}^{<\omega}\omega \,.$$

11.2 Exercise (ZF). *There is a null set N of reals such that ${}^{\omega}\omega - N$ is meager.*

Hint. First, for $j < \omega$ find a $D_j \subseteq {}^{\omega}\omega$ such that ${}^{\omega}\omega - D_j$ is nowhere dense and $m_L(D_j) < \frac{1}{j+1}$. For example, for any $s \in {}^{<\omega}\omega$ let $s * n$ be s concatenated with n zero's, and set $D_j = \bigcup_{i \in \omega} O(s_i * f_j(i))$ for some $f_j \in {}^{\omega}\omega$ sufficiently fast-growing so that $m_L(D_j) < \frac{1}{j+1}$. Then set $N = \bigcap_{j<\omega} D_j$. ⊣

The perfect set property dates back to Cantor's original efforts to establish his Continuum Hypothesis. For $x \in {}^{\omega}\omega$ and $A \subseteq {}^{\omega}\omega$, x is a *limit point of A iff* for any $O(s)$ with $x \in O(s)$, $(A \cap O(s)) - \{x\} \neq \emptyset$. x is an *isolated point of A iff* $x \in A$ and is not a limit point of A, i.e. there is an $O(s)$ such that $A \cap O(s) = \{x\}$. For $A \subseteq {}^{\omega}\omega$,

A is *perfect iff* it is nonempty, closed, and has no isolated points .

Cantor formulated this concept in his topological investigations of $\mathbb{R}$. Its relevance to the Continuum Hypothesis can be seen from the following observations:

11.3 Proposition (ZF).

(a) (Cantor [83: 575][84: 471], Bendixson [83]) *For any uncountable closed set C of reals, there is a perfect set $P \subseteq C$ such that $C - P$ is at most countable.*

(b) (Cantor [84a]) *Any perfect set of reals has cardinality $2^{\aleph_0}$.*

Proof. (a) For any $A \subseteq {}^{\omega}\omega$, let A' = the limit points of A (Cantor's [72] concept of the *derived set* of A). It is simple to check that A' is always a closed set, and for A closed, that $A' \subseteq A$. Starting with an uncountable closed set C of reals, define $\langle C_\alpha \mid \alpha < \omega_1 \rangle$ by: $C_0 = C$, $C_{\alpha+1} = C'_\alpha$, and for limit $\delta > 0$, $C_\delta = \bigcap_{\alpha<\delta} C_\alpha$. Then $X = \bigcup_{\alpha<\omega_1} (C_\alpha - C_{\alpha+1})$ is at most countable: If $x \in X$, noting that x is an isolated point of exactly one C_α let $\chi(x)$ be the least i such that $O(\mathbf{s}_i) \cap C_\alpha = \{x\}$. It is simple to see that $\chi \colon X \to \omega$ is injective.

Since X is at most countable, there must be a $\gamma < \omega_1$ such that $C_{\gamma+1} = C_\gamma \neq \emptyset$. Then $P = C_\gamma$ is perfect, and $C - P = X$.

(b) Suppose that $P \subseteq {}^{\omega}\omega$ is perfect. Then there are $p_t \in P$ for every $t \in {}^{<\omega}2$ and a strictly increasing sequence $\langle n_i \mid i \in \omega \rangle$ of integers such that for any $t, u \in {}^{<\omega}2$,

(i) if $t \subseteq u$, then $p_t | n_{|t|} = p_u | n_{|t|}$, and
(ii) if $|t| = |u|$ yet $t \neq u$, then $p_t | n_{|t|} \neq p_u | n_{|t|}$.

This can be established by a straightforward recursion on $|t|$. For each $x \in {}^{\omega}2$, let $p_x = \lim_n \{p_{x|n} \mid n \in \omega\}$ in the natural sense based on (i). Then $p_x \in P$ since P is closed, and $x \neq y$ implies that $p_x \neq p_y$ by (ii). ⊣

Cantor thus verified the Continuum Hypothesis for closed sets: every closed set is either countable or has cardinality $2^{\aleph_0}$. Put another way, he had reduced his efforts to finding a closed set of reals of cardinality $\aleph_1$. Through Cohen's work it is now known that this is impossible in ZFC, but Cantor had initiated a programmatic approach to the Continuum Problem based on the topological complexity of sets which was to be vigorously pursued in descriptive set theory.

For any $A \subseteq {}^{\omega}\omega$,

$$A \text{ has the } \textit{perfect set property} \textit{ iff } A \text{ is countable or else has a perfect subset}.$$

Like Lebesgue measurability and the Baire property, the perfect set property can be regarded as a property indicative of well-behaved sets, and eventually all Borel sets and more were shown to possess it. However, Bernstein, who was initially a student of Cantor at Halle, showed that a modest form of the Axiom of Choice already confronts this approach to the Continuum Problem. In 1897 when just 19, Bernstein had established, with a readily avoidable use of AC, what is usually known as the Schröder-Bernstein Theorem: if $|A| \leq |B|$ and $|B| \leq |A|$, then there is a bijection between A and B. The sense of this derives from $\leq$ for "cardinals" in the ZF setting denoting injective embeddability. In particular, $\omega_1 \leq 2^{\aleph_0}$ means there is an uncountable, well-orderable set of reals. Bernstein's argument for the perfect set property can be analyzed as follows:

11.4 Proposition (ZF)(Bernstein [08]).

(a) If the reals are well-orderable, then there is a set A of reals of cardinality $2^{\aleph_0}$ such that $P \cap A \neq \emptyset$ and $P - A \neq \emptyset$ for any perfect set P of reals.

(b) If $\omega_1 \leq 2^{\aleph_0}$, then there is a set of reals without the perfect set property.

Proof. (a) By a straightforward counting argument using the Schröder-Bernstein Theorem there are exactly $2^{\aleph_0}$ perfect sets (the easier $\leq 2^{\aleph_0}$ suffices for the proof). Using a fixed well-ordering of ${}^{\omega}\omega$ let $\langle P_\alpha \mid \alpha < 2^{\aleph_0}\rangle$ be an enumeration of the perfect sets and define $a_\alpha \neq b_\alpha$ for $\alpha < 2^{\aleph_0}$ as follows: If a_β, b_β have been defined for $\beta < \alpha$, let a_α be the least a in the well-ordering such that

$$a \in P_\alpha - (\{a_\beta \mid \beta < \alpha\} \cup \{b_\beta \mid \beta < \alpha\})$$

and then let b_α be the least b in the well-ordering such that

$$b \in P_\alpha - (\{a_\beta \mid \beta \leq \alpha\} \cup \{b_\beta \mid \beta < \alpha\}) .$$

This is possible by 11.3(b). The desired A can be taken to be $\{a_\alpha \mid \alpha < 2^{\aleph_0}\}$.

(b) If $2^{\aleph_0} \not\leq \omega_1$, then by the hypothesis and 11.3(b) there is a well-orderable set of reals without the perfect set property. If $2^{\aleph_0} \leq \omega_1$, then by hypothesis ${}^{\omega}\omega$ is bijective with ω_1 by the Schröder-Bernstein Theorem, and so the conclusion follows from (a). ⊣

A remark of Bernstein's (p. 330) is interpretable as asserting that the denial of the perfect set property in the strong sense of (a) leads to a non-Lebesgue measurable set. This is simple to see, since if A were measurable, either $m_L(A) > 0$ or $m_L({}^\omega\omega - A) > 0$, say the first. By 0.9(a) there would be a closed set $C \subseteq A$ such that $m_L(C) > 0$, so by 11.3(a) a perfect set $P \subseteq C$, which is a contradiction.

Moreover, the set A cannot have the Baire property either: Suppose otherwise; by taking its complement if necessary it can assumed that A is not meager. By 0.12(b) there is a G_δ set $X \subseteq A$ such that $A - X$ is meager. But X is uncountable as A is not meager, yet any uncountable Borel set contains a perfect subset (by 12.2(c)), which is a contradiction.

The perfect set property led to the first instance of an important phenomenon in set theory: the derivation of equiconsistency results based on the complementary methods of forcing and inner models. A large cardinal hypothesis is typically transformed into a proposition about sets of reals by forcing that collapses the cardinal to ω_1 or enlarges the power of the continuum to the cardinal. Conversely, the proposition entails the same large cardinal hypothesis in the clarity of an inner model. Thus, in a subtle synthesis, hypotheses of length concerning the extent of transfinite are correlated with hypotheses of width concerning the fullness of power sets. Solovay's 11.1 provided the forcing direction from an inaccessible cardinal to the proposition that every set of reals has the perfect set property. Notably, Specker [57] had already made the conceptual move to inner models; through a sequence of implications he had in effect established in ZF that if every set of reals has the perfect set property and ω_1 is regular, then ω_1 is inaccessible in L (cf. Mycielski [64: 213]). Thus, Solovay's use of an inaccessible cardinal was necessary for getting 11.1(c)(d), and its collapse to ω_1 complemented Specker's observation.

Specker's conclusion can be readily strengthened to ω_1 *is inaccessible from reals:*

$$\forall a \in {}^\omega\omega(\omega_1^{L[a]} < \omega_1) .$$

This soon became a focal hypothesis, and the description is justified by a simple self-refinement:

11.5 Proposition (ZF). *If ω_1 is regular and $\forall a \in {}^\omega\omega(\omega_1^{L[a]} < \omega_1)$, then*

$$\forall a \in {}^\omega\omega(\omega_1 \text{ is inaccessible in } L[a]) .$$

Proof. Assume to the contrary that for some $a \in {}^\omega\omega$, ω_1 is not inaccessible in $L[a]$. Since $L[a]$ satisfies GCH by the same proof as for L, ω_1 must be a successor cardinal in $L[a]$, say $\omega_1 = \rho^{+L[a]}$. Since ρ is countable, let $b \in {}^\omega\omega$ code a well-ordering of ω of ordertype ρ. But if $c \in {}^\omega\omega$ codes a and b, then $\omega_1^{L[c]} = \omega_1$, which is a contradiction. ⊣

Finally, Levy's initial result with his collapse $\mathrm{Col}(\omega, \kappa)$ can be brought into the scheme of things; he essentially observed the equiconsistency of (a) and (c) below:

11.6 Theorem (Solovay [65b, 70], Specker [57], Levy [63, 70]). *The following theories are equiconsistent:*

(a) ZFC + $\exists\kappa(\kappa$ is inaccessible).

(b) ZF + DC + ⌜every set of reals has the perfect set property⌝.

(c) ZF + ω_1 is regular + $\omega_1 \not\leq 2^{\aleph_0}$.

(d) ZF + ω_1 is regular + $\forall a \in {}^\omega\omega(\omega_1^{L[a]} < \omega_1)$.

Proof. Con(d) → Con(a) follows from 11.5 and Con(a) → Con(b) from 11.1 once it is established. (b) → (c) and (c) → (d) are direct, the first from 11.4(b) and the fact that DC implies that ω_1 is regular, and the second as follows: For any $a \in {}^\omega\omega$, $\mathcal{P}(\omega) \cap L[a]$ has a well-ordering in $L[a]$ of ordertype $\omega_1^{L[a]}$, so $\omega_1^{L[a]} = \omega_1$ would imply that $\omega_1 \leq 2^{\aleph_0}$. ⊣

The question of whether an inaccessible cardinal is necessary in 11.1 for Lebesgue measurability or the Baire property was left unresolved, and considerable effort was eventually expended into trying to eliminate the large cardinal hypothesis. In one of those twists that makes set theory so intriguing Shelah [84] in 1979 put these issues to rest by establishing that indeed an inaccessible cardinal is not necessary for the Baire property, but rather unexpectedly, it *is* necessary for Lebesgue measurability (see volume II).

After this extended preamble, we finally begin the ascent to 11.1 with the preliminaries toward the Lebesgue measurability result.

Random Reals

Solovay took the first step beyond the p.o.'s like Col(λ, S) of §10 that arose initially with the creation of forcing and investigated forcing with the familiar σ-algebra of Borel sets. With the §0 terminology, consider

$$\mathcal{B}^* = \{X \in \mathcal{B} \mid X \text{ is not null}\},$$

ordered by

$$p \leq q \textit{ iff } p \subseteq q$$

as a p.o. for forcing. The following is simple:

11.7 Exercise.

(a) For $p, q \in \mathcal{B}^*$, $p \perp q$ iff $p \cap q$ is null.

(b) $\mathcal{B}^*$ has the ω_1-c.c.

Hint. (b) is like 2.1. ⊣

It is simple to see that $\mathcal{B}^*$ is not separative, and that its separative quotient $\mathcal{B}^*/\!\approx$ can be formulated with $p \approx q$ *iff* $p \,\triangle\, q$ is null. Using this as a definition $\approx$ can be extended to all of $\mathcal{M}_L$, the σ-algebra of Lebesgue measurable sets. The quotient $\mathcal{M}_L/\!\approx$, which coincides with $\mathcal{B}/\!\approx$, is a Boolean algebra under the set-theoretic operations modulated by $\approx$ called the *Lebesgue measure algebra.*

Solovay worked in terms of this algebra, but carrying on with $\mathcal{B}^*$ avoids $\approx$-equivalence classes and makes no difference by 10.3.

Toward Solovay's characterizations of $\mathcal{B}^*$-genericity, *codes* for the open, closed, and G_δ sets are developed: For each $c \in {}^\omega\omega$, set

$$A_c = \begin{cases} \bigcup\{O(\mathbf{s}_i) \mid c(i+1) = 1\} & \text{if } c(0) = 0\ , \\ {}^\omega\omega \ - \ \bigcup\{O(\mathbf{s}_i) \mid c(i+1) = 1\} & \text{if } c(0) = 1\ , \text{ and} \\ \bigcap_n(\bigcup\{O(\mathbf{s}_i) \mid c(2^n 3^{i+1}) = 1\}) & \text{if } c(0) > 1\ . \end{cases}$$

c is an *open code* if $c(0) = 0$, c is a *closed code* if $c(0) = 1$, and c a G_δ *code* if $c(0) > 1$. For every open, closed, or G_δ set X there is a respective code c such that $X = A_c$. All the Borel sets can be coded as in Solovay [70], but that complexity is not necessary because of the approximation result 0.9. While the extensions of these sets may vary from model to model, their Boolean and measure-theoretic essence can be described absolutely in terms of their codes.

11.8 Lemma. *Suppose that M be a transitive $\in$-model of* ZF. *Then:*

(a) The following are absolute for M:

$$x \in A_c\ ;\ A_c \neq \emptyset\ ;\ A_c \subseteq A_d\ ;\ A_c \subseteq ({}^\omega\omega - A_d)\ ;\ \text{and } A_c \cap A_d = A_e.$$

(That is, if $x, c \in M$ and c is a code, then $x \in A_c^M$ iff $x \in A_c$; and so forth.)

(b) If $c \in M$ is a code, then $m_L^M(A_c^M) = m_L(A_c)$.

Proof. (a) This is straightforward since the codes determine the interplay of the $O(\mathbf{s}_i)$'s, which in turn only depends on the $\mathbf{s}_i$'s.

(b) Suppose first that c is an open code. A sequence $\langle i_j \mid j \in \omega\rangle$ can be defined from c such that A_c is a disjoint union of the $O(\mathbf{s}_{i_j})$'s, as in the proof of 0.8(d). Thus, $m_L(A_c)$ is determined absolutely.

For a closed code c, let d be an open code such that $A_d = {}^\omega\omega - A_c$ and use (a).

Finally, suppose that c is a G_δ code. A sequence $\langle c_n \mid n \in \omega\rangle$ of open codes can be defined from c such that $A_{c_m} \supseteq A_{c_n}$ if $m < n$ and $A_c = \bigcap_n A_{c_n}$. $m_L(A_c) = \inf(\{m_L(A_{c_n}) \mid n \in \omega\})$, and thus is determined absolutely. ⊣

Forcing with $\mathcal{B}^*$ can now be characterized in terms of adding a real:

11.9 Theorem (Solovay [65b, 70]). *Suppose that G is $\mathcal{B}^*$-generic. Then there is a unique $x \in {}^\omega\omega$ such that for any closed code $c \in V$,*

$$x \in A_c^{V[G]} \textit{ iff } A_c^V \in G\ ,$$

and $V[x] = V[G]$.

Proof. The "C part" of 0.9(a) has the following consequences:

(i) For any $n \in \omega$,

$$\{C \in \mathcal{B}^* \mid C \text{ is closed } \wedge \ \exists k(C \subseteq \{f \in {}^\omega\omega \mid f(n) = k\})\}$$

is dense in $\mathcal{B}^*$.

(ii) For any Lebesgue measurable $A \subseteq {}^\omega\omega$ with $A \in V$,

$$\{C \in \mathcal{B}^* \mid C \text{ is closed } \wedge \ (C \subseteq A \ \vee \ C \cap A = \emptyset)\}$$

is dense in $\mathcal{B}^*$.

For (ii), suppose that $p \in \mathcal{B}^*$ is arbitrary. If $m_L(p \cap A) > 0$, there is a closed $C \in \mathcal{B}^*$ such that $C \subseteq p \cap A$. Otherwise $m_L(p - A) > 0$ and there is a closed $C \in \mathcal{B}^*$ such that $C \subseteq p - A$.

Arguing in $V[G]$ there is a unique $x \in {}^\omega\omega$ specified by

$$\{x\} = \bigcap\{A_c^{V[G]} \mid c \in V \text{ is a closed code } \wedge \ A_c^V \in G\}$$

since this is an intersection of closed sets with the finite intersection property and (i) holds.

Suppose now that $c \in V$ is an arbitrary closed code. If $A_c^V \in G$, then $x \in A_c^{V[G]}$ by definition. Conversely, suppose that $x \in A_c^{V[G]}$. To show that $A_c^V \in G$, it suffices by (ii) to show that for any closed code $d \in V$ with $A_d^V \in G$, $A_c^V \cap A_d^V \neq \emptyset$. But this follows from $x \in A_c^{V[G]} \cap A_d^{V[G]}$ and 11.8(a).

Finally, G is definable in $V[x]$ by

$$G = \{p \in B^* \mid \exists c(c \in V \text{ is a closed code } \wedge \ x \in A_c^{V[x]} \ \wedge \ A_c^{V[x]} \subseteq p)\}\,.$$

Hence, $V[x] = V[G]$. ⊣

Solovay called such an x a *random real*; more precisely, for M a transitive $\in$-model of ZFC and $x \in {}^\omega\omega$,

x is *random over* M *iff* x is as in 11.9 for some $(\mathcal{B}^*)^M$-generic G over M .

The next result is an informative characterization of random reals in terms of codes, and Solovay [70: 33] observed how it reflects the intuitive concept of randomness. Ironically, as most of the generic reals were eventually named after their discoverers, Solovay squandered an opportunity by his choice of *le mot juste*! In what follows, being null is unambiguous by 11.8(b), and 11.8(a) is used without further mention.

11.10 Theorem (Solovay [65b, 70]). *Suppose that M is a transitive $\in$-model of ZFC. Then $x \in {}^\omega\omega$ is random over M iff $x \notin A_c$ for any $c \in M$ which is a G_δ code for a null set.*

Proof. Suppose first that x is random over M with G the corresponding $(\mathcal{B}^*)^M$-generic over M, and $c \in M$ is a G_δ code for a null set. Then

$$M \models \ulcorner D = \{C \in \mathcal{B}^* \mid C \text{ is closed } \wedge \ C \subseteq {}^\omega\omega - A_c\} \text{ is dense in } \mathcal{B}^{*}\urcorner\,.$$

Hence, there is a closed code $d \in M$ such that $A_d^M \in G \cap D$ and so $x \in A_d^{M[G]}$. But then, $x \in A_d$ and so $x \notin A_c$.

Conversely, suppose that $x \notin A_c$ whenever $c \in M$ is a G_δ code for a null set. Considering how x can be used to define a corresponding $G \subseteq (\mathcal{B}^*)^M$ as in the proof of 11.9, to show that x is random over M it suffices to show that: whenever $D \in M$ is dense in $(\mathcal{B}^*)^M$, there is a closed code $c \in M$ and a $p \in D$ such that $A_c^M \subseteq p$ and $x \in A_c$.

To do this, let $D \in M$ be dense and first argue in M: Let W be a maximal antichain satisfying $W \subseteq \{C \mid C \text{ is closed} \wedge \exists p \in D(C \subseteq p)\}$. W is countable by the ω_1-c.c., so using closed codes let $\langle A_{c_n} \mid n \in \omega\rangle$ enumerate it. Now let $c \in {}^\omega\omega$ be any function satisfying $c(0) > 1$ and

$$c(2^n 3^{i+1}) = 1 \ \textit{ iff } \ O(\mathbf{s}_i) \cap A_{c_n} = \emptyset\ .$$

Since each A_{c_n} is closed, c is a G_δ code for $\bigcap_n({}^\omega\omega - A_{c_n})$, a null set by the maximality of W.

Stepping out of M, by hypothesis $x \notin A_c$. Hence, $x \in A_{c_n}$ for some $c_n \in M$ to complete the proof. ⊣

Random reals enter the coming consistency results through this characterization. Beyond this initial use they have found a variety of applications in set theory.

Proof of Solovay's Theorem

The main theorem 11.1 is proved by first establishing an auxiliary result about the Levy collapse and then passing to an inner model. This result is itself quite significant in that it asserts in a ZFC context that a large collection of sets of reals is well-behaved.

A set X is ${}^\omega$On-*definable iff* for some $a \in {}^\omega\text{On}$ and formula $\varphi(v_1, v_2)$,

$$y \in X \ \textit{ iff } \ \varphi[a, y]\ .$$

Of course, using one parameter in ${}^\omega$On is the same as using finitely many from On $\cup$ ${}^\omega$On by coding. Like the well-known concept of ordinal definability (see Kunen [80: V§2]) this concept is formally definable within set theory through the Reflection Principle for ZF: X is ${}^\omega$On-definable *iff*

$$\exists\alpha\exists a\exists\varphi(a \in {}^\omega\text{On} \cap V_\alpha \ \wedge\ \forall y(y \in X \ \leftrightarrow\ y \in V_\alpha \wedge V_\alpha \models \varphi[a, y]))\ .$$

11.11 Theorem (Solovay [65b, 70]). *Suppose that κ is an inaccessible cardinal and G is* Col(ω, κ)*-generic. Then in $V[G]$, every ${}^\omega$On-definable set of reals is Lebesgue measurable, has the Baire property, and has the perfect set property.*

The proof hinges on a key lemma that uses previous facts developed about the Levy collapse Col(ω, κ):

11.12 Lemma. *For each formula $\varphi(v_1)$ there is a corresponding formula $\tilde{\varphi}(v_1)$ such that: for any $x \in {}^{\omega}\mathrm{On} \cap V[G]$,*

$$V[G] \models \varphi[x] \ \textit{ iff } \ V[x] \models \tilde{\varphi}[x] .$$

Proof. Set $P = \mathrm{Col}(\omega, \kappa)$. For any $x \in {}^{\omega}\mathrm{On} \cap V[G]$, by 10.21 there is a P-generic H over $V[x]$ such that $V[G] = V[x][H]$. However, by the weak homogeneity of P (10.19) it is not necessary to know H to decide whether or not $\varphi[x]$ is satisfied:

$$V[x][H] \models \varphi[x] \ \textit{ iff } \ V[x] \models \ulcorner \Vdash_P \varphi(\check{x}) \urcorner .$$

Hence, $\tilde{\varphi}$ can be taken to be a formula in one free variable corresponding to this last forcing assertion about x and P. ⊣

Proof of 11.11. Argue in $V[G]$. Note first that

for any $a \in {}^{\omega}\mathrm{On}$, ${}^{\omega}\omega \cap V[a]$ is countable

since κ is still inaccessible in $V[a]$ by 10.12 and 10.17(e).

Suppose now that $A \subseteq {}^{\omega}\omega$ is ${}^{\omega}\mathrm{On}$-definable, so that for some $a \in {}^{\omega}\mathrm{On}$ and $\varphi(v_1, v_2)$,

$$x \in A \ \textit{ iff } \ \varphi[a, x] .$$

Using a two-variable version of 11.12 there is a formula $\tilde{\varphi}(v_1, v_2)$ such that

$$x \in A \ \textit{ iff } \ V[a][x] \models \tilde{\varphi}[a, x] .$$

We first verify A's Lebesgue measurability: Note that

$$\{x \in {}^{\omega}\omega \mid x \text{ is not random over } V[a]\}$$

is null, being equal by 11.10 to

$$\bigcup\{A_c \mid c \in V[a] \text{ is a } G_\delta \text{ code for a null set}\} ,$$

a countable union of null sets as ${}^{\omega}\omega \cap V[a]$ is countable. Thus, it suffices to find a Borel set X such that $A \triangle X$ consists of reals not random over $V[a]$:

For $\mathcal{B}^*$ forcing in the sense of $V[a]$ with $\dot{r}$ a canonical name for the random real, let $Y \in V[a]$ be a maximal antichain consisting of closed sets each deciding $\tilde{\varphi}(\check{a}, \dot{r})$, all in the sense of $V[a]$. If x is random over $V[a]$, by genericity

$$x \in A \ \textit{ iff } \ x \in \bigcup\{A_c \mid A_c^{V[a]} \in Y \ \wedge \ A_c^{V[a]} \Vdash \tilde{\varphi}(\check{a}, \dot{r})\} .$$

Since Y is countable by the ω_1-c.c., the set on the right is a Borel (in fact an F_σ) set, and hence can serve as our desired set X.

The verification of the Baire property is only outlined since it parallels the Lebesgue measurability argument. (As mentioned before, Shelah [84] has shown that an inaccessible cardinal is not necessary to get the consistency of every set of reals having the Baire property.) The idea is to force instead with

$$\mathcal{B}^{\dagger} = \{X \in \mathcal{B} \mid X \text{ is not meager}\} ,$$

ordered by inclusion. The separative quotient this time has a countable dense subset, namely the equivalence classes of the $O(s)$'s, and so by 10.20 forcing with $\mathcal{B}^{\dagger}$ coincides with forcing with $\mathrm{Col}(\omega, \{\omega\})$, i.e. adding a Cohen real. Using 0.12(b) and introducing codes for the $F_{\sigma\delta}$ sets (countable intersections of F_σ sets), analogues of the characterizations 11.9 and 11.10 for Cohen reals can be established with G_δ replacing closed and $F_{\sigma\delta}$ replacing G_δ. The rest of the argument goes through, modulated only by these changes.

Finally, we verify the perfect set property for A. It can be assumed that A is uncountable, so that because ${}^{\omega}\omega \cap V[a]$ is countable, there must be a $y \in A - V[a]$. By 10.21 there is a $\mathrm{Col}(\omega, \kappa)$-generic H over $V[a]$ such that $V[a][H] = V[G]$ and so by 10.17(e) there is a p.o. $Q \in V[a]$ satisfying $Q \subseteq \mathrm{Col}(\omega, \kappa)$ and $|Q| < \kappa$ such that $y \in V[a][H \cap Q]$. Temporarily let $\Vdash$ denote the forcing relation for Q in $V[a]$. There is a condition p and a name $\dot{y}$ such that

$$p \Vdash \dot{y} \in {}^{\omega}\omega - \check{V}[a] \ \wedge \ \check{V}[a][\dot{y}] \models \tilde{\varphi}(\check{a}, \dot{y}) .$$

Also, $\mathcal{P}^{V[a]}(Q)$ is countable in $V[G]$ as κ is inaccessible in $V[a]$, so there is an enumeration $\langle D_n \mid n \in \omega \rangle$ of the dense subsets of Q in $V[a]$. To each $t \in {}^{<\omega}2$ can now be associated a $p_t \in Q$ and a $u_t \in {}^{<\omega}2$ as follows:

Let $p_\emptyset \leq p$ be such that $p_\emptyset \in D_0$. Given $p_t \leq p$ with $|t| = n$, define u_t, $p_{t^\frown 0}$, and $p_{t^\frown 1}$ as follows: Because $p \Vdash \dot{y} \notin \check{V}[a]$ there must be a least $k \in \omega$ such that p_t does not decide a value for $\dot{y}(k)$. Let u_t satisfy $p_t \Vdash \dot{y}|k = \check{u}_t$, and choose $p_{t^\frown 0}, p_{t^\frown 1}$ both in D_{n+1} so that $p_{t^\frown 0}$ and $p_{t^\frown 1}$ decide different values for $\dot{y}(k)$.

For each $x \in {}^{\omega}2$, set

$$G_x = \{q \in Q \mid \exists t (p_t \leq q \ \wedge \ t \subseteq x)\} .$$

By the construction each G_x is Q-generic over $V[a]$. Next, let

$$C = \{\dot{y}^{G_x} \mid x \in {}^{\omega}2\} .$$

Then $C \subseteq A$ since $p \in G_x$ for every $x \in {}^{\omega}2$. Finally, C is perfect, since it contains no isolated points and

$$C = \bigcap_n \bigcup_t \{O(u_t) \mid |t| = n\}$$

where each finite union is closed by 0.8(c) so that overall intersection is closed.

This completes the proof of 11.11. ⊣

Solovay established the main result 11.1 by passing in $V[G]$ to the inner model arising naturally from the previous proof, the class of *hereditarily* ${}^{\omega}$*On-definable sets*, i.e. those sets x such that every member of the transitive closure $\mathrm{tc}(\{x\})$ is ${}^{\omega}$On-definable. Here, we shall pass to the constructible closure (§3) of the reals, $L(\mathbb{R})$. This is the standard name for the model, and $\mathbb{R}$ was taken in §0 to be the Dedekind completion of the rationals, but in any case

$$L(\mathbb{R}) = L({}^{\omega}\omega) = L(\mathcal{P}(\omega))$$

by interdefinability. Although $L(\mathbb{R})$ is in general a smaller model than Solovay's, it can be shown that they are the same had L been the ground model for the initial Levy collapse. Moreover, $L(\mathbb{R})$ has figured prominently in the investigation of strong hypotheses (§§27, 30). The verification of DC is less immediate than for Solovay's original inner model, but is of intrinsic interest.

Proof of 11.1. Building on 11.11, continue to work in $V[G]$. ${}^{\omega}\omega \cap L(\mathbb{R}) = {}^{\omega}\omega$, so through codes every open, closed, G_δ and hence F_σ set is in $L(\mathbb{R})$ and satisfies these respective properties in $L(\mathbb{R})$. (In fact, every Borel set is in $L(\mathbb{R})$.)

Suppose now that $A \subseteq {}^{\omega}\omega$ with $A \in L(\mathbb{R})$. Then A is ${}^{\omega}$On-definable, and so by the proof of 11.11 there is an F_σ set X such that $A \triangle X$ is null. It follows (by 0.9(a)) that for each $n \in \omega$ there is an open set $O_n \supseteq A \triangle X$ such that $m_L(O_n) < \frac{1}{n+1}$. By the previous paragraph and absoluteness (11.8), all this holds in $L(\mathbb{R})$ as well, and so $(A \text{ is Lebesgue measurable})^{L(\mathbb{R})}$. The argument for the Baire property is analogous.

For the perfect set property, it can be assumed in addition that A is uncountable in $L(\mathbb{R})$. Then A is (really) uncountable, since every countable well-ordering is coded by a real and hence in $L(\mathbb{R})$. Thus, A has a perfect subset C by 11.11, and it is simple to see that C is perfect in the sense of $L(\mathbb{R})$.

The following proposition completes the proof of 11.1.

11.13 Proposition (ZF + DC). $(\mathrm{DC})^{L(\mathbb{R})}$.

Proof. First note that by straightforward means as for the definition of a well-ordering of L, a surjection $\Phi\colon \mathrm{On} \times \mathbb{R} \to L(\mathbb{R})$ can be defined by transfinite recursion so that for any α, $\Phi|(\alpha \times \mathbb{R}) \in L(\mathbb{R})$.

Suppose now that $X, R \in L(\mathbb{R})$, $X \neq \emptyset$, $R \subseteq X \times X$, and

$$\forall x \in X \exists y \in X(\langle x, y\rangle \in R) .$$

Arguing in V, by DC there is an $f \in {}^{\omega}X$ such that $\langle f(n), f(n+1)\rangle \in R$ for every $n \in \omega$. Let δ be such that $\mathrm{ran}(f) \subseteq \Phi``(\delta \times \mathbb{R})$, and using AC_ω, a consequence of DC, let $\langle a_n \mid n \in \omega\rangle$ be such that for every $n \in \omega$ there is a $\xi < \delta$ such that $\Phi(\xi, a_n) = f(n)$. Now define a relation $\prec$ on $\delta \times \omega$ by:

$$\langle \eta, i\rangle \prec \langle \zeta, j\rangle \ \ \textit{iff} \ \ i = j + 1 \ \wedge \ \langle \Phi(\zeta, a_j), \Phi(\eta, a_i)\rangle \in R .$$

$\Phi|(\delta \times \mathbb{R}) \in L(\mathbb{R})$, and the sequence of reals $\langle a_n \mid n \in \omega\rangle \in L(\mathbb{R})$ since it can be coded by one real. Hence, $\prec \in L(\mathbb{R})$. Now $\prec$ is ill-founded in $L(\mathbb{R})$ because of $\mathrm{ran}(f)$ by absoluteness (0.3). Let $E \subseteq \delta \times \omega$ with $E \in L(\mathbb{R})$ be non-empty and without a $\prec$-minimal element. Then starting with any $\langle \eta, i_0\rangle \in$

E a sequence $\langle \eta_n \mid n \in \omega \rangle$ can be defined by recursion so that $\eta_0 = \eta$ and $\langle \Phi(\eta_n, a_{i_0+n}), \Phi(\eta_{n+1}, a_{i_0+n+1}) \rangle \in R$ for every $n \in \omega$, specifying η_{n+1} to be the minimal possible given η_n. By definability $\langle \Phi(\eta_n, a_{i_0+n}) \mid n \in \omega \rangle \in L(\mathbb{R})$, and the proof is complete. $\dashv$

Having finally completed the proof of 11.1, this section is concluded with a compact discussion of one further problem resolved in Solovay's model. An early observation in the study of partition relations was that if there is a well-ordering of the reals, then $\omega \longrightarrow (\omega)^{\omega}_{2}$ fails (7.1). By devising a new generic real, Mathias [68, 77] established that this partition property holds in Solovay's model.

The Mathias forcing p.o. is

$$P = [\omega]^{<\omega} \times [\omega]^{\omega}$$

ordered by:

$$\langle s, A \rangle \leq \langle t, B \rangle \textit{ iff } t \text{ is an initial segment of } s \ \wedge\ A \cup (s - t) \subseteq B\,.$$

The idea is that a condition $\langle s, A \rangle$ determines an initial segment s of a new subset of ω, whose further members are to be restricted to A. Such a forcing was first formulated by Prikry for measurable cardinals (§18). Note that if $\langle s, A \rangle \parallel \langle t, B \rangle$, then $A \cap B$ is infinite and s is an initial segment of t or *vice versa*. Thus, if G is P-generic, then G determines a subset of ω,

$$x_G = \bigcup \{ s \mid \exists A (\langle s, A \rangle \in G) \}\,.$$

x_G in turn determines G, since if

$$G' = \{ \langle s, A \rangle \in P \mid s \text{ is an initial segment of } x_G \text{ and } x_G - s \subseteq A \}\,,$$

then $G = G'$: It is simple to check that $G \subseteq G'$ and G' is P-generic. But then $G = G'$ on general grounds: For any $\langle s, A \rangle \in P$,

$$\{ \langle t, B \rangle \in P \mid \langle t, B \rangle \leq \langle s, A \rangle \ \vee\ \langle t, B \rangle \perp \langle s, A \rangle \}$$

is dense in P, and so $\langle s, A \rangle \in G'$ readily implies that $\langle s, A \rangle \in G$.

Thus, $V[x_G] = V[G]$. Such an x_G is called a *Mathias real*; more precisely, for M a transitive $\in$-model of ZFC and $x \subseteq \omega$,

$$x \text{ is a } \textit{Mathias real} \text{ over } M \textit{ iff } x = x_G \text{ for some } P^M\text{-generic } G \text{ over } M\,.$$

(Here, it is more convenient to consider reals as subsets of ω.) P does not have the ω_1-c.c., but it preserves ω_1 and has other useful features that have led to several applications. Mathias himself established some crucial properties:

11.14 Proposition (Mathias [77: 86]).

(a) For any $\langle s, A \rangle \in P$ and formula φ in the forcing language, there is a $B \in [A]^{\omega}$ such that $\langle s, B \rangle \parallel \varphi$.

(b) If x is a Mathias real over M, then any $y \in [x]^{\omega}$ is also a Mathias real over M. $\dashv$

The point of (a) is that the s need not be extended to decide formulas; simpler proofs for analogous results for measurable cardinals will be provided later (§18).

For discussing versions of $\omega \longrightarrow (\omega)^{\omega}_{2}$ depending on the complexity of partitions,

$$Y \subseteq [\omega]^{\omega} \text{ is } \textit{Ramsey} \;\; \textit{iff} \;\; \exists x \in [\omega]^{\omega}([x]^{\omega} \subseteq Y \;\vee\; [x]^{\omega} \cap Y = \emptyset) \,.$$

Thus, $\omega \longrightarrow (\omega)^{\omega}_{2}$ *iff every* $Y \subseteq [\omega]^{\omega}$ *is Ramsey.* See the end of §27 for more about this property.

11.15 Exercise (Mathias [68, 77]). *Suppose that there is an inaccessible cardinal* κ *and* G *is* $\mathrm{Col}(\omega, \kappa)$*-generic. Then*

(a) In $V[G]$*, every* ${}^{\omega}\mathrm{On}$*-definable subset of* $[\omega]^{\omega}$ *is Ramsey.*

(b) In $L(\mathbb{R})^{V[G]}$*,* $\omega \longrightarrow (\omega)^{\omega}_{2}$ *holds.*

Hint. For (a), let $Y \subseteq [\omega]^{\omega}$ satisfy

$$x \in Y \;\; \textit{iff} \;\; V[a][x] \models \varphi[a, x]$$

with $a \in {}^{\omega}\omega$. By 11.14(a), there is an $\langle \emptyset, A \rangle \in P^{V[a]}$ such that $\langle \emptyset, A \rangle \parallel \varphi(\check{a}, \dot{r})$, where $\dot{r}$ is a canonical name for the Mathias real. Now note that there is a Mathias real x over $V[a]$ satisfying $x \subseteq A$, and conclude by 11.14(b) that such an x verifies that Y is Ramsey. ⊣

Unlike the other properties considered, it is not known whether an inaccessible cardinal is necessary for $\omega \longrightarrow (\omega)^{\omega}_{2}$:

11.16 Question. *Does* Con(ZFC) *imply* Con(ZF + DC + $\omega \longrightarrow (\omega)^{\omega}_{2}$)?

Solovay's fruitful ideas, first developed to establish the global result 11.1, were to lead to important hierarchical extensions of classical results in descriptive set theory (§14).

12. Descriptive Set Theory

Descriptive set theory is the definability theory of the continuum, the study of the structural properties of definable sets of reals. Motivated initially by constructivist concerns, a major incentive for the subject has been to investigate the extent of the *regularity properties*, properties indicative of well-behaved sets of reals of which Lebesgue measurability, the Baire property, and the perfect set property are the prominent examples. The subject developed progressively from Suslin's results on analytic sets in 1916, until Gödel established delimitative results showing under $V = L$ that there are simply defined sets of reals that do not have the regularity properties. In the ensuing years Kleene developed what turned out to be the effective version of the theory as a generalization of his foundational work in recursion theory, and considerably refined the classical results. Then in 1965 Solovay reactivated the classical program by building on his 11.1 to provide characterizations for the regularity properties at the level of Gödel's $V = L$ delimitation. While 11.1 had established the relative consistency of every set of reals having the regularity properties, Solovay's further results began the use of large cardinal hypotheses to derive directly the regularity properties for sets of reals, an approach that eventually met with remarkable success by the late 1980's as investigations with forcing and inner models established unexpectedly close connections between large cardinal hypotheses and definable sets of reals.

This and the next section serve as a basis for succeeding sections by reviewing the historical development and the basic theory, and §§14, 15 discuss work of Solovay and then Martin incorporating large cardinal hypotheses. Moschovakis [80] provides a comprehensive treatment of descriptive set theory in a general context, Rogers [80] a compendium of recent results on analytic sets, Kechris [95] an exposition of the classical theory in light of recent work, and Hinman [78] an approach emphasizing hierarchies. Mansfield-Weitkamp [85] and Jech [03: §25] are shorter accounts that incorporate modern developments using forcing. This section begins by recounting the early junctures of the subject, deferring proofs to the next section.

Descriptive set theory has its origins in the work of the French analysts Borel, Baire, and Lebesgue at the turn of the century. Following important contributions to the theory of integration by Cantor as part of his study of sets (see Hawkins [75: 3.3]), Borel [98: 46-47] isolated the Borel sets through an axiomatically presented measure. Taking a concrete approach to the general concept of function, Baire in his thesis [99] distinguished and provided a hierarchical analysis of the *Baire functions*, those functions in the smallest collection containing the continuous functions of several real variables and closed under the taking of pointwise limits. While Lebesgue's thesis [02] is fundamental for measure theory as the source of his concept of measurability, his first major work in a distinctive direction was to be the seminal paper of descriptive set theory: In the memoir [05] Lebesgue investigated the Baire functions, stressing that they are exactly the functions definable via analytic expressions (in a sense made precise). He correlated the Baire functions with the Borel sets by showing that the latter are exactly the pre-images of open sets by the former. With this he had introduced the first hierarchy for the Borel sets, differing in minor details from the now standard one from Hausdorff [14].

In §0 ${}^{\omega}\omega$ served as the reals for formulations and the extension of concepts to ${}^{k}({}^{\omega}\omega)$ for $k \in \omega$ was assumed through homeomorphisms. It is now incumbent to state our further definitions explicitly for ${}^{k}({}^{\omega}\omega)$, with k implicitly ranging over ω. The classes $\mathbf{\Sigma}^0_\alpha$ and $\mathbf{\Pi}^0_\alpha$ for $0 < \alpha < \omega_1$ of the Borel hierarchy are defined by stipulating membership for each $A \subseteq {}^{k}({}^{\omega}\omega)$ as follows:

$$A \in \mathbf{\Sigma}^0_1 \text{ \emph{iff} } A \text{ is open in } {}^{k}({}^{\omega}\omega)\text{ , and}$$
$$A \in \mathbf{\Pi}^0_1 \text{ \emph{iff} } A \text{ is closed in } {}^{k}({}^{\omega}\omega)\text{ ,}$$

and recursively for $\alpha > 1$,

$$A \in \mathbf{\Sigma}^0_\alpha \text{ \emph{iff} } A \text{ is the union of a countable subset of } \bigcup_{0<\beta<\alpha} \mathbf{\Pi}^0_\beta\text{ , and}$$
$$A \in \mathbf{\Pi}^0_\alpha \text{ \emph{iff} } A \text{ is the intersection of a countable subset of } \bigcup_{0<\beta<\alpha} \mathbf{\Sigma}^0_\beta\text{ .}$$

Each $\mathbf{\Sigma}^0_\alpha \cap \mathcal{P}({}^{k}({}^{\omega}\omega))$ is closed under the taking of countable unions, and each $\mathbf{\Pi}^0_\alpha \cap \mathcal{P}({}^{k}({}^{\omega}\omega))$, the taking of countable intersections. It is readily seen by induction that $\mathbf{\Sigma}^0_\alpha$ and $\mathbf{\Pi}^0_\alpha$ are *dual* collections, i.e. $A \in \mathbf{\Sigma}^0_\alpha$ *iff* ${}^{k}({}^{\omega}\omega) - A \in \mathbf{\Pi}^0_\alpha$ for $A \subseteq {}^{k}({}^{\omega}\omega)$. Moreover, $\mathbf{\Sigma}^0_1 \subseteq \mathbf{\Sigma}^0_2$ by 0.8(c)(d), and so by induction $0 < \beta < \alpha$ implies that $\mathbf{\Sigma}^0_\beta \cup \mathbf{\Pi}^0_\beta \subseteq \mathbf{\Sigma}^0_\alpha \cap \mathbf{\Pi}^0_\alpha$. By a cofinality argument the subsets of ${}^{k}({}^{\omega}\omega)$ in

$$\bigcup_{0<\alpha<\omega_1} \mathbf{\Sigma}^0_\alpha = \bigcup_{0<\alpha<\omega_1} \mathbf{\Pi}^0_\alpha$$

comprise a σ-algebra, and hence are the Borel subsets of ${}^{k}({}^{\omega}\omega)$. In §§0, 11 Hausdorff's terms F_σ and G_δ were used for the subsets of ${}^{\omega}\omega$ in $\mathbf{\Sigma}^0_2$ and $\mathbf{\Pi}^0_2$ respectively; the now standard $\mathbf{\Sigma}^0_\alpha$ and $\mathbf{\Pi}^0_\alpha$ notation is from Addison [58].

The main results of Lebesgue [05] are that this is a *proper* hierarchy since he showed in effect that $\mathbf{\Sigma}^0_\alpha \not\subseteq \mathbf{\Pi}^0_\alpha$ for any countable $\alpha > 0$, and that there is a Lebesgue measurable set which is not a Borel set. In fact, this paper is remarkable for anticipating central concerns and techniques of mathematical logic: the systematic study of definable sets and the use of Cantor's universal enumeration and diagonalization argument to effect transcendence. Although Lebesgue's reservations about arbitrary functions had turned to open skepticism with the appearance of Zermelo's Well-Ordering Theorem [04], it is ironic that his paper contained tacit but essential uses of the Axiom of Choice (see Moore [82: 2.3]).

Descriptive set theory emerged as a distinct discipline through the initiatives of the Russian mathematician Nikolai Luzin. Through a focal seminar that he began in 1914 at the University of Moscow, he was to establish a prominent school in the theory of functions of a real variable (see Phillips [78]; Keldysh [74], Uspenskii [85], and Kanovei [85] survey the contributions to descriptive set theory). Luzin had become acquainted with the work and views of the French analysts while he was in Paris as a student, and from the beginning a major topic of his seminar was the "descriptive theory of functions". Significantly, the young

Sierpiński was an early participant; he had been interned in Moscow in 1915, and Luzin and his teacher Dmitry Egorov interceded on his behalf to let him live freely until his repatriation to Poland a year later. Not only did this lead to a decade-long collaboration between Luzin and Sierpiński, but it undoubtedly encouraged the latter in his founding of the Polish school (§2) and laid the basis for its interest in descriptive set theory.

Luzin described the cardinality problem for Borel sets (operatively becoming whether they have the perfect set property) to Pavel Aleksandrov, an early member of the seminar and later a pioneer of modern topology. That the Borel sets are Lebesgue measurable and have the Baire property are intrinsic to these concepts, but little was known about the perfect set property beyond the Cantor-Bendixson result 11.3(a). The property was implicit in a scheme of William Young [03] clarified by Hausdorff [14: 465ff], who used it to show that every $\mathbf{\Sigma}^0_4$ set is either countable or has cardinality $2^{\aleph_0}$. Hausdorff [16] then extended this conclusion to all Borel sets. Aleksandrov [16] was able to establish that *the Borel sets have the perfect set property*. The proof required a new way of comprehending the Borel sets, as it turns out that a set having the perfect set property does not imply that its complement does, and so an inductive proof through the Borel hierarchy is not possible. The new, more direct analysis of Borel sets broke the ground for a dramatic development:

Soon afterwards another student of Luzin's, Mikhail Suslin (often rendered "Souslin" in the French transliteration), began reading Lebesgue [05]. Memoirs of Sierpiński [50: 28ff] recalled how Suslin then made a crucial discovery. For any set B, the *projection of B* is

$$pB = \{x \mid \exists y(\langle x, y\rangle \in B)\} .$$

In particular, for $B \subseteq {}^{k+1}({}^{\omega}\omega) = {}^{k}({}^{\omega}\omega) \times {}^{\omega}\omega$,

$$pB = \{\langle x_1, \ldots, x_k\rangle \mid \exists y(\langle x_1, \ldots, x_k, y\rangle \in B)\} .$$

Suslin noticed that at one point Lebesgue asserted ([05: 191-192]) that the projection of a Borel subset of the plane (${}^{2}({}^{\omega}\omega)$ in our formulation) is also a Borel set. This was based on the mistaken claim that given a countable collection of subsets of the plane the projection of their intersection is equal to the intersection of their projections – in terms of quantifiers, an unjustifiable $\forall\exists$ to $\exists\forall$ switch. Suslin soon found a counterexample to Lebesgue's assertion, and this led to his inspired investigation of the *analytic sets*. For $A \subseteq {}^{k}({}^{\omega}\omega)$,

$$A \text{ is } \textit{analytic} \textit{ iff } A = pB \text{ for some closed subset } B \text{ of } {}^{k+1}({}^{\omega}\omega) .$$

(To proceed classically with $\mathbb{R}$ instead of ${}^{\omega}\omega$, "closed" here must be replaced by "G_δ", i.e. countable intersections of open subsets of $\mathbb{R}$.) Suslin [17] first formulated the analytic sets as the A-sets (les ensembles (A)), sets resulting from an explicit operation, the Operation (A), and then characterized them as projections. He announced three main results, that *every Borel set is analytic,* that *there is an analytic set which is not Borel*, and the following characterization (established by 13.5):

12.1 Theorem (Suslin [17]). *For $A \subseteq {}^k({}^\omega\omega)$, A is Borel iff A and ${}^k({}^\omega\omega) - A$ are both analytic.*

Thus, the Borel sets can be characterized in simple terms from above, as well as analyzed through a hierarchy from below. Paradigmatic for later hierarchy results, this theorem really began the subject of descriptive set theory. [17] was to be Suslin's sole publication, for he succumbed to typhus in a Moscow epidemic in 1919 at the age of 25. (The whole episode recalls a pivotal mistake by Augustin-Louis Cauchy and the clarification due to the young Niels Abel that led to the concept of uniform convergence, even to an unjustifiable $\forall\exists$ to $\exists\forall$ switch and Abel's untimely death.)

In an accompanying note, Luzin announced the regularity properties for analytic sets ((a) and (b) are established by 14.4 and (c) by 14.8):

12.2 Theorem (Luzin [17]; (c) is attributed there to Suslin).

(a) Every analytic set is Lebesgue measurable.

(b) Every analytic set has the Baire property.

(c) Every analytic subset of ${}^\omega\omega$ has the perfect set property.

(The attribution of (c) to Suslin is actually a faint echo of a question of priority. According to memoirs of Aleksandrov [79: 284-286] it was he who had defined the A-sets, and Suslin proposed the name, as well as "Operation (A)" for the corresponding operation, in Aleksandrov's honor. This eponymy is not mentioned in Suslin [17] but is supported by recollections of e.g. Keldysh [74: 180] and Kuratowski [80: 69]. Aleksandrov recalled that he had shown that every Borel set is an A-set and that every A-set has the perfect set property, although this is not explicit in his [16]. He then tried hard in 1916 to show that every A-set is Borel, only discontinuing when it became known that in the summer Suslin had found a non-Borel A-set. According to Aleksandrov: "Many years later Luzin started to call A-sets analytic sets and began, contrary to the facts, which he knew well, to assert that the term 'A-set' is only an abbreviation for 'analytic set'. But by this time my personal relations with Luzin, at one time close and sincere, were estranged." Luzin [25, 27] did go to some pains to trace the term analytic back to Lebesgue [05] and pointed out that the original example there of a non-Borel Lebesgue measurable set is in fact the first example of a non-Borel analytic set. See also the text Luzin [30: 186ff], in which the Operation (A) is conspicuous by its absence.)

The notes Suslin [17] and Luzin [17] were considerably elaborated in the ensuing years, as the investigation of analytic sets was continued by Luzin and his circle in Moscow and by Sierpiński in Warsaw. With the Borel sets having been properly extended through the new operation of projection, Luzin [25a] and Sierpiński [25] formulated the *projective sets* as those sets obtainable from the Borel sets by the iterated application of projection and complementation. There is the corresponding *projective hierarchy*, formulated in modern notation for ${}^\omega\omega$:

For $A \subseteq {}^k({}^\omega\omega)$,

$$A \in \mathbf{\Sigma}_0^1 \ \textit{iff} \ A \in \mathbf{\Sigma}_1^0, \text{ i.e. } A \text{ is open },$$

and recursively for $n \in \omega$,

$$A \in \mathbf{\Pi}_n^1 \ \textit{iff} \ {}^k({}^\omega\omega) - A \in \mathbf{\Sigma}_n^1 \text{ , and}$$
$$A \in \mathbf{\Sigma}_{n+1}^1 \ \textit{iff} \ A = pB \text{ for some } \mathbf{\Pi}_n^1 \text{ subset } B \text{ of } {}^{k+1}({}^\omega\omega) \text{ .}$$

Thus, $A \in \mathbf{\Sigma}_1^1$ *iff* A is analytic. Also define

$$\mathbf{\Delta}_n^1 = \mathbf{\Sigma}_n^1 \cap \mathbf{\Pi}_n^1 \text{ .}$$

These terms are also used as adjectives, as in "$\mathbf{\Sigma}_n^1$ set" or "A is $\mathbf{\Pi}_n^1$". $\mathbf{\Sigma}_n^1$ and $\mathbf{\Pi}_n^1$ are dual collections, i.e. $A \in \mathbf{\Sigma}_n^1$ *iff* ${}^k({}^\omega\omega) - A \in \mathbf{\Pi}_n^1$ for $A \subseteq {}^k({}^\omega\omega)$. Moreover, $\mathbf{\Sigma}_n^1 \cup \mathbf{\Pi}_n^1 \subseteq \mathbf{\Delta}_{n+1}^1$. This follows readily by induction from $\mathbf{\Pi}_0^1 \subseteq \mathbf{\Pi}_1^1$, which itself is seen as follows:

Suppose that $A \in \mathbf{\Pi}_0^1$, i.e. A is closed, so that $A = \bigcap_{n\in\omega} G_n$ for open sets G_n (by 0.8(c)(d)). Set $G = \bigcup_{n\in\omega} G_n \times O(\langle n\rangle) \subseteq {}^2({}^\omega\omega)$. Then G is open, and for any $x \in {}^\omega\omega$,

$$\begin{aligned} x \in A &\leftrightarrow \forall n(x \in G_n) \\ &\leftrightarrow \forall y(\langle x, y\rangle \in G) \end{aligned}$$

so that $A \in \mathbf{\Pi}_1^1$.

Suslin's result 12.1 is the assertion that $\mathbf{\Delta}_1^1$ is exactly the collection of Borel sets. Luzin [25c] and Sierpiński [25] observed that like the Borel hierarchy the projective hierarchy is a proper hierarchy (12.8), and soon its basic properties were established. For example, *each* $\mathbf{\Sigma}_n^1 \cap \mathcal{P}({}^k({}^\omega\omega)))$ *and* $\mathbf{\Pi}_n^1 \cap \mathcal{P}({}^k({}^\omega\omega))$ *is closed under countable unions and intersections.* (However, the projective subsets of $\mathcal{P}({}^k({}^\omega\omega))$ as a whole are not so closed; to obtain this closure, one can continue the projective hierarchy into the transfinite incorporating countable unions at limit stages.)

On the other hand, the investigation of projective sets encountered basic obstacles from the beginning. For one thing, unlike for $\mathbf{\Sigma}_1^1$ sets the cardinality problem for $\mathbf{\Pi}_1^1$ sets (operatively, whether they have the perfect set property) could not be resolved. Luzin [17: 94] had already noted this difficulty, and it was emphasized as a major problem in Luzin [25a]. In a confident and remarkably prophetic passage he declared that his efforts towards its resolution led him to a conclusion "totally unexpected", that "one does not know *and one will never know*" of the family of projective sets, although it has cardinality $2^{\aleph_0}$ and consists of "effective sets", whether every member has cardinality $2^{\aleph_0}$ if uncountable, has the Baire property, or is even Lebesgue measurable. Luzin [25b] pointed out the specific problem of establishing the Lebesgue measurability of $\mathbf{\Sigma}_2^1$ sets. Both these difficulties at $\mathbf{\Pi}_1^1$ and $\mathbf{\Sigma}_2^1$ were also observed by Sierpiński [25: 242], although he was able to show that *every* $\mathbf{\Sigma}_2^1$ *set is a union of* $\aleph_1$ *Borel sets* (13.7).

The first wave of progress from Suslin's results having worked itself out, Luzin provided systematic accounts, especially in a classic text [30]. He aired the constructivist views of his French predecessors, and not only did he contrive somewhat self-effacingly to establish definite precedents for his own work in theirs, but he also espoused their distrust of the unbridled Axiom of Choice and advocated their views on definability. Lebesgue [05] had ended with the question, "Can one name a non-measurable set?" Taking this as a starting point of his own work, Luzin [30: 323] wrote: "... the author considers the question of whether all projective sets are measurable or not to be unsolvable [insoluble], since in his view the methods of defining the projective sets and Lebesgue measure are not comparable, and consequently, not logically related."

If the projective sets proved intractable with respect to the regularity properties, significant progress was nonetheless made in other directions. In Luzin [30a] the investigation of *uniformization* was proposed. For $A, B \subseteq {}^2({}^\omega\omega)$,

$$A \text{ is \textit{uniformized} by } B \text{ \textit{iff} } \quad B \subseteq A \ \wedge\ \forall x(\exists y(\langle x, y\rangle \in A) \leftrightarrow \exists! y(\langle x, y\rangle \in B)) .$$

(As usual, $\exists!$ abbreviates the formalizable "there exists exactly one".) Since B is in effect a choice function for an indexed family of sets, asserting the uniformizability of arbitrary subsets of ${}^2({}^\omega\omega)$ is a version of the Axiom of Choice. Taking this approach to the problem of definable choices, Luzin announced several results about the uniformizability of $\mathbf{\Sigma}^1_1$ sets by like sets. One was affirmed in a sharp form by Petr Novikov [31], who showed that there is a closed set that cannot be uniformized by any $\mathbf{\Sigma}^1_1$ set. It was eventually shown by Vladimir Yankov [41] that any $\mathbf{\Sigma}^1_1$ set can be uniformized by a set that is a countable intersection of countable unions of differences of $\mathbf{\Sigma}^1_1$ sets. Interestingly enough, von Neumann [49: 448ff] also established a less structured uniformization result for $\mathbf{\Sigma}^1_1$ sets as part of an extensive study of rings of operators. Presumably because of the difficulties encountered in the study of $\mathbf{\Pi}^1_1$ sets Luzin [30a: 351] claimed that there are $\mathbf{\Pi}^1_1$ sets that cannot be uniformized by any "distinguishable" set and gave a purported example. Notwithstanding, Sierpiński [30: 139] asked whether every $\mathbf{\Pi}^1_1$ set can be uniformized by a $\mathbf{\Pi}^1_2$, or even any projective, set, and a result of Novikov in Luzin-Novikov [35] did lead to the conclusion that every $\mathbf{\Pi}^1_1$ set can be uniformized by a $\mathbf{\Sigma}^1_2$ set. Building on this, the Japanese mathematician Motokiti Kondô established (see 13.17) the $\mathbf{\Pi}^1_1$ *Uniformization Theorem*:

12.3 Theorem (Kondô [37, 39]). *Every $\mathbf{\Pi}^1_1$ subset of ${}^2({}^\omega\omega)$ can be uniformized by a $\mathbf{\Pi}^1_1$ set.*

This was the culminating result of the early period. As Kondô noted, his result implies through projections that every $\mathbf{\Sigma}^1_2$ set can be uniformized by a $\mathbf{\Sigma}^1_2$ set, but the question of whether every $\mathbf{\Pi}^1_2$ set can be uniformized by any projective set was left open.

The total impasse in descriptive set theory with respect to the regularity properties was to be explained by Gödel's work on the constructible hierarchy. In his initial article [38] on L, he announced that *if $V = L$, then there is a Δ^1_2 set of reals which is not Lebesgue measurable and a $\mathbf{\Pi}^1_1$ set of reals which does not have the perfect set property.* (See 13.10 and 13.12 for proofs.) Thus, the early descriptive set theorists were confronting an obstacle formalizable in ZFC. The importance that Gödel attached to these results can be evinced from his listing of each of them on equal footing with his AC and GCH results. However, he did not publish proofs, and more than a decade was to pass before proofs first appeared in Novikov [51]. In the meantime, Gödel [51:67] (see Gödel [90:33]) had sketched a more basic result that readily implies under $V = L$ that there is a non-measurable $\mathbf{\Delta}^1_2$ set of reals: *if $V = L$, then there is a $\mathbf{\Sigma}^1_2$ well-ordering of the reals*. (According to Kreisel [80:197], "... according to Gödel's notes, not he, but S. Ulam, steeped in the Polish tradition of descriptive set theory, noticed that the definition of the well-ordering ... of subsets of ω was so simple that it supplied a non-measurable PCA [i.e. $\mathbf{\Sigma}^1_2$] set of real numbers ...") Mostowski had also established the result in a manuscript destroyed during the war, but it is not apparent in Novikov [51]. Details were eventually provided by Addison [59] who built on this work to settle the uniformization problem for projective sets and related issues in L. Addison was a student of Kleene, who through initiatives from mathematical logic had been developing what is now called *effective descriptive set theory*.

After his fundamental work on recursive function theory in the 1930's, Kleene expanded his investigations of effectiveness and developed a general theory of definability for relations on ω. In [43] he considered the *arithmetical relations*, those relations obtainable from the recursive relations by applications of number quantifiers. Developing canonical representations he classified these relations into a hierarchy according to quantifier complexity and showed that it is a proper hierarchy. In [55, 55a, 55b] he studied the *analytical relations*, those relations obtainable from the arithmetical relations by applications of function quantifiers. Again he worked out representation and hierarchy results, and moreover he established an elegant theorem that turned out to be an effective version of Suslin's characterization 12.1 of Borel sets.

Kleene was developing what amounted to the effective content of the classical theory, unaware that his work had direct antecedents in the papers of Lebesgue, Luzin, and Sierpiński. Certainly, he had very different motivations: With the arithmetical relations he wanted to extend the Incompleteness Theorem, and analytical relations grew out of his investigations of notations for recursive ordinals. Moreover, his work featured refinements made possible by his Recursion Theorem and its sometimes miraculous applications in the method of effective transfinite recursion. On the other hand, already in [43:50] he did make elliptic remarks about possible analogies with the classical theory. Pursuing one such analogy, Mostowski [47] had also derived the results on arithmetical relations. But further investigations exposed shortcomings with this analogy, and it was Addison who established the proper correlation: it is the analytical relations that are analogous to the projective sets, with the arithmetical relations being analogous to the sets in $\bigcup_{0<n<\omega} \mathbf{\Sigma}^0_n$, the finite levels of the Borel hierarchy. As he would put it,

$$\frac{\text{analytical}}{\text{projective}} = \frac{\text{arithmetical}}{\text{finite level Borel}} \, .$$

The conceptual move in Kleene's work from relations on ω to relations on ${}^{\omega}\omega$ revealed that this correlation was exact: as pointed out by Addison [58: 126] the clopen subsets of ${}^{\omega}\omega$ are exactly those sets *recursive in some real* (12.6(a)). Had Kleene seen his work in this light from the beginning, it is doubtful that he would have initially used "analytic" as he did for "analytical"; the latter term may have evolved to draw a distinction from the classical analytic sets, but is still a source of confusion.

The development of effective descriptive set theory considerably clarified the classical context, injected recursion-theoretic techniques into the subject, and placed definability considerations squarely at its forefront. Not only were new approaches to classical problems provided, but results and questions could now be formulated in a refined setting. That there is a basic connection between the projective sets and logical definability had been observed in Kuratowski-Tarski [31] and Kuratowski [31]; it is the careful attention to real parameters that distinguishes the modern approach. The effective context is forthwith established via definability in what follows.

The development of descriptive set theory can be carried out in ZF + DC, DC being the Principle of Dependent Choices from 11.1. The principle becomes crucial for the later use of strong hypotheses (Chapter 6); it is then needed for the characterization of well-foundedness in terms of the lack of infinite descending chains. Recalling that DC implies the countable choice principle AC_{ω}, consider its further restriction to sets of reals:

($\mathrm{AC}_{\omega}({}^{\omega}\omega)$) Every countable set consisting of non-empty sets of reals has a choice function .

This principle, applied to choose countably many reals coding well-orderings, suffices to show that ω_1 is regular. An important observation for later is that:

The main development through this and the next section can be carried out in ZF + $\mathrm{AC}_{\omega}({}^{\omega}\omega)$.

Those few results and remarks involving more substantial uses of AC will serve mainly to frame the discussion and will be noted explicitly.

The Definability Context

Second-order arithmetic is taken to be the two-sorted structure

$$\mathcal{A}^2 = \langle \omega, {}^{\omega}\omega, ap, +, \times, exp, <, 0, 1 \rangle \, ,$$

where ω and ${}^{\omega}\omega$ are two separate domains connected by the binary operation $ap \colon {}^{\omega}\omega \times \omega \to \omega$ of *application*,

$$ap(x, m) = x(m) \, ,$$

and $+, \times, exp, <, 0, 1$ impose the usual arithmetical structure on ω, with exp being exponentiation. (exp is traditionally not taken as a basic operation, but its inclusion simplifies our presentation without altering the significant concepts of descriptive set theory.) In addition to symbols for these features the underlying language has two sorts of variables, those ranging over ω to be denoted by $v_0^0, v_1^0, v_2^0, \ldots$ and those ranging over ${}^\omega\omega$ to be denoted by $v_0^1, v_1^1, v_2^1, \ldots$. Corresponding are the *number quantifiers* and *function quantifiers*, to be denoted by $\exists^0, \forall^0$ and $\exists^1, \forall^1$ respectively. *Terms for numbers* are defined recursively by closing off $0, 1, v_0^0, v_1^0, v_2^0, \ldots$ under $+, \times, exp$, and $ap(v_i^1, \cdot)$. For any such term τ, $(\exists^0 v_i^0 < \tau)\varphi$ is written for $\exists^0 v_i^0(v_i^0 < \tau \wedge \varphi)$ and $(\forall^0 v_i^0 < \tau)\varphi$ for $\forall^0 v_i^0(v_i^0 < \tau \rightarrow \varphi)$ whenever possible; these are the *bounded quantifiers*. $\mathcal{A}^2$ is a full second-order structure with ${}^\omega\omega$ playing the role of $\mathcal{P}(\omega)$.

Generally speaking, questions of definability in this structure can pertain to any relations $A \subseteq {}^r\omega \times {}^k({}^\omega\omega)$ for some $r, k \in \omega$. The case $A \subseteq {}^k({}^\omega\omega)$ was the preoccupation of classical descriptive set theory, but the general case is intrinsic to the effective theory. Regarding the members of ${}^r\omega \times {}^k({}^\omega\omega)$ as $(r+k)$-tuples, to be written interchangeably are

$$A(m_1, \ldots, m_r, x_1, \ldots, x_k) \text{ iff } \langle m_1, \ldots, m_r, x_1, \ldots, x_k\rangle \in A ,$$

with A regarded as a relation on the one hand and a set on the other. Notational variations like

$$A(m, \mathbf{w})$$

when $r > 0$, $m \in \omega$, and $\mathbf{w} \in {}^{r-1}\omega \times {}^k({}^\omega\omega)$ have the expected meaning. Such an A is *definable in $\mathcal{A}^2$ by a formula $\varphi(v_1^0, \ldots, v_r^0, v_1^1, \ldots, v_k^1)$ iff* for any $m_1, \ldots, m_r \in \omega$ and $x_1, \ldots, x_k \in {}^\omega\omega$,

$$A(m_1, \ldots, m_r, x_1, \ldots, x_k) \text{ iff } \mathcal{A}^2 \models \varphi[m_1, \ldots, m_r, x_1, \ldots, x_k] .$$

Henceforth, *the syntax and semantics of $\mathcal{A}^2$ are informally conflated* in such travesties as $\forall^0 m \exists^1 x A(m, x)$ and $\neg A(\mathbf{w})$, focusing primarily on relations rather than their defining formulas.

Classifying relations according to quantifier complexity, stipulate for $r, k \in \omega$ and $A \subseteq {}^r\omega \times {}^k({}^\omega\omega)$ that:

$$A \in \Delta_0^0 \text{ iff } A \text{ is definable in } \mathcal{A}^2 \text{ by a formula whose only quantifiers are bounded} .$$

Δ_0^0 is also used as an adjective.

Δ_0^0 codings of sequences will often be used. For definiteness, let p_i denote the $(i+1)$st prime, and for $m_0, \ldots, m_r, m, i \in \omega$ set

$$\langle m_0, \ldots, m_r\rangle = p_0^{m_0+1} \cdot p_1^{m_1+1} \cdots p_r^{m_r+1} , \text{ and}$$
$$(m)_i = \min(\{e \mid p_i^{e+2} \text{ does not divide } m\}) .$$

These functions are Δ^0_0 (and it is for this that exp was included in $\mathcal{A}^2$). Moreover, $(m)_i$ is always defined, and $(\langle\!\langle m_0, \ldots, m_r \rangle\!\rangle)_i = m_i$ for $i \le r$. In §0 an enumeration $\langle \mathbf{s}_i \mid i \in \omega \rangle$ of ${}^{<\omega}\omega$ had been fixed; henceforth, *the decoding of* $\mathbf{s}_i$ *from i will be taken to be as* Δ^0_0, using

$$\mathbf{s}_i = \begin{cases} \langle m_0, \ldots, m_r \rangle & \text{if } i = \langle\!\langle m_0, \ldots, m_r \rangle\!\rangle \text{ , and} \\ \emptyset & \text{otherwise .} \end{cases}$$

For $x \in {}^{\omega}\omega$ and $m \in \omega$ set

$$\overline{x}(m) = \langle\!\langle x(0), \ldots, x(m-1) \rangle\!\rangle \in \omega$$

and define $(x)_i \in {}^{\omega}\omega$ for $i \in \omega$ by

$$(x)_i(m) = x(\langle\!\langle i, m \rangle\!\rangle) \ .$$

The $\overline{x}$ notation is extended to $\mathbf{w} \in {}^{r}\omega \times {}^{k}({}^{\omega}\omega)$ by stipulating that *for* $m \in \omega$, $\overline{\mathbf{w}}(m) \in {}^{r+k}\omega$ *is like* $\mathbf{w}$ *except that each occurrence of any* $x \in {}^{\omega}\omega$ *has been replaced by* $\overline{x}(m)$.

Continuing,

$$A \text{ is } \textit{arithmetical} \textit{ iff } A \text{ is definable in } \mathcal{A}^2 \text{ by a formula without function quantifiers .}$$

Bounded quantifiers can be shifted to the right and like quantifiers contracted using the following equivalences and their duals (obtained by negating both sides):

$$(\forall^0 p < q)\exists^0 m R(p, q, m, \mathbf{w}) \leftrightarrow \exists^0 m (\forall^0 p < q) R(p, q, (m)_p, \mathbf{w})$$
$$\exists^0 m \exists^0 p R(m, p, \mathbf{w}) \leftrightarrow \exists^0 m R((m)_0, (m)_1, \mathbf{w}) \ .$$

Hence, the arithmetical relations can be classified into a hierarchy, the *arithmetical hierarchy*, by defining the (lightface!) Σ^0_n and Π^0_n classes for $n \in \omega$ as follows:

$$A \in \Sigma^0_n \textit{ iff } \forall \mathbf{w}(A(\mathbf{w}) \leftrightarrow \exists^0 m_1 \forall^0 m_2 \ldots Q m_n R(m_1, \ldots, m_n, \mathbf{w})) \text{ , and}$$
$$A \in \Pi^0_n \textit{ iff } \forall \mathbf{w}(A(\mathbf{w}) \leftrightarrow \forall^0 m_1 \exists^0 m_2 \ldots Q m_n R(m_1, \ldots, m_n, \mathbf{w}))$$

for some Δ^0_0 $R \subseteq {}^{r+n}\omega \times {}^{k}({}^{\omega}\omega)$, where Q is $\exists^0$ if n is odd and $\forall^0$ if n is even in the first and *vice versa* in the second. (It is notationally incumbent that $\Sigma^0_0 = \Pi^0_0 = \Delta^0_0$, but this will be irrelevant.) Σ^0_n and Π^0_n are dual collections, i.e. $A \in \Sigma^0_n$ *iff* ${}^{r}\omega \times {}^{k}({}^{\omega}\omega) - A \in \Pi^0_n$; $\Sigma^0_n \cup \Pi^0_n \subseteq \Sigma^0_{n+1} \cap \Pi^0_{n+1}$; and these terms are also used as adjectives.

$$A \text{ is } \textit{recursive} \textit{ iff } A \in \Sigma^0_1 \cap \Pi^0_1 \ .$$

This conforms for $A \subseteq {}^{r}\omega$ to the original 1930's concept of recursive relation (although this by no means is immediate from our formulation). Kleene extended the theory in the 1950's to include $A \subseteq {}^{r}\omega \times {}^{k}({}^{\omega}\omega)$. There are recursive relations that are not Δ^0_0, i.e. $\Sigma^0_1 \cap \Pi^0_1$ properly extends Δ^0_0.

Finally,

$$A \text{ is } \textit{analytical} \textit{ iff } A \text{ is definable in } \mathcal{A}^2 .$$

Number quantifiers can be shifted to the right and like function quantifiers contracted using the following equivalences and their duals:

$$(*) \qquad \begin{aligned} \forall^0 m \exists^1 x R(m, \mathbf{w}, x) &\leftrightarrow \exists^1 x \forall^0 m R(m, \mathbf{w}, (x)_m) \\ \exists^0 m \exists^1 x R(m, \mathbf{w}, x) &\leftrightarrow \exists^1 x R((x)_0(0), \mathbf{w}, (x)_1) \\ \exists^1 x \exists^1 y R(\mathbf{w}, x, y) &\leftrightarrow \exists^1 x R(\mathbf{w}, (x)_0, (x)_1) . \end{aligned}$$

(The first makes use of $\mathrm{AC}_\omega({}^\omega\omega)$; $(*)$ will be seen to be necessary for the classification of relations, leading to the coming 12.7 and other basic results.) It follows that the analytical relations can be classified into a hierarchy, the *analytical hierarchy*, by defining the (lightface!) Σ^1_n and Π^1_n classes for $n \in \omega$ as follows: First, for later inductive schemes set

$$\Sigma^1_0 = \Sigma^0_1 \text{ and } \Pi^1_0 = \Pi^0_1 .$$

Then for $n > 0$, define

$$\begin{aligned} A \in \Sigma^1_n &\textit{ iff } \forall \mathbf{w}(A(\mathbf{w}) \leftrightarrow \exists^1 x_1 \forall^1 x_2 \ldots Q x_n R(\mathbf{w}, x_1, \ldots, x_n)) , \text{ and} \\ A \in \Pi^1_n &\textit{ iff } \forall \mathbf{w}(A(\mathbf{w}) \leftrightarrow \forall^1 x_1 \exists^1 x_2 \ldots Q x_n R(\mathbf{w}, x_1, \ldots, x_n)) \end{aligned}$$

for some arithmetical $R \subseteq {}^r\omega \times {}^{k+n}({}^\omega\omega)$, where Q is $\exists^1$ if n is odd and $\forall^1$ if n is even in the first and *vice versa* in the second. Also define

$$\Delta^1_n = \Sigma^1_n \cap \Pi^1_n .$$

As before, Σ^1_n and Π^1_n are dual collections, $\Sigma^1_n \cup \Pi^1_n \subseteq \Delta^1_{n+1}$, and these terms are also used as adjectives. The terminology is consistent with the classification of formulas in §0, but the focus on relations and the use of ${}^\omega\omega$ rather than $\mathcal{P}(\omega)$ makes the present context distinctive.

The quantifier equivalences $(*)$ above show that *the classification of any relation in the analytical hierarchy only depends on the function quantifiers and their alternations*; this is frequently relied upon without further mention. Although the foregoing may be the natural formulation of the hierarchy, there is an important self-refinement that leads to simple normal forms (and vindicates our definitions of Σ^1_0 and Π^1_0):

12.4 Exercise. *If $n > 0$ is even, there is a Δ^0_0 $\hat{R} \subseteq {}^{r+k+n+1}\omega$ such that the arithmetical $R(\mathbf{w}, x_1, \ldots, x_n)$ above can be replaced by*

$$\exists^0 m \hat{R}(m, \overline{\mathbf{w}}(m), \overline{x}_1(m), \ldots, \overline{x}_n(m)) .$$

An analogous assertion holds for n odd, with $\forall^0 m$ replacing $\exists^0 m$.

Hint. Take $n > 0$ even, so that the rightmost function quantifier Q is $\forall^1$; the other case follows by taking negations. First change each number quantifier $\forall^0 p$ in a prenex arithmetical definition of R to a function quantifier $\forall^1 x$ by replacing occurrences of p by $x(0)$. Using the equivalences $(*)$, shift these new quantifiers to the left and contract like quantifiers. Hence, all the $\forall^0$'s of R have been absorbed into Q, and what remains is of form $\exists^0 p R_0(p, \mathbf{w}, x_1, \ldots, x_n)$ where R_0 is Δ^0_0. Being definable in $\mathcal{A}^2$ with only bounded quantifiers, for each p, $R_0(p, \mathbf{w}, x_1, \ldots, x_n)$ only depends on $\overline{\mathbf{w}}(q), \overline{x}_1(q), \ldots, \overline{x}_n(q)$ for some $q \in \omega$ sufficiently large. Thus,

$$\exists^0 p R_0(p, \mathbf{w}, x_1, \ldots, x_n) \leftrightarrow \exists^0 p \exists^0 q R_1(p, \overline{\mathbf{w}}(q), \overline{x}_1(q), \ldots, \overline{x}_n(q))$$

for some Δ^0_0 R_1, and the quantifiers can be contracted to complete the argument. ⊣

Having defined the effective ("lightface") hierarchies the correlation with the Borel and projective ("boldface") hierarchies can be made through the simple expedient of *relativization* to real parameters. For $a \in {}^\omega\omega$, *second-order arithmetic in a* is the expanded structure

$$\mathcal{A}^2(a) = \langle \omega, {}^\omega\omega, ap, +, \times, exp, <, 0, 1, a \rangle$$

where a is regarded as a binary relation on ω. Replacing $\mathcal{A}^2$ by $\mathcal{A}^2(a)$ in the preceding there are the corresponding relativized notions:

Δ^0_0 *in a*
recursive in a
arithmetical in a
$\Sigma^0_n(a)$, $\Pi^0_n(a)$
analytical in a
$\Sigma^1_n(a)$, $\Pi^1_n(a)$, $\Delta^1_n(a)$.

The following is a simple observation that will be used without further mention:

12.5 Exercise. *For $a \in {}^\omega\omega$, $R \subseteq {}^r\omega \times {}^k({}^\omega\omega)$ is $\Sigma^0_1(a)$ iff there is a Σ^0_1 $S \subseteq {}^r\omega \times {}^{k+1}({}^\omega\omega)$ such that $\forall \mathbf{w}(R(\mathbf{w}) \leftrightarrow S(\mathbf{w}, a))$. Hence, analogous statements hold for Σ^0_n, Π^0_n, Σ^1_n, and Π^1_n.* ⊣

This provides a way of defining the relativized hierarchies in $\mathcal{A}^2$, but one has to be careful: although $\Delta^1_n(a) = \Sigma^1_n(a) \cap \Pi^1_n(a)$, 12.5 does not hold with the Σ^0_1's replaced by Δ^1_n.

Here is the correlation with the classical hierarchies:

12.6 Proposition. *Suppose that $A \subseteq {}^k({}^\omega\omega)$ and $n > 0$. Then:*

(a) $A \in \mathbf{\Sigma}^0_n$ iff $A \in \Sigma^0_n(a)$ for some $a \in {}^\omega\omega$, and similarly for $\mathbf{\Pi}^0_n$.

(b) $A \in \mathbf{\Sigma}^1_n$ iff $A \in \Sigma^1_n(a)$ for some $a \in {}^\omega\omega$, and similarly for $\mathbf{\Pi}^1_n$.

Proof. (a) For $n = 1$, suppose first that $A \in \mathbf{\Sigma}^0_1$. Using our Δ^0_0 decoding of $\mathbf{s}_i \in {}^{<\omega}\omega$ from i and laying out the codes for the basic open sets that make up A in blocks of length k, there is an $a \in {}^\omega\omega$ such that for any $\mathbf{w} \in {}^k({}^\omega\omega)$,

$$A(\mathbf{w}) \leftrightarrow \exists^0 m(\mathbf{w} \in O(\mathbf{s}_{a(mk)}) \times O(\mathbf{s}_{a(mk+1)}) \times \ldots \times O(\mathbf{s}_{a(mk+k-1)}))\ .$$

Hence, it is readily seen that $A \in \Sigma^0_1(a)$. Conversely, if $A \in \Sigma^0_1(a)$ for some $a \in {}^\omega\omega$, then for any $\mathbf{w} \in {}^k({}^\omega\omega)$,

$$A(\mathbf{w}) \leftrightarrow \exists^0 m R(m, \mathbf{w})$$

where R is recursive in a. For each m, $R(m, \mathbf{w})$ only depends on $\overline{\mathbf{w}}(q)$ for some q. Consequently each $R_m = \{\mathbf{w} \mid R(m, \mathbf{w})\}$ is open, and so is $A = \bigcup_m R_m$.

The $\mathbf{\Pi}^0_n$ case is an immediate consequence of the $\mathbf{\Sigma}^0_n$ case, and the $\mathbf{\Sigma}^0_{n+1}$ case follows from the $\mathbf{\Pi}^0_n$ case by an argument similar to the base case.

(b) This is a direct consequence of the definitions, 12.4, and the $n = 1$ case of (a). ⊣

Thus, via $\Sigma^0_1(a) \cap \Pi^0_1(a)$ being the collection of relations recursive in a we have Addison's dictum mentioned earlier that the clopen subsets of ${}^\omega\omega$ are exactly those sets recursive in some real. The uniform notation of $\mathbf{\Sigma}$, Σ, and so forth that has been used all along was advocated by Addison [58] to reflect the correlations. Results will often be stated in terms of the $\Sigma^1_n(a)$'s for the effective content; these renderings imply of course the corresponding results about the classical $\mathbf{\Sigma}^1_n$'s.

There is also an effective analogue of the transfinite Borel levels $\mathbf{\Sigma}^0_\alpha$, $\mathbf{\Pi}^0_\alpha$ for $\alpha \geq \omega$ (see Hinman [78: IV§4]). For relations $\subseteq {}^r\omega$, the original formulation through integer notations for ordinals is Kleene's *hyperarithmetic hierarchy*; indeed, its study was the major motivation of his formulation of the analytical hierarchy. In the classical context, extending the hierarchy $\bigcup_{0<n<\omega} \mathbf{\Sigma}^0_n$ of the finite level Borel sets into the transfinite results in the Borel sets, which by Suslin's fundamental result 12.1 coincides with the $\mathbf{\Delta}^1_1$ sets. In the effective context, extending the hierarchy $\bigcup_{n\in\omega} \Sigma^0_n$ of arithmetical sets into the transfinite through analogous but recursive procedures results in the *hyperarithmetic sets*, which by Kleene's analogue for 12.1 coincides with the Δ^1_1 sets. (For more on hyperarithmetic theory, see Sacks [90: A§II].)

Having established a working context for descriptive set theory, we now pursue the development of the subject through its classical results. The emphasis will be on the traditional relations $\subseteq {}^k({}^\omega\omega)$, although relations $\subseteq {}^r\omega \times {}^k({}^\omega\omega)$ can be easily accommodated with only notational complications.

The first important idea of descriptive set theory was the use of Cantor's universal enumeration and diagonalization argument to effect a hierarchical transcendence. Borel [05: note III] had mentioned the idea, and Lebesgue [05] incorporated definability considerations to achieve his hierarchy results. An application is given here in set form to show that the projective hierarchy is proper. For any collection Γ of sets,

U is *universal for* Γ *iff* (i) $U \subseteq {}^2({}^\omega\omega)$ and $U \in \Gamma$, and
(ii) for any $A \subseteq {}^\omega\omega$, $A \in \Gamma$ *iff*
$A = \{x \mid \langle a, x\rangle \in U\}$ for some $a \in {}^\omega\omega$.

For $0 < k \in \omega$ any homeomorphic identification of ${}^k({}^\omega\omega)$ with ${}^\omega\omega$ yields subsets of ${}^{k+1}({}^\omega\omega)$ in Γ analogously universal for the subsets of ${}^k({}^\omega\omega)$ in Γ, a concept that occurs in the following proposition.

12.7 Proposition (Luzin [25c]; Sierpiński [25]). *For each $n > 0$, there is a Σ^1_n set universal for $\mathbf{\Sigma}^1_n$ (and so a Π^1_n set universal for $\mathbf{\Pi}^1_n$).*

Proof. First note that for $0 < k \in \omega$ there is a Σ^0_1 set $W_k \subseteq {}^{k+1}({}^\omega\omega)$ universal for $\mathbf{\Sigma}^0_1$ subsets of ${}^k({}^\omega\omega)$, i.e. for any $A \subseteq {}^k({}^\omega\omega)$,

$$A \in \mathbf{\Sigma}^0_1 \;\; \textit{iff} \;\; A = \{\mathbf{w} \mid W_k(a, \mathbf{w})\} \text{ for some } a \in {}^\omega\omega \, .$$

Simply set

$$W_k(a, \mathbf{w}) \;\textit{iff}\; \exists^0 m(\mathbf{w} \in O(\mathbf{s}_{a(mk)}) \times O(\mathbf{s}_{a(mk+1)}) \times \ldots \times O(\mathbf{s}_{a(mk+k-1)}))$$

as in the proof of 12.6(a). It is now straightforward to check, recalling 12.4, that if U_n is defined by

$$U_n(a, x) \;\textit{iff}\; \exists^1 x_1 \forall^1 x_2 \ldots \forall^1 x_n W_{n+1}(a, x, x_1, \ldots, x_n)$$

when n is even, and with $\forall^1 x_n$ replaced by $\exists^1 x_n$ and W_{n+1} by its complement when n is odd, then U_n is universal for $\mathbf{\Sigma}^1_n$. The parenthetical remark about $\mathbf{\Pi}^1_n$ follows from taking the complement of U_n. ⊣

12.8 Corollary. *For each $n > 0$, $\Sigma^1_n \not\subseteq \mathbf{\Delta}^1_n$, $\Pi^1_n \not\subseteq \mathbf{\Delta}^1_n$, and $\Delta^1_{n+1} \not\subseteq \mathbf{\Sigma}^1_n \cup \mathbf{\Pi}^1_n$.*

Proof. This uses Cantor's diagonalization argument. With U_n as in the previous proof, set $D_n(x) \leftrightarrow U_n(x, x)$ for any $x \in {}^\omega\omega$. Then $D_n \in \Sigma^1_n$, and $D_n \notin \mathbf{\Pi}^1_n$: Otherwise, $\neg D_n = {}^\omega\omega - D_n \in \mathbf{\Sigma}^1_n$ and there would be an $a \in {}^\omega\omega$ such that for any $x \in {}^\omega\omega$, $\neg D_n(x) \leftrightarrow U_n(a, x)$. But then, there is the paradigmatic contradiction

$$\neg D_n(a) \;\leftrightarrow\; U_n(a, a) \;\leftrightarrow\; D_n(a) \, .$$

That $\Pi^1_n \not\subseteq \mathbf{\Delta}^1_n$ follows by considering $\neg D_n$, and that $\Delta^1_{n+1} \not\subseteq \mathbf{\Sigma}^1_n \cup \mathbf{\Pi}^1_n$ follows by considering the recursive union

$$E_n = \{(x)^0 \mid x \in D_n\} \cup \{(x)^1 \mid x \notin D_n\}$$

where for $i < 2$, $(x)^i \in {}^\omega\omega$ is given by

$$(x)^i(m) = \begin{cases} i & \text{if } m = 0 \text{ , and} \\ x(m-1) & \text{if } m > 0 \, . \end{cases}$$

$E_n \in \Delta^1_{n+1}$ since $D_n \in \Sigma^1_n$ and $\neg D_n \in \Pi^1_n$, yet $E_n \notin \mathbf{\Sigma}^1_n \cup \mathbf{\Pi}^1_n$ since both D_n and $\neg D_n$ can be recovered from it. ⊣

Note that the diagonalization argument shows that there can be no $\mathbf{\Delta}^1_n$ set universal for $\mathbf{\Delta}^1_n$. An analogous argument with number quantifiers shows that the finite levels of the Borel hierarchy are proper. Lebesgue [05] similarly showed that all levels of the hierarchy are proper (and bearing on his skepticism of the Axiom of Choice, $\mathrm{AC}_\omega({}^\omega\omega)$ is nonetheless seen to be necessary here to choose cofinal sequences for limit ordinals).

The next important idea of descriptive set theory was to analyze sets by associating ordinals to their members. With roots in the Lebesgue paper and first applications in Luzin-Sierpiński [18], the approach became basic in Luzin-Sierpiński [23], a fundamental paper on analytic sets. Establishing some terminology in sufficient generality for later use, stipulate for any $k \in \omega$ and set Y that

$$\begin{aligned} T \text{ is a } \textit{tree on } {}^k\omega \times Y \ \textit{ iff } \ &\text{(i) } T \subseteq \textstyle\bigcup_{m\in\omega} {}^k({}^m\omega) \times {}^mY \text{ , and} \\ &\text{(ii) if } \langle t_0, \dots, t_k\rangle \in T \text{, then} \\ &\qquad \langle t_0|m, \dots, t_k|m\rangle \in T \text{ for any } m \leq |t_0| \ . \end{aligned}$$

(When $k = 0$, T is a *tree on* Y *iff* $T \subseteq {}^{<\omega}Y$ and is closed under the taking of initial segments; analogous conventions apply when $Y = \emptyset$.) Thus, T consists of $(k+1)$-tuples of finite sequences *of the same length* and is a tree in the sense of §7 under a natural initial segment ordering. $[T]$ denotes the collection of infinite branches of T:

$$\begin{aligned} \langle x_0, \dots, x_{k-1}, f\rangle \in [T] \ \textit{ iff } \ &\langle x_0, \dots, x_{k-1}, f\rangle \in {}^k({}^\omega\omega) \times {}^\omega Y \text{ and} \\ &\forall^0 m(\langle x_0|m, \dots, x_{k-1}|m, f|m\rangle \in T) \ . \end{aligned}$$

Also, for $\mathbf{w} = \langle x_0, \dots, x_{l-1}\rangle \in {}^l({}^\omega\omega)$ with $l \leq k$, $T_{\mathbf{w}}$ is that tree on ${}^{k-l}\omega \times Y$ defined by

$$\langle t_0, \dots, t_{k-l}\rangle \in T_{\mathbf{w}} \ \textit{ iff } \ \exists^0 m(\langle x_0|m, \dots, x_{l-1}|m, t_0, \dots, t_{k-l}\rangle \in T) \ .$$

T_x is written for $T_{\langle x\rangle}$.

Next define $\subseteq^*$ to be coordinate-wise inclusion for n-tuples, i.e.

$$\langle x_1, \dots, x_n\rangle \subseteq^* \langle y_1, \dots, y_n\rangle \ \textit{ iff } \ x_i \subseteq y_i \text{ for } 1 \leq i \leq n \ ,$$

and let $\subset^*$, $\supseteq^*$, and $\supset^*$ have the derived meanings. Also define $^\frown{}^*$ to be coordinate-wise concatenation for n-tuples of sequences, i.e.

$$\langle x_1, \dots, x_n\rangle {}^\frown{}^* \langle y_1, \dots, y_n\rangle = \langle x_1{}^\frown y_1, \dots, x_n{}^\frown y_n\rangle \ .$$

When $k = 0$ or $Y = \emptyset$ in the definition of T, $\supset^*$ is regarded as reverting back to $\supset$ and so forth.

In these terms, some further terminology for a T as above is developed: First, for any set t,

$$T/t = \{s \mid t{}^\frown{}^* s \in T\} \ .$$

Next,

$$T \text{ is } \textit{well-founded } \ \textit{iff } \ [T] = \emptyset \ ,$$

i.e. *reverse* coordinate-wise inclusion $\supset^*$ is a well-founded relation on T. (In this context, salient is the lack of infinite descending chains. The equivalence of this with the usual formulation of well-foundedness requires DC in general, but the focus will be on well-orderable Y's and hence well-orderable T's, for which DC is not needed.) For such T, the *rank function* $\rho_T\colon T \to \mathrm{On}$ is defined in the usual way (cf. 0.3) by:

$$\rho_T(t) = \sup(\{\rho_T(s) + 1 \mid s \supset^* t\}) ,$$

noting that $\rho_T(t) = \sup(\emptyset) = 0$ *iff* t is maximal in $\supset^*$. The *height* of T is then defined by

$$\|T\| = \mathrm{ran}(\rho_T) .$$

Note that $\|T\|$ is an ordinal, indeed $\|T\| = \rho_T(\langle\emptyset, \ldots, \emptyset\rangle)$, and generally $\rho_T(t) = \|T/t\|$. ρ_T is a basic example of an *order-preserving map* $\rho\colon \langle T, \supset^*\rangle \to \langle \mathrm{On}, <\rangle$, i.e. if $s, t \in T$ with $s \supset^* t$, then $\rho(s) < \rho(t)$. There is also the concept of an *order-preserving map* $\rho\colon T_1 \to T_2$ between two trees T_1 and T_2, i.e. if $s, t \in T_1$ with $s \supset^* t$, then $\rho(s) \supset^* \rho(t)$. The following are simple but crucial observations; no appeal will be made to (a) unless Y is well-orderable.

12.9 Exercise.

(a) For a tree T on ${}^k\omega \times Y$, T is well-founded iff there is an order-preserving map $\rho\colon T \to \mathrm{On}$. If such a ρ exists, then $\rho_T(t) \leq \rho(t)$ for every $t \in T$, and $\rho_T\colon T \to \max(\{\omega_1, |Y|^+\})$.

(b) For trees S and T, there is an order-preserving map: $S \to T$ iff either T is ill-founded or else $\|S\| \leq \|T\|$.

Hint. For (a), recall 0.3.

For (b), if $f\colon S \to T$ is order-preserving and T is well-founded, then $\rho_T \circ f\colon S \to \mathrm{On}$ is order-preserving, so by (a),

$$\|S\| = \rho_S(\emptyset) \leq \rho_T(f(\emptyset)) \leq \|T\| .$$

For the converse, if T is ill-founded, then fixing an infinite branch through T, S can be mapped in an order-preserving manner into T by assigning all the elements in a level to the single node on the branch at the corresponding level. If T is well-founded and $\|S\| \leq \|T\|$, then proceed by induction on $\|T\|$: For each $\langle i\rangle \in S$, $\|S/\langle i\rangle\| < \|T\|$, and so there is a $t_i \in T$ such that $t_i \neq \emptyset$ and $\|S/\langle i\rangle\| \leq \|T/t_i\|$. Let $f_i\colon S/\langle i\rangle \to T/t_i$ be order-preserving and define $f\colon S \to T$ by

$$f(s) = \begin{cases} \emptyset & \text{if } s = \emptyset \text{ , and} \\ t_i{}^\frown f_i(\hat{s}) & \text{if } s = \langle i\rangle^\frown \hat{s} \text{ .} \end{cases}$$

⊣

Trees provide a simple analysis of closed sets:

12.10 Proposition. *For $C \subseteq {}^k({}^\omega\omega)$, C is closed iff there is a tree T on ${}^k\omega$ such that $C = [T]$.*

Proof. Take $k = 1$ for notational simplicity. Given C, set

$$T = \{x|m \mid x \in C \ \wedge \ m \in \omega\}$$

so that $C \subseteq [T]$. If C is closed, then for any $y \notin C$ there is an $m \in \omega$ such that $O(y|m) \cap C = \emptyset$, so $y \notin [T]$. Hence, $C = [T]$.

Conversely, if T is a tree on ω and $x \notin [T]$, then for some $m \in \omega$, $x|m \notin T$ and consequently $O(x|m) \cap [T] = \emptyset$. Thus, ${}^{\omega}\omega - [T]$ is open. ⊣

The next section begins with a tree representation of $\mathbf{\Pi}^1_1$ sets and proceeds to the consequent classical results.

13. $\mathbf{\Pi}^1_1$ Sets and $\mathbf{\Sigma}^1_2$ Sets

This section is devoted to the classical analysis of $\mathbf{\Pi}^1_1$ sets and $\mathbf{\Sigma}^1_2$ sets, building on the groundwork laid in the previous section. As described there, the development can be carried out in ZF + $\mathrm{AC}_\omega({}^\omega\omega)$.

Well-founded trees as articulated at the end of the previous section provide a basic representation of $\mathbf{\Pi}^1_1$ sets, the classical result later given effective content:

13.1 Theorem (Luzin-Sierpiński [23], Kleene [55]). *Suppose that $a \in {}^\omega\omega$ and $A \subseteq {}^k({}^\omega\omega)$. Then A is $\Pi^1_1(a)$ iff there is a tree T on ${}^k\omega \times \omega$ such that for any $\mathbf{w} \in {}^k({}^\omega\omega)$,*

$$A(\mathbf{w}) \leftrightarrow T_{\mathbf{w}} \text{ is well-founded },$$

and the relation $\{\langle i, \mathbf{w}\rangle \mid \mathbf{s}_i \in T_{\mathbf{w}}\}$ is recursive in a.

Proof. This can be seen as a corollary to 12.10, but is established anew. Given such a representation, $A(\mathbf{w})$ is clearly $\Pi^1_1(a)$:

$$A(\mathbf{w}) \leftrightarrow \forall^1 y \exists^0 i \exists^0 p(\mathbf{s}_i = y|p \ \wedge \ \mathbf{s}_i \notin T_{\mathbf{w}}) .$$

For the converse, take $k = 1$ for notational simplicity. By 12.4, there is an $R \subseteq {}^3\omega$ recursive in a such that for any $x \in {}^\omega\omega$,

$$A(x) \leftrightarrow \forall^1 y \exists^0 m R(m, \overline{x}(m), \overline{y}(m)) .$$

Let T be that tree on $\omega \times \omega$ consisting of the "unsecured" sequences:

$$T = \{\langle s, t\rangle \mid \forall^0 p < |s| + 1(\neg R(p, \langle\!\langle s(0), \ldots, s(p-1)\rangle\!\rangle, \langle\!\langle t(0), \ldots, t(p-1)\rangle\!\rangle))\} .$$

Then $A(x)$ *iff* T_x is well-founded, and $\{\langle i, x\rangle \mid \mathbf{s}_i \in T_x\}$ is a relation recursive in a. ⊣

By this means, for any $\mathbf{\Pi}^1_1$ set A with corresponding tree T an ordinal $\|T_{\mathbf{w}}\|$ can be associated to each $\mathbf{w} \in A$. Actually, Luzin-Sierpiński [23] focused on well-orderings, and Kleene [55] on integer notations for ordinals. Both in effect linearized trees using an ordering that first appeared in the earlier paper, although it is now known as the *Kleene-Brouwer ordering*: For $s, t \in {}^{<\omega}\mathrm{On}$,

$$s <_{\mathrm{KB}} t \ \textit{ iff } \text{ either } s \supset t \text{ or else for the least } i \text{ such that } s(i) \neq t(i),\ s(i) < t(i) .$$

$<_{\mathrm{KB}}$ is a strict linear ordering of ${}^{<\omega}\mathrm{On}$. Note that it is $s \supset t$ in the definition rather than $s \subset t$. As a simple way to remember the ordering, $s <_{\mathrm{KB}} t$ *iff* $s^\frown\langle\infty, \infty, \infty, \ldots\rangle$ is lexicographically less than $t^\frown\langle\infty, \infty, \infty, \ldots\rangle$ where ∞ is formally greater than all the ordinals.

13.2 Exercise (Luzin-Sierpiński [23]). *If T is a tree on γ, then T is well-founded iff T is well-ordered by $<_{\mathrm{KB}}$. Hence, "well-founded" can be replaced in 13.1 by "well-ordered by $<_{\mathrm{KB}}$."* ⊣

Hint. Suppose that $\langle t_i \mid i \in \omega \rangle$ is an infinite $<_{\mathrm{KB}}$ descending chain in T. The t_i's for $i > 0$ cannot be the empty sequence, and since $\langle t_i(0) \mid 0 < i < \omega \rangle$ is a non-increasing sequence of ordinals, there is a $k_0 \in \omega$ such that $t_i(0)$ has a fixed value $y(0)$ for $i \geq k_0$. The t_i's for $i > k_0$ cannot have length 1, and so there is a $k_1 \in \omega$ with $k_0 < k_1$ such that $t_i(1)$ has a fixed value $y(1)$ for $i \geq k_1$. Continuing, the resulting $y \in {}^{\omega}\omega$ verifies that T is ill-founded. ⊣

The development is carried out in the more immediate terms of well-founded relations. Not only have they become as basic in set theory as well-orderings, but later generalizations were to establish direct connections with well-founded models.

The representation 13.1 provides a basic stratification:

13.3 Proposition (Luzin-Sierpiński [18, 23]). *Every $\mathbf{\Pi}_1^1$ (and hence $\mathbf{\Sigma}_1^1$) set is both a union and an intersection of $\aleph_1$ Borel sets.*

Proof. Suppose that $A \subseteq {}^k({}^{\omega}\omega)$ is $\mathbf{\Pi}_1^1$, and T is a tree on ${}^k\omega \times \omega$ as given by 13.1 so that for any $\mathbf{w} \in {}^k({}^{\omega}\omega)$,

$$A(\mathbf{w}) \;\leftrightarrow\; T_{\mathbf{w}} \text{ is well-founded} .$$

For $s \in {}^{<\omega}\omega$ and $\alpha < \omega_1$, set

$$B_\alpha^s = \{\mathbf{w} \in {}^k({}^{\omega}\omega) \mid T_{\mathbf{w}}/s \text{ is well-founded with } \|T_{\mathbf{w}}/s\| \leq \alpha\} .$$

Then it follows by induction on α that these sets are Borel:

$$B_0^s = \bigcap\nolimits_{i\in\omega}\{\mathbf{w} \in {}^k({}^{\omega}\omega) \mid s^\frown\langle i\rangle \notin T_{\mathbf{w}}\} ,$$

and since $s^\frown\langle i\rangle \notin T_{\mathbf{w}}$ depends only on the initial segment $\overline{\mathbf{w}}(|s|+1)$, B_0^s is Borel, in fact $\mathbf{\Pi}_2^0$. For $\alpha > 0$, note that a tree U on ω is well-founded with $\|U\| \leq \alpha$ exactly when for any $i \in \omega$ there is a $\beta < \alpha$ such that $U/\langle i\rangle$ is well-founded with $\|U/\langle i\rangle\| \leq \beta$. Hence,

$$B_\alpha^s = \bigcap\nolimits_{i\in\omega}\bigcup\nolimits_{\beta<\alpha} B_\beta^{s^\frown\langle i\rangle}$$

is Borel by induction. A is consequently a union of $\aleph_1$ Borel sets:

$$A = \bigcup\nolimits_{\alpha<\omega_1} B_\alpha^{\emptyset} .$$

To show that A is also an intersection of $\aleph_1$ Borel sets, for $\alpha < \omega_1$ set

$$C_\alpha = \{\mathbf{w} \in {}^k({}^{\omega}\omega) \mid (T_{\mathbf{w}} \text{ is well-founded with } \|T_{\mathbf{w}}\| \leq \alpha) \vee \exists s \in {}^{<\omega}\omega(T_{\mathbf{w}}/s \text{ is well-founded with } \|T_{\mathbf{w}}/s\| = \alpha)\} .$$

That the B_α^s's are Borel implies in a straightforward manner that the C_α's are Borel. We shall verify that

$$A = \bigcap\nolimits_{\alpha<\omega_1} C_\alpha .$$

If $\mathbf{w} \in A$, then $T_{\mathbf{w}}$ is well-founded, so that for any $\alpha < \omega_1$ either $\|T_{\mathbf{w}}\| \leq \alpha$ or else there is an $s \in {}^{<\omega}\omega$ such that $\|T_{\mathbf{w}}/s\| = \alpha$, i.e. $\mathbf{w} \in C_\alpha$. On the other hand, if $\mathbf{w} \notin A$, then $T_{\mathbf{w}}$ is ill-founded, and also for any $s \in {}^{<\omega}\omega$ there is at most one $\beta < \omega_1$ such that $T_{\mathbf{w}}/s$ is well-founded with $\|T_{\mathbf{w}}/s\| = \beta$. But since ${}^{<\omega}\omega$ is countable, there must be some $\alpha < \omega_1$ such that $\mathbf{w} \notin C_\alpha$. ⊣

The Borel sets $B_\alpha^\emptyset$ figure in a general bounding principle for analytic sets:

13.4 Proposition (Luzin-Sierpiński [18, 23]). *Suppose that $A \subseteq {}^k({}^\omega\omega)$ is $\mathbf{\Pi}_1^1$ with a corresponding tree T on ${}^k\omega \times \omega$ as given by 13.1, and $B \subseteq A$ is $\mathbf{\Sigma}_1^1$. Then there is an $\alpha < \omega_1$ such that*

$$B \subseteq \{\mathbf{w} \in {}^k({}^\omega\omega) \mid T_{\mathbf{w}} \text{ is well-founded with } \|T_{\mathbf{w}}\| \leq \alpha\} .$$

Proof. Taking $k = 1$ for simplicity, by 12.4 let U be a tree on $\omega \times \omega$ so that $B = p[U]$, i.e. for any $w \in {}^\omega\omega$,

$$w \in B \leftrightarrow \exists^1 y(\langle w, y\rangle \in [U]) .$$

For $n \in \omega$ and $s, t, u \in {}^n\omega$, set

$$B_{s,t,u} = \{w \in B \mid \langle s, t\rangle \in T \;\wedge\; s \subseteq w \;\wedge\; \exists^1 y(\langle w, y\rangle \in [U] \;\wedge\; u \subseteq y)\} .$$

Assume now to the contrary that $\sup(\{\|T_w\| \mid w \in B\}) = \omega_1$. Define $s_n, t_n, u_n \in {}^n\omega$ for every $n \in \omega$ such that $s_n \subseteq s_{n+1}$, $t_n \subseteq t_{n+1}$, $u_n \subseteq u_{n+1}$, and

$$(*) \qquad \sup(\{\|T_w/t_n\| \mid w \in B_{s_n,t_n,u_n}\}) = \omega_1$$

by recursion on n as follows: Set $s_0 = t_0 = u_0 = \emptyset$, so that $B_{s_0,t_0,u_0} = B$ and $(*)$ holds for $n = 0$ by assumption. Given s_n, t_n, u_n satisfying $(*)$, since there are only countably many one-step extensions of s_n, t_n, u_n, and since $\|T_w/t_n\| = \sup(\{\|T_w/t_n^\frown\langle i\rangle\| + 1 \mid i \in \omega\})$, recursively define $s_{n+1} \supseteq s_n$, $u_{n+1} \supseteq u_n$, and $t_{n+1} \supseteq t_n$ so that $(*)$ is satisfied with n replaced by $n + 1$.

Set $w = \bigcup_n s_n$, $x = \bigcup_n t_n$, and $y = \bigcup_n u_n$. Then it is easy to see that $\langle w, y\rangle \in [U]$, i.e. $w \in B$, yet T_w is ill-founded since $x \in [T_w]$. Contradiction! ⊣

The foregoing proof was drawn from the original publications. Curiously, recent texts give the following less direct proof:

Second proof of 13.4. Assume to the contrary that for any $\alpha < \omega_1$ there is a $\mathbf{w} \in B$ with $\|T_{\mathbf{w}}\| > \alpha$. It will be shown that any $\mathbf{\Pi}_1^1$ set of reals is also $\mathbf{\Sigma}_1^1$, contradicting the hierarchy result 12.7. So, suppose that $C \subseteq {}^\omega\omega$ is $\mathbf{\Pi}_1^1$ with a corresponding tree U on $\omega \times \omega$ as given by 13.1 so that for any $x \in {}^\omega\omega$,

$$C(x) \leftrightarrow U_x \text{ is well-founded} .$$

By 12.9(b) and the assumption on B, it follows that

$$C(x) \leftrightarrow \exists \mathbf{w}(B(\mathbf{w}) \wedge \exists f(f\colon U_x \to T_{\mathbf{w}} \text{ is an order-preserving map}))\ .$$

But the $\exists f \ldots$ can be rendered in $\mathbf{\Sigma}^1_1$ form as

$$\exists^1 y \forall^0 i \forall^0 j(\mathbf{s}_i \in U_x \ \wedge\ \mathbf{s}_j \in U_x \ \wedge\ \mathbf{s}_i \supset \mathbf{s}_j \to$$
$$\mathbf{s}_{y(i)} \in T_{\mathbf{w}} \ \wedge\ \mathbf{s}_{y(j)} \in T_{\mathbf{w}} \ \wedge\ \mathbf{s}_{y(i)} \supset \mathbf{s}_{y(j)})\ .$$

Hence, C is $\mathbf{\Sigma}^1_1$. ⊣

A direct consequence of 13.4 is *the Borel separability of analytic sets* (Luzin [27: 50ff]): If $X, Y \subseteq {}^k({}^\omega\omega)$ are disjoint $\mathbf{\Sigma}^1_1$ sets, there is a Borel set $Z \subseteq {}^k({}^\omega\omega)$ such that $X \subseteq Z$ and $Y \cap Z = \emptyset$. This in turn leads directly to the fundamental result that began descriptive set theory:

13.5 Exercise (Suslin [17]). *A set is Borel iff it is* $\mathbf{\Delta}^1_1$.

Hint. For the forward direction, show that the $\mathbf{\Delta}^1_1$ subsets of ${}^k({}^\omega\omega)$ form a σ-algebra; closure under countable unions follows from the quantifier equivalences $(*)$ before 12.4. ⊣

This is essentially the first published proof of Suslin's result, which appeared in Luzin-Sierpiński [18]. Luzin-Sierpiński [23] started with a formulation of analytic sets differing from Suslin's, one that essentially regarded reals as codes for relations. In our context, any $x \in {}^\omega\omega$ encodes a binary relation E_x on ω given by

$$\langle m, n\rangle \in E_x \ \textit{ iff } \ x(\langle\!\langle m, n\rangle\!\rangle) = 0\ .$$

If E_x is well-founded, it has a rank function ρ_x defined recursively on ω by:

$$\rho_x(n) = \sup(\{\rho_x(m) + 1 \mid \langle m, n\rangle \in E_x\})\ ,$$

so that $\rho_x(n) = \sup(\emptyset) = 0$ *iff* n is minimal in E_x or else n is not in its field. (The latter is allowed for the convenience of having $\mathrm{dom}(\rho_x) = \omega$.) Define

$$\|x\| = \sup(\{\rho_x(n) \mid n \in \omega\})\ .$$

Now set

$$\mathrm{WF} = \{x \in {}^\omega\omega \mid E_x \text{ is well-founded}\}\ , \text{ and}$$
$$\mathrm{WF}_\alpha = \{x \in \mathrm{WF} \mid \|x\| \leq \alpha\}$$

for $\alpha < \omega_1$. WF is Π^1_1, since for any $x \in {}^\omega\omega$,

$$x \in \mathrm{WF} \ \leftrightarrow\ \forall^1 y \exists^0 m(x(\langle\!\langle y(m+1), y(m)\rangle\!\rangle) \neq 0)\ .$$

In terms of 13.1, this is already in normal form, with the corresponding tree on $\omega \times \omega$ of "unsecured" sequences being

$$T^{\mathrm{WF}} = \{\langle s, t\rangle \in \textstyle\bigcup_m ({}^m\omega \times {}^m\omega) \mid$$
$$\forall^0 i < |s| - 1(\langle\!\langle t(i+1), t(i)\rangle\!\rangle < |s| \to s(\langle\!\langle t(i+1), t(i)\rangle\!\rangle) = 0)\}\ .$$

For $x \in \mathrm{WF}$, $\|x\| = \|T_x^{\mathrm{WF}}\|$ and so

$$\mathrm{WF}_\alpha = \{x \in {}^\omega\omega \mid T_x^{\mathrm{WF}} \text{ is well-founded and } \|T_x^{\mathrm{WF}}\| \leq \alpha\} .$$

The following is now a simple consequence of the argument for 13.3 and of 13.4:

13.6 Exercise (Luzin-Sierpiński [23]).

(a) For any $\alpha < \omega_1$, WF_α *is Borel.*

(b) If $B \subseteq \mathrm{WF}$ *is* $\mathbf{\Sigma}^1_1$, *then there is an* $\alpha < \omega_1$ *such that* $B \subseteq \mathrm{WF}_\alpha$.

(c) $\mathrm{WF} \in \mathbf{\Pi}^1_1 - \mathbf{\Sigma}^1_1$. ⊣

Spector [55] established an effective version of (b) in terms of well-orderings for (lightface) Σ^1_1 sets and with ω_1 replaced by

$$\omega_1^{\mathrm{CK}} = \sup\{\|x\| \mid x \in \mathrm{WF} \ \wedge \ (\text{the graph of) } x \text{ is recursive}\} .$$

This Σ^1_1 boundedness theorem is a cornerstone of hyperarithmetic theory (see Sacks [90: A§II]).

(c) provides a natural example of a member of $\mathbf{\Pi}^1_1 - \mathbf{\Sigma}^1_1$. Note that with the first proof of 13.4, this approach to $\mathbf{\Pi}^1_1 - \mathbf{\Sigma}^1_1 \neq \emptyset$ does not depend on the diagonalization argument for 12.8. By 13.1, for any $\Pi^1_1(a)$ set $A \subseteq {}^k({}^\omega\omega)$ there is a function $f\colon {}^k({}^\omega\omega) \to {}^\omega\omega$ (whose graph is) recursive in a such that $A = f^{-1}(\mathrm{WF})$. Hence, WF is "complete" for this sort of reducibility for the $\mathbf{\Pi}^1_1$ sets. Versions of WF and (a) had already occurred in Lebesgue [05], but he could not get to (c) because of a crucial mistake, discussed earlier, that spurred Suslin's discovery of the analytic sets.

Mazurkiewicz [36] provided a more mathematical example of a member of $\mathbf{\Pi}^1_1 - \mathbf{\Sigma}^1_1$. Since a continuous function: $\mathbb{R} \to \mathbb{R}$ is determined by its values on the rationals, for present purposes such functions can be identified with members of ${}^\omega\omega$ in a systematic way much as codes for open sets were developed (see before 11.8). Mazurkiewicz showed that

$$\{f\colon \mathbb{R} \to \mathbb{R} \mid f \text{ is differentiable}\} \in \mathbf{\Pi}^1_1 - \mathbf{\Sigma}^1_1 .$$

See Becker [92] for an account of recent work along these lines.

The first significant result about the second level of the projective hierarchy was the following extension of part of 13.3:

13.7 Proposition (Sierpiński [25]). *Every* $\mathbf{\Sigma}^1_2$ *set is a union of* $\aleph_1$ *Borel sets.*

Two proofs are provided; the first is the original one and uses AC, and the second (see after 13.14) does not.

First proof of 13.7. Suppose that $A \subseteq {}^k({}^\omega\omega)$ is $\mathbf{\Sigma}^1_2$, with $\mathbf{\Pi}^1_1$ $B \subseteq {}^{k+1}({}^\omega\omega)$ such that for any $\mathbf{w} \in {}^k({}^\omega\omega)$,

$$A(\mathbf{w}) \leftrightarrow \exists^1 y B(\mathbf{w}, y) .$$

By 13.3, $B = \bigcup_{\alpha<\omega_1} C_\alpha$ where each C_α is Borel, and so

$$A(\mathbf{w}) \leftrightarrow \exists^1 y \exists \alpha < \omega_1 C_\alpha(\mathbf{w}, y)$$
$$\leftrightarrow \exists \alpha < \omega_1 \exists^1 y C_\alpha(\mathbf{w}, y) \; .$$

For each $\alpha < \omega_1$, $D_\alpha = \{\mathbf{w} \in {}^k({}^\omega\omega) \mid \exists^1 y C_\alpha(\mathbf{w}, y)\}$ is $\mathbf{\Sigma}^1_1$, and so again by 13.3 $D_\alpha = \bigcup_{\xi<\omega_1} E^\alpha_\xi$ where each E^α_ξ is Borel. (The choice for each $\alpha < \omega_1$ of a family $\{E^\alpha_\xi \mid \xi < \omega_1\}$ uses AC.) Hence,

$$A = \bigcup_{\alpha<\omega_1} \bigcup_{\xi<\omega_1} E^\alpha_\xi \; . \qquad \dashv$$

With the perfect set property for Borel sets (see 12.2(c)) Sierpiński was able to conclude that *every* $\mathbf{\Sigma}^1_2$ *set has cardinality at most* $\aleph_1$ *or a perfect subset and hence cardinality* $2^{\aleph_0}$. (This result is improved by 14.9 in a modern setting.)

The main classical results about $\mathbf{\Pi}^1_1$ and $\mathbf{\Sigma}^1_2$ were established using these ideas (without the effective content, of course), and Kondô's $\mathbf{\Pi}^1_1$ Uniformization theorem (see 13.17), considered difficult for some time, was a culminating result of the ordinal analysis of $\mathbf{\Pi}^1_1$ sets. 13.7 was to become paradigmatic for reducing the analysis of projective sets to unions of Borel sets; digressing somewhat, we briefly take a look at work in this vein in the post-Cohen era:

The converse to 13.7,

$$\text{any union of } \aleph_1 \text{ Borel sets is } \mathbf{\Sigma}^1_2 \; ,$$

fails in ZFC + CH, since any set of reals of cardinality $\aleph_1$ is a union of $\aleph_1$ Borel sets, namely the singletons of its members. Notwithstanding, Luzin [35] considered the possibility of that proposition holding, and it is actually a consequence of ZFC + MA + $\neg$CH + $\exists a \in {}^\omega\omega(\omega_1^{L[a]} = \omega_1)$ (Martin-Solovay [70: §3]). This incidentally showed that one cannot prove in ZFC that every $\mathbf{\Pi}^1_2$ set is a union of $\aleph_1$ Borel sets. Solovay showed that it is consistent to have $2^{\aleph_0}$ large and a $\mathbf{\Pi}^1_2$ set that is not the union of fewer than $2^{\aleph_0}$ Borel sets.

Considering arbitrary sets of reals, Levy-Solovay [72] first observed that ZFC + $\neg$CH implies that there is a set of reals which is not the union of any $\aleph_1$ Borel sets, and went on to show that some substantial use of AC is necessary: They established using Solovay's Lebesgue measure arguments that Con(ZFC + $\exists\kappa$(κ is inaccessible)) implies Con(ZFC + $\neg$CH + ⌜every ${}^\omega$On-definable set of reals is a union of $\aleph_1$ Borel sets⌝) (cf. 11.11), and passing to an inner model, Con(DC + $\omega_1 \not\leq 2^{\aleph_0}$ + ⌜every set of reals is a union of $\aleph_1$ Borel sets⌝) (cf. 11.1). Finally, in the rather different setting of ZF + AD, the Axiom of Determinacy, the converse to 13.7 does hold, and moreover, elegant generalizations through the projective hierarchy have been obtained (30.20).

Regularity Properties in L

We next discuss Gödel's delimitative results with $V = L$. A conspicuous feature of the modern theory is the study of reals as codes for countable models. Any $x \in {}^\omega\omega$ encodes the binary relation $E_x = \{\langle m, n\rangle \mid x(\langle m, n\rangle) = 0\}$ as specified before, and consequently a structure

$$M_x = \langle \omega, E_x \rangle$$

for the language $\mathcal{L}_\in$ of set theory. If M_x is well-founded and extensional, then by the Collapsing Lemma 0.4 it has a unique transitive collapse $\mathrm{tr}(M_x)$ and an isomorphism

$$\pi_x \colon M_x \to \mathrm{tr}(M_x)\ .$$

For φ a formula and $s \in {}^{<\omega}\omega$,

$$M_x \models \varphi[s]$$

is written with the understanding that φ has at most the variables $v_0, \ldots, v_{|s|-1}$ free and is satisfied in M_x with the variable assignment of $s(i)$ to v_i. Temporarily specify that $\ulcorner\varphi\urcorner \in \omega$ is the Gödel number of formula φ in some fixed recursive arithmetization of $\mathcal{L}_\in$, and for any collection Φ of formulas, define $\ulcorner\Phi\urcorner \colon \omega \to 2$ by $\ulcorner\Phi\urcorner(m) = 1$ *iff* $m = \ulcorner\varphi\urcorner$ for some $\varphi \in \Phi$.

13.8 Proposition.
(a) For any $n \in \omega$,

$$\{\langle m, i, x\rangle \in {}^2\omega \times {}^\omega\omega \mid m = \ulcorner\varphi\urcorner \ \wedge\ \varphi \text{ is } \Sigma_n \ \wedge\ M_x \models \varphi[\mathbf{s}_i]\}$$

is arithmetical, in fact Σ^0_n when $n > 0$.
(b) For any collection Φ of $\mathcal{L}_\in$ sentences, $\{x \in {}^\omega\omega \mid M_x \text{ models } \Phi\}$ is $\Delta^1_1(\ulcorner\Phi\urcorner)$.
(c) If $a \in {}^\omega\omega$ and M_a is well-founded and extensional, then

$$\{\langle m, x\rangle \in \omega \times {}^\omega\omega \mid \pi_a(m) = x\}$$

is arithmetical in a.

Proof. (a) and (b) follow from the arithmetization of the satisfaction relation. The first can be established inductively following the definition of $\models^n_M$ given in §0. The second recasts the Δ^{ZF}_1 definability of the satisfaction relation for set structures (see e.g. Devlin [84: 41] or Drake [74: 92]).

(c) Using (a) and the definition of each member of ω in set theory, it is simple to see that

$$\{\langle i, p\rangle \ \in {}^2\omega \mid \pi_a(i) = p\}$$

is arithmetical in a. Consequently, if ϕ is the formula $\langle v_0, v_1\rangle \in v_2$, then for any $m \in \omega$ and $x \in {}^\omega\omega$,

$$\begin{aligned}&\pi_a(m) = x \ \leftrightarrow \\ &\quad \forall^0 p \forall^0 q (x(p) = q \ \leftrightarrow\ \exists^0 i \exists^0 j (\pi_a(i) = p \ \wedge\ \pi_a(j) = q \ \wedge\ M_a \models \phi[i, j, m]))\ ,\end{aligned}$$

which shows that the relation is arithmetical in a. ⊣

It can be checked that the relation in (c) is $\Pi^0_2(a)$.

By 3.3(a), there is a sentence σ_0 such that for any transitive set model N of σ_0, $N = L_\delta$ for some limit $\delta > \omega$, and a formula $\varphi_0(v_1, v_2)$ that defines in L a well-ordering $<_L$ of L such that for any limit $\delta > \omega$ and $x, y \in L_\delta$, $x <_L y$ *iff* $L_\delta \models \varphi_0[x, y]$. By the proof of CH in L the well-ordering

$$^2({}^\omega\omega) \cap <_L = \{\langle x, y\rangle \mid x, y \in {}^\omega\omega \cap L \ \wedge \ x <_L y\}$$

has ordertype ω_1^L.

13.9 Theorem (Gödel [51:67]). *${}^\omega\omega \cap L$ and ${}^2({}^\omega\omega) \cap <_L$ are Σ_2^1.*

Proof. Proceeding directly to the well-ordering, for any $x, y \in {}^\omega\omega \cap L$,

$$\begin{aligned} x \ <_L y \ \leftrightarrow \ & \exists^1 z \exists^0 p \exists^0 q (M_z \text{ is well-founded and extensional} \\ & \wedge \ \pi_z(p) = x \ \wedge \ \pi_z(q) = y \ \wedge \ M_z \ \models \sigma_0 \wedge \varphi_0[p, q]) \ . \end{aligned}$$

By 13.8 this is Σ_2^1. ⊣

13.10 Corollary. *If ${}^\omega\omega \subseteq L$, then ${}^2({}^\omega\omega) \cap <_L$ is a Δ_2^1 set which is not Lebesgue measurable and does not have the Baire property.*

Proof. With ${}^\omega\omega \subseteq L$, ${}^2({}^\omega\omega) \cap <_L$ is a well-ordering of all the reals, and so it must also be Δ_2^1, since for any $x, y \in {}^\omega\omega$,

$$x <_L y \ \leftrightarrow \ x \neq y \ \wedge \ \neg(y <_L x) \ .$$

Assume now that ${}^2({}^\omega\omega) \cap <_L$ is Lebesgue measurable. For any $y \in {}^\omega\omega \cap L$, $\{x \in {}^\omega\omega \cap L \mid x <_L y\}$ is countable and hence null, and so ${}^2({}^\omega\omega) \cap <_L$ is null by Fubini's Theorem 0.10. However, the same argument could be applied to its complement $\{\langle x, y\rangle \in {}^2({}^\omega\omega) \cap L \mid y <_L x \vee x = y\}$ to show that it too must be null, reaching a contradiction. An analogous argument using the Kuratowski-Ulam Theorem 0.13 shows that ${}^2({}^\omega\omega) \cap <_L$ does not have the Baire property. ⊣

Gödel [38] announced the Lebesgue measure result; the Baire property result seems to be first explicit in Mycielski [64:216]. The following refinement of 13.9, establishing in the parlance that in L, ${}^2({}^\omega\omega) \cap <_L$ is a Σ_2^1-*good* well-ordering, led to the further results of descriptive set theory in L.

13.11 Exercise (Addison [59]). *The relation $\mathrm{IS}_L \subseteq {}^2({}^\omega\omega)$ given by*

$$\mathrm{IS}_L(x, y) \ \leftrightarrow \ \{(x)_i \mid i \in \omega\} = \{z \in {}^\omega\omega \mid z <_L y\}$$

is Σ_2^1.

Hint. This is similar to 13.9. ⊣

The delimitative result for the perfect set property is more involved. Gödel never published a proof, and those published have relied on $\mathbf{\Pi}_1^1$ Uniformization. He once described to Levy the following incisive argument, one that has a wider applicability (20.21); Solovay also came to essentially the same argument (cf. Silver [71b: 440]). It can be rendered with just the 3.3(a) properties of σ_0 and φ_0 reviewed before 13.8.

13.12 Theorem. *Suppose that $\omega_1^L = \omega_1$. Then there is a Π_1^1 set of reals without the perfect set property.*

Proof. The idea is to get a Π_1^1 set of unique codes for ordinals less than ω_1^L. For such an ordinal α, there is an $x_0 \in {}^\omega\omega \cap L$ such that M_{x_0} is a well-ordered set with ordertype α. To specify the $<_L$-least such x_0 is not Π_1^1, but it is Π_1^1 to say of some $x_1 \in {}^\omega\omega \cap L$: $M_{x_1} \cong \langle L_\delta, \in\rangle$ for the least limit ordinal $\delta > \omega$ such that some $x_0 \in L_\delta$ is the $<_L|(L_\delta \times L_\delta)$-least possible as above. Again, to specify the $<_L$-least such x_1 is not Π_1^1, but it can be similarly specified in terms of some M_{x_2}, and so forth. The resulting $\langle x_i \mid i \in \omega\rangle$ can then serve as a unique code for α.

For $k \in \omega$ and $x \in {}^\omega\omega$, let $M_x|k$ be the initial segment of M_x determined by k, i.e. set

$$D_{x,k} = \{n \mid x(\langle\!\langle n, k\rangle\!\rangle) = 0\}\ , \text{ and}$$
$$M_x|k = \langle D_{x,k}, E_x|(D_{x,k} \times D_{x,k})\rangle\ .$$

Recall that each $x \in {}^\omega\omega$ encodes a countable sequence $\langle (x)_i \mid i \in \omega\rangle$ with each $(x)_i \in {}^\omega\omega$. The description above then leads to the following relation on ${}^\omega\omega$:

$$\begin{aligned} A(x) \;\leftrightarrow\; & M_{(x)_0} \text{ is well-ordered} \wedge \forall^0 i[M_{(x)_{i+1}} \text{ is well-founded} \\ & \text{and extensional} \wedge M_{(x)_{i+1}} \models \sigma_0 \wedge (x)_i \in \mathrm{tr}(M_{(x)_{i+1}}) \\ & \wedge \forall^0 k(M_{(x)_{i+1}}|k \models \sigma_0 \to (x)_i \notin \mathrm{tr}(M_{(x)_{i+1}}|k)) \\ & \wedge \forall^1 y \forall^0 p \forall^0 q[\pi_{(x)_{i+1}}(p) = y \wedge \pi_{(x)_{i+1}}(q) = (x)_i \\ & \wedge M_{(x)_{i+1}} \models \varphi_0[p,q] \to M_y \not\cong M_{(x)_i}]]\ . \end{aligned}$$

That $M_y \not\cong M_{(x)_i}$ can be cast as a Π_1^1 assertion $\forall^1 f(f$ does not code an isomorphism: $M_y \cong M_{(x)_i})$ (cf. the second proof of 13.4), and so applications of 13.8(a)(c) and a quantifier analysis show that A is Π_1^1.

If $A(x)$, there is a corresponding sequence $\langle \alpha_i^x \mid i \in \omega\rangle$ of ordinals less than ω_1^L such that, using the 3.3(a) properties of σ_0 and φ_0,

(a) $M_{(x)_0}$ has ordertype α_0^x, and for each $i \in \omega$, $M_{(x)_{i+1}} \cong \langle L_{\alpha_{i+1}^x}, \in\rangle$ and α_{i+1}^x is the least limit ordinal $> \omega$ such that $(x)_i \in L_{\alpha_{i+1}^x}$.
(b) For each $i \in \omega$, if $y <_L (x)_i$ (in $L_{\alpha_{i+1}^x}$ and hence in L), then $M_y \not\cong M_{(x)_i}$.

Moreover, it is simple to see that for any $\alpha < \omega_1^L$ there is a unique $x \in {}^\omega\omega$ such that $A(x)$ and $\alpha_0^x = \alpha$. Hence, A is a Π_1^1 set of cardinality $|\omega_1^L|$.

A cannot have a perfect subset; in fact, any $\mathbf{\Sigma}_1^1$ subset B of A is countable: $B' = \{(x)_0 \mid x \in B\}$ is also $\mathbf{\Sigma}_1^1$, and in terms of 13.6, $B' \subseteq \mathrm{WF}$ so that $B' \subseteq \mathrm{WF}_\beta$ for some $\beta < \omega_1$. This means that $\|(x)_0\| = \alpha_0^x \leq \beta$ for $x \in B$, and so by the uniqueness of the coding, $|B| \leq |\beta|$.

Finally, the assumption $\omega_1^L = \omega_1$ implies that A is uncountable, and so the proof is complete. ⊣

Guaspari [73], Kechris [75], and Sacks [76] independently showed that

$$C_1 = \{x \in {}^\omega\omega \mid x \in L_{\omega_1^x}\} ,$$

where ω_1^x is the least ordinal not recursive in x (i.e. not the ordertype of any well-ordering $\subseteq {}^2({}^\omega\omega)$ recursive in x), is also a Π^1_1 set of cardinality $|\omega_1^L|$ without a perfect subset. Moreover, they showed that *any* Π^1_1 set without a perfect subset is a subset of C_1, and provided detailed analyses which are generalized through the projective hierarchy by Kechris [75] under strong hypotheses.

For any $a \in {}^\omega\omega$, $L[a]$ has a canonical well-ordering $<_{L[a]}$ with properties analogous to $<_L$. Hence 13.8-13.12 all relativize to produce corresponding results about $L[a]$ and $\Pi^1_1(a)$, $\Delta^1_2(a)$, and $\Sigma^1_2(a)$ sets. Moreover, this relativization is uniform, so that for example $\{\langle x, a\rangle \in {}^2({}^\omega\omega) \mid x \in L[a]\}$ is Σ^1_2.

Analysis of $\mathbf{\Sigma}^1_2$ Sets

A basic tree representation of $\mathbf{\Sigma}^1_2$ sets implicit in Shoenfield [61] provided the backdrop for the analysis of their regularity properties. As this foreshadowed the development of similar representations of all projective sets under strong hypotheses, a general setting is developed. Recall that the projection of a set B is $pB = \{x \mid \exists y(\langle x, y\rangle \in B)\}$. For $A \subseteq {}^k({}^\omega\omega)$ and set Y,

$$A \text{ is } Y\text{-}Suslin \textit{ iff } \text{there is a tree } T \text{ on } {}^k\omega \times Y \text{ such that } A = p[T] .$$

This last can be further expressed in our tree terminology by: For $\mathbf{w} \in {}^k({}^\omega\omega)$,

$$\begin{aligned} A(\mathbf{w}) &\leftrightarrow [T_\mathbf{w}] \neq \emptyset \\ &\leftrightarrow T_\mathbf{w} \text{ is ill-founded} . \end{aligned}$$

In terms of a natural topology imposed on ${}^k({}^\omega\omega) \times {}^\omega Y$, a subset C is closed *iff* there is a tree on ${}^k\omega \times Y$ such that $C = [T]$ (cf. the proof of 12.10). Thus, being Y-Suslin is a generalization of being analytic. While it may be necessary to replace ω by other Y, such representations turn out to have significant consequences (see e.g. §30), and so in anticipation,

$$A \text{ is } Suslin \textit{ iff } A \text{ is } Y\text{-Suslin for some set } Y .$$

The following serves to frame the discussion, with all but (c) and (d) making substantial uses of AC.

13.13 Proposition. *Suppose that $A \subseteq {}^k({}^\omega\omega)$. Then:*

(a) A is Y-Suslin iff A is $|Y|$-Suslin.

(b) A is $2^{\aleph_0}$-Suslin.

(c) A is ω-Suslin iff A is $\mathbf{\Sigma}^1_1$.

(d) If $k > 1$ and A is κ-Suslin, then pA is also κ-Suslin. Consequently, if $n \in \omega$ and every $\mathbf{\Pi}^1_n$ set is κ-Suslin, then so is every $\mathbf{\Sigma}^1_{n+1}$ set.

(e) If A_ξ is a κ-Suslin subset of ${}^k({}^\omega\omega)$ for each $\xi < \kappa$, then so is $\bigcup_{\xi<\kappa} A_\xi$.

(f) (Martin) *For any $n > 0$, A is ω_n-Suslin iff A is a union of $\aleph_n$ Borel sets.*

(g) Suppose that $B \subseteq {}^l({}^\omega\omega)$ is Y-Suslin, $f\colon {}^k({}^\omega\omega) \to {}^l({}^\omega\omega)$ is continuous, and $A = f^{-1}(B)$. Then A is Y-Suslin.

Proof. (a) Only the cardinality of Y matters; a bijection: $Y \to |Y|$ induces an isomorphism between the corresponding trees.

(b) Assuming that $A \neq \emptyset$, let $A = \{a_\xi \mid \xi < 2^{\aleph_0}\}$. Then if

$$T = \{\langle a_\xi|m, m \times \{\xi\}\rangle \mid m \in \omega\} ,$$

it follows that $A = p[T]$.

(c) This follows from 12.10 and the definition of analytic sets; cf. 13.1.

(d) If a tree T on ${}^k\omega \times \kappa$ is such that $A = p[T]$, then that tree recast as T' on ${}^{k-1}\omega \times (\omega \times \kappa)$ is such that $pA = p[T']$. The result now follows as for (a) by applying a bijection: $\omega \times \kappa \to \kappa$.

(e) Suppose that for each $\xi < \kappa$ there is a tree T^ξ on ${}^k\omega \times \kappa$ such that $A_\xi = p[T^\xi]$. For $z \in {}^\omega\kappa$ define $z^+ \in {}^\omega\kappa$ by: $z^+(i) = z(i+1)$. Let T be that tree on ${}^k\omega \times \kappa$ such that

$$[T] = \{\langle \mathbf{w}, y\rangle \mid \langle \mathbf{w}, y^+\rangle \in T^{y(0)}\} .$$

Then

$$\begin{aligned} \mathbf{w} \in \textstyle\bigcup_{\xi<\kappa} A_\xi &\leftrightarrow \exists\xi < \kappa\exists y(\langle \mathbf{w}, y\rangle \in [T^\xi]) \\ &\leftrightarrow \exists z(\langle \mathbf{w}, z^+\rangle \in [T^{z(0)}]) \\ &\leftrightarrow \exists z(\langle \mathbf{w}, z\rangle \in [T]) . \end{aligned}$$

(f) For one direction, (c) and (e) imply that a union of κ Borel sets is κ-Suslin. For the converse, proceed by induction on n: Suppose that T is a tree on ${}^k\omega \times \omega_n$ with $n > 0$ and $A = p[T]$. For each $\gamma < \omega_n$, let $T^\gamma = T \cap ({}^k({}^{<\omega}\omega) \times {}^{<\omega}\gamma)$. Then $A = \bigcup_{\gamma<\omega_n} p[T^\gamma]$ since $\mathrm{cf}(\omega_n) > \omega$. But by (a), each $p[T^\gamma]$ is ω_{n-1}-Suslin. Consequently, if $n = 1$ then by (c) and 13.3, $p[T^\gamma]$ is a union of $\aleph_1$ Borel sets, and if $n > 1$, then $p[T^\gamma]$ is a union of $\aleph_{n-1}$ Borel sets by induction. In either case, it follows that A is a union of $\aleph_n$ Borel sets.

(g) Take $k = l = 1$ for notational simplicity. By the continuity of f, for any $t \in {}^{<\omega}\omega$ there is an $s \in {}^{<\omega}\omega$ such that $O(s) \subseteq f^{-1}(O(t))$. A sort of converse is needed, so first note that for any $s \in {}^{<\omega}\omega$,

$$W_s = \{t \in {}^{<\omega}\omega \mid O(s) \subseteq f^{-1}(O(t))\}$$

is nonempty as it contains $\emptyset$, and for $t, \bar{t} \in W_s$, either $t \subseteq \bar{t}$ or $\bar{t} \subseteq t$. So, let $\theta(s)$ be the largest $t \in W_s$ satisfying $|t| \leq |s|$.

For any $x \in {}^\omega\omega$,

$$f(x) = \textstyle\bigcup_{m\in\omega}\theta(x|m) .$$

The $\supseteq$ direction follows directly from the definition of θ. For the $\subseteq$ direction, suppose that $n \in \omega$. Since f is continuous and $x \in f^{-1}(O(f(x)|n))$, there must be an $m \in \omega$ such that $O(x|m) \subseteq f^{-1}(O(f(x)|n))$. Taking such an $m \geq n$, it follows that $f(x)|n \subseteq \theta(x|m)$.

Suppose now that $B = p[T]$ with T a tree on $\omega \times Y$, and let S be a tree on $\omega \times Y$ defined by:

$$S = \{\langle s, u\rangle \mid |s| = |u| \ \wedge \ \langle \theta(s), u||\theta(s)|\rangle \in T\} .$$

Then for any $x \in {}^{\omega}\omega$,

$$\begin{aligned} x \in p[S] &\leftrightarrow \exists g \in {}^{\omega}Y(\langle x, g\rangle \in [S]) \\ &\leftrightarrow \exists g \in {}^{\omega}Y\forall^0 m(\langle\theta(x|m), g||\theta(x|m)|\rangle \in T) \\ &\leftrightarrow \exists g \in {}^{\omega}Y(\langle f(x), g\rangle \in [T]) \\ &\leftrightarrow f(x) \in p[T] \\ &\leftrightarrow x \in A . \end{aligned}$$

⊣

With ZF the ambient theory the uses of AC can be made explicit by adding hypotheses, e.g. that Y is well-orderable in (a) and that the choice of trees $\langle T^{\xi} \mid \xi < \kappa\rangle$ is given beforehand in (e). (f) is a ZFC construction principle which will be applied later (15.13). (b) is merely a foil; the main interest is in Suslin representations established in ZF with trees having definability and homogeneity properties (as will become clearer in §15). With (c) as the starting point Shoenfield provided such a representation for $\mathbf{\Sigma}^1_2$ sets:

13.14 Theorem (Shoenfield [61]). *Every $\mathbf{\Sigma}^1_2$ set is ω_1-Suslin. In fact, if $a \in {}^{\omega}\omega$ and $A \subseteq {}^{k}({}^{\omega}\omega)$ is $\Sigma^1_2(a)$, there is a tree $U \in L[a]$ on ${}^{k}\omega \times \omega_1$ such that $A = p[U]$.*

Proof. The result is first established assuming that A is $\Pi^1_1(a)$. Taking $k = 1$ for notational simplicity, by 13.1 there is a tree $T \in L[a]$ on $\omega \times \omega$ such that for any $x \in {}^{\omega}\omega$,

$$A(x) \leftrightarrow T_x \text{ is well-founded} .$$

Consequently,

$$A(x) \leftrightarrow \exists g(g\colon T_x \to \omega_1 \text{ is an order-preserving map})$$

by 12.9(a).

This can be recast in terms of getting an infinite branch through a tree. Remembering that there is a fixed recursive enumeration $\langle \mathbf{s}_i \mid i \in \omega\rangle$ of ${}^{<\omega}\omega$ with $|\mathbf{s}_i| \leq i$, define a tree $\hat{T}$ on $\omega \times \omega_1$ by

$$\hat{T} = \{\langle s, u\rangle \mid \forall i, j < |s|(\mathbf{s}_i \supset \mathbf{s}_j \ \wedge \ \langle s||\mathbf{s}_i|, \mathbf{s}_i\rangle \in T \to u(i) < u(j))\} .$$

It is straightforward to see that $\hat{T} \in L[a]$, being definable from T and ω_1, and that for any $x \in {}^{\omega}\omega$,

$$\begin{aligned} A(x) &\leftrightarrow \exists g \in {}^{\omega}\omega_1(\langle x, g\rangle \in [\hat{T}]) \\ &\leftrightarrow x \in p[\hat{T}] . \end{aligned}$$

The full result for $\Sigma^1_2(a)$ sets now follows by the argument for 13.13(d), using a bijection: $\omega \times \omega_1 \leftrightarrow \omega_1$ in $L[a]$ at one point. $\dashv$

The ω_1 of this result is emphatically in the sense of V and could well differ from $\omega_1^{L[a]}$. Note that in the full AC context the converse,

$$\text{any } \omega_1\text{-Suslin set is } \mathbf{\Sigma}^1_2 ,$$

is independent by the discussion (after 13.7) of the proposition "any union of $\aleph_1$ Borel sets is $\mathbf{\Sigma}^1_2$" and the connecting 13.13(f).

The 13.14 representation leads to a proof of the classical 13.7 avoiding AC:

Second Proof of 13.7. Suppose that $A \subseteq {}^k({}^\omega\omega)$ is $\mathbf{\Sigma}^1_2$, and let U be a tree on ${}^k\omega \times \omega_1$ such that $A = p[U]$. For each $\gamma < \omega_1$ let $U^\gamma = U \cap ({}^k({}^{<\omega}\omega) \times {}^{<\omega}\gamma)$, so that $A = \bigcup_{\gamma<\omega_1} p[U^\gamma]$ as $\mathrm{cf}(\omega_1) > \omega$. This follows the proof of 13.13(f); by other parts of 13.13 not dependent on AC, each $p[U^\gamma]$ is ω-Suslin and hence $\mathbf{\Sigma}^1_1$. In any case, we can now proceed as in the proof of 13.3 to show, uniformly in γ, that for any $\alpha < \omega_1$,

$$\begin{aligned} C^\gamma_\alpha = \{\mathbf{w} \in {}^k({}^\omega\omega) \mid (U^\gamma_\mathbf{w} \text{ is well-founded and } \|U^\gamma_\mathbf{w}\| \leq \alpha) \vee \\ \exists s \in {}^{<\omega}\omega_1 (U^\gamma_\mathbf{w}/s \text{ is well-founded and } \|U^\gamma_\mathbf{w}/s\| = \alpha)\} \end{aligned}$$

is Borel, and arguing as in that proof but in terms of complements, $p[U^\gamma] = \bigcup_{\alpha<\omega_1}({}^k({}^\omega\omega) - C^\gamma_\alpha)$. Consequently,

$$A = \bigcup\nolimits_{\gamma<\omega_1} \bigcup\nolimits_{\alpha<\omega_1} ({}^k({}^\omega\omega) - C^\gamma_\alpha) . \qquad \dashv$$

Shoenfield established an important absoluteness result with this analysis of $\mathbf{\Sigma}^1_2$ sets. To discuss the absoluteness of analytical relations, first associate to each analytical A a definite normal form as given by 12.4. For instance, if $A \subseteq {}^\omega\omega$ is Σ^1_2, then fix a recursive $R \subseteq {}^4\omega$ such that for any $x \in {}^\omega\omega$,

$$A(x) \leftrightarrow \exists^1 y \forall^1 z \exists^0 m R(m, \overline{x}(m), \overline{y}(m), \overline{z}(m)) .$$

Let M be a transitive class such that for any $r \in \omega$ and recursive $S \subseteq {}^r\omega$, $S \in M$; this certainly holds if M is an $\in$-model of ZF. Then A^M denotes the relativization of (the normal form of) A to M. For the Σ^1_2 A above and $x \in {}^\omega\omega \cap M$,

$$A^M(x) \leftrightarrow (\exists y \in {}^\omega\omega \cap M)(\forall z \in {}^\omega\omega \cap M)\exists^0 m R(m, \overline{x}(m), \overline{y}(m), \overline{z}(m)) .$$

Finally,

$$A \text{ is } \textit{absolute for } M \textit{ iff } \text{for any } \mathbf{u} \in M,\ A(\mathbf{u}) \leftrightarrow A^M(\mathbf{u}) .$$

For any $a \in {}^\omega\omega$, these concepts relativize directly to relations analytical in a and M such that $a \in M$. Here is Shoenfield's Σ^1_2 *Absoluteness Theorem*, parsimoniously stated; as before, ω_1 is in the sense of V.

13.15 Theorem (Shoenfield [61]). *There is a theorem $\hat{\sigma}$ of* ZF *such that for any $a \in {}^\omega\omega$ and M a transitive $\in$-model of $\hat{\sigma}$ with $\omega_1 \cup \{a\} \subseteq M$, every $\Sigma^1_2(a)$ (and hence $\Pi^1_2(a)$) relation is absolute for M.*

Proof. The proof is carried out assuming that M is an inner model of ZFC with $a \in M$, the necessary refinements articulated later. For notational simplicity consider a $\Sigma^1_2(a)$ relation $A \subseteq {}^\omega\omega$.

First, let T be a tree on ${}^2\omega \times \omega$ such that for any $x \in {}^\omega\omega$,

$$A(x) \leftrightarrow \exists^1 y(T_{\langle x,y\rangle} \text{ is well-founded}) ,$$

where by 13.1 it can be assumed that $T^M = T$. Now if $x \in M$ and $A^M(x)$, there is a $y \in M$ such that $T_{\langle x,y\rangle}$ is well-founded in the sense of M. It follows by 12.9(a) that in M there is an order-preserving map: $T_{\langle x,y\rangle} \to$ On. Of course, the map also exists in V, so $A(x)$ holds.

The converse uses the 13.14 representation of $\Sigma^1_2(a)$ sets. Let $U \in L[a] \subseteq M$ be a tree on $\omega \times \omega_1$ such that for any $x \in {}^\omega\omega$,

$$A(x) \leftrightarrow U_x \text{ is ill-founded} .$$

Now if $x \in M$ and $A(x)$, the ill-foundedness of U_x in V implies by 12.9(a) that there is no order-preserving map: $U_x \to \omega_1$, and so no such map can exist in M either. Thus, U_x is ill-founded in the sense of M, and using 13.14 there $A^M(x)$ holds.

To complete the proof, note first that M need only model a sufficiently strong theorem $\hat{\sigma}$ of ZF. Neither the uses of 12.9(a) nor of 13.14 require DC, as it is not needed in the equivalence between well-foundedness and the lack of infinite descending chains for well-orderable trees. Also, it can be checked that U need only be a definable class in M, for which $\omega_1 \subseteq M$ suffices. ⊣

The mathematical significance of this result lies in the fact that many propositions of classical analysis are $\mathbf{\Sigma}^1_2$, and so if they can be established assuming for example $V = L$ (and hence CH), then they can be established in ZF alone!

The result is the best possible in the projective hierarchy, since $\exists^1 x(x \notin L)$ is Σ^1_3 and cannot be absolute for $M = L$. Inner models involving large cardinals were to be devised that possess stronger absoluteness properties (see volume II).

13.16 Corollary. *If $r \in \omega$ and $A \subseteq {}^r\omega$ is $\Sigma^1_2(a)$, then $A \in L[a]$.*

Proof. Although we have been working with tree representations of subsets of ${}^k({}^\omega\omega)$, the preceding argument works with only notational changes for $A \subseteq {}^r\omega$. Consequently, $A^{L[a]} = A \cap L[a] = A$, so $A \in L[a]$. ⊣

The idea behind 13.14 also provides a straightforward approach to the proof of Kondô's $\mathbf{\Pi}^1_1$ Uniformization Theorem 12.3, which as Addison observed can be effectivized.

13.17 Theorem (Kondô [37, 39]; Addison). *For $a \in {}^{\omega}\omega$, any $\Pi^1_1(a)$ subset of ${}^2({}^{\omega}\omega)$ can be uniformized by a $\Pi^1_1(a)$ set.*

Proof. Suppose that $A \subseteq {}^2({}^{\omega}\omega)$ is $\Pi^1_1(a)$. The task is to find a $\Pi^1_1(a)$ set $A_0 \subseteq A$ such that for any $x \in {}^{\omega}\omega$, if there is a y such that $A(x, y)$, then there is exactly one y such that $A_0(x, y)$. Let T be a tree over ${}^2\omega \times \omega$ as in 13.1 such that

$$A(x, y) \leftrightarrow T_{x,y} \text{ is well-founded },$$

where $T_{x,y}$ is written for $T_{\langle x,y\rangle}$ from now on for the sake of convenience. The argument for 13.14 provides a corresponding tree $\hat{T}$ on ${}^2\omega \times \omega_1$ such that

$$A(x, y) \leftrightarrow \exists g \in {}^{\omega}\omega_1(\langle x, y, g\rangle \in [\hat{T}]) ,$$

where g provides an order-preserving map: $T_{x,y} \to \omega_1$ verifying the well-foundedness of $T_{x,y}$.

For any $x, y \in {}^{\omega}\omega$, if there is a g such that $\langle x, y, g\rangle \in [\hat{T}]$ at all, then as for any ill-founded tree, there is a lexicographically least ("leftmost") one, i.e. a g_0 such that for any other g, if i is least such that $g_0(i) \neq g(i)$, then $g_0(i) < g(i)$. However, for $\hat{T}$ there is an "honest leftmost" one, i.e. a $g_{x,y}$ such that for any other g, $g_{x,y}(i) \leq g(i)$ for *every* $i \in \omega$: By 12.9(a) and the definition of $\hat{T}$, we can take

$$g_{x,y}(i) = \begin{cases} \rho_{T_{x,y}}(\mathbf{s}_i) & \text{if } \mathbf{s}_i \in T_{x,y} \text{ , and} \\ 0 & \text{otherwise .} \end{cases}$$

Next, for any $x \in {}^{\omega}\omega$, suppose that there is a y such that $A(x, y)$, i.e. a pair $\langle y, g\rangle$ such that $\langle x, y, g\rangle \in [\hat{T}]$. Then there is a lexicographically least ("leftmost") one, i.e. a pair $\langle y_x, g_x\rangle$ such that for any other pair $\langle y, g\rangle$, if i is least such that $\langle y_x(i), g_x(i)\rangle \neq \langle y(i), g(i)\rangle$, then either $y_x(i) < y(i)$ or else $y_x(i) = y(i)$ and $g_x(i) < g(i)$. The uniformization is carried out by choosing y_x. That this can be done in a $\Pi^1_1(a)$ fashion relies on the observation that g_x must be the honest leftmost g_{x,y_x}.

Turning to formalities, because of how honest leftmost branches can be defined let $R^{\leq}$ and $R^{<}$ be relations on $\omega \times {}^3({}^{\omega}\omega)$ given by

$$R^{\leq}(i, x, y, z) \leftrightarrow T_{x,z}/\mathbf{s}_i \text{ is ill-founded} \vee \rho_{T_{x,y}}(\mathbf{s}_i) \leq \rho_{T_{x,z}}(\mathbf{s}_i) \text{ , and}$$
$$R^{<}(i, x, y, z) \leftrightarrow T_{x,z}/\mathbf{s}_i \text{ is ill-founded} \vee \rho_{T_{x,y}}(\mathbf{s}_i) < \rho_{T_{x,z}}(\mathbf{s}_i) \text{ .}$$

Remembering that $\rho_T(t) = \|T/t\|$ for well-founded trees T and $t \in T$, by 12.9,

$$R^{\leq}(i, x, y, z) \leftrightarrow \exists f(f\colon T_{x,y}/\mathbf{s}_i \to T_{x,z}/\mathbf{s}_i \text{ is order-preserving}) \text{ , and}$$
$$R^{<}(i, x, y, z) \leftrightarrow \exists f \exists^0 k(f\colon T_{x,y}/\mathbf{s}_i \to T_{x,z}/\mathbf{s}_i{}^\frown\langle k\rangle \text{ is order-preserving}) \text{ .}$$

Since by 13.1, $\{\langle i, x, y\rangle \mid \mathbf{s}_i \in T_{x,y}\}$ is recursive in a, $R^{\leq}$ and $R^{<}$ are thus seen to be $\Sigma^1_1(a)$ (cf. the second proof of 13.4).

Now define $A_0 \subseteq {}^2({}^\omega\omega)$ by

$$\begin{aligned} A_0(x, y) \quad \leftrightarrow \quad & A(x, y) \;\wedge\; \forall^1 z \forall^0 m[(\overline{y}(m) = \overline{z}(m) \\ & \wedge\; \forall^0 i < m(R^{\leq}(i, x, y, z) \;\wedge\; R^{\leq}(i, x, z, y)) \\ & \rightarrow (y(m) < z(m) \;\vee\; (y(m) = z(m) \;\wedge\; \neg R^{<}(m, x, z, y))))] \,. \end{aligned}$$

This is essentially of form $\Pi^1_1(a) \wedge \forall^1[\Sigma^1_1(a) \rightarrow \neg\Sigma^1_1(a)]$, and hence $\Pi^1_1(a)$. To decipher it, note that $R^{\leq}(i, x, y, z) \wedge R^{\leq}(i, x, z, y)$ is equivalent to

$$(T_{x,y}/\mathbf{s}_i \text{ and } T_{x,z}/\mathbf{s}_i \text{ are both ill-founded } \vee \; \rho_{T_{x,y}}(\mathbf{s}_i) = \rho_{T_{x,z}}(\mathbf{s}_i)) \,.$$

$\neg R^{<}(m, x, z, y)$ is equivalent to

$$(T_{x,y} \text{ is well-founded } \wedge \; \rho_{T_{x,z}}(\mathbf{s}_m) \not< \rho_{T_{x,y}}(\mathbf{s}_m)) \,,$$

which in turn is equivalent to

$$(T_{x,y} \text{ is well-founded } \wedge \; \rho_{T_{x,y}}(\mathbf{s}_m) \leq \rho_{T_{x,z}}(\mathbf{s}_m)) \,.$$

Hence, for any $x \in {}^\omega\omega$, through honest leftmost branches $A_0(x, y)$ does indeed pick out that y of the leftmost pair $\langle y, g\rangle$ such that $\langle x, y, g\rangle \in [\hat{T}]$ if there is one, and so the proof is complete. ⊣

Elements of this proof, particularly the simple definability of the honest leftmost branch, were later abstracted in the concept of a *scale* (cf. 29.2). Adjoining an $\exists^1$ to the result is straightforward:

13.18 Exercise (Kondô [39]). *For $a \in {}^\omega\omega$, any $\Sigma^1_2(a)$ subset of ${}^2({}^\omega\omega)$ can be uniformized by a $\Sigma^1_2(a)$ set.* ⊣

This is the limit of directly derivable results about uniformization. Addison settled the issue in L:

13.19 Exercise (Addison [59]). *If ${}^\omega\omega \subseteq L$, then for any $a \in {}^\omega\omega$ and $2 \leq n \in \omega$, every $\Sigma^1_n(a)$ subset of ${}^2({}^\omega\omega)$ can be uniformized by a $\Sigma^1_n(a)$ set.*

Hint. Note that if $A \subseteq {}^2({}^\omega\omega)$ is $\Pi^1_{n-1}(a)$, then in terms of 13.11,

$$A_0(x, y) \;\leftrightarrow\; A(x, y) \;\wedge\; \exists^1 u(\mathrm{IS}_L(u, y) \wedge \forall^0 i \neg A(x, (u)_i))$$

is $\Sigma^1_n(a)$, and $\exists y A(x, y) \leftrightarrow \exists! y A_0(x, y)$. ⊣

However, Levy [65a] observed that if one adjoins a Cohen real to L then in the resulting extension there is a Π^1_2 set that cannot be uniformized by any analytical set (see also Levy [70]). Moreover, if one adjoins uncountably many Cohen reals to L, then in the resulting extension there is a Π^1_2 set (namely $\{\langle x, y\rangle \mid y \notin L[x]\}$) that cannot be uniformized by any projective set. Strong hypotheses were later to settle these issues in different ways (15.14, 29.9); in fact, they breathed new life into the study of structural properties of classical vintage (§§28, 29).

14. $\mathbf{\Sigma}^1_2$ Sets and Sharps

In 1965 Solovay reactivated the classical program of investigating the extent of the regularity properties by providing characterizations for the $\mathbf{\Sigma}^1_2$ sets, the level of Gödel's delimitative results. He also investigated the definability of the transcendent $0^{\#}$ and made several related conjectures bearing on the scope of the forcing method that inspired new avenues of research. This section features these developments, which served to establish the relevance of large cardinal hypotheses to the study of the projective hierarchy.

Building on his Lebesgue result 11.1, Solovay used forcing to establish characterizations for the measurability of the $\mathbf{\Sigma}^1_2$ sets. Through this and other results, forcing soon became established as a basic *construction* technique within set theory.

14.1 Theorem (Solovay). *The following are equivalent for* $a \in {}^{\omega}\omega$:
(a) *Every* $\Sigma^1_2(a)$ *set of reals is Lebesgue measurable.*
(b) $\bigcup\{A_c \mid c \in L[a]$ is a G_δ code for a null set$\}$ *is null.*
(c) $\{x \mid x$ is not random over $L[a]\}$ *is null.*
(d) *Every set of form* $\{x \in {}^{\omega}\omega \mid L[a][x] \models \varphi[p, x]\}$ *for some formula* φ *and* $p \in L[a]$ *is Lebesgue measurable.*

Proof. By 11.10, (b) and (c) are synoptic and imply (d) by the argument for 11.11 with V replaced by L. (d) $\to$ (a) follows from $\mathbf{\Sigma}^1_2$ absoluteness 13.15.

Finally, for (a) $\to$ (b), let B be the set described in (b). If $C \subseteq {}^2({}^{\omega}\omega)$ is given by

$$C(x, c) \leftrightarrow c \text{ is a } G_\delta \text{ code for a null set } \land\ x \in A_c\ ,$$

then it is straightforward to check that C is arithmetical. Since

$$x \in B \leftrightarrow \exists^1 c(c \in L[a] \land C(x, c))\ ,$$

B is $\Sigma^1_2(a)$ by 13.9 relativized. B is therefore measurable by assumption, and it remains to verify that it is null. For this purpose, an elaboration of the well-ordering argument for 13.10 is used:

First define $D \subseteq {}^2({}^{\omega}\omega)$ by

$$D(x, c) \leftrightarrow c \in L[a] \land C(x, c) \land \forall^1 d(d <_{L[a]} c \to \neg C(x, d))\ .$$

Then define $R \subseteq {}^2({}^{\omega}\omega)$ by

$$R(x, y) \leftrightarrow \exists^1 c \exists^1 d(D(x, c) \land D(y, d) \land (c <_{L[a]} d \lor c = d))\ .$$

Thus, R orders the x's in B according to the $<_{L[a]}$-least G_δ code c for a null set such that $x \in A_c$. R is a *prewellordering*, i.e. a well-ordering except that there may be distinct x and y such that $R(x, y)$ and $R(y, x)$.

By 13.11 relativized D is $\Sigma^1_2(a)$, and so also is R. Hence, R is measurable by assumption. Since ${}^2({}^{\omega}\omega) \cap <_{L[a]}$ has ordertype $\omega_1^{L[a]} \leq \omega_1$, for any y,

$R_y = \{x \mid R(x, y)\}$ is null, being a countable union of null sets. Thus, by Fubini's Theorem 0.10, R is null. However, the same argument can be applied to the complement $\{\langle x, y\rangle \in B \times B \mid \neg R(x, y)\}$, a measurable set since B is, to show that it too must be null. But then, $B \times B$ is null, and so also must be B. ⊣

An analogous argument based on remarks in the proof of 11.11 about the Baire property establishes the following:

14.2 Theorem (Solovay). *The following are equivalent for $a \in {}^\omega\omega$:*

(a) Every $\Sigma^1_2(a)$ set of reals has the Baire property.

(b) $\bigcup\{A_c \mid c \in L[a]$ is an $F_{\sigma\delta}$ code for a meager set$\}$ *is meager.*

(c) $\{x \mid x$ not a Cohen real over $L[a]\}$ *is meager.*

(d) Every set of form $\{x \in {}^\omega\omega \mid L[a][x] \models \varphi[p, x]\}$ for some formula φ and $p \in L[a]$ has the Baire property. ⊣

14.3 Corollary. *If $\omega_1^{L[a]} < \omega_1$, then every $\Sigma^1_2(a)$ set of reals is Lebesgue measurable and has the Baire property. In particular, if there is a measurable cardinal, then every $\mathbf{\Sigma}^1_2$ set of reals is Lebesgue measurable and has the Baire property.*

Proof. If ${}^\omega\omega \cap L[a]$ is countable, then the (b)'s of the theorems obtain. The second assertion follows from the first, since by 9.19, ω_1 is inaccessible in $L[a]$ for every $a \in {}^\omega\omega$. ⊣

Solovay's ideas were recast by Moschovakis [80: 8G] to provide a forcing-free proof of this corollary. While the statement about measurable cardinals is not very sharp, it is a direct implication between classical concepts that provided a counterweight to Gödel's $V = L$ delimitations for definable sets of reals. In this way, Solovay established that although measurable cardinals do not decide the Continuum Hypothesis, they do have a direct mathematical effect on the reals. On the other hand, Silver was to show that even measurable cardinals do not entail the Lebesgue measurability of Δ^1_3 sets (20.19). One of the prominent themes of the pursuit of stronger hypotheses is the measurability of more and more projective sets.

The *consistency* of ZFC plus the proposition that every $\mathbf{\Sigma}^1_2$ set is Lebesgue measurable and has the Baire property does not require large cardinal hypotheses. Martin-Solovay [70: §4] showed that it follows from Martin's Axiom (MA) + ¬CH, and moreover, by an ω_1-length iteration of the forcing described there, it can be made consistent with CH. On the other hand, Shelah [84] was to show that the Lebesgue measurability of every $\mathbf{\Sigma}^1_3$ set does require the consistency strength of an inaccessible cardinal (see volume II), complementing Solovay's result 11.1.

Solovay also established outright the Lebesgue measurability of sets in a substantial subclass of $\mathbf{\Delta}^1_2$. To describe this class, we have to look more carefully at the defining formulas for relations.

For $a \in {}^\omega\omega$ and $A \subseteq {}^\omega\omega$, A is *provably* $\Delta^1_2(a)$ *iff* there are $\mathcal{A}^2$ formulas $\psi(v^1_1, v^1_2, v^1_3, v^1_4)$ and $\varphi(v^1_1, v^1_2, v^1_3, v^1_4)$ without function quantifiers such that for any $x \in {}^{\omega}\omega$, $A(x)$ *iff* $\mathcal{A}^2 \models \exists^1 v_3^1 \forall^1 v_4^1 \psi[x, a]$ *iff* $\mathcal{A}^2 \models \forall^1 v_3^1 \exists^1 v_4^1 \varphi[x, a]$, and moreover, provable in ZFC is:

$$\mathcal{A}^2 \models \forall^1 v_1^1 \forall^1 v_2^1 (\exists^1 v_3^1 \forall^1 v_4^1 \psi \leftrightarrow \forall^1 v_3^1 \exists^1 v_4^1 \varphi) \; .$$

(The definition extends analogously to $A \subseteq {}^r\omega \times {}^k({}^{\omega}\omega)$ and also to the corresponding concepts *provably* Δ_2^1 and *provably* $\mathbf{\Delta}_2^1$.) Thus, A is $\Delta_2^1(a)$ in the strong sense that there are $\Sigma_2^1(a)$ and $\Pi_2^1(a)$ formulations whose equivalence is a theorem of ZFC. For example, $V = L$ implies that ${}^2({}^{\omega}\omega) \cap <_L$ is Δ_2^1, but it is not provably Δ_2^1. By introducing vacuous quantifiers, every $\Sigma_1^1(a)$ relation is provably $\Delta_2^1(a)$. Andrei Kolmogorov [28] introduced the *R-sets* by generalizing the Operation (A) of Suslin [17] into a transfinite hierarchy, and showed that they are Lebesgue measurable. Recently John Burgess [82, 83] and Rana Barua [84] have redeveloped and extended the theory of these sets. It can be shown that the R-sets are provably $\mathbf{\Delta}_2^1$. Thus, the following subsumes 12.2(a)(b) and more:

14.4 Exercise (Solovay – cf. Fenstad-Normann [74]). *Every provably* $\mathbf{\Delta}_2^1$ *set of reals is Lebesgue measurable and has the Baire property.*

Hint. Suppose that A is provably $\Delta_2^1(a)$ for some $a \in {}^{\omega}\omega$, and let τ be a theorem of ZFC establishing the equivalence between a Σ_2^1 formulation and a Π_2^1 formulation of A. If $\hat{\sigma}$ is the theorem of ZF of 13.15, then any transitive $\in$-model M of $\hat{\sigma} \wedge \tau$ with $a \in M$ and containing the recursive relations is absolute for A. (The point is that $\omega_1 \subseteq M$ as in 13.15 is no longer required, as the Π_2^1 formulation can be used contrapositively in place of the argument with the tree over $\omega \times \omega_1$.) Now work with a countable such M satisfying finitely more axioms of ZFC so that the arguments for (b) $\rightarrow$ (a) of 14.1 and 14.2 can be used with M in place of $L[a]$.⊣

Finally, recent characterizations are provided that further attest to the intrinsic importance of random and Cohen reals in descriptive set theory: They complement 14.1 and 14.2, inform on Gödel's 13.10, and could well have been noticed much earlier. First, a preparatory exercise:

14.5 Exercise. *Suppose that* M *is an inner model of* ZFC. *Then:*

(a) If there is a real random over M*, then such reals are dense: For any closed code* $c \in M$ *for a non-null set there is a real in* A_c *random over* M.

(b) If there is a Cohen real over M*, then such reals are dense: For any* $s \in {}^{<\omega}\omega$ *there is a Cohen real* $\supseteq s$ *over* M.

Hint. (b) is much easier than (a), and both can be deduced from the fact that the notions of forcing are weakly homogeneous (cf. 10.19). However, this fact for random real forcing depends on involved results about Lebesgue measure, so a direct approach is outlined for (a):

Let x be random over M. The idea is to develop a permutation $p\colon \omega \rightarrow \omega$ so that the induced homeomorphism $\overline{p}\colon {}^{\omega}\omega \rightarrow {}^{\omega}\omega$ (defined by $\overline{p}(f)(n) = f(p(n))$

for $n \in \omega$) satisfies $\overline{p}(x) \in A_c$. Such induced homeomorphisms preserve measure, so $\overline{p}(x)$ will also be random over M. To begin with, it can be assumed that

$$A_c = {}^\omega\omega - \bigcup\{O(\mathbf{s}_{i_j}) \mid j \in \omega\}$$

where the $O(\mathbf{s}_{i_j})$'s are pairwise disjoint. p will be comprised of transpositions $p(a_m) = b_m$ and $p(b_m) = a_m$ for a_m's and b_m's all distinct and exhausting ω, defined in pairs $\langle a_m, b_m\rangle$ by recursion as follows:

Suppose that the a_m's and b_m's for $m < n$ have been defined, recursively satisfying

$$m_L(\bigcup\{O(\mathbf{s}_{i_j}) \mid \mathbf{s}_{i_j} \supseteq \{\langle a_0, x(b_0)\rangle, \dots, \langle a_{n-1}, x(b_{n-1})\rangle\}\}) < \prod_{m<n} 2^{-(x(b_m)+1)} .$$

(The basis $n = 0$ is included vacuously here with the value on the right being 1, a consequence of the assumption that A_c is not null.) Let a_n be the least member of ω different from all the a_m's and b_m's for $m < n$. (So $a_0 = 0$.) There must be a $k \in \omega$ such that

$$\begin{aligned} &m_L(\bigcup\{O(\mathbf{s}_{i_j}) \mid \mathbf{s}_{i_j} \supseteq \{\langle a_0, x(b_0)\rangle, \dots, \langle a_{n-1}, x(b_{n-1})\rangle, \langle a_n, k\rangle\}\}) \\ &\quad < 2^{-(k+1)} \cdot \prod_{m<n} 2^{-(x(b_m)+1)} . \end{aligned}$$

Since x is random, choose $b_n \neq a_n$ and different from all the a_m's and b_m's for $m < n$ such that $x(b_n) = k$. Thus, the recursion assumption is perpetuated. It can be checked that this construction insures that $\overline{p}(x) \in A_c$. ⊣

14.6 Theorem (Ihoda-Shelah [89]). *Suppose that $a \in {}^\omega\omega$. Then:*

(a) Every $\Delta^1_2(a)$ set of reals is Lebesgue measurable iff there is a real random over $L[a]$.

(b) Every $\Delta^1_2(a)$ set of reals has the Baire property iff there is a Cohen real over $L[a]$.

Proof. As usual, only the measure result is established. The forward direction follows the proof of (a) → (b) of 14.1, noting that the R in that proof would actually be $\Delta^1_2(a)$ if there were no reals random over $L[a]$ (cf. 13.10).

The converse is a variation of the argument for 14.4. Suppose that there is a real random over $L[a]$ and that $A \subseteq {}^k({}^\omega\omega)$ is $\Delta^1_2(a)$. Taking $k = 1$ for simplicity, there are consequently $\Sigma^1_2(a)$ relations $R_0, R_1 \subseteq {}^\omega\omega$ such that for any $x \in {}^\omega\omega$,

$$A(x) \leftrightarrow R_0(x) \leftrightarrow \neg R_1(x) .$$

As usual, if M is a transitive class containing all recursive relations, then R_i^M denotes the relativization of (the normal form of) R_i to M. Let $P = (\mathcal{B}^*)^{L[a]}$, the random real forcing in the sense of $L[a]$, and $\dot{r}$ a P-name for the random real. Then

$$D = \{p \in P \mid p \Vdash R_i^{L[a][\dot{r}]}(\dot{r}) \text{ for exactly one } i < 2\} \in L[a]$$

is dense in P by the following argument in V: Starting with an arbitrary member of P, which for demonstrating density can be taken to be of form $A_c^{L[a]}$ for some closed code $c \in L[a]$, there is an $x \in A_c$ random over $L[a]$ by 14.5(a). There is exactly one $i < 2$ such that $R_i(x)$, whereupon $\mathbf{\Sigma}^1_2$ Absoluteness 13.15 implies that $R_i^{L[a][x]}(x)$. Hence, there is a $p \leq A_c^{L[a]}$ such that $p \in D$.

Since D is dense *open* in P, there is in $L[a]$ a maximal antichain $E \subseteq D$ consisting of closed sets, which by the ω_1-c.c. is countable. For $i < 2$, set

$$\begin{aligned} Y_i &= \{c \in {}^\omega\omega \mid c \in L[a] \text{ is a closed code } \wedge \\ &\qquad\qquad A_c^{L[a]} \in E \ \wedge \ A_c^{L[a]} \Vdash R_i^{L[a][\dot{r}]}(\dot{r})\}\ , \text{ and} \\ Z_i &= \bigcup\{A_c \mid c \in Y_i\}\ . \end{aligned}$$

It now suffices to verify that $Z_i - R_i$ is null for $i < 2$. Granted this and invoking the absoluteness 11.8 of codes, that

(i) $Z_0 \cap Z_1$ is null since $Y_0 \cap Y_1 = \emptyset$, and
(ii) $m_L(Z_0) + m_L(Z_1) = 1$ by the maximality of E

will together imply that $Z_0 \triangle R_0$ is null. Hence, it can be concluded that $R_0 = A$ is measurable.

To verify that $Z_i - R_i$ is null for $i < 2$, let $M \in L[a]$ be a countable transitive $\in$-model of enough of ZFC with $Y_0, Y_1 \in M$. (It will suffice to take $M \prec L_\lambda[a]$ for some sufficiently large regular λ, and such M model ZFC except possibly for the Power Set Axiom. In what follows, pretend that $M \models$ ZFC for the use of established terminology.) Since M is countable, the set of reals *not* random over M is null (cf. 11.10). If on the other hand $x \in {}^\omega\omega$ is random over M and $x \in Z_i$, then $x \in A_c$ for some $c \in Y_i$ and so $x \in A_c^{M[x]}$. Hence, $R_i^{M[x]}(x)$ by genericity, and so by the upward direction of $\mathbf{\Sigma}^1_2$ Absoluteness 13.15 (which does not require $\omega_1 \subseteq M$) it follows that $R_i(x)$. This confirms what was to be verified, and so the proof is complete. ⊣

As observed by Solovay, if $\aleph_1$ many random reals are generically added to $L[a]$, then the $\Sigma^1_2(a)$ set ${}^\omega\omega \cap L[a]$ is not Lebesgue measurable in the extension (see Jech [03: 535ff]). Thus, the measurability of every $\Delta^1_2(a)$ set does not imply the measurability of every $\Sigma^1_2(a)$ set. Analogous remarks hold for the Baire property. As to the difference in the proofs of 14.4 and 14.6, the countable M of the former is absolute for A by provability, while it was necessary to incorporate a maximal antichain into the M of the latter, which in turn depended on the property that $R_0(x) \leftrightarrow \neg R_1(x)$ holds for all x random over $L[a]$.

Solovay also characterized the perfect set property for $\mathbf{\Sigma}^1_2$ sets by a forcing argument as in 11.11. Then an elegant result of Richard Mansfield revealed the structural essence. It too was proved initially by forcing, but Solovay provided a proof that can be seen as an elaboration of the Cantor-Bendixson argument (11.3(a)):

14.7 Theorem (Mansfield [70]). *Suppose that for some infinite Y, T is a tree on $\omega \times Y$ such that $T \in L[T]$, $A \subseteq {}^\omega\omega$, and $A = p[T]$, i.e. for any $x \in {}^\omega\omega$,*

$$A(x) \leftrightarrow \exists f \in {}^{\omega}Y(\langle x, f\rangle \in [T]) .$$

Then either $A \subseteq L[T]$ and $|A| \leq |T|$, or else A has a perfect subset.

Proof. For any tree U on $\omega \times Y$, let U' consist of those members of U having extensions in U with incompatible first coordinates, i.e.

$$U' = \{\langle s, h\rangle \in U \mid \exists \langle s_0, h_0\rangle, \langle s_1, h_1\rangle \in U(\langle s_0, h_0\rangle \supseteq^* \langle s, h\rangle \\ \wedge \langle s_1, h_1\rangle \supseteq^* \langle s, h\rangle \wedge |s_0| = |s_1| \wedge s_0 \neq s_1)\} .$$

Now set $T_0 = T$, $T_{\alpha+1} = T'_\alpha$, and for limit ordinals $\delta > 0$, $T_\delta = \bigcap_{\alpha<\delta} T_\alpha$. It is clear that each $T_\alpha \in L[T]$. There is a $\beta < |Y|^+$ such that $T_{\beta+1} = T_\beta$; set $T^* = T_\beta$ for the least such β.

If $T^* = \emptyset$, then $A \subseteq L[T]$ and $|A| \leq |T|$: Suppose that $x \in {}^{\omega}\omega$ satisfies $A(x)$, so that there is some $f \in {}^{\omega}Y$ with $\langle x, f\rangle \in [T]$. Clearly there is an $\alpha < \beta$ such that $\langle x, f\rangle \in [T_\alpha] - [T_{\alpha+1}]$. But then, there is an $m \in \omega$ such that $\langle x|m, f|m\rangle \in T_\alpha - T_{\alpha+1}$, so that any $\langle s, h\rangle \in T_\alpha$ with $s \supseteq x|m$ and $h \supseteq f|m$ satisfies $s \subseteq x$. Hence,

$$x = \bigcup\{s \mid \exists h(s \supseteq x|m \wedge h \supseteq f|m \wedge \langle s, h\rangle \in T_\alpha)\} ,$$

and so $x \in L[T]$. Thus, $A \subseteq L[T]$. This analysis shows that any $x \in {}^{\omega}\omega$ satisfying $A(x)$ is determined by an $\alpha < \beta$ and a pair $\langle x|m, f|m\rangle \in T_\alpha$. Hence also $|A| \leq |\beta|\cdot|T| = |T|$.

The proof is completed by showing that if $T^* \neq \emptyset$, then A has a perfect subset: Starting with a member of T^* and using $T^{*\prime} = T^*$, recursively define $s_t \in {}^{<\omega}\omega$ for each $t \in {}^{<\omega}2$ such that for any $t, u \in {}^{<\omega}2$,

(i) $\exists h(\langle s_t, h\rangle \in T^*)$,
(ii) if $t \subset u$, then $s_t \subset s_u$, and
(iii) if $|t| = |u|$ yet $t \neq u$, then $s_t(i) \neq s_u(i)$ for some $i \in \mathrm{dom}(s_t) \cap \mathrm{dom}(s_u)$.

Then $S = \{s \in {}^{<\omega}\omega \mid \exists t \in {}^{<\omega}2(s \subseteq s_t)\}$ is a tree on ω, $[S] \subseteq A$, and building on 12.10, $[S]$ is perfect. ⊣

The following corollaries are immediate from the tree representations of $\mathbf{\Sigma}^1_1$ sets (13.13(c)) and $\mathbf{\Sigma}^1_2$ sets (13.14).

14.8 Corollary (Suslin – Luzin [17]). *Every $\mathbf{\Sigma}^1_1$ set of reals has the perfect set property.* ⊣

14.9 Corollary (Solovay [69]). *If $a \in {}^{\omega}\omega$ and A is a $\Sigma^1_2(a)$ set of reals, then either $A \subseteq L[a]$ or A has a perfect subset.* ⊣

Recalling the set C_1 defined after 13.12, 14.9 implies that if $\omega_1^{L[a]} < \omega_1$, then the countable set

$$C_2^a = {}^{\omega}\omega \cap L[a]$$

is a $\Sigma^1_2(a)$ set such that any $\Sigma^1_2(a)$ set without a perfect subset is a subset of C_2^a. Thus:

14.10 Theorem (Solovay [69]). *The following are equivalent for any $a \in {}^{\omega}\omega$:*
(a) Every $\Sigma^1_2(a)$ set of reals has the perfect set property.
(b) Every $\Pi^1_1(a)$ set of reals has the perfect set property.
(c) $\omega_1^{L[a]} < \omega_1$.

Proof. (a) $\to$ (b) is immediate, (b) $\to$ (c) is 13.12 relativized, and (c) $\to$ (a) follows from the above remarks. ⊣

This result refines 11.6 and shows that the perfect set property for all $\mathbf{\Pi}^1_1$ sets requires the consistency strength of an inaccessible cardinal. As observed by Lyubeckij [70], 14.3 and 14.10 have the curious consequence that if every $\mathbf{\Sigma}^1_2$ set has the perfect set property, then every $\mathbf{\Sigma}^1_2$ set is Lebesgue measurable – with the implication between these classical properties proceeding through $\forall a \in {}^{\omega}\omega(\omega_1^{L[a]} < \omega_1)$. Lyubeckij [71] rederived several of Solovay's results about $\mathbf{\Sigma}^1_2$ sets.

Definability of Sharps

The main motivation for Solovay's study of $0^{\#}$ was to analyze its definability. By 13.16 every Σ^1_2 subset of ω is constructible. Using methods of Gaifman and Rowbottom, Solovay showed in the autumn of 1965 that if there is a measurable cardinal, then there is a non-constructible Δ^1_3 subset of ω. Then in the context of Silver's work on indiscernibles, Solovay and Silver independently saw that $0^{\#}$ has stronger properties. For present purposes, temporarily specify that if φ is a formula in the language $\mathcal{L}^*_{\in}$ of §9, then $\ulcorner\varphi\urcorner \in \omega$ is its Gödel number in some recursive arithmetization of $\mathcal{L}^*_{\in}$. Recall that by this means, $0^{\#}$ is regarded as a subset of ω.

14.11 Theorem (ZF)(Solovay [67], Silver [71]). *The relation $R \subseteq {}^{\omega}\omega$ defined by*

$$R(x) \leftrightarrow 0^{\#} \textit{ exists} \wedge x \in {}^{\omega}2 \wedge \{m \mid x(m) = 1\} = 0^{\#}$$

is Π^1_2.

Proof. We analyze the definition from §9 that $0^{\#}$ is the unique EM blueprint satisfying conditions (I)-(III); this uniqueness is a consequence of ZF, since as observed in §9 the $0^{\#}$ theory does not require AC. For any $x \in {}^{\omega}\omega$, let $\mathrm{Th}_x = \{\varphi \mid x(\ulcorner\varphi\urcorner) = 1\}$. Now define $S \subseteq {}^{\omega}\omega$ by: $S(x)$ *iff* $x \in {}^{\omega}2$; for any $m \in \omega$, $x(m) = 1$ implies that $m = \ulcorner\varphi\urcorner$ for some sentence φ of $\mathcal{L}^*_{\in}$; and Th_x is a complete and consistent theory of $\mathcal{L}^*_{\in}$ containing

(i) the sentence σ_0 of 3.3(a) ,

(ii) the sentences asserting that the c_i's are indiscernibles , and
(iii) the sentences for satisfying conditions (II) and (III) .

Then it is straightforward to check that S is arithmetical; to assert that Th_x is consistent, it is enough to say that there is no integer coding a proof from it of $v_0 \neq v_0$. Note that if $S(x)$, then by the Completeness Theorem Th_x has a countable model. But technically speaking, (i) and (ii) do not suffice to show that Th_x is an EM blueprint, for only if the model is well-founded can its transitive collapse be taken and 3.3(a) invoked. However, this will be taken care of when condition (I) is incorporated.

For this purpose, recall the terminology of 13.8; in particular, every countable well-ordering is an E_y for some $y \in {}^\omega\omega$. The $R \subseteq {}^\omega\omega$ of the theorem is then seen to be equivalent to:

$$S(x) \wedge \forall^1 y \forall^1 z[(E_y \text{ is a well-ordering } \wedge\ M_z \text{ satisfies 9.4 with } \mathrm{Th}_x \text{ the theory and indiscernibles of ordertype } E_y) \rightarrow M_z \text{ is well-founded}]\ .$$

That M_z satisfies 9.4 etc. is Σ^1_1: Beyond what M_z is to model, one must assert that there is an order-preserving injection of E_y into On^{M_z} such that its image is a set of indiscernibles for M_z, and every element of M_z results from a Skolem term applied to these indiscernibles. Thus, R has form $\forall^1 y \forall^1 z[\Pi^1_1 \wedge \Sigma^1_1 \rightarrow \Pi^1_1]$ and so is Π^1_2. ⊣

14.12 Corollary.

(a) $0^\#$ is absolute for transitive $\in$-models M of ZF *such that $\omega_1 \subseteq M$ in the following sense:*

$$M \models \ulcorner\text{There is an EM blueprint satisfying (I)-(III)}\urcorner \quad \textit{iff} \quad 0^\# \in M\ ,$$

in which case $M \models \ulcorner 0^\#$ is the unique EM blueprint satisfying (I)-(III)$\urcorner$.

(b) $0^\#$ is a non-constructible Δ^1_3 subset of ω.

(c) Assume $0^\#$ exists. Then every $x \in \mathcal{P}(\omega) \cap L$ is one-one reducible to $0^\#$, i.e. there is an injective, total recursive function $f\colon \omega \to \omega$ such that $x = f^{-1}(0^\#)$, and so in particular, x is Δ^1_3.

Proof. (a) This follows from Σ^1_2 Absoluteness 13.15. With R as in 14.11, if $M \models$ There is an EM blueprint satisfying (I)-(III), then there is an $x \in M$ such that $(R(x))^M$. Hence $R(x)$, and so $\{m \mid x(m) = 1\} = 0^\# \in M$.

(b) By (a), $0^\# \notin L$. Also, for any $m \in \omega$,

$$\begin{aligned} m \in 0^\# &\leftrightarrow \exists x(R(x) \wedge x(m) = 1) \\ &\leftrightarrow \forall x(R(x) \rightarrow x(m) = 1)\ . \end{aligned}$$

The first expression shows that $0^\#$ is Σ^1_3, and the second, that it is Π^1_3.

(c) $x = t^L(\iota_{\xi_1}, \ldots, \iota_{\xi_n})$ for some term t and indiscernibles $\iota_{\xi_1}, \ldots, \iota_{\xi_n}$. As a recursive arithmetization had been assumed, the function $f \in {}^\omega\omega$ defined by

$$f(m) = \ulcorner m \in t(c_1, \ldots, c_n)\urcorner$$

is an injective total recursive function. But clearly, $x = f^{-1}(0^\#)$. The last assertion of (c) follows, as $0^\#$ is Δ^1_3. ⊣

(c) implies by remarks after 9.12 that *if* $0^\#$ *exists, then the set of (Gödel numbers of) sentences true in* L *is one-one reducible to* $0^\#$.

These properties of $0^\#$ relativize directly for any $a \subseteq \omega$ to $a^\#$ (as defined at the end of §9) over $L[a]$. Consequently, the assumption that $a^\#$ exists for every $a \subseteq \omega$ provides a rich landscape for the study of relative constructibility, akin to the study of Turing reducibility in recursion theory. For any sets a and b,

$$a \leq_c b \;\; \textit{iff} \;\; a \in L[b] \,,$$

and let $<_c$ and $\equiv_c$ have the derived meanings. Restricting the discussion to $\mathcal{P}(\omega)$, the equivalence classes under $\equiv_c$ are called the *constructibility degrees* of reals, and the orderings $<_c$ and $\leq_c$ carry over to these degrees. $\mathcal{P}(\omega) \cap L$ is the minimum degree and the degrees form an upper semi-lattice under $<_c$. Because of the transcendence properties of $a <_c a^\#$, # can be regarded as the analogue of the jump operation on Turing degrees. Jeffrey Paris [74] established a significant result on patterns of indiscernibles which implies that if $a \leq_c b <_c a^\#$, then $a^\# \equiv_c b^\#$; the analogous assertion about Turing degrees with jump is well-known to be false. Paris's result was a basic ingredient for further results: Leo Harrington and Kechris in their [77] established partial results toward a conjecture of Solovay (see below). Also, Farrington [82] showed that the first-order theory of the upper semi-lattice of constructibility degrees with # has the same sort of recursive isomorphism to the truth set of second-order arithmetic as was established by Simpson [77] for the Turing degrees. In evident analogy with the Turing degrees, much more can still be done in this investigation of #.

The remarkable properties of $0^\#$ prompted Solovay to make sweeping conjectures in the late 1960's that raised broad issues about the scope of the forcing method. The first was the *Genericity Conjecture*:

If $a \subseteq \mathrm{On}$ and $0^\# \notin L[a]$, then a is generic over L .

This conjecture turns on what is meant by genericity over L, and $a \in L$ are allowed to be generic over L in a trivial sense. If M is an inner model, a set $a \subseteq \mathrm{On}$ is *set-generic over* M *iff* there is a p.o. $P \in M$ and a G P-generic over M such that $M[a] = M[G]$. Since forcing with any (set) p.o. P preserves cardinals $\geq |P|^+$, $0^\#$ *is not set-generic over* L by 9.17(c) and 14.12(a). Confident in the apparent generality of forcing, Solovay originally meant his conjecture to be an assertion of minimal transcendence for $0^\#$ from the set forcing point of view. However, in 1975-6 Jensen refuted this version of the conjecture using his impressive "coding the universe in a real" technique, with which he established the following result.

14.13 Theorem (Jensen – Beller-Jensen-Welch [82]). *Assuming* GCH, *there is a proper class notion of forcing* P *such that if* G *is* P*-generic, then* V *and* $V[G]$

have the same cardinals and cofinalities, and for some $a \in \mathcal{P}(\omega) \cap V[G]$,

$$V[G] \models V = L[a] . \qquad \dashv$$

Forcing with P moreover preserves various large cardinal properties consistent with $V = L$ like weak compactness. In the mid-1980's Sy Friedman [94] refined and simplified Jensen's proof toward the development of the much stronger "strong coding" technique of S. Friedman [87, 87a]. In 1989 Shelah and Lee Stanley (see their [90, 95, 95a]) found a short proof of Jensen's original coding result assuming that $0^\#$ does not exist, and subsequently so did S. Friedman [97].

Jensen also showed that it is consistent relative to ZFC that there is a real a as in 14.13 so that $0^\# \notin L[a]$ yet a is not set-generic over L, and that such an a can be constructed outright from $0^\#$:

14.14 Theorem (Jensen – Beller-Jensen-Welch [82]). *Assume that* $0^\#$ *exists. Then there is an* $a \subseteq \omega$ *such that* $a <_c 0^\#$ *yet* a *is not set-generic over* L. $\dashv$

The more liberal interpretations of genericity over L involve proper class notions of forcing; see Beller-Jensen-Welch [82: 155] for several formulations. Suppose that M is an inner model. For $P \subseteq M$, $\langle M, P\rangle$ is *amenable iff* $P \cap x \in M$ for every $x \in M$. This is a weaker condition than requiring that P be a (definable) class in the sense of M. $a \subseteq \mathrm{On}$ is *amenable-generic* over M *iff* there is a class p.o. P such that $\langle M, P\rangle$ is amenable and the forcing relation $\Vdash_P$ restricted to formulas of any bounded complexity is definable in $\langle M, P\rangle$, and there is a G P-generic over M (i.e. G meets every dense subclass of P definable in M in addition to the usual genericity requirements) such that $a \in M[G]$. $0^\#$ is not amenable-generic over L. S. Friedman developed a notion of forcing with hyperclasses (classes of classes) to show that if $0^\#$ exists, then there is an $a \subseteq \omega$ such that $a <_c 0^\#$ yet a is not amenable-generic over L. Mack Stanley [94] established the consistency of the existence of a real not amenable-generic over any inner model to which it does not belong. (However, the existence of Stanley's real is incompatible with that of $0^\#$.)

A second conjecture of Solovay's had to do more directly with the definability of $0^\#$. For $x \in {}^\omega\omega$, or $x \subseteq \omega$ regarded as a member of ${}^\omega 2$ through its characteristic function,

$$x \text{ is a } \Pi^1_n \textit{ singleton iff } \{x\} \text{ is a } \Pi^1_n \text{ subset of } {}^\omega\omega .$$

By $\mathbf{\Sigma}^1_2$ Absoluteness 13.15, a Π^1_2 singleton is definable by a formula that defines it in every inner model that contains it, and by 14.11, $0^\#$ is such a singleton. Because this absolute definability seemed a structurally stringent constraint, Solovay made the Π^1_2 *Singleton Conjecture:*

There are no Π^1_2 singletons a such that $a \notin L$ yet $a <_c 0^\#$.

That is, $0^\#$ is a minimal, non-constructible Π^1_2 singleton with respect to relative constructibility. The first related work was the discovery of several forcing notions for adjoining a Π^1_2 singleton to a model of ZFC without large cardinal hypotheses: Jensen-Solovay [70], Jensen [70], and Jensen-Johnsbråten [74]. These are particularly elegant constructions, but they do not yield Π^1_2 singletons if $0^\#$ exists. The Jensen-Johnsbråten real was the most absolute: if $\omega_1^L = \omega_1$, then it is a Π^1_2 singleton. Building on Jensen's coding techniques, several people then established results more directly toward the conjecture. For instance, René David [82] showed that there is a non-constructible real a satisfying $a <_c 0^\#$ such that for any set-generic $b \subseteq \mathrm{On}$ over $L[a]$, a is a Π^1_2 singleton in $L[a][b]$ (through the same Π^1_2 formula for all such b). See David [84, 89] for various properties of $0^\#$ simulated by reals in generic extensions. Then in culminating work, Friedman finally refuted the conjecture outright:

14.15 Theorem (S. Friedman [90]). *Assume that $0^\#$ exists. Then there is a Π^1_2 singleton a such that $a \notin L$ yet $a <_c 0^\#$.* ⊣

Solovay apparently conjectured the negation of his original conjecture soon after learning of Jensen's 14.14, and this is the orientation of S. Friedman [90].

Solovay had made a third conjecture asserting in part that all constructibility degrees containing Π^1_2 singletons are produced by the iterative application of #. Before this was refuted by Friedman's work, Harrington-Kechris [77], building on Paris [74], had made observations that tended to support the conjecture. They showed that if $x \subseteq \omega$ is a Π^1_2 singleton, then either $0^\# \leq_c x$ or $x^\# \equiv_c 0^\#$. Relativizing this, they observed that the constructibility degrees of $0^\#$, $0^{\#\#}$, $0^{\#\#\#}, \ldots$ are the first ω degrees of *sharps* of Π^1_2 singletons.

Uniform Indiscernibles

Solovay's conjectures generated results that have led to a considerable understanding of sharps, forcing, and definability at the second level of the projective hierarchy. Another important chapter was begun in the late 1960's when, as discussed in the next section, sharps were used to analyze the sets in the third level of the projective hierarchy. In preparation, some results are established here about the interplay of the sharps, building on the §9 theory in relativized form.

Up to the discussion above of Π^1_n singletons it was convenient to regard reals for the theory of sharps as subsets of ω, since the classes $L[A]$ had only been defined for $A \subseteq \mathrm{On}$; henceforth, it is incumbent to consider reals as members of ${}^\omega\omega$. For $a \in {}^\omega\omega$, set

$$a^\# = \{\langle n, a(n)\rangle \mid n \in \omega\}^\# ,$$

i.e. the sharp of the graph of a coded as a subset of ω, and stipulate that

$$L[a] = L[\{\langle n, a(n)\rangle \mid n \in \omega\}] \text{ , and}$$

I_a is the corresponding class of indiscernibles for $L[a]$.

Moreover, for either $a \subseteq \omega$ or $a \in {}^\omega\omega$, construe $a^\# \subseteq \omega$ also as

$$a^\# \in {}^\omega 2 \text{ via the characteristic function}$$

as befits the context.

The uniformity of the relativization of 14.11 is first recorded:

14.16 Exercise (ZF). *The relation* $S \subseteq {}^2({}^\omega\omega)$ *defined by*

$$S(a, x) \leftrightarrow a^\# \text{ exists} \wedge x = a^\#$$

is Π^1_2. ⊣

For $a \in {}^\omega\omega$, the relativization of 14.12(a) leading to the absoluteness of $a^\#$ is direct. Taken for granted in what follows is the straightforward consequence that the definition from $a^\#$ of the class I_a of indiscernibles for $\langle L[a], \in, a\rangle$, in terms of Skolem hulls and so forth as in §9, is absolute for any inner model M containing $a^\#$. Properties of I_a that follow from the relativization of 9.14 will also be assumed, e.g. I_a is a closed unbounded class containing every uncountable cardinal, and if $\delta \in I_a$, then $\langle L_\delta[a], \in, a\rangle \prec \langle L[a], \in, a\rangle$.

When assuming $\forall a \in {}^\omega\omega(a^\# \text{ exists})$, set

$$I^* = \bigcap\nolimits_{a\in{}^\omega\omega} I_a \ ,$$

the class of *uniform indiscernibles*.

14.17 Exercise (ZF).

(a) Suppose that $a, b \in {}^\omega\omega$ *with* $a \in L[b]$, *and* $b^\#$ *exists. Then* $a^\#$ *exists and* $I_b \subseteq I_a$.

(b) Assume $\forall a \in {}^\omega\omega(a^\# \text{ exists})$. *Then* I^* *is a closed unbounded class of ordinals containing every uncountable cardinal, and for any* $a \in {}^\omega\omega$ *and* $u \in I^*$, $I_a \cap u$ *has ordertype* u.

Hint. (a) For the existence of $a^\#$, note that for any $\lambda > \omega$, $(I_b \cap \lambda) - \eta$ is a set of indiscernibles for $L_\lambda[a] \in L[b]$ for some $\eta < \lambda$ (thrown in to get above indiscernibles figuring in some definition of a).

Assuming next that $\gamma \notin I_a$, we show that $\gamma \notin I_b$: First,

$$\gamma = t^{\langle L[a], \in, a\rangle}(x_1, \ldots, x_m, y_1, \ldots, y_n)$$

for some Skolem term t (of $\mathcal{L}_\in(\dot{a})$) and $x_1 < \ldots < x_m < y_1 < \ldots < y_n$ all in I_a with $x_m < \gamma < y_1$. By the remarkable condition (III) of §9 relativized, it can be assumed that the y_i's are in the closed unbounded class $I_a \cap I_b$. Also, by further applications of (III) each x_i is definable in $L[b]$ in terms of members of $I_b - \{\gamma\}$ and a is definable in $L[b]$ in terms of members of $I_b - (\gamma + 1)$. Consequently, γ is definable in $L[b]$ in terms of members of $I_b - \{\gamma\}$, and hence $\gamma \notin I_b$.

(b) For the last assertion, suppose that $a \in {}^\omega\omega$, $u \in I^*$, and $\xi < u$. By the remarkable condition (III) and the simple observation that u must be a limit member of I_b for every $b \in {}^\omega\omega$, ξ is definable in $\langle L[a^\#], \in, a^\# \rangle$ in terms of members of $I_{a^\#} \cap u$. By absoluteness of $a^\#$ it follows that the ξth member of I_a is definable in $L[a^\#]$ in terms of members of $I_{a^\#} \cap u$, and so that member is less than u (by condition (II) of §9). But this holds for any $\xi < u$, so it follows that the ordertype of $I_a \cap u$ is u. ⊣

Assuming $\forall a \in {}^\omega\omega(a^\#\text{exists})$, set

$$I^* = \{u_\xi \mid 0 < \xi \in \text{On}\}$$

with the indexing in increasing order; then

$$u_1 = \omega_1 \quad \text{and} \quad \forall \xi > 0(u_\xi \leq \omega_\xi) ,$$

the first as every countable ordinal is coded by a real, and the second as I^* contains every uncountable cardinal.

14.18 Proposition (ZF)(Solovay). *Assume* $\forall a \in {}^\omega\omega(a^\# \text{ exists})$ *and* $\xi > 0$. *Then:*

(a) $\text{cf}(u_{\xi+1}) = \text{cf}(u_2) \leq \omega_2$.

(b) For any $\gamma < u_\xi$, *there is an* $a \in {}^\omega\omega$, *a Skolem term* t, *and* $0 < \xi_1 < \ldots < \xi_n < \xi$ *such that*

$$\gamma = t^{\langle L[a], \in, a \rangle}(u_{\xi_1}, \ldots, u_{\xi_n}) .$$

Proof. Define n: On $\times\, {}^\omega\omega \to$ On by

$$n(\alpha, a) = \text{least member of } I_a - (\alpha + 1) .$$

Then for any $\xi > 0$,

$$(*) \qquad \sup(\{n(u_\xi, a) \mid a \in {}^\omega\omega\}) = u_{\xi+1} .$$

To show this, note that we have $\leq$, and so it suffices to show that the supremum is in I^*. But for any $b \in {}^\omega\omega$,

$$\sup(\{n(u_\xi, a) \mid a \in {}^\omega\omega\}) = \sup(\{n(u_\xi, a) \mid a \in {}^\omega\omega \ \wedge \ b \in L[a]\})$$

by 14.17(a) since for any $a \in {}^\omega\omega$ there is an $\overline{a} \in {}^\omega\omega$ such that $\{a, b\} \subseteq L[\overline{a}]$, and the latter supremum is in I_b as it is closed.

To establish (a), note that

$$(**) \qquad \text{If } n(u_1, a) = n(u_1, b), \text{ then } n(u_\xi, a) = n(u_\xi, b) .$$

To see this, let $c \in {}^\omega\omega$ code a and b, e.g. $(c)_0 = a$ and $(c)_1 = b$. Then it is straightforward to see by 14.17(a) and absoluteness of sharps that there is a formula $\psi(v_0)$ such that for any $x \in I_{c^\#}$,

$$n(x, a) = n(x, b) \ \textit{iff} \ \langle L[c^\#], \in, c^\# \rangle \models \psi[x] .$$

Hence, (**) follows by indiscernibility.

By (∗), $\{n(u_1, a) \mid a \in {}^\omega\omega\}$ is cofinal in u_2, and by (∗∗) the map sending $n(u_1, a)$ to $n(u_\xi, a)$ is well-defined. Hence (a) follows.

To establish (b), proceed by induction on ξ and then on γ. So, it can be assumed that $\xi = \zeta + 1$ for some $\zeta > 0$ and $u_\zeta < \gamma < u_{\zeta+1}$. Since $\gamma \notin I^*$, there is an $a \in {}^\omega\omega$ such that

$$\gamma = t_0^{\langle L[a], \in, a\rangle}(x_1, \ldots, x_m, y_1, \ldots, y_n)$$

for some Skolem term t_0 and $x_1 < \ldots < x_m < y_1 < \ldots < y_n$ all in I_a with $x_m < \gamma < y_1$. The x_i's can be replaced by members of I^*: Inductively, each x_i is definable in $\langle L[a_i], \in, a_i\rangle$ for some $a_i \in {}^\omega\omega$ in terms of members of I^* less than $u_{\zeta+1}$, and hence so definable in $\langle L_\delta[a_i], \in, a_i\rangle \prec \langle L[a_i], \in, a_i\rangle$ for some (any) δ in $I^* - u_{\zeta+1}$. So, if necessary, a can be replaced by a real coding $a, a_1, \ldots, a_n$ and each x_i by its definition, subsuming the δ into the y_i's by 14.17(a). Thus,

$$(***) \qquad \gamma = t_1^{\langle L[a], \in, a\rangle}(u_{\xi_1}, \ldots, u_{\xi_r}, y_1, \ldots, y_n)$$

for some new Skolem term t_1, with $\xi_1 < \ldots < \xi_r \leq \zeta$.

Next, by our initial observation (∗), $\gamma < u_{\zeta+1}$ implies that there is a $b \in {}^\omega\omega$ such that $\gamma < n(u_\zeta, b)$. As before, replacing a by a real coding a and b if necessary, it can be assumed by 14.17(a) that $b = a$.

Finally, by the absoluteness of sharps, (∗∗∗) also holds in the sense of $L[a^\#]$. Arguing there with the remarkable condition (III) applied to $\gamma < n(u_\zeta, a)$ and the fact that I_a contains every uncountable cardinal, the y_i's can be replaced by the first n cardinals in the sense of $L[a^\#]$ greater than u_ζ. But these cardinals are definable in $L[a^\#]$ from u_ζ, and hence

$$\gamma = t_2^{\langle L[a^\#], \in, a^\#\rangle}(u_{\xi_1}, \ldots, u_{\xi_s}, u_\zeta)$$

for some Skolem term t_2 and $\xi_1 < \ldots < \xi_s < \zeta$ to complete the proof. ⊣

That 14.18 is provable in ZF was noted for a later purpose. With the Axiom of Choice ω_3 is of course regular, and that $\omega_3 \in I^*$ has the following consequence in ZFC:

14.19 Corollary. $u_n < \omega_3$ *for every* $n \leq \omega$. ⊣

These results will be applied in the forthcoming section.

15. Sharps and $\mathbf{\Sigma}^1_3$ Sets

Martin and Solovay in their [69] established results about $\mathbf{\Sigma}^1_3$ sets under the existence of sharps that began the use of large cardinal hypotheses to extend the methods of classical descriptive set theory. Martin went on to provide a sharp tree representation of $\mathbf{\Sigma}^1_3$ sets, and these first explorations broke the ground for major developments over a decade later: the analysis of the projective sets under the Axiom of Determinacy (§§29, 30), and the consistency of that hypothesis relative to large cardinals (§32). Against this larger backdrop the results for $\mathbf{\Sigma}^1_3$ have a transitional flavor, but they were to form the bases of the grander edifices to follow and, being close enough to the bedrock of $\mathbf{\Sigma}^1_2$, to figure in important absoluteness arguments.

Unlike previous sections this one is largely devoted to a single technical development – one that serves to motivate later generalizations and substantiate how large cardinals provide contextually optimal analyses of the projective sets. A tree representation for $\mathbf{\Sigma}^1_3$ sets implicit in Martin-Solovay [69] was brought to light by Mansfield [71] using a measurable cardinal; this approach is first taken because of its own significance and simpler context, and then the context refined to follow from the existence of sharps, as was done by Martin. Kechris [81] is a good reference.

The idea is to develop a tree representation for $\mathbf{\Sigma}^1_3$ sets based on one for $\mathbf{\Sigma}^1_2$ sets, emulating how a tree representation for $\mathbf{\Sigma}^1_2$ sets was developed based on one for $\mathbf{\Sigma}^1_1$ sets. Recapitulating from 13.13 and 13.14, every $\mathbf{\Sigma}^1_1$ set is ω-Suslin, and to show that every $\mathbf{\Sigma}^1_2$ set is ω_1-Suslin it sufficed to show that every $\mathbf{\Pi}^1_1$ set is ω_1-Suslin. But if $A \subseteq {}^k({}^\omega\omega)$ is $\mathbf{\Pi}^1_1$, then there is a tree T on ${}^k\omega \times \omega$ such that for any $\mathbf{w} \in {}^k({}^\omega\omega)$,

$$\begin{aligned} A(\mathbf{w}) &\leftrightarrow \mathbf{w} \notin p[T] \\ &\leftrightarrow T_{\mathbf{w}} \text{ is well-founded} \\ &\leftrightarrow \exists g(g\colon T_{\mathbf{w}} \to \omega_1 \text{ is an order-preserving map}) \\ &\leftrightarrow \mathbf{w} \in p[\hat{T}]\,, \end{aligned}$$

where $\hat{T}$ is a tree defined on ${}^k\omega \times \omega_1$ from T as in the proof of 13.14. Thus, a tree representation for $\mathbf{\Sigma}^1_1$ sets was converted into one for $\mathbf{\Pi}^1_1$ sets by appealing to the equivalence between the Π^{ZF}_1 property of well-foundedness and the Σ^{ZF}_1 property of having an order-preserving map and formulating a new tree of approximations to such a map.

The tree representation for $\mathbf{\Sigma}^1_2$ sets is to be similarly "dualized" to get one for $\mathbf{\Pi}^1_2$ sets and hence for $\mathbf{\Sigma}^1_3$ sets. For this purpose, a more specific representation for $\mathbf{\Sigma}^1_2$ sets is first derived:

Suppose that $B \subseteq {}^k({}^\omega\omega)$ is $\mathbf{\Sigma}^1_2$. Taking $k = 1$ for simplicity, let T be a tree on ${}^2\omega \times \omega$ such that for any $x \in {}^\omega\omega$,

$$B(x) \leftrightarrow \exists^1 y(T_{\langle x,y\rangle} \text{ is well-founded})\,.$$

T can be defined from a relation recursive in a real (13.1), and by slightly altering that relation if necessary *it can be assumed that* $\langle\emptyset,\emptyset,\emptyset\rangle \in T$.

In terms of the fixed recursive enumeration $\langle \mathbf{s}_i \mid i \in \omega\rangle$ of ${}^{<\omega}\omega$ such that $|\mathbf{s}_i| \leq i$, for any $x, y \in {}^{\omega}\omega$ let $<_{x,y}$ be that strict linear ordering of ω defined by:

$$\begin{aligned} i <_{x,y} j \;\leftrightarrow\; & \mathbf{s}_i \notin T_{\langle x,y\rangle} \;\wedge\; \mathbf{s}_j \notin T_{\langle x,y\rangle} \;\wedge\; i < j \;;\text{ else} \\ & \mathbf{s}_i \notin T_{\langle x,y\rangle} \;\wedge\; \mathbf{s}_j \in T_{\langle x,y\rangle} \;;\text{ else} \\ & \mathbf{s}_i \in T_{\langle x,y\rangle} \;\wedge\; \mathbf{s}_j \in T_{\langle x,y\rangle} \;\wedge\; \mathbf{s}_i <_{\mathrm{KB}} \mathbf{s}_j \end{aligned}$$

where $<_{\mathrm{KB}}$ is the Kleene-Brouwer ordering, defined before 13.2. Thus, those i's so that $\mathbf{s}_i$ is not in $T_{\langle x,y\rangle}$ appear first in their indexing ordering, followed by those i's so that $\mathbf{s}_i$ is in $T_{\langle x,y\rangle}$ ordered according to $<_{\mathrm{KB}}$. Note that since $\langle\emptyset,\emptyset,\emptyset\rangle \in T$ and $\mathbf{s}_0 = \emptyset$ (as $|s_i| \leq i$), 0 is the maximum of $<_{x,y}$ by the definition of $<_{\mathrm{KB}}$. The following is a consequence of 13.2.

15.1 Lemma. *For any* $x \in {}^{\omega}\omega$, $B(x) \leftrightarrow \exists^1 y(<_{x,y}$ is a well-ordering). ⊣

For $s, t \in {}^{<\omega}\omega$ with $|s| = |t|$, temporarily setting

$$T_{\langle s,t\rangle} = \{w \mid \exists m < |s|(\langle s|m, t|m, w\rangle \in T)\}$$

let

$$\begin{aligned} <_{s,t} \;=\; & \text{that strict linear ordering of } |s| = \{i \mid i < |s|\} \\ & \text{defined like } <_{x,y} \text{ but with } T_{\langle x,y\rangle} \text{ replaced by } T_{\langle s,t\rangle}\,. \end{aligned}$$

For $|s| > 0$, note that 0 is again the maximum of $<_{s,t}$. If $s, t, \bar{s}, \bar{t} \in {}^{<\omega}\omega$ with $|\bar{s}| = |\bar{t}|$ and for some $m \leq |\bar{s}|$, $s = \bar{s}|m$ and $t = \bar{t}|m$, then $\mathbf{s}_i \in T_{\langle s,t\rangle}$ *iff* $\mathbf{s}_i \in T_{\langle \bar{s},\bar{t}\rangle}$ for $i < m$ since $|\mathbf{s}_i| \leq i$. Consequently, $<_{s,t} \subseteq <_{\bar{s},\bar{t}}$ as relations. Similarly, for $x, y \in {}^{\omega}\omega$

$$i <_{x,y} j \;\; \textit{iff} \;\; i <_{x|m,y|m} j$$

for any $m \in \omega$ such that $i, j < m$, and so

$$(*) \qquad <_{x,y} = \bigcup_{m\in\omega} <_{x|m,y|m} \;.$$

In terms of a regular $\kappa > \omega$ to be further specified later, a tree T^* on ${}^2\omega \times \kappa$ is now defined by

$$T^* = \{\langle s,t,u\rangle \in \bigcup_{m\in\omega}({}^m\omega \times {}^m\omega \times {}^m\kappa) \mid \forall i, j < |s|(u(i) < u(j) \leftrightarrow i <_{s,t} j)\}\,.$$

T^* is like the $\hat{T}$ defined in the proof of 13.14 with $\kappa = \omega_1$, except that the linearity of $<_{s,t}$ affords the possibility of *iff*. The following is a consequence of 15.1 and $(*)$.

15.2 Lemma. *For any* $x \in {}^{\omega}\omega$, $B(x) \leftrightarrow \exists^1 y \exists g \in {}^{\omega}\kappa(\langle x, y, g\rangle \in [T^*])$. ⊣

Toward a tree representation for $\mathbf{\Pi}^1_2$ sets, suppose that $A \subseteq {}^k({}^{\omega}\omega)$ is such a set. Again taking $k = 1$, A can be considered to be the complement of B above so that for any $x \in {}^{\omega}\omega$,

$$\begin{aligned} A(x) &\leftrightarrow \forall^1 y \forall g \in {}^{\omega}\kappa \exists^0 m(\langle x|m, y|m, g|m\rangle \notin T^*) \\ &\leftrightarrow T^*_x \text{ is well-founded}. \end{aligned}$$

To convert as in the $\mathbf{\Pi}^1_1$ case, note that this in turn is equivalent (in ZF) to the existence of an order-preserving map: $T^*_x \to \kappa^+$ by 12.9(a), as T^*_x is injectible into κ. A nicety for now but of consequence later, the κ^+ here can in effect be replaced by κ:

First, in the definition of B the maximum of $<_{s,t}$ is $\emptyset$ by arrangement so that if $\langle s, t, u\rangle \in T^*$ with $|s| > 0$ then $u(0) = \max(\mathrm{ran}(u))$. Next, define a strict ordering $<_{\mathrm{KB}^2}$ on $T^*_x(= \{\langle t, u\rangle \mid \exists^0 m(\langle x|m, t, u\rangle \in T^*)\})$ as follows:

$$\begin{aligned} \langle t, u\rangle <_{\mathrm{KB}^2} \langle \bar{t}, \bar{u}\rangle \ \textit{ iff } \ & t = u = \emptyset \ \wedge \ |\bar{t}| = |\bar{u}| > 0 \ ; \ \text{else} \\ & t \supset \bar{t} \ \wedge \ u \supset \bar{u} \ ; \ \text{else} \\ & \exists m < |t|(t|m = \bar{t}|m \ \wedge \ u|m = \bar{u}|m \ \wedge \\ & (u(m) < \bar{u}(m) \ \vee \ (u(m) = \bar{u}(m) \ \wedge \ t(m) < \bar{t}(m)))) \ . \end{aligned}$$

Thus, $\langle \emptyset, \emptyset\rangle$ is the minimum, and the other $\langle t, u\rangle$ are ordered according to an antilexicographical version of $<_{\mathrm{KB}}$. If $\langle t, u\rangle <_{\mathrm{KB}^2} \langle \bar{t}, \bar{u}\rangle$ then either $u = \emptyset$ or else $u(0) \leq \bar{u}(0)$, and since $u(0) = \max(\mathrm{ran}(u))$ as noted above, it follows that the set of predecessors of $\langle \bar{t}, \bar{u}\rangle$ has cardinality $|{}^{<\omega}\omega \times {}^{<\omega}(\bar{u}(0) + 1)| < \kappa$. On the other hand, $\{\langle\langle 0\rangle, \langle \xi\rangle\rangle \mid \xi < \kappa\}$ is a subset of T^*_x of ordertype κ under $<_{\mathrm{KB}^2}$. Hence, by arguments as for $<_{\mathrm{KB}}$,

$$T^*_x \text{ is well-founded } \textit{ iff } \ T^*_x \text{ is well-ordered by } <_{\mathrm{KB}^2} \text{ in ordertype } \kappa \ .$$

$\langle \emptyset, \emptyset\rangle$ was made the minimum of $<_{\mathrm{KB}^2}$ to render the ordertype κ (rather than $\kappa+1$), whereas the well-foundedness of T^*_x is with respect to $\supset^*$ (as described before 12.9) for which $\langle \emptyset, \emptyset\rangle$ is the maximum. Accepting this, we have the following:

15.3 Lemma. *For any $x \in {}^{\omega}\omega$,*

$$A(x) \leftrightarrow \exists f(f\colon T^*_x - \{\langle \emptyset, \emptyset\rangle\} \to \kappa \text{ is an order-preserving map}) \ . \qquad \dashv$$

This is recast (cf. 13.14) in terms of getting an infinite branch through a tree: For $s, t \in {}^{<\omega}\omega$ with $|s| = |t|$, set

$$T^*_{s,t} = \{u \mid \langle s, t, u\rangle \in T^*\} \ .$$

Then stipulate for $s \in {}^{<\omega}\omega$ that

$$\begin{aligned} \langle h_i \mid i < |s|\rangle \text{ is } \textit{orderly for } s \ \textit{ iff } \ & \text{for } i, j < |s| \ , \\ & \text{(a) } h_i\colon T^*_{s||\mathbf{s}_i|, \mathbf{s}_i} \to \kappa \ , \text{ and} \\ & \text{(b) if } \mathbf{s}_i \supset \mathbf{s}_j \neq \emptyset \text{ and } u \in T^*_{s||\mathbf{s}_i|, \mathbf{s}_i}, \text{ then } h_i(u) < h_j(u||\mathbf{s}_j|) \ . \end{aligned}$$

((b) imposes no conditions on $h_0\colon \{\emptyset\} \to \kappa$, which corresponds to $\langle \emptyset, \emptyset\rangle \in T^*_x$; it is only maintained for notational convenience.) Now let

$$T^+ = \{\langle s, \langle h_i \mid i < |s|\rangle\rangle \mid s \in {}^{<\omega}\omega \ \wedge \ \langle h_i \mid i < |s|\rangle \text{ is orderly for } s\} \ .$$

The following is then straightforward.

15.4 Lemma. *For any* $x \in {}^{\omega}\omega$, $A(x) \leftrightarrow \exists h \forall^0 m(\langle x|m, h|m\rangle \in T^+)$. ⊣

This is a sort of Suslin property, but one with ordinal-valued functions h_i composing the tree rather than ordinals. The final step involves ranking the h_i's in a coherent way, and the crucial idea is *to assign to the* h_i*'s ordinals that they represent in well-founded ultrapowers.*

Note first that the tree T^* has the following *homogeneity property*: For $s, t \in {}^{<\omega}\omega$ with $|s| = |t|$, $<_{s,t}$ is a strict linear ordering, and so for any $u \in T^*_{s,t}$, $\operatorname{ran}(u) \in [\kappa]^{|s|}$. Moreover, for any $w \in [\kappa]^{|s|}$ there is a unique $u \in T^*_{s,t}$ with $\operatorname{ran}(u) = w$, the one that orders the members of w to satisfy the definition of T^*. Hence, *the range function* ran: $T^*_{s,t} \to [\kappa]^{|s|}$ *is a bijection.*

κ *is now specified to be a measurable cardinal*, and the homogeneity of T^* is used to define ultrafilters over $T^*_{s,t}$ as follows: Let U be a normal ultrafilter over κ. For $s, t \in {}^{<\omega}\omega$ with $|s| = |t|$, define $U_{s,t}$ by:

$$X \in U_{s,t} \ \text{ iff } \ X \subseteq T^*_{s,t} \ \wedge \ \exists H \in U([H]^{|s|} \subseteq \text{ran“}X) \ .$$

By Rowbottom's result 7.17, for any $n \in \omega$ and $Y \subseteq [\kappa]^n$ there is an $H \in U$ such that either $[H]^n \subseteq Y$ or else $[H]^n \cap Y = \emptyset$. It follows that $U_{s,t}$ *is a* κ*-complete ultrafilter over* $T^*_{s,t}$. (Here, the principal ultrafilter $U_{\emptyset,\emptyset} = \{\{\emptyset\}\}$ is being allowed for notational convenience, contravening a convention.) Let

$$j_{s,t}\colon V \prec M_{s,t} \cong \text{Ult}(V, U_{s,t})$$

(so that $j_{\emptyset,\emptyset}$ is the identity and $M_{\emptyset,\emptyset} = V$). For each $\langle s, \langle h_i \mid i < |s|\rangle\rangle \in T^+$, h_i is to be replaced by the ordinal $[h_i]$ that it represents in the ultrapower $M_{s||\mathbf{s}_i|,\mathbf{s}_i}$. Since $[h_i] < j_{s||\mathbf{s}_i|,\mathbf{s}_i}(\kappa)$ by condition (a) of orderliness, the result will be a tree on $\omega \times \gamma_U$, where

$$\gamma_U = \sup(\{j_{s,t}(\kappa) \mid s, t \in {}^{<\omega}\omega \ \wedge \ |s| = |t|\}) \ .$$

(γ_U has a simple description in terms of U, but this will not be needed.)

Toward expressing the condition (b) of orderliness in terms of ordinals, note the following *coherence property*: If $s, t, \bar{s}, \bar{t} \in {}^{<\omega}\omega$ with $|\bar{s}| = |\bar{t}| = n$ say and $s = \bar{s}|m$ and $t = \bar{t}|m$ for some $m \leq n$, then for $X \subseteq T^*_{s,t}$,

$$X \in U_{s,t} \ \text{ iff } \ \{u \in T^*_{\bar{s},\bar{t}} \mid u|m \in X\} \in U_{\bar{s},\bar{t}}$$

since for $H \in U$, $[H]^m \subseteq \text{ran“}X$ *iff* $[H]^n \subseteq \{w \in [\kappa]^n \mid [w]^m \subseteq \text{ran“}X\}$. As is readily checked, there is a corresponding elementary embedding

$$j_{\langle s,t\rangle,\langle\bar{s},\bar{t}\rangle}\colon M_{s,t} \prec M_{\bar{s},\bar{t}}$$

defined by

$$j_{\langle s,t\rangle,\langle \bar{s},\bar{t}\rangle}([f]_{U_{s,t}}) = [\langle f(u|m) \mid u \in T^*_{\bar{s},\bar{t}}\rangle]_{U_{\bar{s},\bar{t}}}$$

satisfying $j_{\langle s,t\rangle,\langle \bar{s},\bar{t}\rangle} \circ j_{s,t} = j_{\bar{s},\bar{t}}$. Condition (b) of orderliness can now be suitably rendered, recasting T^+ as the following tree on $\omega \times \gamma_U$:

$$\begin{aligned}\tilde{T} = \{\langle s, \langle \delta_i \mid i < |s|\rangle\rangle \in \textstyle\bigcup_{m\in\omega}({}^m\omega \times {}^m\gamma_U) \mid \\ \forall i, j < |s|(\mathbf{s}_i \supset \mathbf{s}_j \neq \emptyset \;\rightarrow\; \delta_i < j_{\langle s||\mathbf{s}_j|,\mathbf{s}_j\rangle,\langle s||\mathbf{s}_i|,\mathbf{s}_i\rangle}(\delta_j))\}\ .\end{aligned}$$

The coherence property of the $U_{s,t}$'s provides a real Suslin property:

15.5 Proposition. $A = p[\tilde{T}]$.

Proof. The characterization 15.4 of A in terms of branches through T^+ and how $\tilde{T}$ was defined from T^+ readily implies that $A \subseteq p[\tilde{T}]$.

Conversely, suppose that $x \in p[\tilde{T}]$, say with $\langle x, g\rangle \in [\tilde{T}]$. For $i \in \omega$ let $h_i\colon T^*_{x||\mathbf{s}_i|,\mathbf{s}_i} \rightarrow \mathrm{On}$ be such that $g(i) = [h_i]$ in $M_{x||\mathbf{s}_i|,\mathbf{s}_i}$. Then by the definition of $\tilde{T}$, for $\mathbf{s}_i \supset \mathbf{s}_j \neq \emptyset$,

$$\{u \in T^*_{x||\mathbf{s}_i|,\mathbf{s}_i} \mid h_i(u) < h_j(u||\mathbf{s}_j|)\} \in U_{x||\mathbf{s}_i|,\mathbf{s}_i}\ ,$$

and so there is an $X_{ij} \in U$ such that

$$(\dagger) \qquad \mathrm{ran}(u) \in [X_{ij}]^{|\mathbf{s}_i|} \rightarrow h_i(u) < h_j(u||\mathbf{s}_j|)\ .$$

Set $Y = \bigcap\{X_{ij} \mid i, j \in \omega \;\wedge\; \mathbf{s}_i \supset \mathbf{s}_j \neq \emptyset\} \in U$.

Assume now to the contrary that $\neg A(x)$. Then going back to the earlier juncture 15.1 in terms of B, there is a $y \in {}^\omega\omega$ such that $<_{x,y}$ is a well-ordering. As Y is uncountable, there is a function $d\colon \omega \rightarrow Y$ such that for $i, j \in \omega$,

$$d(i) < d(j) \;\; \textit{iff} \;\; i <_{x,y} j\ .$$

This implies that for every $m \in \omega$, $d|m \in T^*_{x|m,y|m}$, and hence by $(\dagger)$ that whenever $y \supset \mathbf{s}_i \supset \mathbf{s}_j \neq \emptyset$, $h_i(d||\mathbf{s}_i|) < h_j(d||\mathbf{s}_j|)$. This leads to an infinite descending sequence of ordinals and thus a contradiction. ⊣

The foregoing analysis will soon be adapted to a hypothesis weaker than measurability, but we draw on its strength to derive an interesting and useful conclusion based on the definability of $\tilde{T}$. This result extends $\mathbf{\Sigma}^1_2$ Absoluteness 13.15, and applicable is the discussion before it of the relativization of analytical relations in terms of their normal forms and the corresponding formulation of absoluteness.

15.6 Theorem (Martin-Solovay [69: 152]). *Suppose that A is the $\mathbf{\Pi}^1_2$ set as above with associated tree $\tilde{T}$ defined from the normal ultrafilter U over κ, and P is a p.o. satisfying $|P| < \kappa$. Then*

$$\exists^1 x A(x) \;\; \textit{iff} \;\; \Vdash_P \exists^1 x A(x)\ .$$

Hence, if there is a measurable cardinal, $\mathbf{\Sigma}^1_3$ relations are absolute for V in any generic extension via a p.o. P of cardinality less than that cardinal.

Proof. By the argument for the Levy-Solovay result 10.15,

$$\Vdash \dot{U} = \{Y \subseteq \kappa \mid \exists X \in \check{U}(X \subseteq Y)\}$$

is a κ-complete ultrafilter over κ. Moreover, as for $(*)$ in that argument, but incorporating Rowbottom's 7.17, we have the following: If $n \in \omega$ and $p \Vdash \dot{t}\colon [\kappa]^n \to \check{V}$, then there is a $q \leq p$, a $Z \in U$, and an $f\colon [Z]^n \to V$ such that $q \Vdash \dot{t}|[\check{Z}] = \check{f}$.

Suppose now that G is P-generic. Then it is straightforward to check with the property just established that for $s, t \in {}^{<\omega}\omega$ with $|s| = |t|$, if $j^G_{s,t}$ is defined in $V[G]$ from $\dot{U}^G$ just as $j_{s,t}$ had been defined in V from U, then $j^G_{s,t}$ and $j_{s,t}$ agree on the ordinals. Consequently, if $\tilde{T}^G$ is defined in $V[G]$ as $\tilde{T}$ had been defined in V, then $\tilde{T}^G = \tilde{T}$.

With this absoluteness property the argument can now be concluded as for 13.15: If in V there is an $x \in {}^\omega\omega$ such that $A(x)$, then $x \in p[\tilde{T}] \subseteq (p[\tilde{T}])^{V[G]}$ and so $A^{V[G]}(x)$. Conversely, if in V there is no $x \in {}^\omega\omega$ such that $A(x)$, then $[\tilde{T}] = \emptyset$, i.e. $\tilde{T}$ is well-founded and so there is an order-preserving map: $\tilde{T} \to \mathrm{On}$. Such a map also exists in $V[G]$, and so $[\tilde{T}]^{V[G]} = \emptyset$, i.e. in $V[G]$ there is no $x \in {}^\omega\omega$ such that $A^{V[G]}(x)$. ⊣

Hence, *if there is a proper class of measurable cardinals, then $\mathbf{\Sigma}^1_3$ relations are absolute for V in any generic extension.* Generic absoluteness would become an important consideration in the study of strong hypotheses.

Another analysis of $\mathbf{\Sigma}^1_2$ sets is provided next, one not needed for present purposes but through which 15.5 could have been established. This approach was to be the basis of a generalization through the projective hierarchy using strong hypotheses (see 32.6 about weakly homogeneous trees).

For $x, y \in {}^\omega\omega$ it is simple to check that

$$\langle\langle M_{x|m,y|m} \mid m \in \omega\rangle, \langle j_{\langle x|m,y|m\rangle,\langle x|n,y|n\rangle} \mid m \leq n\rangle\rangle$$

is a directed system (§0), so stipulate that

$$\langle M_{x,y}, E_{x,y}\rangle \text{ is the direct limit, and}$$
$$k_{x|m,y|m}\colon \langle M_{x|m,y|m}, \in\rangle \prec \langle M_{x,y}, E_{x,y}\rangle$$

the corresponding embeddings so that for $m \leq n$,

$$k_{x|n,y|n} \circ j_{\langle x|m,y|m\rangle,\langle x|n,y|n\rangle} = k_{x|m,y|m}\ .$$

In what follows, B is the representative $\mathbf{\Sigma}^1_2$ set of which we have been viewing A as the complement.

15.7 Exercise.

(a) For any $x, y \in {}^{\omega}\omega$, $\langle M_{x,y}, E_{x,y}\rangle$ *is well-founded iff*

(∗) *whenever* $X_m \in U_{x|m,y|m}$ *for* $m \in \omega$ *there is a* $d \in {}^{\omega}\kappa$ *such that* $d|m \in X_m$ *for every* $m \in \omega$.

(b) For any $x \in {}^{\omega}\omega$,

$$B(x) \leftrightarrow \exists^1 y(\langle M_{x,y}, E_{x,y}\rangle \text{ is well-founded}) .$$

Hint. (a) Suppose that (∗) holds, and assume to the contrary that $\langle M_{x,y}, E_{x,y}\rangle$ is ill-founded, i.e. there are $\{z_i \mid i \in \omega\} \subseteq M_{x,y}$ such that $z_{i+1}\ E_{x,y}\ z_i$ for every $i \in \omega$. For each such i, let $m_i \in \omega$ and $h_i\colon T^*_{x|m_i,y|m_i} \to V$ be such that $k_{x|m_i,y|m_i}([h_i]) = z_i$, with $m_i \leq m_{i+1}$. It follows that

$$[h_{i+1}] \in j_{\langle x|m_i,y|m_i\rangle,\langle x|m_{i+1},y|m_{i+1}\rangle}[h_i] \text{ for every } i \in \omega .$$

This is similar to a situation in the argument for 15.5, so invoking (∗) get an appropriate $d \in {}^{\omega}\kappa$ as there to get a contradiction.

Suppose next that (∗) fails, i.e. there are $X_m \in U_{x|m,y|m}$ for $m \in \omega$ with no $d \in {}^{\omega}\kappa$ such that $d|m \in X_m$ for every such m. Set

$$W = \{w \in {}^{<\omega}\kappa \mid \forall m \leq |w|(w|m \in X_m)\} .$$

Then $\langle W, \supset\rangle$ is a well-founded tree on κ and so has a rank function ρ_W (see before 12.9). By the coherence property of the $U_{s,t}$'s, $W \cap X_m \in U_{x|m,y|m}$ for every $m \in \omega$. Now show that for such m,

$$k_{x|m+1,y|m+1}([\rho_W|(W \cap X_{m+1})])\ \ E_{x,y}\ \ k_{x|m,y|m}([\rho_W|(W \cap X_m)])$$

to confirm that $M_{x,y}$ is ill-founded.

(b) If there is a $y \in {}^{\omega}\omega$ such that $\langle M_{x,y}, E_{x,y}\rangle$ is well-founded, then by (∗) there is a $d \in {}^{\omega}\kappa$ such that $\langle x, y, d\rangle \in [T^*]$, so that $B(x)$. For the converse, argue as for 15.5 to establish (∗) for a $y \in {}^{\omega}\omega$ such that $<_{x,y}$ is a well-ordering by finding a $d\colon \omega \to Y$ for an appropriate $Y \in U$. ⊣

The Analysis with Sharps

Having developed a tree representation for $\mathbf{\Pi}^1_2$ (and hence $\mathbf{\Sigma}^1_3$) sets with the broad strokes afforded by a measurable cardinal, the analysis is now refined to establish results in ZF assuming only

$$\forall a \in {}^{\omega}\omega(a^{\#} \text{ exists}) .$$

This hypothesis was discussed at the end of §14, and the results established there will soon be applied. In effect, the approach of Mansfield [71] is incorporated into the original Martin-Solovay [69] context, as was done by Martin.

Following the previous theory for our representative $\mathbf{\Pi}^1_2$ set A up to the formulation of T^+, the first task is to replace the normal ultrafilter U by an adequate analogue. For this purpose we simply take

$$\kappa = \omega_1$$

and the closed unbounded filter $\mathcal{C}_{\omega_1}$ over ω_1 restricted to a suitable class. Recalling that for $a \in {}^\omega\omega$, $L[a] = L[\{\langle n, a(n)\rangle \mid n \in \omega\}]$ as stipulated in §14, set

$$\tilde{L} = \bigcup_{a \in {}^\omega\omega} L[a] ,$$

the class of *sets constructible from some real*. $\tilde{L} \subseteq L(\mathbb{R}) = L({}^\omega\omega)$, and the equality $\tilde{L} = L(\mathbb{R})$ holds *iff* $\mathbb{R} \in L[a]$ for some $a \in {}^\omega\omega$ *iff* $\tilde{L}$ is an inner model. $\tilde{L}$ serves as a convenient class, rich enough for the full play of the reals, but restricted enough to be constrained by the sharps.

15.8 Lemma. *For any $n \in \omega$ and $Y \in \mathcal{P}([\omega_1]^n) \cap \tilde{L}$, there is an H closed unbounded in ω_1 such that either $[H]^n \subseteq Y$ or else $[H]^n \cap Y = \emptyset$.*

Proof. Suppose that $Y \in \mathcal{P}([\omega_1]^n) \cap L[a]$ where $a \in {}^\omega\omega$. Then

$$Y = t^{\langle L[a], \in, a\rangle}(x_1, \ldots, x_m, y_1, \ldots, y_p)$$

for some Skolem term t and $x_1 < \ldots < x_m < y_1 < \ldots < y_p$ all in I_a, the class of $a^\#$ indiscernibles, with $x_m < \omega_1 \leq y_1$. If $H = I_a \cap (\omega_1 - (x_m + 1))$, then H is closed unbounded in ω_1, and either $[H]^n \subseteq Y$ or else $[H]^n \cap Y = \emptyset$ by indiscernibility. ⊣

With this, $\mathcal{C}_{\omega_1} \cap \tilde{L}$ can be used in place of U to define analogues $\overline{U}_{s,t}$ to $U_{s,t}$ for $s, t \in {}^{<\omega}\omega$ with $|s| = |t|$ by

$$X \in \overline{U}_{s,t} \;\; \textit{iff} \;\; X \in \mathcal{P}(T^*_{s,t}) \cap \tilde{L} \;\wedge\; (\exists H \in \mathcal{C}_{\omega_1} \cap \tilde{L})([H]^{|s|} \subseteq \text{ran``}X) .$$

($\overline{U}_{\emptyset,\emptyset} = \{\{\emptyset\}\}$ is allowed as before.) Note that $T^*_{s,t} \in \tilde{L}$, since it is simple to check that $T^* \in L[b] \subseteq \tilde{L}$ for any $b \in {}^\omega\omega$ such that A is $\Pi^1_2(b)$ (cf. 13.14; as there, ω_1 may differ from $\omega_1^{L[b]}$). It follows by 15.8 that *$\overline{U}_{s,t}$ is an ω_1-complete ultrafilter on $\mathcal{P}(T^*_{s,t}) \cap \tilde{L}$.* (In our approach through a motivating analogy, some use of AC is needed to choose countably many H's in $\mathcal{C}_{\omega_1} \cap \tilde{L}$ here, and to get an order-preserving map into On from well-foundedness below. DC suffices, but as we note before the summarizing result 15.12, all uses of choice principles can be eliminated from its proof.)

As before there is a coherence property: if $s, t, \bar{s}, \bar{t} \in {}^{<\omega}\omega$ with $|\bar{s}| = |\bar{t}|$ and $s = \bar{s}|m$ and $t = \bar{t}|m$ for some $m \leq |\bar{s}|$, then for $X \in \mathcal{P}(T^*_{s,t}) \cap \tilde{L}$,

$$X \in \overline{U}_{s,t} \;\; \textit{iff} \;\; \{u \in T^*_{\bar{s},\bar{t}} \mid u|m \in X\} \in \overline{U}_{\bar{s},\bar{t}} .$$

What about the analogues for the maps $j_{\langle s,t\rangle,\langle \bar{s},\bar{t}\rangle}$? Note that only their restrictions to On, indeed to γ_U, had been needed. So, first consider for $s, t \in {}^{<\omega}\omega$ with $|s| = |t|$ the ultrapower of $\langle \text{On}, <\rangle$ by $\overline{U}_{s,t}$ using only functions: $T^*_{s,t} \to \text{On}$ that are members of $\tilde{L}$:

For such a function f, set

$$(f)_{\overline{U}_{s,t}} = \{g \in \tilde{L} \mid g\colon T^*_{s,t} \to \mathrm{On} \wedge \{u \in T^*_{s,t} \mid f(u) = g(u)\} \in \overline{U}_{s,t}\} .$$

(Note that if $f \in L[a_0]$ and $g \in L[a_1]$ for some $a_0, a_1 \in {}^\omega\omega$, then $\{f, g\} \subseteq L[a]$ for any $a \in {}^\omega\omega$ coding a_0 and a_1, so that $\{u \in T^*_{s,t} \mid f(u) = g(u)\} \in L[a] \subseteq \tilde{L}$.) Now define $(f)^0_{\overline{U}_{s,t}}$ as in §5 to consist of those members of $(f)_{\overline{U}_{s,t}}$ of minimal rank, so that $(f)^0_{\overline{U}_{s,t}}$ is a set. The domain of the ultrapower can then be formulated as

$${}^{T^*_{s,t}}\mathrm{On}/\overline{U}_{s,t} = \{(f)^0_{\overline{U}_{s,t}} \mid f\colon T^*_{s,t} \to \mathrm{On} \wedge f \in \tilde{L}\}$$

with its ordering $<_{\overline{U}_{s,t}}$ defined by

$$(f)^0_{\overline{U}_{s,t}} <_{\overline{U}_{s,t}} (g)^0_{\overline{U}_{s,t}} \quad \textit{iff} \quad \{u \in T^*_{s,t} \mid f(u) < g(u)\} \in \overline{U}_{s,t} .$$

(Note that $\{u \in T^*_{s,t} \mid f(u) < g(u)\} \in \tilde{L}$ as before.) Since $\overline{U}_{s,t}$ is ω_1-complete, this ultrapower is well-ordered (cf. 5.3), and hence there is a unique isomorphism

$$\pi_{\overline{U}_{s,t}}\colon \langle {}^{T^*_{s,t}}\mathrm{On}/\overline{U}_{s,t}, <_{\overline{U}_{s,t}}\rangle \to \langle \mathrm{On}, <\rangle .$$

As in §5, set

$$[f]_{\overline{U}_{s,t}} = \pi_{\overline{U}_{s,t}}((f)^0_{\overline{U}_{s,t}}) .$$

There is a corresponding ultrapower embedding $\overline{j}_{s,t}\colon \mathrm{On} \to \mathrm{On}$ given by $\overline{j}_{s,t}(\alpha) = [f_\alpha]_{\overline{U}_{s,t}}$, where f_α is the constant function: $T^*_{s,t} \to \{\alpha\}$.

For $s, t, \overline{s}, \overline{t} \in {}^{<\omega}\omega$ with $|s| = |t|$ and $s = \overline{s}|m$ and $t = \overline{t}|m$ for some $m \leq |s|$, define analogues

$$\overline{j}_{\langle s,t\rangle,\langle \overline{s},\overline{t}\rangle}\colon \mathrm{On} \to \mathrm{On}$$

to $j_{\langle s,t\rangle,\langle \overline{s},\overline{t}\rangle}$ by:

$$\overline{j}_{\langle s,t\rangle,\langle \overline{s},\overline{t}\rangle}([f]_{\overline{U}_{s,t}}) = [\langle f(u|m) \mid u \in T^*_{\overline{s},\overline{t}}\rangle]_{\overline{U}_{\overline{s},\overline{t}}} .$$

The coherence property for the ultrafilters insures that these maps are order-preserving.

From the tree T^+, a new tree analogous to the former $\tilde{T}$ is now defined. This time, for $\langle s, \langle h_i \mid i < |s|\rangle\rangle \in T^+$, h_i is to be replaced by the ordinal $[h_i]$ that it represents in the ultrapower of On by $\overline{U}_{s||\mathbf{s}_i|,\mathbf{s}_i}$. Since $[h_i] < \overline{j}_{s||\mathbf{s}_i|,\mathbf{s}_i}(\omega_1)$ by condition (a) of orderliness, the result will be a tree on $\omega \times \overline{\gamma}$, where

$$\overline{\gamma} = \sup(\{\overline{j}_{s,t}(\omega_1) \mid s, t \in {}^{<\omega}\omega \wedge |s| = |t|\}) .$$

Important conclusions will soon be drawn from a detailed analysis of $\overline{\gamma}$, but for now the condition (b) of orderliness is cast in the formulation of the new tree, which again will be called $\tilde{T}$:

$$\tilde{T} = \{\langle s, \langle \delta_i \mid i < |s| \rangle\rangle \in \bigcup_{m\in\omega}({}^m\omega \times {}^m\overline{\gamma}) \mid \\ \forall i, j < |s|(\mathbf{s}_i \supset \mathbf{s}_j \neq \emptyset \ \rightarrow \ \delta_i < \overline{j}_{\langle s||\mathbf{s}_j|, \mathbf{s}_j\rangle, \langle s||\mathbf{s}_i, \mathbf{s}_i\rangle}(\delta_j))\}$$

The analogue of 15.5 holds:

15.9 Exercise. $A = p[\tilde{T}]$ *(with the current definition of* $\tilde{T}$*).*

Hint. To show that $A \subseteq p[\tilde{T}]$, suppose that $x \in {}^\omega\omega$ and $A(x)$, i.e. T^*_x is well-founded. $T^* \in L[b]$ for $b \in {}^\omega\omega$ such that A is $\Pi^1_2(b)$, so if $c \in {}^\omega\omega$ codes b and x, then $T^*_x \in L[c]$. Arguing as for $\mathbf{\Sigma}^1_2$ Absoluteness 13.15, T^*_x is also well-founded in $L[c]$, and so applying 15.3 in $L[c]$ there is an order-preserving map $f\colon T^*_x - \{\langle\emptyset, \emptyset\rangle\} \to \omega_1$ with $f \in L[c]$. Such a map can be elaborated as a set $\{h_i \mid i \in \omega\} \subseteq L[c] \subseteq \tilde{L}$ such that $\langle x|m, \langle h_0, \ldots, h_{m-1}\rangle\rangle$ is orderly for $x|m$ (as in the definition of T^+) for every $m \in \omega$. Consequently, if $g\colon \omega \to \overline{\gamma}$ is defined by $g(i) = [h_i]$, the equivalence class in the sense of $\overline{U}_{x||\mathbf{s}_i|, \mathbf{s}_i}$, then $\langle x, g\rangle \in [\tilde{T}]$ to confirm that $x \in p[\tilde{T}]$.

The converse is just as for 15.5. ⊣

Whenever there are Suslin representations for the $\mathbf{\Pi}^1_2$ sets, there are corresponding representations for the $\mathbf{\Sigma}^1_3$ sets, and some observations can be made about their absoluteness and regularity properties. Suppose that for $x \in {}^\omega\omega$,

$$C(x) \leftrightarrow \exists^1 y D(x, y)$$

where $D \subseteq {}^2({}^\omega\omega)$ is $\Pi^1_2(a)$ for some $a \in \omega$, and there is a tree T^D on ${}^2\omega \times \kappa$ for some κ such that $D = p[T^D]$. Then

$$C(x) \leftrightarrow T^D_x \text{ is ill-founded} .$$

Recall the discussion before 13.15 about the relativization of analytical relations in terms of their normal forms and the corresponding formulation of absoluteness.

15.10 Exercise (Mansfield [71]). *Suppose that* M *is an inner model with* $\{a, T^D\} \subseteq M$. *Then:*

(a) C *is absolute for* M.

(b) If ${}^\omega\omega \cap M$ *is countable, then* C *is Lebesgue measurable, has the Baire Property, and has the perfect set property.*

Hint. (a) For any $x \in {}^\omega\omega \cap M$,

$$\begin{aligned} C(x) &\leftrightarrow T^D_x \text{ is ill-founded} \\ &\leftrightarrow (T^D_x \text{ is ill-founded})^M \\ &\leftrightarrow (\exists y \in {}^\omega\omega \cap M)(T^D_{\langle x,y\rangle} \text{ is ill-founded})^M \\ &\leftrightarrow (\exists y \in {}^\omega\omega \cap M)(T^D_{\langle x,y\rangle} \text{ is ill-founded}) \\ &\leftrightarrow (\exists y \in {}^\omega\omega \cap M)D(x, y) \\ &\leftrightarrow (\exists y \in {}^\omega\omega \cap M)D^M(x, y) \ \ \text{(by } \mathbf{\Sigma}^1_2 \text{ Absoluteness)} \\ &\leftrightarrow C^M(x) . \end{aligned}$$

(b) For the first two properties, note that by (a),

$$C = \{x \in {}^{\omega}\omega \mid M[x] \models \varphi[x]\}$$

for some formula φ, and use Solovay's arguments for 14.1 and 14.2. For the third property, apply 14.7. ⊣

The significance of the tree $\tilde{T}$ on $\omega \times \overline{\gamma}$ lies not only in such properties, but also in the possibility afforded by the sharps of establishing a contextually optimal Suslin property low in the cumulative hierarchy. This is a consequence of a calculation of $\overline{\gamma}$ in terms of the class $\{u_\xi \mid 0 < \xi \in \mathrm{On}\}$ of uniform indiscernibles discussed at the end of §14. The results established there lead to the following connection between the indiscernibles and the ultrapower embeddings that figured in the formulation of $\tilde{T}$:

15.11 Lemma. *Suppose that $s, t \in {}^{<\omega}\omega$ with $|s| = |t|$. Then*

$$\overline{j}_{s,t}(\omega_1) = u_{|s|+1} \ .$$

Proof. Since $\overline{j}_{s,t}(\omega_1)$ is

$$\{[h]_{\overline{U}_{s,t}} \mid h\colon T^*_{s,t} \to \omega_1 \ \wedge \ h \in \tilde{L}\} \ ,$$

we must show that the uniform indiscernible $u_{|s|+1}$ coincides with this set of ordinals.

Suppose that $h\colon T^*_{s,t} \to \omega_1$ with $h \in \tilde{L}$. Then for some $a \in {}^{\omega}\omega$, $h \in L[a]$, and moreover by the proof of GCH in $L[a]$, h is the γth element in the canonical well-ordering of $L[a]$ (3.3(b)) for some $\gamma < u_2$ (as $\omega_1 = u_1 < u_2$). Using 14.18(b) for this γ and coding more into a if necessary (or proceeding directly as in its proof), it follows that

$$h = t_1^{\langle L[a], \in, a\rangle}(\omega_1)$$

for some Skolem term t_1.

It can henceforth be assumed that $|s| > 0$, as $\overline{j}_{\emptyset,\emptyset}(\omega_1) = \omega_1 = u_1$. By definition of T^* there is a unique permutation $\pi\colon |s| \to |s|$ such that for any $\xi_0 < \ldots < \xi_{|s|-1} < \omega_1$, $\langle \xi_{\pi(0)}, \ldots, \xi_{\pi(|s|-1)}\rangle \in T^*_{s,t}$. A small adjustment of t_1 leads to a new Skolem term t such that for any $\xi_0 < \ldots < \xi_{|s|-1} < \omega_1$,

$$h(\langle \xi_{\pi(0)}, \ldots, \xi_{\pi(|s|-1)}\rangle) = t^{\langle L[a], \in, a\rangle}(\xi_0, \ldots, \xi_{|s|-1}, \omega_1) < \omega_1 \ .$$

Cast this way, h can be naturally extended. In particular, consider

$$\eta(h, a, t) = t^{\langle L[a], \in, a\rangle}(u_1, \ldots, u_{|s|}, u_{|s|+1}) < u_{|s|+1} \ .$$

(By an argument used in the proof of 14.18(b), $\eta(h, a, t) = \overline{t}^{\langle L[a^\#], \in, a^\#\rangle}(u_1, \ldots, u_{|s|})$ for some Skolem term $\overline{t}$, i.e. the dependence on $u_{|s|+1}$ can be eliminated at the cost of replacing a by $a^\#$. However, while resonating with a surjectivity argument at the end of this proof, this further transformation is not needed here.)

First observe that if $\overline{h} \in [h]_{\overline{U}_{s,t}}$ and $\eta(\overline{h}, \overline{a}, \overline{t})$ is derived from some $\overline{a}$ and $\overline{t}$ for $\overline{h}$, then $\eta(h, a, t) = \eta(\overline{h}, \overline{a}, \overline{t})$: By definition of $\overline{U}_{s,t}$, there is a C closed unbounded in ω_1 such that for $\xi_0 < \ldots < \xi_{|s|-1}$ all in C, $h(\langle \xi_{\pi(0)}, \ldots, \xi_{\pi(|s|-1)} \rangle) = \overline{h}(\langle \xi_{\pi(0)}, \ldots, \xi_{\pi(|s|-1)} \rangle)$. Since $I_a \cap I_{\overline{a}} \cap \omega_1$ is also closed unbounded in ω_1, there are such $\xi_0 < \ldots < \xi_{|s|-1}$ all in $I_a \cap I_{\overline{a}}$, and so

$$\begin{aligned} t^{\langle L[a], \in, a \rangle}(\xi_0, \ldots, \xi_{|s|-1}) &= h(\langle \xi_{\pi(0)}, \ldots, \xi_{\pi(|s|-1)} \rangle, \omega_1) \\ &= \overline{h}(\langle \xi_{\pi(0)}, \ldots, \xi_{\pi(|s|-1)} \rangle) \\ &= \overline{t}^{\langle L[\overline{a}], \in, \overline{a} \rangle}(\xi_0, \ldots, \xi_{|s|-1}, \omega_1) \ . \end{aligned}$$

Hence, if $c \in {}^\omega\omega$ codes a and $\overline{a}$, then by definability and indiscernibility applied in $L[c]$,

$$\begin{aligned} \eta(h, a, t) &= t^{\langle L[a], \in, a \rangle}(u_1, \ldots, u_{|s|}, u_{|s|+1}) \\ &= \overline{t}^{\langle L[\overline{a}], \in, \overline{a} \rangle}(u_1, \ldots, u_{|s|}, u_{|s|+1}) \\ &= \eta(\overline{h}, \overline{a}, \overline{t}) \ . \end{aligned}$$

It now follows that

$$e\colon \{[h]_{\overline{U}_{s,t}} \mid h\colon T^*_{s,t} \to \omega_1 \ \wedge \ h \in \tilde{L}\} \ \to \ u_{|s|+1}$$

is well-defined by: $e([h]_{\overline{U}_{s,t}}) = \eta(h, a, t)$ for some a and t for h. An analogous argument shows that e is order-preserving. Finally, any $\gamma < u_{|s|+1}$ is of form $t^{\langle L[a], \in, a \rangle}(u_1, \ldots, u_{|s|})$ for some $a \in {}^\omega\omega$ by 14.18(b), so it follows that e is surjective. Hence, e is a bijection (and consequently the identity map), and the proof is complete. ⊣

That the map e is the identity implies that $\tilde{T}$ could have been formulated in terms of a well-ordering of Skolem terms involving the u_n's without reference to the ultrafilters $\overline{U}_{s,t}$. Consequently, the foregoing analysis can be carried out in ZF without relying on a choice principle (DC) to ensure well-foundedness of ultrapowers (cf. Moschovakis [80: 596ff]). On the other hand, the ultrafilter formulation motivates $\tilde{T}$ by embedding it into a larger context.

It follows from 15.11 that

$$\begin{aligned} \overline{\gamma} &= \sup(\{\overline{j}_{s,t}(\omega_1) \mid s, t \in {}^{<\omega}\omega \ \wedge \ |s| = |t|\}) \\ &= \sup(\{u_n \mid 0 < n \in \omega\}) \\ &= u_\omega \ , \end{aligned}$$

and this leads to the following result, incorporating the previous remark about provability in ZF.

15.12 Theorem (ZF)(Martin). *Assume that* $\forall a \in {}^\omega\omega(a^\#$ exists). *Then every* $\mathbf{\Sigma}^1_3$ *set is* u_ω*-Suslin.* ⊣

The Axiom of Determinacy actually implies that $u_n = \omega_n$ for every $n \leq \omega$ (28.9). In the ZFC context we have the following remarkable structure result for $\mathbf{\Sigma}^1_3$ sets:

15.13 Corollary (Martin). *Assume that* $\forall a \in {}^\omega\omega(a^\#$ exists). *Then every* $\mathbf{\Sigma}^1_3$ *set is* ω_2*-Suslin, and hence a union of* $\aleph_2$ *Borel sets.*

Proof. By 14.19, $u_\omega < \omega_3$. The latter assertion follows from 13.13(f). ⊣

This is analogous to the classical result 13.7, and implies under the existence of sharps that *every* $\mathbf{\Sigma}^1_3$ *set has cardinality at most* $\aleph_2$ *or a perfect subset and hence cardinality* $2^{\aleph_0}$.

The definability properties of $\tilde{T}$ also lead to a uniformization result for $\mathbf{\Pi}^1_2$ sets. As discussed at the end of §13, every $\mathbf{\Sigma}^1_2$ set can be uniformized by a $\mathbf{\Sigma}^1_2$ set, and but it is consistent that there is a Π^1_2 set that cannot be uniformized by any projective set. The tree analysis of Mansfield [71] together with the sharps analysis of Martin-Solovay [69] leads to the following:

15.14 Theorem (Mansfield [71], Martin-Solovay [69]). *Assume that* $\forall a \in {}^\omega\omega(a^\#$ exists). *Then for any* $b \in {}^\omega\omega$, *every* $\Pi^1_2(b)$ *set can be uniformized by a* $\Pi^1_3(b)$ *set.*

Proof. The structure of the proof is as for the $\mathbf{\Pi}^1_1$ Uniformization 13.17, which established: If $A \subseteq {}^2({}^\omega\omega)$ is $\Pi^1_1(a)$, then there is a $\Pi^1_1(a)$ set $A_0 \subseteq A$ such that for any $x \in {}^\omega\omega$, if there is a y such that $A(x, y)$ then there is exactly one y such that $A_0(x, y)$. Availing ourselves of our representative $\mathbf{\Pi}^1_2$ set $A \subseteq {}^\omega\omega$ and the theory developed for it through 15.9, we shall establish the following "basis" theorem: For $b \in {}^\omega\omega$ such that A is $\Pi^1_2(b)$, there is a $\Pi^1_3(b)$ set $A_0 \subseteq A$ such that if there is a y such that $A(y)$, then there is exactly one y such that $A_0(y)$. With only notational changes A and A_0 can be recast as subsets of ${}^2({}^\omega\omega)$ and a parametrized version of the proof can be given to establish the usual uniformization assertion.

Let T^* on ${}^2\omega \times \omega_1$ and $\tilde{T}$ on $\omega \times \overline{\gamma}$ be as before so that for any $y \in {}^\omega\omega$,

$$\begin{aligned} A(y) &\leftrightarrow T^*_y \text{ is well-founded} \\ &\leftrightarrow \exists g \in {}^\omega\overline{\gamma}(\langle y, g\rangle \in [\tilde{T}]) . \end{aligned}$$

First observe that for any $y \in {}^\omega\omega$ if there is a g such that $\langle y, g\rangle \in [\tilde{T}]$ at all, then there is an honest leftmost one, i.e. a g_y such that for any other g, $g_y(i) \leq g(i)$ for *every* $i \in \omega$:

Suppose that $A(y)$, so that T^*_y is well-founded. By 12.9(a) and 15.3 the restriction of its rank function $\rho_{T^*_y}$ to $T^*_y - \{\langle\emptyset, \emptyset\rangle\}$ is an order-preserving map into ω_1. Moreover, by the relativization argument for 15.9, for any $c \in {}^\omega\omega$ that codes b and y, $\rho_{T^*_y} \in L[c] \subseteq \tilde{L}$. Consequently, $\rho_{T^*_y}$ can be elaborated as $\{h^y_i \mid i \in \omega\} \subseteq \tilde{L}$ where $h^y_i \colon T^*_{y||\mathbf{s}_i|,\mathbf{s}_i} \to \omega_1$ is specified by $h^y_i(u) = \rho_{T^*_y}(\langle \mathbf{s}_i, u\rangle)$ for $i \neq 0$ and

$h^y_0(\emptyset) = 0$. Then if $g_y \in {}^{\omega}\overline{\gamma}$ is defined by $g_y(i) = [h^y_i]$, the equivalence class in the sense of $\overline{U}_{y||\mathbf{s}_i|,\mathbf{s}_i}$, it follows that $\langle y, g_y \rangle \in [\tilde{T}]$.

Suppose now that $\langle y, g \rangle \in [\tilde{T}]$ with g arbitrary. For $i \in \omega$ let $h_i \colon T^*_{y||\mathbf{s}_i|,\mathbf{s}_i} \to$ On be such that $g(i) = [h_i]$, the equivalence class in the sense of $\overline{U}_{y||\mathbf{s}_i|,\mathbf{s}_i}$. Then for $\mathbf{s}_i \supset \mathbf{s}_j \neq \emptyset$,

$$\{u \in T^*_{y||\mathbf{s}_i|,\mathbf{s}_i} \mid h_i(u) < h_j(u||\mathbf{s}_j|)\} \in \overline{U}_{y||\mathbf{s}_i|,\mathbf{s}_i} ,$$

and so there is a C_{ij} closed unbounded in ω_1 such that

$$(\dagger) \qquad \operatorname{ran}(u) \in [C_{ij}]^{|\mathbf{s}_i|} \to h_i(u) < h_j(u||\mathbf{s}_j|) .$$

Set $C = \bigcap\{C_{ij} \mid i, j \in \omega \ \wedge \ \mathbf{s}_i \supset \mathbf{s}_j \neq \emptyset\}$, a set closed unbounded in ω_1.

Next, let $e \colon C \to \omega_1$ be the order-preserving bijection. Then e induces an isomorphism between T^*_y and a tree T^C on $\omega \times C$ defined by

$$\langle t, u \rangle \in T^C \quad \textit{iff} \quad \langle t, e \circ u \rangle \in T^*_y$$

since the order relationships for the second coordinates are preserved. As $A(y)$, both T^*_y and T^C are well-founded, and their rank functions are related by

$$\rho_{T^C}(\langle t, u \rangle) = \rho_{T^*_y}(\langle t, e \circ u \rangle) .$$

Hence, by 12.9(a) and how the h_i's make up an order-preserving map: $T^C \to$ On by $(\dagger)$,

$$\rho_{T^*_y}(\langle \mathbf{s}_i, e \circ u \rangle) \leq h_i(u) \text{ for } 0 < i \in \omega \text{ and } u \in {}^{|\mathbf{s}_i|}C .$$

Finally, it is simple to see that there is a $D \subseteq C$ closed unbounded in ω_1 on which e is the identity, so that

$$\rho_{T^*_y}(\langle \mathbf{s}_i, u \rangle) \leq h_i(u) \text{ for } 0 < i \in \omega \text{ and } u \in {}^{|\mathbf{s}_i|}D .$$

But how the h^y_i's were defined from $\rho_{T^*_y}$ implies that

$$g_y(i) = [h^y_i] \leq [h_i] = g(i) \ \text{ for } \textit{every } i \in \omega .$$

Proceeding now as in the argument for $\mathbf{\Pi}^1_1$ Uniformization, the idea is to uniformize by choosing that y of the leftmost $\langle y, g \rangle$ in $[\tilde{T}]$ if there is one. With A being $\Pi^1_2(b)$, that this can be done in a $\Pi^1_3(b)$ fashion relies on the observation that g would have to be the honest leftmost g_y as described above.

Let $R^{\leq}$ and $R^{<}$ be relations on $\omega \times {}^2({}^{\omega}\omega)$ given by

$$R^{\leq}(i, y, z) \leftrightarrow \neg A(y) \ \vee \ \neg A(z) \ \vee \ g_y(i) \leq g_z(i) , \text{ and}$$
$$R^{<}(i, y, z) \leftrightarrow \neg A(y) \ \vee \ \neg A(z) \ \vee \ g_y(i) < g_z(i) .$$

It suffices to show that these are $\Sigma^1_3(b)$. If $A_0 \subseteq A$ were then defined following the proof for 13.17, it would uniformize as desired and be essentially of form $\Pi^1_2(b) \ \wedge \ \forall^1[\Sigma^1_3(b) \to \neg\Sigma^1_3(b)]$ and hence $\Pi^1_3(b)$ to complete the proof.

To show that $R^{\leq}$ is $\Sigma^1_3(b)$, the argument for $R^<$ being analogous, note first that $\neg A$ is $\Sigma^1_2(b)$, so it remains to show assuming $A(y)$ and $A(z)$ that $g_y(i) \leq g_z(i)$ is $\Sigma^1_3(b)$:

For this purpose, recall that any $x \in {}^\omega\omega$ codes $\langle (x)_k \mid k \in \omega\rangle \in {}^\omega({}^\omega\omega)$ by $(x)_k(m) = x(\langle k, m\rangle)$, and let $a(y,z) \in {}^\omega\omega$ be such that $(a(y,z))_0 = b$, $(a(y,z))_1 = y$, $(a(y,z))_2 = z$, and $(a(y,z))_n =$ the constant function: $\omega \to \{\emptyset\}$ for $n > 2$. Now by the characterization of the honest leftmost g_y, for $i \in \omega$, $g_y(i) = [h^y_i]$, the equivalence class in the sense of $\overline{U}_{y||\mathbf{s}_i|,\mathbf{s}_i}$, where in turn $h^y_i \colon T^*_{y||\mathbf{s}_i|,\mathbf{s}_i} \to \omega_1$ is given by $h^y_i(u) = \rho_{T^*_y}(\langle \mathbf{s}_i, u\rangle)$. Applying a previous remark, $\rho_{T^*_y} \in L[a(y,z)]$ as $a(y,z)$ codes b and y, and in fact a closer look shows that $\rho_{T^*_y}$ is definable in $\langle L[a(y,z)], \in, a(y,z)\rangle$, via the same formula for all y and z, in terms of the one parameter needed to define T^*_y there, namely (the real) ω_1.

We next incorporate aspects of the proof of 15.11. For $0 < i \in \omega$, there is a unique permutation $\pi_{i,y} \colon |\mathbf{s}_i| \to |\mathbf{s}_i|$ such that for any $\xi_0 < \ldots < \xi_{|\mathbf{s}_i|-1} < \omega_1$, $\langle \xi_{\pi_{i,y}(0)}, \ldots, \xi_{\pi_{i,y}(|\mathbf{s}_i|-1)}\rangle \in T^*_{y||\mathbf{s}_i|,\mathbf{s}_i}$. It follows from the aforementioned definability of $\rho_{T^*_y}$ in terms of ω_1 that for $0 < i \in \omega$ Skolem terms t_i can be defined uniformly, not depending on y or z, such that

$$\begin{aligned} h^y_i(\langle \xi_{\pi_{i,y}(0)}, \ldots, \xi_{\pi_{i,y}(|\mathbf{s}_i|-1)}\rangle) &= \rho_{T^*_y}(\langle \mathbf{s}_i, \langle \xi_{\pi_{i,y}(0)}, \ldots, \xi_{\pi_{i,y}(|\mathbf{s}_i|-1)}\rangle\rangle) \\ &= t_i^{\langle L[a(y,z)], \in, a(y,z)\rangle}(\xi_0, \ldots, \xi_{|\mathbf{s}_i|-1}, \omega_1)\ . \end{aligned}$$

The proof of 15.11 now shows (through its mapping e being the identity) that

$$[h^y_i] = t_i^{\langle L[a(y,z)], \in, a(y,z)\rangle}(u_1, \ldots, u_{|\mathbf{s}_i|}, u_{|\mathbf{s}_i|+1})\ .$$

All of the foregoing applies, of course, to get an analogous expression for $[h^z_i]$ in terms of some $\bar{t}_i$. Consequently, for each $i \in \omega$ there is a formula φ_i such that

$$\begin{aligned} g_y(i) \leq g_z(i) &\leftrightarrow [h^y_i] \leq [h^z_i] \\ &\leftrightarrow \langle L[a(y,z)], \in, a(y,z)\rangle \models \varphi_i[u_1, \ldots, u_{|\mathbf{s}_i|+1}] \\ &\leftrightarrow \ulcorner \varphi_i(c_0, \ldots, c_{|\mathbf{s}_i|})\urcorner \in a(x,y)^\# \end{aligned}$$

where the c_i's are constants for indiscernibles in the §9 language $\mathcal{L}^*_\in$ for the theory of sharps. By the uniformity of the definitions it is straightforward to see that the function taking i to $\ulcorner \varphi_i(c_0, \ldots, c_{|\mathbf{s}_i|})\urcorner$ can be taken to be recursive. Hence, with b coded in $a(x,y)$, 14.16 and the argument for 14.12(c) implies that the $g_y(i) \leq g_z(i)$ is $\Sigma^1_3(b)$ (in fact, $\Delta^1_3(b)$) to complete the proof. ⊣

Adjoining an $\exists$ to the result is straightforward (cf. 13.18).

15.15 Exercise. *Assume that* $\forall a \in {}^\omega\omega(a^\#$ exists). *Then for any* $b \in {}^\omega\omega$, *every* $\Sigma^1_3(b)$ *set can be uniformized by a* $\Sigma^1_4(b)$ *set.* ⊣

Thus, the methods and results of classical descriptive set theory were extended to the third level of the projective hierarchy. What about the fourth level?

Having developed a tree representation for $\mathbf{\Pi}^1_2$ sets and hence for $\mathbf{\Sigma}^1_3$ sets, the next step would be to convert the latter to a tree representation for $\mathbf{\Pi}^1_3$ sets by again appealing to the basic equivalence between well-foundedness and having an order-preserving map into the ordinals. Suppose that $E \subseteq {}^\omega\omega$ is $\mathbf{\Pi}^1_3$, say

$$E(x) \;\leftrightarrow\; \neg\exists y A(x, y)$$

where $A \subseteq {}^2({}^\omega\omega)$ is $\mathbf{\Pi}^1_2$. This A can be regarded as a binary version of our previous, representative A; recalling that $\overline{\gamma} = u_\omega$ (as noted before 15.12), the new A is $p[\tilde{T}]$ for a corresponding $\tilde{T}$ on ${}^2\omega \times u_\omega$. Consequently,

$$\begin{aligned} E(x) \;&\leftrightarrow\; \neg\exists y\exists g \in {}^\omega u_\omega(\langle x, y, g\rangle \in [\tilde{T}]) \\ &\leftrightarrow\; \tilde{T}_x \text{ is well-founded} \\ &\leftrightarrow\; \exists f(f\colon \tilde{T}_x \to u^+_\omega \text{ is an order-preserving map})\,. \end{aligned}$$

A tree analogous to T^* could now be developed of finite approximations to order-preserving maps. This done, certain ordinal-valued functions would have to be ranked, requiring ω_1-complete ultrafilters $U_{s,t}$ on $\mathcal{P}(X_{s,t})$ for some subsets $X_{s,t}$ of ${}^{|s|}u_\omega$. These ultrafilters would again have to satisfy coherence and well-foundedness properties, and these would be derived most directly from homogeneity properties of $\tilde{T}$. How can such properties be established?

Looking further into the construction of $\tilde{T}$, if $\langle s, \langle \delta_i \mid i < |s|\rangle\rangle \in \tilde{T}$, there is a sequence $\langle h_i \mid i < |s|\rangle$ orderly for s such that $\delta_i = [h_i]$. Such a sequence in turn corresponds to a function on $Z_s = \{\langle \mathbf{s}_i, u\rangle \mid \exists i < |s|(\langle s||\mathbf{s}_i|, \mathbf{s}_i, u\rangle) \in T^*\}$ given by $h(\langle \mathbf{s}_i, u\rangle) = h_i(u)$. Moreover, for $s \neq \emptyset$, Z_s is seen to be well-ordered by $<_{\mathrm{KB}^2}$ in ordertype ω_1 by remarks before 15.3. Thus, Z_s can be identified with a member of $[\omega_1]^{\omega_1}$, which would now play the role of $[\omega_1]^n$. But now, we would need a partition property for $[\omega_1]^{\omega_1}$ analogous to 15.8, i.e. some substantive version of the full $\omega_1 \longrightarrow (\omega_1)^{\omega_1}_\omega$. Such properties contradict AC, but actually follow from the Axiom of Determinacy (cf. 28.13), and moreover the ZFC context could be maintained by only assuming that axiom in the inner model $L(\mathbb{R})$, which could play the role of $\tilde{L}$.

In 1971 Kunen thus proceeded to derive a tree representation for $\mathbf{\Pi}^1_3$ sets, and forging ahead, similarly comprehended the $\mathbf{\Pi}^1_4$ sets. This work was to stand as the high point of the contextually optimal analysis of the projective sets for well over a decade, further progress being stymied by the inability to gain sufficient information about an ordinal known as δ^1_5. How this work was resoundingly completed in the mid-1980's is discussed in §30.

Chapter 4

Aspects of Measurability

This chapter describes the wide-ranging investigation of measurability carried out in the late 1960's with the complementary methods of forcing and inner models. These developments not only provided a coherent and illuminating structural analysis, but suggested new questions and provided paradigms for the investigation of stronger hypotheses. §16 and §17 discuss saturated ideals, a subject brought to the forefront by work of Solovay. The first section presents his combinatorial analysis, and the second, his consistency result on real-valued measurability and then the cycle of Kunen-Paris consistency results based on the extendibility of elementary embeddings. §18 describes Prikry forcing, a notion that provided yet another understanding of measurability, and a generalization, one of several that were to become crucial to the careful study of consistency strength. The rest of the chapter is based on Kunen's work on iterated ultrapowers, inner models of measurability, and elementary embeddings: With the general concept of M-ultrafilter, his iterated ultrapower theory is developed in §19. §20 applies it to derive the structure theorems of Kunen and Silver on inner models of measurability. Finally, §21 gives Kunen's result that the existence of $0^{\#}$ is equivalent to the existence of a $j\colon L \prec L$, and a detailed analysis of Solovay's $0^{\dagger}$, the existence of which is a principle of transcendence over inner models of measurability.

16. Saturated Ideals I

Addressing another aspect of Lebesgue measure, Solovay derived the principal results on its extendibility, in the presence of AC, to a measure on all sets of reals. This is real-valued measurability $\leq 2^{\aleph_0}$, the other branch of the bifurcation found by Ulam in his fundamental paper [30] on Banach's measure problem (§2).

16.1 Theorem (Solovay [66, 71]).

(a) If κ is a measurable cardinal, then there is a p.o. P such that

$$\Vdash_P \kappa = 2^{\aleph_0} \wedge \kappa \text{ is real-valued measurable} .$$

(b) If κ is a real-valued measurable cardinal, then there is an inner model of ZFC *in which κ is measurable.*

(a) is established in the next section (17.5); the original proof of (b) is sketched in the setting of inner models of measurability (20.4); and a sharp version of (b) is established much later (see Precipitous Ideals in volume II). Solovay's paper [71] was particularly influential, for in establishing 16.1 not only did it broaden the study of large cardinal properties from ultrafilters to ideals, but it also described how forcing and ultrapowers can be combined in a useful technique now known as *generic ultrapowers*. Owing to various developments in the 1970's generic ultrapowers was to emerge as a standard technique of wide applicability, and it is taken up against a broader backdrop in volume II.

What played a key role in this area and soon became a staple feature of large cardinal theory is the concept of *saturated ideal*, formulated and studied by Tarski [45]. Building on the §0 preliminaries about ideals, for I an ideal over a set S, set

$$I^* = \{X \subseteq S \mid S - X \in I\} , \text{ and}$$
$$I^+ = \mathcal{P}(S) - I .$$

I^* is the *dual filter of* I, and I^+ the collection of I^*-stationary sets, i.e. those sets $X \subseteq S$ such that $X \cap Z \neq \emptyset$ for any $Z \in I^*$. $A \in I^+$ *iff*

$$I|A = \{X \subseteq S \mid X \cap A \in I\}$$

is a (proper) ideal, the *restriction of* I *to* A or the *ideal generated by* $I \cup \{S - A\}$. Intuitively, I, I^+, and I^* are the collections of negligible, non-negligible, and all but negligible subsets of S with respect to I. For S a regular cardinal $\kappa > \omega$, the main example is $I = \text{NS}_\kappa$, the ideal of non-stationary subsets, when I^+ is the collection of stationary subsets and I^* the closed unbounded filter $\mathcal{C}_\kappa$. Of course, the theory of ideals can be recast entirely in terms of filters, an approach consistent with the modern preference for ultrafilters over maximal ideals. However, like the choice of $p \leq q$ in forcing for p is stronger than q, the ideal formulation is more natural because of the interplay with Boolean algebras.

Turning to Tarski's concept, for an ideal I,

$$I \text{ is } \lambda\text{-}saturated \;\; iff \;\; \text{for any } \{X_\alpha \mid \alpha < \lambda\} \subseteq I^+ \text{ there are } \beta < \gamma < \lambda \text{ such that } X_\beta \cap X_\gamma \in I^+ ,$$

and

$$\text{sat}(I) = \text{the least } \lambda \text{ such that } I \text{ is } \lambda\text{-saturated} .$$

sat(I) is a measure of how close I is to being a maximal ideal, with sat$(I) = 2$ *iff* I is maximal. Erdős-Tarski [43] established that sat(I), if infinite, is always a regular uncountable cardinal.

Only ideals I over cardinals κ will be considered in this section, for which by a §0 convention $\alpha = \{\xi \mid \xi < \alpha\} \in I$ for every $\alpha < \kappa$. Furthermore, for the study of large cardinal properties analogous to measurability only ideals of requisite completeness will be considered:

$$\text{By } \textit{ideal over } \kappa \text{ is meant } \kappa\text{-complete ideal over } \kappa .$$

It is simply to see that *for such ideals to exist* κ *must be regular.*

Saturation of ideals is a rather general concept; for example, the ideal $[\kappa]^{<\kappa}$ over κ is not λ-saturated *iff* there is a family F of λ *almost disjoint* subsets of κ: $F \subseteq [\kappa]^\kappa$ with $|F| = \lambda$ such that $X \neq Y$ both in F implies that $|X \cap Y| < \kappa$. An early result of combinatorial set theory is that such a family exists for $\lambda = \kappa^+$ by a simple recursive construction (Sierpiński [28]), so that $[\kappa]^{<\kappa}$ is not κ^+-saturated. The connection with real-valued measurability is provided by 2.1:

16.2 Exercise. *If m is a κ-additive measure over κ, then $\{X \subseteq \kappa \mid m(X) = 0\}$ is an ω_1-saturated ideal over κ.* ⊣

Solovay showed that only this property is all that is needed to establish the large cardinal consequences of real-valued measurability. Beyond this, the study of saturated ideals led to a basic result about stationary sets and soon developed into an important area of research in its own right. This section establishes the basic combinatorial results about saturated ideals and the next, the first relative consistency results.

Initial observations about the saturation of ideals were based on Ulam's fundamental work on measurability.

16.3 Exercise (Ulam [30]). *For any λ, there is no λ^+-saturated ideal over λ^+.*

Hint. Check that the proof of 2.4 with Ulam matrices works. ⊣

The following proposition is established using transfinite versions of Ulam's tree argument for 2.5(a); its (c) affirms 7.12 by 16.2.

16.4 Proposition. *Suppose that I is an ideal over κ.*

(a) (Tarski [45: 53]) *If I is λ-saturated where $2^{<\lambda} < \kappa$, then κ is measurable.*

(b) (Levy, Silver) *If I is κ-saturated and κ is weakly compact, then κ is measurable.*

(c) If I is λ-saturated for some $\lambda < \kappa$, then κ has the tree property.

Proof. For (a) or (b), suppose first that I has an atom, i.e. a set $A \in I^+$ such that whenever $A = B \cup C$ is a disjoint union, either $B \in I$ or $C \in I$. Then $\{X \subseteq \kappa \mid X \cap A \in I^+\}$ is a κ-complete ultrafilter, and hence κ is a measurable cardinal. So, we can assume to the contrary that I has no atoms, and derive a contradiction.

For this purpose, build a tree T ordered by $\supset$ and consisting of sets indexed by some of the members of ${}^{<\kappa}2$ as follows: Set $X_\emptyset = \kappa$. If X_s has been defined for some $s \in {}^{<\kappa}2$, then define $X_{s^\frown\langle 0\rangle}$ and $X_{s^\frown\langle 1\rangle}$ exactly when $X_s \in I^+$, in which case they are to satisfy

$$X_{s^\frown\langle 0\rangle}, X_{s^\frown\langle 1\rangle} \in I^+ \ , \ X_s = X_{s^\frown\langle 0\rangle} \cup X_{s^\frown\langle 1\rangle} \ , \ \text{and } X_{s^\frown\langle 0\rangle} \cap X_{s^\frown\langle 1\rangle} = \emptyset \ .$$

That I has no atoms makes this possible. Finally, if $\delta < \kappa$ is a limit and $s \in {}^{\delta}2$, define $X_s = \bigcap_{\alpha<\delta} X_{s|\alpha}$ exactly when $X_{s|\alpha}$ has been defined for each $\alpha < \delta$. The tree T thus built has the following property:

(∗) If $\gamma \leq \kappa$ and $s \in {}^{\gamma}2$, then the collection of "offshoots"
$\{X_{s|\alpha^\frown\langle i\rangle} \mid \alpha < \gamma \ \wedge \ X_{s|\alpha+1} \text{ is defined} \wedge s(\alpha) \neq i\}$ is pairwise disjoint .

For (a), (∗) and λ-saturation imply that T has height at most λ. But then, κ is a union of $2^{<\lambda} < \kappa$ sets in I, namely those X_s's without tree successors. This contradicts the κ-completeness of I.

For (b), note that if T had height less than κ, then by the inaccessibility of κ and the above argument, the κ-completeness of I would again be violated. Thus, it can be assumed that T is a κ-tree, and so by weak compactness it has a cofinal branch. But by (∗) this contradicts the κ-saturation of I.

To establish (c), suppose that $\langle T, <_T\rangle$ is a κ-tree; it can be assumed that $T = \kappa$. For $\xi < \kappa$ set $S_\xi = \{\zeta < \kappa \mid \xi <_T \zeta\}$, and let $U = \{\xi < \kappa \mid S_\xi \in I^+\}$. Then $\xi \in U$ and $\xi' <_T \xi$ implies that $\xi' \in U$. It is simple to see that the subtree determined by U has height κ because of the κ-completeness of I, and that each of its levels has cardinality less than λ by λ-saturation. 7.9 then implies that this subtree has a cofinal branch, which is also a cofinal branch through $\langle \kappa, <_T\rangle$. ⊣

The hypothesis of (c) cannot be weakened to I is κ-saturated (Kunen [78]). It follows from (a) that the theory of λ-saturation over κ essentially reduces to that of measurability unless $\lambda > \omega$ and $2^{<\lambda} \geq \kappa$. Having saturated ideals over κ does not require the strong inaccessibility of κ as evidenced by 16.2, and subject to some restrictions like 16.3 there could be saturated ideals low in the cumulative hierarchy.

Since an ideal I over κ is trivially $(2^\kappa)^+$-saturated, the substantive cases for λ-saturation are for λ in the range $\omega < \lambda \leq 2^\kappa$. In fact, the theory quickly devolved into three cases: κ^+, κ, and $<\kappa$ saturation. In what follows, Solovay's results are analyzed by progressively considering these cases in turn, deriving consequences of measurability which are really consequences of these restricted hypotheses. Although restricted as combinatorial properties, it should be mentioned at the outset that even κ^+-saturation has the consistency strength of measurability (see remarks at the beginning of §17). λ-saturation for $\kappa^+ < \lambda \leq 2^\kappa$ has been studied in Baumgartner [76: §5] and more extensively in Baumgartner-Taylor [82]; interesting questions remain, but large cardinals are generally not involved.

While κ-saturation amounts to the assertion that no subset of κ can be partitioned into κ disjoint I^+ sets, κ^+-saturation has no such intuitive feel. What is significant is the completeness of the quotient Boolean algebra $\mathcal{P}(\kappa)/I$. This is the Boolean algebra consisting of equivalence classes $[X]_I$ for $X \subseteq \kappa$, where

$$[X]_I = [Y]_I \ \textit{ iff } \ X \triangle Y \in I \ .$$

As usual, the subscripts are suppressed when clear from the context. The Boolean operations are the set-theoretic ones modulated by ideal properties:

$$[X] + [Y] = [X \cup Y]\ , \ [X] \cdot [Y] = [X \cap Y]\ , \ \text{and} \ -[X] = [\kappa - X]\ .$$

Consequently,

$$[X] \leq [Y] \ \textit{ iff } \ X - Y \in I\ ,$$

and

$$\mathbf{0} = [\emptyset] = I \ \text{ and } \ \mathbf{1} = [\kappa] = I^*$$

are the zero and one elements of the Boolean algebra. $\langle I^+, \subseteq \rangle$ is later considered as a notion of forcing (see Precipitous Ideals in volume II). Since then $p \perp q$ *iff* $p \cap q \in I$, the separative quotient $I^+/{\approx}$ is just $\mathcal{P}(\kappa)/I - \{\mathbf{0}\}$. In this guise *the* λ*-saturation of* I *is equivalent to the* λ*-c.c. of the forcing notion.*

The following was the first significant result about κ^+-saturation. Recall that a Boolean algebra B is *complete iff* for any $S \subseteq B$, the least upper bound $\sum S$ exists in the sense of B.

16.5 Proposition (Smith-Tarski [57: 254]). *If I is a κ^+-saturated ideal over κ, then $\mathcal{P}(\kappa)/I$ is a complete Boolean algebra.*

Proof. Set $B = \mathcal{P}(\kappa)/I$ and let $S \subseteq B$ be arbitrary. The proof that $\sum S$ exists proceeds progressively through three cases depending on $|S|$:

Case (i): $|S| < \kappa$. With $S = \{[X_\alpha] \mid \alpha < \gamma\}$ where $\gamma < \kappa$ it follows that $\sum S = [\bigcup_{\alpha<\gamma} X_\alpha]$: If $[Y] \geq [X_\alpha]$ for every $\alpha < \gamma$, then $(\bigcup_{\alpha<\gamma} X_\alpha) - Y = \bigcup_{\alpha<\gamma}(X_\alpha - Y) \in I$ by κ-completeness, and so $[Y] \geq [\bigcup_{\alpha<\gamma} X_\alpha]$.

Case (ii): $|S| = \kappa$. Let $S = \{b_\alpha \mid \alpha < \kappa\}$, and using (i) set $a_\alpha = b_\alpha - \sum_{\xi<\alpha} b_\xi$ for $\alpha < \kappa$. Then straightforward arguments show that:

(a) if $\alpha < \beta < \kappa$, then $a_\alpha \cdot a_\beta = \mathbf{0}$;
(b) if $\alpha < \kappa$, then $\sum_{\xi<\alpha} a_\xi = \sum_{\xi<\alpha} b_\xi$; and hence
(c) if $\sum_{\alpha<\kappa} a_\alpha$ exists, then so does $\sum_{\alpha<\kappa} b_\alpha$ and $\sum_{\alpha<\kappa} a_\alpha = \sum_{\alpha<\kappa} b_\alpha$.

We can now work with the a_α's instead, which can be assumed to be nonzero. Extend $\{a_\alpha \mid \alpha < \kappa\}$ to a maximal family $A \subseteq B$ such that $x \neq y$ both in A implies that $x \neq \mathbf{0}$ yet $x \cdot y = \mathbf{0}$. By κ^+-saturation $|A| \leq \kappa$, so let $A = \{[Y_\delta] \mid \delta < \kappa\}$. The Y_δ's can be taken to be pairwise disjoint. (Replace each Y_δ by $Y_\delta - \bigcup_{\xi<\delta} Y_\xi = Y_\delta - \bigcup_{\xi<\delta}(Y_\xi \cap Y_\delta)$, noting that the subtracted set is in I by κ-completeness.) Finally, set $D = \bigcup\{Y_\delta \mid [Y_\delta] = a_\alpha$ for some $\alpha < \kappa\}$; then $[D] = \sum_{\alpha<\kappa} a_\alpha$:

Suppose that $[E] \geq a_\alpha$ for every $\alpha < \kappa$. To show that $[E] \geq [D]$, set $b = [D] - [E]$. Certainly $b \cdot a_\alpha = \mathbf{0}$ for every $\alpha < \kappa$ by hypothesis on $[E]$. Also, for $[Y_\delta]$ not an a_α, $Y_\delta \cap D = \emptyset$, and so $b \cdot [Y_\delta] = \mathbf{0}$. Hence, $b = \mathbf{0}$ by the maximality of A.

Case (iii): $|S| > \kappa$. It can be assumed inductively that if $T \subseteq S$ with $|T| < |S|$, then $\sum T$ exists. Let $S = \{b_\alpha \mid \alpha < \lambda\}$, and using the inductive hypothesis, set $a_\alpha = b_\alpha - \sum_{\xi<\alpha} b_\xi$ for $\alpha < \lambda$. Then (a)–(c) of (ii) hold with κ replaced by λ. But by κ^+-saturation at most κ of the a_α's are non-zero, and so by (ii) $\sum_{\alpha<\lambda} a_\alpha = \sum_{\alpha<\lambda} b_\alpha$ exists. ⊣

This result has a partial converse:

16.6 Exercise (Solovay [71: 427]). *Suppose that I is an ideal over κ such that $\mathcal{P}(\kappa)/I$ is complete. Then for any λ such that I is not λ-saturated, $2^\lambda \leq 2^\kappa$. In particular:*

(a) I is 2^κ-saturated.

(b) If $2^\kappa < 2^{\kappa^+}$, then I is κ^+-saturated.

Hint. Suppose that $\{b_\alpha \mid \alpha < \lambda\} \subseteq \mathcal{P}(\kappa)/I$ is a counterexample to λ-saturation. For $X \subseteq \lambda$ set $a_X = \sum_{\alpha \in X} b_\alpha$, and note that if $X \neq Y$, then $a_X \neq a_Y$. But $|\mathcal{P}(\kappa)/I| \leq 2^\kappa$. ⊣

The completeness of $\mathcal{P}(\kappa)/I$ is itself a substantial hypothesis; see Kanamori-Shelah [95].

Solovay used 16.5 to show that κ^+-saturation is enough to "normalize" ideals in analogy to 5.11 for measurability. To establish some further terminology, for I an ideal over κ, $X \subseteq \kappa$, and $f\colon X \to \kappa$,

f is *I-small on X iff* $\{\xi \in X \mid f(\xi) = \alpha\} \in I$ for every $\alpha < \kappa$,

f is *I-incompressible on X iff* f is I-small on X yet for any $Y \subseteq X$ with $Y \in I^+$ and $g\colon Y \to \kappa$ with $g(\xi) < f(\xi)$ for $\xi \in Y$, g is not I-small on Y, and

I is *normal iff* I^* is a normal filter,

i.e. for any $X \in I^+$ and f regressive on X, there is a $Y \subseteq X$ with $Y \in I^+$ such that f is constant on Y.

16.7 Proposition (Solovay [71]). *Suppose that I is a κ^+-saturated ideal over κ. Then:*

(a) For any $X \in I^+$ there is an I-incompressible function on some $Y \subseteq X$ with $Y \in I^+$.

(b) There is an I-incompressible function on κ.

(c) If f is an I-incompressible function on κ, then

$$f_*(I) = \{X \subseteq \kappa \mid f^{-1}(X) \in I\}$$

is a normal κ^+-saturated ideal over κ.

Proof. (a) Assume to the contrary that this fails for some $X \in I^+$. The following is first established:

(∗) For any f I-small on X, there is an $\overline{f}$ I-small on X such that $[\{\xi \in X \mid \overline{f}(\xi) < f(\xi)\}] = [X]$.

Given such an f, note that by assumption whenever $Y \subseteq X$ with $Y \in I^+$, there is a $Z \subseteq Y$ with $Z \in I^+$ and a g I-small on Z such that $g(\xi) < f(\xi)$ for $\xi \in Z$. Hence, by taking an appropriately maximal family $\subseteq \mathcal{P}(X) \cap I^+$ and invoking κ^+-saturation, there is a collection $\{\langle Z_\alpha, g_\alpha\rangle \mid \alpha < \kappa\}$ satisfying:

(i) $\alpha < \beta < \kappa$ implies that $Z_\alpha \subseteq X$ with $Z_\alpha \in I^+$ yet $Z_\alpha \cap Z_\beta \in I$;
(ii) g_α is I-small on Z_α and $g_\alpha(\xi) < f(\xi)$ for $\xi \in Z_\alpha$; and
(iii) $\sum_{\alpha<\kappa}[Z_\alpha] = [X]$.

As for 16.5, the Z_α's can be taken to be pairwise disjoint (by replacing Z_α with $Z_\alpha - \bigcup_{\xi<\alpha} Z_\xi$). Now let $\overline{f}$ be any extension of $\bigcup_{\alpha<\kappa} g_\alpha$ to all of X. Then it is simple to see that $\overline{f}$ satisfies the conclusion of (∗). For example, $\overline{f}$ must be I-small on X, else if $\{\xi \in X \mid \overline{f}(\xi) = \gamma\} \in I^+$ for some $\gamma < \kappa$, then by (iii) $\{\xi \in Z_\alpha \mid g_\alpha(\xi) = \overline{f}(\xi) = \gamma\} \in I^+$ for some $\alpha < \kappa$, contradicting (ii).

With (∗) we can readily derive a contradiction as follows: Let f_0 be the identity map on X, and for $n \in \omega$ inductively set $f_{n+1} = \overline{f}_n$ as provided by (∗). Setting

$$X_n = \{\xi \in X \mid f_{n+1}(\xi) < f_n(\xi)\} ,$$

$[X_n] = [X]$ implies that $X - X_n \in I$ and hence $X - \bigcap_n X_n \in I$, so that $\bigcap_n X_n \in I^+$. But any ξ in this set corresponds to a descending sequence of ordinals

$$f_0(\xi) > f_1(\xi) > f_2(\xi) > \cdots .$$

(b) By taking an appropriately maximal family and invoking κ^+-saturation, it follows from (a) that there is a collection $\{\langle Y_\alpha, f_\alpha\rangle \mid \alpha < \kappa\}$ satisfying

(i) $\alpha < \beta < \kappa$ implies that $Y_\alpha \in I^+$ yet $Y_\alpha \cap Y_\beta \in I$,
(ii) f_α is I-incompressible on Y_α , and
(iii) $\sum_{\alpha<\kappa}[Y_\alpha] = \mathbf{1}$.

Again the Y_α's can be taken to be pairwise disjoint. Then any extension of $\bigcup_{\alpha<\kappa} f_\alpha$ to all of κ is an I-incompressible function on κ.

(c) This is straightforward. ⊣

The strengthening to κ-saturation opens the door to results about the large size of κ analogous to those for measurability. The proof of (c) below was based on the original Keisler-Tarski one of 5.14. Recall that

$$M(X) = \{\alpha \in X \mid X \cap \alpha \text{ is stationary in } \alpha\} ,$$

Mahlo's operation. (e) was also obtained by Jensen and in part by Fremlin (cf. Solovay [71: 397]).

16.8 Proposition (Solovay [71]). *Suppose that I is a normal κ-saturated ideal over κ. Then:*

(a) If $X \in I^+$ and f is regressive on X, then there is a $Y \subseteq X$ with $[Y] = [X]$ and a $\gamma < \kappa$ such that $f``Y \subseteq \gamma$.

(b) $\{\alpha < \kappa \mid \alpha$ is regular$\} \in I^*$.

(c) If $S \subseteq \kappa$ is stationary, then $\{\alpha < \kappa \mid S \cap \alpha$ is stationary in $\alpha\} \in I^*$.

(d) For any $X \subseteq \kappa$, $[M(X)] = [X]$.

(e) κ is κ-weakly Mahlo.

Proof. (a) Set $X_\alpha = \{\xi \in X \mid f(\xi) = \alpha\}$ for $\alpha < \kappa$, and $E = \{\alpha < \kappa \mid X_\alpha \in I^+\}$. By κ-saturation $|E| < \kappa$, and so $\gamma = \sup(E) < \kappa$. Hence, we can take $Y = \bigcup_{\alpha \in E} X_\alpha$ since $X - Y \in I$ by a simple application of normality.

(b) Assume to the contrary that $\{\alpha < \kappa \mid \mathrm{cf}(\alpha) < \alpha\} \in I^+$, so that by normality $A = \{\alpha < \kappa \mid \mathrm{cf}(\alpha) = \delta\} \in I^+$ for some $\delta < \kappa$. For each $\alpha \in A$, let $\langle \eta^\alpha_\xi \mid \xi < \delta \rangle$ be cofinal in α, and then for each $\xi < \delta$ define regressive functions g_ξ on A by: $g_\xi(\alpha) = \eta^\alpha_\xi$. By (a), for each $\xi < \delta$ there is an $A_\xi \subseteq A$ with $[A_\xi] = [A]$ and a $\gamma_\xi < \kappa$ such that $g_\xi``A_\xi \subseteq \gamma_\xi$. Set $\gamma = \sup(\{\gamma_\xi \mid \xi < \delta\}) < \kappa$. By κ-completeness $\bigcap_{\xi<\delta} A_\xi \in I^+$ as in previous arguments yet $\bigcap_{\xi<\delta} A_\xi \subseteq \gamma$, which is a contradiction.

(c) Assume to the contrary that $W = \{\alpha < \kappa \mid S \cap \alpha$ is not stationary in $\alpha\} \in I^+$. Then for each $\alpha \in W$ there is a C_α closed unbounded in α such that $C_\alpha \cap S = \emptyset$. Define $f_\xi \colon W \to \kappa$ for $\xi < \kappa$ by:

$$f_\xi(\alpha) = \begin{cases} \xi th \text{ element of } C_\alpha & \text{if there is one , and} \\ 0 & \text{otherwise .} \end{cases}$$

Then f_ξ is regressive on W, so by (a) there is a $W_\xi \subseteq W$ with $[W_\xi] = [W]$ and a $\gamma_\xi < \kappa$ such that $f_\xi``W_\xi \subseteq \gamma_\xi$.

Let D be the diagonal intersection of the W_ξ's, i.e.

$$D = \{\alpha \in W \mid \alpha \in \bigcap_{\xi<\alpha} W_\xi\} .$$

Then $[D] = [W]$: If not, $\{\alpha \in W \mid \exists h(\alpha) < \alpha(\alpha \notin W_{h(\alpha)})\} \in I^+$, and so by normality $\{\alpha \in W \mid \alpha \notin W_\xi\} \in I^+$ for some $\xi < \kappa$; but this contradicts $[W_\xi] = [W]$. Next set $C = \{\alpha < \kappa \mid \alpha$ is a limit $\wedge\ \forall \xi < \alpha(\gamma_\xi < \alpha)\}$, a closed unbounded set.

To conclude the argument, since S is stationary, let $\eta \in S \cap C$, and $\zeta \in D$ with $\zeta > \eta$. By (b) we can assume that ζ is regular, so that $f_\eta(\zeta)$ is the ηth element of C_ζ. In fact, $f_\eta(\zeta) = \eta$ since

$$\eta \leq f_\eta(\zeta) = \sup(\{f_\xi(\zeta) \mid \xi < \eta\}) \leq \sup(\{\gamma_\xi \mid \xi < \eta\}) \leq \eta \, .$$

The equality holds since η is a limit ordinal, the next inequality since $\zeta \in D$, and the last since $\eta \in C$. Thus, $\eta = f_\eta(\zeta) \in C_\zeta$ yet $\eta \in S$, which is a contradiction.

(d) We can assume that $X \in I^+$; X is then stationary since I^* is normal and thus contains every closed unbounded subset. As

$$M(X) = X \cap \{\alpha < \kappa \mid X \cap \alpha \text{ is stationary in } \alpha\} \, ,$$

the result now follows from (c).

(e) This follows by induction from (b),(d), and κ-completeness. ⊣

As with weakly compact cardinals (e) can be extended using diagonal intersections: $\{\alpha < \kappa \mid \alpha \text{ is } \alpha\text{-weakly Mahlo}\} \in I^*$ and so forth.

Solovay used 16.8(d) to establish a basic result about stationary sets that does not involve large cardinal hypotheses:

16.9 Theorem (Solovay [71]). *Suppose that $\kappa > \omega$ is regular and $X \subseteq \kappa$ is stationary. Then X can be partitioned into κ disjoint stationary sets, i.e. there are pairwise disjoint stationary sets X_ξ for $\xi < \kappa$ such that $X = \bigcup_{\xi<\kappa} X_\xi$.*

Proof. The crux of the matter is:

(∗) If $S \subseteq \kappa$ is stationary and $\kappa - S$ is unbounded in κ, then $S - M(S) = \{\alpha \in S \mid S \cap \alpha \text{ is not stationary in } \alpha\}$ is stationary in κ .

To show this, suppose that C closed unbounded in κ. Let D be the set of limit points of C less than κ, and E the set of limit points of $\kappa - S$ less than κ. Then D and E are both closed unbounded in κ, so $S \cap D \cap E \neq \emptyset$. Let α be the least member of $S \cap D \cap E$. If $\mathrm{cf}(\alpha) = \omega$, then since $\alpha \in E$, there is an unbounded subset of α of ordertype ω disjoint from $S \cap \alpha$. If $\mathrm{cf}(\alpha) > \omega$, then it is simple to see that $D \cap E \cap \alpha$ is a closed unbounded subset of α disjoint from $S \cap \alpha$. In either case, $\alpha \in (S - M(S)) \cap C$.

To establish the theorem, assume by taking a subset of X if necessary that $\kappa - X$ is unbounded in κ. If X cannot be partitioned into κ disjoint stationary subsets, then it is simple to check that $\mathrm{NS}_\kappa | X$, the ideal of non-stationary subsets of κ restricted to X, would be a normal κ-saturated ideal. But then, (∗) for $S = X$ would contradict 16.8(d). ⊣

This result has become a standard tool of combinatorial set theory, and proofs have been devised that avoid 16.8(d) (see e.g. Jech [03: 95]). Nonetheless, it is a prominent example of a basic result of set theory first found in the context of large cardinals. Another is Silver's 1974 result on the Singular Cardinals Problem (see volume II). These results further attest to the "pragmatic success" of large cardinals; techniques developed in their study have led to crucial results in set theory not involving large cardinal hypotheses.

A basic combinatorial question about the ideal NS_κ of non-stationary subsets of κ sets was soon raised:

Can NS_κ, or even $\mathrm{NS}_\kappa|S$ for some stationary $S \subseteq \kappa$, ever be κ^+-saturated for some regular $\kappa > \omega$?

16.9 had denied κ-saturation, but the question of κ^+-saturation, particularly for the focal case $\kappa = \omega_1$, could not be readily resolved. Results soon showed that having any κ^+-saturated ideal over a *successor* cardinal κ was a very substantial hypothesis, and the possibility at ω_1 was particularly enticing. Progressive research steadily clarified the relationship between ideals and large cardinal hypotheses, and eventually led in the mid-1980's to a positive answer to the question (see §32 and volume II).

What about partition properties? Using $<\kappa$ saturation, Solovay established a generalization of Rowbottom's result 7.17 on normal ultrafilters. For a filter F over a cardinal κ and $\omega < \nu < \kappa$,

F is ν-*Rowbottom* *iff* for any $f\colon [\kappa]^{<\omega} \to \gamma$ with $\gamma < \kappa$ there is an $X \in F$ such that $|f``[X]^{<\omega}| < \nu$.

F is *Rowbottom* *iff* F is ω_1-Rowbottom.

This filter amplification of ν-Rowbottom cardinals is analogous to the relationship of normal ultrafilters to Ramsey cardinals. A simple observation is made before turning to the theorem:

16.10 Exercise. *Suppose that I is a ν-saturated ideal over κ where $\nu < \kappa$. If $f\colon \kappa \to \gamma$ where $\gamma < \kappa$ and $E = \{\xi < \gamma \mid f^{-1}(\{\xi\}) \in I^+\}$, then $E \neq \emptyset$, $|E| < \nu$, and $\bigcup_{\xi \in E} f^{-1}(\{\xi\}) \in I^*$.* ⊣

16.11 Theorem (Solovay [71]). *Suppose that I is a normal ν-saturated ideal over κ where ν is regular and $\omega < \nu < \kappa$. Then I^* is a ν-Rowbottom filter.*

Proof. It suffices to establish the following for every $n \in \omega$: for any $f\colon [\kappa]^n \to \gamma$ with $\gamma < \kappa$ there is an $X \in I^*$ such that $|f``[X]^n| < \nu$.

Proceeding by induction, the case $n = 1$ follows from 16.10. So, assume that the statement holds for n, and suppose that $f\colon [\kappa]^{n+1} \to \gamma$ with $\gamma < \kappa$. Define $f_\alpha\colon [\kappa]^n \to \gamma$ for each $\alpha < \kappa$ as follows:

$$f_\alpha(s) = \begin{cases} f(\{\alpha\} \cup s) & \text{if } \alpha < \min(s)\text{ , and} \\ 0 & \text{otherwise .} \end{cases}$$

By the inductive hypothesis, for each $\alpha < \kappa$ let $H_\alpha \in I^*$ be such that $f_\alpha``[H_\alpha]^n = S_\alpha$ with $|S_\alpha| < \nu$.

We now whittle down further. First, let $g\colon \kappa \to \nu$ be given by $g(\alpha) = |S_\alpha|$. By 16.10, let $W \in I^*$ such that $|g``W| < \nu$. If $\delta = \sup(g``W)$, then $\delta < \nu$ by the regularity of ν. For each $\alpha \in W$, let $h_\alpha: \delta \to S_\alpha$ be a surjection. Then define $g_\xi: W \to \gamma$ for each $\xi < \delta$ by: $g_\xi(\alpha) = h_\alpha(\xi)$. Again, for each $\xi < \delta$ there is a $W_\xi \subseteq W$ with $W_\xi \in I^*$ such that $|g_\xi``W_\xi| < \nu$. By the regularity of $\nu$, if $E = \bigcup_{\xi<\delta} g_\xi``W_\xi$, then $|E| < \nu$.

Finally, set $X = \bigcap_{\xi<\delta} W_\xi \cap \Delta_{\alpha<\kappa} H_\alpha$. $X \in I^*$ by normality, so to complete the proof it suffices to show that $f``[X]^{n+1} \subseteq E$. Suppose then that $t \in [X]^{n+1}$, written $t = \{\alpha\} \cup s$ where $\alpha < \min(s)$. Then $f(t) = f_\alpha(s) \in S_\alpha$ as $s \in [H_\alpha]^n$, and as $\alpha \in \bigcap_{\xi<\delta} W_\xi$ it follows that $S_\alpha \subseteq E$. ⊣

16.12 Corollary (Prikry [75]).

(a) Suppose that there is a ν-saturated ideal over κ, where $\omega < \nu < \kappa$ and $2^{<\nu} = \kappa$. Then for $\nu \leq \lambda < \kappa$, $2^\lambda = \kappa$.

(b) If $2^{\aleph_0}$ is real-valued measurable, then $2^\lambda = 2^{\aleph_0}$ for every $\lambda < 2^{\aleph_0}$.

Proof. (a) If ν is regular, the result follows from 16.11 and 8.9. If ν is singular, then $2^\nu = \kappa$ (see the first paragraph of the proof of 8.9); but $\nu < \kappa$ implies $\nu^+ < \kappa$ by 16.3, and so just ν^+-saturation implies the result by 16.11 and 8.9. (There is another argument appealing to the Erdős-Tarski [43] result that the least λ such that some ideal is λ-saturated is always regular.)

(b) This follows from (a). ⊣

(b) was Prikry's main objective; his argument was factored through 8.9. The conclusion of (b) is also a consequence of Martin's Axiom, a hypothesis known to be inconsistent with the real-valued measurability of $2^{\aleph_0}$.

The analogy with measurable cardinals was further extended by Kunen (see Solovay [71]), who established that a cardinal κ carrying a ν-saturated ideal for some $\nu < \kappa$ has a kind of Π^2_1-indescribability property.

17. Saturated Ideals II

This section is devoted to the first relative consistency results about saturated ideals obtained by forcing. Solovay was led to 16.11 so that he could establish 16.1(b) by adapting an argument of Silver's from measurability – see 20.3 and 20.4. Soon afterwards Kunen [70] using iterated ultrapowers showed that κ^+-saturation sufficed for a comprehensive result: *If I is a normal κ^+-saturated ideal over κ, then κ is measurable in $L[I]$, the class of sets relatively constructible from I. Moreover, in $L[I]$ the dual filter to $I \cap L[I]$ is a normal ultrafilter over κ.* A refined version of this result will be established later (see Precipitous Ideals in volume II). What is pertinent for us now is that the forcing arguments below all assume the existence of a measurable cardinal in the ground model, an assumption both necessary as well as sufficient.

An early forcing result of Prikry spurred the initial work in this area. He showed that merely chain conditions suffice to preserve saturation, and thus saturated ideals are resilient under a variety of generic extensions. The following is an extension of his result.

17.1 Theorem (Prikry [66,70] for $\lambda < \kappa$, Solovay [71:411] for $\lambda = \kappa$, and Kakuda [72] for $\lambda = \kappa^+$). *Suppose that I is a λ-saturated ideal over κ where $\lambda \le \kappa^+$ is regular, and P is a p.o. with the ν-c.c. where $\nu < \kappa$ and $\nu \le \lambda$. Then*

$$\Vdash_P \check{I} \text{ generates a } \lambda\text{-saturated ideal over } \kappa\ .$$

Proof. Let $\Vdash \dot{I} = \{X \subseteq \kappa \mid \exists Y \in \check{I}(X \subseteq Y)\}$, the ideal generated by I. Only $\nu \le \kappa$ is needed to show that $\Vdash \dot{I}$ is κ-complete: It suffices to show that if $\rho < \kappa$ and $p \Vdash \{\dot{X}_\alpha \mid \alpha < \rho\} \subseteq \dot{I}$, then there is an $A \in I$ such that $p \Vdash \bigcup_{\alpha<\rho} \dot{X}_\alpha \subseteq \check{A}$. To do this, for each $\alpha < \rho$ let Q_α be a maximal antichain below p such that for each $q \in Q_\alpha$ there is an $A^q_\alpha \in I$ such that $q \Vdash \dot{X}_\alpha = \check{A}^q_\alpha$. Since $|Q_\alpha| < \nu \le \kappa$ by the ν-c.c., $A_\alpha = \bigcup\{A^q_\alpha \mid q \in Q_\alpha\} \in I$. Thus also $A = \bigcup_{\alpha<\rho} A_\alpha \in I$, and it is simple to see that $p \Vdash \bigcup_{\alpha<\rho} \dot{X}_\alpha \subseteq \check{A}$.

Turning to λ-saturation, suppose that $p \Vdash \{\dot{Y}_\alpha \mid \alpha < \lambda\} \subseteq \dot{I}^+$. The proof is completed by finding a $q \le p$ and $\alpha < \beta < \lambda$ so that $q \Vdash \dot{Y}_\alpha \cap \dot{Y}_\beta \in \dot{I}^+$:

For each $\alpha < \lambda$ let

$$A_\alpha = \{\xi < \kappa \mid \exists q \le p(q \Vdash \xi \in \dot{Y}_\alpha)\}\ .$$

Then $A_\alpha \in I^+$ since $p \Vdash \dot{Y}_\alpha \subseteq \check{A}_\alpha$. Hence, there must be an $A \in I^+$ satisfying:

$$(*) \qquad \forall W(W \subseteq A \ \wedge\ W \in I^+ \ \rightarrow\ |\{\alpha < \lambda \mid W \cap A_\alpha \in I^+\}| = \lambda)\ .$$

This follows directly from λ-saturation and the regularity of λ: Let $\{W_\gamma \mid \gamma < \rho\} \subseteq I^+$ be maximal such that $\gamma < \delta < \rho$ implies that $W_\gamma \cap W_\delta \in I$ and $|\{\alpha < \lambda \mid W_\gamma \cap A_\alpha \in I^+\}| < \lambda$. Then $\rho < \lambda$, and so it is readily seen that $A = A_\beta$ must satisfy $(*)$ for any sufficiently large $\beta < \lambda$.

With $(*)$, sets $\{S_\eta \mid \eta < \nu\} \subseteq [\lambda]^{<\lambda}$ and $\{B_\alpha \mid \alpha \in S_\eta\}$ can be defined recursively so that for $\eta < \zeta < \nu$,

(i) $S_\eta \cap S_\zeta = \emptyset$,
(ii) if $\{\alpha, \beta\} \in [S_\eta]^2$, then $B_\alpha \subseteq A_\alpha \cap A$ and $B_\alpha \in I^+$ yet $B_\alpha \cap B_\beta = \emptyset$, and
(iii) $[A] = \sum_{\alpha \in S_\eta} [B_\alpha]$.

At stage $\eta < \nu$ of the recursion, $|\bigcup_{\xi<\eta} S_\xi| < \lambda$ since $\nu \leq \lambda$ and λ is regular, and so with $(*)$ find $S_\eta \subseteq \lambda - \bigcup_{\xi<\eta} S_\xi$ and $\{B_\alpha \mid \alpha \in S_\eta\}$ satisfying (iii) such that for $\{\alpha, \beta\} \in [S_\eta]^2$, $B_\alpha \subseteq A_\alpha \cap A$ and $B_\alpha \in I^+$ yet $B_\alpha \cap B_\beta \in I$. $|S_\eta| < \lambda$ by λ-saturation, and so the B_α's can be taken to be pairwise disjoint by replacing B_α with $B_\alpha - \bigcup\{B_\beta \mid \beta < \alpha \ \wedge \ \beta \in S_\eta\}$ as usual.

Let $A' = \bigcap_{\eta<\nu} \bigcup_{\alpha \in S_\eta} B_\alpha$, so that by (iii) and κ-completeness, $[A'] = [A]$ and thus $A' \in I^+$. By (ii), for each $\xi \in A'$ and $\eta < \nu$ there is exactly one $\alpha \in S_\eta$ such that $\xi \in B_\alpha \subseteq A_\alpha$. By definition of A_α let $q_\xi^\eta \leq p$ such that $q_\xi^\eta \Vdash \xi \in \dot{Y}_\alpha$. Note that by the ν-c.c., for each $\xi \in A'$ there must be $\eta_\xi < \zeta_\xi < \nu$ and an $r_\xi \in P$ such that $r_\xi \leq q_\xi^{\eta_\xi}, q_\xi^{\zeta_\xi}$. (Note that if $\lambda \leq \kappa$, we could have taken the $\dot{Y}_\alpha$'s to be forced disjoint and finish the argument here with a contradiction, since r_ξ forces ξ to be in two different $\dot{Y}_\alpha$'s.) By κ-completeness and $\nu < \kappa$, there must be fixed $\eta < \zeta < \nu$ and a $B \subseteq A'$ with $B \in I^+$ such that for $\xi \in B$, $\langle \eta_\xi, \zeta_\xi \rangle = \langle \eta, \zeta \rangle$. Finally, by (i), (iii), and the completeness of $\mathcal{P}(\kappa)/I$ there are $\alpha < \beta$ such that $\alpha \in S_\eta$ and $\beta \in S_\zeta$ and $Z = B_\alpha \cap B_\beta \cap B \in I^+$.

All of the foregoing was to establish that there are $Z \in I^+$ and $\alpha < \beta < \lambda$ satisfying:

$$\text{for each } \xi \in Z, \text{ there is an } r_\xi \leq p \text{ such that } r_\xi \Vdash \xi \in \dot{Y}_\alpha \cap \dot{Y}_\beta \ .$$

The rest is simple. Let $Q \subseteq P$ be a maximal antichain below p such that for each $q \in Q$ there is an $Z_q \in I$ such that $q \Vdash \dot{Y}_\alpha \cap \dot{Y}_\beta \subseteq \check{Z}_q$. Since $|Q| < \nu \leq \kappa$, $Z - \bigcup_{q \in Q} Z_q \in I^+$. For any ξ in this set, note that r_ξ must be incompatible with every member of Q. Thus, $r_\xi \Vdash \dot{Y}_\alpha \cap \dot{Y}_\beta \in \dot{I}^+$ to complete the proof. ⊣

This result does not extend to the κ-c.c. since the Levy collapse $\mathrm{Col}(\omega, \kappa)$ of an uncountable regular cardinal κ to ω_1 has the κ-c.c. yet there are no ω_1-saturated ideals over ω_1 by 16.3. For κ^+-saturated ideals over κ, Laver [82] and later Baumgartner-Taylor [82a] characterized those κ-c.c. P for which the conclusion holds. 17.1 is subsumed by, and its proof derived from, this characterization. The case $\lambda < \kappa$ led to the first consistency result on saturated ideals:

17.2 Corollary (Prikry [66, 70]). *Suppose that κ is a measurable cardinal. Then there is an ω_1-c.c. p.o. P such that*

$$\Vdash_P 2^{\aleph_0} = \kappa \ \wedge \ \exists I (I \text{ is an } \omega_1\text{-saturated ideal over } \kappa) \ .$$

Proof. Let P be the p.o. for adding κ Cohen reals, i.e.

$$P = \{p \mid p \text{ is a finite function } \wedge \text{ dom}(p) \subseteq \kappa \times \omega \wedge \text{ ran}(p) \subseteq 2\} .$$

It is well-known that P has the ω_1-c.c. (by the Δ-system Lemma 10.4) and that $\Vdash_P 2^{\aleph_0} = \kappa$ (using 10.5 in one direction). By 17.1 any ideal dual to a κ-complete ultrafilter over κ generates an ω_1-saturated ideal over κ in the generic extension. ⊣

Prikry [70] verified that in this generic extension, κ is not real-valued measurable. Solovay had originally been inspired by Prikry's work to show that after "adding κ random reals" instead, κ does remain real-valued measurable. This is the result established next, after a discussion of the forcing.

Product Measure Forcing

For any infinite cardinal κ, we consider a product measure for ${}^{\kappa}2$ that recalls Lebesgue measure as formulated in §0. For each function s with $\text{dom}(s) \in [\kappa]^{<\omega}$ and $\text{ran}(s) \subseteq 2$, set

$$O(s) = \{f \in {}^{\kappa}2 \mid f \supseteq s\} .$$

${}^{\kappa}2$ is topologized by taking these $O(s)$'s as the basic open sets. Specify that

$$\mathcal{B}_\kappa \text{ is the } \sigma\text{-algebra on } {}^{\kappa}2 \text{ generated by the } O(s)\text{'s.}$$

To endow $\mathcal{B}_\kappa$ with a measure, let m_2 be the measure on $\mathcal{P}(2)$ specified by: $m_2(\emptyset) = 0$, $m_2(\{0\}) = m_2(\{1\}) = \frac{1}{2}$, and $m_2(\{0, 1\}) = 1$. Then stipulate that

$$m_\kappa \text{ is the product measure on } \mathcal{B}_\kappa \text{ induced by } m_2.$$

$m_\kappa(O(s)) = 2^{-|s|}$ for every s, and m_κ is the unique measure on $\mathcal{B}_\kappa$ with this property.

Consider

$$\mathcal{B}_\kappa^* = \{X \in \mathcal{B}_\kappa \mid m_\kappa(X) > 0\}$$

ordered by

$$p \leq q \ \textit{ iff } \ p \subseteq q$$

as a p.o. for forcing. The following recalls 11.7.

17.3 Exercise.
(a) If $p, q \in \mathcal{B}_\kappa^*$, then $p \perp q$ iff $m_\kappa(p \cap q) = 0$.
(b) $\mathcal{B}_\kappa^*$ has the ω_1-c.c. ⊣

$\mathcal{B}_\kappa^*$ is not separative, and its separative quotient $\mathcal{B}_\kappa^*/\approx$ is formulated with $p \approx q$ *iff* $m_\kappa(p \,\triangle\, q) = 0$. Using this as a definition to extend $\approx$ to all of $\mathcal{B}_\kappa$, $\mathcal{B}_\kappa/\approx$ is a Boolean algebra with the Boolean operations being the set-theoretic ones modulated by $\approx$. This algebra is analogous to $\mathcal{P}(\kappa)/I$ discussed previously, with the corresponding ideal being $\{X \in \mathcal{B}_\kappa \mid m_\kappa(X) = 0\}$. By a simple version of the proof of 16.5, the ω_1-c.c. implies that *$\mathcal{B}_\kappa/\!\approx$ is a complete Boolean algebra.*

Proceeding as in §11, we can develop codes for appropriate $\mathcal{B}^*_\kappa$ sets, show that forcing with $\mathcal{B}^*_\kappa$ amounts to adding a new function: $\kappa \to 2$ analogous to a random real, and provide characterizations like 11.9 and 11.10. See Kunen [84] for details. The next proposition shows how forcing with $\mathcal{B}^*_\kappa$ affects the size of the continuum.

17.4 Proposition. $\Vdash_{\mathcal{B}^*_\kappa} \ulcorner 2^{\aleph_0} \geq \kappa \urcorner$, *and if* $\kappa^{\aleph_0} = \kappa$, *then* $\Vdash_{\mathcal{B}^*_\kappa} \ulcorner 2^{\aleph_0} = \kappa \urcorner$.

Proof. For each *limit* ordinal $\gamma < \kappa$ let $\dot{f}_\gamma$ be a name satisfying $\Vdash \dot{f}_\gamma \colon \omega \to 2$ such that for any $n \in \omega$ and $i < 2$,

$$\Vdash \dot{f}_\gamma(n) = i \;\leftrightarrow\; \check{O}(\{\langle \gamma + n, i \rangle\}) \in \dot{G}$$

(where $\dot{G}$ is the canonical name for the generic object). To show that $\Vdash 2^{\aleph_0} \geq \kappa$, it suffices to show that $\Vdash \dot{f}_\gamma \neq \dot{f}_\delta$ for limit ordinals $\gamma < \delta < \kappa$:

Let $p \in \mathcal{B}^*_\kappa$ be arbitrary. Choose $k \in \omega$ so large that $2^{-k} < m_\kappa(p)$. For each $t \in {}^k2$, let

$$\bar{t} \colon \{\gamma, \gamma + 1, \ldots, \gamma + k - 1, \delta, \delta + 1, \ldots, \delta + k - 1\} \to 2$$

satisfy $\bar{t}(\gamma + i) = \bar{t}(\delta + i) = t(i)$ for each $i < k$. If $q = \bigcup_{t \in {}^k2} O(\bar{t})$, then $m_\kappa(q) = 2^k(2^{-2k}) = 2^{-k}$. Thus, $m_\kappa(p - q) > 0$, and a standard forcing argument shows that $p - q \Vdash \dot{f}_\gamma | k \neq \dot{f}_\delta | k$.

If $\kappa^{\aleph_0} = \kappa$, note that an analysis of $\mathcal{B}_\kappa$ in a hierarchy of length ω_1 as for the Borel hierarchy for ${}^\omega\omega$ shows that $|\mathcal{B}_\kappa| = \aleph_1 \cdot \kappa^{\aleph_0} = \kappa$. Hence by the ω_1-c.c. and 10.5, $\Vdash 2^{\aleph_0} \leq \kappa$. ⊣

For any $\gamma < \kappa$, the map $e_\gamma \colon \omega \to \kappa$ given by $e_\gamma(n) = \gamma + n$ induces an injective, complete embedding: $\mathcal{B}^*_\omega \to \mathcal{B}^*_\kappa$. Hence by 10.1(a), if G is $\mathcal{B}^*_\kappa$-generic, then $e_\gamma^{-1}(G)$ is $\mathcal{B}^*_\omega$-generic. Moreover, for limit γ these generics are determined by the $\dot{f}_\gamma$ of the previous proof, which can thus be considered random reals. (In §11, random reals were defined via $\mathcal{B}^* \subseteq \mathcal{P}({}^\omega\omega)$ whereas $\mathcal{B}^*_\omega \subseteq \mathcal{P}({}^\omega 2)$. However, there is a natural homeomorphism between ${}^\omega\omega$ and ${}^\omega 2$, if countably many points are ignored, that preserves measure (see e.g. Levy [79: VII§3]) and hence randomness.) It is in this sense that forcing with $\mathcal{B}^*_\kappa$ is often referred to as "adding κ random reals", although it is not the same as adding κ random reals "side-by-side" with product forcing because of the measure-theoretic overlay.

Solovay established his result 16.1(a) with $\mathcal{B}^*_\kappa$ using Boolean-valued models. Towards a proof of his result, some of that approach is adopted in the assignment of Boolean values in a weak sense to formulas of the forcing language for $\mathcal{B}^*_\kappa$: For any formula φ, choose a maximal antichain $A \subseteq \{p \in \mathcal{B}^*_\kappa \mid p \parallel \varphi\}$. A is countable by the ω_1-c.c., so set

$$[\![\varphi]\!] = \bigcup\{p \in A \mid p \Vdash \varphi\} \in \mathcal{B}_\kappa \,.$$

$[\![\varphi]\!]$ has two crucial properties: if $[\![\varphi]\!] \neq \emptyset$, then $[\![\varphi]\!] \Vdash \varphi$; and for any $q \in \mathcal{B}^*_\kappa$, if $q \Vdash \varphi$, then $m_\kappa(q - [\![\varphi]\!]) = 0$. Although $[\![\varphi]\!]$ depends on the choice of A, only $m_\kappa([\![\varphi]\!])$ and $m_\kappa([\![\varphi]\!] \cap p)$ for $p \in \mathcal{B}^*_\kappa$ will be used, and it is simple to check that these values do *not* depend on A.

The following result establishes 16.1(a):

17.5 Theorem (Solovay [66, 71]). *Suppose that κ is measurable. Then*

$$\Vdash_{\mathcal{B}^*_\kappa} 2^{\aleph_0} = \kappa \ \wedge \ \kappa \text{ is real-valued measurable}.$$

Proof. By 17.4 it suffices to establish real-valued measurability. Let U be a κ-complete ultrafilter over κ. First define for each $p \in \mathcal{B}^*_\kappa$ a "measure" μ_p on names $\dot{X}$ such that $p \Vdash \dot{X} \subseteq \kappa$ by:

$$\mu_p(\dot{X}) = r \ \textit{ iff } \ \{\xi < \kappa \mid \frac{m_\kappa([\![\xi \in \dot{X}]\!] \cap p)}{m_\kappa(p)} = r\} \in U .$$

This is well-defined, since $2^{\aleph_0} < \kappa$ and U is κ-complete. It is straightforward to check that

(i) If $X \in U$, then $\mu_p(\check{X}) = 1$; if $X \notin U$, then $\mu_p(\check{X}) = 0$.
(ii) If $p \Vdash \dot{X} \subseteq \dot{Y} \subseteq \kappa$, then $\mu_p(\dot{X}) \leq \mu_p(\dot{Y})$.
(iii) If $\gamma < \kappa$ and $p \Vdash \ulcorner \dot{X} = \bigcup\{\dot{X}_\alpha \mid \alpha < \gamma\} \subseteq \kappa$ is a pairwise disjoint union$\urcorner$, then $\mu_p(\dot{X}) = \sum_{\alpha<\gamma} \mu_p(\dot{X}_\alpha)$.

For the last, note that for any $\xi < \kappa$,

$$m_\kappa([\![\xi \in \dot{X}]\!] \cap p) = m_\kappa(\bigcup_{\alpha<\gamma} [\![\xi \in \dot{X}_\alpha]\!] \cap p) = \sum_{\alpha<\gamma} m_\kappa([\![\xi \in \dot{X}_\alpha]\!] \cap p)$$

since the union is really a countable union of $\mathcal{B}^*_\kappa$ sets by the ω_1-c.c..

A crucial observation is that for any $p \in \mathcal{B}^*_\kappa$, name $\dot{X}$ such that $p \Vdash \dot{X} \subseteq \kappa$, and $r \in \mathbb{R}$ with $0 \leq r \leq 1$,

$$(*) \qquad \text{if } \forall q \leq p \exists s \leq q(\mu_s(\dot{X}) \leq r), \text{ then } \mu_p(\dot{X}) \leq r$$

(and an analogous assertion holds with the last two $\leq$'s replaced by $\geq$'s). To show this, let $A = \{p_n \mid n \in \omega\}$ be a maximal antichain below p so that $\mu_{p_n}(\dot{X}) \leq r$ for $n \in \omega$. By the ω_1-completeness of U, there is an $E \in U$ such that for each $\xi \in E$, $m_\kappa([\![\xi \in \dot{X}]\!] \cap p_n) \leq r \cdot m_\kappa(p_n)$ for every $n \in \omega$. For such ξ, the maximality of the antichain implies that

$$m_\kappa([\![\xi \in \dot{X}]\!] \cap p) = \sum_n m_\kappa([\![\xi \in \dot{X}]\!] \cap p_n) \leq r \cdot \sum_n m_\kappa(p_n) = r \cdot m_\kappa(p),$$

and hence $\mu_p(\dot{X}) \leq r$.

The theorem is established by showing that for any $\mathcal{B}^*_\kappa$-generic G, $\mu \colon \mathcal{P}(\kappa) \to [0, 1]$ given by

$$\mu(X) = \lim(\{\mu_p(\dot{X}) \mid p \in G\})$$

where $\dot{X}$ is any name for X, is a κ-additive measure over κ in $V[G]$. To verify that the limit exists, does not depend on the name $\dot{X}$, and has the requisite properties, Solovay used the Radon-Nikodym Theorem of measure theory; proceeding more gingerly, a lim inf is taken instead, following Jech [03: 416ff]:

First define for each $p \in \mathcal{B}^*_\kappa$ a μ^*_p on names $\dot{X}$ such that $p \Vdash \dot{X} \subseteq \kappa$ by

$$\mu^*_p(\dot{X}) = \inf(\{\mu_q(\dot{X}) \mid q \leq p\}) .$$

Then let $\dot{\mu}$ be a name such that $\Vdash \dot{\mu}\colon \mathcal{P}(\kappa) \to [0, 1]$ and satisfying

$$\Vdash \dot{X} \subseteq \kappa \to \dot{\mu}(\dot{X}) = \sup(\{\check{\mu}^*_p(\dot{X}) \mid p \in \dot{G}\})$$

(where $\dot{G}$ is the canonical name for the generic object). There is a basic connection between μ^*_p and $\dot{\mu}$: For any $p \in \mathcal{B}^*_\kappa$, name $\dot{X}$ such that $p \Vdash \dot{X} \subseteq \kappa$, and $r \in [0, 1]$,

$$(**) \qquad \mu^*_p(\dot{X}) \geq r \ \textit{ iff } \ p \Vdash \dot{\mu}(\dot{X}) \geq \check{r} .$$

The forward direction is clear. For the converse, suppose that $p \Vdash \dot{\mu}(\dot{X}) \geq \check{r}$, and let $r_0 \in \mathbb{R}$ satisfy $0 \leq r_0 < r$. It suffices to show that $\mu^*_p(\dot{X}) \geq r_0$: By definition of $\dot{\mu}$, $p \Vdash \exists s \in \dot{G}(\check{\mu}^*_s(\dot{X}) \geq r_0)$. This means that any $q \leq p$ is compatible with some s such that $\mu^*_s(\dot{X}) \geq r_0$, and since generally speaking $p_1 \leq p_2$ implies that $\mu^*_{p_1}(\dot{X}) \geq \mu^*_{p_2}(\dot{X})$, it follows that

$$\forall q \leq p \exists s \leq q(\mu^*_s(\dot{X}) \geq r_0) .$$

With any $p_0 \leq p$ playing the role of p here, it follows from the $\geq$-variant of $(*)$ that $\mu_{p_0}(\dot{X}) \geq r_0$. Hence, $\mu^*_p(\dot{X}) \geq r_0$.

The proof is completed by establishing that $\Vdash \dot{\mu}$ is a well-defined κ-additive measure. By (i) above, if $X \in U$, then $\Vdash \dot{\mu}(\check{X}) = 1$; and if $X \notin U$, then $\Vdash \dot{\mu}(\check{X}) = 0$. Also, if $p \Vdash \dot{X} \subseteq \dot{Y} \subseteq \kappa$, then for any $q \leq p$, $\mu^*_q(\dot{X}) \leq \mu^*_q(\dot{Y})$ by (ii) above, and hence $p \Vdash \dot{\mu}(\dot{X}) \leq \dot{\mu}(\dot{Y})$. (Otherwise, there would be a $q \leq p$ and a rational r such that $q \Vdash \dot{\mu}(\dot{X}) \geq \check{r} > \dot{\mu}(\dot{Y})$, which by $(**)$ leads to the contradictory $\mu^*_q(\dot{X}) \geq r > \mu^*_q(\dot{Y})$.) In particular, if $p \Vdash \dot{X} = \dot{Y}$, then $p \Vdash \dot{\mu}(\dot{X}) = \dot{\mu}(\dot{Y})$, and so for any $\mathcal{B}^*_\kappa$-generic G, $\dot{\mu}^G$ is well-defined on elements in $V[G]$. The rest of the proof is devoted to showing that $\Vdash \dot{\mu}$ is κ-additive:

We first verify that $\Vdash \dot{\mu}$ is finitely additive, for which it suffices to show that if $\Vdash \dot{X}, \dot{Y} \subseteq \kappa$ are disjoint, then $\Vdash \dot{\mu}(\dot{X} \cup \dot{Y}) = \dot{\mu}(\dot{X}) + \dot{\mu}(\dot{Y})$: Suppose that r_1 and r_2 are rationals and $p \Vdash \dot{\mu}(\dot{X}) \geq \check{r}_1 \ \wedge \ \dot{\mu}(\dot{Y}) \geq \check{r}_2$. Then

$$\mu^*_p(\dot{X} \cup \dot{Y}) \geq \mu^*_p(\dot{X}) + \mu^*_p(\dot{Y}) \geq r_1 + r_2 .$$

Here, the first inequality follows from (iii) above and the definition of μ^*_p, and the second, from $(**)$. So again by $(**)$, $p \Vdash \dot{\mu}(\dot{X} \cup \dot{Y}) \geq \check{r}_1 + \check{r}_2$. Since p, r_1, and r_2 were arbitrary, it follows that $\Vdash \dot{\mu}(\dot{X} \cup \dot{Y}) \geq \dot{\mu}(\dot{X}) + \dot{\mu}(\dot{Y})$.

In the other direction, assume to the contrary that $\Vdash \dot{\mu}(\dot{X} \cup \dot{Y}) > \dot{\mu}(\dot{X}) + \dot{\mu}(\dot{Y})$. Then it is simple to see that for some p and rationals r_3, r_4, r_5, and r_6,

$$p \Vdash \dot{\mu}(\dot{X}) < \check{r}_3 < \check{r}_4 \ \wedge\ \dot{\mu}(\dot{Y}) < \check{r}_5 < \check{r}_6 \ \wedge\ \dot{\mu}(\dot{X} \cup \dot{Y}) \geq \check{r}_4 + \check{r}_6 \, .$$

For any $q \leq p$, $\mu^*_q(\dot{X}) < r_3$ by (∗∗), and so there is an $s \leq q$ such that $\mu_s(\dot{X}) < r_3$. By (∗), it follows that $\mu_p(\dot{X}) \leq r_3$. Similarly, $\mu_p(\dot{Y}) \leq r_5$, and consequently

$$\mu^*_p(\dot{X} \cup \dot{Y}) \leq \mu_p(\dot{X} \cup \dot{Y}) = \mu_p(\dot{X}) + \mu_p(\dot{Y}) \leq r_3 + r_5 < r_4 + r_6 \, ,$$

which by (∗∗) contradicts $p \Vdash \dot{\mu}(\dot{X}) \cup (\dot{Y}) \geq r_4 + r_6$.

To verify κ-additivity, suppose that $\gamma < \kappa$ and $\Vdash \ulcorner \dot{X} = \bigcup\{\dot{X}_\alpha \mid \alpha < \gamma\} \subseteq \kappa$ is a pairwise disjoint union$\urcorner$. By finite additivity, $\Vdash \dot{\mu}(\dot{X}) \geq \sum_\alpha \dot{\mu}(\dot{X}_\alpha)$. In the other direction, assume to the contrary that for some p and rationals r_1, r_2,

$$p \Vdash \textstyle\sum_\alpha \dot{\mu}(\dot{X}_\alpha) < \check{r}_1 < \check{r}_2 \leq \dot{\mu}(\dot{X}) \, .$$

For any $t \in [\gamma]^{<\omega}$, let $\Vdash \dot{X}_t = \bigcup_{\alpha \in t} \dot{X}_\alpha$. Then $p \Vdash \dot{\mu}(\dot{X}_t) < \check{r}_1$ by finite additivity, and so an application of (∗) as in the previous paragraph shows that $\mu_p(\dot{X}_t) \leq r_1$. Since t was arbitrary, $\mu_p(\dot{X}) \leq r_1$ by (iii) above. Hence, $\mu^*_p(\dot{X}) < r_2$, which by (∗∗) contradicts $p \Vdash \dot{\mu}(\dot{X}) \geq \check{r}_2$. This completes the proof. ⊣

Kunen-Paris Forcing

Kunen and Paris in their [71] provided the first widely applicable method for preserving properties of measurable cardinals κ in forcing extensions by p.o.'s P such that $|P| \geq \kappa$, and used it to establish consistency results about saturated ideals I over κ such that $\mathrm{sat}(I) \geq \kappa$. Their basic scheme is presented in a form that anticipates later developments; the rest of the section is then devoted to its applications.

Suppose that κ is a measurable cardinal, U a normal ultrafilter over κ, and $j\colon V \prec M \cong \mathrm{Ult}(V, U)$. Some general conditions are imposed on a p.o. so that j can be extended, in a weak sense, in an intermediate generic extension. Assume that P is a p.o. so that, regarding $j(P)$ as a p.o. in V,

(i) $j(P) \cong P * \dot{Q}$ under an identification through which
(ii) $j(p) = \langle p, \mathbb{1}_Q \rangle$ for every $p \in P$.

These conditions are met by a variety of p.o.'s that can be analyzed as iterations.

Let G be $j(P)$-generic over V. Then by (i) and 10.9 there are corresponding G_0 and G_1 such that $G = G_0 * G_1$, G_0 is P-generic over V, G_1 is $\dot{Q}^{G_0}$-generic over $V[G_0]$, and $V[G] = V[G_0][G_1]$. Moreover by (ii),

$$(*) \qquad \text{if } p \in G_0, \text{ then } j(p) \in G \, .$$

Since $M \subseteq V$, G is also $j(P)$-generic over M, and we can consider the generic extension $M[G]$ of M. The key observation is that with (∗) the embedding j can be extended to an elementary embedding

$$\bar{j}\colon V[G_0] \prec M[G]$$

definable in $V[G]$. The idea is to apply j to forcing names:

Suppose that $x \in V[G_0]$ and $\dot{x}$ is any P-name such that $\dot{x}^{G_0} = x$. Then $j(\dot{x})$ is a $j(P)$-name, so we can set

$$\overline{j}(x) = j(\dot{x})^{M[G]} .$$

This is well-defined, i.e. does not depend on the choice of name for x: Suppose that $\dot{y}$ is another name and $p \in G_0$ is such that

$$p \Vdash_P \dot{x} = \dot{y} .$$

Then $j(p) \in G$ by (*), and in M,

$$j(p) \Vdash_{j(P)} j(\dot{x}) = j(\dot{y}) .$$

Hence, $j(\dot{x})^{M[G]} = j(\dot{y})^{M[G]}$. The verification that $\overline{j}$ is elementary is analogous, with the atomic formula $\dot{x} = \dot{y}$ replaced by a general formula. Finally, $\overline{j}$ extends j, since if $x \in V$, then $x = \check{x}^{G_0}$, $j(\check{x}) = \check{j}(x)$ (under any reasonable formalization of the forcing language), and so $\overline{j}(x) = j(x)$.

The embedding $\overline{j}$ generates an ultrafilter. For N an inner model,

> W is an *N-normal ultrafilter over κ iff*
> (i) W is an ultrafilter *on* $\mathcal{P}(\kappa) \cap N$, and
> (ii) For any $f \in {}^{\kappa}\kappa \cap N$ with $\{\xi < \kappa \mid f(\xi) < \xi\} \in W$ there is an $\alpha < \kappa$ such that $\{\xi < \kappa \mid f(\xi) = \alpha\} \in W$.

Note that (ii) *subsumes what would be called N-κ-completeness*. Of course, the point is that $W \in N$ is not required. The following recalls 5.6:

17.6 Exercise. *In $V[G]$ with $\overline{j}$ as above, define $\overline{U}$ by:*

$$X \in \overline{U} \quad \textit{iff} \quad X \in \mathcal{P}(\kappa) \cap V[G_0] \ \wedge \ \kappa \in \overline{j}(X) .$$

Then $\overline{U}$ is a $V[G_0]$-normal ultrafilter over κ such that $\overline{U} \supseteq U$. ⊣

This in sum is the basic Kunen-Paris scheme; its applications are based on those P such that the $\dot{Q}$ part of $j(P) \cong P * \dot{Q}$ is tame. For example, if forcing with $\dot{Q}$ adds no new subsets of κ, then the $\overline{U}$ of 17.6 verifies the measurability of κ in $V[G]$. Also, the following lemma may be applicable in $V[G]$ to the dual of $\overline{U}$ to yield saturated ideals in intermediate models N, $V[G_0] \subseteq N \subseteq V[G]$. For N an inner model,

> J is a *N-λ-saturated ideal over κ iff*
> (i) J is an ideal on $\mathcal{P}(\kappa) \cap N$ such that if $\gamma < \kappa$ and $f \in {}^{\gamma}J \cap N$, then $\bigcup_{\alpha<\gamma} f(\alpha) \in J$, and
> (ii) For any $g \in {}^{\lambda}J^{+} \cap N$ there are $\alpha < \beta < \lambda$ such that $g(\alpha) \cap g(\beta) \in J^{+}$.

17.7 Lemma. *Suppose that κ and λ are regular and P is a p.o. with the λ-c.c.. If*

$$\Vdash_P \dot{J} \text{ is a } \check{V}\text{-}\lambda\text{-saturated ideal over } \kappa\ ,$$

then

$$I = \{X \subseteq \kappa \mid \Vdash_P \check{X} \in \dot{J}\}$$

is a λ-saturated ideal over κ.

Proof. Suppose that $\{X_\alpha \mid \alpha < \lambda\} \subseteq I^+$. Then

$$(*) \qquad \Vdash |\{\alpha < \lambda \mid \check{X}_\alpha \in \dot{J}^+\}| = \lambda\ .$$

If not, by the regularity of λ and the λ-c.c. of P, there would be a $\gamma < \lambda$ such that

$$\Vdash \{\alpha < \lambda \mid \check{X}_\alpha \in \dot{J}^+\} \subseteq \gamma\ .$$

But then, $X_\gamma \in I$, which is a contradiction. With $(*)$, the $\check{V}$-λ-saturation of $\dot{J}$ implies that there is a $p \in P$ and $\alpha < \beta < \lambda$ such that $p \Vdash \check{X}_\alpha \cap \check{X}_\beta \in \dot{J}^+$, i.e. $X_\alpha \cap X_\beta \in I^+$. ⊣

Kunen and Paris showed in applications of their scheme that straightforward Easton products (as described before 10.7) preserve large cardinal properties. The first application preserved measurability, and showed that while normal ultrafilters may be special, it is consistent to have the maximum possible number of them.

17.8 Theorem (Kunen-Paris [71]). *Suppose that κ is measurable. Then there is a κ-c.c. p.o. E such that*

$$\Vdash_E \text{There are } 2^{2^\kappa} \text{ normal ultrafilters over } \kappa\ .$$

Proof. Let P be the Easton product of the p.o.'s for adding a generic subset to each *successor* cardinal less than κ. P can be regarded as consisting of those functions p with $\mathrm{dom}(p) \subseteq \kappa \times \kappa$ and $\mathrm{ran}(p) \subseteq 2$ such that

(i) for any $\langle \xi, \zeta \rangle \in \mathrm{dom}(p)$, ξ is a successor cardinal less than κ and $\zeta < \xi$, and

(ii) for any regular ν, $|\{\langle \xi, \zeta \rangle \in \mathrm{dom}(p) \mid \xi \leq \nu\}| < \nu$.

Here, the $\xi \leq \nu$ comprises the case $\xi < \nu$ for the Easton product, and the case $\xi = \nu$ (only for successor cardinals ν) for adding a subset at ν. Note that $|p| < \kappa$ for every $p \in P$ by taking $\nu = \kappa$. P is ordered by: $p \leq_P q$ *iff* $p \supseteq q$.

Some observations generally applicable to Easton products are now made, partly to establish a context for later proofs. First, there is a standard product analysis: For any γ, set

$$(P)_\gamma = \{p \in P \mid p \subseteq \gamma \times \kappa \times 2\}\ , \text{ and}$$
$$(P)^\gamma = \{p \in P \mid p \cap (\gamma \times \kappa \times 2) = \emptyset\}\ .$$

Then $P \cong (P)_\gamma \times (P)^\gamma$ under the natural identification of $p \in P$ with $\langle p \cap \gamma \times \kappa \times 2,\ p - \gamma \times \kappa \times 2 \rangle$. $(P)^\gamma$ is γ^+-closed, and if $\gamma^{<\gamma} = \gamma$, then $(P)_\gamma$ has the γ^+-c.c. by a typical application of the Δ-system Lemma 10.4. Hence by 10.8, forcing with $(P)^\gamma$ after $(P)_\gamma$ adds no further subsets of γ and preserves γ^+ as a cardinal. Since there are arbitrarily large such $\gamma < \kappa$, it follows that κ remains inaccessible after forcing with P. We next observe that because κ is Mahlo, P has the κ-c.c.:

Suppose that $A \subseteq P$ is a maximal antichain. By 6.2(b), let $\delta < \kappa$ be inaccessible such that

$$\langle V_\delta, \in, P \cap V_\delta, A \cap V_\delta \rangle \prec \langle V_\kappa, \in, P, A \rangle .$$

For any $p \in P$, $p \cap V_\delta \in V_\delta$ by (ii), so by the maximality of $A \cap V_\delta$ in V_δ, $p \cap V_\delta$ is compatible with some element of $A \cap V_\delta$. But then, it is readily seen that p itself is compatible with that element. Hence, $A \cap V_\delta$ is a maximal antichain of P, so that $A = A \cap V_\delta$ and $|A| < \kappa$.

Suppose now that U is a normal ultrafilter over κ, and $j\colon V \prec M \cong \mathrm{Ult}(V, U)$. Then

$$j(P) \cong P \times Q\ ,\ \text{ where } Q = (j(P))^\kappa\ ,$$

since $(j(P))_\kappa = P$. Here, for any $p \in P$, $j(p)$ is identified with $\langle p, \emptyset \rangle = \langle p, 1\!\!1_Q \rangle$, since $|p| < \kappa$ and so $j(p) = p$. Hence, the Kunen-Paris conditions are satisfied.

Generally speaking, the $\overline{U}$ of 17.6 is seen by the definition of $\overline{j}$ to correspond to a $P \times Q$-name $\dot{U}$ such that for any P-name $\dot{X}$ for a subset of κ and $\langle p, q \rangle \in P \times Q$,

$$\langle p, q \rangle \Vdash \dot{X} \in \dot{U} \leftrightarrow \kappa \in j(\dot{X})\ .$$

In the present case, because $P \times Q$ is identified with $j(P)$ by taking $\langle p, q \rangle$ to $p \cup q$, if $q = [f_q]_U$ where $f_q\colon \kappa \to P$, it follows that

$$(*) \qquad \langle p, q \rangle \Vdash \dot{X} \in \dot{U} \ \textit{ iff } \ \{\xi < \kappa \mid p \cup f_q(\xi) \Vdash \xi \in \dot{X}\} \in U\ .$$

Furthermore, Q is κ^+-closed (since κ is not a successor cardinal). Because P has the κ^+-c.c., it follows from 10.8 that forcing with Q after P adds no further subsets of κ. Hence,

$$\Vdash_{P \times Q} \dot{U} \text{ is a normal ultrafilter over } \kappa\ .$$

To get 2^{2^κ} normal ultrafilters, proceed as follows: In V, set $\lambda = 2^{2^\kappa}$ and let R be the κ^+-product of λ copies of Q, i.e. the p.o. consisting of functions $r \in {}^\lambda Q$ such that $|\{\alpha < \lambda \mid r(\alpha) \neq \emptyset\}| \leq \kappa$, ordered by: $r \leq_R s$ *iff* $\forall \alpha < \lambda (r(\alpha) \leq_Q s(\alpha))$. Then force with $P \times R$ instead. It is readily seen that $P \times R$ is also κ^+-closed, so for each $\alpha < \lambda$ there is a $P \times R$-name $\dot{U}_\alpha$ corresponding to $\dot{U}$ satisfying

$$\Vdash_{P \times R} \dot{U}_\alpha \text{ is a normal ultrafilter over } \kappa\ ,$$

such that for any $\langle p, r\rangle \in P \times R$ and P-name $\dot{X}$,

$$(**) \qquad \langle p, r\rangle \Vdash_{P\times R} \dot{X} \in \dot{U}_\alpha \ \textit{ iff } \ \langle p, r(\alpha)\rangle \Vdash_{P\times Q} \dot{X} \in \dot{U} .$$

A genericity argument shows that for $\alpha < \beta < \lambda$, $\Vdash_{P\times R} \dot{U}_\alpha \neq \dot{U}_\beta$:

Given any $\langle p, r\rangle \in P \times R$, let $\langle \overline{p}, \overline{r}\rangle \leq \langle p, r\rangle$ be such that $\overline{r}(\alpha) \perp \overline{r}(\beta)$ in Q. If $[g]_U = \overline{r}(\alpha)$ and $[h]_U = \overline{r}(\beta)$, it follows that

$$Y = \{\xi < \kappa \mid g(\xi) \perp h(\xi) \text{ in } P\} \in U .$$

Let $\dot{X}$ be a P-name for a subset satisfying

$$\Vdash_P \forall\xi(\xi \in \dot{X} \leftrightarrow \xi \in \check{Y} \wedge g(\check{\xi}) \in \dot{G})$$

(where $\dot{G}$ is the canonical name for the generic object), i.e. for any $z \in P$, $z \Vdash \xi \in \dot{X}$ *iff* $\xi \in Y \wedge z \leq_P g(\xi)$. Then using $(*)$ and $(**)$ it is straightforward to verify that

$$\langle \overline{p}, \overline{r}\rangle \Vdash \dot{X} \in \dot{U}_\alpha - \dot{U}_\beta .$$

Finally, $\lambda = 2^{2^\kappa}$ in the sense of the generic extension: Since there are λ distinct ultrafilters over κ, it remains to verify that $2^{2^\kappa} \leq \lambda$: Clearly $P \times R$ has cardinality λ, and a use of the Δ-system lemma 10.4 shows that it has the $(2^\kappa)^+$-c.c.. Hence, the computation of 10.5 yields the desired result. (Actually, it is seen that 2^{2^κ} can be made arbitrarily large in the generic extension.) ⊣

The next two theorems establish consistency results about ideals I over κ such that $\text{sat}(I) = \kappa$ and $\text{sat}(I) = \kappa^+$.

17.9 Theorem (Kunen-Paris [71]). *Suppose that κ is measurable. Then there is a p.o. E such that $\Vdash_E$ (a) + (b) + (c), where:*

(a) $2^{\aleph_0} = \kappa$.

(b) There is a κ-saturated ideal over κ.

(c) For no $\lambda < \kappa$ is there a λ-saturated ideal over κ.

Proof. Let P be the Easton product of the p.o.'s for adding κ subsets of ω and of each *successor* cardinal less than κ. P can be regarded as consisting of those functions p with $\text{dom}(p) \subseteq \kappa \times \kappa \times \kappa$ and $\text{ran}(p) \subseteq 2$ such that

(i) for any $\langle \xi, \zeta, \eta\rangle \in \text{dom}(p)$, ξ is ω or a successor cardinal less than κ, and $\zeta < \xi$, and

(ii) for any regular ν, $|\{\langle \xi, \zeta, \eta\rangle \in \text{dom}(p) \mid \xi \in \nu\}| < \nu$,

intending $p(\langle \xi, \zeta, \eta\rangle) = 1$ *iff* "the ηth new subset of ξ contains ζ". Note that $|p| < \kappa$ for every $p \in P$ by taking $\nu = \kappa$. P is ordered by: $p \leq_P q$ *iff* $p \supseteq q$. As for 17.8, P has the κ-c.c..

Suppose now that U is a normal ultrafilter over κ and $j\colon V \prec M \cong \text{Ult}(V, U)$. Let

$$P_1 = \{p \in j(P) \mid p \subseteq (j(\kappa) - \kappa) \times j(\kappa) \times j(\kappa) \times 2\}\ , \text{ and}$$
$$P_2 = \{p \in j(P) \mid p \subseteq \kappa \times \kappa \times (j(\kappa) - \kappa) \times 2\}\ .$$

Then

$$j(P) \cong P \times P_1 \times P_2$$

through a natural identification. Here, for any $p \in P$, $j(p)$ is identified with $\langle p, \emptyset, \emptyset \rangle = \langle p, \mathbb{1}_{P_1}, \mathbb{1}_{P_2} \rangle$ since $|p| < \kappa$ and so $j(p) = p$. Hence, the Kunen-Paris conditions are satisfied.

We now proceed to show that $P \times P_1$ is the desired p.o.. Let G_0 be P-generic over V and G_1 P_1-generic over $V[G_0]$. To invoke the Kunen-Paris scheme let G_2 be P_2-generic over $V[G_0][G_1]$, and set $G = G_0 \times G_1 \times G_2$. Since P_1 is κ^+-closed (as κ is not a successor cardinal), it follows from 10.8 that forcing with P_1 after P adds no further subsets of κ. Hence,

(∗) the $\overline{U}$ of 17.6 is a $V[G_0][G_1]$-normal ultrafilter over κ .

Observe next that $\Vdash_{P \times P_1} \check{P}_2$ has the κ-c.c.: Given $\{p_\alpha \mid \alpha < \kappa\} \subseteq P_2$, let

$$X = \{\eta \mid \exists\alpha\exists\xi\exists\zeta(\langle \xi, \zeta, \eta \rangle \in \mathrm{dom}(p_\alpha))\} \subseteq j(\kappa) - \kappa\ .$$

Then $|X| \leq \kappa$, and so an injection: $X \to \kappa$ can be used to associate to each p_α a $q_\alpha \in P$ so that: $p_\alpha \perp p_\beta$ *iff* $q_\alpha \perp q_\beta$. Thus, the κ-c.c. of P implies the κ-c.c. of P_2. By a standard antichain argument, another appeal to the κ-c.c. of P further shows that $\Vdash_P \check{P}_2$ has the κ-c.c.. Finally, P_1 is κ^+-closed, so 10.8 ensures that $\Vdash_{P \times P_1} \check{P}_2$ has the κ-c.c..

Using this and (∗), 17.7 can be applied to the dual of $\overline{U}$ to conclude that in $V[G_0][G_1]$ there is a κ-saturated ideal over κ. That $2^{\aleph_0} = \kappa$ also holds follows from standard arguments (using 10.5 and the κ-c.c. in one direction). To complete the proof, it is verified that in $V[G_0][G_1]$ there are no λ-saturated ideals over κ for any $\lambda < \kappa$:

Assume to the contrary that there is such an ideal for some $\lambda < \kappa$, which can be taken to satisfy $\lambda^{<\lambda} = \lambda$. Then by a product analysis as in 17.8, $P \cong (P)_\lambda \times (P)^\lambda$ where $(P)_\lambda$ has the λ^+-c.c. and $(P)^\lambda$ is λ^+-closed. G_0 can thus be considered as $H_0 \times H_1$ where $H_0 \subseteq (P)_\lambda$ and $H_1 \subseteq (P)^\lambda$. If $V[G_0][G_1]$ is regarded as a generic extension of $V[H_1][G_1]$ using $(P)_\lambda$, then by 17.7 there is a λ^+-saturated ideal in $V[H_1][G_1]$. Now $2^\lambda < \kappa$ holds in this model as $(P)^\lambda \times P_1$ is seen to be λ^+-closed, so it follows from 16.4(a) that κ is measurable there. However, $2^{\lambda^+} = \kappa$ also holds in the model because of $(P)^\lambda$. This is a contradiction. ⊣

17.10 Theorem (Kunen-Paris [71]). *Suppose that κ is measurable. Then there is a p.o. E such that $\Vdash_E$ (a) + (b) + (c), where:*

(a) κ is weakly compact.

(b) There is a κ^+-saturated ideal over κ.

(c) There is no κ-saturated ideal over κ.

Proof. Let P be the Easton product of the p.o.'s for adding a generic subset to each *regular* cardinal less than κ, i.e. P is as in the proof of 17.8 except that "successor" is replaced by "regular" in its (i). Again, P has the κ-c.c..

Suppose that U is a normal ultrafilter over κ and $j\colon V \prec M \cong \mathrm{Ult}(V, U)$. Let

$$P_1 = \{p \in j(P) \mid p \subseteq (j(\kappa) - \kappa + 1) \times j(\kappa) \times 2\}\ , \text{ and}$$
$$P_2 = \{p \in j(P) \mid p \subseteq \{\kappa\} \times \kappa \times 2\}\ .$$

Then P_1 is κ^+-closed, P_2 is κ-closed and has the κ^+-c.c., and since

$$j(P) \cong P \times P_1 \times P_2$$

through a natural identification which for any $p \in P$ identifies $j(p)$ with $\langle p, 1\!\!1_{P_1}, 1\!\!1_{P_2}\rangle$, the Kunen-Paris conditions are again satisfied.

We now proceed to show that $P \times P_1$ is the desired p.o.. Let G_0 be P-generic over V and G_1 P_1-generic over $V[G_0]$. To invoke the Kunen-Paris scheme, let G_2 be P_2-generic over $V[G_0][G_1]$, and set $G = G_0 \times G_1 \times G_2$. As in the previous proof, the κ^+-closure of P_1 implies that in $V[G]$, there is a $V[G_0][G_1]$-normal ultrafilter over κ. Since P_2 has the κ^+-c.c., it follows from 17.7 that there is a κ^+-saturated ideal over κ in $V[G_0][G_1]$.

We next verify that κ is weakly compact in $V[G_0][G_1]$: By 7.8, κ is weakly compact *iff* κ is inaccessible and has the tree property, and by 10.8 it suffices to establish these properties in $V[G_0]$. First, κ is inaccessible in $V[G_0]$ by remarks at the beginning of the proof of 17.8. Next, Suppose that $\langle T, <_T\rangle$ is a κ-tree in $V[G_0]$ with $T \subseteq \kappa$. Applying the Kunen-Paris $\overline{j}\colon V[G_0] \prec M[G]$ extending j, $\langle \overline{j}(T), \overline{j}(<_T)\rangle$ is a $\overline{j}(\kappa)$-tree in $M[G]$. Hence, any γ at its κth level determines a κ-branch through T since $\overline{j}(T) \cap \kappa = T$. It remains to show that such a branch already exists in $V[G_0]$. This is not immediate since $\mathcal{P}(\kappa) \cap V[G_0] = \mathcal{P}(\kappa) \cap V[G]$ fails; we work in $V[G_0]$ and rely on the fact that $V[G]$ is a generic extension by $P_1 \times P_2$, a κ-closed notion of forcing:

By the above remarks, there is a $P_1 \times P_2$-name τ such that

$$\Vdash_{P_1 \times P_2} \tau \text{ is a } \kappa\text{-branch through } \check{T}\ .$$

Using κ-closure, define conditions $p_\alpha \in P_1 \times P_2$ and sets b_α for $\alpha < \kappa$ by recursion such that for $\alpha < \beta < \kappa$: (i) $p_\alpha \Vdash \tau \cap \alpha = \check{b}_\alpha$, and (ii) $p_\beta \leq p_\alpha$. But then, $b = \bigcup_{\alpha<\kappa} b_\alpha$ is a κ-branch. (Note that we did not need any p such that $p \Vdash \tau = \check{b}$.)

To complete the proof, it remains only to verify that there are no κ-saturated ideals over κ in $V[G_0][G_1]$. To do this, it suffices by 16.4(b) to argue in $V[G_0][G_1]$ that κ cannot be measurable there: For any regular $\alpha < \kappa$, P had added a generic set $g(\alpha) \subseteq \alpha$ such that $g(\alpha) \cap \xi \in V$ for any $\xi < \alpha$. If κ were measurable and W a normal ultrafilter over κ, let $x = [g]_W$. As usual, $x = j_W(x) \cap \kappa$, i.e. $\{\alpha \mid g(\alpha) = x \cap \alpha\} \in W$. However, for any $\alpha < \beta$ both in this set $g(\alpha) = g(\beta) \cap \alpha \in V$, which is a contradiction. ⊣

17.11 Exercise. *17.10 holds with "κ is weakly compact" replaced by "κ is weakly inaccessible and $2^{\aleph_0} = \kappa$".*

Hint. One approach is to force first with the p.o. E of 17.10, and then do the ω_1-c.c. forcing for adding κ Cohen reals. 17.1 verifies (b), and 17.7, (c). $\dashv$

Some further consistency results are cited as the possibilities for non-measurable κ and $\text{sat}(I) \leq \kappa^+$ for some ideal I over κ: If κ is weakly compact, the only possibility by 16.4(b) is $\text{sat}(I) = \kappa^+$, shown consistent by 17.10. If κ is strongly inaccessible but not weakly compact, the only possibilities are $\text{sat}(I) = \kappa$, shown consistent by Kunen [78], and $\text{sat}(I) = \kappa^+$, shown consistent by Boos [74]. If κ is weakly but not strongly inaccessible, then $\omega < \text{sat}(I) < \kappa$ is consistent by 17.1, $\text{sat}(I) = \kappa$ by 17.9, and $\text{sat}(I) = \kappa^+$ by 17.11. Finally, if κ is a successor cardinal, the only possibility by 16.3 is $\text{sat}(I) = \kappa^+$, shown consistent by Kunen [78] in 1972.

This last result generated considerable interest. The others had all assumed the consistency strength of measurability, the exact strength required. However, Kunen [70] had previously shown through inner models that the consistency strength of having a κ^+-saturated ideal over a *successor* cardinal κ was strictly stronger than that of measurability. He then proceeded to establish that, assuming a strong hypothesis, the existence of a huge cardinal (§24), there is a generic extension in which there is an ω_2-saturated ideal over ω_1. Furthermore, he suggested *prima facie* evidence for the necessity of some such hypothesis. This is an early instance of an emerging approach to the analysis of combinatorial propositions about the low orders of the cumulative hierarchy: By using the methods of forcing and inner models, their consistency strength is located between two hypotheses in the hierarchy of large cardinals. Short of achieving an equiconsistency result, their strength can then be compared to others, and suggested possibilities pursued to refine the large cardinal correlations. As for getting ω_2-saturated ideals over ω_1, far less than huge cardinals are now known to suffice (see §32, especially 32.10).

18. Prikry Forcing

This section describes a simple but elegant notion of forcing for measurable cardinals devised by Karel Prikry, as well as a useful generalization. These together with his forcing results discussed in §17 appeared in his 1968 Berkeley dissertation with Silver. Prikry and also Thomas Jech were originally members of Vopěnka's productive seminar, and their emigration was a Prague spring of sorts for set theory in the United States. Although Prikry's forcing may have seemed a bit of a curiosity at first, it soon became an integral part of the theory of measurable cardinals. Not only did it add to the growing store of intuitions about them, but it provided a basic paradigm for later generalizations.

Maintaining some generality, let F be a filter over a cardinal κ. Then Prikry's forcing p.o. for F is

$$P_F = [\kappa]^{<\omega} \times F$$

ordered by

$$\langle s, A\rangle \leq \langle t, B\rangle \ \ \textit{iff} \ \ t \text{ is an initial segment of } s \text{ and } A \cup (s - t) \subseteq B\ .$$

The idea is that a condition $\langle s, A\rangle$ determines a finite initial segment s of a new subset of κ whose further members are to be restricted to a set A, kept large in the sense of being in F. Note that P_F is not separative, since for any $\langle s, A\rangle, \langle t, B\rangle \in P_F$, $\langle s, A\rangle \leq \langle t, B\rangle$ exactly when $\langle s, A\rangle \leq \langle t, B - (\max(t) + 1)\rangle$. It could be required of conditions $\langle s, A\rangle$ that $A \cap (\max(s) + 1) = \emptyset$ with only notational complications.

18.1 Lemma.

(a) If $\langle s, A\rangle \parallel \langle t, B\rangle$, then s is an initial segment of t or vice versa.

(b) P_F has the κ^+-c.c..

(c) If G is P_F-generic and

$$x = \bigcup\{s \mid \exists A(\langle s, A\rangle \in G)\}\ ,$$

then x is a unbounded subset of κ of ordertype ω, and $V[x] = V[G]$.

Proof. (a) is clear, and (b) follows from the observation that any two conditions with the same first component are compatible.

For (c), that x is unbounded in κ with ordertype $\geq \omega$ follows from a density argument (by a §0 convention, $\{\xi \mid \alpha \leq \xi < \kappa\} \in F$ for every $\alpha < \kappa$). On the other hand, for any $\langle s, A\rangle \in G$, s must be an initial segment of x by (a), and so x must have ordertype $\leq \omega$. Finally, G can be recovered from x since if

$$G_x = \{\langle s, A\rangle \in P_F \mid s \text{ is an initial segment of } x \text{ and } x - s \subseteq A\}\ ,$$

then $G = G_x$ (cf. before 11.14). ⊣

Because of this last point, for M a transitive $\in$-model of ZFC such that $M \models \ulcorner F$ is a filter over a cardinal $\kappa\urcorner$ and $x \in [\kappa]^\omega$,

$$x \text{ is } P_F\text{-generic over } M \text{ iff } G_x \text{ is } P_F\text{-generic over } M\ ,$$

and identifying x with its increasing enumeration, that

$$x \text{ is a } \textit{Prikry sequence over } M \textit{ iff } x \text{ is } P_F\text{-generic over } M \text{ for some such } F\ .$$

Since the cofinality of κ is ω in any forcing extension by P_F, κ may be collapsed. The following key lemma is how large cardinal ideas enter the picture in the case by which Prikry forcing is primarily known.

18.2 Lemma (Prikry [70]). *Suppose that κ is measurable and U is a normal ultrafilter over κ. Then for any $\langle s, A\rangle \in P_U$ and formula φ in the forcing language, there is a $B \subseteq A$ with $B \in U$ such that $\langle s, B\rangle \parallel \varphi$.*

Proof. Define a function $f\colon [A - (\max(s)+1)]^{<\omega} \to 2$ by

$$f(t) = \begin{cases} 0 & \text{if } \exists X(\langle s \cup t, X\rangle \Vdash \varphi)\ , \text{ and} \\ 1 & \text{otherwise}\ . \end{cases}$$

By Rowbottom's 7.17, there is a $B \subseteq A-(\max(s)+1)$ with $B \in U$ homogeneous for f. It follows that $\langle s, B\rangle \parallel \varphi$: Otherwise, there would be a $\langle s\cup t_0, B_0\rangle \leq \langle s, B\rangle$ and a $\langle s \cup t_1, B_1\rangle \leq \langle s, B\rangle$ such that $\langle s \cup t_0, B_0\rangle \Vdash \varphi$ and $\langle s \cup t_1, B_1\rangle \Vdash \neg\varphi$. By extending one condition if necessary, we can assume that $|t_0| = |t_1|$. But then $f(t_0) = 0$ and $f(t_1) = 1$, contradicting the homogeneity of B. ⊣

For any $\langle s, A\rangle \in P_U$ and formula φ, there is always a $\langle t, B\rangle \leq \langle s, A\rangle$ that decides φ; the point of the lemma is that we can take $t = s$. This leads to a strong conclusion usually drawn from the κ-closure of a p.o.:

18.3 Lemma. *With the hypotheses of 18.2, for any P_U-name $\dot{\tau}$ and set y with $|y| < \kappa$, there is a $z \subseteq y$ such that $\Vdash \dot{\tau} \subseteq \check{y} \to \dot{\tau} = \check{z}$.*

Proof. By a standard forcing argument (as for the maximal principle), it suffices to assume that $\langle s, A\rangle \Vdash \dot{\tau} \subseteq \check{y}$ and find a $\langle t, B\rangle \leq \langle s, A\rangle$ and a $z \in V$ such that $\langle t, B\rangle \Vdash \dot{\tau} = \check{z}$. For each $a \in y$, by 18.2 let $\langle s, B_a\rangle \parallel \check{a} \in \dot{\tau}$. Since $|y| < \kappa$, $B = \bigcap_{a\in y} B_a \in U$. If z is then defined by $a \in z$ *iff* $\langle s, B\rangle \Vdash \check{a} \in \dot{\tau}$, it follows that $\langle s, B\rangle \Vdash \dot{\tau} = \check{z}$. ⊣

The following summarizes the situation:

18.4 Theorem (Prikry [70]). *Suppose that κ is measurable and U is a normal ultrafilter over κ. If G is P_U-generic, then $(V_\kappa)^V = (V_\kappa)^{V[G]}$ and the cardinals of V and $V[G]$ coincide, yet $\mathrm{cf}^{V[G]}(\kappa) = \omega$.*

Proof. That $(V_\kappa)^V = (V_\kappa)^{V[G]}$ can be established by a straightforward induction on rank using 18.3, so that the cardinals below κ are preserved. κ too remains a cardinal since it is a limit cardinal, and the cardinals above κ are preserved by the κ^+-c.c.. ⊣

So, without disturbing the universe below κ, P_U introduces an ω-sequence overlay on κ and the crux of the matter is Prikry's lemma 18.2. 18.4 first solved the problem of getting an extension of a model of ZFC with the same ordinals and cardinals but a different cofinality. Remarkably, later developments in inner model theory would demonstrate that the consistency strength of a measurable cardinal is necessary for this purpose (see volume II). In any case, just for the preservation of cardinals alone, a Rowbottom filter (as defined before 16.10) suffices:

18.5 Exercise (Prikry [70]). *Suppose that F is a Rowbottom filter over κ. Then forcing with P_F preserves all cardinals.*

Hint. It suffices to show that no *successor* cardinal $\lambda^+ < \kappa$ in V is collapsed by P_F. So, set $\mu = \lambda^+$ and suppose that $\langle s, A\rangle \Vdash \dot{\tau}\colon \lambda \to \mu$. For each $t \in [A - (\max(s)+1)]^{<\omega}$ let

$$E_t = \{\eta \mid \exists\xi\exists X(\langle s \cup t, X\rangle \Vdash \dot{\tau}(\xi) = \eta)\}\,.$$

$|E_t| \leq \lambda$ as $\dot{\tau}$ is forced to be a function, so set $f(t) = \sup(E_t) < \mu$. Let $B \subseteq A - (\max(s)+1)$ with $B \in F$ be such that $|f``[B]^{<\omega}| \leq \omega$, so that $\delta = \sup(f``[B]^{<\omega}) < \mu$. Finally, show that $\langle s, B\rangle \Vdash \dot{\tau}``\lambda \subseteq \delta$. ⊣

Devlin [74] characterized those F over κ such that P_F preserves cardinals by a property slightly weaker than being Rowbottom which is implicit in the preceding proof. On the other hand, he and Paris (Devlin [74: 30]) established the full converse for P_F rendering $(V_\kappa)^V = (V_\kappa)^{V[G]}$ for any P_F-generic G: F must be a κ-complete ultrafilter satisfying the Rowbottom partition property 7.17 – all that 18.2 required of normality.

It is not surprising that P_F should preserve large cardinal properties; the following result provided an example of a Rowbottom cardinal of cofinality ω.

18.6 Exercise (Prikry [70]). *Suppose that F is a Rowbottom filter over κ. Then in any generic extension by P_F, F generates a Rowbottom filter over κ.*

Hint. If $\lambda < \kappa$ and $\langle s, A\rangle \Vdash \dot{\tau}\colon [\kappa]^{<\omega} \to \lambda$, it suffices to find a $B \subseteq A$ with $B \in F$ such that $\langle s, B\rangle \Vdash |\dot{\tau}``[\check{B}]^{<\omega}| \leq \omega$. To do this, for each $t \in [A - (\max(s)+1)]^{<\omega}$ let

$$f(t) = \{\eta \mid \exists u,v\exists X(t = u \cup v \ \wedge\ \langle s \cup u, X\rangle \Vdash \dot{\tau}(v) = \eta)\}\,.$$

$f(t) \in [\lambda]^{<\omega}$ since $\dot{\tau}$ is forced to be a function. Let $B \subseteq A - (\max(s)+1)$ with $B \in F$ such that $|f``[B]^{<\omega}| \leq \omega$, and set $W = \bigcup f``[B]^{<\omega}$. Finally, show that $\langle s, B\rangle \Vdash \dot{\tau}``[\check{B}]^{<\omega} \subseteq \check{W}$. ⊣

A revealing characterization of P_U-genericity for normal ultrafilters U is established next. If F is any ultrafilter over κ (in V) and $x \in [\kappa]^\omega$ is P_F-generic over V, then since x manages to burrow its way into every $X \in F$ by genericity, it is simple to see that x *generates* F in the following sense:

$$\forall X \in V \cap \mathcal{P}(\kappa)(X \in F \leftrightarrow x - X \text{ is finite}) .$$

Mathias provided a basic structural rationale for Prikry forcing by establishing the converse for normal ultrafilters. Generalizing a familiar concept, for $\langle A_s \mid s \in [\kappa]^{<\omega} \rangle$ with each $A_s \subseteq \kappa$, its *diagonal intersection* is

$$\triangle_s A_s = \{\xi < \kappa \mid \xi \in \bigcap \{A_s \mid \max(s) < \xi\}\} .$$

If U is a normal ultrafilter over κ and each $A_s \in U$, then it is simple to see that $\triangle_s A_s \in U$.

18.7 Theorem (Mathias [73]). *Suppose that M is a transitive $\in$-model of* ZFC *in which κ is a measurable cardinal and U a normal ultrafilter over κ. Then for any $x \in [\kappa]^\omega$,*

$$x \text{ is } P_U\text{-generic over } M \quad \textit{iff} \quad \forall X \in U(x - X \text{ is finite}) .$$

18.8 Corollary. *With M as above, if x is P_U-generic over M and $y \in [x]^\omega$, then y is also P_U-generic over M.* ⊣

Proof of 18.7. In the substantive direction, suppose that $x \in [\kappa]^\omega$ and $x - X$ is finite for every $X \in U$. We must show that

$$G_x = \{\langle s, A \rangle \in P_U \mid s \text{ is an initial segment of } x \text{ and } x - s \subseteq A\}$$

is P_U-generic over M, which reduces to showing that if $D \in M$ is dense open in P_U, then $G_x \cap D \neq \emptyset$:

For each $s \in [\kappa]^{<\omega}$, define $f_s \colon [\kappa - (\max(s) + 1)]^{<\omega} \to 2$ by

$$f_s(t) = \begin{cases} 0 & \text{if } \exists X(\langle s \cup t, X \rangle \in D) \text{ , and} \\ 1 & \text{otherwise .} \end{cases}$$

By 7.17, there is an $A_s \subseteq \kappa - (\max(s) + 1)$ with $A_s \in U$ homogeneous for f_s. Set $A = \triangle_s A_s \in U$.

Next, for each $s \in [\kappa]^{<\omega}$, if there is an X such that $\langle s, X \rangle \in D$, let B_s be such an X; otherwise, let $B_s = \kappa$. Set $B = A \cap \triangle_s B_s \in U$. Then

$$(*) \qquad \forall s \in [\kappa]^{<\omega}(\exists X(\langle s, X \rangle \in D \to \langle s, B - (\max(s) + 1) \rangle \in D)) ,$$

noting that D is dense open.

By hypothesis, there is an initial segment s_0 of x such that $x - s_0 \subseteq B$. Now by density there is a $t \in [B - (\max(s_0) + 1)]^{<\omega}$ such that $\langle s_0 \cup t, X \rangle \in D$ for some X. Consequently, if t_0 consists of the first $|t|$ members of $x - s_0 \subseteq B$, then the homogeneity of $B - (\max(s_0) + 1) \subseteq A_{s_0}$ for f_{s_0} implies that $\langle s_0 \cup t_0, X \rangle \in D$ for some X. But then by $(*)$, $\langle s_0 \cup t_0, B - (\max(t_0) + 1) \rangle \in D$. This condition is also a member of G_x, and so the proof is complete. ⊣

18.8 is analogous to 11.14(b) for Mathias reals, and indeed the two were proved together by Mathias. The analogies extend to the circumstance that Mathias forcing is equivalent to first adjoining an ultrafilter U over ω with strong properties by an ω_1-closed forcing, and then forcing with P_U (Mathias [77: §4]). An external ω sequence generating an internal normal ultrafilter also occurs in the theory of iterated ultrapowers as described in the next section. To conclude it, we establish a connection with Prikry forcing extensively studied by Patrick Dehornoy. Claude Sureson [89] has studied the interplay of various Prikry extensions in terms of degrees of constructibility.

As inner model theory developed through the 1970's, intriguing results were to provide intimate connections among the forcing and inner model phenomena and the concept of generating indiscernibles. With this synthesis, Prikry forcing has become a integral part of the study of measurability.

A Generalization

Prikry [70: 17, 18, 38] also considered a generalization of his forcing for which a version of his crucial lemma 18.2 can be established without assuming the partition property 7.17. The generalization being a natural and useful one for a variety of purposes, it has since been rediscovered by Dehornoy [78], who established the analogue of 18.7 for it, and variants by Moti Gitik [80, 86], Andreas Blass [88], and James Henle [90] (see also the work of Arthur Apter and Henle in their [91, 92]). A general formulation is provided, the main properties established, and then several examples given.

Let $\langle I, <_I \rangle$ be a non-empty directed set (§0) with no maximal elements, and with $<_I$ understood let ${}^{<\omega}I_+$ denote that subset of ${}^{<\omega}I$ consisting of increasing functions, i.e.

$${}^{<\omega}I_+ = \{s \mid \exists n \in \omega(s\colon n \to I \ \wedge\ \forall i \forall j(i < j < n \to s(i) <_I s(j)))\} .$$

Next, let $\mathcal{F}$ be a function on ${}^{<\omega}I_+$ such that for $s \in {}^{<\omega}I_+$,

(i) $\mathcal{F}(s)$ is an ultrafilter over I , and
(ii) for any $i \in I$, $\{j \mid i <_I j\} \in \mathcal{F}(s)$.

For such $\mathcal{F}$,

$$\prod \mathcal{F} = \{\mathcal{A} \mid \mathcal{A} \text{ is a function} \wedge \operatorname{dom}(\mathcal{A}) = {}^{<\omega}I_+ \wedge \forall s \in {}^{<\omega}I_+(\mathcal{A}(s) \in \mathcal{F}(s))\} .$$

For $\mathcal{A}, \mathcal{B} \in \prod \mathcal{F}$ set

$$\mathcal{A} \subseteq^* \mathcal{B} \ \ \textit{iff} \ \ \forall s \in {}^{<\omega}I_+(\mathcal{A}(s) \subseteq \mathcal{B}(s)) .$$

Finally, define a notion of forcing with

$$P_{\mathcal{F}} = {}^{<\omega}I_+ \times \prod \mathcal{F}$$

ordered by

$$\langle s, \mathcal{A}\rangle \leq \langle t, \mathcal{B}\rangle \ \ \textit{iff} \ \ \text{(i) } t \subseteq s \ , \quad \text{(ii) } \mathcal{A} \subseteq^* \mathcal{B} \ , \text{ and} \quad \text{(iii) for } |t| \leq i < |s|, \ s(i) \in \mathcal{B}(s|i) \ .$$

($t \subseteq s$ is equivalent of course to $s||t| = t$, i.e. t is an initial segment of s.)

When $I = \kappa$, $<_I$ is the membership relation on κ, and $\mathcal{F}$ is the constant function with range $\{U\}$ where U is a normal ultrafilter over κ, then $P_{\mathcal{F}}$ and P_U as defined earlier are equivalent forcing notions: Temporarily ignoring the difference between members of ${}^{<\omega}[\kappa]_+$ and their ranges in $[\kappa]^{<\omega}$, define $i_0\colon P_U \to P_{\mathcal{F}}$ by setting $i_0(\langle s, A\rangle) = \langle s, \mathcal{A}\rangle$ where $\mathcal{A}$ is the constant function: $[\kappa]^{<\omega} \to \{A\}$, and $i_1\colon P_{\mathcal{F}} \to P_U$ by setting $i_1(\langle t, \mathcal{B}\rangle) = \langle t, B\rangle$ where $B = \triangle_s \mathcal{B}(s)$, so that $B - (\max(s)+1) \subseteq \mathcal{B}(s)$ for each s. Noting that for any $\langle s, \mathcal{A}\rangle, \langle t, \mathcal{B}\rangle \in P_{\mathcal{F}}$, $\langle s, \mathcal{A}\rangle \leq i_0(i_1(\langle t, \mathcal{B}\rangle))$ implies that $\langle s, \mathcal{A}\rangle \leq \langle t, \mathcal{B}\rangle$, it is straightforward to check that i_0 and i_1 are dense embeddings (cf. 10.1).

The idea of the generalization is that for a finite s, instead of considering extensions drawn from a single large set, we consider those that are chains in a tree, the possible one-point extensions at each node being large in a sense specific to that node.

The following is the analogue of 18.1:

18.9 Exercise.

(a) If $\langle s, \mathcal{A}\rangle \parallel \langle t, \mathcal{B}\rangle$, then $s \subseteq t$ or vice versa.

(b) $P_{\mathcal{F}}$ has the $|I|^+$-c.c..

(c) If G is $P_{\mathcal{F}}$-generic and

$$x = \bigcup \{s \mid \exists \mathcal{A}(\langle s, \mathcal{A}\rangle \in G)\} \ ,$$

then $x\colon \omega \to I$ and is unbounded in $\langle I, <_I\rangle$, i.e. for any $i \in I$ there is an $n \in \omega$ such that $i <_I x(n)$, and $V[x] = V[G]$. ⊣

The last assertion of (c) is established by showing that if

$$G_x = \{\langle s, \mathcal{A}\rangle \in P_{\mathcal{F}} \mid s \subseteq x \ \wedge \ \forall i(|s| \leq i < \omega \to s(i) \in \mathcal{A}(s|i))\} \ ,$$

then $G = G_x$. Because of this, for M a transitive $\in$-model of ZFC in which I and $\mathcal{F}$ are as above and $x\colon \omega \to I$,

$$x \text{ is } (P_{\mathcal{F}})^M\text{-generic over } M \ \ \textit{iff} \ \ G_x \cap M \text{ is } (P_{\mathcal{F}})^M\text{-generic over } M \ .$$

The following is the analogue to 18.2, and it is crucial for this context just as 18.2 was for its.

18.10 Lemma. *Suppose that $\langle s, \mathcal{A}\rangle \in P_{\mathcal{F}}$ and φ is a formula of the forcing language. Then there is a $\mathcal{B} \in \prod \mathcal{F}$ with $\mathcal{B} \subseteq^* \mathcal{A}$ such that $\langle s, \mathcal{B}\rangle \parallel \varphi$.*

Proof. The following is first established:

(∗) For any $t \in {}^{<\omega}I_+$, there is a $\mathcal{X} \in \prod \mathcal{F}$ such that $\langle t, \mathcal{X}\rangle \Vdash \varphi$ *iff* $\{i \in I \mid \exists \mathcal{Y}(\langle t^\frown\langle i\rangle, \mathcal{Y}\rangle \in P_{\mathcal{F}} \wedge \langle t^\frown\langle i\rangle, \mathcal{Y}\rangle \Vdash \varphi)\} \in \mathcal{F}(t)$.

The forward direction is clear, since for any $i \in \mathcal{X}(t)$ with $t(|t|-1) <_I i$, $\langle t^\frown\langle i\rangle, \mathcal{X}\rangle \leq \langle t, \mathcal{X}\rangle$. For the converse, let W be the hypothesized member of $\mathcal{F}(t)$, and for each $i \in W$ with $t(|t|-1) <_I i$ let $\mathcal{Y}_i$ be such that $\langle t^\frown\langle i\rangle, \mathcal{Y}_i\rangle \Vdash \varphi$. Define $\mathcal{X} \in \prod \mathcal{F}$ by:

$$\mathcal{X}(u) = \begin{cases} \mathcal{Y}_i(u) & \text{if } i \in W \text{ and } t^\frown\langle i\rangle \subseteq u \ , \\ W & \text{if } u = t \ , \text{ and} \\ I & \text{otherwise} \ . \end{cases}$$

Then it is simple to check that $\langle t, \mathcal{X}\rangle \Vdash \varphi$.

To establish the lemma assume to the contrary that for no $\mathcal{B} \subseteq^* \mathcal{A}$ does $\langle s, \mathcal{B}\rangle \parallel \varphi$. For $t \in {}^{<\omega}I_+$ with $s \subseteq t$, set

$$\mathcal{A}'(t) = \{i \in \mathcal{A}(t) \mid \neg\exists \mathcal{Y}(\langle t^\frown\langle i\rangle, \mathcal{Y}\rangle \in P_{\mathcal{F}} \wedge \langle t^\frown\langle i\rangle, \mathcal{Y}\rangle \parallel \varphi)\} \ .$$

For such t satisfying $\forall i(|s| \leq i < |t| \rightarrow t(i) \in \mathcal{A}'(t|i))$, it is straightforward to check that $\mathcal{A}'(t) \in \mathcal{F}(t)$ by induction on $|t|$, repeatedly applying (∗) for φ and $\neg\varphi$ "relativized" to $\mathcal{A}$. Hence, if we further set $\mathcal{A}'(t) = \mathcal{A}(t)$ for those t such that $s \not\subseteq t$, then $\mathcal{A}' \in \prod \mathcal{F}$ with $\mathcal{A}' \subseteq^* \mathcal{A}$. But then, no $\langle t, \mathcal{X}\rangle \leq \langle s, \mathcal{A}'\rangle$ can ever decide φ, which is a contradiction. ⊣

The following is the analogue of 18.3:

18.11 Exercise. *Suppose that λ is regular and $\mathcal{F}(s)$ is λ-complete for every $s \in {}^{<\omega}I_+$. Then for any $P_{\mathcal{F}}$-name $\dot{\tau}$ and set y with $|y| < \lambda$, there is a $z \subseteq y$ such that $\Vdash \dot{\tau} \subseteq \check{y} \rightarrow \dot{\tau} = \check{z}$.* ⊣

All this was known to Prikry [70]; Dehornoy [78] went on to establish the following analogue of the characterization 18.7 in his context. For the general result, the partition property argument for 18.7 is replaced by a well-foundedness argument.

18.12 Theorem. *Suppose that M is a transitive $\in$-model of* ZFC *in which I and $\mathcal{F}$ are as above. Then for any $<_I$-increasing $x\colon \omega \rightarrow I$,*

$$x \text{ is } (P_{\mathcal{F}})^M\text{-generic over } M \textit{ iff}$$
$$\forall \mathcal{A} \in (\textstyle\prod\mathcal{F})^M \exists n \in \omega \forall i (n \leq i < \omega \rightarrow x(i) \in \mathcal{A}(x|i)) \ .$$

Proof. The forward direction is clear since for any $\mathcal{A} \in (\prod \mathcal{F})^M$, $\{\langle t, \mathcal{B}\rangle \in (P_{\mathcal{F}})^M \mid \mathcal{B} \subseteq^* \mathcal{A}\}$ is dense.

For the converse direction, we must show that

$$G_x = \{\langle s, \mathcal{A}\rangle \in (P_{\mathcal{F}})^M \mid s \subseteq x \wedge \forall i(|s| \leq i < \omega \rightarrow s(i) \in \mathcal{A}(s|i))\}$$

is $(P_{\mathcal{F}})^M$-generic over M, which soon reduces to showing that it meets every dense open set in M.

To this end, argue in M: Assume that D is dense open. For $\langle s, \mathcal{A}\rangle \in P_{\mathcal{F}}$ set

$$\tau(s, \mathcal{A}) = \{t \mid s \subseteq t \ \wedge \ \forall i(|s| \leq i < |t| \to t(i) \in \mathcal{A}(t|i)) \\ \wedge \ \exists \mathcal{B} \in \prod \mathcal{F}(\langle t, \mathcal{B}\rangle \notin D)\} ,$$

i.e. the set of possible first coordinates t of extensions of $\langle s, \mathcal{A}\rangle$ "unsecured" in the sense that there is still a $\mathcal{B}$ such that $\langle t, \mathcal{B}\rangle \notin D$. As D is open, if $t_0 \subseteq t_1$ and $t_1 \in \tau(s, \mathcal{A})$, then $t_0 \in \tau(s, \mathcal{A})$. Define

$$\langle s, \mathcal{A}\rangle \textit{ snares } D \textit{ iff } \text{there is no } y\colon \omega \to I \text{ such that} \\ \text{for every } n \in \omega,\ y|n \in \tau(s, \mathcal{A})$$

(i.e. $\langle \tau(s, \mathcal{A}), \supset\rangle$ is well-founded). Then for any $s \in {}^{<\omega}I_+$, there is an $\mathcal{A} \in \prod \mathcal{F}$ such that $\langle s, \mathcal{A}\rangle$ snares D:

Arguing as for 18.10, the following can be established:

$$\text{For any } t \in {}^{<\omega}I_+ \text{ there is a } \mathcal{X} \in \prod \mathcal{F} \text{ such that } \langle t, \mathcal{X}\rangle \text{ snares } D \\ \textit{iff } \{i \in I \mid \exists \mathcal{Y}(\langle t^\frown\langle i\rangle, \mathcal{Y}\rangle \in P_{\mathcal{F}} \ \wedge \ \langle t^\frown\langle i\rangle, \mathcal{Y}\rangle \text{ snares } D)\} \in \mathcal{F}(t) .$$

Using this, assume to the contrary that there is no $\mathcal{A} \in \prod \mathcal{F}$ such that $\langle s, \mathcal{A}\rangle$ snares D and get an $\mathcal{A}' \in \prod \mathcal{F}$ such that for any $t \in {}^{<\omega}I_+$ with $s \subseteq t$,

$$\mathcal{A}'(t) = \{i \in I \mid \neg\exists \mathcal{Y}(\langle t^\frown\langle i\rangle, \mathcal{Y}\rangle \in P_{\mathcal{F}} \ \wedge \ \langle t^\frown\langle i\rangle, \mathcal{Y}\rangle \text{ snares } D)\} .$$

However, any $\langle t, \mathcal{X}\rangle \leq \langle s, \mathcal{A}'\rangle$ with $\langle t, \mathcal{X}\rangle \in D$ certainly snares D, which is a contradiction.

To complete the proof of the theorem, for each $s \in {}^{<\omega}I_+$ let $\mathcal{A}_s \in \prod \mathcal{F}$ be such that $\langle s, \mathcal{A}_s\rangle$ snares D. Define a "diagonal intersection" of the $\mathcal{A}_s$'s by setting

$$\mathcal{B}(t) = \bigcap\nolimits_{s \subseteq t} \mathcal{A}_s(t) \quad \text{for } t \in {}^{<\omega}I_+ .$$

$\mathcal{B} \in \prod \mathcal{F}$, and it is simple to check that for any $s \in {}^{<\omega}I_+$, $\tau(s, \mathcal{B}) \subseteq \tau(s, \mathcal{A}_s)$, so that $\langle s, \mathcal{B}\rangle$ snares D.

By hypothesis, there is an $n \in \omega$ such that for $n \leq i < \omega$, $x(i) \in \mathcal{B}(x|i)$. Now $\langle x|n, \mathcal{B}\rangle$ snares D, i.e. $\langle \tau(x|n, \mathcal{B}), \supset\rangle$ is well-founded, so there is (0.3) an order-preserving function $\rho\colon \tau(x|n, \mathcal{B}) \to \text{On}$. Finally stepping out of M, such a ρ also verifies that $\tau(x|n, \mathcal{B})^M$ is well-founded in V, and consequently there must be an $m \geq n$ such that $\langle x|m, \mathcal{B}\rangle \in D$. But $\langle x|m, \mathcal{B}\rangle \in G_x$, and so the proof is complete. ⊣

18.13 Corollary. *With M as in 18.12, if x is $(P_{\mathcal{F}})^M$-generic over M and $y\colon \omega \to I$ is $<_I$-increasing with* $\text{ran}(y) \subseteq \text{ran}(x)$, *then y is also $(P_{\mathcal{F}})^M$-generic over M.*⊣

Having duly established that the main features of Prikry forcing carry over to the general setting, this section is concluded with several examples:

1. (Prikry [70: 38]) I is κ^+, $<_I$ the membership relation on κ^+ and $\mathcal{F}$ a constant function with range $\{U\}$ where U is a κ-complete ultrafilter over κ^+. Then forcing with $P_\mathcal{F}$ adds no new bounded subsets of κ and preserves all cardinals except κ^+, whose cofinality is changed to ω.

2. (Prikry [70: 38]) I is κ^+, $<_I$ the membership relation on κ^+, and there is an increasing sequence $\langle \kappa_n \mid n \in \omega \rangle$ of cardinals cofinal in κ such that U_n is a κ_n-complete ultrafilter over κ^+ for $n \in \omega$ and $\mathcal{F}(s) = U_{|s|}$ for $s \in {}^{<\omega}I_+$. Then again forcing with $P_\mathcal{F}$ adds no new bounded subsets of κ and preserves all cardinals except κ^+, whose cofinality is changed to ω. (But here, we must first note that the $\mathcal{B}$ of 18.10 can also be required to satisfy $\mathcal{B}(t) = \mathcal{A}(t)$ for $s \not\subseteq t$ by suitably adjusting the proof, and using this, that for 18.11 $\mathcal{F}(s)$ need only be λ-complete for $|s|$ sufficiently large.)

3. (Dehornoy [78]) I is $\kappa > \omega$, $<_I$ the membership relation on κ, and $\mathcal{F}$ a constant function with range $\{U\}$ where U is a (not necessarily normal) κ-complete ultrafilter over κ. In this case, forcing with $P_\mathcal{F}$ has an equivalent formulation which is more like Prikry forcing:

Let $f \in {}^\kappa\kappa$ be such that $[f]_U = \kappa$, i.e. for any $\alpha < \kappa$, $\{\xi < \kappa \mid \alpha < f(\xi)\} \in U$, yet for any $g \in {}^\kappa\kappa$ such that $\{\xi < \kappa \mid g(\xi) < f(\xi)\} \in U$ there is an $\alpha < \kappa$ satisfying $\{\xi < \kappa \mid g(\xi) = \alpha\} \in U$. For $\langle A_s \mid s \in [\kappa]^{<\omega} \rangle$ with each $A_s \subseteq \kappa$, its *f-diagonal intersection* is

$$\Delta_s^f A_s = \{\xi < \kappa \mid \xi \in \bigcap\{A_s \mid \max(s) < f(\xi)\}\} .$$

As for the case when f is the identity map: $\kappa \to \kappa$, i.e. U is normal, it is simple to check that if each $A_s \in U$, then $\Delta_s^f A_s \in U$.

Setting

$${}^{<\omega}[\kappa]_f = \{s \in {}^{<\omega}[\kappa] \mid \forall i < |s| - 1(s(i) < f(s(i+1)))\} ,$$

let

$$P_U^f = {}^{<\omega}[\kappa]_f \times U$$

ordered by

$$\langle s, A \rangle \leq \langle t, B \rangle \ \textit{iff} \ \ t \text{ is an initial segment of } s \text{ and } A \cap (\mathrm{ran}(s) - \mathrm{ran}(t)) \subseteq B .$$

It can be assumed that $f(\xi) \leq \xi$ for each $\xi < \kappa$, and so this is a stringent version of Prikry forcing where beyond $s(i) < s(i+1)$ we require $s(i) < f(s(i+1)) \leq s(i+1)$.

Echoing remarks before 18.9 and again temporarily ignoring the difference between members of ${}^{<\omega}[\kappa]_+$ and their ranges in $[\kappa]^{<\omega}$, define i_0: $P_U^f \to P_\mathcal{F}$ by setting $i_0(\langle s, A \rangle) = \langle s, \mathcal{A} \rangle$ where $\mathcal{A}$ is the constant function $[\kappa]^{<\omega} \to \{A\}$, and i_1: $P_\mathcal{F} \to P_U^f$ by setting $i_1(\langle t, \mathcal{B} \rangle) = \langle t, B \rangle$ where $B = \Delta_s^f \mathcal{B}(s)$. Note that for any $\langle t, \mathcal{B} \rangle \in P_\mathcal{F}$, if $\mathcal{B}'$: $[\kappa]^{<\omega} \to U$ is defined by

$$\mathcal{B}'(s) = \mathcal{B}(s) \cap \{\xi < \kappa \mid \max(s) < f(\xi)\} \in U ,$$

then for any $\langle s, \mathcal{A}\rangle \leq \langle t, \mathcal{B}'\rangle$ and $i+1 \in \mathrm{dom}(s) - \mathrm{dom}(t)$, $s(i) < f(s(i+1))$. Consequently, $\langle s, \mathcal{A}\rangle \leq i_0(i_1(\langle t, \mathcal{B}'\rangle))$ implies that $\langle s, \mathcal{A}\rangle \leq \langle t, \mathcal{B}'\rangle$. It is now straightforward to check that i_0 and i_1 are dense embeddings.

Gitik-Magidor [92] used this forcing in the P_U version.

4. With $\kappa \leq \gamma$, I is $\mathcal{P}_\kappa\gamma = \{x \subseteq \gamma \mid |x| < \kappa\}$, $<_I$ the proper inclusion relation on $\mathcal{P}_\kappa\gamma$, and $\mathcal{F}$ a constant function with range $\{U\}$ where U is a κ-complete ultrafilter over $\mathcal{P}_\kappa\gamma$ such that $\{x \in \mathcal{P}_\kappa\gamma \mid \alpha \in x\} \in U$ for every $\alpha < \gamma$. Then forcing with $P_\mathcal{F}$ adds no new bounded subsets of κ, and for every regular δ with $\kappa \leq \delta \leq \gamma$ renders its cofinality ω: if $x\colon \omega \to \mathcal{P}_\kappa\gamma$ is $P_\mathcal{F}$-generic, then $\bigcup_{n\in\omega} x(n) \cap \delta = \delta$. Since the cardinal successor of κ in the generic extension must be regular there as well as in V, it follows that no δ with $\kappa < \delta \leq \gamma$ remains a cardinal.

5. (Gitik) The same as 4., except that $<_I$ is the relation $<$ defined on $\mathcal{P}_\kappa\gamma$ by:

$$x < y \ \textit{ iff } \ x \subseteq y \ \text{ and } \ |x| < |y \cap \kappa| \,.$$

(See the end of §25 for more on this relation.) Then forcing with $P_\mathcal{F}$ has the properties described in 4. and moreover has an alternate formulation as in 3. See Gitik [80, 86] and Apter-Henle [91, 92] for applications.

6. I is ω and $<_I$ the membership relation on ω. Blass [88] considered various possibilities for $\mathcal{F}$ and extended results of Mathias [77] established using Mathias forcing (see the end of §11). Woodin and also Judah-Shelah [91] showed assuming CH how to get an $\mathcal{F}$ so that $P_\mathcal{F}$ (and hence an ω_1-c.c. notion of forcing) adjoins a real of minimal constructibility degree.

As a large part of the increasing use of strong hypotheses to establish relative consistency results, Prikry forcing was to be generalized in various directions for a variety of purposes. The possibility of adjoining a new unbounded subset of some large κ with minimal disturbance to the universe, as in the basic paradigm, came to be considered a prominent feature of large cardinals. In the decades to follow, generalizations for changing cofinality were to become central to the study of the possible values of 2^λ for singular strong limit λ, and equiconsistency results were to be established using an elegant and technically sharp generalization for adjoining closed unbounded subsets of some κ while preserving its large cardinal properties, *Radin forcing* (see volume II).

19. Iterated Ultrapowers

Kenneth Kunen was a student with Scott at Stanford University and went on to become a prominent set theorist and topologist at the University of Wisconsin. Results of his have already been discussed in §17 and elsewhere. His 1968 dissertation [68] is rivaled only by Silver's of a couple of years earlier for its impact on the development of set theory. Sophisticated and wide-ranging, it featured elegant results about inner models and elementary embeddings that considerably clarified the structure theory of measurable cardinals and $0^{\#}$. This section develops the techniques that led to these results, which are then established in the two subsequent sections.

Like Silver, Kunen was inspired by Gaifman's seminal work. Gaifman had systematized the process of iterating the ultrapower construction into the transfinite and revealed its potentialities by establishing 9.1. Kunen realized that ultrapowers of inner models can be taken by an ultrafilter not necessarily a member of the model, and the process iterated if an "amenability" condition is satisfied. His approach is formulated in a way that anticipates later developments, turning first to a generalization of the ultrapower construction of §5.

Suppose that M is a transitive $\in$-model of ZFC$^-$; i.e. ZFC with the Power Set Axiom deleted. Suppose also that κ is an infinite cardinal in the sense of M; *κ need not be a cardinal in V*. (To the end of this chapter especially germane is the liberal convention that $\kappa, \lambda, \ldots$ may denote cardinals only in the sense of some transitive $\in$-model.) Then

U is an *M-ultrafilter over κ iff*
(i) $\langle M, \in, U\rangle \models U$ is a normal ultrafilter over κ ; and
(ii) $\langle M, U\rangle$ is *weakly amenable*: for any
$F \in {}^{\kappa}M \cap M$, $\{\xi < \kappa \mid F(\xi) \in U\} \in M$.

An equivalent formulation of (i) was used for a definition before 17.6, and as noted there, *normality subsumes κ-completeness in the sense of M, and $U \in M$ is not required.* The present formulation in terms of $\langle M, \in, U\rangle$ anticipates the use of such structures below, so it should be pointed out that when M is a proper class, the usual caveats about the formalizability of satisfaction apply (§0). The new condition (ii) is a weak form of Jensen's concept of amenability: For $R \subseteq A$, $\langle A, R\rangle$ is *amenable iff* for any $x \in A$, $R \cap x \in A$. (ii) imposes this condition on $\langle M, U\rangle$ only for those $x \in M$ of cardinality at most κ in the sense of M. It should thus be clear that in considering $\langle M, \in, U\rangle$, U is not to play a role in the Replacement Schema for $M \models$ ZFC$^-$.

The current definition of M-ultrafilter over κ differs from the original one in Kunen [70] in two respects: First, transitive $\in$-models M not only of ZFC but of ZFC$^-$ are allowed; examples are L_λ and H_λ for regular $\lambda > \omega$ and their elementary substructures, and so such models abound. (On the other hand, such an M, if it contains all the ordinals, must be an inner model of ZFC (cf. before 3.1).) Second, for the sake of simplicity normality is imposed on U at the outset, whereas Kunen had began only with κ-completeness in the sense of M. Kunen [70: 188] observed that for countable $M \models$ ZFC, having an M-ultrafilter over κ in his weaker sense is equivalent to κ being weakly compact in M. With normality Kleinberg [78] established an analogous equivalence with a stronger large cardinal property (see Subtle Properties in volume II). This property is also compatible with $V = L$, but in any case external measurability with the weak amenability condition (b) does lead to strength within M, at least if $M \models$ ZFC.

With U an M-ultrafilter over κ, an ultrapower of $\langle M, \in, U\rangle$ can be formed: For any $f \in {}^{\kappa}M \cap M$, set

$$(f)_U = \{g \mid g \in {}^{\kappa}M \cap M \ \wedge \ \{\xi < \kappa \mid f(\xi) = g(\xi)\} \in U\} .$$

Thus, only functions in M are considered and equivalence classes of these taken. To accommodate proper classes M set

$$(f)_U^0 = \{g \mid g \in (f)_U \ \wedge \ \forall h(h \in (f)_U \rightarrow \ \text{rank}(g) \leq \text{rank}(h))\}$$

as in §5 so that the $(f)_U^0$'s are sets. The domain of the ultrapower can then be formulated as

$${}^{\kappa}M/U = \{(f)_U^0 \mid f \in {}^{\kappa}M \cap M\}$$

with its membership relation E_U defined by

$$(f)_U^0 \ E_U \ (g)_U^0 \quad \textit{iff} \quad \{\xi < \kappa \mid f(\xi) \in g(\xi)\} \in U .$$

Furthermore, taking $\mathcal{L}_\in(\dot{A})$ to be the ambient language of $\langle M, \in, U\rangle$, the new interpretation $\dot{A}_U$ of the unary predicate symbol $\dot{A}$ can be defined appropriately:

$$(f)_U^0 \ E_U \ \dot{A}_U \quad \textit{iff} \quad \{\xi < \kappa \mid f(\xi) \in U\} \in U .$$

This motivates weak amenability: that $\{\xi < \kappa \mid f(\xi) \in U\} \in M$ makes this definition possible.

Proceeding as in §5, a crucial use of $M \models \text{ZFC}^-$ is to check that Łoś's Theorem holds for $\langle {}^{\kappa}M/U, E_U\rangle$; the Separation Axiom is needed for its very statement, and the Axiom of Choice for the existential quantifier step. However, E_U may not be well-founded, as U is κ-complete only for sequences in M. *If* it is well-founded, by the Collapsing Lemma 0.4 there is a transitive N, a $W \subseteq N$, and an isomorphism

$$\pi\colon \langle {}^{\kappa}M/U, E_U, \dot{A}_U\rangle \ \rightarrow \ \langle N, \in, W\rangle .$$

We can then set

$$[f]_U = \pi((f)_U^0) \quad \text{for } f \in {}^{\kappa}M \cap M$$

and

$$j(x) = [f_x]_U \text{ for } x \in M$$

where f_x is the constant function: $\kappa \rightarrow \{x\}$. Because of Łoś's Theorem,

$$j\colon \langle M, \in\rangle \prec \langle N, \in\rangle$$

and so $N \models \mathrm{ZFC}^-$ and $\mathrm{On} \cap M \subseteq \mathrm{On} \cap N$, and

$$\mathrm{crit}(j) = \kappa \quad\text{and}\quad [\mathrm{id}]_U = \kappa$$

by normality, where id: $\kappa \to \kappa$ is the identity function. When M is an inner model of ZFC and $U \in M$, the foregoing is just the §5 ultrapower construction carried out in M, so that $W \in N \subseteq M$ and N is a definable class in the sense of M. However, it can well be that $\mathrm{On} \cap M < \mathrm{On} \cap N$: Suppose that κ is a measurable cardinal and U is a normal ultrafilter over κ, and consider $M = H_{\kappa^+}$. Then $\langle {}^\kappa M/U, E_U\rangle$ is well-founded, and there is an N and j as above. Since this j is the restriction to M of the usual embedding defined on all of V,

$$\mathrm{On} \cap M = \kappa^+ \leq 2^\kappa < j(\kappa) < \mathrm{On} \cap N\ .$$

By taking a countable elementary substructure of an amalgamation of M, U, and N and forming its transitive collapse, we can get countable M_0, U_0, and N_0 such that $\mathrm{On} \cap M_0 < \mathrm{On} \cap N_0$.

Parts of the next lemma recall 5.7; as usual, the subscript U is suppressed.

19.1 Lemma.

(a) j is cofinal: for any $y \in N$ there is an $x \in M$ such that $y \in j(x)$. Moreover, if y is an ordinal, then x can be taken to be an ordinal as well.

(b) If M is a set, then $|M| = |N|$.

(c) $j(x) = x$ for every $x \in V_\kappa \cap M$, $V_\kappa \cap M = V_\kappa \cap N$, $\mathcal{P}(\kappa) \cap M = \mathcal{P}(\kappa) \cap N$, and $\kappa^{+M} = \kappa^{+N}$.

(d) $U \notin N$.

(e) W is an N-ultrafilter over $j(\kappa)$.

Proof. (a) If $y = [f]$ say, take $x = \mathrm{ran}(f)$. If also $y \in \mathrm{On}$, then we can assume that $\mathrm{ran}(f) \subseteq \mathrm{On}$ and take $x = \sup(\mathrm{ran}(f)) + 1$.

(b) $|N| \leq |{}^\kappa M \cap M| \leq |M|$, but also $|M| \leq |N|$ because of j.

(c) The rank function can be defined in ZFC^- by the usual recursion (0.2). Using this, we shall show that for each $\alpha \leq \kappa$,

$$(*) \qquad j(x) = x \ \text{ for every } \ x \in V_\alpha \cap M, \ \text{ and } V_\alpha \cap M = V_\alpha \cap N\ .$$

This proceeds by induction on α, the limit stage being immediate. For the successor stage, suppose that $(*)$ holds, and let $x \in M$ with $\mathrm{rank}(x) = \alpha$. Then $\mathrm{rank}(j(x)) = j(\alpha) = \alpha$ by definability of rank, so by induction

$$\begin{aligned} j(x) &= \{y \in V_\alpha \cap N \mid y \in j(x)\} \\ &= \{y \in V_\alpha \cap M \mid y \in j(x)\} \\ &= \{y \in V_\alpha \cap M \mid j(y) \in j(x)\} \\ &= \{y \mid y \in x\} \\ &= x\ . \end{aligned}$$

This also implies that $V_{\alpha+1} \cap M \subseteq V_{\alpha+1} \cap N$, so it remains to establish the converse to complete the inductive proof of (∗):

Suppose that $x \in N$ with $\mathrm{rank}(x) = \alpha$, say $x = [f]$. Since $j(\alpha) = \alpha$, it can be assumed that $\mathrm{rank}(f(\xi)) = \alpha$ for every $\xi < \kappa$. Set $u = \bigcup \mathrm{ran}(f) \in M$. Assume to the contrary that $|u| \geq \kappa$ in M, so that there is a surjection $s\colon u \to \kappa$ with $s \in M$. Let $g\colon \kappa \to u$ with $g \in M$ such that $s(g(\xi)) = \xi$ for every $\xi < \kappa$. Since $\mathrm{rank}(g(\xi)) < \alpha$ for every $\xi < \kappa$, $[g] \in V_\alpha \cap N = V_\alpha \cap M$, say $[g] = z \in M$, so that $j(z) = z = [g]$. But this is a contradiction, since g through its definition must be injective.

It follows that $|u| \leq \kappa$ in M. But then,

$$x = \{y \in u \mid \{\xi < \kappa \mid y \in f(\xi)\} \in U\} \in M\ .$$

The equality holds since for any $y \in x$, $y \in V_\alpha \cap N = V_\alpha \cap M$ and $j(y) = y$ by induction, and membership in M follows from the weak amenability of $\langle M, U\rangle$.

For the final parts of (c), note first that if $X \in \mathcal{P}(\kappa) \cap M$, then $j(X) \cap \kappa = X \in N$. Conversely, if $Y \in \mathcal{P}(\kappa) \cap N$, say $Y = [f]$, then

$$Y = \{\alpha \in \kappa \mid \{\xi < \kappa \mid \alpha \in f(\xi)\} \in U\}\ ,$$

which is in M by the weak amenability of $\langle M, U\rangle$. That $\kappa^{+M} = \kappa^{+N}$ follows, as M and N contain the same well-orderings of κ.

(d) Assume to the contrary that $U \in N$. Then

$$\mathcal{P}(\kappa) \cap M = U \cup \{\kappa - X \mid X \in U\} \in N\ .$$

Since $\mathcal{P}(\kappa)^N = \mathcal{P}(\kappa) \cap N = \mathcal{P}(\kappa) \cap M$ by (c) and $j(\kappa) = \{[f] \mid f \in {}^\kappa\kappa \cap M\}$, it is not difficult to see that

$$N \models \exists\alpha < j(\kappa)\exists y\exists g(y = \mathcal{P}(\alpha)\ \wedge\ g\colon y \to j(\kappa) \text{ is surjective})$$

since we can take $\alpha = \kappa$ and define an appropriate g from $U \in N$ using Replacement. But then,

$$M \models \exists\alpha < \kappa\exists y\exists g(y = \mathcal{P}(\alpha)\ \wedge\ g\colon y \to \kappa \text{ is surjective})\ ,$$

and this is enough to derive a contradiction by the argument of 2.8. (Alternately, let $h\colon \kappa \to y$ with $h \in M$ satisfy $g(h(\xi)) = \xi$ for every $\xi < \kappa$. Then $[h] \in V_{\alpha+1} \cap N = V_{\alpha+1} \cap M$ by (c), and as in its proof, the injectivity of h leads to a contradiction.)

(e) To verify the weak amenability of $\langle N, W\rangle$, suppose that $F \in {}^{j(\kappa)}N \cap N$, say $F = [f]$. For each $\xi < \kappa$ it can be assumed that $f(\xi) \in {}^\kappa M$, and so a function $\overline{f}\colon \kappa \times \kappa \to M$ can be defined by: $\overline{f}(\xi, \eta) = f(\xi)(\eta)$. Since $\overline{f} \in M$, $X = \{\langle \xi, \eta\rangle \mid \overline{f}(\xi, \eta) \in U\} \in M$ by the weak amenability of $\langle M, U\rangle$. Now define $g \in {}^\kappa M$ by: $g(\xi) = \{\eta \mid \langle \xi, \eta\rangle \in X\}$. Then $g \in M$, and for any $h \in {}^\kappa M \cap M$,

$$
\begin{aligned}
[h] \in [g] \ & \textit{iff} \ \{\xi < \kappa \mid h(\xi) \in g(\xi)\} \in U \\
& \textit{iff} \ \{\xi < \kappa \mid f(\xi)(h(\xi)) \in U\} \in U \\
& \textit{iff} \ F([h]) \in W \ ,
\end{aligned}
$$

i.e. $[g] = \{\alpha < j(\kappa) \mid F(\alpha) \in W\}$.

It is similarly straightforward to show that $\langle N, \in, W\rangle \models W$ is a normal ultrafilter. ⊣

The import of (d) is that whether or not $U \in M$, it is definitely the case that $U \notin N$. Assertions like (a) become pertinent since it is not necessarily the case that $\mathrm{On} \cap M = \mathrm{On} \cap N$.

Recalling 5.6, elementary embeddings also give rise to ultrafilters in the present context:

19.2 Exercise. *Suppose that M and N are transitive $\in$-models of* ZFC$^-$, $k\colon M \prec N$ *with* $\mathrm{crit}(k) = \kappa$, *and* $\mathcal{P}(\kappa) \cap M = \mathcal{P}(\kappa) \cap N$. *Then*

$$\{X \in \mathcal{P}(\kappa) \cap M \mid \kappa \in k(X)\}$$

is an M-ultrafilter over κ. ⊣

The process of iterating the ultrapower construction is now described. Starting with an M-ultrafilter over κ, if the ultrapower of M by U is well-founded, we can define N, W, and j as above and by 19.1(e) repeat the process with N and W, continuing through successor stages as long as ultrapowers are well-founded. At limit stages we will have engendered a directed system of models and embeddings as reviewed in §0, and so we can take a direct limit. The following observation shows that the process can be continued if the result is a well-founded structure. For convenience, let

$$e\colon \langle X, \in, R\rangle \prec^- \langle Y, E, S\rangle$$

indicate that (i) $e\colon \langle X, \in\rangle \prec \langle Y, E\rangle$, i.e. e is elementary for $\mathcal{L}_\in$ formulas; and (ii) e also preserves the predicate, i.e. e is an embedding of $\langle X, \in, R\rangle$ into $\langle Y, E, S\rangle$. With regard to (i), recall our convention emanating from 5.1(c) that for transitive proper class $\in$-structures, elementary is Σ_1-elementary. For (ii), note that if R is definable in $\langle X, \in\rangle$ using parameters, then $\prec^-$ is equivalent to $\prec$, full elementary embedding.

19.3 Exercise. *Suppose that for each $\alpha < \delta$, W_α is an N_α-ultrafilter over λ_α and that*

$$\langle\langle\langle N_\alpha, \in, W_\alpha\rangle \mid \alpha < \delta\rangle, \langle j_{\alpha\beta} \mid \alpha \leq \beta\rangle\rangle$$

is a directed system of $\prec^-$ embeddings which has a well-founded limit. Then it has a transitive collapse $\langle N, \in, W\rangle$, and if for $\alpha < \delta$, $j_{\alpha\delta}\colon \langle N_\alpha, \in, W_\alpha\rangle \prec^- \langle N, \in, W\rangle$ is the direct limit embedding modulated by the transitive collapse, then W is an N-ultrafilter over $\lambda = j_{\alpha\delta}(\lambda_\alpha)$ for some (and so any) $\alpha < \delta$.

Hint. Toward showing that the Collapsing Lemma 0.4 is applicable to the direct limit, note that it is set-like: Let $\langle D, E, A\rangle$ denote the direct limit and $i_\alpha\colon \langle N_\alpha, \in, W_\alpha\rangle \to \langle D, E, A\rangle$ the corresponding embeddings for $\alpha < \delta$. Suppose that $a \in D$, so that for some $\alpha < \delta$ there is a $b \in N_\alpha$ such that $i_\alpha(b) = a$. Then for any $x \in D$,

$$x \ E \ a \ \textit{ iff } \ \exists\beta\exists y(\alpha \leq \beta < \delta \ \wedge \ i_\beta(y) = x \ \wedge \ y \in j_{\alpha\beta}(b)) ,$$

and so $\{x \in D \mid x \ E \ a\}$ is a set by Replacement.

For the latter assertion, the verification of weak amenability is typical: Suppose that $F \in {}^\lambda N \cap N$, so that $F = j_{\alpha\delta}(\overline{F})$ for some $\alpha < \delta$ and $\overline{F} \in {}^{\lambda_\alpha} N_\alpha \cap N_\alpha$. By the weak amenability of $\langle N_\alpha, W_\alpha\rangle$,

$$X = \{\xi < \lambda_\alpha \mid \overline{F}(\xi) \in W_\alpha\} \in N_\alpha ,$$

and hence

$$j_{\alpha\delta}(X) = \{\xi < \lambda \mid F(\xi) \in W\} \in N . \qquad \dashv$$

For any M-ultrafilter U over κ, we can thus proceed recursively to define $\mathcal{L}_\in(\dot{A})$ structures $\langle M_\alpha, \in, U_\alpha\rangle$ for $\alpha < \tau$ where U_α is an M_α-ultrafilter over κ_α, and embeddings $i_{\alpha\beta}\colon \langle M_\alpha, \in, U_\alpha\rangle \prec^- \langle M_\beta, \in, U_\beta\rangle$ for $\alpha \leq \beta < \tau$ as follows: Set $M_0 = M$, $U_0 = U$, $\kappa_0 = \kappa$, and i_{00} the identity on M. Having defined $M_\alpha, U_\alpha, \kappa_\alpha$, and $i_{\alpha\beta}$ for $\alpha < \beta < \delta$, there are two cases:

(i) δ is a successor ordinal, say $\delta = \gamma + 1$. If the ultrapower of M_γ by U_γ is well-founded, let M_δ be its transitive collapse and $U_\delta \subseteq M_\delta$ such that

$$j\colon \langle M_\gamma, \in, U_\gamma\rangle \prec^- \langle M_\delta, \in, U_\delta\rangle$$

is the corresponding embedding. Set $\kappa_\delta = j(\kappa_\gamma)$, $i_{\gamma\delta} = j$, $i_{\alpha\delta} = j \circ i_{\alpha\gamma}$ for $\alpha < \gamma$, and $i_{\delta\delta}$ the identity on M_δ. If on the other hand the ultrapower is ill-founded, set $\delta = \tau$.

(ii) δ is a limit ordinal. If the direct limit of

$$\langle\langle\langle M_\alpha, \in, U_\alpha\rangle \mid \alpha < \delta\rangle, \langle i_{\alpha\beta} \mid \alpha \leq \beta\rangle\rangle$$

is well-founded, let M_δ be its transitive collapse and $U_\delta \subseteq M_\delta$ such that for each $\alpha < \delta$ there is a direct limit embedding:

$$\langle M_\alpha, \in, U_\alpha\rangle \prec^- \langle M_\delta, \in, U_\delta\rangle$$

modulated by the transitive collapse. Call this embedding $i_{\alpha\delta}$ for $\alpha < \delta$, set $\kappa_\delta = i_{\alpha\delta}(\kappa_\alpha)$ for some (and hence, any) $\alpha < \delta$ and $i_{\delta\delta}$ the identity on M_δ. If on the other hand the direct limit is ill-founded, set $\delta = \tau$.

If this definition proceeds through all the ordinals, set $\tau = \mathrm{On}$. Otherwise, $\tau \in \mathrm{On}$ denotes the least stage at which the iteration process encounters ill-foundedness. We could have trudged on through ill-founded structures with only notational complications, but such toil would be gratuitous for present purposes.

To summarize the notation,

$$\langle M_\alpha, U_\alpha, \kappa_\alpha, i_{\alpha\beta}\rangle_{\alpha\le\beta\in\tau} \text{ is the } \textit{iteration} \text{ of } \langle M, \in, U\rangle\ ;$$

τ is the *length* of the iteration; and for $\alpha \in \tau$, $\langle M_\alpha, \in, U_\alpha\rangle$ is an *iterate*, specifically the *αth iterate*, of $\langle M, \in, U\rangle$. Also,

$$\langle M, \in, U\rangle \text{ is } \textit{iterable} \text{ (and } U \text{ is an } \textit{iterable } M\text{-ultrafilter)} \ \textit{iff}\ \ \tau = \text{On}\ .$$

Concerning formalizability, note that even if M were a proper class, the iteration of $\langle M, \in, U\rangle$ can be regarded as a single class definable from M and U. In particular, if M is an inner model of ZFC and $U \in M$, then the iteration is definable entirely in M.

Iterability is a substantial hypothesis, one that will soon be considered in detail. But for now, we fix $\langle M, \in, U\rangle$ and its iteration as given above, and proceed through several lemmata to establish some basic properties.

19.4 Lemma. *Suppose that $\alpha < \beta \in \tau$.*

(a) $\text{crit}(i_{\alpha\beta}) = \kappa_\alpha$ and $i_{\alpha\beta}(\kappa_\alpha) = \kappa_\beta$.

(b) $i_{\alpha\beta}(x) = x$ for every $x \in V_{\kappa_\alpha} \cap M_\alpha$, $V_{\kappa_\alpha} \cap M_\alpha = V_{\kappa_\alpha} \cap M_\beta$, and $\mathcal{P}(\kappa_\alpha) \cap M_\alpha = \mathcal{P}(\kappa_\alpha) \cap M_\beta$.

(c) If β is a limit ordinal, then $\kappa_\beta = \sup(\{\kappa_\gamma \mid \gamma < \beta\})$.

(d) If M is a set, $|M_\alpha| = |M| \cdot |\alpha|$.

Proof. (a) follows directly from the construction, and (b) is a consequence of 19.1(c). For (c), if $\xi < \kappa_\beta$, then by definition of direct limit $\xi = i_{\gamma\beta}(\overline{\xi})$ for some $\gamma < \beta$ and $\overline{\xi} < \kappa_\gamma$. Hence, $\xi = \overline{\xi} < \kappa_\gamma$.

For (d), $|M_\alpha| \le |M| \cdot |\alpha|$ follows by induction on α, using 19.1(b) and bounding direct limits. Conversely, M is injectible into M_α and $\{\kappa_\gamma \mid \gamma < \alpha\} \subseteq M_\alpha$, and so $|M| \cdot |\alpha| \le |M_\alpha|$. ⊣

For any $\alpha + 1 \in \tau$ and $X \in \mathcal{P}(\kappa_\alpha) \cap M_\alpha$,

$$X \in U_\alpha \ \ \textit{iff}\ \ \{\xi \mid \xi \in X\} \in U_\alpha \ \ \textit{iff}\ \ \kappa_\alpha \in i_{\alpha,\alpha+1}(X)$$

by normality. This leads to a further characterization at limit ordinals crucial to later definability results – another testimonial to the efficacy of the concept of normality.

19.5 Lemma (Kunen [68, 70]). *Suppose that β is a limit ordinal less than τ. Then for any $X \in \mathcal{P}(\kappa_\beta) \cap M_\beta$,*

$$X \in U_\beta \ \ \textit{iff}\ \ \exists\alpha < \beta(\{\kappa_\gamma \mid \alpha \le \gamma < \beta\} \subseteq X)\ .$$

Proof. Suppose that $\gamma < \beta$ and $X = i_{\gamma\beta}(\overline{X})$ for some $\overline{X} \in \mathcal{P}(\kappa_\gamma) \cap M_\gamma$. Then

$$\begin{aligned} X \in U_\beta \ \ &\textit{iff}\ \ \overline{X} \in U_\gamma \\ &\textit{iff}\ \ \kappa_\gamma \in i_{\gamma,\gamma+1}(\overline{X}) \\ &\textit{iff}\ \ \kappa_\gamma \in i_{\gamma\beta}(\overline{X}) = X\ , \end{aligned}$$

since $i_{\gamma+1,\beta}(\kappa_\gamma) = \kappa_\gamma$. Thus, for any $\gamma < \beta$, if $X \in \text{ran}(i_{\gamma\beta})$, then $X \in U_\beta$ *iff* $\kappa_\gamma \in X$. This suffices, by the definition of direct limit. $\dashv$

Thus, U_β is generated by the closed unbounded set $\{\kappa_\gamma \mid \gamma < \beta\} \subseteq \kappa_\beta$, and so in particular, this set cannot belong to M_β. The following result provides a representation of iterated ultrapowers that generalizes 5.13(a) and plays a basic role in the structure theory.

19.6 Lemma. *For any $\alpha \in \tau$ and $x \in M_\alpha$, there are $n \in \omega$, $f \in {}^{[\kappa]^n}M \cap M$, and $\gamma_1 < \ldots < \gamma_n < \alpha$ such that $x = i_{0\alpha}(f)(\kappa_{\gamma_1}, \ldots, \kappa_{\gamma_n})$.*

Proof. Proceed by induction on α. Suppose that the proposition holds for α, $\alpha+1 \in \tau$, and $x \in M_{\alpha+1}$. By 5.13(a), $x = i_{\alpha,\alpha+1}(g)(\kappa_\alpha)$ for some $g \in {}^{\kappa_\alpha}M_\alpha \cap M_\alpha$, so we are done if $\alpha = 0$. Otherwise, by induction $g = i_{0\alpha}(h)(\kappa_{\gamma_1}, \ldots, \kappa_{\gamma_n})$ for some $n \in \omega$, $h \in {}^{[\kappa]^n}M \cap M$, and $\gamma_1 < \ldots < \gamma_n < \alpha$, where it can be assumed that $\text{ran}(h)$ consists of functions. Define $f \in {}^{[\kappa]^{n+1}}M \cap M$ by: $f(\xi_1, \ldots, \xi_n, \xi_{n+1}) = h(\xi_1, \ldots, \xi_n)(\xi_{n+1})$. Then:

$$\begin{aligned} i_{0,\alpha+1}(f)(\kappa_{\gamma_1}, \ldots, \kappa_{\gamma_n}, \kappa_\alpha) &= i_{0,\alpha+1}(h)(\kappa_{\gamma_1}, \ldots, \kappa_{\gamma_n})(\kappa_\alpha) \\ &= i_{\alpha,\alpha+1}(i_{0\alpha}(h)(\kappa_{\gamma_1}, \ldots, \kappa_{\gamma_n}))(\kappa_\alpha) \\ &= i_{\alpha,\alpha+1}(g)(\kappa_\alpha) \\ &= x \ . \end{aligned}$$

If $\delta < \tau$ is a limit ordinal and $x \in M_\delta$, then $x = i_{\alpha\delta}(\overline{x})$ for some $\alpha < \delta$ and $\overline{x} \in M_\alpha$. By induction $\overline{x} = i_{0\alpha}(f)(\kappa_{\gamma_1}, \ldots, \kappa_{\gamma_n})$ for some $n \in \omega$, $f \in {}^{[\kappa]^n}M \cap M$, and $\gamma_1 < \ldots \gamma_n < \alpha$. Hence, $x = i_{0\delta}(f)(\kappa_{\gamma_1}, \ldots, \kappa_{\gamma_n})$ since $i_{\alpha\delta}(\kappa_{\gamma_i}) = \kappa_{\gamma_i}$ for $1 \le i \le n$. $\dashv$

M_α is thus generated in a concrete fashion from $\text{ran}(i_{0\alpha})$ and $\{\kappa_\gamma \mid \gamma < \alpha\}$. This leads the following corollary on the action of the $i_{0\alpha}$'s on ordinals; the cardinals are in the sense of V as usual.

19.7 Corollary.

(a) If $\xi \in \text{On} \cap M$ and $\alpha \in \tau$, then

$$i_{0\alpha}(\xi) < (|{}^\kappa\xi \cap M| \cdot |\alpha|)^+ \ .$$

(b) If ν is a cardinal such that $|{}^\kappa\kappa \cap M| < \nu \in \tau$, then

$$\kappa_\nu = i_{0\nu}(\kappa_0) = \nu \ .$$

(c) If θ is a cardinal, $M \models \ulcorner\text{ZFC} \wedge \theta$ is a strong limit of cofinality $> \kappa\urcorner$, *and $\alpha < \min(\theta, \tau)$, then*

$$i_{0\alpha}(\theta) = \theta \ .$$

Proof. (a) This follows from cardinality considerations, since 19.6 implies that $\eta < i_{0\alpha}(\xi)$ *iff* $\eta = i_{0\alpha}(f)(\kappa_{\gamma_1}, \ldots, \kappa_{\gamma_n})$ for some $n \in \omega$, $f \in {}^{[\kappa]^n}\xi \cap M$, and $\gamma_1 < \ldots < \gamma_n < \alpha$.

(b) We have

$$\nu \leq \kappa_\nu = \sup(\{\kappa_\alpha \mid \alpha < \nu\}) \leq \sup(\{(|{}^\kappa\kappa \cap M| \cdot |\alpha|)^+ \mid \alpha < \nu\}) \leq \nu .$$

The equality follows from 19.4(c), and the middle inequality from (a) above.

(c) It suffices to show assuming $\eta < i_{0\alpha}(\theta)$ that $\eta < \theta$. By 19.6, $\eta = i_{0\alpha}(f)(\kappa_{\gamma_1}, \ldots, \kappa_{\gamma_n})$ for some $n \in \omega$, $f \in {}^{[\kappa]^n}\theta \cap M$, and $\gamma_1 < \ldots < \gamma_n < \alpha$. Since $\mathrm{cf}(\theta) > \kappa$ in M, there is a $\xi < \theta$ such that $f \in {}^{[\kappa]^n}\xi$. Hence,

$$\eta < i_{0\alpha}(\xi) < (|{}^\kappa\xi \cap M| \cdot |\alpha|)^+ \leq \theta .$$

The second inequality is from (a), and the third follows from the hypotheses on θ. ⊣

19.6 veers toward Kunen's original formulation of iterated ultrapowers; the beginnings of that formulation are broached as part of this development: For $n \in \omega$, set

$$U^n = \{X \in \mathcal{P}([\kappa]^n) \cap M \mid \exists H \in U([H]^n \subseteq X)\} .$$

Thus, U^n is a (definable) class in the sense of $\langle M, \in, U\rangle$. The normality of U together with weak amenability shows that this provides the appropriate notion of n-fold product:

19.8 Exercise.

(a) For $n \in \omega$ and $X \in \mathcal{P}([\kappa]^{n+1}) \cap M$,

$$X \in U^{n+1} \text{ iff } \{s \in [\kappa]^n \mid \{\xi < \kappa \mid s \cup \{\xi\} \in X\} \in U\} \in U^n .$$

(b) For $0 < n \in \omega$,

$$\langle M, \in, U\rangle \models U^n \text{ is a } \kappa\text{-complete ultrafilter over } [\kappa]^n .$$

Hint. Verify (a) along the way to (b); Rowbottom's 7.17 holds for U and those $f\colon [\kappa]^n \to 2$ such that $f \in M$ since U is an M-ultrafilter. ⊣

Temporarily liberalizing conventions, for $0 < n \in \omega$ we can proceed to form the ultrapower

$${}^{[\kappa]^n}M/U^n = \{(f)^0_{U^n} \mid f \in {}^{[\kappa]^n}M \cap M\}$$

and a corresponding embedding

$$k_n\colon \langle M, \in, U\rangle \prec^- \langle {}^{[\kappa]^n}M/U^n, E_{U^n}, \dot{A}_{U^n}\rangle$$

given by

$$k_n(x) = (f_x)^0_{U^n} \quad \text{for } x \in M$$

where f_x is the constant function: $[\kappa]^n \to \{x\}$.

The next result, together with the representation 19.6, provides a reduction of the satisfaction relation for an iterate to that of $\langle M, \in, U\rangle$, and shows in particular that the ultrapowers just defined are isomorphic to the finite iterates of $\langle M, \in, U\rangle$. Note that $U^0 = \{0\}$ by definition. For proper classes M, (a) below and the coming corollary should be taken to be schemas of theorems, one for each class and formula.

19.9 Lemma.
(a) For any formula $\varphi(v_0, \ldots, v_n)$ of $\mathcal{L}_\in$, $x \in M$, and $\gamma_1 < \ldots < \gamma_n < \alpha \in \tau$,

$$\langle M_\alpha, \in\rangle \models \varphi[i_{0\alpha}(x), \kappa_{\gamma_1}, \ldots, \kappa_{\gamma_n}]$$
$$\text{iff } \langle M, \in, U\rangle \models \{\{\xi_1, \ldots, \xi_n\} \in [\kappa]^n \mid \varphi[x, \xi_1, \ldots, \xi_n]\} \in U^n .$$

(b) For $n \in \omega$ with $0 < n \in \tau$, π_n: $\langle {}^{[\kappa]^n}M/U^n, E_{U^n}, \dot{A}_{U^n}\rangle \to \langle M_n, \in, U_n\rangle$ given by

$$\pi_n((f)^0_{U^n}) = i_{0n}(f)(\kappa_0, \ldots, \kappa_{n-1})$$

is a well-defined isomorphism. Hence, it is the collapsing isomorphism onto the transitive collapse. Moreover, $i_{0n} = \pi_n \circ k_n$.

Proof. (a) Proceed by induction on n; the case $n = 0$ is immediate. Generally,

$$\begin{aligned}
&\langle M_\alpha, \in\rangle \models \varphi[i_{0\alpha}(x), \kappa_{\gamma_1}, \ldots, \kappa_{\gamma_n}]\\
&\quad \text{iff } \langle M_{\gamma_n+1}, \in\rangle \models \varphi[i_{0,\gamma_n+1}(x), \kappa_{\gamma_1}, \ldots, \kappa_{\gamma_n}]\\
&\quad \text{iff } \langle M_{\gamma_n}, \in, U_{\gamma_n}\rangle \models \{\xi < \textstyle\bigcup U_{\gamma_n} \mid \varphi[i_{0\gamma_n}(x), \kappa_{\gamma_1}, \ldots, \kappa_{\gamma_{n-1}}, \xi]\} \in U_{\gamma_n}\\
&\quad \text{iff } \langle M, \in, U\rangle \models \{\{\xi_1, \ldots, \xi_{n-1}\} \in [\kappa]^{n-1} \mid\\
&\qquad\qquad \{\xi < \textstyle\bigcup U \mid \varphi[x, \xi_1, \ldots, \xi_{n-1}, \xi]\} \in U\} \in U^{n-1}\\
&\quad \text{iff } \langle M, \in, U\rangle \models \{\{\xi_1, \ldots, \xi_n\} \in [\kappa]^n \mid \varphi[x, \xi_1, \ldots, \xi_n]\} \in U^n .
\end{aligned}$$

(Of course, $\bigcup U_{\gamma_n}$ is κ_{γ_n} and $\bigcup U$ is κ.)
(b) This is an application of (a) and 19.6 with $\alpha = n$. ⊣

19.8(a) suggests the general notion of product ultrafilter. In general model-theoretic terms, that finite iterations of ultraproducts can be rendered as a single ultraproduct by a product ultrafilter had been observed by Frayne-Morel-Scott [62]. For technical reasons obviated by the representation 19.6 for normal ultrafilters, Kunen carried through a similar analysis into the transfinite, representing all iterates as single ultrapowers of sorts. Independently, Keisler also developed such a representation (Chang-Keisler [90: 6.5]).

The following immediate corollary of 19.9(a) shows how iterated ultrapowers lead to indiscernibles. This insight of Kunen's led to a new structural understanding of $0^\#$ (21.1).

19.10 Corollary (Kunen [68, 70]). *For any $\varphi(v_0, \ldots, v_n)$ of $\mathcal{L}_\in$, $x \in M$, $\gamma_1 < \ldots < \gamma_n < \alpha \in \tau$, and $\delta_1 < \ldots < \delta_n < \beta \in \tau$,*

$$\langle M_\alpha, \in\rangle \models \varphi[i_{0\alpha}(x), \kappa_{\gamma_1}, \ldots, \kappa_{\gamma_n}] \textit{ iff } \langle M_\beta, \in\rangle \models \varphi[i_{0\beta}(x), \kappa_{\delta_1}, \ldots, \kappa_{\delta_n}]\ .$$

In particular, $\{\kappa_\gamma \mid \gamma < \alpha\}$ *is a set of indiscernibles for* $\langle M_\alpha, \in, i_{0\alpha}(x)\rangle_{x\in M}$. ⊣

Iterability

Continuing in terms of $\langle M, \in, U\rangle$ and its iteration $\langle M_\alpha, U_\alpha, \kappa_\alpha, i_{\alpha\beta}\rangle_{\alpha\leq\beta\in\tau}$, several characterizations of iterability are established next. For convenience, On *is treated as an ordinal when discussing lengths of iterations.* For example, $\tau \geq \omega_1$ is understood to include the possibility $\tau = \text{On}$. Indeed, the historically first result about iterability, due to Gaifman, was the assertion that $\tau = \text{On}$ *iff* $\tau \geq \omega_1$. This is established as part of a larger framework, one that incorporates a sufficient condition due to Kunen:

U is *countably complete iff* for any $\{X_n \mid n \in \omega\} \subseteq U$, $\bigcap_n X_n \neq \emptyset$.

This suffices for the forward direction of 5.3; $\{X_n \mid n \in \omega\}$ may not be a member of M, and for that reason requiring $\bigcap_n X_n \in U$ as in ω_1-completeness would be too stringent. Kunen showed that this external completeness implies iterability. For later characterizations, the implication is factored through a property attributable to Jensen and implicit in Dodd [82: 65]:

$\langle M, \in, U\rangle$ is *countably iterable iff* for any N-ultrafilter W with $\langle N, \in, W\rangle$ countable and $\prec^-$ embeddable into $\langle M, \in, U\rangle$, the length of the iteration of $\langle N, \in, W\rangle$ is $\geq \omega_1$.

19.11 Lemma (Jensen). *If* U *is countably complete, then* $\langle M, \in, U\rangle$ *is countably iterable.*

Proof. Suppose that W is an N-ultrafilter, $\langle N, \in, W\rangle$ is countable, and

$$e\colon \langle N, \in, W\rangle \prec^- \langle M, \in, U\rangle\ .$$

Letting $\langle N_\alpha, W_\alpha, \lambda_\alpha, j_{\alpha\beta}\rangle_{\alpha\leq\beta\in\sigma}$ be the iteration of $\langle N, \in, W\rangle$ it must be verified that $\sigma \geq \omega_1$. To do this, proceed by induction on $\alpha < \omega_1$ to establish that:

(i) $\langle N_\alpha, \in, W_\alpha\rangle$ is defined, and for inductive purposes ,
(ii) there is an $e_\alpha\colon \langle N_\alpha, \in, W_\alpha\rangle \prec^- \langle M, \in, U\rangle$ such that $e_\gamma = e_\alpha \circ j_{\gamma\alpha}$ for $\gamma < \alpha$.

Initially set $e_0 = e$. For the induction argument at $\alpha + 1$, suppose that (i) and (ii) have been satisfied through stages $\leq \alpha$. In particular, for any $X \in W_\alpha$, $e_\alpha(X) \in U$. Also, since N is countable and $\alpha < \omega_1$, N_α is countable by 19.4(d). Hence, there is an $\eta \in \bigcap\{e_\alpha(X) \mid X \in W_\alpha\}$ by the countable completeness of U. Now η defines an embedding of the ultrapower $\langle {}^{\lambda_\alpha}N_\alpha/W_\alpha, E_{W_\alpha}, \dot{A}_{W_\alpha}\rangle$ of $\langle N_\alpha, \in, W_\alpha\rangle$ into $\langle M, \in, U\rangle$ by: $j((f)^0_{W_\alpha}) = e_\alpha(f)(\eta)$. Moreover, j is $\mathcal{L}_\in$ elementary: Using a one-variable formula $\varphi(v_0)$ for simplicity,

$$\begin{aligned}\langle {}^{\lambda_\alpha}N_\alpha/W_\alpha, E_{W_\alpha}\rangle \models \varphi[(f)^0_{W_\alpha}] &\\ \textit{iff}\ \ &\{\xi < \lambda_\alpha \mid \langle N_\alpha, \in\rangle \models \varphi[f(\xi)]\} \in W_\alpha\\ \textit{iff}\ \ &\eta \in \{\xi < \kappa \mid \langle M, \in\rangle \models \varphi[e_\alpha(f)(\xi)]\}\\ \textit{iff}\ \ &\langle M, \in\rangle \models \varphi[e_\alpha(f)(\eta)]\ .\end{aligned}$$

But, $\langle {}^{\lambda_\alpha}N_\alpha/W_\alpha, E_{W_\alpha}, \dot{A}_{W_\alpha}\rangle$ is well-founded, being embeddable into a transitive structure, and so $\langle N_{\alpha+1}, \in, W_{\alpha+1}\rangle$ is duly defined as its transitive collapse. Consequently, j modulated by the collapsing map yields an embedding

$$e_{\alpha+1}\colon \langle N_{\alpha+1}, \in, W_{\alpha+1}\rangle \prec^- \langle M, \in, U\rangle\ .$$

Furthermore, it is simple to check that $e_\alpha = e_{\alpha+1} \circ j_{\alpha,\alpha+1}$, and hence that for any $\gamma < \alpha,\ e_\gamma = e_\alpha \circ j_{\gamma\alpha} = e_{\alpha+1} \circ j_{\gamma,\alpha+1}$.

Suppose now that $\delta < \omega_1$ is a limit ordinal and (i) and (ii) have been satisfied through stages $< \delta$. In particular, the direct limit of

$$\langle\langle\langle N_\alpha, \in, W_\alpha\rangle \mid \alpha < \delta\rangle, \langle j_{\alpha\beta} \mid \alpha \leq \beta\rangle\rangle$$

is $\prec^-$ embeddable into $\langle M, \in, U\rangle$ by 0.7 because of the e_α's. Hence, this direct limit is well-founded, and so $\langle N_\delta, \in, W_\delta\rangle$ is duly defined as its transitive collapse. Moreover, that embedding of the direct limit into $\langle M, \in, U\rangle$ modulated by the collapsing map yields an embedding

$$e_\delta\colon \langle N_\delta, \in, W_\delta\rangle \prec^- \langle M, \in, W\rangle$$

such that $e_\gamma = e_\delta \circ j_{\gamma\delta}$ for $\gamma < \delta$. ⊣

19.12 Lemma. *If $\langle M, \in, U\rangle$ is countably iterable, then it is iterable.*

Proof. Assume to the contrary that $\tau \in$ On. If τ is a successor, say $\tau = \gamma + 1$, then temporarily let $\langle M_\tau, E_\tau, U_\tau\rangle$ be the ultrapower $\langle {}^{\kappa_\gamma}M_\gamma/U_\gamma, E_{U_\gamma}, \dot{A}_{U_\gamma}\rangle$ of $\langle M_\gamma, \in, U_\gamma\rangle$, and

$$i_\tau\colon \langle M, \in, U\rangle \prec^- \langle M_\tau, E_\tau, U_\tau\rangle$$

the composition of $i_{0\gamma}$ followed by the natural embedding into that ultrapower. If τ is a limit ordinal, let $\langle M_\tau, E_\tau, U_\tau\rangle$ be the direct limit of the iteration of $\langle M, \in, U\rangle$, and

$$i_\tau\colon \langle M, \in, U\rangle \prec^- \langle M_\tau, E_\tau, U_\tau\rangle$$

the corresponding embedding. In either case $\langle M_\tau, E_\tau, U_\tau\rangle$ is ill-founded by assumption. Nonetheless, a look at the inductive proofs of 19.6 and 19.9(a) shows that they also hold "at τ", i.e. with $\alpha = \tau$, $\langle M_\alpha, \in, U_\alpha\rangle$ replaced by $\langle M_\tau, E_\tau, U_\tau\rangle$, and $i_{0\alpha}$ replaced by i_τ – *if* in the case $\tau = \gamma + 1$, κ_γ is replaced by $[\mathrm{id}]_{U_\gamma}$ where id: $\kappa_\gamma \to \kappa_\gamma$ is the identity. This is how 19.6 and 19.9(a) are applied "at τ" in what follows.

Let $\{x_k \mid k \in \omega\} \subseteq M_\tau$ such that $x_{k+1}\ E_\tau\ x_k$ for each $k \in \omega$. By 19.6 "at τ",

$$x_k = i_\tau(f_k)(\kappa_{\gamma_1^k}, \ldots, \kappa_{\gamma_{n(k)}^k}) \text{ for some } n(k) \in \omega\ ,$$
$$f_k \in {}^{[\kappa]^{n(k)}}M \cap M\ , \text{ and } \gamma_1^k < \ldots < \gamma_{n(k)}^k < \tau\ .$$

Take a Skolem hull of $\{f_k \mid k \in \omega\}$ in $\langle M, \in, U\rangle$, and then its transitive collapse. This produces a countable $\langle N, \in, W\rangle$ where W is an N-ultrafilter, and the inverse of the collapsing map is an embedding

$$e\colon \langle N, \in W\rangle \prec^- \langle M, \in, U\rangle\ .$$

Let $\overline{f}_k$ satisfy $e(\overline{f}_k) = f_k$ for each $k \in \omega$. By hypothesis, the iteration $\langle N_\alpha, W_\alpha, \lambda_\alpha, j_{\alpha\beta}\rangle_{\alpha\le\beta\in\sigma}$ of $\langle N, \in, W\rangle$ satisfies $\sigma \ge \omega_1$.

Let $\zeta < \omega_1$ be the ordertype of the set $Z = \{\gamma_m^k \mid k \in \omega \wedge 1 \le m \le n(k)\}$ and $h\colon Z \to \zeta$ the unique order-preserving function. For each $k \in \omega$, set

$$\delta_m^k = h(\gamma_m^k) \text{ for } 1 \le m \le n(k)\ , \text{ and}$$
$$\overline{x}_k = j_{0\zeta}(\overline{f}_k)(\lambda_{\delta_1^k}, \ldots, \lambda_{\delta_{n(k)}^k})\ .$$

Furthermore, let $\varphi_k(v_0, v_1, \ldots)$ be a formula such that $\varphi_k[i_\tau(f_k), i_\tau(f_{k+1}), \ldots]$ asserts that $x_{k+1}\ E_\tau\ x_k$ in $\langle M_\tau, E_\tau, U_\tau\rangle$, where $\ldots$ denotes the listing of the set $\{\kappa_{\gamma_1^k}, \ldots, \kappa_{\gamma_{n(k)}^k}, \kappa_{\gamma_1^{k+1}}, \ldots, \kappa_{\gamma_{n(k+1)}^{k+1}}\}$ in increasing order. Finally, let $\varphi_k^*(v_0, v_1)$ be the formula to the right of $\langle M, \in, U\rangle \models$ in the application of 19.9(a) "at τ" to φ_k. Then for each $k \in \omega$,

$$\begin{aligned} x_{k+1}\ E_\tau\ x_k\ &\textit{iff}\ \ \langle M, \in, U\rangle \models \varphi_k^*[f_k, f_{k+1}] \\ &\textit{iff}\ \ \langle N, \in, W\rangle \models \varphi_k^*[\overline{f}_k, \overline{f}_{k+1}] \\ &\textit{iff}\ \ \overline{x}_{k+1} \in \overline{x}_k\ , \end{aligned}$$

by using φ_k appropriately in $\langle N_\zeta, \in, W_\zeta\rangle$. This of course is a contradiction, and so the proof is complete. ⊣

19.13 Corollary (Kunen [68, 70]). *If U is countably complete, then $\langle M, \in, U\rangle$ is iterable.* ⊣

The converse is false, since there are iterable $\langle N, \in, W\rangle$ that are countable (19.15(e)) in which case W cannot be countably complete. However, we do have the following:

19.14 Lemma. *If $\langle M, \in, U\rangle$ is iterable, then it has an iterate $\langle M_\alpha, \in, U_\alpha\rangle$ such that U_α is countably complete.*

Proof. This is affirmed with $\alpha = \omega_1$: Suppose that $\{X_n \mid n \in \omega\} \subseteq U_{\omega_1}$. By 19.5, for each $n \in \omega$ there is an $\alpha_n < \omega_1$ such that $\{\kappa_\gamma \mid \alpha_n \le \gamma < \omega_1\} \subseteq X_n$. Let $\beta = \sup(\{\alpha_n \mid n \in \omega\})$. Then $\kappa_\beta \in \bigcap_n X_n$. ⊣

The arguments for these lemmata can now be used to concoct a daisy chain of implications that provides several characterizations of iterability:

19.15 Theorem (Gaifman for (b), Jensen, Dodd [82]). *The following are equivalent:*

(a) $\langle M, \in, U\rangle$ *is iterable, i.e.* $\tau = \mathrm{On}$.

(b) $\tau \geq \omega_1$.

(c) There is an iterate $\langle M_\alpha, \in, U_\alpha\rangle$ *such that* U_α *is countably complete.*

(d) $\langle M, \in, U\rangle$ *is countably iterable.*

(e) Any $\langle N, \in, W\rangle \prec^-$ *embeddable into* $\langle M, \in, U\rangle$ *is iterable.*

Proof. (a) $\to$ (b) is immediate. For (b) $\to$ (c), note first that $\tau \geq \omega_1$ implies that $\tau > \omega_1$, i.e. there is an ω_1-iterate, since the corresponding direct limit is readily seen to be well-founded. Hence, (c) follows by the argument for 19.14. For (c) $\to$ (d), note that $\langle M_\alpha, \in, U_\alpha\rangle$ is countably iterable by 19.11, and so $\langle M, \in, U\rangle$ is countably iterable since it is embeddable into $\langle M_\alpha, \in, U_\alpha\rangle$. For (d) $\to$ (e), note that $\langle N, \in, W\rangle$ is countably iterable, and use 19.12. (e) $\to$ (a) is immediate. ⊣

The following absoluteness result is a consequence:

19.16 Proposition. *Suppose that* S *is a transitive* $\in$*-model of* ZFC, $\omega_1 \subseteq S$, *and* $M, U \subseteq S$ *are classes in the sense of* S *such that* $S \models U$ is an M-ultrafilter. *Then*

$$\langle M, \in, U\rangle \text{ is iterable } \textit{ iff } \; S \models \langle M, \in, U\rangle \text{ is iterable}\,.$$

In particular, if U *is an* M*-ultrafilter,* $M \models$ ZFC, *and* $\omega_1 \cup \{U\} \subseteq M$, *then* $\langle M, \in, U\rangle$ *is iterable.*

Proof. Clearly U really is an M-ultrafilter, and in particular M really models ZFC$^-$ – assertions implicit in the conclusion. The hypotheses further imply by absoluteness of satisfaction that for any $\alpha < \omega_1$, $(\langle M_\alpha, \in, U_\alpha\rangle)^S = \langle M_\alpha, \in, U_\alpha\rangle$ when both are defined. Moreover, for class structures in the sense of S, well-foundedness is absolute (0.3). Hence, if $S \models \langle M, \in, U\rangle$ is iterable, then the (real) iteration of $\langle M, \in, U\rangle$ has length $\geq \omega_1$, and so by 19.15 it is iterable. The converse is clear.

The last assertion follows from taking $S = M$ and noting that $M \models U$ is countably complete. ⊣

A Lemma on Measurable Cardinals

Tucked in next is a striking result of Kunen's, a dividend of his work on iterated ultrapowers. Chang [71] formulated for cardinals κ the class C^κ of sets constructible using the infinitary language $L_{\kappa\kappa}$ (described in §4) and observed various generalizations of properties of $L = C^\omega$. For example, C^κ is the least inner model M closed under the taking of arbitrary $<\kappa$-sequences, i.e. ${}^{<\kappa}M \subseteq M$. Also, Scott's result 5.5 can be extended to show that if there is a measurable cardinal κ, then $V \neq C^{\kappa^+}$. Sureson [86] studied transcendence over these models in terms of indiscernibles, and showed in particular that transcendence over C^{ω_1} is equiconsistent with the existence of a measurable cardinal. Using iterated ultrapowers Kunen [73] showed that unlike L these models do not generally have intrinsic well-orderings: *if κ is regular and there are κ^+ measurable cardinals then the Axiom of Choice fails in C^{κ^+}.* Toward this result he established a striking lemma that revealed an unexpected global constraint on measurable cardinals imposed by ZFC. Here, a proof due to William Fleissner is provided that does not use iterated ultrapowers. As usual, if κ is measurable and U is a κ-complete ultrafilter over κ, then $j_U \colon V \prec M_U \cong \mathrm{Ult}(V, U)$ and $[f]_U \in M_U$ corresponds to $f \colon \kappa \to V$.

19.17 Theorem (Kunen [73]). *For any ξ, the following set is finite:*

$$\{\kappa > \omega \mid \text{there is a } \kappa\text{-complete ultrafilter } U \text{ over } \kappa \text{ such that } \xi < j_U(\xi)\}\ .$$

Proof (Fleissner [75]). First, stipulate for U a κ-complete ultrafilter over $\kappa > \omega$ that a half-open interval of ordinals $[\alpha, \beta) = \{\xi \mid \alpha \leq \xi < \beta\}$ is a *moving interval for U iff* j_U fixes cofinally many ordinals less than α as well as β, yet it moves every ordinal in $[\alpha, \beta)$. Note that if these conditions are satisfied, then

(i) $\mathrm{cf}(\alpha) = \kappa$, and
(ii) $\beta = \sup(\{j_U^n(\xi) \mid n \in \omega\})$ for any $\xi \in [\alpha, \beta)$,

where j_U^n indicates n iterative applications of j_U. (For (i), $\sup(\{j_U(\eta) \mid \eta < \alpha\}) = \alpha$ since cofinally many ordinals less than α are fixed by j_U, so a standard argument shows that $j_U(\alpha) > \alpha$ necessitates $\mathrm{cf}(\alpha) = \kappa$.)

Next, observe that if U is a κ-complete ultrafilter over $\kappa > \omega$ and W is a λ-complete ultrafilter over $\lambda > \kappa$, then j_U and j_W commute: For any set x, $j_U(x) = \{[f]_U \mid f \in {}^{\kappa}x\}$ and since j_W fixes κ and U,

$$j_W(j_U(x)) = \{[f]_U \mid f \in {}^{\kappa}j_W(x)\} = j_U(j_W(x))\ .$$

It follows that with U and W as above if, $[\alpha, \beta)$ is a moving interval for U and $[\overline{\alpha}, \overline{\beta})$ is a moving interval for W, then either they are disjoint or strictly nested, i.e. if $\alpha < \overline{\alpha} < \beta$, then $\overline{\beta} < \beta$. To show this, first choose an η such that $\alpha \leq \eta < \overline{\alpha}$ and $j_W(\eta) = \eta$; this is possible by hypothesis on $\overline{\alpha}$. By (ii) above, there is an $n \in \omega$ such that $\overline{\alpha} < j_U^n(\eta) < \beta$. Since j_U and j_W commute,

$$j_W(j_U^n(\eta)) = j_U^n(j_W(\eta)) = j_U^n(\eta)\ .$$

Hence, $\overline{\beta} \leq j_U^n(\eta)$ since $[\overline{\alpha}, \overline{\beta})$ is a moving interval for W, and so $\overline{\beta} < \beta$.

The proof can now be completed: Assume to the contrary that for some ξ there is an increasing sequence $\langle \kappa_i \mid i \in \omega \rangle$ of measurable cardinals such that for each $i \in \omega$ there is a κ_i-complete ultrafilter U_i over κ_i with $\xi < j_{U_i}(\xi)$. It follows that for each $i \in \omega$, ξ is in a moving interval $[\alpha_i, \beta_i)$ for U_i. By (i) above, $i \neq j$ implies that $\alpha_i \neq \alpha_j$, so that we can assume by taking a subsequence that $i < j$ implies that $\alpha_i < \alpha_j$. But then, as the $[\alpha_i, \beta_i)$'s all contain ξ, they are nested intervals, and so $\beta_0 > \beta_1 > \beta_2 \dots$ is a descending sequence of ordinals. Contradiction! ⊣

Friedrich Wehrung [89] gave another proof of Kunen's result and established related results.

A Connection with Prikry Forcing

This section is concluded by establishing an interesting connection between iterated ultrapowers and Prikry forcing. Suppose that M is an inner model of ZFC, $U \in M$, and $M \models U$ is a normal ultrafilter over κ. Then $\langle M, \in, U\rangle$ is iterable by 19.16, so let $\langle M_\alpha, U_\alpha, \kappa_\alpha, i_{\alpha\beta}\rangle_{\alpha\le\beta\in\mathrm{On}}$ be its iteration. Then for any γ, $i_{0\gamma}(U) = U_\gamma \in M_\gamma$, and by absoluteness of satisfaction $\langle\langle M_\alpha, \in, U_\alpha\rangle \mid \gamma \le \alpha \in \mathrm{On}\rangle$ is definable in $\langle M_\gamma, \in, U_\gamma\rangle$ as its sequence of iterated ultrapowers. In particular, $\alpha \le \beta$ implies that $M_\beta \subseteq M_\alpha$.

For limit ordinals $\delta > 0$, set

$$M_\delta^+ = \bigcap\nolimits_{\alpha<\delta} M_\alpha \ .$$

Then $M_\delta \subseteq M_\delta^+$, and M_δ^+ *is an inner model*: in terms of our formulation before 3.1,

$$\bigcup M_\delta^+ \subseteq M_\delta^+ \quad \text{and} \quad \forall\xi(\mathrm{def}(M_\delta^+ \cap V_\xi) \subseteq M_\delta^+) \ .$$

For the latter, if $x \in \mathrm{def}(M_\delta^+ \cap V_\xi)$ for some ξ, then for any $\alpha < \delta$, by definability and absoluteness of M_δ^+ and rank in M_α we have $x \in M_\alpha$, and hence $x \in \bigcap_{\alpha<\delta} M_\alpha = M_\delta^+$.

Focusing on $\delta = \omega$, note that $\{\kappa_n \mid n \in \omega\} \in M_\omega^+ - M_\omega$: For each $i \in \omega$, $\{\kappa_n \mid i \le n \in \omega\} \in M_i$ because of the definability of the iteration process from M_i on within M_i, and so $\{\kappa_n \mid n \in \omega\} \in M_i$ as it is only a finite modification. However, $\{\kappa_n \mid n \in \omega\} \notin M_\omega$ since $\kappa_\omega = \sup(\{\kappa_n \mid n \in \omega\})$ is measurable in M_ω. Solovay noted that $\{\kappa_n \mid n \in \omega\}$ is in fact Prikry P_{U_ω}-generic over M_ω; Bukovský [73] observed on general grounds that M_ω^+ is a generic extension of M_ω; and then Dehornoy and Bukovský made the exact connection:

19.18 Theorem.
(a) (Solovay) $\{\kappa_n \mid n \in \omega\}$ *is Prikry* P_{U_ω}*-generic over* M_ω.
(b) (Dehornoy [75, 78]; Bukovský) $M_\omega[\{\kappa_n \mid n \in \omega\}] = M_\omega^+$.

Proof. (a) This is a direct consequence of 18.7 and 19.5.

(b) Set $g = \{\kappa_n \mid n \in \omega\}$; it remains to establish that $M_\omega^+ \subseteq M_\omega[g]$. First,

(∗) if $x \in M_\omega$, then for sufficiently large $n \in \omega$, $i_{n\omega}(x) = i_{\omega,\omega2}(x)$.

This is a consequence of the following for sufficiently large $n \in \omega$: There is a $x_n \in M_n$ such that $i_{n\omega}(x_n) = x$, i.e. $\langle M_n, \in, U_n\rangle \models \ulcorner$the image of x_n under the embedding into the ωth iterate is $x\urcorner$. Applying $i_{n\omega}$, $\langle M_\omega, \in, U_\omega\rangle \models \ulcorner$the image of $i_{n\omega}(x_n)$ under the embedding into the ωth iterate is $i_{n\omega}(x)\urcorner$. But as the embedding in question is $i_{\omega,\omega2}$, $i_{n\omega}(x) = i_{\omega,\omega2}(i_{n\omega}(x_n)) = i_{\omega,\omega2}(x)$.

Next, every set of *ordinals* in M_ω^+ is also in $M_\omega[g]$: Suppose that $x \in \mathcal{P}(\gamma) \cap M_\omega^+$. By 19.6 and definability there is a sequence $\langle f_n \mid n \in \omega \rangle \in M$ such that for each $n \in \omega$, $f_n\colon [\kappa]^n \to M$ and $x = i_{0n}(f_n)(\kappa_0, \ldots, \kappa_{n-1})$, so that $i_{n\omega}(x) = i_{0\omega}(f_n)(\kappa_0, \ldots, \kappa_{n-1})$ as $i_{n\omega}(\kappa_i) = \kappa_i$ for $i < n$. Since

$$\langle i_{0\omega}(f_n) \mid n \in \omega \rangle = i_{0\omega}(\langle f_n \mid n \in \omega \rangle) \in M_\omega ,$$

it follows by definability from g that $\langle i_{n\omega}(x) \mid n \in \omega \rangle \in M_\omega[g]$. Now for any $\xi < \gamma$, $\xi \in x$ *iff* for sufficiently large $n \in \omega$, $i_{n\omega}(\xi) \in i_{n\omega}(x)$ *iff* for sufficiently large $n \in \omega$, $i_{\omega,\omega 2}(\xi) \in i_{n\omega}(x)$ by $(*)$ above. Hence, by the definability of $i_{\omega,\omega 2}$ in M_ω it follows that $x \in M_\omega[g]$.

Finally, $M_\omega^+ \subseteq M_\omega[g]$ can be established by induction on rank: Suppose that $x \in M_\omega^+$ and inductively, $x \subseteq M_\omega[g]$. Then $x \subseteq (V_{\text{rank}(x)})^{M_\omega[g]}$, and since $M_\omega[g]$ satisfies AC (being a generic extension by (a) of a model of AC), let $f \in M_\omega[g]$ be an injection: $(V_{\text{rank}(x)})^{M_\omega[g]} \to \text{On}$. Then $f \in M_\omega^+ \supseteq M_\omega[g]$, and $f``x$ is a set of ordinals in $M_\omega^+$. Hence, $f``x \in M_\omega[g]$ by the previous paragraph, and so $x = f^{-1}(f``x) \in M_\omega[g]$. ⊣

This last is part of the argument for showing that if two transitive $\in$-models N_0 and N_1 of ZF have the same sets of ordinals and at least one satisfies AC, then $N_0 = N_1$ (see Jech [03: 196ff]).

Dehornoy [76, 78] completely described the inner models M_δ^+ in terms of the M_α's by cases as follows: (i) $\delta = \alpha + \omega$ for some $\alpha < \delta$. Then as for 19.18, $\{\kappa_{\alpha+n} \mid n \in \omega\}$ is Prikry P_{U_δ}-generic over M_δ and $M_\delta^+ = M_\delta[\{\kappa_{\alpha+n} \mid n \in \omega\}]$. (ii) For some $\alpha < \delta$, $\langle M_\alpha, \in, U_\alpha \rangle \models \text{cf}(\delta) > \omega$. Then $M_\delta^+ = M_\delta$. For instance, $\langle M_\omega, \in, U_\omega \rangle \models \kappa_\omega$ is regular, so $M_{\kappa_\omega}^+ = M_{\kappa_\omega}$. (iii) Neither (i) nor (ii) obtain for δ. Then $M_\delta^+ = M_\delta[G_\delta]$, where G_δ is a specified set of Prikry generic sequences not well-orderable in M_δ^+. In particular, AC fails in M_δ^+. Dehornoy [83] extended the scheme to inner models of stronger hypotheses and forcing for changing to an uncountable cofinality.

The basic theory of this section is extended in the next to derive the main structure theorems for inner models of measurability.

20. Inner Models of Measurability

One of the emerging themes of the theory of large cardinals in the late 1960's was the investigation of the smallest inner models in which they maintain their defining property. Like L for the theory ZFC these canonical inner models were found to crystallize the necessary consequences of the corresponding theories as well as to establish the relative consistency of minimal principles like GCH. Silver and Kunen initiated this direction of research with their structure theorems on inner models of measurability, which are established in this section. As attention shifted to strong hypotheses, the development of inner model theory was to become a full-fledged program. Guiding structural initiatives were thus introduced beyond the combinatorial theory of large cardinals, and this was to lead to remarkable equiconsistency results by the complementary methods of forcing and inner models.

Inner model theory began, of course, with Gödel's fundamental work on L. For modest large cardinals like inaccessible, Mahlo, and weakly compact cardinals, the corresponding inner model is L itself (3.1, 6.7). Jensen's systematic investigations beginning in the late 1960's revealed a fine structure for L which is thus compatible with these cardinals, and led moreover to informative characterizations in this sharpened context. He also isolated combinatorial principles obtaining in L whose *negations* turned out to entail these cardinals in L (see volume II). Thus, the careful investigation of L resulted in a new structural understanding of modest large cardinals.

Before these developments, the relativity introduced by Scott's result that measurable cardinals contradict $V = L$ had set the stage for the development of inner models for measurability, first considered by Solovay. Let U be a κ-complete ultrafilter over $\kappa > \omega$. Since the measurability of κ amounts to the existential postulation of such a set U, Solovay looked at $L[U]$, Levy's inner model of sets constructible relative to U (§3). $\overline{U} = U \cap L[U] \in L[U]$ and so $L[\overline{U}] = L[U]$, and the following is simple to check:

20.1 Exercise.

$$L[U] \models \overline{U} \text{ is a } \kappa\text{-complete ultrafilter over } \kappa ,$$

and if U is normal, then

$$L[U] \models \overline{U} \text{ is normal} . \qquad \dashv$$

Thus, κ is measurable in $L[U]$, and like L with respect to ZF, it is consistent with κ being measurable that $V = L[U]$, so that U could have been $\overline{U}$ all along. Focusing on these *inner models of measurability,*

$\langle L[U], \in, U\rangle$ is a *κ-model iff* $\langle L[U], \in, U\rangle \models U$ is a normal ultrafilter over κ.

Thus, $U \in L[U]$ is incorporated from the beginning as a unary relation. That $U \in L[U]$ implies that $L[U]^{L[U]} = L[U]$ by 3.2(b), and hence that $L[U] \models \ulcorner V = L[U]\urcorner$. *It is not assumed that κ is a cardinal in V.* (This section continues the extensive use of $\kappa, \lambda, \ldots$ to denote cardinals only in the sense of some transitive $\in$-model.) Initial observations were promising:

20.2 Lemma (Solovay). *Suppose that $\langle L[U], \in, U\rangle$ is a κ-model. Then the following hold in $L[U]$:*

(a) $\forall\lambda \geq \kappa(2^\lambda = \lambda^+)$.

(b) κ is the only measurable cardinal.

Proof. (a) Gödel's argument for GCH in L works in any $L[A]$ for sufficiently large cardinals; in the present case, that $U \subseteq \mathcal{P}(\kappa)$ allows it to work for any $\lambda \geq \kappa$: For such λ we shall verify that $\mathcal{P}(\lambda) \cap L[U] \subseteq L_{\lambda^+}[U]$, drawing the conclusion from $|L_{\lambda^+}[U]| = \lambda^+$. Since $L[U] \models V = L[U]$, the argument carried out inside $L[U]$ then gives the desired result.

Suppose then that $x \in \mathcal{P}(\lambda) \cap L[U]$. Let $\gamma > \lambda$ be a limit ordinal such that $x, U \in L_\gamma[U]$, and $\langle H, \in, U \cap H\rangle \prec \langle L_\gamma[U], \in, U\rangle$ such that $\lambda \cup \{x, U\} \subseteq H$ and $|H| = \lambda$. Let $\langle N, \in, W\rangle$ be the transitive collapse of $\langle H, \in, U \cap H\rangle$ and π the collapsing isomorphism. Since $\lambda \subseteq H$, $\pi(y) = y$ for every $y \in \mathcal{P}(\lambda) \cap H$. Consequently, $\pi(x) = x$ and $W = \pi``(U \cap H) = U \cap N$. By elementarity $\langle N, \in, U \cap N\rangle$ satisfies the sentence σ_1 of 3.3(b), so that $N = L_\delta[U]$ for some δ. Finally, $|N| = \lambda$ so that $\delta < \lambda^+$, and the result follows since $x \in L_\delta[U] \subseteq L_{\lambda^+}[U]$.

(b) Assuming throughout that $V = L[U]$, suppose to the contrary that there is a measurable cardinal $\lambda \neq \kappa$. Let W be a λ-complete ultrafilter over λ and

$$j\colon V \prec M \cong \mathrm{Ult}(V, W)\ .$$

Since V is the class $L[U]$ definable from U, by elementarity and 3.2(b), $M = L[j(U)]^M = L[j(U)]$. It will in fact be established that $M = L[U]$, and this suffices for a contradiction since by 5.7(e), $W \notin M$.

If $\lambda > \kappa$, then $j(U) = U$, and so we are done. Suppose then that $\lambda < \kappa$. As U is normal,

$$E = \{\alpha < \kappa \mid \alpha > \lambda\ \wedge\ \alpha \text{ is inaccessible}\} \in U\ ,$$

and by 5.7(c), $j(\alpha) = \alpha$ for every $\alpha \in E$. It follows from this that $j(U) = U \cap M$: First, suppose that $X \in j(U)$, say $X = [f]_W$ with $f \in {}^\lambda U$. Then $Y = \bigcap_{\xi<\lambda} f(\xi) \in U$ by κ-completeness, and clearly $j(Y) \subseteq X$. But $j(Y) \supseteq j``(Y \cap E) = Y \cap E \in U$ since j is the identity on E, and so $X \in U \cap M$. With $j(U) \subseteq U \cap M$ verified, it follows that $j(U) = U \cap M$ since $j(U)$ is an ultrafilter in M. 3.2(b) now implies that

$$M = L[j(U)] = L[U \cap M] = L[U]\ . \qquad \dashv$$

Extending 20.2(a), Silver obtained the first substantial result about κ-models, that full GCH holds:

20.3 Theorem (Silver [66a, 71a]). *Suppose that* $\langle L[U], \in, U\rangle$ *is a* κ*-model. Then* $L[U] \models$ GCH.

Proof. Keeping in mind that $L[U] \models V = L[U]$, we argue in $L[U]$ to verify GCH below κ there: Suppose that $\lambda < \kappa$. Recall the well-ordering $<_{L[U]}$ of $L[U]$ given by the formula $\varphi_1(v_0, v_1)$ of 3.3(b). Although there may be new subsets of λ appearing late in the $L[U] = \bigcup_\alpha L_\alpha[U]$ hierarchy, we shall establish a property that clearly suffices, that $<_{L[U]} |(\mathcal{P}(\lambda) \times \mathcal{P}(\lambda))$ has ordertype $\leq \lambda^+$. In other words, for any $y \in \mathcal{P}(\lambda) \cap L[U]$,

$$|\{x \in \mathcal{P}(\lambda) \cap L[U] \mid x <_{L[U]} y\}| \leq \lambda .$$

Assume to the contrary that this fails. Then there is a $y \in \mathcal{P}(\lambda) \cap L[U]$ such that with

$$R = \{x \in \mathcal{P}(\lambda) \cap L[U] \mid x <_{L[U]} y\} ,$$

$|R| = \lambda^+$.

Let γ be a limit ordinal $> \kappa$ with $y, U \in L_\gamma[U]$, and consider the structure

$$\mathcal{A} = \langle L_\gamma[U], \in, U, R, \{y\}\rangle .$$

Since U is λ^+-Rowbottom (as defined before 16.10), by the argument for 8.4(b) using a language with constants for each ordinal less than λ, there is a

$$\mathcal{B} = \langle B, \in, U \cap B, R \cap B, \{y\}\rangle \prec \mathcal{A}$$

such that $|R \cap B| \leq \lambda$, $\lambda \subseteq B$, and $B \cap \kappa \in U$. Let $\langle N, \in, W\rangle$ be the transitive collapse of $\langle B, \in, U \cap B\rangle$, and π the collapsing isomorphism. Since $\lambda \subseteq B$, $\pi(x) = x$ for any $x \in \mathcal{P}(\lambda) \cap B$. In particular, $y \in N$ and $R \cap N = R \cap B$. We shall verify that $W = U \cap N$. By elementarity and 3.3(b) this would imply that $N = L_\rho[U]$ for some limit ρ, and $R \subseteq N$ since $y \in N$, leading to the contradiction $|N| = |R \cap N| = |R \cap B| \leq \lambda$.

To verify that $W = U \cap N$, note that: $B \cap \kappa \in U$, $\pi(\xi) \leq \xi$ for any $\xi \in B$, and π is injective. It follows from the normality of U that

$$E = \{\xi \in B \cap \kappa \mid \pi(\xi) = \xi\} \in U .$$

For any $S \in N$, say $S = \pi(\overline{S})$ with $\overline{S} \in B$, it follows that: $S \in W$ *iff* $\exists D \in U(D \cap E \subseteq \overline{S})$ (taking $D = \overline{S}$ in one direction since $W = \pi``(U \cap B))$ *iff* $\exists D \in U(D \cap E \subseteq S)$)(since π is the identity on $D \cap E$) *iff* $S \in U$. Hence, $W = U \cap N$ and the proof is complete. ⊣

Instead of going into inner models Jensen [74] established the relative consistency of GCH with a measurable cardinal by forcing. Relying on a combinatorial contingency, he showed that a measurable cardinal in the ground model remains measurable upon forcing with an Easton product of *collapsing* p.o.'s. This method first established the relative consistency of GCH with Ramsey cardinals, for which an inner model theory became available only later.

Digressing somewhat, the following sketches Solovay's original proof of 16.1(b), that real-valued measurable cardinals are measurable in an inner model.

20.4 Exercise (Solovay [71]). *Suppose that I is a normal λ-saturated ideal over κ where $\lambda < \kappa$. Then*

$$L[I] \models \kappa \text{ is measurable .}$$

Hint. Check that if $\overline{I} = I \cap L[I]$, then $L[I] \models \overline{I}$ is a normal λ-saturated ideal over κ. Now work in $L[I]$: $\lambda^+ < \kappa$ by 16.3, and $\overline{I}^*$ is λ^+-Rowbottom by 16.11, so check that the argument for 20.3 works to show that $2^\lambda = \lambda^+$. Now apply 16.4(a). ⊣

The foregoing results about $L[U]$ depended on U being preserved as a predicate in transitive substructures, i.e. if N is the domain, what corresponds to U is just $U \cap N$. The further results required a more sophisticated analysis of relative constructibility. Although just the Rowbottom property of normal ultrafilters suffices for the GCH result 20.3, Silver saw that the Ramsey property leading to substructures generated by indiscernibles provides a Δ^1_3 well-ordering of the reals in $L[U]$, sharpening the analogy with L. Kunen established that a κ-model $\langle L[U], \in, U\rangle$ only depends on κ, and in fact, if W is any κ-complete ultrafilter over κ then $L[U] = L[W]$. Moreover, he showed that all the κ-models for different κ's are generated by taking iterated ultrapowers! In particular, all these inner models of measurability have the same theory.

With the preparations of §19, these and further structure theorems will be established in a terminology and approach that anticipates later developments. In particular, iterated ultrapower proofs will be provided for Silver's results on GCH and the Δ^1_3 well-ordering of the reals. The clear internal structure and striking global coherence of inner models of measurability revealed by these results provide a forceful argument for the *consistency* of the theory: ZFC plus there is a measurable cardinal.

We begin by considering structures that approximate κ-models and sharpen the §19 context:

$\langle M, \in, U\rangle$ is a ZFC$^-$ *premouse (at κ) iff*
U is an M-ultrafilter over κ and
$M = L_\zeta[U]$ for some ζ (allowing $M = L[U]$) .

The terminology is a capitulation to historical development, and anticipates a rodent infestation in a finer context (the Core Model – see volume II). The ZFC$^-$ refers to the implicit assumption in the term M-ultrafilter that $M \models$ ZFC$^-$, but for the sake of brevity

by *premouse* is meant ZFC$^-$ premouse in this section

with the caveat that the term will have a different meaning in volume II. Some simple observations:

(i) The weak amenability condition for M-ultrafilters U is redundant for $M = L_\zeta[U]$.
(ii) Any iterate of a premouse is also a premouse by 3.3(b) and and elementarity.
(iii) A κ-model $\langle L[U], \in, U\rangle$ is a premouse such that $U \in L[U]$.

To allow this inner model case implicitly, we stipulate that in this context that

$$L_\zeta[U] \text{ includes the possibility “}\zeta = \text{On”, i.e. } L_\zeta[U] = L[U]\,.$$

Iterability in such cases is an immediate consequence of 19.16:

20.5 Lemma. *Suppose that $\langle L_\zeta[U], \in, U\rangle$ is a premouse, $L_\zeta[U] \models$ ZFC, and $\omega_1 \cup \{U\} \subseteq L_\zeta[U]$. Then $\langle L_\zeta[U], \in, U\rangle$ is iterable.* ⊣

In particular, any κ-model is iterable (and its iterates are definable in it).

The following is a crucial observation about the definability of certain iterates of a premouse. Recall that if $\nu > \omega$ is regular, then $\mathcal{C}_\nu$ denotes the closed unbounded filter over ν.

20.6 Lemma. *Suppose that $\langle M, \in, U\rangle$ is a premouse at κ with iteration $\langle M_\alpha, U_\alpha, \kappa_\alpha, i_{\alpha\beta}\rangle_{\alpha\le\beta\in\tau}$, and $\nu \in \tau$ is a regular cardinal $> |{}^\kappa\kappa \cap M|$. Then for some ζ,*

$$M_\nu = L_\zeta[\mathcal{C}_\nu] \text{ and } U_\nu = \mathcal{C}_\nu \cap L_\zeta[\mathcal{C}_\nu]\,.$$

Proof. $\kappa_\nu = \nu$ by 19.7(b), and $U_\nu \subseteq \mathcal{C}_\nu \cap M_\nu$ by 19.5. Hence, $U_\nu = \mathcal{C}_\nu \cap M_\nu$ since U_ν is an ultrafilter on $\mathcal{P}(\nu) \cap M_\nu$. Since $\langle M_\nu, \in, \mathcal{C}_\nu \cap M_\nu\rangle$ satisfies the sentence σ_1 of 3.3(b), the result follows. ⊣

20.7 Corollary (Kunen [68, 70]). *Suppose that there is a κ-model, and ν is a regular cardinal greater than κ^+. Set $\overline{\mathcal{C}_\nu} = \mathcal{C}_\nu \cap L[\mathcal{C}_\nu]$. Then $L[\overline{\mathcal{C}_\nu}]$ is a ν-model, i.e.*

$$\langle L[\overline{\mathcal{C}_\nu}], \in, \overline{\mathcal{C}_\nu}\rangle \models \overline{\mathcal{C}_\nu} \text{ is a normal ultrafilter over } \nu\,.$$

Proof. Let $\langle L[U], \in, U\rangle$ be a κ-model. By 20.5, $\langle L[U], \in, U\rangle$ is iterable and by 20.2(a), $\kappa^+ \ge |{}^\kappa\kappa \cap L[U]|$. Hence, the result follows from 20.6, since $L[\mathcal{C}_\nu] = L[\overline{\mathcal{C}_\nu}]$ is the νth iterate of $L[U]$. ⊣

Thus, constructing relative to filters definable in ZFC can result in internal measurability! This compelling insight of Kunen's became a recurring theme in inner model theory, bringing Ulam's concept of measurability further into the fold of set theory.

The next result is the main technical lemma from which the structure theorems flow.

$\langle M, \in, U\rangle$ and $\langle N, \in, W\rangle$ are *comparable iff*
either $U = W \cap M$ or $W = U \cap N$.

In other words, there is an F such that $U = F \cap M$ and $W = F \cap N$ with $M = L_\zeta[F]$ and $N = L_\eta[F]$ for some ζ and η, so that M and N are initial segments of the same relative constructible hierarchy. Of course, if M and N are inner models, then $M = N$. The following direct consequence of 20.6 is the *Comparison Lemma for* ZFC$^-$ *Premice.*

20.8 Lemma. *Suppose that $\langle M, \in, U\rangle$ and $\langle N, \in, W\rangle$ are iterable premice. Then there is an iterate of $\langle M, \in, U\rangle$ and an iterate of $\langle N, \in, W\rangle$ that are comparable.* ⊣

We need one further preliminary result on definability. The 3.3(b) formula $\varphi_1(v_0, v_1)$ of $\mathcal{L}_\in(\dot{A})$ defines in any $\langle L[A], \in, A \cap L[A]\rangle$ a well-ordering $<_{L[A]}$ of $L[A]$ such that: for any limit $\delta > \omega$ and $x, y \in L_\delta[A]$,

$$x <_{L[A]} y \quad \textit{iff} \quad \langle L_\delta[A], \in, A \cap L_\delta[A]\rangle \models \varphi_1[x, y]\,.$$

Following a discussion after 9.3, for any formula $\varphi(v_0, \ldots, v_m)$ of $\mathcal{L}_\in(\dot{A})$ the *canonical Skolem term* t_φ for φ can be defined that chooses the least witness to v_0 according to φ_1 whenever possible. Consequently, for any limit ordinal $\delta > \omega$ and $X \subseteq L_\delta[A]$, *the* Skolem hull of X in $\langle L_\delta[A], \in, A \cap L_\delta[A]\rangle$ can be taken to be well-defined, with domain

$$\{t_\varphi^{\langle L_\delta[A], \in, A\cap L_\delta[A]\rangle}(x_1, \ldots, x_n) \mid \varphi \text{ is a formula of } \mathcal{L}_\in(\dot{A}) \ \wedge \ x_1, \ldots, x_n \in X\}\,.$$

This coincides with the collection of those $x \in L_\delta[A]$ such that $\{x\}$ is definable over $\langle L_\delta[A], \in, A \cap L_\delta[A]\rangle$ using parameters from X.

By *Skolem term* is meant one of these t_φ's in this section .

20.9 Lemma. *Suppose that $\langle L[U], \in, U\rangle$ is a κ-model, $B \subseteq$ On with $|B| \geq \kappa^+$, and δ is a limit ordinal such that $B \subseteq L_\delta[U]$. Then $\mathcal{P}(\kappa) \cap L[U]$ is a subset of the Skolem hull of $\kappa \cup B$ in $\langle L_\delta[U], \in, U\rangle$, so that for any $X \in \mathcal{P}(\kappa) \cap L[U]$,*

$$X = t^{\langle L_\delta[U], \in, U\rangle}(\xi_1, \ldots, \xi_m, \eta_1, \ldots, \eta_n)$$

for some Skolem term t, $\xi_1, \ldots, \xi_m \in \kappa$, and $\eta_1, \ldots, \eta_n \in B$.

Proof. Let $\langle H, \in, U \cap H\rangle$ be the stated Skolem hull, $\langle N, \in, W\rangle$ its transitive collapse, and π the collapsing isomorphism. Then $\pi(\alpha) = \alpha$ for every $\alpha < \kappa$, so $\pi(X) = X$ for every $X \in \mathcal{P}(\kappa) \cap H$. Consequently, $W = U \cap N$, so that using the sentence σ_1 of 3.3(b), $N = L_\zeta[U]$ for some ζ. $\zeta \geq \kappa^+$ since $|B| \geq \kappa^+$, so by the proof of 20.2(a), $\mathcal{P}(\kappa) \cap L[U] \subseteq N$. Hence, $\mathcal{P}(\kappa) \cap L[U] \subseteq H$ as π is the identity on subsets of κ. ⊣

The Main Results

We are now in a position to prove the main structure theorems on κ-models.

20.10 Theorem (Kunen [68, 70]). *Suppose that* $\langle L[U], \in, U\rangle$ *and* $\langle L[W], \in, W\rangle$ *are both* κ*-models. Then* $U = W$ *and so* $L[U] = L[W]$.

Proof. By 20.5 $\langle L[U], \in, U\rangle$ and $\langle L[W], \in, W\rangle$ are iterable, so by 20.8 they have comparable iterates and hence a common iterate $\langle L[F], \in, F\rangle$ since $L[U]$ and $L[W]$ are inner models. Let

$$i: \langle L[U], \in, U\rangle \prec \langle L[F], \in, F\rangle \quad \text{and}$$
$$j: \langle L[W], \in, W\rangle \prec \langle L[F], \in, F\rangle$$

be the corresponding embeddings. By 19.7(c),

$$\{\theta > \kappa \mid \theta = |\theta| = i(\theta) = j(\theta)\}$$

is a proper class, so let $B \cup \{\delta\}$ consist of members of this class so that $|B| = \kappa^+$ and $\delta > \sup(B)$.

For any $X \in U$, by 20.9

$$X = t^{\langle L_\delta[U], \in, U\rangle}(\xi_1, \ldots, \xi_m, \eta_1, \ldots, \eta_n)$$

for some Skolem term t, $\xi_1, \ldots, \xi_m \in \kappa$, and $\eta_1, \ldots, \eta_n \in B$. Set

$$Y = t^{\langle L_\delta[W], \in, W\rangle}(\xi_1, \ldots, \xi_m, \eta_1, \ldots, \eta_n) .$$

Since Skolem terms are definable,

$$i(X) = t^{\langle L_\delta[F], \in, F\rangle}(\xi_1, \ldots, \xi_m, \eta_1, \ldots, \eta_n) = j(Y)$$

noting that all the parameters are fixed by i and j. $X \in U$, so $i(X) = j(Y) \in F$, and hence $Y \in W$. But also

$$X = i(X) \cap \kappa = i(Y) \cap \kappa = Y .$$

This argument establishes that $U \subseteq W$, and the converse follows *mutatis mutandis*. ⊣

20.11 Corollary. *If* $\langle L[U], \in, U\rangle$ *is the* κ*-model, then* U *is the only normal ultrafilter over* κ *in* $L[U]$.

Proof. Suppose that W is a normal ultrafilter over κ in $L[U]$. Setting $\overline{W} = W \cap L[W]$, $L[\overline{W}]$ is a κ-model. Hence $\overline{W} = U$, and since W is an ultrafilter in $L[U]$, $W = U$. ⊣

Thus, if there is a κ-model $\langle L[U], \in, U\rangle$, then it is unique, κ is the only measurable cardinal in $L[U]$ (20.2(b)), and U is the only normal ultrafilter over κ in $L[U]$. This last contrasts with the consistency of having the maximum possible number 2^{2^κ} of normal ultrafilters (17.8). (The possible number of normal ultrafilters is further discussed before 22.13.)

The next theorem shows that all the κ-models for various κ are produced by iterating the $\overline{\kappa}$-model where $\overline{\kappa}$ is least possible; consequently, all κ-models for various κ are definable subclasses of this $\overline{\kappa}$-model and are elementarily equivalent to each other.

20.12 Theorem (Kunen [70]). *Suppose that $\langle L[U], \in, U\rangle$ is the κ-model and $\langle L[W], \in, W\rangle$ is the λ-model with $\kappa < \lambda$. Then $\langle L[W], \in, W\rangle$ is an iterate of $\langle L[U], \in, U\rangle$. In particular, $L[W] \subseteq L[U]$ and $\langle L[W], \in, W\rangle$ is definable in $\langle L[U], \in, U\rangle$ from λ.*

Proof. Let $\langle L[U_\alpha], U_\alpha, \kappa_\alpha, i_{\alpha\beta}\rangle_{\alpha\leq\beta\in\mathrm{On}}$ be the iteration of $\langle L[U], \in, U\rangle$, so that in particular each $\langle L[U_\alpha], \in, U_\alpha\rangle$ is definable in $\langle L[U], \in, U\rangle$. Since the κ_α's form a closed unbounded class of ordinals, there is a unique β such that $\kappa_\beta \leq \lambda < \kappa_{\beta+1}$. If $\kappa_\beta = \lambda$, we are done by 20.10. So, assume toward a contradiction that $\kappa_\beta < \lambda < \kappa_{\beta+1}$:

Let $\langle L[F], \in, F\rangle$ be a common iterate of $\langle L[U], \in, U\rangle$ and $\langle L[W], \in, W\rangle$, with

$$i\colon \langle L[U], \in, U\rangle \prec \langle L[F], \in, F\rangle \ \text{ and}$$
$$j\colon \langle L[W], \in, W\rangle \prec \langle L[F], \in, F\rangle$$

the corresponding embeddings. Hence, $i = i_{0\gamma}$ for some $\gamma > \beta$. Arguing as for 20.10, let $B \cup \{\delta\}$ consist of cardinals greater than λ fixed by both $i_{\beta\gamma}$ and j, with $|B| = \kappa_\beta^+$ and $\delta > \sup(B)$.

By 5.13(a) there is an $f \in {}^{\kappa_\beta}\kappa_\beta \cap L[U_\beta]$ such that $\lambda = i_{\beta,\beta+1}(f)(\kappa_\beta)$. By 20.9,

$$f = t^{\langle L_\delta[U_\beta], \in, U_\beta\rangle}(\xi_1, \dots, \xi_m, \eta_1, \dots, \eta_n)$$

for some Skolem term t, $\xi_1, \dots, \xi_m \in \kappa_\beta$, and $\eta_1, \dots, \eta_n \in B$. Hence,

$$\lambda = i_{\beta,\beta+1}(f)(\kappa_\beta) = u^{\langle L_\delta[U_{\beta+1}], \in, U_{\beta+1}\rangle}(\xi_1, \dots, \xi_m, \kappa_\beta, \eta_1, \dots, \eta_n)$$

for some Skolem term u. By definability of Skolem terms,

$$\lambda = u^{\langle L_\delta[F], \in, F\rangle}(\xi_1, \dots, \xi_m, \kappa_\beta, \eta_1, \dots, \eta_n)$$

since $\lambda < \kappa_{\beta+1}$ and so every parameter as well as λ is fixed by $i_{\beta+1,\gamma}$. But since $\kappa_\beta < \lambda$, the expression on the right also equals

$$j(u^{\langle L_\delta[W], \in, W\rangle}(\xi_1, \dots, \xi_m, \kappa_\beta, \eta_1, \dots, \eta_n))$$

for similar reasons. Thus $\lambda \in \mathrm{ran}(j)$, contradicting $\lambda = \mathrm{crit}(j)$. ⊣

These results have emphasized normal ultrafilters. Kunen also showed that constructing relative to an arbitrary κ-complete ultrafilter over κ again yields the κ-model, and analyzed the structure of the κ-complete ultrafilters over κ in the κ-model. The following technical result is the basis:

20.13 Exercise (Kunen [70]). *Suppose that W is any κ-complete ultrafilter over $\kappa > \omega$, and $j\colon V \prec M_W \cong \mathrm{Ult}(V, W)$. Let $\langle L[U], \in, U\rangle$ be the κ-model with iteration $\langle L[U_\alpha], U_\alpha, \kappa_\alpha, i_{\alpha\beta}\rangle_{\alpha\leq\beta\in\mathrm{On}}$. Then for some β, j and $i_{0\beta}$ agree on $\mathcal{P}(\kappa)\cap L[U]$.*

Hint. $L[j(U)]^{M_W} = L[j(U)]$ is *the* $j(\kappa)$-model, so by 20.12 there is a β such that $\kappa_\beta = j(\kappa)$ and $U_\beta = j(U)$. Show that this β is as desired with an appropriate use of 20.9. ⊣

20.14 Corollary. *Under the above hypotheses, $L[W] = L[U]$.*

Proof. Since κ is measurable in $L[W]$, there is a κ-model which is a subclass of $L[W]$. Hence, by 20.10, $L[U] \subseteq L[W]$. For the converse, it suffices to show that $W \cap L[U] \in L[U]$, for then $L[U] \supseteq L[W \cap L[U]] = L[W]$. For this purpose, let $\rho = [\mathrm{id}]_W$, where id: $\kappa \to \kappa$ is the identity, so that for any $X \subseteq \kappa$, $X \in W$ *iff* $\rho \in j(X)$. Then using 20.13,

$$\begin{aligned} W \cap L[U] &= \{X \in \mathcal{P}(\kappa) \cap L[U] \mid \rho \in j(X)\} \\ &= \{X \in \mathcal{P}(\kappa) \cap L[U] \mid \rho \in i_{0\beta}(X)\} \in L[U] \end{aligned}$$

since $i_{0\beta}$ is definable in $\langle L[U], \in, U\rangle$. ⊣

Proceeding toward a description of the κ-complete ultrafilters over κ in the κ-model, the rudiments of a structure theory for ultrafilters are discussed next, a subject which is pursued further in volume II:

Suppose that S and T are sets, $f\colon S \to T$, and D is an ultrafilter over S. Then

$$f_*(D) = \{Y \subseteq T \mid f^{-1}(Y) \in D\}\,.$$

If principal ultrafilters were allowed, it can be checked that $f_*(D)$ is always an ultrafilter over T, the natural projection of D via f. Note that if $X \in D$, then $f``X \in f_*(D)$. If moreover E is an ultrafilter over T,

D and E are *equivalent iff* there is a bijection $f\colon S \to T$ such that $E = f_*(D)$.

Thus, equivalent ultrafilters are the same up to a relabeling of their underlying sets. The following is well-known.

20.15 Proposition. *Suppose that D is an ultrafilter over S and $f\colon S \to T$.*

(a) If $g\colon S \to T$ and $\{i \in S \mid f(i) = g(i)\} \in D$, then $g_*(D) = f_*(D)$.

(b) If $h\colon S \to S$ and $h_*(D) = D$, then $\{i \in S \mid h(i) = i\} \in D$.

(c) If E is an ultrafilter over T, $|T| = |S|$, $f_*(D) = E$, and $g\colon T \to S$ is such that $g_*(E) = D$, then D and E are equivalent.

Proof. (a) This is clear.

(b) Assume to the contrary that $\{i \in S \mid h(i) > i\} \in D$; the $<$ case is analogous. By (a), it can be assumed that $h(i) > i$ for *every* $i \in S$. Let $h^0(i) = i$, $h^{n+1}(i) = h(h^n(i))$ for $n \in \omega$, and define a relation $\sim$ on S by:

$$i \sim j \ \textit{ iff } \ \exists m \exists n (h^m(i) = h^n(j)) \, .$$

It can be checked that $\sim$ is an equivalence relation, so fix a representative for each equivalence class. Let $S_0 \subseteq S$ consist of those i such that if r is the representative of the equivalence class of i and $h^m(i) = h^n(r)$, then $|m - n|$ is an even number. (Note that this is independent of m and n as h is increasing on S.) Set $S_1 = S - S_0$. Then for some $e < 2$, $S_e \in D$, and so $h``S_e \in h_*(D) = D$. But it is simple to see that $h``S_e \subseteq S_{1-e}$, leading to the contradictory $S_e \cap h``S_e = \emptyset \in D$.

(c) $(g \circ f)_*(D) = D$, so that $\{i \in S \mid g(f(i)) = i\} \in D$ by (b), and f is injective on this set. Since $|S| = |T|$, there is a bijection $\overline{f}\colon S \to T$ such that $\{i \in S \mid f(i) = \overline{f}(i)\} \in D$, and so the proof is complete by (a). ⊣

In $L[U]$, where $\langle L[U], \in, U\rangle$ is the κ-model, U is the only normal ultrafilter over κ, and from U the product ultrafilters U^n over $[\kappa]^n$ for $n \in \omega$ can be defined (as after 19.7). (b) below asserts that every κ-complete ultrafilter over κ in $L[U]$ is essentially one of these.

20.16 Theorem (Kunen [70]; Paris [69] also for (a)). *Suppose that* $V = L[U]$ *where* $\langle L[U], \in, U\rangle$ *is the* κ*-model with iteration* $\langle L[U_\alpha], U_\alpha, \kappa_\alpha, i_{\alpha\beta}\rangle_{\alpha\leq\beta\in\mathrm{On}}$. *Let* W *be any* κ*-complete ultrafilter over* κ. *Then:*

(a) $W = \{X \subseteq \kappa \mid \rho \in i_{0\omega}(X)\}$ *for some* $\rho < \kappa_\omega$.

(b) W *is equivalent to* U^n *for some* n *with* $0 < n \in \omega$.

Proof. Letting $j\colon V \prec M_W \cong \mathrm{Ult}(V, W)$, first observe that $V = L[U]$ implies that the β of 20.13 must be finite: By 20.14, $V = L[W]$ and $M_W = L[j(W)] = L[U_\beta]$ by the uniqueness of $j(\kappa)$-models. If to the contrary $\beta \geq \omega$, then κ_ω is inaccessible in $L[U_\beta]$ since it is measurable in $L[U_\omega]$ and $\mathcal{P}(\kappa_\omega) \cap L[U_\omega] = \mathcal{P}(\kappa_\omega) \cap L[U_\beta]$. However, since M_W contains every countable sequence of ordinals by 5.7(d), κ_ω has cofinality ω there, which is a contradiction. Hence, $\beta = n$ for some n with $0 < n \in \omega$.

(a) now follows from the proof of 20.14, since $i_{n\omega}(\rho) = \rho$ for any $\rho < j(\kappa) = i_{0n}(\kappa)$.

For (b), if

$$e_n\colon V \prec M_{U^n} \cong {}^{[\kappa]^n}V/U^n \, ,$$

then $M_{U^n} = L[U_n]$ and $e_n = i_{0n}$ by the analysis of 19.9(b). Let id: $\kappa \to \kappa$ and $\overline{\mathrm{id}}\colon [\kappa]^n \to [\kappa]^n$ be the identity maps on their respective domains, and $f\colon \kappa \to [\kappa]^n$ and $g\colon [\kappa]^n \to \kappa$ such that $[f]_W = [\overline{\mathrm{id}}]_{U^n}$ and $[g]_{U^n} = [\mathrm{id}]_W$. Because j and e_n agree on $\mathcal{P}(\kappa)$ and hence on $\mathcal{P}([\kappa]^n)$, for any $X \subseteq [\kappa]^n$,

$$\begin{aligned} X \in U^n \ &\textit{iff} \ [\overline{\mathrm{id}}]_{U^n} \in e_n(X) \\ &\textit{iff} \ [f]_W \in j(X) \\ &\textit{iff} \ f^{-1}(X) \in W. \end{aligned}$$

Hence, $f_*(W) = U^n$. Similarly, $g_*(U^n) = W$, and so the proof is complete by 20.15(c). ⊣

The next goal is to establish Silver's result that there is a Δ^1_3 well-ordering of the reals in $L[U]$. The argument is in terms of premice, and to motivate it, another proof of Silver's GCH result 20.3 is first given, one that has the advantage of having generalizations to inner models of stronger hypotheses.

Second Proof of 20.3. As before, it suffices to show for $\lambda < \kappa$ that for any $y \in \mathcal{P}(\lambda) \cap L[U]$,

$$|\{x \in \mathcal{P}(\lambda) \cap L[U] \mid x <_{L[U]} y\}| \leq \lambda .$$

Suppose then that $y \in \mathcal{P}(\lambda) \cap L[U]$, and let $x \in \mathcal{P}(\lambda) \cap L[U]$ with $x <_{L[U]} y$, i.e. $\langle L[U], \in, U\rangle \models \varphi_1[x, y]$ for $\varphi_1(v_0, v_1)$ of 3.3(b). Then:

(∗) If $\langle N, \in, W\rangle$ is an iterate premouse at some $\nu > \lambda$ and $y \in N$, then $x \in N$.

To show this, by 20.8 let $\langle L[F], \in, F\rangle$ and $\langle L_\zeta[F], \in, F \cap L_\zeta[F]\rangle$ be comparable iterates of $\langle L[U], \in, U\rangle$ and $\langle N, \in, W\rangle$ respectively. Then $y \in V_\nu \cap N = V_\nu \cap L_\zeta[F]$ by 19.4(b). Remembering that $\lambda < \kappa$, $\langle L[F], \in, F\rangle \models \varphi_1[x, y]$ by elementarity, and so by 3.3(b), $y \in L_\zeta[F]$ implies that $x \in L_\zeta[F]$. Hence, $x \in V_\nu \cap L_\zeta[F] = V_\nu \cap N$. (Note for the next theorem that $\langle N, \in, W\rangle \models \varphi_1[x, y]$, again by elementarity.)

With (∗), the proof devolves to finding an iterable premouse $\langle N, \in, W\rangle$ at some $\nu > \lambda$ with $|N| \leq \lambda$. But this is simple: Let θ be a sufficiently large regular cardinal such that $y, U \in L_\theta[U]$. Then by the absoluteness 19.16 applied with $S = L[U]$, $\langle L_\theta[U], \in, U\rangle$ is an iterable premouse. Take the Skolem hull of $\lambda \cup \{y\}$ in $\langle L_\theta[U], \in, U\rangle$; then its transitive collapse is a premouse as desired, with iterability a consequence of 19.15(e). ⊣

The next results are about the set of reals in the κ-model. An early intimation of the Core Model (see volume II) is that this set is independent even of κ; this follows by a simpler version of the previous argument depending only on 20.12:

20.17 Exercise. *Suppose that $\langle L[U], \in, U\rangle$ is the κ-model and $\langle L[W], \in, W\rangle$ is the λ-model for some κ and λ. Then*

$${}^\omega\omega \cap L[U] = {}^\omega\omega \cap L[W] \text{ and } {}^2({}^\omega\omega) \cap <_{L[U]} = {}^2({}^\omega\omega) \cap <_{L[W]} . \qquad ⊣$$

In referring to these sets in what follows, $L[U]$ is used generically since the sets do not depend on any particular choice. Recall from before 13.8 that any $z \in {}^\omega\omega$ encodes a binary relation $E_z = \{\langle m, n\rangle \mid z(\langle\!\langle m, n\rangle\!\rangle) = 0\}$ and consequently a structure $M_z = \langle\omega, E_z\rangle$ for the language $\mathcal{L}_\in$ of set theory. Defining $A_z = \{n \mid z(n) = 1\}$, z also encodes a structure

$$N_z = \langle\omega, E_z, A_z\rangle$$

for the augmented language $\mathcal{L}_\in(\dot{A})$. As before, if N_z is well-founded and extensional, then it has a transitive collapse $\text{tr}(N_z)$ and an isomorphism $\pi_z\colon N_z \to \text{tr}(N_z)$.

The following is the analogue of Gödel's result 13.9 for L:

20.18 Theorem (Silver [71b]). *${}^{\omega}\omega \cap L[U]$ and ${}^{2}({}^{\omega}\omega) \cap <_{L[U]}$ are Σ^1_3.*

Proof. Proceeding directly to the well-ordering, for any $x, y \in {}^{\omega}\omega \cap L[U]$,

$$(**) \qquad x <_{L[U]} y \ \ \textit{iff} \ \ \text{there is a countable, iterable premouse } \langle N, \in, W\rangle \text{ with } x, y \in N \text{ and } \langle N, \in, W\rangle \models \varphi_1[x, y]\,.$$

This follows from the argument for $(*)$ in the proof of 20.16 together with the Skolem hull argument at its end. It remains to verify that this gives a Σ^1_3 description:

Building on 3.3(b) and 13.8(a)(c),

$$Z = \{z \in {}^{\omega}\omega \mid N_z \text{ is well-founded and extensional } \wedge \ \text{tr}(N_z) \text{ is a premouse}\}$$

is Π^1_1. It now suffices to show that ($z \in Z \ \wedge \ \text{tr}(N_z)$ is iterable) is Π^1_2, for then $(**)$ can be rendered as $\exists z(\Pi^1_2)$.

By 19.15(b) a premouse is iterable *iff* its iteration has length $\geq \omega_1$. Recalling the analysis 14.11 of $0^{\#}$, this can be asserted through codes for countable well-orderings. First, with $\prec^-$ as before 19.3 let $R \subseteq {}^{3}({}^{\omega}\omega)$ be given by $R(a, b, z)$ *iff*

(i) $E_a = \{\langle m, n\rangle \mid a(\langle m, n\rangle) = 0\}$ is a well-ordering with field ω.
(ii) b codes a set $\{z^b_n \mid n \in \omega\} \subseteq Z$ together with codes for embeddings $i_{mn}\colon \text{tr}(N_{z^b_m}) \prec^- \text{tr}(N_{z^b_n})$ for $\langle m, n\rangle \in E_a$ such that:
(iii) if n is the minimum in terms of E_a, then $z^b_n = z$;
(iv) if n is the immediate successor of m in terms of E_a, then $\text{tr}(N_{z^b_n})$ is the ultrapower of $\text{tr}(N_{z^b_m})$ and i_{mn} the corresponding embedding; and
(v) if n is a limit point of E_a, then $\text{tr}(N_{z^b_n})$ is the direct limit of the corresponding structures and embeddings, and i_{mn} for $\langle m, n\rangle \in E_a$ the corresponding embedding into the direct limit.

It is tedious but straightforward to check that R is Π^1_1 since Z is Π^1_1 and satisfaction for N_{z_n} is Δ^1_1 (cf. 13.8(b)(c)).

Next, let $S \subseteq {}^{2}({}^{\omega}\omega)$ be given by $S(a, b)$ *iff*

(i), (ii), (iv), and (v) above.
(vi) if E_a has a maximum element n, then the ultrapower of $\text{tr}(N_{z_n})$ is well-founded.
(vii) if E_a has no maximum element, then the direct limit of the structures and embeddings coded by b is well-founded.

S is also Π^1_1.

Finally,

$$(z \in Z \ \wedge \ \mathrm{tr}(N_z) \text{ is iterable}) \ \mathit{iff} \ \ z \in Z \ \wedge \ \forall^1 a \forall^1 b(R(a,b,z) \to S(a,b)) \,,$$

which is of form $\Pi^1_1 \wedge \forall^1 a \forall^1 b(\Pi^1_1 \to \Pi^1_1)$, and hence Π^1_2. ⊣

20.19 Corollary. *If* $^\omega\omega \subseteq L[U]$, *then* $^2(^\omega\omega) \cap <_{L[U]}$ *is a* Δ^1_3 *set which is not Lebesgue measurable and does not have the Baire property.*

Proof. As for 13.10. ⊣

As for L (cf. 13.11) there is a refinement of 20.19, establishing in the parlance that in $L[U]$, $^2(^\omega\omega) \cap <_{L[U]}$ is a Σ^1_3-good well-ordering:

20.20 Exercise (Silver [71b: 440]). *The relation* $\mathrm{IS}_{L[U]} \subseteq {}^2(^\omega\omega)$ *given by*

$$\mathrm{IS}_{L[U]}(x,y) \ \leftrightarrow \ \{(x)_i \mid i \in \omega\} = \{z \in {}^\omega\omega \mid z <_{L[U]} y\}$$

is Σ^1_3. ⊣

For the perfect set property, the argument for L (13.12) can be adapted. For this purpose, first observe that there is a natural ordering of iterable premice:

$\langle M, \in, U\rangle <_{\mathrm{ip}} \langle N, \in, W\rangle$ *iff* there is an F such that $\langle L_\zeta[F], \in, F \cap L_\zeta[F]\rangle$ is an iterate of $\langle M, \in, U\rangle$, $\langle L_\eta[F], \in, F \cap L_\eta[F]\rangle$ is an iterate of $\langle N, \in, W\rangle$, and $\zeta < \eta$.

It is straightforward to check using the Comparison Lemma 20.8 that $<_{\mathrm{ip}}$ is a well-defined well-ordering of iterable set premice. Next observe that countable such $\langle M, \in, U\rangle$ and $\langle N, \in, W\rangle$ have *countable* iterates that are comparable: Suppose that

$$\langle M_\alpha, U_\alpha, \kappa_\alpha, i_{\alpha\beta}\rangle_{\alpha\le\beta\in\mathrm{On}} \text{ is the iteration of } \langle M, \in, U\rangle \text{ , and}$$
$$\langle N_\alpha, W_\alpha, \lambda_\alpha, j_{\alpha\beta}\rangle_{\alpha\le\beta\in\mathrm{On}} \text{ is the iteration of } \langle N, \in, W\rangle \text{ .}$$

Noting that $\kappa_{\omega_1} = \omega_1 = \lambda_{\omega_1}$ by 19.7(c) so that $\{\kappa_\alpha \mid \alpha < \omega_1\}$ and $\{\lambda_\alpha \mid \alpha < \omega_1\}$ are both closed unbounded in ω_1, let $\langle \gamma_n \mid n \in \omega\rangle$ be an increasing sequence of ordinals drawn from their intersection and set $\gamma = \sup(\{\gamma_n \mid n \in \omega\})$. Then $\gamma = \kappa_\alpha = \lambda_\beta$ for some $\alpha, \beta < \omega_1$, and with F the filter generated by $\langle \gamma_n \mid n \in \omega\rangle$ (i.e. $X \in F$ *iff* $\exists m \in \omega(\{\gamma_n \mid m \le n \in \omega\} \subseteq X \subseteq \gamma)$),

$$\langle M_\alpha, \in, U_\alpha\rangle = \langle L_\zeta[F], \in, F \cap L_\zeta[F]\rangle \text{ , and}$$
$$\langle N_\beta, \in, W_\beta\rangle = \langle L_\eta[F], \in, F \cap L_\eta[F]\rangle$$

for some ζ and η.

20.21 Theorem (Silver [71b: 440]; Solovay). *If* $\omega_1^{L[U]} = \omega_1$, *then there is a* Π^1_2 *set without the perfect set property.*

Proof. Adapting the argument for 13.12, the idea is to get a Π^1_2 set of unique codes for ordinals less than $\omega_1^{L[U]}$. For such an ordinal α, there is an $x_0 \in {}^\omega\omega \cap L[U]$ such that $M_{x_0} = \langle \omega, E_{x_0} \rangle$ is a well-ordered set with ordertype α. To specify the $<_{L[U]}$-least such x_0 is not Π^1_2, but it turns out to be Π^1_2 to say of some $x_1 \in {}^\omega\omega \cap L[U]$: $N_{x_1} \cong \langle N, \in, W \rangle$ where the latter is the $<_{\mathrm{ip}}$-least countable iterable premouse such that some $x_0 \in N$ is the $<_{L[U]} |(N \times N)$-least possible as above. Again, to specify the $<_{L[U]}$-least such x_1 is not Π^1_2, but it can be similarly specified it in terms of some N_{x_2}, and so forth. The resulting $\langle x_i \mid i \in \omega \rangle$ can then serve as a unique code for α.

For any $x, y \in {}^\omega\omega$, that $\mathrm{tr}(N_y)$, if defined, is the $<_{\mathrm{ip}}$-least possible countable iterable premouse containing x is equivalent by the remark preceding the theorem to: $\forall^1 z(\mathrm{tr}(N_z)$ is defined and is a premouse with a countable iterate isomorphic to a proper initial segment of a countable iterate of $\mathrm{tr}(N_y) \rightarrow x \notin \mathrm{tr}(N_z))$, noting that the antecedent to this implication implies that $\mathrm{tr}(N_z)$ is iterable.

Recalling the relation A of the proof of 13.12 and availing ourselves of the Π^1_1 Z and R of the proof of 20.18, we are led to the following relation on ${}^\omega\omega$:

$$\begin{aligned}
A(x) \quad \leftrightarrow \quad & M_{(x)_0} \text{ is well-ordered } \wedge\ \forall^0 i[(x)_{i+1} \in Z \\
& \wedge\ \mathrm{tr}(N_{(x)_{i+1}}) \text{ is iterable } \wedge\ (x)_i \in \mathrm{tr}(N_{(x)_{i+1}}) \\
& \wedge\ \forall^1 z \forall^1 a \forall^1 b \forall^1 c \forall^1 d[z \in Z\ \wedge\ R(a, b, z)\ \wedge\ R(c, d, (x)_{i+1}) \\
& \wedge\ m \text{ is the maximum element of } E_a \\
& \wedge\ n \text{ is the maximum element of } E_c \\
& \wedge\ \exists^1 f(f \text{ codes an isomorphism of } N_{z^b_m} \text{ into a} \\
& \text{proper initial segment of } N_{z^d_n}) \rightarrow\ (x)_i \notin \mathrm{tr}(N_{z^d_n})] \\
& \wedge\ \forall^1 y \forall^0 p \forall^0 q[\pi_{(x)_{i+1}}(p) = y\ \wedge\ \pi_{(x)_{i+1}}(q) = (x)_i \\
& \wedge\ (N_{(x)_{i+1}} \models \varphi_1[p, q]\ \rightarrow\ (i = 0\ \rightarrow\ M_y \not\cong M_{(x)_0}) \\
& \wedge\ (i > 0\ \rightarrow\ N_y \not\cong N_{(x)_i}))]]
\end{aligned}$$

where $\varphi_1(v_0, v_1)$ is the well-ordering formula of 3.3(b). Now getting on track and following the 13.12 argument, it is straightforward to check that A is a Π^1_2 set of cardinality $|\omega_1^{L[U]}|$, and that A does not have a perfect (even uncountable $\mathbf{\Sigma}^1_1$) subset. ⊣

These results are the best possible in the projective hierarchy since Solovay had shown that if there is a measurable cardinal, then every $\mathbf{\Sigma}^1_2$ set is Lebesgue measurable, has the Baire property, and has the perfect set property (14.3, 14.10). Silver had been motivated in part by this, and his results established a new delimitation at Δ^1_3. This in turn led to an investigation of stronger large cardinal principles that might similarly correlate with regularity properties higher in the projective hierarchy. Unlike Gödel's hopes for the Continuum Problem (§3), the relative concreteness of the projective hierarchy afforded the possibility of a systematic study, one that was to meet with remarkable success by the late 1980's (§32).

The concluding results of this section concerns other large cardinals and inner models of measurability. Of course, in the presence of a measurable cardinal more modest large cardinals exist in great profusion by previous reflection results. The Jónsson and Rowbottom cardinals (§8) are somewhat anomalous in that although they have substantial consistency strength (21.4) they themselves may not be very large. Kunen on the other hand established that in the κ-model these cardinals coincide with the ostensibly stronger Ramsey cardinals. Thus, one cannot establish in the theory ZFC plus there is a measurable cardinal that there are, e.g. singular Jónsson cardinals. An outright limitation reminiscent of 8.13 is first noted:

20.22 Exercise. *Suppose that $\langle L[U], \in, U\rangle$ is the κ-model and there is a Jónsson cardinal $\lambda > \kappa$. Then $V \neq L[U]$.*

Hint. Working in $L[U]$, argue by contradiction and take λ to be the least Jónsson cardinal $> \kappa$. By 8.14(c) and 8.15(a)(b) there is a regular ν satisfying $\kappa < \nu < \lambda$, such that $\lambda \to [\lambda]^{<\omega}_{\nu,<\nu}$. By 8.4(b) let

$$\langle H, \in, U \cap H, \nu \cap H\rangle \prec \langle L_\lambda[U], \in, U, \nu\rangle$$

be such that $|H| = \lambda$, $\kappa + 1 \subseteq H$, and $|\nu \cap H| < \nu$. The transitive collapse of $\langle H, \in, U \cap H\rangle$ is again $\langle L_\lambda[U], \in, U\rangle$, so if $i\colon \langle L_\lambda[U], \in, U\rangle \prec \langle L_\lambda[U], \in, U\rangle$ is the inverse of the collapsing map, crit(i) is a measurable cardinal greater than κ in $L[U]$, which is a contradiction. ⊣

This result is extended by 21.23.

20.23 Theorem (Kunen [70]). *Suppose that $\langle L[U], \in, U\rangle$ is the κ-model for some κ. Then*

$$L[U] \models \forall\lambda(\lambda \text{ is Jónsson } \leftrightarrow \lambda \text{ is Ramsey})\,.$$

Proof. Working in $L[U]$, suppose that λ is Jónsson, and $f\colon [\lambda]^{<\omega} \to 2$. We must find a member of $[\lambda]^\lambda$ homogeneous for f, and this is first done under an additional assumption:

(∗) There is a $\delta < \lambda$ and an $x \subseteq \delta$ such that $f <_{L[U]} x$.

This uses arguments employed in the proof of GCH: Noting that $\lambda < \kappa$ by 20.22, there is an iterable premouse $\langle N, \in, W\rangle$ at some $\rho > \delta$ with $x \in N$ and $|N| = |\delta|$. If $\langle N_\lambda, \in, W_\lambda\rangle$ is then the λth iterate of $\langle N, \in, W\rangle$ with $j_{0\lambda}$ the corresponding embedding, $j_{0\lambda}(\rho) = \lambda$ and $x \in N_\lambda$. By Comparison, let $\langle L[F], \in, F\rangle$ and $\langle L_\zeta[F], \in, F \cap L_\zeta[F]\rangle$ be comparable iterates of $\langle L[U], \in, U\rangle$ and $\langle N_\lambda, \in, W_\lambda\rangle$ respectively. Since $f <_{L[U]} x$, by elementarity, $f <_{L[F]} x$. Also $\mathcal{P}(\lambda) \cap N_\lambda = \mathcal{P}(\lambda) \cap L_\zeta[F]$, so that $x \in N_\lambda$ implies that $f \in N_\lambda$. But λ is measurable in N_λ, so f has a homogeneous set as required.

The general case yields to similar arguments. First, there is an iterable pre-mouse $\langle M, \in, Y\rangle$ at some $\sigma > \lambda$ with $f \in M$ and $|M| = \lambda$. Let $g\colon \lambda \to M$ be a bijection. Since λ is Jónsson, there is a proper substructure

$$\langle H, \in, Y \cap H, g|H, \{f\}\rangle \prec \langle M, \in, Y, g, \{f\}\rangle$$

with $|H| = \lambda$. Let $\langle T, \in, Z, \overline{g}, \{\overline{f}\}\rangle$ be the transitive collapse and

$$j\colon \langle T, \in, Z, \overline{g}, \{\overline{f}\}\rangle \prec \langle M, \in, Y, g, \{f\}\rangle$$

the inverse of the collapsing map. Using g, $|H \cap \lambda| = \lambda$ yet $H \cap \lambda \neq \lambda$, so that $j(\lambda) = \lambda$ yet j has a critical point $\delta < \lambda$. The usual

$$\{X \in \mathcal{P}(\delta) \cap T \mid \delta \in j(X)\}$$

cannot witness the measurability of δ in $L[U]$, so there must be an $x \in \mathcal{P}(\delta) - T$.

It is clear that $\langle T, \in, Z\rangle$ is an iterable premouse at λ and $\overline{f} \in T$, so an application of Comparison as before shows that $\overline{f} <_{L[U]} x$. Finally, by the argument from $(*)$ there is an $S \in [\lambda]^\lambda$ homogeneous for $\overline{f}$. But then, j"$S \in [\lambda]^\lambda$ is homogeneous for f, and the proof is complete. ⊣

Kunen [68] generalized his structure results to models $M = L[\langle U_\alpha \mid \alpha < \gamma\rangle]$ where in M, U_α is a normal ultrafilter over some λ_α, $\alpha < \beta < \gamma$ implies that $\lambda_\alpha < \lambda_\beta$, and the sequence is short: $\gamma < \lambda_0$. For example, he showed that in M the U_α's are the only normal ultrafilters (over any cardinal). Silver [71b] showed that M satisfies GCH, and there is a Δ^1_3 well-ordering of its reals. Far beyond this first generalization, inner model theory was to develop into a sophisticated area of set theory in the hands of Mitchell, Jensen, Dodd, Steel, and others, encompassing stronger and stronger hypotheses and providing forceful arguments for their consistency.

21. Embeddings, $0^\#$, and $0^\dagger$

Iterated ultrapowers is not only the appropriate technique for investigating inner models, but also their embeddings. Using it Kunen established an elegant characterization of $0^\#$: it exists exactly when there is an elementary embedding: $L \prec L$. Reminiscent of the elementary embedding characterization of measurable cardinals, the initial isolation of $0^\#$ was bolstered by Kunen's result. Moreover, it provided a simple paradigm for transcendence over inner models M in general in terms of their *non-rigidity*, i.e. the existence of an elementary embedding: $M \prec M$. For inner models of measurability Solovay had already isolated a real $0^\dagger$ analogous to $0^\#$ for L. The existence of $0^\dagger$ is a canonical principle for transcendence over *all* inner models of measurability and also has characterizations in terms of their non-rigidity. This section explores embeddings of L and develops the theory of $0^\dagger$ in considerable detail.

The following theorem analyzes Kunen's result about the non-rigidity of L through several equivalences. Strictly speaking, the quantification $\exists j(j\colon L \prec L)$ cannot be formalized, and (b) articulates what is needed.

21.1 Theorem (Kunen). *The following are equivalent:*

(a) There is an elementary embedding: $L \prec L$.

(b) For some α and β there is an elementary embedding: $L_\alpha \prec L_\beta$ with a critical point less than $|\alpha|$.

(c) There is an L-ultrafilter U such that the ultrapower of L by U is well-founded.

(d) There is an iterable L-ultrafilter.

(e) $0^\#$ exists.

Proof. (d) $\to$ (e) The domain of each iterate is again L by elementarity, so 19.10 can be applied to get an uncountable set of indiscernibles for some L_α.

(e) $\to$ (a) This is 9.17(a).

(a) $\to$ (b) Suppose that $k\colon L \prec L$ with critical point κ. For any α with $\kappa < |\alpha|$, the satisfaction relation for L_α can be used to show that $k|L_\alpha$ satisfies (b).

(b) $\to$ (c) Let j be an elementary embedding as hypothesized, say with critical point κ, and consider the usual

$$U = \{X \in \mathcal{P}(\kappa) \cap L_\alpha \mid \kappa \in j(X)\}\ .$$

$\mathcal{P}(\kappa) \cap L \subseteq L_\alpha$ since $\alpha \geq \kappa^{+L}$, so U is an L-ultrafilter (cf. 19.2). It remains to verify that the ultrapower of L by U is well-founded:

Assume to the contrary that for each $n \in \omega$ there is a function $f_n \in L$ with domain κ such that

$$X_n = \{\xi < \kappa \mid f_{n+1}(\xi) \in f_n(\xi)\} \in U\ .$$

Let $\gamma > \kappa$ be a limit ordinal such that $\{f_n \mid n \in \omega\} \subseteq L_\gamma$, $H \prec L_\gamma$ a Skolem hull of $\kappa \cup \{f_n \mid n \in \omega\}$, and N its transitive collapse with $\pi\colon H \to N$ the collapsing isomorphism. Then $N = L_\delta$ for some $\delta < \alpha$ (this is where $\alpha \geq \kappa^+$ is needed, not just $\alpha \geq \kappa^{+L}$). Letting $\pi(f_n) = \overline{f}_n$ for $n \in \omega$ and noting that $\pi(\xi) = \xi$ for $\xi < \kappa$, for such n

$$X_n = \{\xi < \kappa \mid \overline{f}_{n+1}(\xi) \in \overline{f}_n(\xi)\} \in U \ .$$

Since $\{\overline{f}_n \mid n \in \omega\} \subseteq L_\alpha$, it follows from the definition of U that

$$j(\overline{f}_{n+1})(\kappa) \in j(\overline{f}_n)(\kappa)$$

for every $n \in \omega$, which is a contradiction.

(c) $\to$ (d) Suppose that U is an L-ultrafilter with corresponding iteration

$$\langle M_\alpha, U_\alpha, \kappa_\alpha, i_{\alpha\beta}\rangle_{\alpha \leq \beta \in \tau}$$

where $\tau > 1$; note that each $M_\alpha = L$. It will be established that in fact $\tau = \mathrm{On}$. This is the heart of the matter, and uses the canonicity of L to derive full iterability from just the well-foundedness of the first ultrapower. The following two lemmata, showing successively that τ is neither a successor ordinal nor a limit ordinal, complete the proof of the theorem.

21.2 Lemma. *τ is not a successor ordinal.*

Proof. Assuming that $\alpha < \tau$ we must show that $\alpha + 1 < \tau$, i.e. the ultrapower of M_α by U_α is well-founded. By 19.6,

$$(*) \qquad M_\alpha = L = \{i_{0\alpha}(g)(\kappa_{\gamma_1}, \ldots, \kappa_{\gamma_n}) \mid g \in {}^{[\kappa]^n}L \cap L \text{ for some } n \in \omega \text{ and } \gamma_1 < \ldots < \gamma_n < \alpha\} \ .$$

Since $\tau > 1$, $M_1 \cong {}^{\kappa}L/U$ is again L, so a g as in $(*)$ is of form $g = [f]_U$, where it can be assumed that for every $\xi < \kappa$, $f(\xi)$ is a function with domain $[\xi]^n$. For such f, define $\overline{f}\colon [\kappa]^n \to {}^{\kappa}L$ by:

$$\overline{f}(\xi_1, \ldots, \xi_n)(\xi) = \begin{cases} f(\xi)(\xi_1, \ldots, \xi_n) & \text{if } \xi_1 < \ldots < \xi_n < \xi \ , \text{ and} \\ 0 & \text{otherwise} \ . \end{cases}$$

Incorporating this into the representation $(*)$ of L, we show that the map $e\colon \langle L, \in\rangle \to \langle {}^{\kappa_\alpha}M_\alpha/U_\alpha, E_{U_\alpha}\rangle$ given by:

$$e(i_{0\alpha}([f]_U)(\kappa_{\gamma_1}, \ldots, \kappa_{\gamma_n})) = (i_{0\alpha}(\overline{f})(\kappa_{\gamma_1}, \ldots, \kappa_{\gamma_n}))^0_{U_\alpha}$$

is a well-defined isomorphism, thereby confirming that the latter structure is well-founded:

Taking for simplicity a one-variable formula $\varphi(v_0)$ of $\mathcal{L}_\in$, note that as a schema,

$\langle L, \in\rangle \models \varphi[i_{0\alpha}([f]_U)(\kappa_{\gamma_1}, \ldots, \kappa_{\gamma_n})]$
iff $\langle L, \in, U\rangle \models \{\{\xi_1, \ldots, \xi_n\} \in [\kappa]^n \mid \varphi[[f]_U(\xi_1, \ldots, \xi_n)]\} \in U^n$ (by 19.9(a))
iff $\langle L, \in, U\rangle \models \{\{\xi_1, \ldots, \xi_n\} \in [\kappa]^n \mid \{\xi < \kappa \mid \varphi[\overline{f}(\xi_1, \ldots, \xi_n)(\xi)]\} \in U\} \in U^n$
iff $\langle M_\alpha, \in, U_\alpha\rangle \models \{\xi < \kappa_\alpha \mid \varphi[i_{0\alpha}(\overline{f})(\kappa_{\gamma_1}, \ldots, \kappa_{\gamma_n})(\xi)]\} \in U^\alpha$ (by 19.9(a))
iff $\langle {M_\alpha}^{\kappa_\alpha}/U_\alpha, E_{U_\alpha}\rangle \models \varphi[(i_{0\alpha}(\overline{f})(\kappa_{\gamma_1}, \ldots, \kappa_{\gamma_n}))^0_{U_\alpha}]$.

Using atomic formulas φ, it is simple to check that e is well-defined (i.e. $e(x)$ for $x \in L$ does not depend on the representation of x), that it is injective, and that it preserves membership. Also, e is surjective: every member of ${M_\alpha}^{\kappa_\alpha}/U_\alpha$ is of form $(i_{0\alpha}(\overline{f})(\kappa_{\gamma_1}, \ldots, \kappa_{\gamma_n}))^0_{U_\alpha}$ for some appropriate f and $\gamma_1 < \ldots < \gamma_n < \alpha$. Hence, e is an isomorphism as desired. ⊣

21.3 Lemma. *τ is not a limit ordinal.*

Proof. Assuming that $\delta \leq \tau$ is a limit ordinal we must show that $\delta < \tau$, i.e. that the direct limit of

$$\langle\langle\langle M_\alpha, \in, U_\alpha\rangle \mid \alpha < \delta\rangle, \langle i_{\alpha\beta} \mid \alpha \leq \beta < \delta\rangle\rangle$$

is well-founded. This is done by developing a different representation of the iteration based on Skolem hulls constructed uniformly in L. In order to formalize the procedure in ZFC generated Σ_1-elementary substructures are used, in keeping with the convention based on 5.1(c) that elementary embeddings between proper classes are formalized as Σ_1-elementary embeddings.

Recall the canonical Skolem terms t_φ for φ a formula of $\mathcal{L}_\in$ as defined after 9.3. Let T consist of the terms in the closure under term composition of the t_φ's for φ a Σ_1 formula. Since the satisfaction relation $\models^1_L$ for L restricted to Σ_1 formulas is definable in ZFC (§0), the collection of L interpretations $\{t^L \mid t \in T\}$, properly coded, is definable in ZFC. Consequently, we can uniformly define for any $X \subseteq L$ *the Σ_1 Skolem hull of X in L*, denoted $H^L_1(X)$, to be that $\in$-substructure of L with domain

$$\{t^L(x_1, \ldots, x_m) \mid t \in T \text{ is } m\text{-ary and } x_1, \ldots, x_m \in X\} .$$

$H^L_1(X)$ is extensional, since if $a, b \in H^L_1(X)$ and $\exists x(x \in a \mathbin{\triangle} b)$, then $\exists x \in H^L_1(X)(x \in a \triangle b)$. It follows (0.4) that $H^L_1(X)$ has a transitive collapse $N(X)$, and by definition of T the inverse of the collapsing isomorphism is a map $e_X\colon N(X) \prec_1 L$, which by 5.1(c) also satisfies $e_X\colon N(X) \prec_n L$ for each particular n. As $N(X)$ satisfies the sentence σ_0 of 3.3(a), it follows that $N(X) = L$ for proper classes X.

Proceeding with the proof, set

$$H = \{x \in L \mid \forall\alpha < \delta(i_{0\alpha}(x) = x)\} .$$

Then $\kappa \subseteq H$ and H is a proper class. Stipulate for $\alpha \leq \delta$ that

$H_\alpha = H^L_1(H \cup \{\kappa_\gamma \mid \gamma < \alpha\})$,
N_α is its transitive collapse , and

$$e_\alpha \colon N_\alpha \prec L \text{ the inverse of the collapsing isomorphism .}$$

Then it is simple to see that $H_0 = H$, and by previous remarks, that each $N_\alpha = L$. For $\alpha \leq \beta < \delta$, keeping in mind that $H_\alpha \subseteq H_\beta$, set

$$e_{\alpha\beta} = e_\beta^{-1} \circ e_\alpha \colon L \prec_1 L \ .$$

Then the system of embeddings $\langle e_{\alpha\beta} \mid \alpha \leq \beta < \delta \rangle$ forms a directed system. Moreover, for any limit ordinal γ with $0 < \gamma \leq \delta$, $H_\gamma = \bigcup_{\alpha<\gamma} H_\alpha$ and so $N_\gamma = L$ is the direct limit of $\langle e_{\alpha\beta} \mid \alpha \leq \beta < \gamma \rangle$. Since this holds for $\gamma = \delta$ itself, it suffices to show that $i_{\alpha\beta} = e_{\alpha\beta}$ for $\alpha \leq \beta < \delta$ to complete the proof.

We first establish the following for $\alpha \leq \beta < \delta$:

(i) $\kappa_\alpha \subseteq H_\alpha$.

(ii) For any $X \in \mathcal{P}(\kappa_\alpha) \cap L$, $i_{\alpha\beta}(X) = e_{\alpha\beta}(X) \cap \kappa_\beta$.

(iii) $\mathrm{crit}(e_{\alpha\beta}) = \kappa_\alpha$ and $e_{\alpha\beta}(\kappa_\alpha) = \kappa_\beta$.

For (i), let $\xi < \kappa_\alpha$ be arbitrary. Then by 19.6, $\xi = i_{0\alpha}(f)(\kappa_{\gamma_1}, \ldots, \kappa_{\gamma_n})$ for some $f \colon [\kappa]^n \to \kappa$ with $f \in L$ and $\gamma_1 < \ldots < \gamma_n < \alpha$. But $f = e_0(f) | [\kappa]^n$ since $\kappa \subseteq H$, and so $i_{0\alpha}(f) = e_0(f) | [\kappa]^n$ since $e_0(f) \in H$. Hence, $\xi = e_0(f)(\kappa_{\gamma_1}, \ldots, \kappa_{\gamma_n}) \in H_\alpha$.

For (ii), note first that $X = e_\alpha(X) \cap \kappa_\alpha$ since e_α fixes every member of κ_α by (i). Hence,

$$i_{\alpha\beta}(X) = i_{\alpha\beta}(e_\alpha(X) \cap \kappa_\alpha) = e_\alpha(X) \cap \kappa_\beta = e_{\alpha\beta}(X) \cap \kappa_\beta \ .$$

The second equality follows from the fact that $i_{\alpha\beta}$ fixes every member of $H \cup \{\kappa_\eta \mid \eta < \alpha\}$ and hence of H_α, and the third equality, that e_β fixes every member of κ_β by (i) for β.

(iii) follows from (i) and by taking $X = \kappa_\alpha$ in (ii).

We can now proceed to establish $i_{\alpha\beta} = e_{\alpha\beta}$ for $\alpha \leq \beta < \delta$ by induction on β: The limit case is immediate since direct limits are taken. If $\beta = \alpha + 1$, only $i_{\alpha,\alpha+1} = e_{\alpha,\alpha+1}$ need be verified. For this purpose, note that

$$M_{\alpha+1} = L = \{ i_{\alpha,\alpha+1}(f)(\kappa_\alpha) \mid f \in {}^{\kappa_\alpha}L \cap L \}$$

by the usual representation of ultrapowers. We shall show that the map $k \colon M_{\alpha+1} \to N_{\alpha+1}$ given by

$$k(i_{\alpha,\alpha+1}(f)(\kappa_\alpha)) = e_{\alpha,\alpha+1}(f)(\kappa_\alpha)$$

is an isomorphism between $M_{\alpha+1} = L$ and $N_{\alpha+1} = L$. This more than suffices, for then k must be the identity, and so for any $x \in L$,

$$i_{\alpha,\alpha+1}(x) = i_{\alpha,\alpha+1}(f_x)(\kappa_\alpha) = e_{\alpha,\alpha+1}(f_x)(\kappa_\alpha) = e_{\alpha,\alpha+1}(x)$$

where f_x is the constant function: $\kappa_\alpha \to \{x\}$.

An appeal to (ii) shows that k is a well-defined injection:

$$\begin{aligned} i_{\alpha,\alpha+1}(f)(\kappa_\alpha) = i_{\alpha,\alpha+1}(g)(\kappa_\alpha) \quad & \textit{iff } \kappa_\alpha \in i_{\alpha,\alpha+1}(\{\xi < \kappa_\alpha \mid f(\xi) = g(\xi)\}) \\ & \textit{iff } \kappa_\alpha \in e_{\alpha,\alpha+1}(\{\xi < \kappa_\alpha \mid f(\xi) = g(\xi)\}) \\ & \textit{iff } e_{\alpha,\alpha+1}(f)(\kappa_\alpha) = e_{\alpha,\alpha+1}(g)(\kappa_\alpha) \; . \end{aligned}$$

Clearly, $=$ can be replaced by $\in$ throughout.

Finally, it remains to show that k is surjective. Suppose that $y \in N_{\alpha+1}$. Then $e_{\alpha+1}(y) \in H_{\alpha+1}$, and so

$$e_{\alpha+1}(y) = t^L(x_1, \ldots, x_m, \kappa_{\gamma_1}, \ldots, \kappa_{\gamma_n}, \kappa_\alpha)$$

for some $t \in T$ (as defined at the beginning of the proof), $x_1, \ldots, x_m \in H$, and $\gamma_1 < \ldots < \gamma_n < \alpha$. Let $f \in {}^{\kappa_\alpha}L \cap L$ be defined by:

$$f(\xi) = t^L(x_1, \ldots, x_m, \kappa_{\gamma_1}, \ldots, \kappa_{\gamma_n}, \xi) \; .$$

Then

$$\begin{aligned} e_{\alpha,\alpha+1}(f)(\kappa_\alpha) &= e_{\alpha+1}^{-1}(e_\alpha(f)(\kappa_\alpha)) \\ &= e_{\alpha+1}^{-1}(t^L(x_1, \ldots, x_m, \kappa_{\gamma_1}, \ldots, \kappa_{\gamma_n}, \kappa_\alpha)) \\ &= y \; . \end{aligned}$$

We should justify the middle equality because of the formalization through Σ_1-elementary embeddings:

Suppose that the well-ordering formula $\varphi_0(v_0, v_1)$ of 3.3(a) is Σ_k where $k > 0$. (Actually, φ_0 can be required to be Σ_1; see Devlin [84: 75].) Then for any $t(v_1, \ldots, v_m) \in T$, $v_0 = t(v_1, \ldots, v_m)$ is definable by a Σ_{k+1} formula: Proceeding by induction on complexity of term composition, if t is a canonical Skolem term t_φ where φ is Σ_1, then the definition after 9.3 is a Boolean combination $(\Sigma_1 \wedge \Pi_k) \wedge \Pi_1$, and hence is Σ_{k+1}. Suppose now that $t \in T$ is of form $t_\varphi(t_1(v_1, \ldots, v_m), \ldots, t_r(v_1, \ldots, v_m))$ where φ is Σ_1 and inductively each $v_0 = t_i(v_1, \ldots, v_m)$ has a Σ_{k+1} definition $\varphi_i(v_0, v_1, \ldots, v_m)$. Then $v_0 = t(v_1, \ldots, v_m)$ *iff*

$$\begin{aligned} & \exists w_1 \ldots \exists w_r (\varphi_1(w_1, v_1, \ldots, v_m) \;\wedge \ldots \\ & \qquad\qquad \wedge\; \varphi_r(w_r, v_1, \ldots, v_m) \;\wedge\; v_0 = t_\varphi(w_1, \ldots, w_r)) \; . \end{aligned}$$

The last conjunct having a Σ_{k+1} definition by the previous argument, the whole formula can be recast as a Σ_{k+1} formula by "pushing" existential quantifiers to the left.

Finally, $e_{\alpha,\alpha+1}$ is Σ_{k+2}-elementary for our particular k by 5.1(c), so applying it to

$$\forall \xi < \kappa_\alpha (f(\xi) = t^L(x_1, \ldots, x_m, \kappa_{\gamma_1}, \ldots, \kappa_{\gamma_n}, \xi))$$

results in

$$\forall \xi < \kappa_{\alpha+1}(e_{\alpha,\alpha+1}(f)(\xi) = t^L(x_1, \ldots, x_m, \kappa_{\gamma_1}, \ldots, \kappa_{\gamma_n}, \xi))$$

since $e_{\alpha,\alpha+1}(\kappa_\alpha) = \kappa_{\alpha+1}$ by (iii) and $x_1, \ldots, x_m, \kappa_{\gamma_1}, \ldots, \kappa_{\gamma_n}$ are all fixed by $e_{\alpha,\alpha+1}$. In particular, we have the desired justification

$$e_{\alpha,\alpha+1}(f)(\kappa_\alpha) = t^L(x_1, \ldots, x_m, \kappa_{\gamma_1}, \ldots, \kappa_{\gamma_n}, \kappa_\alpha) .$$

This completes the proof of 21.3, and with it, the theorem. ⊣

Proofs of 21.1 have been devised by Silver (see Devlin [84: V§4]) and Paris (see Kanamori-Magidor [78: §10]) that generate indiscernibles directly without relying on iterated ultrapowers. Parsimony may be appropriate were the theorem an end in itself, but in volume II generalizations will be considered for which the richer context of iterated ultrapowers provides the natural setting.

As foreshadowed, we can now draw a stronger conclusion from model-theoretic hypotheses discussed in §8:

21.4 Corollary (Kunen). *If Chang's Conjecture holds or there is a Jónsson cardinal, then* $0^\#$ *exists.*

Proof. These hypotheses imply 21.1(b) by the proofs of 8.3 and 8.13 respectively. ⊣

There is another consequence, about the closed unbounded filter being an ultrafilter in a restricted sense:

21.5 Exercise. *The following are equivalent for regular* $\kappa > \omega$:

(a) $0^\#$ *exists.*

(b) Every $X \in \mathcal{P}(\kappa) \cap L$ *either contains or is disjoint from a closed unbounded subset of* κ.

Hint. For (a) → (b), suppose that $X = t^L(x_1, \ldots, x_m, y_1, \ldots, y_n)$ for a Skolem term t and $x_1 < \ldots < x_m < y_1 < \ldots < y_n$ all in I, the class of $0^\#$ indiscernibles, with $x_m < \kappa \leq y_1$. Show that X either contains or is disjoint from $I \cap (\kappa - (x_m + 1))$. For (b) → (a), produce a k: $L \prec L$. ⊣

Finally, as noted in the proof of 9.17(a) any order-preserving injection from the class of indiscernibles given by $0^\#$ into itself induces an elementary embedding: $L \prec L$ via Skolem terms. This is how *all* elementary embeddings: $L \prec L$ are generated:

21.6 Proposition. *Suppose that* $0^\#$ *exists,* j: $L \prec L$, *and* I *is the class of indiscernibles given by* $0^\#$. *Then* crit$(j) \in I$. *Also,* $j|I$: $I \to I$, *and hence* $j|I$ *induces* j *via Skolem terms.*

Proof. First observe that for any k: $L \prec L$, crit$(k) \in I$: Let $\delta = \text{crit}(k)$, and assume to the contrary that $\delta \notin I$. Then

$$\delta = t^L(x_1, \ldots, x_m, y_1, \ldots, y_n)$$

for some Skolem term t and $0^{\#}$ indiscernibles $x_1 < \ldots < x_m < y_1 \ldots < y_n$ with $x_m < \delta < y_1$. It can be assumed that there is a proper class of cardinals fixed by k. (If not, switch from k to j_U: $L \prec L \cong {}^{\delta}L/U$ where U is the L-ultrafilter over δ defined by: $X \in U$ *iff* $X \in \mathcal{P}(\delta) \cap L \wedge \delta \in k(X)$ as in the proof of 21.1, as j_U also has critical point δ.) But then, by the $0^{\#}$ condition (III) it can be assumed that $y_1, \ldots, y_n$ are all fixed by k. Hence, being definable in terms of fixed parameters we have $k(\delta) = \delta$, which is a contradiction.

It remains to show for the given j that if $\iota \in I$, then $j(\iota) \in I$. For this purpose, it suffices by the above observation to show that $j(\iota)$ is the critical point of *some* k: $L \prec L$. Let H be a Skolem hull of $j(\iota) \cup j``(I - (\iota + 1))$. (This is directly formalizable, since the satisfaction relation for L is definable in the presence of $0^{\#}$.) The transitive collapse of H is L, and the inverse of the collapsing isomorphism is an elementary embedding k: $L \prec L$. It will be shown that $\mathrm{crit}(k) = j(\iota)$; since $j(\iota) \subseteq H$, this reduces to checking that $j(\iota) \notin H$:

Assume to the contrary that there are $\eta_1 < \ldots < \eta_m < j(\iota)$ such that

$$j(\iota) = t^L(\eta_1, \ldots, \eta_m, j(y_1), \ldots, j(y_n))$$

for some Skolem term t and $\iota < y_1 < \ldots < y_n$ all in I. Then by the elementarity of j, there are $\xi_1 < \ldots < \xi_m < \iota$ such that

$$\iota = t^L(\xi_1, \ldots \xi_m, y_1, \ldots, y_n) \ .$$

However, by simple applications of the $0^{\#}$ condition (III) each ξ_i is definable in terms of members of $I - \{\iota\}$, and hence ι itself is definable in terms of *other* indiscernibles. This is a contradiction. $\dashv$

The Set $0^{\dagger}$

Having developed a detailed analysis of transcendence over L, we next consider a canonical formulation of transcendence over inner models of measurability. If $\langle L[U], \in, U\rangle$ is the κ-model for some ordinal κ, by the remarks at the end of §9 (slightly amended to allow $U \subsetneq \mathcal{P}(\kappa)$) there exists under sufficient assumptions a set $U^{\#} \subseteq \kappa$ analogous to $0^{\#}$ that generates a closed unbounded class of indiscernibles for $\langle L[U], \in, U, \xi\rangle_{\xi \leq \kappa}$. However, because the κ-models for various κ are merely iterates of each other, one might expect a *unifying* transcendence principle. In fact, soon after the isolation of $0^{\#}$ Solovay formulated such a principle: the existence of the set of integers $0^{\dagger}$ ("zero dagger"). The theory of $0^{\dagger}$ is developed in considerable detail as in Kanamori-Awerbuch-Friedlander [90], first following the analysis leading to $0^{\#}$ in §9 and then providing non-rigidity and other characterizations.

The idea behind $0^{\dagger}$ is to develop a canonical theory for structures of form

$$\langle L_{\zeta}[U], \in, U\rangle \models U \text{ is a normal ultrafilter over } \kappa$$

with *two* sets of indiscernibles, one below κ and one above, that together generate the structure. For $\mathcal{M}$ a structure and X and Y subsets of the domain of $\mathcal{M}$ so that

$X \cup Y$ is linearly ordered by a relation $<$, $\langle X, Y, <\rangle$ (or in context, just $\langle X, Y\rangle$) is a *double set of indiscernibles for* $\mathcal{M}$ *iff* for every formula $\varphi(v_1, \ldots, v_{n+s})$ in the language of $\mathcal{M}$; $x_1 < \ldots < x_n$ and $\overline{x}_1 < \ldots < \overline{x}_n$ all in X; and $y_1 < \ldots < y_s$ and $\overline{y}_1 < \ldots < \overline{y}_s$ all in Y,

$$\mathcal{M} \models \varphi[x_1, \ldots, x_n, y_1, \ldots, y_s] \text{ iff } \mathcal{M} \models \varphi[\overline{x}_1, \ldots, \overline{x}_n, \overline{y}_1, \ldots, \overline{y}_s] .$$

Let $\mathcal{L}_\in(\dot{A})^{**}$ be $\mathcal{L}_\in(\dot{A})$ augmented by constants $\{c_k \mid k \in \omega\} \cup \{d_k \mid k \in \omega\}$. An EM^2 *blueprint* is the theory in $\mathcal{L}_\in(\dot{A})^{**}$ of some structure

$$\langle L_\zeta[U], \in, U, x_k, y_k\rangle_{k\in\omega}$$

where ζ is a limit ordinal greater than ω; for some ordinal κ, $\langle L_\zeta[U], \in, U\rangle \models U$ is a normal ultrafilter over κ; and $\langle\{x_k \mid k \in \omega\}, \{y_k \mid k \in \omega\}\rangle$ is a double set of ordinal indiscernibles for $\langle L_\zeta[U], \in, U\rangle$ such that for every $k \in \omega$,

$$x_k < x_{k+1} < \kappa < y_k < y_{k+1} .$$

The canonical Skolem terms t_φ for φ a formula of $\mathcal{L}_\in(\dot{A})$ and corresponding Skolem hulls were formulated before 20.9.

By *Skolem term* is meant one of these t_φ's for the rest of this section .

We can suppose that if a structure $\langle M, E, R\rangle$ is elementarily equivalent to one of form $\langle L_\delta[A], \in, A\rangle$ for some limit ordinal $\delta > \omega$, then for any $X \subseteq M$ *the* Skolem hull of X in $\langle M, E, R\rangle$ is well-defined and given by these Skolem terms.

The following is the analogue of 9.4. For a theory T in $\mathcal{L}_\in(\dot{A})^{**}$, temporarily let T^- denote its restriction to $\mathcal{L}_\in(\dot{A})$. Note for here and later that if $\dot{A}$ is interpreted by a normal ultrafilter in a structure, then $\bigcup \dot{A}$ is a way of denoting the corresponding measurable cardinal in the structure.

21.7 Exercise. *Suppose that T is an* EM^2 *blueprint. Then for any α and γ, there is a model $\mathcal{M} = \mathcal{M}(T, \alpha, \gamma)$ of T^- unique up to isomorphism such that:*

(a) There is a double set $\langle X, Y\rangle$ of indiscernibles for $\mathcal{M}$ with $X \cup Y \subseteq \mathrm{On}^{\mathcal{M}}$, X of ordertype α and Y of ordertype γ under $<^{\mathcal{M}}$, and $x <^{\mathcal{M}} \bigcup \dot{A}^{\mathcal{M}} <^{\mathcal{M}} y$ for every $x \in X$ and $y \in Y$. Moreover, for any formula $\varphi(v_1, \ldots, v_{n+s})$ of $\mathcal{L}_\in(\dot{A})$, $x_1 <^{\mathcal{M}} \ldots <^{\mathcal{M}} x_n$ all in X, and $y_1 <^{\mathcal{M}} \ldots <^{\mathcal{M}} y_s$ all in Y,

$$\mathcal{M} \models \varphi[x_1, \ldots, x_n, y_1, \ldots, y_s] \text{ iff } \varphi(c_0, \ldots, c_{n-1}, d_0, \ldots, d_{s-1}) \in T .$$

(b) The Skolem hull of $X \cup Y$ in $\mathcal{M}$ is again $\mathcal{M}$. ⊣

If $\mathcal{M}(T, \alpha, \gamma)$ is well-founded, then its transitive collapse is of form $\langle L_\delta[U], \in, U\rangle$ for some limit ordinal $\delta > \omega$. In this case,

$$\mathcal{M}(T, \alpha, \gamma) \text{ is identified with } \langle L_\delta[U], \in, U\rangle .$$

The following is the analogue of 9.5.

21.8 Exercise. *Suppose that T is an* EM^2 *blueprint. Then $\mathcal{M}(T,\alpha,\gamma)$ is well-founded for every α,γ iff*

(I) $\qquad \mathcal{M}(T,\alpha,\gamma)$ is well-founded for every $\alpha,\gamma<\omega_1$. $\dashv$

A sufficient hypothesis leads quickly to EM^2 blueprints with potent properties; the following is an analogue of 9.6.

21.9 Lemma. *Suppose that there is a κ-model for some κ and a Ramsey cardinal greater than κ. Then there is an* EM^2 *blueprint satisfying* (I) *of 21.8.*

Proof. Let $\langle L[U],\in,U\rangle$ be the κ-model, $\langle L[U_\alpha],U_\alpha,\kappa_\alpha,i_{\alpha\beta}\rangle_{\alpha\leq\beta\in\mathrm{On}}$ its iteration, and ν a Ramsey cardinal greater than κ. By the argument for 9.3, there is a set of indiscernibles $H\in[\nu-\kappa]^{\omega_1}$ for the structure $\langle L_\nu[U],\in,U\rangle$. Set $X=\{\kappa_\alpha\mid\alpha<\omega_1\}$ and $Y=i_{0\omega_1}$"H. Then straightforward uses of 19.9(a) with $i_{0\omega_1}(\nu)=\nu$ as a parameter show that $\langle X,Y\rangle$ is a double set of indiscernibles for $\langle L_\nu[U_{\omega_1}],\in,U_{\omega_1}\rangle$. Hence, $\langle X,Y\rangle$ determines an EM^2 blueprint, and since X and Y are uncountable, this EM^2 blueprint satisfies (I) by the argument for 9.6. $\dashv$

By 20.22, the hypothesis of 21.9 already implies that V is not the κ-model $L[U]$. Assuming that hypothesis, we can deduce the existence of an EM^2 blueprint fully analogous to $0^{\#}$. Evidently, a weaker partition property than Ramsey suffices for the proof of 21.9; in any case, on the basis of that proof, stipulate that

(i) $\rho<\zeta$ are limit ordinals and $\langle L_\zeta[U],\in,U\rangle\models U$ is a normal ultrafilter over ρ;
(ii) $\langle X,Y\rangle$ is a double set of indiscernibles with X and Y both of ordertype ω_1 and $\sup(X)=\rho$;
(iii) ζ is the least possible satisfying (i) and (ii);
(iv) X and Y have the least possible ωth elements; and
(v) T_0 is the corresponding EM^2 blueprint.

21.10 Exercise. *The following conditions hold for $T=T_0$:*

(IIa) *For any $(n+s)$-ary Skolem term t, T contains the sentence:*

$$t(c_0,\ldots,c_{n-1},d_0,\ldots,d_{s-1})\in\bigcup\dot{A}\rightarrow t(c_0,\ldots,c_{n-1},d_0,\ldots,d_{s-1})<c_n\ .$$

(IIb) *For any $(n+s)$-ary Skolem term t, T contains the sentence:*

$$t(c_0,\ldots,c_{n-1},d_0,\ldots,d_{s-1})\in\mathrm{On}\rightarrow t(c_0,\ldots,c_{n-1},d_0,\ldots,d_{s-1})<d_s\ .$$

(IIIa) *For any $(m+n+s+1)$-ary Skolem term t, T contains the sentence:*

$$t(c_0,\ldots,c_{m+n},d_0,\ldots,d_{s-1})<c_m\rightarrow t(c_0,\ldots,c_{m+n},d_0,\ldots,d_{s-1})= t(c_0,\ldots,c_{m-1},c_{m+n+1},\ldots,c_{m+2n+1},d_0,\ldots,d_{s-1})\ .$$

(IIIb) *For any $(n+r+s+1)$-ary Skolem term t, T contains the sentence:*

$$t(c_0,\ldots,c_{n-1},d_0,\ldots,d_{r+s})<d_r\rightarrow t(c_0,\ldots,c_{n-1},d_0,\ldots,d_{r+s})= t(c_0,\ldots,c_{n-1},d_0,\ldots,d_{r-1},d_{r+s+1},\ldots,d_{r+2s+1})\ .$$

Hint. A simple argument by contradiction establishes (IIa) from $\sup(X) = \rho$ and the argument for 9.8 establishes (IIb) by the minimality of ζ. The argument for 9.10 establishes (IIIa) and (IIIb) from the minimality of the ωth elements of X and Y respectively. ⊣

For an EM^2 blueprint satisfying (I) and for any α, γ, temporarily let

$$\langle\langle \chi_\eta^{T,\alpha,\gamma} \mid \eta < \alpha\rangle, \langle \iota_\xi^{T,\alpha,\gamma} \mid \xi < \gamma\rangle\rangle \text{ and } \chi_\alpha^{T,\alpha,\gamma}$$

denote the double set of indiscernibles and the measurable cardinal of $\mathcal{M}(T, \alpha, \gamma)$ respectively. The χ notation anticipates the following:

21.11 Lemma. *Suppose that T is an* EM^2 *blueprint satisfying* (I)-(III). *Then for any β, δ:*

(a) $\{\chi_\eta^{T,\beta,\delta} \mid \eta < \beta\}$ is a closed set of ordinals, unbounded in $\chi_\beta^{T,\beta,\delta}$ if β is an infinite limit ordinal.

(b) If $\omega \le \alpha \le \beta$ and $\omega \le \gamma \le \delta$ with α and γ limit ordinals, then for every $\eta \le \alpha$,

$$\chi_\eta^{T,\alpha,\gamma} = \chi_\eta^{T,\beta,\delta} .$$

Proof. (a) (IIa) implies that the set is unbounded in $\chi_\beta^{T,\beta,\delta}$ if β is a limit ordinal. (IIIa) implies that the set is closed (by the argument for 9.11, also used below).

(b) Let $\mathcal{H}$ be the Skolem hull of

$$\{\chi_\eta^{T,\beta,\delta} \mid \eta < \alpha\} \cup \{\iota_\xi^{T,\beta,\delta} \mid \xi < \gamma\}$$

in $\mathcal{M}(T, \beta, \delta)$. Its transitive collapse is $\mathcal{M}(T, \alpha, \gamma)$ by uniqueness. We shall show that $\chi_\alpha^{T,\beta,\delta}$ is a subset of the domain of $\mathcal{H}$. This suffices, since the collapsing isomorphism consequently fixes every member of $\{\chi_\eta^{T,\beta,\delta} \mid \eta < \alpha\}$ making this set *the* lower set of the double set of indiscernibles for $\mathcal{M}(T, \alpha, \gamma)$ (and thus also $\chi_\alpha^{T,\alpha,\gamma} = \chi_\alpha^{T,\beta,\delta}$ by (a) for α as well as for β).

To show that $\chi_\alpha^{T,\beta,\delta}$ is a subset of the domain of $\mathcal{H}$, let $\sigma < \chi_\alpha^{T,\beta,\delta}$ be arbitrary. Suppressing the superscript ${}^{T,\beta,\delta}$ from the indiscernibles for convenience,

$$\sigma = t^{\mathcal{M}(T,\beta,\delta)}(\chi_{\eta_0}, \ldots, \chi_{\eta_{m-1}}, \chi_{\zeta_0}, \ldots, \chi_{\zeta_n}, \iota_{\xi_0}, \ldots, \iota_{\xi_{s-1}})$$

for some Skolem term t and the indiscernibles listed in increasing order with $\eta_{m-1} < \alpha \le \zeta_0$. Applying (IIIb) with $r = 0$, ξ_i can be replaced by i for $i < s$, and then applying (IIIa), ζ_i can be replaced by $\eta_{m-1} + i$ for $i < n + 1$. Since α and γ are limit ordinals, the resulting expression shows that σ is in the domain of $\mathcal{H}$. ⊣

21.12 Lemma. *Suppose that T is an* EM^2 *blueprint satisfying* (I)-(III), $\omega \le \gamma \le \delta$, *and γ and α are limit ordinals (allowing $\alpha = 0$). Then with $\mathcal{M}(T, \alpha, \delta) = \langle L_\zeta[D], \in, D\rangle$ say, the transitive collapse of the Skolem hull of*

$$\{\chi_\eta^{T,\alpha,\delta} \mid \eta < \alpha\} \cup \{\iota_\xi^{T,\alpha,\delta} \mid \xi < \gamma\}$$

in $\mathcal{M}(T,\alpha,\delta)$ *is* $\langle L_\iota[D], \in, D\cap L_\iota[D]\rangle$, *where* $\iota = \iota_\gamma^{T,\alpha,\delta}$. *Consequently,*

$$\mathcal{M}(T,\alpha,\gamma) = \langle L_\iota[D], \in, D\cap L_\iota[D]\rangle \text{ and } \iota_\xi^{T,\alpha,\gamma} = \iota_\xi^{T,\alpha,\delta} \text{ for every } \xi \le \gamma .$$

Proof. Let $\mathcal{H}$ be the stated Skolem hull. Then its transitive collapse is $\mathcal{M}(T,\alpha,\gamma)$ by uniqueness. By the argument for 21.11(b) with $\alpha = \beta$ (and a simple version for $\alpha = 0$), $\chi_\alpha^{T,\alpha,\delta}$ is a subset of the domain of $\mathcal{H}$, so that $\mathcal{M}(T,\alpha,\gamma)$ must be of form $\langle L_\iota[D], \in, D\cap L_\iota[D]\rangle$ for some ι. Checking that $\iota = \iota_\gamma^{T,\alpha,\delta}$ by the argument for 9.11, the proof is complete. ⊣

By 21.11 and 21.12, if T is an EM^2 blueprint satisfying (I)-(III), then for any η, ξ, and α with α a limit ordinal (possibly 0), we can unambiguously set

$$\begin{aligned} \chi_\eta^T &= \chi_\eta^{T,\beta,\gamma} \text{ for any limit ordinals } \beta \text{ and } \gamma \text{ with } \beta > \eta\,,\\ \iota_\xi^{T,\alpha} &= \iota_\xi^{T,\alpha,\gamma} \text{ for any limit ordinal } \gamma > \xi\ , \text{ and}\\ Y^{T,\alpha} &= \{\iota_\xi^{T,\alpha} \mid \xi \in \mathrm{On}\}\,. \end{aligned}$$

Finally, stipulate that if α is a limit ordinal,

$$D_\alpha^T \text{ is the normal ultrafilter over } \chi_\alpha^T \text{ in the sense of } \mathcal{M}(T,\alpha,\omega)\,.$$

The following is the analogue of 9.12.

21.13 Exercise. *Suppose that* T *is an* EM^2 *blueprint satisfying* (I)-(III) *and* α *is a limit ordinal. Then:*

(a) $\langle L[D_\alpha^T], \in, D_\alpha^T\rangle$ *is the* χ_α^T*-model, and for* $\xi \le \zeta$,

$$\langle L_{\iota_\xi^{T,\alpha}}[D_\alpha^T], \in, D_\alpha^T\rangle \prec \langle L_{\iota_\zeta^{T,\alpha}}[D_\alpha^T], \in, D_\alpha^T\rangle\,.$$

(b) $|\chi_\eta^T| = |\eta| + \aleph_0$ *for every* η, *and* $|\iota_\xi^{T,\alpha}| = |\xi| + |\alpha|$ *for every* ξ.

(c) $\{\chi_\eta^T \mid \eta \in \mathrm{On}\}$ *and* $Y^{T,\alpha}$ *are closed unbounded classes of ordinals.*

(d) For any cardinal $\lambda > \omega$, $\chi_\lambda^T = \lambda$ *and if* $\lambda > \alpha$, $\iota_\lambda^{T,\alpha} = \lambda$ *and so* $\mathcal{M}(T,\alpha,\lambda) = \langle L_\lambda[D_\alpha^T], \in, D_\alpha^T\rangle$.

(e) T *is the only* EM^2 *blueprint satisfying* (I)–(III).

Hint. For (a), note that 21.12 implies that there is a D such that

$$\mathcal{M}(T,\alpha,\gamma) = \langle L_{\iota_\gamma^{T,\alpha}}[D], \in, D\cap L_{\iota_\gamma^{T,\alpha}}[D]\rangle$$

for any limit ordinal $\gamma \ge \omega$. Hence, $\langle L[D], \in, D\rangle$ is the χ_α^T-model. By indiscernibility, $(\chi_\alpha^T)^{+L[D]} < \iota_\xi^{T,\alpha}$ for any ξ. In particular, $D \subseteq L_{\iota_\omega^{T,\alpha}}[D]$ by 20.2(a), and so $D = D_\alpha^T$. The rest of (a) and the exercise is just like 9.12. ⊣

We shall soon derive more information about the χ_η^T's and $\iota_\xi^{T,\alpha}$'s, incorporating successor α's into the scheme using iterated ultrapowers. As with $0^{\#}$, the hypothesis of 21.13 implies through its (a) and (d) that for any infinite limit ordinal α,

$$\text{the satisfaction relation for } \langle L[D_\alpha^T], \in, D_\alpha^T\rangle \text{ is definable in ZFC}\,.$$

Because of this, various upcoming assertions like the following about inner models are directly formalizable.

21.14 Lemma. *For any limit ordinal α, $\langle\{\chi_\eta^T \mid \eta < \alpha\}, Y^{T,\alpha}\rangle$ is a double class of indiscernibles for the χ_α^T-model and the Skolem hull of $\{\chi_\eta^T \mid \eta < \alpha\} \cup Y^{T,\alpha}$ in the model is again the model.* ⊣

With 21.13(e) in hand, stipulate that

$$0^\dagger \text{ is the unique EM}^2 \text{ blueprint satisfying (I)-(III)}$$

if there is one, and use the accepted solecism

$$0^\dagger \text{ exists}$$

for ⌜there is an EM^2 blueprint satisfying (I)-(III)⌝. Through a recursive arithmetization of $\mathcal{L}_\in(\dot{A})^{**}$, $0^\dagger$ is regarded as a subset of ω. The following summarizing theorem gives the main features (with one direction of its (a) given by the proof of 21.9):

21.15 Theorem (Solovay).

(a) $0^\dagger$ exists iff there is a κ-model for some ordinal κ that has an uncountable set of indiscernibles whose minimum element is greater than κ. Hence, if there is a κ-model for some κ and a Ramsey cardinal greater than κ (e.g. if there are two measurable cardinals), then $0^\dagger$ exists.

(b) $0^\dagger$ exists iff for every uncountable cardinal λ there is a λ-model and a double class $\langle X, Y\rangle$ of indiscernibles for it such that: $X \subseteq \lambda$ is closed unbounded, $Y \subseteq \mathrm{On} - (\lambda + 1)$ is a closed unbounded class, $X \cup \{\lambda\} \cup Y$ contains every uncountable cardinal and the Skolem hull of $X \cup Y$ in the λ-model is again the model. ⊣

In this last situation, since there are regular cardinals and there are strong limit cardinals, every member of Y is inaccessible in the λ-model by indiscernibility. Hence, as Solovay noted early on, the existence of $0^\dagger$ implies the existence of set, transitive $\in$-models of measurability. To be sure, much more can be said about the size of the indiscernibles, along the lines of 9.17.

Solovay also established the direct analogue of 14.11 on the definability of $0^\dagger$:

21.16 Theorem (ZF)(Solovay). *The relation $R \subseteq {}^\omega\omega$ defined by*

$$R(x) \leftrightarrow 0^\dagger \text{ exists} \wedge x \in {}^\omega 2 \wedge \{m \mid x(m) = 1\} = 0^\dagger$$

is Π_2^1. ⊣

21.17 Corollary.

(a) $0^\dagger$ *is absolute for transitive* $\in$*-models of* ZF *such that* $\omega_1 \subseteq M$ *as follows:*

$$M \models \ulcorner\text{There is an EM}^2 \text{ blueprint satisfying (I)-(III)}\urcorner \;\; \textit{iff} \;\; 0^\dagger \in M ,$$

in which case $M \models \ulcorner 0^\dagger$ is the unique EM^2 blueprint satisfying (I)-(III)$\urcorner$.

(b) $0^\dagger$ *is a* Δ^1_3 *subset of* ω *which is not a member of the* κ*-model for any* κ.

Proof. For (b), if to the contrary $0^\dagger$ were a member of the κ-model for some κ, then by iterating ultrapowers it can be assumed that κ is a cardinal. But this is a contradiction by (a) and 21.15(b). ⊣

Having developed the analogue of the $0^\#$ theory in §9, we now proceed to derive more information with iterated ultrapowers that sharpens the focus. But first, a respite from the rigors: Instead of yet another recipe, we offer the following chess problem (M. Henneberger, first and second prize, "Revista de Sah" 1928):

White. King on b1, Rooks on b7 and c7, and Bishop on b5.
Black. King on a8, Rook on a3, and Pawn on f2.
White to play and win.

Send complete solutions to the author for a small prize.

For the rest of this section, stipulate that

$$\langle L[U_\alpha], U_\alpha, \kappa_\alpha, i_{\alpha\beta}\rangle_{\alpha\leq\beta\in\text{On}} \text{ is the iteration of the least } \kappa\text{-model} ,$$

i.e. with κ the least possible. All the κ-models for various κ appear in this iteration (20.12). The results quickly follow from the basic uniqueness property of 21.7 tailored for $0^\dagger$ and κ-models and stated here for emphasis.

21.18 Lemma. *For any* α*, there is at most one double class* $\langle X, Y\rangle$ *of indiscernibles for the* κ_α*-model with* $X \subseteq \kappa_\alpha$ *of ordertype* α *and* $Y \subseteq \text{On} - (\kappa_\alpha + 1)$ *such that:*

(a) For any formula $\varphi(v_1, \ldots, v_{n+s})$ *of* $\mathcal{L}_\in(\dot{A})$, $x_1 < \ldots < x_n$ *all in* X*, and* $y_1 < \ldots < y_s$ *all in* Y*,*

$$\text{the } \kappa_\alpha\text{-model satisfies } \varphi[x_1, \ldots, x_n, y_1, \ldots, y_s] \;\; \textit{iff} \;\; \varphi(c_0, \ldots, c_{n-1}, d_0, \ldots, d_{s-1}) \in 0^\dagger .$$

(b) The Skolem hull of $X \cup Y$ *in the* κ_α*-model is again the model.*

Proof. If $\langle X, Y\rangle$ and $\langle \overline{X}, \overline{Y}\rangle$ were two such classes, then the order-preserving injection: $X \cup Y \to \overline{X} \cup \overline{Y}$ extends to an isomorphism of the κ_α-model into itself and hence must be the identity. ⊣

Dropping the superscript ^{T} by uniqueness, in the presence of $0^\dagger$ there are corresponding classes

$$\{\chi_\eta \mid \eta \in \mathrm{On}\} \text{ and } Y^\alpha = \{\iota_\xi^\alpha \mid \xi \in \mathrm{On}\}$$

for every limit α as defined before 21.13 and with the properties ascribed by it. By 21.14, for every limit ordinal α, $\langle\{\chi_\eta \mid \eta < \alpha\}, Y^\alpha\rangle$ satisfies 21.18(a)(b) for the χ_α-model. In particular, the $\langle X, Y\rangle$ of 21.18 need not determine an EM^2 blueprint: X could be finite, with $\langle\emptyset, Y^0\rangle$ for the χ_0-model being an example. In fact, Y^0 is the basis of the next lemma, one that is not surprising in view of the uniqueness properties of $0^\dagger$.

21.19 Lemma.

(a) For every α, $\langle X, Y\rangle = \langle\{\kappa_\eta \mid \eta < \alpha\}, i_{0\alpha}\text{“}Y^0\rangle$ satisfies 21.18(a)(b) for the κ_α-model.

(b) $\chi_\eta = \kappa_\eta$ for every η, and $i_{0\alpha}\text{“}Y^0 = Y^\alpha$ for every limit ordinal α.

Proof. We first argue that $\chi_0 = \kappa_0$. Clearly, $\kappa_0 \le \chi_0$ by definition of κ_0. For the converse, we work in some generality for a later inference: Let λ be a regular cardinal such that $\kappa_\lambda = \lambda = \chi_\lambda$; there are arbitrarily large such λ by 19.7(b) and 21.13(d). Note for what follows that by 21.14, $\langle\{\chi_\eta \mid \eta < \lambda\}, Y^\lambda\rangle$ satisfies 21.18(a)(b) for the λ-model. By 19.10, if E is the proper class of cardinals greater than λ fixed by $i_{0\lambda}$, then $\langle\{\kappa_\eta \mid \eta < \lambda\}, E\rangle$ is a double class of indiscernibles for the λ-model. Note that this pair satisfies 21.18(a) for this model: There are infinitely many χ_η's in the set $\{\kappa_\eta \mid \eta < \lambda\}$, and $E \subseteq Y^\lambda$ as Y^λ contains every cardinal greater than λ. Let $\mathcal{N}$ be the transitive collapse of the Skolem hull of $\{\kappa_\eta \mid \eta < \lambda\} \cup E$ in the λ-model, and π the collapsing isomorphism. Then $\mathcal{N}$ is again the λ-model since $\pi(\lambda) = \lambda$. Hence, $\pi(\kappa_\eta) = \chi_\eta$ for every $\eta < \lambda$ by 21.18. Consequently, $\chi_\eta \le \kappa_\eta$ for every $\eta < \lambda$ since π is collapsing.

Having deduced that $\kappa_0 = \chi_0$, argue as follows for any α: Remembering that $\langle L[U_\alpha], \in, U_\alpha\rangle$ is the κ_α-model, by 19.6 any $x \in L[U_\alpha]$ is of form $x = i_{0\alpha}(f)(\kappa_{\gamma_1}, \ldots, \kappa_{\gamma_n})$ for some function $f \in L[U_0]$ and $\gamma_1 < \ldots < \gamma_n < \alpha$. Since $\kappa_0 = \chi_0$, f is definable in the κ_0-model from elements in Y^0, and so $i_{0\alpha}(f)$ is definable in the κ_α-model from elements in $i_{0\alpha}\text{“}Y^0$. Hence, $\langle\{\kappa_\eta \mid \eta < \alpha\}, i_{0\alpha}\text{“}Y^0\rangle$ satisfies 21.18(b) for the κ_α-model.

Next, take any regular cardinal $\lambda \ge \alpha$ such that $\kappa_\lambda = \lambda = \chi_\lambda$. It follows from a previous remark that $\langle\{\kappa_\eta \mid \eta < \lambda\}, i_{0\lambda}\text{“}Y^0\rangle$ satisfies 21.18(a) for the λ-model. Hence, the elementarity of $i_{\alpha\lambda}$ shows that $\langle\{\kappa_\eta \mid \eta < \alpha\}, i_{0\alpha}\text{“}Y^0\rangle$ satisfies 21.18(a) for the κ_α-model.

(b) of the lemma is now a direct consequence of 21.14 and 21.18. ⊣

We can now define Y^α for *successor* ordinals α by:

$$Y^\alpha = i_{0\alpha}\text{“}Y^0 .$$

These Y^α's are also closed unbounded classes by the usual argument from (IIb) and (IIIb) of $0^\dagger$. Save for a small initial shift, the Y^α's coincide:

21.20 Exercise. *For any $\alpha \leq \beta$:*

(a) $Y^{\beta} = i_{\alpha\beta}$“$(Y^{\alpha} \cap (\beta+1)) \cup (Y^{\alpha} - (\beta+1))$, a disjoint union.

(b) If $\beta < \min(Y^{\alpha})$, then $Y^{\beta} = Y^{\alpha}$.

Hint. Any $\iota \in Y^{\alpha}$ is inaccessible in the κ_{α}-model. Since the iteration of inner models of measurability beyond the κ_{α}-model can be defined in the model using U_{α}, 19.7(c) applied there implies that $i_{\alpha\beta}(\iota) = \iota$ for any $\beta < \iota$. ⊣

The following overall theorem describes the global coherence of κ_{α}-models and their indiscernibles.

21.21 Theorem. *Assume that $0^{\dagger}$ exists. Then for every α there is a class Y^{α} of ordinals characterized by:*

(a) Y^{α} is closed unbounded and $\langle\{\kappa_{\eta} \mid \eta < \alpha\}, Y^{\alpha}\rangle$ is a double class of indiscernibles for the κ_{α}-model.

(b) The Skolem hull of $\{\kappa_{\eta} \mid \eta < \alpha\} \cup Y^{\alpha}$ in the κ_{α}-model is again the model. Moreover, for any $\alpha \leq \beta$, $i_{\alpha\beta}$“$Y^{\alpha} = Y^{\beta}$, and if π^{β}_{α} is a collapsing isomorphism from the Skolem hull of $\{\kappa_{\eta} \mid \eta < \alpha\} \cup Y^{\beta}$ in the κ_{β}-model into its transitive collapse, then: $\pi^{\beta}_{\alpha}(\kappa_{\eta}) = \kappa_{\eta}$ for $\eta < \alpha$, $\pi^{\beta}_{\alpha}(\kappa_{\beta}) = \kappa_{\alpha}$, and π^{β}_{α}“$Y^{\beta} = Y^{\alpha}$.

Proof. The characterization of Y^{α} follows from 21.18 and the fact that two closed unbounded classes have many common members.

For the assertion about π^{β}_{α}, the transitive collapse must be the κ-model for some κ with $\langle\{\pi^{\beta}_{\alpha}(\kappa_{\eta}) \mid \eta < \alpha\}, \pi^{\beta}_{\alpha}$“$Y^{\beta}\rangle$ satisfying 21.18(a)(b) for that κ-model. Because of the ordertype α of the lower set of this pair, the only possibility by uniqueness is $\kappa = \kappa_{\alpha}$, and the conclusions follow, also by uniqueness. ⊣

Thus, the existence of $0^{\dagger}$ leads to remarkable conclusions about the simple generation of inner models of measurability and their relation to each other. Taking into account the absoluteness result 21.17(a), for every α the κ_{α}-model is a subclass of $L[0^{\dagger}]$, and is moreover definable in $L[0^{\dagger}]$ as $\bigcup_{\gamma} \mathcal{M}(0^{\dagger}, \alpha, \gamma)$, i.e. by taking the lower set of indiscernibles of ordertype α and "stretching" the upper set. The resulting systems of indiscernibles are closely interrelated by iterated ultrapowers and Skolem hulls as described by 21.21.

As expected, there is a non-rigidity characterization:

21.22 Exercise. *The following are equivalent:*

(a) $0^{\dagger}$ exists.

(b) There is a κ-model for some κ and an elementary embedding of that model into itself with critical point greater than κ.

Hint. In the forward direction, for any α let Y^{α} be the closed unbounded class for the κ_{α}-model as given by $0^{\dagger}$ and h any order-preserving injection of $\{\kappa_{\eta} \mid \eta < \alpha\} \cup Y^{\alpha}$ into itself such that $h(\kappa_{\eta}) = \kappa_{\eta}$ for every $\eta < \alpha$ and $h(\iota) > \iota$ for some $\iota \in Y^{\alpha}$. Then h induces an elementary embedding of the κ_{α}-model into itself as desired.

The converse can be established by following the arguments for 21.1: Suppose that for some α, k is an elementary embedding of the κ_α-model $\langle L[U_\alpha], \in, U_\alpha\rangle$ into itself with $\mathrm{crit}(k) = \rho > \kappa_\alpha$. k determines an $L[U_\alpha]$-ultrafilter

$$W = \{X \in \mathcal{P}(\rho) \cap L[U_\alpha] \mid \rho \in k(X)\}$$

over ρ. It now suffices to show:

(∗) W is an iterable $L[U_\alpha]$-ultrafilter .

The argument could then be completed by iterating with W to get uncountably many indiscernibles for $L[U_\alpha]$ by 19.10, and then proceeding to get $0^\dagger$ by the argument for 21.9 and the ensuing analysis.

To establish (∗), note that the ultrapower of $L[U_\alpha]$ by W is well-founded and check that the arguments for 21.2 and 21.3 can be generalized from L using the following basic observation deriving from 3.3(b): If M and N are transitive, $j\colon M \prec N$, and $j(\xi) = \xi$ for $\xi \le \kappa_\alpha$, then $M = L[U_\alpha]$ *iff* $N = L[U_\alpha]$. The forward direction is for ultrapower arguments, and the latter, for Skolem hull arguments. ⊣

There are further equivalences as for 21.1; one of them figures in the following analogue of 21.4, which has a simple argument if we avail ourselves of the result that if μ is regular, then μ^+ is not Jónsson (8.17).

21.23 Exercise. *Suppose that there is a κ-model for some κ, and a Jónsson cardinal $\lambda > \kappa$. Suppose also that if $|\kappa|$ is singular, then $\lambda > \kappa^+$. Then $0^\dagger$ exists.*

Hint. Building on the outline for 20.22, note that the current hypotheses together with the result that successors of regular cardinals are not Jónsson are enough to show *without assuming* $V = L[U]$ that there is a regular ν, $\kappa < \nu < \lambda$, such that $\lambda \longrightarrow [\lambda]^{<\omega}_{\nu,<\nu}$. Hence, the embedding i can be defined as before, and then a corresponding $L[U]$-ultrafilter W over $\mathrm{crit}(i) > \kappa$ such that the ultrapower of $L[U]$ by W is well-founded by the argument for 21.1(b) → (c). ⊣

With more sophisticated techniques, the conditions on singular $|\kappa|$ can be eliminated (Donder-Koepke [83: 248]). The Jónsson cardinal can also be replaced by a suitable analogue of Chang's Conjecture for uncountable languages to extend the analogy to 21.4. Also, 21.5 has a direct analogue in the present context. Finally, the argument for 21.6 can be adapted to show that *every* elementary embedding as for 21.22 is induced by an h as described in its proof.

The existence of $0^\dagger$ does not imply the existence of measurable cardinals, only inner models of measurability. The following is a slight reformulation of an observation of Kunen in the presence of a measurable cardinal.

21.24 Proposition (Kunen [70]). *Suppose that κ is a measurable cardinal and $\langle L[U], \in, U\rangle$ the κ-model. Then the following are equivalent:*

(a) $0^\dagger$ *exists.*
(b) $\kappa^{++L[U]} \le 2^\kappa$.

Proof. The forward direction is clear; every (real) cardinal greater than κ is in the class Y^κ of indiscernibles for the κ-model given by $0^\dagger$, and hence large in the model by simple indiscernibility arguments.

For the converse, first note that in the continuing terminology, $\kappa = \kappa_\alpha$ and $U = U_\alpha$ for some α. Since the iteration of $\langle L[U], \in, U\rangle$ can be defined in $L[U]$ using U, 19.7(a) applied there implies that $\kappa_{\alpha+\omega} < \kappa^{++L[U]}$. Also, if F is the filter over $\kappa_{\alpha+\omega}$ generated by $\{\kappa_{\alpha+n} \mid n \in \omega\}$, i.e.

$$X \in F \quad \textit{iff} \quad \exists m \in \omega\{\kappa_{\alpha+n} \mid m \le n \in \omega\} \subseteq X\ ,$$

then by 19.5 $L[F]$ is the $\kappa_{\alpha+\omega}$-model.

Suppose now that W is any κ-complete ultrafilter over κ and $j_W \colon V \prec M_W \cong \mathrm{Ult}(V, W)$. Then by assumption

$$\kappa_{\alpha+\omega} < \kappa^{++L[U]} \le 2^\kappa < j_W(\kappa)\ .$$

Also $\{\kappa_{\alpha+n} \mid n \in \omega\} \in M_W$ since ${}^\omega M_W \subseteq M_W$, and hence $F \cap M_W \in M_W$ so that $L[F] = L[F \cap M_W]$ is definable in M_W. Consequently,

$$M_W \models \text{There is a } \rho\text{-model for some } \rho < j_W(\kappa)\ ,$$

so that by elementarity there is a ρ-model for some $\rho < \kappa$. By 21.9 this entails the existence of $0^\dagger$. $\dashv$

As Kunen observed, it follows that *if there is a measurable cardinal* κ *such that* $\kappa^+ < 2^\kappa$, *then* $0^\dagger$ *exists.* This was the first inkling of a genuine impediment to forcing at measurable cardinals: Beyond the restriction from below of 5.17, the measurability of κ imposes sufficient constraints on $\mathcal{P}(\kappa)$ so that achieving $\kappa^+ < 2^\kappa$ requires stronger hypotheses and presumably a new forcing approach. Such an approach was to be discovered by Silver (see volume II).

Kunen soon elaborated his argument to draw a stronger conclusion from several hypotheses:

21.25 Theorem (Kunen [71a]). *Suppose that for some* κ *at least one of the following holds:*

(a) κ *is measurable and* $\kappa^+ < 2^\kappa$.

(b) Every κ*-complete filter over* κ *can be extended to a* κ*-complete ultrafilter over* κ *(e.g.* κ *is strongly compact).*

(c) There is a κ*-complete ultrafilter* D *over* κ^+ *which is uniform, i.e.* $|X| = \kappa^+$ *whenever* $X \in D$.

Then for any λ *there is an inner model of* ZFC *with* λ *measurable cardinals.*

The building store of intuitions indicated that once a conclusion like this about a large cardinal concept can be deduced, presumably a genuine transcendence over that concept can be achieved. This was indeed confirmed by later developments in inner model theory, with an actual equiconsistency result established for 21.25(a) (see volume II).

Daggers

As for $0^\#$, the formulation of $0^\dagger$ can be relativized in a straightforward way. Suppose that a is a set of ordinals with $\alpha = \bigcup a$. An EM^2 *blueprint for a* is the theory of some structure

$$\langle L_\zeta[U], \in, U, x_k, y_k, a, \xi\rangle_{k\in\omega, \xi\leq\alpha}$$

where ζ is a limit ordinal; for some ordinal $\kappa > \alpha$, $\langle L_\zeta[U], \in, U, a, \xi\rangle_{\xi\leq\alpha} \models U$ is a normal ultrafilter over κ; and $\langle\{x_k \mid k \in \omega\}, \{y_k \mid k \in \omega\}\rangle$ is a double set of ordinal indiscernibles for $\langle L_\zeta[U], \in, U, a, \xi\rangle_{\xi\leq\alpha}$ such that for every $k \in \omega$,

$$x_k < x_{k+1} < \kappa < y_k < y_{k+1}\ .$$

Let $L[A, B]$ be the expected inner model constructible relative to A and B (the least inner model M such that for $x \in M$, $A \cap x \in M$ and $B \cap x \in M$) defined through a hierarchy as for $L[A]$ (§3). The theory of inner models of measurability developed in §19 can be adapted in a straightforward way to iterate models $\langle L[U, a], \in, U, a, \xi\rangle_{\xi\leq\alpha}$ that satisfy $\ulcorner U$ is a normal ultrafilter over $\kappa\urcorner$ with $\kappa > \alpha$. Again, such a model only depends on κ (and not on U) and all such models are iterates of each other. The theory for $0^\dagger$ carries over with minor changes, so that from sufficient hypotheses we get a unique EM^2 blueprint $a^\dagger$ for a satisfying (the analogues of) the conditions (I)-(III). $a^\dagger$ through coding can be regarded as a subset of $|\alpha|$. The following analogue of 21.15(a) is typical:

21.26 Theorem. *Suppose that a is a set of ordinals with $\alpha = \bigcup a$. Then $a^\dagger$ exists iff there is an inner model M with $a \in M$ such that for some $U \in M$, $M \models U$ is* a normal ultrafilter over $\kappa > \alpha$, *and $\langle M, \in, U, a, \xi\rangle_{\xi\leq\alpha}$ has an uncountable set of indiscernibles whose minimum element is greater than κ.*

As with sharps, the main interest is in reals a; with our development having shifted from $a \subseteq \omega$ to $a \in {}^\omega\omega$, the foregoing is readily adapted to such a.

The results of this section attest to the breadth and fecundity of Kunen's iterated ultrapowers. Whether they can be proved in other terms, iterated ultrapowers looms large as the framework in which they were first conceived. The uncovered uniformity of inner models and their embeddings offer compelling evidence for the coherence and efficacy of the concepts of measurability, $0^\#$, and $0^\dagger$.

Broadly speaking, this chapter described the first wave of methodical research on Ulam's original concept of measurability using the techniques that became available in the expansive period of the 1960's. Cohen's creation of forcing led to the relative consistency analysis in terms of saturated ideals, real-valued measurability, and Prikry forcing, and Scott's result 5.5 spurred interest in ultrapowers and their iterations, leading to the analysis of definability in the enriched contexts involving elementary embeddings: inner models of measurability, $0^{\#}$, and $0^{\dagger}$. This latter direction of research was to be vigorously pursued by Jensen and Dodd in their investigation of covering properties and the Core Model in the mid-1970's (see volume II). Before then, the forcing analysis of large cardinals was to be advanced by crucial consistency results of Silver and Magidor involving strong hypotheses (see volume II also). In the next venture forth beyond the extended elaboration of measurability, these hypotheses were just beginning to be explored, the most prominent among them being *supercompactness*.

Chapter 5

Strong Hypotheses

This chapter provides the basic theory for the strong hypotheses that generalize measurability. §22 discusses Solovay and Reinhardt's concept of supercompactness, a global reflection property, and the relation of supercompactness to strong compactness. §23 describes the stronger hypotheses that evolved from Reinhardt's proposals: extendibility and a *prima facie* extension shown inconsistent by Kunen. §24 then considers hypotheses on the verge of that inconsistency, and then spanning the expanse, n-hugeness and Vopěnka's Principle. Pursuing offshoots of the theory of supercompactness, §25 describes the combinatorial study of $\mathcal{P}_\kappa\gamma$, and §26 provides the fundamentals of extenders and related large cardinals, refined concepts that were to lead to major advances in inner model theory.

22. Supercompactness

In the late 1960's, even as measurability was being methodically investigated Solovay and William Reinhardt, as a graduate student at Berkeley, were charting out stronger hypotheses using straightforward guiding ideas. Taking the concept of elementary embedding as basic they considered various generalizations of the elementary embedding characterization of measurability. These new hypotheses had succinct formulations and clear motivations, and were soon found to have a simple but elegant basic theory. Prominent features were that the hypotheses form a linear hierarchy in terms of consistency strength, and that an outright inconsistency result of Kunen provided a delimiting upper bound. The basic framework generated several avenues of research inspired by intriguing questions and natural possibilities for structural elaboration, and relative consistency results in terms of the new hypotheses soon began to be established for propositions low in the cumulative hierarchy. With mounting extrinsic evidence in the next two decades, these strong hypotheses together with the hierarchy up through measurability became widely accepted as *the* measuring rod of exhaustive principles against which all possible consistency strengths can be gauged. This section discusses Solovay and Reinhardt's concept of supercompactness, and §§23, 24, the stronger hypotheses arising from Reinhardt's main proposals. The basic theory appeared belatedly in Solovay-Reinhardt-Kanamori [78], which serves as a general reference for these sections; Di Prisco-Marek [85a] is also a useful source.

Strong compactness (§4) was the one concept stronger than measurability that had been considered up to this time, and Solovay and Reinhardt independently devised supercompactness as a further strengthening – hence the term. Recall that if κ is measurable, U a κ-complete ultrafilter over κ, and $j\colon V \prec M \cong \mathrm{Ult}(V, U)$, then ${}^{\kappa}M \subseteq M$. This, especially spiced with normality for U, leads to reflection phenomena at κ (6.5). Generalizing, for $\kappa \leq \gamma$,

κ is *γ-supercompact* *iff* there is a $j\colon V \prec M$ such that
(a) $\mathrm{crit}(j) = \kappa$ and $\gamma < j(\kappa)$, and
(b) ${}^{\gamma}M \subseteq M$.
κ is *supercompact* *iff* κ is γ-supercompact for every $\gamma \geq \kappa$.

Thus, *if $\kappa \leq \gamma \leq \delta$ and κ is δ-supercompact, then κ is γ-supercompact*, and *κ is κ-supercompact iff κ is measurable.* Although $j\colon V \prec M$ is formalized through our convention from 5.1(c) that $\prec$ between inner models is $\prec_1$, these formulations are not directly formalizable because of the $\exists j$ ranging over classes. The developing theory provides a characterization in terms of ultrapowers that provides an *a posteriori* formalization, but nonetheless the *intelligibility* of the concepts comes from their elementary embedding formulations. (b) is the crucial closure assumption, asserting that arbitrary γ-sequences of members of M are again members of M and implying in particular that $H_{\gamma^+} \subseteq M$. Only cardinals γ need be considered, but maintaining the general definition has various procedural advantages. In (a) the condition $\gamma < j(\kappa)$ is carried over from how supercompactness evolved from strong compactness. Its inclusion in the definition is convenient for the development, but it turns out to be superfluous: For any $j\colon V \prec M$ with $\mathrm{crit}(j) = \kappa$ and ${}^{\gamma}M \subseteq M$, some nth iterate j^n has these same properties *and* satisfies $\gamma < j^n(\kappa)$ (23.15 (a)). The relationship between supercompactness and strong compactness is later discussed at some length, but first we develop the basic theory of supercompactness.

Straightforward observations illustrate how closure conditions lead to reflection properties:

22.1 Proposition. *Suppose that κ is 2^κ-supercompact. Then there is a normal ultrafilter U over κ such that*

$$\{\alpha < \kappa \mid \alpha \text{ is measurable}\} \in U\ .$$

In particular, κ is the κth measurable cardinal.

Proof. Let $j\colon V \prec M$ witness the 2^κ-supercompactness of κ. If U is defined from j by

$$X \in U \quad \mathit{iff} \quad X \subseteq \kappa \ \wedge\ \kappa \in j(X)\ ,$$

then U is a normal ultrafilter over κ. But since ${}^{2^\kappa}M \subseteq M$, it is simple to see that every ultrafilter over κ is a member of M. Hence, κ is measurable in M, so that $\{\alpha < \kappa \mid \alpha \text{ is measurable}\} \in U$ by its definition. ⊣

22.2 Proposition. *Suppose that κ is γ-supercompact and $2^\alpha = \alpha^+$ for every $\alpha < \kappa$. Then $2^\beta = \beta^+$ for every $\beta \leq \gamma$.*

Proof. Let $j\colon V \prec M$ witness the γ-supercompactness of κ. Arguing as for 5.8, $\beta \leq \gamma < j(\kappa)$ and elementarity implies that $(2^\beta)^M = \beta^{+M}$. But ${}^{\gamma}M \subseteq M$ implies that $\mathcal{P}(\beta)^M = \mathcal{P}(\beta)$, so that $2^\beta \leq (2^\beta)^M$ and $\beta^{+M} = \beta^+$. ⊣

That supercompactness is a global reflection property can be expressed in terms of the notion $A \prec_n V$, that Σ_n formulas are absolute for A. A well-known result of Levy [65] is that for any $\kappa > \omega$, $H_\kappa \prec_1 V$.

22.3 Proposition. *If κ is supercompact, then $V_\kappa \prec_2 V$.*

Proof. Suppose that φ is Σ_2, say $\exists v_0 \psi(v_0, v_1)$ where ψ is Π_1 for simplicity, and $x \in V_\kappa$. If $V_\kappa \models \varphi[x]$, then for some $y \in V_\kappa$, $V_\kappa \models \psi[y, x]$. But κ is inaccessible so that $V_\kappa = H_\kappa$, and the aforementioned result of Levy's implies that $\psi[y, x]$.

Conversely, suppose that $\varphi[x]$. Let z be such that $\psi[z, x]$. Taking $\alpha > \mathrm{rank}(z)$, the $|V_\alpha|$-supercompactness of κ implies that there is a $j\colon V \prec M$ with $\mathrm{crit}(j) = \kappa$ and $z \in (V_{j(\kappa)})^M$. Note that $j(x) = x \in V_\kappa = (V_\kappa)^M$. Since ψ is Π_1, it is readily seen that $(V_{j(\kappa)} \models \varphi[x])^M$. It now follows by elementarity that $V_\kappa \models \varphi[x]$ as desired. ⊣

The preceding three propositions already follow from a substantially weaker hypothesis on κ; see 26.6.

Generalizing the situation with measurability, γ-supercompactness is next characterized in terms of ultrafilters and ultrapowers. This will provide an adequate formalization of the concept, and hence of supercompactness. A general proposition about closure properties of ultrapowers related to 5.7 (noting that $j``\kappa = \kappa$ there) suggests and frames the later work.

22.4 Proposition. *Suppose that U is an ω_1-complete ultrafilter over a set S, and $j\colon V \prec M \cong \mathrm{Ult}(V, U)$. Then:*

(a) If $j``X \in M$ for some set X and $Y \subseteq M$ is such that $|Y| \leq |X|$, then $Y \in M$.

(b) For any γ, $j``\gamma \in M$ iff ${}^{\gamma}M \subseteq M$.

(c) $j``(|S|^+) \notin M$.

(d) $U \notin M$.

Proof. (a) Argue as for 5.7(d): Construe Y as $\{[f_x] \mid x \in X\}$. Let $h\colon S \to \mathcal{P}(X)$ be so that $[h] = j``X$. Define $g\colon S \to V$ by stipulating that $g(i)$ is that function with domain $h(i)$ satisfying $g(i)(x) = f_x(i)$. Then $[g](j(x)) = [f_x]$ for every $x \in X$, and so $\mathrm{ran}([g]) = Y$.

(b) This follows from (a).

(c) Let $[f] \in M$ be arbitrary. If $A = \{i \in S \mid |f(i)| \leq |S|\} \in U$, then there is an $\alpha \in |S|^+ - \bigcup\{f(i) \mid i \in A\}$, so that $j(\alpha) \notin [f]$. Otherwise, $B = \{i \in S \mid |f(i)| > |S|\} \in U$, in which case there is an injective function h on B satisfying $h(i) \in f(i)$ for every $i \in B$, so that $[h] \in [f] - j``V$. In either case, $[f] \neq j``(|S|^+)$.

(d) Assume to the contrary that $U \in M$. Then $S = \bigcup U \in M$, and so $\mathcal{P}(S) = U \cup \{S - X \mid X \in U\} \in M$, and so also $\mathcal{P}(S \times S) \in M$. This last set contains every well-ordering of S, and so it is straightforward to see that ${}^{S}(|S|^+) \in M$. But for any $\alpha < |S|^+$, $j(\alpha)$ is the ordertype of $\{[f] \mid f \in {}^{S}\alpha\}$, and the set of such ordertypes is definable from U and ${}^{S}(|S|^+)$, i.e. $j``(|S|^+) \in M$. This contradicts (c). ⊣

(b) provides a simple way of gauging the closure of ultrapowers, with (c) establishing an upper bound in this direction and (d) in another.

Suppose now that $j\colon V \prec M$ witnesses the γ-supercompactness of κ. Then $j``\gamma \in M$, and (b) above suggests (recalling 5.6 for measurability) that we try to define an ultrafilter U_j by:

$$X \in U_j \quad \textit{iff} \quad j``\gamma \in j(X)\ .$$

Recall that $\mathcal{P}_\nu S = \{x \subseteq S \mid |x| < \nu\}$. Properties of j and M imply that $|j``\gamma| < j(\kappa)$ in $M$, so that $j``\gamma \in (\mathcal{P}_{j(\kappa)} j(\gamma))^M$. Hence, $\mathcal{P}_\kappa \gamma \in U_j$, and this can be taken to be the underlying index set for U_j. The following properties obtain:

(i) U_j is a κ-complete ultrafilter .

(ii) For any $\alpha \in \gamma$, $\{x \in \mathcal{P}_\kappa\gamma \mid \alpha \in x\} \in U_j$.

(iii) For any $\langle X_\alpha \mid \alpha < \gamma\rangle \in {}^{\gamma}U_j$, $\{x \in \mathcal{P}_\kappa\gamma \mid x \in \bigcap_{\alpha \in x} X_\alpha\} \in U_j$.

By (i) and (ii), for any $y \in \mathcal{P}_\kappa\gamma$, $\{x \in \mathcal{P}_\kappa\gamma \mid y \subseteq x\} \in U_j$. For (iii), note that if $j(\langle X_\alpha \mid \alpha < \gamma\rangle) = \langle Y_\beta \mid \beta < j(\gamma)\rangle$, then $j``\gamma \in \bigcap_{\alpha<\gamma} j(X_\alpha) = \bigcap_{\beta\in j``\gamma} Y_\beta$.

These properties motivate the following definitions, stated for later purposes for an arbitrary κ, set S, and filter F over $\mathcal{P}_\kappa S$. If $\langle X_i \mid i \in S\rangle \in {}^S\mathcal{P}(\mathcal{P}(S))$, then its *diagonal intersection* is $\{x \in \mathcal{P}(S) \mid x \in \bigcap_{i\in x} X_i\}$, denoted by $\Delta_{i\in S}X_i$.

F is *fine* *iff* F is κ-complete, and for any $i \in S$, $\{x \in \mathcal{P}_\kappa S \mid i \in x\} \in F$.
F is *normal* *iff* F is fine, and for any $\langle X_i \mid i \in S\rangle \in {}^S F$, $\Delta_{i\in S}X_i \in F$.

Anticipating a possible confusion, note that members of such F consist of certain *subsets* of a set S, not its elements. Also, we usually deal with sets S of ordinals, in which case $\mathcal{P}_\kappa S = [S]^{<\kappa}$; the former notation has nonetheless become standard in the subject. Finally, for S an infinite ordinal γ there are no normal filters over $\mathcal{P}_\omega\gamma$ since $\Delta_{\alpha\in\gamma}\{x \in \mathcal{P}_\omega\gamma \mid \alpha+1 \in x\}$ is empty.

With fineness providing the requisite breadth, this new concept of normality is a generalization of the old one:

22.5 Exercise.

(a) If F is a normal filter over κ, then $\{X \subseteq \mathcal{P}_\kappa\kappa \mid X \cap \kappa \in F\}$ is a normal filter over $\mathcal{P}_\kappa\kappa$.

(b) If F is a normal filter over $\mathcal{P}_\kappa\kappa$, then $F \cap \mathcal{P}(\kappa)$ is a normal filter over κ.

Hint. For (b) note that if $X_\alpha = \{x \in \mathcal{P}_\kappa\kappa \mid \alpha \subseteq x\}$ for $\alpha < \kappa$, then $\Delta_{\alpha<\kappa}X_\alpha = \kappa \in F$. ⊣

Recall that for a filter F over a set I and $X \subseteq I$, X is F-stationary *iff* $X \cap Z \neq \emptyset$ for every $Z \in F$.

22.6 Exercise. *A fine filter F over $\mathcal{P}_\kappa S$ is normal iff whenever X is F-stationary and f is a choice function on X (i.e. $f(x) \in x$ for $x \in X - \{\emptyset\}$), there is an $s \in S$ such that $f^{-1}(\{s\})$ is F-stationary.*

Hint. Argue as for 5.10. ⊣

Having produced an ultrafilter from an embedding, the process can be reversed by taking an ultrapower. Suppose that $\kappa \leq \gamma$ and U is a normal ultrafilter over $\mathcal{P}_\kappa\gamma$. Since this entails $\kappa > \omega$ so that U is ω_1-complete, let $j_U\colon V \prec M_U \cong \mathrm{Ult}(V, U)$. Let id: $\mathcal{P}_\kappa\gamma \to \mathcal{P}_\kappa\gamma$ be the identity map. Then:

(i) $[\mathrm{id}]_U = j_U``\gamma$ and ${}^\gamma M_U \subseteq M_U$.
(ii) $\mathrm{crit}(j_U) = \kappa$ and $\gamma < j_U(\kappa)$.

For (i), $j_U``\gamma \subseteq [\mathrm{id}]_U$ by fineness, and 22.6 implies that equality holds. That ${}^\gamma M_U \subseteq M_U$ then follows by 22.4(b). For (ii), $\mathrm{crit}(j_U) \geq \kappa$ by κ-completeness.

But also, $\mathrm{ot}([\mathrm{id}]_U) < j_U(\kappa)$ in M_U since $\{x \subseteq \gamma \mid \mathrm{ot}(x) < \kappa\} \in U$, and $\gamma = \mathrm{ot}(j_U``\gamma) = \mathrm{ot}([\mathrm{id}]_U)$.

The following characterization has been established:

22.7 Theorem (Solovay, Reinhardt). *If $\kappa \leq \gamma$, κ is γ-supercompact iff there is a normal ultrafilter over $\mathcal{P}_\kappa\gamma$.* ⊣

One observation that can now be made is that supercompactness does not entail the existence of greater large cardinals:

22.8 Exercise.

(a) If κ is supercompact and $\lambda > \kappa$ is inaccessible, then $V_\lambda \models \kappa$ is supercompact.

(b) If Con(ZFC + $\exists\kappa(\kappa$ is supercompact)), *then* Con(ZFC + $\exists\kappa(\kappa$ is supercompact $\wedge \neg\exists\lambda(\lambda > \kappa \ \wedge \ \lambda$ is inaccessible))). ⊣

It can also be deduced from 22.7 that ⌜κ is γ-supercompact⌝ *is* Δ_2^{ZF}: To show that it is Σ_2^{ZF}, note that it can be rendered as $\exists x(x = V_{\gamma+5} \ \wedge \ \varphi(\kappa, \gamma, x))$ where φ asserts that there is a normal ultrafilter over $\mathcal{P}_\kappa\gamma$. ⌜$y = V_\alpha$⌝ is Π_1^{ZF} (0.2), and φ can be taken to be Σ_0 with all quantifiers bounded to $x = V_{\gamma+5}$. To show that it is Π_2^{ZF}, note that it can also be rendered as $\forall x(x = V_{\gamma+5} \rightarrow \varphi(\kappa, \gamma, x))$ with φ as before.

This has a useful consequence about the persistence of γ-supercompactness:

22.9 Exercise. *Suppose that $\kappa \leq \lambda$, κ is γ-supercompact for every γ with $\kappa \leq \gamma < \lambda$, and λ is supercompact. Then κ is supercompact.*

Hint. Apply 22.3, noting that $V_\lambda \models \kappa$ is supercompact.

Alternately, for $\delta \geq \lambda$ glue together a normal ultrafilter over $\mathcal{P}_\kappa\delta$ as follows: Let U be a normal ultrafilter over $\mathcal{P}_\lambda\delta$, and for each $x \in \mathcal{P}_\lambda\delta$ with $|x| \geq \kappa$ let D_x be a normal ultrafilter over $\mathcal{P}_\kappa x$. Define W by:

$$X \in W \text{ iff } X \subseteq \mathcal{P}_\kappa\delta \ \wedge \ \{x \in \mathcal{P}_\lambda\delta \mid |x| \geq \kappa \ \wedge X \cap \mathcal{P}_\kappa x \in D_x\} \in U .$$

Then W is a normal ultrafilter over $\mathcal{P}_\kappa\delta$. ⊣

This shows that 22.3 is optimal in the sense that the Σ_3^{ZF} sentence ⌜there is a supercompact cardinal⌝ does not hold in V_κ if κ is the least supercompact cardinal.

Supercompactness is characterizable directly as a reflection property:

22.10 Theorem (Magidor [71a]). *κ is supercompact iff for every $\eta > \kappa$ there is an $\alpha < \kappa$ and an e: $V_\alpha \prec V_\eta$ with a critical point δ so that $e(\delta) = \kappa$.*

Proof. For the forward direction, suppose that $\eta > \kappa$. Let $j\colon V \prec M$ witness the $|V_\eta|$-supercompactness of κ. Set $\bar{j} = j|V_\eta$; it is simple to see that $\bar{j}\colon V_\eta \prec (V_{j(\eta)})^M$. By the closure of M under $|V_\eta|$-sequences, $V_\zeta = (V_\zeta)^M \in M$ for $\zeta \leq \eta$

by induction, and so also $\overline{j} \in M$ and $M \models \ulcorner\overline{j}: V_\eta \prec V_{j(\eta)}\urcorner$. Hence, $M \models \ulcorner$there is an $\alpha < j(\kappa)$ and an $e: V_\alpha \prec V_{j(\eta)}$ with a critical point δ so that $e(\delta) = j(\kappa)\urcorner$. The desired result now follows from the elementarity of j.

For the converse, suppose that $\gamma \geq \kappa$. Let $\alpha < \kappa$ and $e: V_\alpha \prec V_{\gamma+\omega}$ with a critical point δ so that $e(\delta) = \kappa$. It follows that $\alpha = \beta + \omega$ for a β satisfying $e(\beta) = \gamma$. Noting that $\mathcal{P}(\mathcal{P}_\delta\beta) \subseteq V_\alpha$ and $e``\beta \in \mathcal{P}_\kappa\gamma$, define U by:

$$X \in U \quad \textit{iff} \quad X \subseteq \mathcal{P}_\delta\beta \ \wedge \ e``\beta \in e(X) .$$

Then U is a normal ultrafilter over $\mathcal{P}_\delta\beta$. But $U \in V_\alpha$, and so $e(U)$ is a normal ultrafilter over $\mathcal{P}_{e(\delta)}e(\beta) = \mathcal{P}_\kappa\gamma$. ⊣

Carrying this further Magidor [71a] characterized the least supercompact cardinal in terms of a second-order Löwenheim-Skolem property. See also Kunen [03], especially its 2.9.

Normal Ultrafilters

Having characterized γ-supercompactness in terms of the existence of normal ultrafilters over $\mathcal{P}_\kappa\gamma$ and derived some initial consequences of this, the basic theory of these ultrafilters is now developed as in Solovay-Reinhardt-Kanamori [78].

The first proposition provides more information about ultrapowers.

22.11 Proposition. *Suppose that $\kappa \leq \gamma$, U is a normal ultrafilter over $\mathcal{P}_\kappa\gamma$, $j: V \prec M \cong \mathrm{Ult}(V, U)$, and $|\gamma| = \lambda$. Then:*

(a) For $\alpha \leq \gamma$, $j``\alpha = [\langle x \cap \alpha \mid x \in \mathcal{P}_\kappa\gamma\rangle]_U$ and $\alpha = [\langle \mathrm{ot}(x \cap \alpha) \mid x \in \mathcal{P}_\kappa\gamma\rangle]_U$.

(b) ${}^{\lambda^{<\kappa}}M \subseteq M$.

(c) $2^{\lambda^{<\kappa}} \leq (2^{\lambda^{<\kappa}})^M < j(\kappa) < (2^{\lambda^{<\kappa}})^+$.

Proof. (a) By normality $j``\gamma = [\mathrm{id}]_U$ where id: $\mathcal{P}_\kappa\gamma \to \mathcal{P}_\kappa\gamma$ is the identity map. Now note that for $\alpha \leq \gamma$, $j``\alpha = j``\gamma \cap j(\alpha)$ and $\alpha = \mathrm{ot}(j``\alpha)$.

(b) By 22.4(a), it suffices to show that $j``\mathcal{P}_\kappa\gamma \in M$. But

$$j``\mathcal{P}_\kappa\gamma = \mathcal{P}_\kappa j``\gamma = (\mathcal{P}_\kappa j``\gamma)^M \in M ;$$

the first equality holds as $j(x) = j``x$ for any $x \in \mathcal{P}_\kappa\gamma$ by $\kappa$-completeness, and the second, as $j``\gamma \in M$ and ${}^\gamma M \subseteq M$.

(c) $2^{\lambda^{<\kappa}} \leq (2^{\lambda^{<\kappa}})^M$ since $(\mathcal{P}(\mathcal{P}_\kappa\lambda))^M = \mathcal{P}(\mathcal{P}_\kappa\lambda)$ by (b) and $M \subseteq V$. $(2^{\lambda^{<\kappa}})^M < j(\kappa)$ since $j(\kappa)$ is inaccessible in M. Finally $j(\kappa) = \{[f]_U \mid f: \mathcal{P}_\kappa\gamma \to \kappa\}$ so that $j(\kappa) < (2^{\lambda^{<\kappa}})^+$. ⊣

(b) asserts that in certain situations there is more closure, for example when $\mathrm{cf}(\lambda) < \kappa$.

We next take a closer look, when a normal ultrafilter U is defined from an embedding, at the relationship between that embedding and j_U: Suppose that

$j\colon V \prec M$ witnesses the γ-supercompactness of κ, and U is the normal ultrafilter over $\mathcal{P}_\kappa\gamma$ defined from j with corresponding $j_U\colon V \prec M_U \cong \mathrm{Ult}(V, U)$. Define $k\colon M_U \to M$ by:

$$k([f]_U) = j(f)(j``\gamma) .$$

22.12 Lemma. *Set* $|\gamma| = \lambda$. *Then:*

(a) k *is elementary, and* $j = k \circ j_U$.

(b) $k(\alpha) = \alpha$ *for every* $\alpha < (2^{\lambda^{<\kappa}})^{+M_U}$, *and* $k(z) = z$ *for every* $z \in \mathcal{P}(\mathcal{P}(\mathcal{P}_\kappa\gamma)) \cap M_U$.

(c) If $(2^{\lambda^{<\kappa}})^+ = (2^{\lambda^{<\kappa}})^{+M}$, *then* $\mathrm{crit}(k) = (2^{\lambda^{<\kappa}})^{+M_U}$.

Proof. (a) This is straightforward (5.13(b)).

(b) For $\alpha \le \gamma$, by 22.11(a)

$$k(\alpha) = k([\langle \mathrm{ot}(x \cap \alpha) \mid x \in \mathcal{P}_\kappa\gamma\rangle]_U) = \mathrm{ot}(j``\gamma \cap j(\alpha)) = \alpha .$$

Noting that $\mathcal{P}(\gamma) \subseteq M_U$, this implies that $k(x) = x$ for every $x \in \mathcal{P}(\gamma)$. (In one direction, if $\alpha \in k(x)$, then $k(x) \subseteq k(\gamma) = \gamma$ implies that $k(\alpha) = \alpha$, so that $k(\alpha) \in k(x)$ and consequently $\alpha \in x$.) Noting that $\mathcal{P}(\mathcal{P}_\kappa\gamma) \subseteq M_U$ by 22.11(b) this in turn implies that $k(y) = y$ for every $y \in \mathcal{P}(\mathcal{P}_\kappa\gamma)$. (In one direction, if $x \in k(y)$, then $k(y) \in \mathcal{P}(\mathcal{P}_{k(\kappa)}k(\gamma))^M = \mathcal{P}(\mathcal{P}_\kappa\gamma) \subseteq M_U$ implies that $k(x) = x$ by the previous assertion, so that $k(x) \in k(y)$ and consequently $x \in y$.) The argument can be repeated one more time to show that $k(z) = z$ for every $z \in \mathcal{P}(\mathcal{P}(\mathcal{P}_\kappa\lambda)) \cap M_U$.

That $k(\alpha) = \alpha$ for $\alpha < (2^{\lambda^{<\kappa}})^{+M_U}$ also follows, by induction on α: Assume that $k(\alpha) = \alpha$ for $\alpha < \beta$ and $\beta < (2^{\lambda^{<\kappa}})^{+M_U}$. Then there is an $z \in \mathcal{P}(\mathcal{P}(\mathcal{P}_\kappa\lambda)) \cap M_U$ and a bijection $f\colon z \to \beta$ such that $f \in M_U$. By elementarity, $M \models k(f)\colon k(z) \to k(\beta)$ is a bijection. By the previous paragraph, $k(z) = z$. Moreover, for any $y \in z$,

$$k(f)(y) = k(f)(k(y)) = k(f(y)) = f(y) ;$$

the first equality holds by the previous paragraph, the second by elementarity, and the third by induction. Hence, $k(f) = f$, and so $k(\beta) = \beta$.

(c) In view of (b), it remains to observe that

$$(2^{\lambda^{<\kappa}})^{+M_U} < (2^{\lambda^{<\kappa}})^+ = (2^{\lambda^{<\kappa}})^{+M} = k((2^{\lambda^{<\kappa}})^{+M_U}) .$$

The inequality follows from 22.11(c), and the last equality, from elementarity. ⊣

Suppose now that j is itself a $j_{\tilde{U}}\colon V \prec M_{\tilde{U}} \cong \mathrm{Ult}(V, \tilde{U})$ where $\tilde{U}$ is a normal ultrafilter over $\mathcal{P}_\kappa\delta$ with $\delta \ge \gamma$. Then

$$X \in \tilde{U} \;\textit{ iff }\; X \subseteq \mathcal{P}_\kappa\delta \,\wedge\, j_{\tilde{U}}``\delta \in j_{\tilde{U}}(X) ,$$

and

$$Y \in U \;\textit{ iff }\; Y \subseteq \mathcal{P}_\kappa\gamma \,\wedge\, j_{\tilde{U}}``\gamma \in j_{\tilde{U}}(Y) .$$

Setting

$$\tilde{U}|\gamma = \{\{x \cap \gamma \mid x \in X\} \mid X \in \tilde{U}\}\,,$$

it readily follows that

$$U = \tilde{U}|\gamma\ .$$

$\tilde{U}|\gamma = f_*(\tilde{U})$ in a former terminology, where $f_*(U) = \{Y \mid f^{-1}(Y) \in U\}$ and $f\colon \mathcal{P}_\kappa\delta \to \mathcal{P}_\kappa\gamma$ is given by $f(x) = x \cap \gamma$. The corresponding $k\colon M_U \prec M_{\tilde{U}}$ can be explicated by

$$k([\langle f(x) \mid x \in \mathcal{P}_\kappa\gamma\rangle]_U) = [\langle f(x \cap \gamma) \mid x \in \mathcal{P}_\kappa\delta\rangle]_{\tilde{U}}\ .$$

Note that for $\delta \geq 2^{\lambda^{<\kappa}}$, ${}^{\delta}M_{\tilde{U}} \subseteq M_{\tilde{U}}$ implies by 22.12(c) that $\mathrm{crit}(k) = (2^{\lambda^{<\kappa}})^{+M_U}$.

We next consider the question of the number of normal ultrafilters, first raised for measurable cardinals. If κ is measurable, then by 20.11 there is an inner model in which there is only one normal ultrafilter over κ, and by 17.8 there is a forcing extension in which there is the maximal possible number 2^{2^κ} of normal ultrafilters over κ. If κ is 2^κ-supercompact, then by 5.16 and 22.1 there are at least two normal ultrafilters over κ, one containing $\{\alpha < \kappa \mid \alpha \text{ is measurable}\}$ and the other not. In 1967 Solovay and also Kanji Namba showed that there are $(2^\kappa)^+$ normal ultrafilters. A few years later, Solovay showed that in fact there are 2^{2^κ} normal ultrafilters over κ. The question of whether the number of normal ultrafilters over some κ can be intermediate between 1 and 2^{2^κ} was settled by Mitchell [74]. In the first development in inner model theory for hypotheses stronger than measurability, he showed that assuming sufficient hypotheses about κ (2^κ-supercompactness suffices), if $\tau \leq \kappa$ or is one of the terms κ^+ or κ^{++}, there is an inner model in which there are exactly τ normal ultrafilters over κ. In the model with exactly two normal ultrafilters over a κ, one contained the set $\{\alpha < \kappa \mid \alpha \text{ is measurable}\}$ and the other did not. Stewart Baldwin [85] built on Mitchell's work to show that assuming similarly strong hypotheses about κ, if $\tau < \kappa$, there is an inner model in which there are exactly τ normal ultrafilters over κ *and* κ is the only measurable cardinal. It is not known how to get such relative consistency results assuming only that κ is measurable at the outset.

Solovay's result mentioned above is now established in a general form; it is a consequence of a structural result that has an argument having aspects of self-reference as well as reflection. Again, for an ω_1-complete ultrafilter W, $j_W\colon V \prec M_W \cong \mathrm{Ult}(V, W)$.

22.13 Theorem (Solovay). *If $\kappa \leq \gamma$ and κ is $2^{|\gamma|^{<\kappa}}$-supercompact, then for any $A \in \mathcal{P}(\mathcal{P}(\mathcal{P}_\kappa\gamma))$ there is a normal ultrafilter W over $\mathcal{P}_\kappa\gamma$ such that $A \in M_W$.*

Proof. Let U be a normal ultrafilter over $\mathcal{P}_\kappa(2^{|\gamma|^{<\kappa}})$ and $j\colon V \prec M \cong \mathrm{Ult}(V, U)$. Let $\varphi(v_0, v_1, v_2)$ be a formula such that $\varphi[A, \kappa, \gamma]$ *iff* $A \in \mathcal{P}(\mathcal{P}(\mathcal{P}_\kappa\gamma))$ but for any normal ultrafilter W over $\mathcal{P}_\kappa\gamma$, $A \notin M_W$. φ can be assumed to have the property that $\exists v_0\varphi[\kappa, \gamma]$ exactly when $M \models \exists v_0\varphi[\kappa, \gamma]$: Since $\mathcal{P}(\mathcal{P}(\mathcal{P}_\kappa\gamma)) \subseteq M$, for any normal ultrafilter W over $\mathcal{P}_\kappa\gamma$, $W \in M$. Moreover, $\mathcal{P}(\mathcal{P}(\mathcal{P}_\kappa\gamma)) \cap M_W$ is definable in M since every function: $\mathcal{P}_\kappa\gamma \to V_\kappa$ is in M, and

$$\mathcal{P}(\mathcal{P}(\mathcal{P}_\kappa\gamma)) \cap M_W \subseteq (V_{\gamma+4})^{M_W} \subseteq (V_{j_W(\kappa)})^{M_W} = j_W(V_\kappa)\,.$$

Assume now to the contrary that $\exists v_0 \varphi[\kappa, \gamma]$. Then $M \models \exists v_0 \varphi[\kappa, \gamma]$. Mindful of 22.12 and the discussion following it, let $U|\gamma$ be the normal ultrafilter over $\mathcal{P}_\kappa\gamma$ corresponding to U with $j_0 \colon V \prec M_0 \cong \mathrm{Ult}(V, U|\gamma)$ and $k \colon M_0 \prec M$ such that $j = k \circ j_0$. Since $k(\kappa) = \kappa$ and $k(\gamma) = \gamma$, by elementarity of k there is an $A_0 \in M_0$ such that $M_0 \models \varphi[A_0, \kappa, \gamma]$. Since $k(A_0) = A_0$, it follows that $M \models \varphi[A_0, \kappa, \gamma]$, and hence $\varphi[A_0, \kappa, \gamma]$. But $A_0 \in M_0 = M_{U|\gamma}$. Contradiction! ⊣

22.14 Corollary. *If $\kappa \leq \gamma$ and κ is $2^{|\gamma|^{<\kappa}}$-supercompact, then there are $2^{2^{|\gamma|^{<\kappa}}}$ normal ultrafilters over $\mathcal{P}_\kappa\gamma$. Hence, if κ is 2^κ-supercompact, then there are 2^{2^κ} normal ultrafilters over κ.*

Proof. Set $|\gamma| = \lambda$. Note first that if W is a normal ultrafilter over $\mathcal{P}_\kappa\gamma$, then

$$|\mathcal{P}(\mathcal{P}(\mathcal{P}_\kappa\lambda)) \cap M_W| \leq (2^{2^{\lambda^{<\kappa}}})^{M_W} < j_W(\kappa) < (2^{\lambda^{<\kappa}})^+$$

by the argument for 22.11(c). By 22.13,

$$\mathcal{P}(\mathcal{P}(\mathcal{P}_\kappa\gamma)) = \bigcup\{\mathcal{P}(\mathcal{P}(\mathcal{P}_\kappa\gamma)) \cap M_W \mid W \text{ is a normal ultrafilter over } \mathcal{P}_\kappa\gamma\}\,.$$

Hence, it follows that there must be $|\mathcal{P}(\mathcal{P}(\mathcal{P}_\kappa\gamma))| = 2^{2^{\lambda^{<\kappa}}}$ normal ultrafilters over $\mathcal{P}_\kappa\gamma$.

The second assertion follows from 22.5(b), noting that $\kappa^{<\kappa} = \kappa$. ⊣

Historically, Magidor [71] had first shown that if κ is supercompact, then there are $(2^{|\gamma|^{<\kappa}})^+$ normal ultrafilters over $\mathcal{P}_\kappa\gamma$. A further corollary provides another measure of the wealth of normal ultrafilters:

22.15 Exercise. *If $\kappa \leq \gamma$ and κ is $2^{|\gamma|^{<\kappa}}$-supercompact, then*

$$(2^{|\gamma|^{<\kappa}})^+ = \sup(\{j_W(\kappa) \mid W \text{ is a normal ultrafilter over } \mathcal{P}_\kappa\gamma\})\,.$$

Hint. One direction follows from 22.11(c). For the converse, suppose that $\xi < (2^{|\gamma|^{<\kappa}})^+$. Let $A \in \mathcal{P}(\mathcal{P}(\mathcal{P}_\kappa\gamma))$ be such that $\xi < (2^{|\gamma|^{<\kappa}})^{+L[A]}$, and by 22.13, W a normal ultrafilter over $\mathcal{P}_\kappa\gamma$ such that $A \in M_W$. Show that $\xi < j_W(\kappa)$. ⊣

The proof of 22.14 shows that if κ is 2^κ-supercompact, then there are 2^{2^κ} normal ultrafilters over κ containing the set $\{\alpha < \kappa \mid \alpha \text{ is measurable}\}$. But paradoxically, the following question remains open:

22.16 Question. *If κ is 2^κ-supercompact, does it follow that there is more than one normal ultrafilter over κ containing the set* $\{\alpha < \kappa \mid \alpha \text{ is not measurable}\}$*?*

The theory of fine and normal ultrafilters over $\mathcal{P}_\kappa\gamma$ has been pursued along these lines by Julius Barbanel [82, 82a, 85] and Yoshihiro Abe [84]. The main avenues of related research concern combinatorial aspects of $\mathcal{P}_\kappa\gamma$ (§25).

On Strong Compactness

Having developed the basic theory of supercompactness, one based on simple reflection arguments, we now turn to a contextual characterization of strong compactness and discuss its relationship to supercompactness, continuing to follow Solovay-Reinhardt-Kanamori [78].

In the study of strong compactness fine ultrafilters over $\mathcal{P}_\kappa\gamma$ come naturally into play as in the proof of 4.1. Define for $\omega < \kappa \leq \gamma$:

$$\kappa \text{ is } \gamma\text{-}compact \textit{ iff } \text{there is a fine ultrafilter over } \mathcal{P}_\kappa\gamma\ .$$

Hence, *if κ is γ-supercompact, then κ is γ-compact;* and *κ is κ-compact iff κ is measurable*. Also, if $\kappa \leq \gamma \leq \delta$ and κ is δ-compact with U a fine ultrafilter over $\mathcal{P}_\kappa\delta$, then $U|\gamma$ is a fine ultrafilter over $\mathcal{P}_\kappa\gamma$, and thus κ is γ-compact. Fineness without normality has the following characterizations. Recall that for a set S and $E \subseteq \mathcal{P}(S)$, the filter over S generated by E is $\{X \subseteq S \mid \exists Y \in E(Y \subseteq X)\}$ provided that it is proper and non-principal.

22.17 Theorem. *If $\kappa \leq \gamma$, the following are equivalent:*

(a) κ is γ-compact.

(b) There is a $j\colon V \prec M$ with $\operatorname{crit}(j) = \kappa$ *such that: for any $X \subseteq M$ with $|X| \leq \gamma$, there is a $Y \in M$ such that $Y \supseteq X$ and $M \models |Y| < j(\kappa)$.*

(c) For any set S, every κ-complete filter over S generated by at most $|\gamma|$ sets can be extended to a κ-complete ultrafilter over S.

Proof. (a) → (b) Let U be a fine ultrafilter over $\mathcal{P}_\kappa\gamma$ and $j\colon V \prec M \cong \mathrm{Ult}(V, U)$. Suppose that $X \subseteq M$ with $|X| \leq \gamma$, say $X = \{[f_\alpha]_U \mid \alpha < \gamma\}$. Define $F\colon \mathcal{P}_\kappa\gamma \to V$ by $F(x) = \{f_\alpha(x) \mid \alpha \in x\}$, and set $Y = [F]_U$. Then $Y \supseteq X$ by fineness, and $M \models |Y| < j(\kappa)$. Finally, note that κ-completeness and the preceeding argument with $X = \kappa$ shows that $\operatorname{crit}(j) = \kappa$.

(b) → (c) Suppose that F is a κ-complete filter over a set S generated by a $E \subseteq \mathcal{P}(S)$ with $|E| \leq \gamma$. By (b) let $Y \supseteq j``E$ with $Y \in M$ and $M \models |Y| < j(\kappa)$. In M, $j(F)$ is $j(\kappa)$-complete and $j(F) \cap Y$ is a subset of cardinality less than $j(\kappa)$. Hence, there is a $c \in M$ so that $c \in \bigcap(j(F) \cap Y)$. c generates an ultrafilter U by: $X \in U$ *iff* $X \subseteq S \wedge c \in j(X)$. It is simple to check that U is a κ-complete ultrafilter over S extending F.

(c) → (a) Extend the κ-complete filter over $\mathcal{P}_\kappa\gamma$ generated by the family $\{\{x \in \mathcal{P}_\kappa\gamma \mid \alpha \in x\} \mid \alpha < \gamma\}$ to a κ-complete ultrafilter over $\mathcal{P}_\kappa\gamma$. ⊣

22.18 Corollary. *κ is strongly compact iff κ is γ-compact for every $\gamma \geq \kappa$.*

Proof. By 4.1. ⊣

We now see how the concept that came after is the mightier: Solovay and Reinhardt independently devised supercompactness to ensure $Y = X$ in 22.17(b) through the normality condition. This led to the simple transcendence as given by 22.1, and provided a provisional answer to a question of Tarski: *Is the least strongly compact cardinal necessarily greater than the least measurable cardinal?* The Vopěnka-Hrbáček result 5.8 together with the existence of inner models for measurability had shown that measurable cardinals need not be strongly compact and Kunen's result 21.25, that strong compactness is strictly stronger than measurability in consistency strength. Yet, Tarski's question had remained unanswered.

Having made the conceptual move to supercompactness, the problem was shifted. While the process of producing a normal ultrafilter over a measurable cardinal κ was a simple one, trying to produce a normal ultrafilter over $\mathcal{P}_\kappa\gamma$ from a fine one met with unexpected difficulties. Still, with so little experience strong compactness and supercompactness had the same feel, and Solovay conjectured that they are equivalent. This conjecture was to be refuted, but it inspired much of the early work in this area. In a few years, with mounting extrinsic evidence provided by relative consistency results supercompactness came to be accepted as the proper generalization of measurability in the emerging hierarchy of large cardinals.

Strong compactness was nonetheless to remain a significant hypothesis, particularly because of its combinatorial consequences. Jussi Ketonen [72] clarified the concept with another characterization: *If $\kappa \leq \lambda$ are both regular, then κ is λ-compact iff for every regular μ such that $\kappa \leq \mu \leq \lambda$ there is a uniform κ-complete ultrafilter over μ.* This confirmed a conjecture of Kunen, and reduced strong compactness to having uniform κ-complete ultrafilters over arbitrarily large cardinals. Before this, Solovay had made an initial connection between supercompactness and the Singular Cardinals Problem, and using an idea in Ketonen's proof, he was able to refine that connection to the following solution from strong compactness (see volume II): *If κ is λ^+-compact and λ is a singular strong limit cardinal $> \kappa$, then $2^\lambda = \lambda^+$.*

Solovay's conjecture that strong compactness and supercompactness are equivalent was refuted by his student Telis Menas in late 1972.

22.19 Theorem (Menas [74]).

(a) If κ is measurable and a limit of strongly compact cardinals, then κ is strongly compact.

(b) If κ is the least cardinal as in (a), then $\{\alpha < \kappa \mid \alpha \text{ is measurable}\}$ *is not a stationary subset of κ. In particular, κ is not 2^κ-supercompact.*

Proof. (a) Let U be a κ-complete ultrafilter over κ with

$$A = \{\alpha < \kappa \mid \alpha \text{ is strongly compact}\} \in U\ .$$

(Such a U is simple to produce: If W is any κ-complete ultrafilter over κ and $f\colon \kappa \to A$ is injective, take $U = \{X \subseteq \kappa \mid f^{-1}(X) \in W\}$.) Suppose now that $\gamma \geq \kappa$. For each $\alpha \in A$ let U_α be a fine ultrafilter over $\mathcal{P}_\alpha\gamma$, and define D by:

$$X \in D \textit{ iff } X \subseteq \mathcal{P}_\kappa\gamma \wedge \{\alpha \in A \mid X \cap \mathcal{P}_\alpha\gamma \in U_\alpha\} \in U\ .$$

Then it is simple to check that D is a fine ultrafilter over $\mathcal{P}_\kappa\gamma$.

(b) Set $A = \{\alpha < \kappa \mid \alpha \text{ is measurable}\}$. Define $f \in {}^{\kappa}\kappa$ by:

$$f(\alpha) = \sup(\{\beta < \alpha \mid \beta \text{ is strongly compact}\}) .$$

Then by hypothesis on κ, f is regressive on A, yet f cannot be constant on any unbounded subset of κ. Hence, A is not stationary. It follows that A is not in any normal ultrafilter over κ, and hence by 22.1 that κ is not 2^{κ}-supercompact. ⊣

The D in the proof of (a) illustrates how fine ultrafilters can be composed with relative ease, and (b), how normality with the consequent reflection properties is another kettle of fish. Abe [85] observed that D is not normal. 23.8 will provide measurable limits of strongly compact cardinals.

Menas's 1973 Berkeley thesis substantiated the gap between fine and normal ultrafilters in various ways, and made contributions to the emerging study of $\mathcal{P}_{\kappa}\gamma$ combinatorics ([74, 76] – see §25). Much of the work was devoted to forcing; he developed methods for preserving supercompactness [76a], and having broached the possibility with 22.19, established the relative consistency of there being exactly one strongly compact cardinal and it failing to be supercompact ([74]).

Menachem Magidor, following upon his 1972 Hebrew University thesis on supercompactness from which [71, 71a] have already been cited, established the principal results on the varying nature of strong compactness. A driving force in large cardinal theory, Magidor quickly became a major set theorist. His early and unabashed use of supercompactness in several relative consistency results confirmed their significance and efficacy and steered set theoretic research upward to the investigation of consistency-wise strong propositions. By mid-1972, he [76] had established:

> If κ is supercompact, then there is a forcing extension in which κ remains supercompact and is also the least strongly compact cardinal .

Then at the beginning of 1973, motivated by Menas's 22.19 and relative consistency result, Magidor [76] established:

> If κ is strongly compact, then there is a forcing extension in which κ remains strongly compact and is also the least measurable cardinal .

This result separated strong compactness from supercompactness in the strongest possible way, and rather unexpectedly answered the aforementioned question of Tarski in the negative. See volume II for proofs of Magidor's results.

Several basic questions about strong compactness have remain unanswered. The following was not resolved by Magidor's consistency results:

22.20 Question. *Are the following theories equiconsistent?*

(a) ZFC + $\exists\kappa(\kappa$ is supercompact).

(b) ZFC + $\exists\kappa(\kappa$ is strongly compact).

That there are many normal ultrafilters over κ was established by a reflection argument whose inflexibility led to the question 22.16; are there other means available?

22.21 Question. *If κ is strongly compact, does it follow that there is more than one normal ultrafilter over κ?*

Finally, the following question is related to 22.2 and concerns the possibilities with forcing:

22.22 Question (Woodin). *If κ is strongly compact and $2^\nu = \nu^+$ for every $\nu < \kappa$, then does* GCH *hold?*

§§25, 26 describe developments in different directions related to the analysis of supercompactness. The first pursues the combinatorial study of $\langle \mathcal{P}_\kappa \gamma, \subset \rangle$ based on analogies with $\langle \kappa, < \rangle$ initially seen in connection with normal ultrafilters. The second develops a basic analysis of elementary embeddings and corresponding weak versions of γ-supercompactness. Unlike for measurability a canonical inner model theory for γ-supercompactness is not available, with naïve attempts failing, e.g. if $\kappa < \lambda$ and U is a normal ultrafilter over $\mathcal{P}_\kappa \lambda$, then $L[U] = L$ (25.8). The approach of §26 does lead to a successful inner model theory for reasonably substantial hypotheses.

As set theory advanced in the 1970's to the study of consistency-wise strong combinatorial propositions, various forcing results soon established supercompactness as a hypothesis of sufficient strength for relative consistency. However, in addition to lacking an inner model theory for the sharper study of consistency strength, supercompactness did not seem to have substantial direct consequences for definable sets of reals beyond those of measurability. An early, optimistic conjecture of Solovay [69: 60] was: if there is a supercompact cardinal, then every set of reals in $L(\mathbb{R})$ has the perfect set property. In remarkable developments in the mid-1980's, this and much more ($\mathrm{AD}^{L(\mathbb{R})}$ – see §32) were confirmed and from the weaker hypotheses of §26. However, it was in the expansive setting provided by supercompactness that the crucial techniques were first developed, with the weaker hypotheses only emerging in the effort to sharpen implications. With equiconsistency results established for these hypotheses and continuing progress in their inner model theory, there is no doubt left as to the intrinsic necessity of strong hypotheses based on elementary embeddings with closure conditions.

23. Extendibility to Inconsistency

The large cardinal hypotheses stronger than supercompactness evolved from the proposals of William Reinhardt. A recurring theme of his work was to investigate the concept of set together with broad notions of class or property. He considered various extensions of ZFC based on such notions, and just as an inaccessible cardinal provides a natural model of ZFC, strong large cardinal hypotheses were motivated as providing natural models for the emerging theories.

In his 1967 Berkeley thesis Reinhardt (see [70]) established results about Ackermann's set theory A. With motivations quite different from those for ZFC, A provides for a universe with extensionally determined entities, classes in a broad sense, together with a predicate for sethood. With that predicate rendered as $x \in V$, the main schema of A is: If $X \subseteq V$ is definable using only parameters from V (and without the predicate V), then $X \in V$. Let A* be A augmented by the Axiom of Foundation for members of V. Levy and Vaught had observed that much as for ZF, A* is consistent relative to A, but also that in A* one can prove the existence of classes like $\{V\}$, $\mathcal{P}(V)$, and $\mathcal{P}(\mathcal{P}(V))$. On the other hand, Levy had shown that if the relativization to V of a sentence σ of $\mathcal{L}_\in$ is provable in A*, then σ is provable in ZF. Reinhardt then established the converse, so that in particular A* and ZF are equiconsistent. He also showed that if $0 < \eta < \alpha < \beta$ and

$$(*) \qquad \langle V_{\alpha+\eta}, \in, x\rangle_{x\in V_\alpha} \equiv \langle V_{\beta+\eta}, \in, x\rangle_{x\in V_\alpha}$$

(where $\equiv$ denotes elementary equivalence), then $\langle V_{\beta+\eta}, \in, V_\alpha\rangle$ is a model of A*, with V_α interpreting the predicate V for sethood. Levy had established this result for $\eta = 1$; Reinhardt observed that the existence of $0 < \eta < \alpha < \beta$ satisfying $(*)$ follows from a sufficient indescribability assumption on β of the sort discussed in §6 for the motivating case $\eta \in \omega$.

At the end of his thesis Reinhardt considered an axiom that was soon proved inconsistent by Kunen, a dramatic turn of events that will be discussed later in this section. That axiom was a *prima facie* extension of a proposal for extending Ackermann's set theory, a proposal which can be cast within set theory as follows: Extend $(*)$ above in the motivating case $\eta = 1$ by incorporating for each "class" $X \subseteq V_\alpha$ a corresponding interpretation $j(X) \subseteq V_\beta$ to get

$$\langle V_{\alpha+1}, \in, x, X\rangle_{x\in V_\alpha, X\subseteq V_\alpha} \equiv \langle V_{\beta+1}, \in, x, j(X)\rangle_{x\in V_\alpha, X\subseteq V_\alpha} .$$

It is simple to see that this amounts to asserting that $j\colon V_{\alpha+1} \prec V_{\beta+1}$ with $\mathrm{crit}(j) = \alpha$. This is how Silver formulated Reinhardt's idea, and then Reinhardt considered arbitrary $\eta > 0$:

κ is *η-extendible* *iff* there is a ζ and a $j\colon V_{\kappa+\eta} \prec V_\zeta$ with $\mathrm{crit}(j) = \kappa$ and $\eta < j(\kappa)$.

κ is *extendible* *iff* κ is η-extendible for every $\eta > 0$.

Thus, embeddings are postulated that are sets, but whose domains and target structures have the ultimate closure property: they are initial segments of the universe. Since $\eta \geq \kappa \cdot \omega$ implies that $\kappa + \eta = \eta$, the specific form of the first definition is pertinent only for small η. For $\eta < \kappa$ it is clear that $\zeta = j(\kappa) + \eta$, so that, generalizing the sense from $\eta < \omega$, η-extendibility asserts that $(\eta + 1)$st-order properties are preserved between V_κ and $V_{j(\kappa)}$ in a strong way. The condition $\eta < j(\kappa)$ evolved from the $\eta = 1$ case; it is convenient to include it in the definition, but it is superfluous for full extendibility (23.15(b)).

Silver's 1-extendibility is already quite strong:

23.1 Proposition. *If κ is 1-extendible, then κ is measurable, and there is a normal ultrafilter U over κ such that*

$$\{\alpha < \kappa \mid \alpha \text{ is measurable}\} \in U\ .$$

Proof. Let $j\colon V_{\kappa+1} \prec V_{j(\kappa)+1}$ with $\mathrm{crit}(j) = \kappa$. Since $\mathcal{P}(\kappa) \subseteq V_{\kappa+1}$, a normal ultrafilter can be defined from j as usual by

$$X \in U \;\; \textit{iff} \;\; X \subseteq \kappa \;\wedge\; \kappa \in j(X)\ .$$

But $U \in V_{j(\kappa)+1}$ so that $V_{j(\kappa)+1} \models \kappa$ is measurable, and consequently $\{\alpha < \kappa \mid \alpha \text{ is measurable}\} \in U$ by its definition. ⊣

We shall soon establish that extendibility implies supercompactness, but also that these concepts are closely intertwined. However, while supercompactness was arrived at rather pragmatically as a strengthening of strong compactness, Reinhardt's motivations for extendibility were based on his analyses of the concept of set. We continue to chart his work in order to describe these motivations in context.

In [74] Reinhardt formulated a set theory S based on a set-existence principle of Shoenfield. S also provided for a universe of extensionally determined entities and had a predicate V, and in effect extended Ackermann's A. For S however, Reinhardt now viewed V as the collection of "existing" sets and the universe as consisting of "imaginable" sets – things and imaginings. Shoenfield's principle was rendered as: If $X \subseteq Y$ for some (imaginable) Y and X is definable using only parameters in V, then $X \in V$. With S* bearing the analogous relation to S that A* does to A, Reinhardt observed that the relativization to V of a sentence σ of $\mathcal{L}_\in$ is provable in S* exactly when σ is provable in ZF.

Reinhardt then extended S* to a further theory S^+ by enriching the notion of property. He first introduced a predicate U for the imaginable sets and a further predicate P for "existing properties of sets". The main new principle that he now proposed was that each imaginable set consisting of existing sets corresponds to an existing property of sets, i.e. for any $w \subseteq V$, there is a $Q \in P$ such that $w = V \cap Q$. Reinhardt gave contextual and other arguments for this broad interpretation of property. He then showed that a structure of form $\langle V_\beta, D, \in, V_\alpha \rangle$, where V_β interprets U, $D \subseteq V_{\beta+1}$ interprets P, and V_α interprets V, is a model of S^+ *iff*

(a) If Q' is definable in $\langle V_\beta, \in, Q\rangle_{Q\in D}$, then $Q' \in D$,
(b) $D \cap V_\beta = V_\alpha$, and
(c) $V_{\alpha+1} = \{Q \cap V_\alpha \mid Q \in D\}$.

(a) provides for the requisite store of properties, (b) expresses Shoenfield's principle in the presence of (a), and (c) expresses Reinhardt's new principle. He showed that these conditions imply that α is measurable with a normal ultrafilter containing the set $\{\xi < \alpha \mid \xi \text{ is measurable}\}$, and as is simple to check, if $j\colon V_{\kappa+1} \prec V_{j(\kappa)+1}$ witnesses the 1-extendibility of κ, then $\langle V_{j(\kappa)}, j``V_{\kappa+1}, \in, V_\kappa\rangle$ satisfies the conditions.

[74a] was Reinhardt's most extensive and explicit statement about motivations for large cardinal axioms based on the concept of set. Proceeding informally this time from the vantage point of ZFC, he advanced ideas of the sort that he had pursued in terms of more elaborate formalizations. After considering the role of the class On in reflection arguments for motivating the inaccessible, Mahlo and indescribable cardinals, Reinhardt introduced V and On as constants in the spirit of an imagined projection of ideas over new possibilities like $V_{\mathrm{On}+1}$ and $V_{\mathrm{On}+\mathrm{On}}$. He argued that since Cantor through his doctrine of the Absolute had intended the universe of sets to comprehend all possibilities, the theory of the new, formally projected universe of imaginary sets, even allowing parameters from V, ought to be the same:

$$\forall x \in V(\varphi^V[x] \leftrightarrow \varphi[x]) .$$

Reinhardt next extended these considerations into class-set theory with arbitrary classes $X \subseteq V$. He argued that a proper class X is essentially different from a set x in that if there were more ordinals, then x would have exactly the same members, whereas X would necessarily have new members. Writing $j(X)$ for the extension of X in the formally projected universe, he came to the schema:

$$(\dagger) \qquad \forall x \in V \forall X \subseteq V(\varphi^{\mathcal{P}(V)}[x, X] \leftrightarrow \varphi[x, j(X)]) .$$

This is the exact counterpart to the aforementioned extension of Ackerman's A in Reinhardt's thesis that led to Silver's formulation of 1-extendibility. It is consistent with the persistent theme that sets are sharply delimited and determined by their extensions, whereas classes in the sense of properties have an essential intensional quality.

Towards full extendibility, Reinhardt not only considered the cumulative hierarchy as proceeding beyond $V = V_{\mathrm{On}}$ in the formally projected universe, but projected versions of these further levels. While the introduction of new objects beyond Cantor's intentions may violate the universality of the concept of set, Reinhardt proposed ([74a: 198]) "to mitigate this sorrow by seeing the universality not in the *extension* of the concept of set but in the applicability of the *theory* of sets." This recalls his observation in [74: 13] that the axioms of ZF follow outright, not just relativized to V, from his S*. Let λ be a formal ordinal greater than On, with corresponding V_λ. Reinhardt argued for a projected realm $V_{\mathrm{On}'}$ corresponding to $V = V_{\mathrm{On}}$, a $V_{\lambda'}$ corresponding to V_λ, and a correspondence $j\colon V_\lambda \to V_{\lambda'}$ so that $j(x) = x$ for $x \in V$, $j(\mathrm{On}) = \mathrm{On}'$, and the theory is preserved:

$$\text{There is a } j\colon V_\lambda \prec V_{\lambda'} \text{ with } \mathrm{crit}(j) = \mathrm{On} \text{ and } j(\mathrm{On}) = \mathrm{On}'\,,$$

and in analogy with (†) for $\lambda = \mathrm{On}+1$, $\lambda < j(\mathrm{On})$. This of course is the assertion of λ-extendibility of On (at least when $\lambda \geq \mathrm{On}\cdot\omega$).

The theory of properties with necessity operator in [80] is the most mature and sophisticated formalization of Reinhardt's ideas. Here, properties are elevated to the status of rich, intensional entities, and modal operators are used to mediate with extensional entities. A professed feature is expressive power adequate to formulate extendibility, although he only indicated the treatment of a simple version.

Extendibility is now brought into the fold by describing its relationships to the other hypotheses, especially supercompactness. As for §22, Solovay-Reinhardt-Kanamori [78] continues to serve as a general reference.

A simple lemma leads to the initial observation.

23.2 Lemma. *There is a formula $\varphi(v_0, v_1)$ of $\mathcal{L}_\in$ such that for any $\alpha < \beta$,*

$$V_\beta \models \varphi[\alpha, x] \text{ iff } x = V_\alpha\,.$$

Proof. Consider the formulas

$$\begin{aligned}
\psi_1 &: f \text{ is a function} \wedge \mathrm{dom}(f) \in \mathrm{On} \wedge \\
&\quad\ \forall\gamma \in \mathrm{dom}(f)(f(\gamma) = \textstyle\bigcup\{\mathcal{P}(f(\xi)) \mid \xi < \gamma\})\,.\\
\psi_2 &: v_0 = \mathrm{dom}(f) \ \wedge\ v_1 = \textstyle\bigcup\{\mathcal{P}(f(\gamma)) \mid \gamma \in \mathrm{dom}(f)\}\,.\\
\psi_3 &: v_0 = \mathrm{dom}(f) + 1 \ \wedge\ v_1 = \mathcal{P}(\textstyle\bigcup\{\mathcal{P}(f(\gamma)) \mid \gamma \in \mathrm{dom}(f)\})\,.\\
\psi_4 &: v_0 = \mathrm{dom}(f) + 2 \ \wedge\ v_1 = \mathcal{P}(\mathcal{P}(\textstyle\bigcup\{\mathcal{P}(f(\gamma)) \mid \gamma \in \mathrm{dom}(f)\}))\,.
\end{aligned}$$

ψ_1 asserts that f must satisfy the usual recursive definition of the rank hierarchy, so that $f(\gamma) = V_\gamma$ for $\gamma \in \mathrm{dom}(f)$. (The proof could alternately have been cast in terms of the rank function (cf. 0.2).)

Let $\varphi(v_0, v_1)$ be $\exists f(\psi_1 \wedge (\psi_2 \vee \psi_3 \vee \psi_4))$. To show that this works, suppose that $\alpha < \beta$. It is simple to see that if $V_\beta \models \varphi[\alpha, x]$, then $x = V_\alpha$. For the converse, assume that $x = V_\alpha$. It is straightforward to check that if f satisfies ψ_1, then $f \in V_{\mathrm{dom}(f)+3}$, and if $\mathrm{dom}(f)$ is a limit ordinal, $f \in V_{\mathrm{dom}(f)+1}$. Hence, if $\alpha + 2 < \beta$ or α is a limit ordinal, then $V_\beta \models \exists f(\psi_1 \wedge \psi_2)[\alpha, x]$. In the remaining cases argue as follows: If α is a successor ordinal and either $\alpha + 2 = \beta$ or else $\alpha + 1 = \beta$ and $\alpha - 1$ is a limit ordinal, then there is an $f \in V_\beta$ with $\mathrm{dom}(f) = \alpha - 1$ satisfying ψ_1, and so $V_\beta \models \exists f(\psi_1 \wedge \psi_3)[\alpha, x]$. Finally, if α is a successor ordinal, $\alpha + 1 = \beta$, and $\alpha - 1$ is a successor ordinal, then there is an $f \in V_\beta$ with $\mathrm{dom}(f) = \alpha - 2$ satisfying ψ_1, and so $V_\beta \models \exists f(\psi_1 \wedge \psi_4)[\alpha, x]$. ⊣

23.3 Proposition. *If κ is η-extendible and $0 < \delta < \eta$, then κ is δ-extendible.*

Proof. Let $j\colon V_{\kappa+\eta} \prec V_\zeta$ witness the η-extendibility of κ. It follows from the lemma that $j(V_{\kappa+\delta}) = V_{j(\kappa+\delta)}$. Hence, formulas can be relativized to show that $j|V_{\kappa+\delta}\colon V_{\kappa+\delta} \prec V_{j(\kappa+\delta)}$, confirming δ-extendibility: For any formula $\varphi(v_1, \ldots, v_n)$ and sets $x_1, \ldots, x_n$ in $V_{\kappa+\delta}$,

$$\begin{aligned} V_{\kappa+\delta} \models \varphi[x_1, \ldots, x_n] \ &\textit{iff}\ \ V_{\kappa+\eta} \models \varphi^{V_{\kappa+\delta}}[x_1, \ldots, x_n] \\ &\textit{iff}\ \ V_\zeta \models \varphi^{V_{j(\kappa+\delta)}}[j(x_1), \ldots, j(x_n)] \\ &\textit{iff}\ \ V_{j(\kappa+\delta)} \models \varphi[j(x_1), \ldots, j(x_n)]\,. \end{aligned}$$ ⊣

While it is consistent with the supercompactness of κ that there is no inaccessible cardinal above κ (22.8), the extendibility of κ implies the existence of many large cardinal above κ: If $j\colon V_{\kappa+1} \prec V_{j(\kappa)+1}$ witnesses the 1-extendibility of κ, then κ is measurable so that $j(\kappa)$ is inaccessible in $V_{j(\kappa)+1}$ by elementarity and hence inaccessible in V as $\mathcal{P}(j(\kappa)) \subseteq V_{j(\kappa)+1}$. Moreover, by 23.1 and elementarity $\{\alpha < j(\kappa) \mid \alpha \text{ is measurable}\}$ is unbounded in $j(\kappa)$. In turn, if $j\colon V_{\kappa+2} \prec V_{j(\kappa)+2}$ witnesses the 2-extendibility of κ, then $j(\kappa)$ is itself measurable in V as $\mathcal{P}(\mathcal{P}(j(\kappa))) \subseteq V_{j(\kappa)+2}$. In these ways, extendibility projects stronger and strong hypotheses upward. These considerations illustrate how strongly the existence of an extendible cardinal affects the higher levels of the cumulative hierarchy, and why η-extendibility cannot be formulated, as γ-supercompactness can, in terms of the existence of ultrafilters.

Extendibility is next characterized in terms of the higher-order analogue of the original formulation of strong compactness. This result of Magidor reinforces extendibility by explicating it as a direct generalization of a prior concept. The $L_{\lambda\mu}$ languages were defined in §4; for $1 \leq n \in \omega$, an $L^n_{\lambda\mu}$ language is an extension of an $L_{\lambda\mu}$ language allowing higher-order variables of type at most n. With the intended interpretation of type $i+1$ variables as ranging over $\mathcal{P}^i(D)$ where D is the domain, there are the corresponding notions of *satisfiability* and *ν-satisfiability* as for $L^1_{\lambda\mu} = L_{\lambda\mu}$.

κ is $L^n_{\lambda\mu}$*-compact* *iff* any collection of $L^n_{\lambda\mu}$ sentences, if κ-satisfiable, is satisfiable .

Thus, $\kappa > \omega$ and κ is $L^1_{\kappa\kappa}$-compact *iff* κ is strongly compact.

23.4 Theorem (Magidor [71a]). *The following are equivalent for $\kappa > \omega$:*

(a) κ is extendible.

(b) κ is $L^n_{\kappa\kappa}$-compact for every $1 \leq n \in \omega$.

(c) κ is $L^2_{\kappa\omega}$-compact.

Proof. Suppose first that κ is extendible, $1 \leq n \in \omega$, and Σ is a κ-satisfiable collection of $L^n_{\kappa\kappa}$ sentences. It can be assumed that Σ is coded as a set via an arithmetization of the language where the logical symbols are coded by members of V_κ. Let $\eta \geq \kappa \cdot \omega$ be a limit ordinal sufficiently large so that $|\Sigma| \leq \eta$ and $V_\eta \models \Sigma$ is κ-satisfiable. By η-extendibility, let $j\colon V_\eta \prec V_\zeta$ with $\mathrm{crit}(j) = \kappa$ and

$\eta < j(\kappa)$. Then $V_\zeta \models j(\Sigma)$ is $j(\kappa)$-satisfiable. Note that $j``\Sigma \subseteq j(\Sigma)$ and so $j``\Sigma \in V_\zeta$, and that $|j``\Sigma| \leq \eta < j(\kappa)$ and so this holds in $V_\zeta$ as $\zeta$ must be a limit ordinal. Hence, $V_\zeta \models j``\Sigma$ has a model A, in which case A really is a model of $j``\Sigma$. But then, $A$ is a model of $\Sigma$, since the formulas in $j``\Sigma$ are like those of Σ with at most the non-logical symbols renamed. It follows that κ is $L^n_{\kappa\kappa}$-compact.

To complete the proof, it remains to show that if κ is $L^2_{\kappa\omega}$-compact, then κ is extendible. Suppose then that $\eta \geq \kappa \cdot \omega$; we seek a ζ and a $j\colon V_\eta \prec V_\zeta$ with $\mathrm{crit}(j) = \kappa$ and $\eta < j(\kappa)$. In terms of distinct constants $\dot{x}$ for $x \in V_\eta$ and c_α where $\alpha \leq \eta$, let Σ be the $L^2_{\kappa\omega}$ theory of

$$\langle V_\eta, \in, x \rangle_{x \in V_\eta}$$

together with the sentences $\ulcorner c_\alpha \in \dot{\kappa} \ \wedge \ c_\alpha \in c_\beta \urcorner$ for $\alpha < \beta \leq \eta$. Then Σ is κ-satisfiable, so by hypothesis it is satisfiable. Now Σ contains a Π_1 sentence σ so that for any transitive set X, $X \models \sigma$ *iff* $X = V_\alpha$ for some α (cf. 0.2), and a Π_1 sentence τ asserting that $\in$ is well-founded. Hence, taking a transitive collapse Σ has a model of form

$$\langle V_\zeta, \in, \overline{x}, \gamma_\alpha \rangle_{x \in V_\eta, \alpha \leq \eta} ,$$

where γ_α interprets c_α. Clearly, the map $j\colon V_\eta \to V_\zeta$ defined by $j(x) = \overline{x}$ is elementary. Also, for any $\alpha < \kappa$, Σ contains the $L_{\kappa\omega}$ sentence

$$\forall v (v \in \dot{\alpha} \leftrightarrow \bigvee_{\xi < \alpha} v = \dot{\xi}) ,$$

so that by induction j is the identity on κ. Finally, $\eta < j(\kappa)$ since $\{\gamma_\alpha \mid \alpha \leq \eta\}$ is a subset of $j(\kappa)$ of ordertype $\eta + 1 > \eta$. ⊣

Although extendibility reflects more ethereal ambitions than supercompactness, the next two results show how their local versions are closely intertwined.

23.5 Exercise. *If κ is $|V_{\kappa+\eta}|$-supercompact and $\eta < \kappa$, then there is a normal ultrafilter U over κ such that*

$$\{\alpha < \kappa \mid \alpha \text{ is } \eta\text{-extendible}\} \in U .$$

Hint. Argue as for the forward direction of 22.10. ⊣

23.6 Proposition. *If κ is η-extendible and $\delta + 2 \leq \eta$, then κ is $|V_{\kappa+\delta}|$-supercompact. Hence, if κ is extendible, then κ is supercompact.*

Proof. Let $j\colon V_{\kappa+\eta} \prec V_\zeta$ witness the η-extendibility of κ. Noting that $\mathcal{P}(\mathcal{P}_\kappa V_{\kappa+\delta}) \subseteq V_{\kappa+\delta+2} \subseteq V_\eta$, define U by:

$$X \in U \ \textit{ iff } \ X \subseteq \mathcal{P}_\kappa V_{\kappa+\delta} \ \wedge \ j``V_{\kappa+\delta} \in j(X) .$$

Then it is straightforward to check that this is a normal ultrafilter over $\mathcal{P}_\kappa V_{\kappa+\delta}$, *provided* that enough apparatus can be coded into the domain of j. This however is routine to check:

To verify, for example, the choice function condition (cf. 22.6), suppose that f is a choice function on $\mathcal{P}_\kappa V_{\kappa+\delta}$. For each $y \in V_{\kappa+\delta}$, set

$$y^+ = (y \cap (V_{\kappa+\delta} - \omega)) \cup \{n+1 \mid n \in y \cap \omega\}$$

so that $\emptyset \notin y$. Then for each $x \in \mathcal{P}_\kappa V_{\kappa+\delta}$, set

$$x_f = \{y^+ \mid y \in x\} \cup \{\{\emptyset\} \cup f(x)^+\}$$

so that $\emptyset$ tags $f(x)$. Clearly $x_f \in V_{\kappa+\delta+1}$. Finally, set

$$X_f = \{x_f \mid x \in \mathcal{P}_\kappa V_{\kappa+\delta}\}$$

so that $X_f \in V_{\kappa+\delta+2}$. Hence, $X_f \in V_\eta$ and faithfully codes f, and can be used in conjunction with the elementarity of j to verify the choice function condition for f. ⊣

That an extendible cardinal is supercompact can also be derived along the lines of 22.10.

The conjunction of even 1-extendibility and supercompactness transcends supercompactness:

23.7 Proposition. *If κ is supercompact and 1-extendible, then there is a normal ultrafilter U over κ such that*

$$\{\alpha < \kappa \mid \alpha \text{ is supercompact}\} \in U\ .$$

In particular, the least supercompact cardinal is not 1-extendible.

Proof. Let $j\colon V_{\kappa+1} \prec V_{j(\kappa)+1}$ witness the 1-extendibility of κ, and U the usual normal ultrafilter over κ defined from j. Since $j(\kappa)$ is inaccessible, $V_{j(\kappa)+1} \models \kappa$ is γ-supercompact for $\kappa \leq \gamma < j(\kappa)$. Hence,

$$A = \{\alpha < \kappa \mid \alpha \text{ is } \gamma\text{-supercompact for } \alpha \leq \gamma < \kappa\} \in U\ .$$

But κ is supercompact, so by 22.9 every $\alpha \in A$ is supercompact. ⊣

This readily leads to a rich supply of cardinals for Menas's result 22.19:

23.8 Exercise. *The conclusion of 23.7 can be strengthened to:*

$$\{\alpha < \kappa \mid \alpha \text{ is supercompact and a limit of supercompact cardinals}\} \in U\ . \quad ⊣$$

On the other hand, supercompactness reins back its stronger cousin in the following sense:

23.9 Exercise.

(a) If $\kappa < \lambda$, κ is extendible, and λ is supercompact, then $V_\lambda \models \kappa$ is extendible.

(b) If Con(ZFC + $\exists\kappa(\kappa$ is extendible)), *then* Con(ZFC + $\exists\kappa(\kappa$ is extendible $\land \neg\exists\lambda(\lambda > \kappa \;\land\; \lambda$ is supercompact))).

Hint. For (a), remembering that $\ulcorner x = V_\alpha \urcorner$ is Π_1^{ZF} check that $\ulcorner\kappa$ is extendible$\urcorner$ is Π_3^{ZF} and apply 22.3. For (b), use (a) and 22.9. ⊣

Hence, while extendibility may imply the existence of proper classes of various large cardinals, this does not encompass supercompact cardinals.

Finally, the next result is analogous to 22.3:

23.10 Proposition. *If κ is extendible, then $V_\kappa \prec_3 V$.*

Proof. Suppose that φ is Σ_3, say $\exists v_0\psi(v_0, v_1)$ where ψ is Π_2 for simplicity, and $x \in V_\kappa$. It follows from 22.3 that if $V_\kappa \models \varphi[x]$, then $\varphi[x]$. For the converse, we only need the simple corollary of extendibility that there are arbitrarily large inaccessible $\lambda > \kappa$ such that $V_\kappa \prec V_\lambda$:

Assume that $\varphi[x]$, and let y be such that $\psi[y, x]$. Let $\lambda > \kappa$ be inaccessible such that $V_\kappa \prec V_\lambda$ and $y \in V_\lambda$. Since ψ is Π_2 and $V_\lambda = H_\lambda$, by the downward direction of Levy's result mentioned before 22.3, $V_\lambda \models \psi[y, x]$. But then $V_\lambda \models \varphi[x]$, and hence $V_\kappa \models \varphi[x]$. ⊣

23.11 Exercise. *If $\kappa < \lambda$, $V_\lambda \models \kappa$ is extendible, and λ is extendible, then κ is extendible.*

Hint. Argue as for 22.9, using $\ulcorner\kappa$ is extendible$\urcorner$ is Π_3^{ZF}. ⊣

This shows that 23.10 is optimal in the sense that the Σ_4^{ZF} sentence $\ulcorner$there is an extendible cardinal$\urcorner$ does not hold in V_κ if κ is the least extendible cardinal.

These various results, drawn mainly from Solovay-Reinhardt-Kanamori [78], show how extendibility extends supercompactness, but also how the two concepts are closely related. Of the two, supercompactness has become a focal hypothesis in set theory, serving as the beginning hypothesis in several relative consistency results, whereas extendibility has receded from view. The simple 23.5, showing how local extendibility is a consequence of local supercompactness, may provide the underlying reason.

Proofs of a Theorem of Kunen

At the end of his thesis Reinhardt briefly considered the following axiom as a strong expression of ideas for extending Ackermann's set theory: There is a $j\colon V \prec V$. As noted in his [74a: 200], this is also suggested as an extension of extendibility: For such a j with $\mathrm{crit}(j) = \kappa$ and any ordinal η, $j|V_{\kappa+\eta}\colon V_{\kappa+\eta} \prec V_{j(\kappa+\eta)}$, and so save for a small quibble about $\eta < j(\kappa)$, it is as if one j works for all η. Although not the original motivation, $j\colon V \prec V$ is most immediately suggested as the ultimate possibility in terms of closure conditions on target inner models, proceeding progressively from measurability through degrees of supercompactness.

Reinhardt's proposal led to a dramatic turn of events. After initial results aroused some suspicion, Kunen established that in ZFC there can be no $j: V \prec V$:

23.12 Theorem (Kunen [71b]). *Suppose that $j: V \prec M$. Then $M \neq V$.*

As the quantification $\forall j$ over classes j cannot be formalized in ZFC, this result can only be regarded as a schema of theorems, one for each j. Kunen's argument does lead to a simple assertion about sets (23.14(a)).

23.12 has delimited the whole large cardinal enterprise. It could have been that $j: V \prec V$ would serve as the culmination of the guiding idea of closure conditions on range models of elementary embeddings; a new guiding idea in some orthogonal direction would have been exploited to formulate still stronger hypotheses; and so on. Rather, in quickly resolving the situation with an appropriately simple statement, Kunen's result sharply defined the context and showed that a completion of ZFC in a specific sense exists. The particular form of the result was intriguing and unexpected, and although the original proof has an *ad hoc* flavor, what it established has not since been superceded by any stronger inconsistency result.

Several proofs of 23.12 are given. The first is that original proof, which applied a combinatorial result of Erdős-Hajnal [66]. The result is of independent interest in connection with Jónsson cardinals (§8) since it shows that if infinitary operations are allowed, then there are Jónsson algebras of every infinite cardinality. For any set x of ordinals and function f,

$$f \text{ is } \omega\text{-}\textit{Jónsson for } x \quad \textit{iff} \quad f: [x]^\omega \to x \text{ and for any } y \subseteq x \text{ with } |y| = |x|,\ f``[y]^\omega = x\,.$$

There is a function ω-Jónsson for x *iff* there is a function ω-Jónsson for $|x|$, and for cardinals κ, there is a function ω-Jónsson for κ *iff* $\kappa \not\rightarrow [\kappa]^\omega_\kappa$ in the notation of §8. The following proof is drawn from Galvin-Prikry [76] and has an antecedent in Galvin [65].

23.13 Theorem (Erdős-Hajnal [66]). *For any λ, there is a function ω-Jónsson for λ.*

Proof. For $x, y \in [\lambda]^\omega$ set $x \sim y$ *iff* they have equal "tails", i.e. for some $\alpha < \bigcup x$, $x - \alpha = y - \alpha$. Then $\sim$ is an equivalence relation on $[\lambda]^\omega$, so for each equivalence class E choose a representative $x_E \in E$. (This is the use of AC in this proof.) Now for any $x \in [\lambda]^\omega$, with E the equivalence class of x and $\alpha \in x_E$ minimal such that $x_E - (\alpha + 1) = x - (\alpha + 1)$, set $g(x) = \alpha$.

To complete the proof, it suffices to find an $A \in [\lambda]^\lambda$ such that for any $B \in [A]^\lambda$, $g``[B]^\omega \supseteq A$, since for such an A a function ω-Jónsson for A can be readily derived from g. So, assume to the contrary that no such A exists. Then for $n \in \omega$ there are $A_n \in [\lambda]^\lambda$ and $a_n \in \lambda$ with $A_n \supseteq A_{n+1}$ and $a_{n+1} \in A_n - (a_n + 1)$ such that

$a_{n+1} \notin g``[A_{n+1}]^\omega$. Set $y = \{a_n \mid n \in \omega\} \in [\lambda]^\omega$. With $E$ the equivalence class of $y$ let $m \in \omega$ be such that for some $\alpha < \bigcup x_E$, $\{a_n \mid m \leq n < \omega\} = x_E - \alpha$. But then, $g(\{a_n \mid m < n < \omega\}) = a_m$ so that $a_m \in g``[A_m]^\omega$, which is a contradiction. ⊣

First Proof of 23.12 (Kunen [71b]). Let $\kappa = \mathrm{crit}(j)$, and j^n denote the nth iteration of j, i.e. for any $x \in V$, $j^0(x) = x$ and $j^{n+1}(x) = j(j^n(x))$ for $n \in \omega$. Set $\lambda = \sup(\{j^n(\kappa) \mid n \in \omega\})$ so that λ is the least ordinal δ above κ such that $j(\delta) = \delta$. To conclude that $M \neq V$, it is established that $j``\lambda \notin M$:

Assume to the contrary that $j``\lambda \in M$, and let $f$ be $\omega$-Jónsson for $\lambda$. In $M$, $j(f)$ is $\omega$-Jónsson for $j(\lambda) = \lambda$, and $j``\lambda \in [\lambda]^\lambda \cap M$. However, we shall check that $j(f)``[j``\lambda]^\omega \subseteq j``\lambda$, which is a contradiction since $j``\lambda \neq \lambda$ (for example, $\kappa \in \lambda - j``\lambda$).

Suppose then that $s \in [j``\lambda]^\omega$. Clearly there is a $t \in [\lambda]^\omega$ such that $j(t) = j``t = s$. But then, $j(f)(s) = j(f)j(t) = j(f(t)) \in j``\lambda$. ⊣

As Kunen observed, since λ is a strong limit cardinal of cofinality ω, $2^\lambda = \lambda^{\aleph_0}$, and to establish 23.13 for such λ there is a simple recursive construction available (albeit one that uses a bit more of AC): Let $\{\langle x_\alpha, y_\alpha\rangle \mid \alpha < 2^\lambda\}$ enumerate $[\lambda]^\lambda \times \lambda$. For $\alpha < 2^\lambda$ recursively choose $s_\alpha \in [x_\alpha]^\omega$ so that $s_\alpha \neq s_\beta$ for $\beta < \alpha$; this is possible since $2^\lambda = \lambda^{\aleph_0}$. Then any function $f\colon [\lambda]^\omega \to \lambda$ such that $f(s_\alpha) = y_\alpha$ is ω-Jónsson for λ.

Other proofs of 23.12 appeared in the late 1980's. One, due to Woodin, depends only on partitioning stationary sets into disjoint stationary subsets:

Second Proof of 23.12 (Woodin). Let $\kappa = \mathrm{crit}(j)$ and $\lambda = \sup(\{j^n(\kappa) \mid n \in \omega\})$ as before. By Solovay's 16.9, there is an $S\colon \kappa \to \mathcal{P}(\lambda^+)$ such that $\mathrm{ran}(S)$ is a partition of $\{\xi < \lambda^+ \mid \mathrm{cf}(\xi) = \omega\}$ into sets stationary in λ^+. (In fact, because of the nature of the set to be partitioned, a recursive definition of such an S is possible.) Since $j(\lambda) = \lambda$, $\lambda^+ \leq j(\lambda^+) = \lambda^{+M} \leq \lambda^+$ so that equality pervades. Hence, by elementarity $j(S)\colon j(\kappa) \to \mathcal{P}(\lambda^+)$ and

$$(j(S)(\kappa) \subseteq \{\xi < \lambda^+ \mid \mathrm{cf}(\xi) = \omega\} \text{ is stationary in } \lambda^+)^M .$$

Assume to the contrary that $j(S)(\kappa)$ is really stationary in λ^+, a consequence of $M = V$. Then by λ^+-completeness $j(S)(\kappa) \cap S(\alpha_0)$ is stationary in λ^+ for some $\alpha_0 < \kappa$. Also,

$$C = \{\xi < \lambda^+ \mid j(\xi) = \xi \ \wedge\ \mathrm{cf}(\xi) = \omega\}$$

is readily seen to be ω-closed unbounded in λ^+. It follows (0.1(d)) that there is a $\xi_0 \in (j(S)(\kappa) \cap S(\alpha_0)) \cap C$. But then, $\xi_0 = j(\xi_0) \in j(S(\alpha_0)) = j(S)(\alpha_0)$ so that $\xi_0 \in j(S)(\kappa) \cap j(S)(\alpha_0)$, which is a contradiction since by elementarity the range of $j(S)$ must consist of pairwise disjoint sets. ⊣

Todorčević observed that the consideration of C is unnecessary: Assume as above that $j(S)(\kappa)$ is really stationary in λ^+. Since $j``\lambda^+$ is unbounded in $j(\lambda^+) = \lambda^+$,

$$\overline{C} = \{\xi < \lambda^+ \mid \exists s(s \in [j``\lambda^+]^\omega \ \wedge\ \xi = \sup(s))\}$$

is ω-closed unbounded, so let $\xi_0 \in j(S)(\kappa) \cap \overline{C}$ with $s \in [j``\lambda^+]^\omega$ such that $\xi_0 = \sup(s)$. There is a $t \in [\lambda^+]^\omega$ such that $j(t) = s$, and $\sup(t) \in S(\alpha_0)$ for some $\alpha_0 < \kappa$. But $j(\sup(t)) = \xi_0 \in j(S)(\kappa)$, and so $j(\alpha_0) = \kappa$ by elementarity, which is a contradiction.

In fact, for regular $\mu > \omega$, with $S\colon \mu \to \mathcal{P}(\mu)$ such that $\mathrm{ran}(S)$ is a partition of $\{\xi < \mu \mid \mathrm{cf}(\xi) = \omega\}$ into sets stationary in μ, $f\colon [\mu]^\omega \to \mu$ defined by $f(s) =$ that unique α such that $\sup(s) \in S(\alpha)$ is a simple example due to Solovay (see Galvin-Prikry [76: 370]) of a function ω-Jónsson for μ. Consequently, we have a reversion to Kunen's argument through an explicit function of ω-Jónsson sort for $\lambda^+ = j(\lambda^+)$.

The following is a simpler version of a proof of 23.12 due to Mikio Harada; rather than depending on combinatorial contingencies it reveals a more structural constraint.

Third Proof of 23.12 (Harada). Letting λ be as before, assume again to the contrary that $j``\lambda \in M$.

Note first that for any bijection $g\colon 2^\lambda \to \mathcal{P}(\lambda)$, $j``\lambda \in \mathcal{P}(\lambda)^M = \mathrm{ran}(j(g))$. So, let $\sigma$ be the *least* ordinal such that for some injective function $F\colon \sigma \to \mathcal{P}(\lambda)$, $j``\lambda \in \mathrm{ran}(j(F))$. Fix such an F, and set $S = \mathrm{ran}(F)$. Now define U by:

$$X \in U \ \ \textit{iff}\ \ X \subseteq S \ \wedge\ j``\lambda \in j(X)\ .$$

Then U is an ω_1-complete ultrafilter over S, so let

$$i\colon V \prec N \cong \mathrm{Ult}(V, U)\ .$$

Recalling the theory of normal ultrafilters from §22, if id: $S \to S$ is the identity map, the definition of U implies successively that:

(i) $i``\lambda = [\mathrm{id}]_U$ and so $i``\lambda \in \mathrm{ran}(i(F))$,
(ii) $\alpha = [\langle \mathrm{ot}(x \cap \alpha) \mid x \in S\rangle]_U$ for $\alpha \leq \lambda$, and
(iii) $i|(\lambda+1) = j|(\lambda+1)$.

(iii) follows by arguments for 22.12, since $k\colon N \to M$ defined by $k([f]_U) = j(f)(j``\lambda)$ is elementary; $j = k \circ i$; and $k(\alpha) = \alpha$ for $\alpha \leq \lambda$ by (ii).

The switch to ultrapowers secures more structure: By 22.4(b), $i``\lambda \in N$ implies that $\mathcal{P}(\lambda)^N = \mathcal{P}(\lambda)$ and also that $i|V_\lambda \in N$ by a simple argument as λ is a strong limit cardinal. Now for any $X \in \mathcal{P}(\lambda)$, $i(X) = \bigcup_{\alpha<\lambda} i(X \cap \alpha)$ since $i(\lambda) = \lambda$, so that in turn $i``\mathcal{P}(\lambda) = \{\bigcup_{\alpha<\lambda} i(X \cap \alpha) \mid X \in \mathcal{P}(\lambda)\} \in N$. Again by 22.4(b) it follows that N has all the enumerations of $\mathcal{P}(\lambda)^N = \mathcal{P}(\lambda)$, and hence

$$(*) \qquad 2^\lambda = (2^\lambda)^N = i((2^\lambda))\ .$$

The properties of σ are now used to derive a contradiction. Clearly the definitions imply that σ is a cardinal and $|S| = \sigma \leq 2^\lambda$. 22.4(c) implies that

$i``\sigma^+ \notin N$; but since $i``\mathcal{P}(\lambda) \in N$ as noted above, it follows from 22.4(a)(b) that $2^\lambda < \sigma^+$. Hence, it must be that $\sigma = 2^\lambda$. But then, $i(\sigma) = \sigma$ from $(*)$, so that $i(\sigma) = \sup(i``\sigma)$. Since $i``\lambda \in \mathrm{ran}(i(F))$ by (i), it follows that $i``\lambda \in \mathrm{ran}(i(F)|i(\alpha)) = \mathrm{ran}(i(F|\alpha))$ for some $\alpha < \sigma$. But then, $j``\lambda \in \mathrm{ran}(j(F|\alpha))$ by definition of U and i, contradicting the minimality of σ. ⊣

These proofs of 23.12 lead to the following refinements:

23.14 Corollary.
(a) For any δ, there is no $j\colon V_{\delta+2} \prec V_{\delta+2}$.
(b) If $j\colon V \prec M$ and δ is the least ordinal above crit(j) *such that $j(\delta) = \delta$, then $j``\delta \notin M$.*

Proof. (a) Assume to the contrary that there is such a j. Clearly $j(\delta) = \delta$, and $\mathrm{crit}(j) < \delta$. (Elementary embeddings are not the identity on their domains by convention, and the argument for 5.1(b) implies that j must move some ordinal.) As before, let $\kappa = \mathrm{crit}(j)$, and noting that each iterate $j^n(\kappa)$ is defined and less than δ, set $\lambda = \sup(\{j^n(\kappa) \mid n \in \omega\}) \leq \delta$. Kunen's contradiction can now be derived since the function ω-Jónsson for λ is in $V_{\lambda+2}$, and hence in the domain of j.

For Woodin's proof, note that subsets of λ^+ can be faithfully coded by members of $V_{\lambda+2}$. For example, for each well-ordering $R \subseteq \lambda \times \lambda$ of λ, let $\alpha_R < \lambda^+$ be its ordertype, and $e_R\colon \langle \alpha_R, <\rangle \to \langle \lambda, R\rangle$ the order-isomorphism. Let $p\colon (\lambda \times \lambda) \times \lambda \to \lambda$ be a bijection, noting that $p \subseteq V_\lambda$. Then for any $X \subseteq \lambda^+$,

$$\{p``(R \times e_R``(X \cap \alpha_R)) \mid R \text{ is a well-ordering of } \lambda\} \in V_{\lambda+2}$$

can serve as a code for X. Short sequences of subsets of λ^+ can be similarly coded, and it is straightforward to see that the relevant assertions like the stationariness of a subset of λ^+ can be expressed in $V_{\lambda+2}$ in terms of codes. Hence, Woodin's contradiction can be derived via codes with $j|V_{\lambda+2}$.

Harada's proof can also be carried out with $j|V_{\lambda+2}$, since every well-ordering of a subset of $\mathcal{P}(\lambda)$ can be coded by a member of $V_{\lambda+2}$, and so the σ and corresponding F and $U \subseteq V_{\lambda+2}$ can be defined via such codes.

(b) Note that if λ is again defined as before from $\mathrm{crit}(j)$, then $\delta = \lambda$. Kunen's and Harada's proofs thus show that $j``\delta \notin M$. To push through Woodin's argument, first switch from $j$ to an ultrapower embedding as in Harada's proof: Assuming that $j``\lambda \in M$, define U over $\mathcal{P}(\lambda)$ by

$$X \in U \text{ iff } X \subseteq \mathcal{P}(\lambda) \wedge j``\lambda \in j(X)\,,$$

and let $i\colon V \prec N \cong \mathrm{Ult}(V, U)$. Then as seen before, $\mathrm{crit}(i) = \mathrm{crit}(j)$, $i(\lambda) = \lambda$, $i``\lambda \in N$, and $i``\mathcal{P}(\lambda) \in N$. Hence ${}^{2^\lambda}N \subseteq N$, and Woodin's proof can be carried out via codes as for (a). ⊣

The following makes good on a small debt concerning the definitions of γ-supercompactness and extendibility.

23.15 Proposition.

(a) κ is γ-supercompact iff there is a $j: V \prec M$ with $\mathrm{crit}(j) = \kappa$ *and* ${}^{\gamma}M \subseteq M$.

(b) κ is extendible iff for any $\eta > \kappa$ there is a $j: V_\eta \prec V_\zeta$ with $\mathrm{crit}(j) = \kappa$.

Proof. (a) Suppose that j is as stated; only the condition $\gamma < j(\kappa)$ is missing. But by 23.14(b), $\gamma < \sup(\{j^n(\kappa) \mid n \in \omega\})$, and hence $\gamma < j^n(\kappa)$ for some $n \in \omega$. Hence, it suffices to establish

(∗) For every $n \in \omega$ with $n > 0$ there is an inner model M_n of ZFC such that $j^n: V \prec M_n$ and ${}^{\gamma}M_n \subseteq M_n$.

To this end first extend j to classes R by stipulating that

$$j^+(R) = \bigcup_\alpha j(R \cap V_\alpha) \,.$$

Note that for classes $R_1, \ldots, R_m$,

$$j\colon \langle V, \in, R_1, \ldots, R_m\rangle \prec \langle M, \in, j^+(R_1), \ldots, j^+(R_m)\rangle$$

(where extending a standing convention, $\prec$ is $\prec_1$ for formalizability) via the Reflection Principle for ZF.

(∗) can now be established by induction on n. For $n = 1$ it holds with $M_1 = M$. Assume now that it holds for n, and set $M_{n+1} = j^+(M_n)$. Since

$$j\colon \langle V, \in, j^n, M_n\rangle \prec \langle M, \in, j^+(j^n), M_{n+1}\rangle \,,$$

M_{n+1} is transitive, and $j^+(j^n)\colon M \prec M_{n+1}$ so that $j^+(j^n) \circ j\colon V \prec M_{n+1}$. But $j^+(j^n) \supseteq \{\langle j(x), j^{n+1}(x)\rangle \mid x \in V\rangle\}$, and so $j^+(j^n) \circ j = j^{n+1}$. Finally, since $\langle V, \in, M_n\rangle \prec \langle M, \in, M_{n+1}\rangle$ and ${}^{\gamma}M_n \subseteq M_n$, ${}^{j(\gamma)}M_{n+1} \cap M \subseteq M_{n+1}$, and so

$${}^{\gamma}M_{n+1} = {}^{\gamma}M_{n+1} \cap M \subseteq M_{n+1} \,.$$

(b) It suffices to show from the latter assertion that, taking $\eta \geq \kappa \cdot \omega$, κ is η-extendible – only the condition $\eta < j(\kappa)$ is missing. For this purpose, let $\gamma > \eta$ be such that

(i) whenever $\beta < \gamma$ and for some ζ there is a $k\colon V_\eta \prec V_\zeta$ with $\mathrm{crit}(k) = \kappa$ and $k(\kappa) = \beta$, there is such a k with a $\zeta < \gamma$.

(ii) $\mathrm{cf}(\gamma) = \omega_1$.

Such a γ exists by a simple closure argument iterated ω_1 times. By hypothesis, let $j\colon V_\gamma \prec V_\rho$ with $\mathrm{crit}(j) = \kappa$. If $j^n(\kappa)$ were defined for every $n \in \omega$, then $\sup(\{j^n(\kappa) \mid n \in \omega\}) < \gamma$ as $\mathrm{cf}(\gamma) = \omega_1$, and we would be able to derive Kunen's contradiction. Hence, it follows that there is an $m \in \omega$ such that $j^m(\kappa) < \gamma \leq j^{m+1}(\kappa)$.

Let $P(i)$ be the assertion:

There is an $e\colon V_\eta \prec V_\zeta$ for some ζ with $\mathrm{crit}(e) = \kappa$ and $e(\kappa) = j^{i+1}(\kappa)$.

It suffices to establish $P(m)$, for then η-extendibility would follow from $\eta < \gamma \leq j^{m+1}(\kappa)$. To this end, $P(i)$ is established for every $i \leq m$ by induction on i:

Set $\overline{j} = j|V_\eta$. Then as for 23.3, $\overline{j}\colon V_\eta \prec V_{j(\eta)}$ with $\mathrm{crit}(\overline{j}) = \kappa$ and $\overline{j}(\kappa) = j(\kappa)$, and $P(0)$ holds. Now assume $P(i)$ where $i < m$. Then because $j^{i+1}(\kappa) < \gamma$ it follows from (i) that for some $\zeta < \gamma$ there is an $k\colon V_\eta \prec V_\zeta$ with $\mathrm{crit}(k) = \kappa$ and $k(\kappa) = j^{i+1}(\kappa)$. j can therefore by applied to k to conclude by 23.2 and elementarity that in V_ρ, and hence in V, there is an $\overline{k}\colon V_{j(\eta)} \prec V_{j(\zeta)}$ with $\mathrm{crit}(\overline{k}) = j(\kappa)$ and $\overline{k}(j(\kappa)) = j^{i+2}(\kappa)$. But then, with $\overline{j}$ as above, $\overline{k} \circ \overline{j}\colon V_\eta \prec V_{j(\zeta)}$ with $\mathrm{crit}(\overline{k} \circ \overline{j}) = \kappa$ and $\overline{k} \circ \overline{j}(\kappa) = j^{i+2}(\kappa)$. Hence, $P(i+1)$ obtains, and the proof is complete. ⊣

Because of its obvious import, Kunen's result quickly came under close scrutiny. A prominent feature of the proofs is the appeal to the Axiom of Choice at the level of a well-ordering of $\mathcal{P}(\lambda)$. This is a crucial factor in getting an ω-Jónsson function for λ in the first proof (actually, a well-ordering of $[\lambda]^\omega$ suffices for the given proof of 23.13), in splitting $\{\xi < \lambda^+ \mid \mathrm{cf}(\xi) = \omega\}$ into κ stationary subsets of λ^+ (through their codes in $\mathcal{P}(\lambda)$), and in arguing for the existence of σ in the third. The following unresolved question is therefore of foundational interest:

23.16 Question. *Is it provable in* ZF *that there is no* $j\colon V \prec V$?

Pending an answer, Kunen's result can best be viewed as an ultimate limitation imposed by the Axiom of Choice on the extent of reflection possible in the universe. ZFC rallies at last to force a veritable Götterdämmerung for large cardinals!

24. The Strongest Hypotheses

This section describes the strongest large cardinal hypotheses not known to be inconsistent, starting with those on the verge of Kunen's result and then n-hugeness and Vopěnka's Principle. Like the development of set theory after the discovery of the antinomies, there was a stepping back from the precipice of Kunen's inconsistency, a charting out of possibilities that remained, and with the passage of time, a growing confidence in the delimited edifice. However, unlike the emergence of the cumulative hierarchy and other guiding ideas that provided intuitive underpinnings for ZFC, it is doubtful that even heuristic arguments can be put forward for the optimality of Kunen's result, because of its basis in a specific mathematical contingency. The strongest hypotheses thus stand on much shakier ground, but their study has a natural appeal owing to the power and simplicity of the concepts involved as well as the possibility of some new apocalyptic inconsistency.

The following propositions just weaker than those denied by 23.14 were considered in Gaifman [74] and Solovay-Reinhardt-Kanamori [78]:

I1. For some δ, there is a $j\colon V_{\delta+1} \prec V_{\delta+1}$.
I2. There is a $j\colon V \prec M$ such that $V_\delta \subseteq M$ for some $\delta > \mathrm{crit}(j)$ satisfying $j(\delta) = \delta$.
I3. For some δ, there is a $j\colon V_\delta \prec V_\delta$.

The class assertion I2 is formalizable through a coming set characterization. A proposition yet stronger than all of these, I0, will be mentioned shortly.

These propositions have thus far defied all attempts at refutation in ZFC. It is clear from the proof of 23.14 that if in each case $\kappa = \mathrm{crit}(j)$ and $\lambda = \sup(\{j^n(\kappa) \mid n \in \omega\})$ as before, then necessarily δ is λ, except that for I3 δ could also be $\lambda + 1$ to yield the stronger I1. One attempt to refute I1 would be to get a function in $V_{\lambda+1}$ that would play the role of the function ω-Jónsson for λ. Since λ is a limit of measurable cardinals, namely the $j^n(\kappa)$'s, λ is Jónsson (8.7) so that $[\lambda]^\omega$ cannot be replaced by $[\lambda]^{<\omega}$ in $f\colon [\lambda]^\omega \to \lambda$. However, a positive answer to the following would refute I1.

24.1 Question (Magidor). *For strong limit λ is there an $f\colon \bigcup_{\nu<\lambda}[\nu]^\omega \to \lambda$ such that whenever $y \in [\lambda]^\lambda$, $f``(\bigcup_{\nu<\lambda}[\nu \cap y]^\omega) = \lambda$?*

That is, is there an analogue of the function ω-Jónsson for λ with domain consisting of the subsets of λ of ordertype ω that are bounded below λ? Of course, such an analogue would be in $V_{\lambda+1}$. Masahiro Shioya [93] has shown that a positive answer to 24.1 would also refute I2.

Some results about I1-I3 are now established, and later some recent work described in light of which these propositions might be approached with some confidence. Recalling the proof of 23.15(a), for $j\colon V_\delta \prec V_\delta$ with δ a limit ordinal define an extension $j^+\colon V_{\delta+1} \to V_{\delta+1}$ by stipulating for $R \subseteq V_\delta$ that

$$j^+(R) = \bigcup_{\alpha<\delta} j(R \cap V_\alpha) .$$

Since $V_\delta \times V_\delta \subseteq V_\delta$, $j^+(R)$ is thus defined for relations R on V_δ, functions $R\colon V_\delta \to V_\delta$, and the like. j^+ is not generally elementary, but significantly, it is the only possibility for an elementary extension of j to domain $V_{\delta+1}$.

The following characterization of I2, evident from Gaifman [74: 89] or Powell [74: 129], clarifies its relationship to I1 and I3:

24.2 Proposition. *The following are equivalent for any κ and δ:*

(a) There is a $j\colon V \prec M$ with $\text{crit}(j) = \kappa < \delta = j(\delta)$ and $V_\delta \subseteq M$.

(b) There is a $j\colon V_\delta \prec V_\delta$ with $\text{crit}(j) = \kappa$ such that whenever R is a well-founded relation on V_δ, so is $j^+(R)$.

Proof. For the forward direction, set $\bar{j} = j|V_\delta$. Then $\bar{j}\colon V_\delta \prec V_\delta$, and for any $R \subseteq V_\delta$, $\bar{j}^+(R) = j(R)$. If moreover R is a well-founded relation, then $M \models j(R)$ is well-founded. But then, by absoluteness (0.3) $j(R)$ is well-founded in V.

For the converse direction, the idea is that although no ultrapower embedding can have the requisite properties, we can take a direct limit of ultrapower embeddings: For $n \in \omega$, set $\kappa_n = j^n(\kappa)$ and define U_n by:

$$X \in U_n \quad \textit{iff} \quad X \subseteq \mathcal{P}(\kappa_n) \ \wedge \ j``\kappa_n \in j(X)\ .$$

Noting that $j``\kappa_n \in \mathcal{P}(\kappa_{n+1})$ and $X \subseteq \mathcal{P}(\kappa_n)$ implies that $j(X) \subseteq \mathcal{P}(\kappa_{n+1})$, it is simple to check that U_n is a κ-complete ultrafilter over $\mathcal{P}(\kappa_n)$. Set

$$i_n\colon V \prec M_n \cong \text{Ult}(V, U_n)\ .$$

Recalling the theory of normal ultrafilters in §22, with $\text{id}_n\colon \mathcal{P}(\kappa_n) \to \mathcal{P}(\kappa_n)$ the identity map,

(i) $i_n``\kappa_n = [\text{id}_n]_{U_n}$, and so

(ii) $\alpha = [\langle \text{ot}(x \cap \alpha) \mid x \in \mathcal{P}(\kappa_n)\rangle]_{U_n}$ for $\alpha \leq \kappa_n$,

(iii) $\text{crit}(i_n) = \kappa$, and

(iv) $V_{\kappa_n} \subseteq M_n$.

For (iii), note that by the representation of κ given by (ii) and the definition of U_n, $\kappa < i_n(\kappa)$ *iff* $\text{ot}(j``\kappa_n \cap j(\kappa)) < j(\kappa)$, and the latter holds since the ordertype is κ. (iv) follows from (i) and 22.4(a) by an inductive argument, since κ_n is inaccessible.

Next, for $n \leq m < \omega$ define $k_{nm}\colon M_n \to M_m$ by:

$$k_{nm}([f]_{U_n}) = [\langle f(x \cap \kappa_n) \mid x \in \mathcal{P}(\kappa_m)\rangle]_{U_m}\ .$$

Then recalling 22.12 and subsequent remarks,

(v) $k_{nm}\colon M_n \prec M_m$, and

(vi) $k_{nm}(x) = x$ for every $x \in V_{\kappa_n}$.

For (vi), $k_{nm}(\alpha) = \alpha$ for every $\alpha \leq \kappa_n$ by (ii), and so with (iv) the rank argument for 5.1(b) can be used.

Finally, as $\langle\langle\langle M_n, \in\rangle \mid n \in \omega\rangle, \langle k_{nm} \mid n \leq m\rangle\rangle$ is seen to be a directed system, let $\langle M, \in_M\rangle$ be its direct limit. For $n \in \omega$ let

$$k_n\colon \langle M_n, \in\rangle \prec \langle M, \in_M\rangle$$

be the corresponding embedding so that $k_m \circ k_{nm} = k_n$ for $n \leq m$, and let

$$i\colon \langle V, \in\rangle \prec \langle M, \in_M\rangle$$

so that $k_n \circ i_n = i$. $\langle M, \in_M\rangle$ is well-founded:

Assume to the contrary that there are $z_r \in M$ such that $z_{r+1} \in_M z_r \in_M \mathrm{On}^M$ for every $r \in \omega$. For such r, let κ_{n_r} be such that $\kappa_{n_r} \leq \kappa_{n_{r+1}}$, and $g_r\colon \mathcal{P}(\kappa_{n_r}) \to \mathrm{On}$ such that $k_{n_r}([g_r]) = z_r$. Since $|\bigcup_{r\in\omega} \mathrm{ran}(g_r)| \leq \delta$, there is a well-ordering R of δ and an order-preserving injection $p\colon \langle\bigcup_{r\in\omega} \mathrm{ran}(g_r), <\rangle \to \langle\delta, R\rangle$. It is now straightforward to check from the definitions of the U_n's and k_{nm}'s that for every $n \in \omega$,

$$\langle j(p \circ g_{r+1})(j``\kappa_{n_{r+1}}), j(p \circ g_r)(j``\kappa_{n_r})\rangle \in j^+(R)\ .$$

But then, $j^+(R)$ is ill-founded, contradicting a key hypothesis on j.

M can now be identified with its transitive collapse, and $\in_M$ with $\in$. The proof is complete, since $\mathrm{crit}(i) = \kappa$ by (iii) and $V_\delta \subseteq M$ by (iv) and (vi). ⊣

This characterization leads to the following transcendence results of I1 over I2 and I2 over I3. For convenience, temporarily let I1(κ, δ) be the proposition that there is a $j\colon V_{\delta+1} \prec V_{\delta+1}$ with $\mathrm{crit}(j) = \kappa$, and analogously define I2(κ, δ) (in terms of the above characterization for formalizability) and I3(κ, δ).

24.3 Proposition. *Suppose that* I1(κ, δ). *Then:*

(a) (Gaifman [74: 89]) I2(κ, δ).

(b) There is a normal ultrafilter U over κ such that

$$\{\alpha < \kappa \mid \mathrm{I2}(\alpha, \delta)\} \in U\ .$$

Proof. Suppose that $j\colon V_{\delta+1} \prec V_{\delta+1}$ with $\mathrm{crit}(j) = \kappa$. For (a), set $\overline{j} = j|V_\delta$. Then $\overline{j}\colon V_\delta \prec V_\delta$ by the argument for 23.3, and for any $R \subseteq V_\delta$, $\overline{j}^+(R) = j(R)$. Moreover, R is a well-founded relation exactly when it is so in the sense of $V_{\delta+1}$, since $V_{\delta+1}$ contains all functions: $\omega \to V_\delta$. Hence, the well-foundedness of R implies the well-foundedness of $j(R)$ by elementarity.

For (b), note that the above argument shows that $V_{\delta+1} \models \mathrm{I2}(\kappa, \delta)$ since there is an $S \in V_{\delta+1}$ coding the satisfaction relation for V_δ. Hence, the result follows by the usual reflection argument using the normal ultrafilter over κ defined from j. ⊣

24.4 Proposition. *Suppose that* I2(κ, δ). *Then:*

(a) I3(κ, δ).

(b) There is a normal ultrafilter U over κ such that

$$\{\alpha < \kappa \mid \mathrm{I3}(\alpha, \delta)\} \in U\ .$$

Proof. (a) is immediate in terms of 24.2. For (b), first let $j\colon V \prec M$ with $\mathrm{crit}(j) = \kappa < \delta = j(\delta)$ and $V_\delta \subseteq M$ as in that characterization. It suffices, by the usual reflection argument using the normal ultrafilter over κ defined from j, to show that

$$M \models \exists i(i\colon V_\delta \prec V_\delta \ \wedge\ \mathrm{crit}(i) = \kappa)\ .$$

It cannot be asserted that $j|V_\delta \in M$, but a well-foundedness argument suffices:

Set

$$I = \{i \mid i\colon V_\alpha \prec V_\beta \text{ for some } \alpha < \beta < \delta \text{ and } \mathrm{crit}(i) = \kappa\}\ .$$

Noting that by elementarity the cofinality of δ in M is also ω, fix a sequence $\langle \gamma_n \mid n \in \omega\rangle \in M$ cofinal in δ. Now define $<$ on I as follows: For $i, i' \in I$, say with $i\colon V_\alpha \prec V_\beta$ and $i'\colon V_{\alpha'} \prec V_{\beta'}$,

$$i < i' \ \text{ iff } \ i \supseteq i' \ \wedge\ \exists n \in \omega(\gamma_n \in \mathrm{dom}(i) - \mathrm{dom}(i'))\ .$$

Then let $\leq^*$ be the reflexive, transitive closure of $<$, i.e. that partial order in which i' is an immediate successor of i exactly when $i < i'$.

Since $V_\delta \subseteq M$ and $\langle \gamma_n \mid n \in \omega\rangle \in M$, $I^M = I$ and $\leq^{*M} = \leq^*$. If $\leq^*$ is ill-founded in M, then the union of any infinite $\leq^*$-descending sequence would be an $i\colon V_\delta \prec V_\delta$ as desired. So, assume to the contrary that $\leq^*$ is well-founded in M. But then, by absoluteness (0.3) $\leq^*$ is well-founded in V. This is a contradiction, since some subsequence of $\langle j|V_{j^{n+1}(\kappa)} \mid n \in \omega\rangle$ is an infinite $\leq^*$-descending chain. ⊣

24.3(b) and 24.4(b) are simple reflection observations focusing on the critical point κ. For reflection on the δ, Solovay (see Gaifman [74: 92]) observed that if I2(κ, δ), then for some $\kappa' < \delta' < \kappa$, I3($\kappa', \delta'$). As part of a latter-day, detailed analysis of I1–I3 through definability, Laver [97] established that if I1(κ, δ), then for some $\kappa' < \delta' < \kappa$, I2($\kappa', \delta'$).

The following is an observation concerning I3. For $j\colon V_\delta \prec V_\delta$ with δ a limit ordinal, j had been extended to a $j^+\colon V_{\delta+1} \to V_{\delta+1}$ by stipulating for $R \subseteq V_\delta$ that $j^+(R) = \bigcup_{\alpha<\delta} j(R \cap V_\alpha)$. What about taking R itself to be a map $k\colon V_\delta \prec V_\delta$?

24.5 Exercise. *Suppose that* $j\colon V_\delta \prec V_\delta$ *and* $k\colon V_\delta \prec V_\delta$. *Then* $j^+(k)\colon V_\delta \prec V_\delta$, *and* $\mathrm{crit}(j^+(k)) = j(\mathrm{crit}(k))$.

Hint. One way to show that $j^+(k)$ is elementary is to replace quantifiers by Skolem functions and to use a reflection argument. ⊣

Hence, while the extendibility of κ may imply the existence of various large cardinals above κ, I3 is a large cardinal property of $\kappa = \mathrm{crit}(j)$ that implies the *very same* large cardinal property for a cardinal above κ.

Attitudes about and expectations concerning I1-I3 have evolved since their formulation, from skepticism toward confidence and acceptance. In Solovay-Reinhardt-Kanamori [78: 109] we had written: "It seems likely that I1, I2, and I3 are all inconsistent since they appear to differ from the proposition proved inconsistent by Kunen only in an inessential technical way." However, this may not be the case. As mentioned earlier, the Axiom of Choice figures prominently in the proof of Kunen's result by providing a well-ordering of $\mathcal{P}(\lambda)$, and such well-orderings first appear, in coded form, in $V_{\lambda+2} - V_{\lambda+1}$. Hence, any refutations of I1-I3 would have to have a different basis. Also, recent work has provided some extrinsic evidence for their coherence and consistency:

Determinacy hypotheses are discussed in Chapter 6. In 1978 Martin [80] showed (see 31.8 and after) that the hypothesis Det($\mathbf{\Pi}^1_2$) follows from a proposition intermediate between I2 and I3, and in 1984 Woodin showed (see 31.9) that $\mathrm{AD}^{L(\mathbb{R})}$ follows from a proposition stronger than I1 and straining the limits of consistency:

I0. For some δ, there is a j: $L(V_{\delta+1}) \prec L(V_{\delta+1})$ with $\mathrm{crit}(j) < \delta$.

(This last condition ensures that $j|V_{\delta+1}$ witnesses I1.) The proofs of 23.12 preclude the possibility here of $L(V_{\delta+1}) \models \mathrm{AC}$. These Martin and Woodin results, discussed in context in §31, provided the first indications that large cardinal hypotheses suffice to establish strong determinacy hypotheses, and further progress was to lead to remarkable equiconsistency results. That propositions at the level of I0–I3 could serve as appropriate hypotheses for the initial implementation of the proof ideas is some extrinsic evidence in their favor. Moreover, Woodin's investigations of I0 was to produce a detailed and coherent structural theory for $L(V_{\delta+1})$. Laver [01, 02] summarizes and develops this theory further.

Work concerning I3 has led to characterizations of corresponding algebras of elementary embeddings. Martin had used the proliferation of elementary embeddings given by 24.5 and composition to establish his determinacy result, and questions soon arose about the supporting algebraic structure. After exploratory work by Laver [86], Dehornoy [88, 89, 89a, 89b], and Randall Dougherty, Laver [92] in 1988 established the freeness of the algebras subject only to the obvious identities.

Set $\mathcal{E}_\delta = \{j \mid j\colon V_\delta \prec V_\delta\}$. In view of 24.5, define a binary operation on $\mathcal{E}_\delta$ by: $j \cdot k = j^+(k)$. Another operation on $\mathcal{E}_\delta$ is composition; if $j, k \in \mathcal{E}_\delta$, then $j \circ k \in \mathcal{E}_\delta$. The following equalities are simple to check; such rules were first applied by Martin [80].

24.6 Exercise. *Suppose that* $i, j, k \in \mathcal{E}_\delta$. *Then:*

(a) $i \circ (j \circ k) = (i \circ j) \circ k$, $(i \circ j) \cdot k = i \cdot (j \cdot k)$, $i \cdot (j \circ k) = (i \cdot j) \circ (i \cdot k)$, *and* $i \circ j = (i \cdot j) \circ i$.

(b) $i \cdot (j \cdot k) = (i \cdot j) \cdot (i \cdot k)$.

Hint. (b) also follows from (a). ⊣

Let Σ denote the set of laws in (a); (b) is the *left distributive law* for $\cdot$. Laver established that these are the only laws:

24.7 Theorem (Laver [92]). *Suppose that* $j \in \mathcal{E}_\delta$. *Then:*

(a) The closure of $\{j\}$ *in* $\langle \mathcal{E}_\delta, \cdot \rangle$ *is the free algebra with one generator satisfying the left distributive law.*

(b) The closure of $\{j\}$ *in* $\langle \mathcal{E}_\delta, \cdot, \circ \rangle$ *is the free algebra with one generator satisfying* Σ. ⊣

Freeness here has the standard meaning. For (b), let W be the set of terms in one constant a in the language of $\cdot$ and $\circ$. Define an equivalence relation $\equiv$ on W by stipulating that $u \equiv v$ *iff* there is a sequence $u = u_0, u_1, \ldots, u_n = v$ with each u_{i+1} obtained from u_i by replacing a subterm of u_i by a term equivalent to it according to one of the laws of Σ. Then $\cdot$ and $\circ$ are well-defined for equivalence classes, and (b) asserts that the resulting structure on $W/{\equiv}$ and the closure of $\{j\}$ in $\langle \mathcal{E}_\delta, \cdot, \circ \rangle$ are isomorphic via the map induced by sending the equivalence class of a to j.

For $u, v \in W$, define $u <_L v$ *iff* for some $w_1, \ldots, w_{n+1} \in W$,

$$v \equiv ((\ldots(u \cdot w_1) \cdot w_2) \ldots \cdot w_n) \cdot w_{n+1}\ , \quad \text{or}$$
$$v \equiv ((\ldots(u \cdot w_1) \cdot w_2) \ldots \cdot w_n) \circ w_{n+1}\ .$$

Laver [92] showed that if $\mathcal{E}_\delta \neq \emptyset$ for some δ, then $<_L$ is irreflexive, and used this result to obtain a normal form theorem for $W/{\equiv}$ showing that $<_L$ actually linearly orders $W/{\equiv}$, from which 24.7 as well as *the solvability of the word problem for* $W/{\equiv}$ follows (this is the problem of effectively deciding whether $u \equiv v$ or not for arbitrary $u, v \in W$). Considerable interest was generated by a hypothesis bordering on the limits of consistency entailing solvability in finitary mathematics, particularly because of the peculiar and enticing possibility that some strong hypothesis may be necessary. Dehornoy [92] provided another approach to the freeness and solvability conclusions from the irreflexivity of $<_L$. A few years later in late 1991, he [92a, 94] settled matters by establishing that irreflexivity outright in ZFC with an elegant argument based on an embedding into an extension of the infinite braid group. Consequently, all the various structure results obtained about $W/{\equiv}$, including the normal form results of Laver [92, 94], are now theorems of ZFC. This is yet another testament to the fecundity of large cardinals: It suggested a simple algebraic structure and inspired its further study in connection with elementary embeddings and critical points. Much of the structure understood, the focus became the question of the irreflexivity of a certain order, and the large cardinals were gracefully retired from the field. See the monograph Dehornoy [00] for developments in braid groups and distributivity.

The study of $\langle \mathcal{E}_\delta, \cdot \rangle$ has raised other interesting finitary issues. For $j \in \mathcal{E}_\delta$, let $\kappa = \mathrm{crit}(j)$ and let $\mathcal{A}_j$ be the closure of $\{j\}$ in $\langle \mathcal{E}_\delta, \cdot \rangle$. Define a corresponding function f on ω by:

$$f(n) = |\{\mathrm{crit}(k) \mid k \in \mathcal{A}_j \ \wedge \ j^n(\kappa) < \mathrm{crit}(k) < j^{n+1}(\kappa)\}|\ .$$

Laver had noted that $f(0) = 0$, $f(1) = 0$, and $f(2) = 1$, but that $f(3)$ is large. Dougherty [93, 96] established a large lower bound for $f(3)$ and showed moreover that f eventually dominates the Ackermann function, and hence cannot be primitive recursive. It was not even clear that each $f(n)$ is finite, but using a result of Steel, Laver [95] duly established this, so that the ordertype of $\{\mathrm{crit}(k) \mid k \in \mathcal{A}_j\}$ is

ω. The argument involved a tower of finite algebras which may be defined without reference to large cardinals. Laver showed that a simple finitistic proposition Φ about the tower is a consequence of I3, and at present it is not known whether Φ is provable in ZFC. Using Dougherty [93], Dougherty and Jech in their [97] showed that Φ cannot be established in Primitive Recursive Arithmetic.

Of course, the extent of structure at the level of I0–I3 may be due to their inconsistency! However, just as in the case of Silver's $0^\#$ investigations for which one motivation was surely the prospect of a possible refutation of measurability, the further elaboration of coherent structure at the level of I0–I3, despite the stalking of possible new inconsistencies, may reinforce confidence in these strong hypotheses.

n-Hugeness

The remainder of this section is devoted to the main hypotheses between extendibility and I0–I3, as charted out in Solovay-Reinhardt-Kanamori [78]. By Kunen's result, if $j\colon V \prec M$ with $\mathrm{crit}(j) = \kappa$, then M is not closed under sequences of length $\lambda = \sup(\{j^n(\kappa) \mid n \in \omega\})$. It is natural to step back and consider weaker closure conditions. For $n \in \omega$,

$$\begin{aligned} \kappa \text{ is } \textit{n-huge} \ \ \textit{iff} \ \ & \text{there is a } j\colon V \prec M \text{ with } \mathrm{crit}(j) = \kappa \\ & \text{and } {}^{j^n(\kappa)}M \subseteq M\,. \\ \kappa \text{ is } \textit{huge} \ \ \textit{iff} \ \ & \kappa \text{ is 1-huge}\,. \\ \kappa \text{ is } \textit{almost huge} \ \ \textit{iff} \ \ & \text{there is a } j\colon V \prec M \text{ with } \mathrm{crit}(j) = \kappa \\ & \text{and } {}^{\gamma}M \subseteq M \text{ for every } \gamma < j(\kappa)\,. \end{aligned}$$

Note that *κ is 0-huge iff κ is measurable*; that *if $j\colon V \prec M$ witnesses the n-hugeness of κ, then $j^n(\kappa)$ is measurable* (in M and hence in V as $\mathcal{P}(j^n(\kappa))^M = \mathcal{P}(j^n(\kappa))$); and similarly, *if $j\colon V \prec M$ witnesses the almost hugeness of κ, then $j(\kappa)$ is inaccessible in V*. We shall see that the existence of an almost huge cardinal is consistency-wise much stronger than the existence of an extendible cardinal. n-huge cardinals are reminiscent of γ-supercompact cardinals and will have a similar characterization in terms of ultrapowers that provides a formalization in ZFC. However, there is a crucial difference: γ-supercompactness is formulated in terms of an *a priori* γ as the degree of closure of M (and requires $\gamma < j(\kappa)$), but n-hugeness posits closure only *a posteriori* in that M is to be closed under $j^n(\kappa)$ sequences, however large $j^n(\kappa)$ may be. Even then, an attempt has been made to motivate n-hugeness as a reflection property by Victoria Marshall [89].

Hugeness was formulated by Kunen in 1972 in order to have a sufficient hypothesis for a relative consistency result. As mentioned at the end of §17, Kunen [78] established that if there is a huge cardinal, then there is a forcing extension in which there is an ω_2-saturated ideal over ω_1. The conclusion was known to have consistency strength stronger than the existence of a measurable cardinal, and this was the first time that a hypothesis stronger than measurability was used in a relative consistency result about the low orders of the cumulative hierarchy.

Since then, supercompactness and, to some extent, hugeness have played the major such roles. Most arguments have depended on the reflection phenomena provided by supercompactness, but on occasion the relative consistency of two-cardinal propositions has made use of the interplay among κ, $j(\kappa)$ and ${}^{j(\kappa)}M \subseteq M$.

Refinements have followed once upper bounds on consistency strength have been established. For example, almost hugeness became more focal when it was realized that it could replace hugeness in Kunen's argument. Then, as part of an important development, Foreman-Magidor-Shelah [88] further reduced the hypothesis in Kunen's result to supercompactness, and then Shelah reduced the hypothesis to an essentially equiconsistent one (see 32.10 and before). Almost hugeness is still needed, however, to get an ω_3-saturated ideal over ω_2. While it is unlikely that equiconsistency results for supercompactness and hugeness lie in this direction of combinatorial propositions, these hypotheses have considerably clarified their consistency strength.

The following characterization is implicit in the proof of 24.2. Adapting a previous definition, for a filter F over some $\mathcal{P}(\lambda)$ (i.e. $F \subseteq \mathcal{P}(\mathcal{P}(\lambda))$,

$$F \text{ is } \textit{normal iff } \text{(i) for any } \alpha < \lambda,\ \{x \in \mathcal{P}(\lambda) \mid \alpha \in x\} \in F \text{ , and}$$
$$\text{(ii) for any } \langle X_\alpha \mid \alpha < \lambda\rangle \in {}^{\lambda}F,\ \triangle_{\alpha<\lambda} X_\alpha \in F \text{ .}$$

24.8 Theorem. *κ is n-huge iff $\kappa > \omega$ and there is a κ-complete normal ultrafilter U over some $\mathcal{P}(\lambda)$ and cardinals $\kappa = \lambda_0 < \lambda_1 < \ldots < \lambda_n = \lambda$ so that for each $i < n$,*

$$\{x \in \mathcal{P}(\lambda) \mid \mathrm{ot}(x \cap \lambda_{i+1}) = \lambda_i\} \in U \text{ .}$$

Proof. Suppose first that $j\colon V \prec M$ witnesses the n-hugeness of κ. Set $\lambda_i = j^i(\kappa)$ for $i \leq n$, and define U over $\mathcal{P}(\lambda_n)$ by:

$$X \in U \ \textit{ iff } \ X \subseteq \mathcal{P}(\lambda_n) \ \wedge \ j``\lambda_n \in j(X) \text{ .}$$

Then it is simple to check that U is a κ-complete normal ultrafilter. Now note that for $i < n$, $\mathrm{ot}(j``\lambda_n \cap j(\lambda_{i+1})) = j(\lambda_i)$.

For the converse, take $j\colon V \prec M \cong \mathrm{Ult}(V, U)$. Then $[\mathrm{id}]_U = j``\lambda$ where id: $\mathcal{P}(\lambda) \to \mathcal{P}(\lambda)$ is the identity map, and so ${}^{\lambda}M \subseteq M$. Also, for $i < n$

$$\begin{aligned}\lambda_{i+1} &= \mathrm{ot}(j``\lambda \cap j(\lambda_{i+1}))\\ &= [\langle \mathrm{ot}(x \cap \lambda_{i+1}) \mid x \in \mathcal{P}(\lambda)\rangle]_U\\ &= [\langle \lambda_i \mid x \in \mathcal{P}(\lambda)\rangle]_U\\ &= j(\lambda_i) \text{ .}\end{aligned}$$

Finally, the case $i = 0$ and κ-completeness implies that $\mathrm{crit}(j) = \kappa$. ⊣

This leads to the expected transcendence results by the usual reflection argument:

24.9 Exercise.

(a) If there is a $j\colon V_\delta \prec V_\delta$ as in I3 *with* $\mathrm{crit}(j) = \kappa$, *then there is a normal ultrafilter U over κ such that*

$$\{\alpha < \kappa \mid \alpha \text{ is } n\text{-huge for every } n\} \in U \ .$$

(b) If κ *is* $(n+1)$*-huge, then there is a normal ultrafilter* U *over* κ *such that*

$$\{\alpha < \kappa \mid \alpha \text{ is } n\text{-huge}\} \in U \ . \qquad \dashv$$

24.8 also shows that $\ulcorner\kappa$ is n-huge$\urcorner$ *is* Σ_2^{ZF} by an argument (before 22.9) for γ-supercompactness. This leads to the following observation.

24.10 Exercise (Morgenstern [79]). *If* κ *is supercompact and there is an n-huge cardinal above* κ*, then there are* κ *n-huge cardinals below* κ*. In particular, if there are supercompact and n-huge cardinals, then the least n-huge cardinal is less than the least supercompact cardinal.*

Hint. For any $\alpha < \kappa$, consider the Σ_2^{ZF} proposition $\ulcorner$there is an n-huge cardinal above $\alpha\urcorner$ and apply 22.3. $\dashv$

It will soon be shown that the existence of a huge cardinal implies the consistency of the existence of supercompact cardinals in a strong sense. 24.10 is a consequence of supercompactness as a global reflection property and is unrelated to the consistency strength of hugeness.

Almost hugeness has a characterization in terms of sequences of ultrafilters. Suppose that $\kappa \leq \lambda$, and for $\kappa \leq \gamma < \lambda$, U_γ is a normal ultrafilter over $\mathcal{P}_\kappa\gamma$. Then $\langle U_\gamma \mid \kappa \leq \gamma < \lambda\rangle$ is a *coherent* sequence *iff* whenever $\kappa \leq \gamma \leq \delta < \lambda$, $U_\delta|\gamma = U_\gamma$. This refers to the projection process described after 22.12; setting $j_\gamma\colon V \prec M_\gamma \cong \mathrm{Ult}(V, U_\gamma)$ there are corresponding embeddings $k_{\gamma\delta}\colon M_\gamma \prec M_\delta$ such that $j_\delta = k_{\gamma\delta} \circ j_\gamma$, given by

$$k_{\gamma\delta}([\langle f(x) \mid x \in \mathcal{P}_\kappa\gamma\rangle]_{U_\gamma}) = [\langle f(x \cap \gamma) \mid x \in \mathcal{P}_\kappa\delta\rangle]_{U_\delta} \ .$$

24.11 Theorem. κ *is almost huge iff there are an inaccessible* $\lambda > \kappa$ *and normal ultrafilters* U_γ *over* $\mathcal{P}_\kappa\gamma$ *for* $\kappa \leq \gamma < \lambda$ *such that* $\langle U_\gamma \mid \kappa \leq \gamma < \lambda\rangle$ *is coherent, and the corresponding* j_γ*'s and* $k_{\gamma\delta}$*'s satisfy: if* $\kappa \leq \gamma < \lambda$ *and* $\gamma \leq \alpha < j_\gamma(\kappa)$*, then there is a* δ *such that* $\gamma \leq \delta < \lambda$ *and* $k_{\gamma\delta}(\alpha) = \delta$.

Proof. For the forward direction, let $j\colon V \prec M$ witness the almost hugeness of κ. Set $\lambda = j(\kappa)$, so that λ is inaccessible. For $\kappa \leq \gamma < \lambda$, define U_γ by:

$$X \in U_\gamma \ \ \textit{iff} \ \ X \subseteq \mathcal{P}_\kappa\gamma \ \wedge \ j``\gamma \in j(X) \ .$$

As readily seen, each U_γ is a normal ultrafilter over $\mathcal{P}_\kappa\gamma$ and $\langle U_\gamma \mid \kappa \leq \gamma < \lambda\rangle$ is a coherent sequence.

Suppose next that $\kappa \leq \gamma < \lambda$ and $f\colon \mathcal{P}_\kappa\gamma \to \kappa$ is such that $\gamma \leq [f]_{U_\gamma} < j_\gamma(\kappa)$. Noting that $\gamma = \mathrm{ot}([\mathrm{id}]_{U_\gamma})$ where id: $\mathcal{P}_\kappa\gamma \to \mathcal{P}_\kappa\gamma$ is the identity, the definition of U_γ implies that $\gamma = \mathrm{ot}(j``\gamma) \leq j(f)(j``\gamma) < j(\kappa) = \lambda$. Set $\delta = j(f)(j``\gamma)$. Then $\mathrm{ot}(j``\delta) = \delta = j(f)(j``\delta \cap j(\gamma))$, and so

$$\{x \in \mathcal{P}_\kappa\delta \mid \mathrm{ot}(x) = f(x \cap \gamma)\} \in U_\delta \ .$$

Hence, $k_{\gamma\delta}([f]_{U_\gamma}) = \delta$ by the definition of $k_{\gamma\delta}$.

For the converse, note that

$$\langle\langle M_\gamma \mid \kappa \leq \gamma < \lambda\rangle, \langle k_{\gamma\delta} \mid \kappa \leq \gamma \leq \delta < \lambda\rangle\rangle$$

is a directed system, so let $\overline{M}$ be its direct limit. $\overline{M}$ is well-founded since λ is inaccessible, so it can be assumed that $\overline{M}$ is an inner model. There are corresponding $\overline{j}\colon V \prec \overline{M}$ and $k_\gamma\colon M_\gamma \prec \overline{M}$ for $\kappa \leq \gamma < \lambda$ such that $\overline{j} = k_\gamma \circ j_\gamma$. Note for what follows that if $\alpha \leq \gamma \leq \delta < \lambda$, then $k_{\gamma\delta}(\alpha) = \alpha$ (22.12(b)), and thus $k_\gamma(\alpha) = \alpha$.

To complete the proof, we show that ${}^\beta\overline{M} \subseteq \overline{M}$ for every $\beta < \lambda$ and that $\overline{j}(\kappa) = \lambda$. If $\beta < \lambda$ and $s = \langle x_\alpha \mid \alpha < \beta\rangle \in {}^\beta\overline{M}$, then by the regularity of λ there are γ with $\beta \leq \gamma < \lambda$ and $\{y_\alpha \mid \alpha < \beta\} \subseteq M_\gamma$ such that $k_\gamma(y_\alpha) = x_\alpha$ for $\alpha < \beta$. $t = \langle y_\alpha \mid \alpha < \beta\rangle \in M_\gamma$ since ${}^\gamma M_\gamma \subseteq M_\gamma$, and noting that $k_\gamma(\beta) = \beta$, it follows that $k_\gamma(t) = s \in \overline{M}$.

To show that $\overline{j}(\kappa) = \lambda$, note first that if $\gamma < \lambda$, then since $\gamma < j_\gamma(\kappa)$,

$$\gamma \leq k_\gamma(\gamma) < k_\gamma(j_\gamma(\kappa)) = \overline{j}(\kappa)\ .$$

Hence, $\overline{j}(\kappa) \geq \lambda$. For the other direction, suppose that $\beta < \overline{j}(\kappa)$. There is a γ and an $\alpha < j_\gamma(\kappa)$ such that $k_\gamma(\alpha) = \beta$. If $\alpha \leq \gamma$, then $\beta = k_\gamma(\alpha) = \alpha < j_\gamma(\kappa)$. If $\alpha > \gamma$, then by hypothesis there is a δ such that $\gamma \leq \delta < \lambda$ and $k_{\gamma\delta}(\alpha) = \delta$; but $k_\delta(\delta) = \delta$, and so it follows that $\beta = k_\delta(\delta) = \delta < j_\delta(\kappa)$. In either case, there is some $\zeta < \lambda$ such that $\beta < j_\zeta(\kappa)$, and so by 22.11(c)

$$\beta < j_\zeta(\kappa) < (2^{|\zeta|^{<\kappa}})^+ < \lambda$$

as λ is inaccessible. Hence, $\overline{j}(\kappa) = \lambda$. ⊣

24.12 Exercise. *If κ is huge, then there is a normal ultrafilter U over κ such that*

$$\{\alpha < \kappa \mid \alpha \text{ is almost huge}\} \in U\ . \qquad ⊣$$

Beyond these results from Solovay-Reinhardt-Kanamori [78], the theory of huge cardinals has been pursued in several papers. Robert Mignone [84] and Carlos Di Prisco and Wiktor Marek in their [84, 85] analyze the ultrafilter characterization, and It Beng Tan [81], Barbanel-Di Prisco-Tan [84], Di Prisco-Henle [85], Abe [86], Di Prisco-Marek [88], and Barbanel [89, 91, 91a] consider various generalizations and elaborations. The following is a global generalization of hugeness in the spirit of supercompactness and extendibility:

κ is *superhuge* *iff* for any γ there is a $j\colon V \prec M$ with $\mathrm{crit}(j) = \kappa$, ${}^{j(\kappa)}M \subseteq M$, and $\gamma < j(\kappa)$.

The following shows how this concept fits into the overall scheme:

24.13 Theorem (Barbanel-Di Prisco-Tan [84]).

(a) If κ is superhuge, then κ is extendible and there is a normal ultrafilter U over κ such that

$$\{\alpha < \kappa \mid \alpha \text{ is extendible}\} \in U\ .$$

(b) If κ is 2-huge, then there is a normal ultrafilter U over κ such that

$$\{\alpha < \kappa \mid V_\kappa \models \alpha \text{ is superhuge}\} \in U \ .$$

Proof. (a) First, it is simple to produce normal ultrafilters over $\mathcal{P}_\kappa\gamma$ for every $\gamma \geq \kappa$ to show that κ is supercompact. With η arbitrary, we next show that κ is η-extendible: Let $j\colon V \prec M$ witness the hugeness of κ with $\eta < j(\kappa)$. Set $e = j|V_{\kappa+\eta}$; since $j(\kappa)$ is inaccessible, $e \in M$ and so $M \models \kappa$ is η-extendible. But by elementarity $M \models j(\kappa)$ is supercompact, and so by the argument for 23.9(a), $M \models (V_{j(\kappa)} \models \kappa$ is η-extendible). Finally, $V^M_{j(\kappa)} = V_{j(\kappa)}$, and so κ really is η-extendible.

To complete the proof of (a), note that the above argument shows that if $j\colon V \prec M$ is *any* witness to the hugeness of κ, then $M \models (V_{j(\kappa)} \models \kappa$ is extendible). Hence, if U is the normal ultrafilter over κ defined from j, then $A = \{\alpha < \kappa \mid V_\kappa \models \alpha \text{ is extendible}\} \in U$. But κ is extendible, so by 23.11 every $\alpha \in A$ is extendible.

(b) Temporarily denote by $\alpha \rightarrow (\beta)$ the following: there is a $j\colon V \prec M$ witnessing the hugeness of α with $j(\alpha) = \beta$. By 24.8, this amounts to asserting the existence of a certain ultrafilter over $\mathcal{P}(\beta)$. Let $j\colon V \prec M$ witness the 2-hugeness of κ. Then $M \models \kappa \rightarrow (j(\kappa))$ since ${}^{j^2(\kappa)}M \subseteq M$, and so if U is the normal ultrafilter over κ defined from j, then $A = \{\alpha < \kappa \mid \alpha \rightarrow (\kappa)\} \in U$. For any $\alpha \in A$, $M \models \alpha \rightarrow (\kappa)$, and so $\{\xi < \kappa \mid \alpha \rightarrow (\xi)\} \in U$ and hence is unbounded in κ. Consequently, $A = \{\alpha < \kappa \mid V_\kappa \models \alpha \text{ is superhuge}\}$. ⊣

Vopěnka's Principle

Finally, a circle is completed by going back to a model-theoretic hypothesis that was first considered about the time that extendible cardinals were formulated. This is *Vopěnka's Principle:*

For any proper class of structures for the same language,
there is one that is elementarily embeddable into another .

This hypothesis has an immediate appeal owing to its unencumbered statement and its readily discernible motivation in the set vs. class distinction. Given a proper class $\{\mathcal{M}_\alpha \mid \alpha \in \text{On}\}$ of structures for the same language, there must be two, in fact a proper subclass, that have the same theory. Vopěnka's Principle can be motivated in terms of a further resemblance based on the elements of the structures: Considerations of the richness of proper classes and the uniformity enforced by the class definition suggest that in cases like $\mathcal{M}_\alpha = \langle V_\alpha, \in\rangle$ one structure ought to be an elementary substructure of another. On the other hand, if $\mathcal{M}_\alpha = \langle V_{\gamma_\alpha}, \in, \{\alpha\}\rangle$ where $\gamma_\alpha > \alpha$, $\mathcal{M}_\alpha \prec \mathcal{M}_\beta$ fails for $\alpha < \beta$ because of the singleton predicates. However, this can be rectified by positing an association up to a renaming of the elements, i.e. an elementary *embedding* $j\colon \mathcal{M}_\alpha \prec \mathcal{M}_\beta$, which must necessarily satisfy $j(\alpha) = \beta$. This step from elementary substructure to elementary embeddability, necessary to state a general resemblance principle involving elements, is a dramatic strengthening. Several people quickly saw what was suggested, that Vopěnka's Principle implies the existence of extendible cardinals.

Vopěnka's Principle cannot be formulated in ZFC. Its consequences were pursued in class theory in Solovay-Reinhardt-Kanamori [78], but this work will be cast here in terms of inaccessible κ so that $V_\kappa \models$ Vopěnka's Principle with arbitrary rather than definable "classes" $X \subseteq V_\kappa$, as in Kanamori [78]. To focus the discussion, a sequence of structures $\langle \mathcal{M}_\alpha \mid \alpha < \kappa \rangle$ is *natural iff* each $\mathcal{M}_\alpha$ has form $\langle V_{f(\alpha)}, \in, \{\alpha\}, R_\alpha \rangle$ where $R_\alpha \subseteq V_{f(\alpha)}$ and $\alpha < \beta < \kappa$ implies that $\alpha < f(\alpha) \leq f(\beta) < \kappa$. Through coding we can construe as natural those sequences where R_α is replaced by a finite number of relations. The specification of $\{\alpha\}$ in $\mathcal{M}_\alpha$ ensures that whenever $\alpha < \beta$ and $j \colon \mathcal{M}_\alpha \prec \mathcal{M}_\beta$, j must move some ordinal since $j(\alpha) = \beta$.

In the spirit of the study of indescribability in terms of corresponding filters (§6), define for inaccessible κ the contextual notion of non-negligible subset of κ: For $X \subseteq \kappa$,

X is *Vopěnka in* κ *iff* for any natural sequence $\langle \mathcal{M}_\alpha \mid \alpha < \kappa \rangle$ there is a $j \colon \mathcal{M}_\alpha \prec \mathcal{M}_\beta$ for some $\alpha < \beta < \kappa$ with critical point in X ,

and

κ is *Vopěnka iff* κ is Vopěnka in κ .

This leads to the consideration of the collection of the all but negligible subsets,

$$F = \{X \subseteq \kappa \mid \kappa - X \text{ is not Vopěnka in } \kappa\} .$$

Note that $X \in F$ *iff* there is a natural sequence $\langle \mathcal{M}_\alpha \mid \alpha < \kappa \rangle$ such that for any $\alpha < \beta < \kappa$ and $j \colon \mathcal{M}_\alpha \prec \mathcal{M}_\beta$, its critical point is in X. Also, if $X, Y \in F$, then $X \cap Y \in F$ by pointwise amalgamating the two corresponding natural sequences into one. Finally, if $\langle \mathcal{A}_\alpha \mid \alpha < \kappa \rangle \in {}^\kappa V_\kappa$ is any sequence of structures for the same language, then each $\mathcal{A}_\alpha$ can be encoded into an $\mathcal{M}_\alpha$ so that $\langle \mathcal{M}_\alpha \mid \alpha < \kappa \rangle$ a natural sequence, and so if X is Vopěnka in κ, there are $\alpha < \beta < \kappa$ and $j \colon \mathcal{A}_\alpha \prec \mathcal{A}_\beta$ such that if it has a critical point, then it belongs to X. These remarks show that

F is a (proper) filter *iff* $V_\kappa \models$ Vopěnka's Principle *iff* κ is Vopěnka .

F is the *Vopěnka filter* over κ, and that it is indeed a (proper) filter will be implicit in the use of this terminology. Although in terms of §6, κ being Vopěnka is a Π^1_1 property of V_κ and hence does not even imply the weak compactness of κ, the following propositions provide structural information which shows that it implies the existence of extendible cardinals in V_κ in a strong sense.

24.14 Proposition. *The Vopěnka filter over κ is normal.*

Proof. Suppose that X is Vopěnka in κ and $f \colon X \to \kappa$ is regressive. Assume to the contrary that for each $\gamma < \kappa$, $f^{-1}(\{\gamma\})$ is not Vopěnka in κ, i.e. there is a natural sequence $\langle \mathcal{M}^\gamma_\alpha \mid \alpha < \kappa \rangle$ such that for any elementary embedding of one structure into another with critical point ρ, $f(\rho) \neq \gamma$. Define a natural sequence $\langle \mathcal{N}_\alpha \mid \alpha < \kappa \rangle$ by setting

$$\mathcal{N}_\alpha = \langle V_{g(\alpha)+\omega}, \in, \{\alpha\}, \langle \mathcal{M}_\alpha^\gamma \mid \gamma < \alpha \rangle, f|\alpha \rangle$$

where $V_{g(\alpha)}$ is the union of the domains of $\mathcal{M}_\alpha^\gamma$ for $\gamma < \alpha$. Since X is Vopěnka, there is a j: $\mathcal{N}_\alpha \prec \mathcal{N}_\beta$ with a critical point $\eta \in X$. But then $f(\eta) < \eta$ implies that $j(f(\eta)) = f(\eta)$ so that $j|\mathcal{M}_\alpha^{f(\eta)}$: $\mathcal{M}_\alpha^{f(\eta)} \prec \mathcal{M}_\beta^{f(\eta)}$ with critical point η, which is a contradiction. ⊣

This result is the $n = 1$ case of a larger scheme in Kanamori [78], which analyzes hypotheses cascading up alongside the n-huge cardinals.

24.15 Proposition. *$\{\xi < \kappa \mid V_\kappa \models \xi$ is extendible$\}$ is in the Vopěnka filter over κ.*

Proof. Define $g \in {}^\kappa\kappa$ by:

$$g(\xi) = \begin{cases} \xi & \text{if } V_\kappa \models \xi \text{ is extendible , else} \\ \xi + \eta & \text{where } \eta \text{ is the least ordinal such} \\ & \text{that } V_\kappa \models \xi \text{ is not } \eta\text{-extendible .} \end{cases}$$

Set $C = \{\rho < \kappa \mid g|\rho \colon \rho \to \rho\}$. Then C is closed unbounded, so that C is in the Vopěnka filter over κ by 24.14. So, let $\langle \mathcal{M}_\alpha \mid \alpha < \kappa \rangle$ be a natural sequence such that for any elementary embedding of one structure into another with a critical point ρ, $\rho \in C$. Now define a natural sequence $\langle \mathcal{N}_\alpha \mid \alpha < \kappa \rangle$ by setting

$$\mathcal{N}_\alpha = \langle V_{\gamma_\alpha}, \in, \{\alpha\}, \mathcal{M}_\alpha, C \cap \gamma_\alpha \rangle$$

where γ_α is the least limit point of C greater than every ordinal in the domain of $\mathcal{M}_\alpha$. To complete the proof, it suffices to show that if j: $\mathcal{N}_\alpha \prec \mathcal{N}_\beta$ with a critical point ξ, then ξ is extendible:

Assume to the contrary that this fails, so that $g(\xi) > \xi$. Since $\xi < \gamma_\alpha$ and $\gamma_\alpha \in C$, $g(\xi) < \gamma_\alpha$. It follows that $j|V_{g(\xi)}$: $V_{g(\xi)} \prec V_{j(g(\xi))}$ with critical point ξ. Also, note that $\xi \in C$ as $\mathcal{M}_\alpha$ is encoded in $\mathcal{N}_\alpha$. Hence, $j(\xi) \in C$ as $C \cap \gamma_\alpha$ is encoded in $\mathcal{N}_\alpha$. It then follows from $\xi < j(\xi)$ that $g(\xi) < j(\xi)$. But then, if $g(\xi) = \xi + \eta$, these properties show that $V_\kappa \models \xi$ is η-extendible, contradicting the definition of $g(\xi)$. ⊣

The following proposition shows that the Vopěnka filter is closed under a strong analogue of Mahlo's operation.

24.16 Proposition. *If X is in the Vopěnka filter over κ, then so also is the set $\{\xi < \kappa \mid \xi$ is measurable and there is a normal ultrafilter over ξ containing $X \cap \xi\}$.*

Proof. Define a natural sequence $\langle \mathcal{M}_\alpha \mid \alpha < \kappa \rangle$ by setting

$$\mathcal{M}_\alpha = \langle V_{\alpha+\omega}, \in, \{\alpha\}, \{X \cap \alpha\} \rangle .$$

It suffices to show that if $j\colon \mathcal{M}_\alpha \prec \mathcal{M}_\beta$ with a critical point $\xi \in X$, then there is a normal ultrafilter over ξ containing $X \cap \xi$. But if U is the usual normal ultrafilter over ξ defined from j, then

$$j(X \cap \xi) = j(X \cap \alpha \cap \xi) = j(X \cap \alpha) \cap j(\xi) = (X \cap \beta) \cap j(\xi) = X \cap j(\xi)$$

and $\xi \in X \cap j(\xi)$ so that $X \cap \xi \in U$. ⊣

24.17 Corollary. *If κ is Vopěnka, then κ is κ-Mahlo and the set*

$$\{\xi < \kappa \mid V_\kappa \models \xi \text{ is extendible and there is a normal ultrafilter over } \xi \text{ containing } \{\zeta < \xi \mid V_\kappa \models \zeta \text{ is extendible}\}\}$$

is stationary in κ.

Proof. By 24.14, 24.15, and 24.16. ⊣

William Powell [72] and others situated Vopěnka's Principle into the large cardinal hierarchy below almost hugeness. Typically, one can establish much more than is needed:

24.18 Theorem. *Suppose that κ is almost huge. Then there is a normal ultrafilter U over κ such that:*

(a) For any natural sequence $\langle \mathcal{M}_\alpha \mid \alpha < \kappa\rangle$ there is a $Y \in U$ such that $\alpha < \beta$ both in Y implies that there is an elementary embedding: $\mathcal{M}_\alpha \prec \mathcal{M}_\beta$ with critical point α.

(b) $\{\alpha < \kappa \mid \alpha \text{ is Vopěnka}\} \in U$.

Proof. Let $j\colon V \prec M$ witness the almost hugeness of κ and U the normal ultrafilter defined from j. To establish (a), let $\langle \mathcal{M}_\alpha \mid \alpha < \kappa\rangle$ be a natural sequence. Setting $X_\alpha = \{\xi < \kappa \mid$ there is an elementary embedding: $\mathcal{M}_\alpha \prec \mathcal{M}_\xi$ with critical point $\alpha\}$ for $\alpha < \kappa$ and $T = \{\alpha < \kappa \mid X_\alpha \in U\}$, we shall show that $T \in U$. Granted this, the normality of U would imply that

$$Y = \{\alpha \in T \mid \alpha \in \textstyle\bigcap_{\beta \in \alpha \cap T} X_\beta\} \in U\ ,$$

and clearly this Y would be as desired.

To show that $T \in U$, set

$$j(\langle \mathcal{M}_\alpha \mid \alpha < \kappa\rangle) = \langle \mathcal{M}^*_\alpha \mid \alpha < j(\kappa)\rangle\ ,$$
$$j(\langle \mathcal{M}^*_\alpha \mid \alpha < j(\kappa)\rangle) = \langle \mathcal{M}^{**}_\alpha \mid \alpha < j^2(\kappa)\rangle\ , \text{ and}$$
$$j(\langle X_\alpha \mid \alpha < \kappa\rangle) = \langle X^*_\alpha \mid \alpha < j(\kappa)\rangle\ .$$

For any $\alpha < \kappa$, noting that $\mathcal{M}^*_\alpha = \mathcal{M}_\alpha$ we have: $X_\alpha \in U$ *iff* $\kappa \in j(X_\alpha)$ *iff* there is an elementary embedding: $\mathcal{M}_\alpha \prec \mathcal{M}^*_\kappa$ with critical point α. So if $\alpha < j(\kappa)$, by elementarity we have: $X^*_\alpha \in j(U)$ *iff* in M, there is an elementary embedding: $\mathcal{M}^*_\alpha \prec \mathcal{M}^{**}_{j(\kappa)}$ with critical point α. Hence, $T \in U$ *iff* $\kappa \in j(T)$ *iff* $X^*_\kappa \in j(U)$ *iff* in M, there is an elementary embedding: $\mathcal{M}^*_\kappa \prec \mathcal{M}^{**}_{j(\kappa)}$ with critical point κ. But setting $\overline{j} = j|\mathcal{M}^*_\kappa$, $\overline{j}$: $\mathcal{M}^*_\kappa \prec \mathcal{M}^{**}_{j(\kappa)}$. Also, $\overline{j}$ is a subset of M with cardinality that of the domain of $\mathcal{M}^*_\kappa$, which in turn is less than $j(\kappa)$. Hence, $\overline{j} \in M$, and so $T \in U$.

For (b), just note that by (a), $M \models \kappa$ is Vopěnka. ⊣

Finally, characterizations of Vopěnka's Principle are drawn from Solovay-Reinhardt-Kanamori [78] in terms of "boldface" versions of extendibility and supercompactness. Characterizations in terms of model-theoretic logics were provided by Edward Fisher [77] and Johann Makowsky [85]. The following definitions serve present purposes: For any set A and $\kappa \leq \eta$,

κ is *η-extendible for A* *iff* there is a ζ and a $j\colon \langle V_\eta, \in, A \cap V_\eta\rangle \prec \langle V_\zeta, \in, A \cap V_\zeta\rangle$ with $\mathrm{crit}(j) = \kappa$ and $\eta < j(\kappa)$,

κ is *η-supercompact for A* *iff* there is an $\alpha < \kappa$ and an $e\colon \langle V_\alpha, \in, A \cap V_\alpha\rangle \prec \langle V_\eta, \in, A \cap V_\eta\rangle$ with $e(\delta) = \kappa$, where $\delta = \mathrm{crit}(e)$.

The second definition is based on the characterization 22.10. Injecting A's into previous arguments lead to characterizations:

24.19 Exercise. *The following are equivalent for inaccessible κ:*

(a) κ is Vopěnka.

(b) For any $A \subseteq V_\kappa$ there is an $\alpha < \kappa$ η-extendible for A for every $\eta < \kappa$.

(c) For any $A \subseteq V_\kappa$ there is an $\alpha < \kappa$ η-supercompact for A for every $\eta < \kappa$.

Hint. The argument for 24.15 adapts to establish (a) → (b).

For (b) → (c), show that for any $A \subseteq V_\kappa$, if α is η-extendible for A for every $\eta < \kappa$, then α is η-supercompact for A for every $\eta < \kappa$: For such an α let

$$j\colon \langle V_{\eta+\omega}, \in, A \cap V_{\eta+\omega}\rangle \prec \langle V_\zeta, \in, A \cap V_\zeta\rangle$$

with $\mathrm{crit}(j) = \alpha$ and $\eta < j(\alpha)$. Set $\overline{j} = j|V_\eta$. Then $\overline{j} \in V_\zeta$, and noting that $A \cap V_\eta = (A \cap V_{j(\eta)}) \cap V_\eta = j(A \cap V_\eta) \cap V_\eta$,

$$\overline{j}\colon \langle V_\eta, \in, j(A \cap V_\eta) \cap V_\eta\rangle \prec \langle V_{j(\eta)}, \in, j(A \cap V_\eta)\rangle \ .$$

Now apply the elementarity of j (cf. 22.10).

Arguing with natural sequences as A's establishes (c) → (a). ⊣

The strong hypotheses from Reinhardt's extendibility to Kunen's inconsistency were formulated and systematized into a linear hierarchy in a relatively short time. Since then, this edifice has proved resilient, reinforced by relative consistency results from hugeness and even I0–I3, and the further elucidation of structure at these hypotheses. Of course, a new inconsistency result would be an exciting development, but as time goes by, the further analysis and application of these hypotheses suggest that they may be approached with increasing confidence if not acceptance.

25. Combinatorics of $\mathcal{P}_\kappa\gamma$

The emergence of $\mathcal{P}_\kappa\gamma$ in the study of supercompactness led to several combinatorial problems about this set that stimulated considerable research; this section is devoted to the work through the 1980's in this offshoot direction. For convenience,

κ denotes a regular uncountable cardinal and $\kappa \leq \gamma$ in this section .

The seminal paper in this area was Jech [73], which in the spirit of the generalization from measurability to γ-supercompactness advanced the idea of generalizing combinatorial properties of and problems about $\langle \kappa, < \rangle$ to $\langle \mathcal{P}_\kappa\gamma, \subset \rangle$. Basic to the subsequent work was the Jech [71] generalization of the concepts of closed unbounded set and stationary set: For $X \subseteq \mathcal{P}_\kappa\gamma$,

X is *unbounded in* $\mathcal{P}_\kappa\gamma$ *iff*	for any $y \in \mathcal{P}_\kappa\gamma$ there is an $x \supseteq y$ such that $x \in X$.
X is *closed in* $\mathcal{P}_\kappa\gamma$ *iff*	for any $\{x_\xi \mid \xi < \beta\} \subseteq X$ with $\beta < \kappa$ and $x_\xi \subseteq x_\zeta$ for $\xi \leq \zeta < \beta$, $\bigcup_{\xi<\beta} x_\xi \in X$.
X is *closed unbounded in* $\mathcal{P}_\kappa\gamma$ *iff*	X is closed in $\mathcal{P}_\kappa\gamma$ and unbounded in $\mathcal{P}_\kappa\gamma$.
X is *stationary in* $\mathcal{P}_\kappa\gamma$ *iff*	$X \cap C \neq \emptyset$ for any C closed unbounded in $\mathcal{P}_\kappa\gamma$.

The "in $\mathcal{P}_\kappa\gamma$" is deleted when clear from the context. Note that *for any* $s \in \mathcal{P}_\kappa\gamma$, $\{x \in \mathcal{P}_\kappa\gamma \mid s \subseteq x\}$ *is closed unbounded*, and so *if* $X \subseteq \mathcal{P}_\kappa\gamma$ *is stationary, then it is unbounded*. David Kueker [72] also formulated these concepts for $\kappa = \omega_1$ for use in model theory, and his [77] provides a full account of his work along these lines.

Jech's original formulation of closed in $\mathcal{P}_\kappa\gamma$ was different and Solovay observed that the current one is equivalent. Because of its usefulness, that first formulation is recovered through the following lemma. A set D is $\subseteq$*-directed iff* for any $x, y \in D$ there is a $z \in D$ such that $x \cup y \subseteq z$. Solovay's observation turns out to be the $\mathcal{P}_\kappa\gamma$ case of a result about transfinite directed sets:

25.1 Lemma (Cohn [65: 33]; Isbell [66]). *X is closed in $\mathcal{P}_\kappa\gamma$ iff for any $\subseteq$-directed $D \subseteq X$ with $|D| < \kappa$, $\bigcup D \in X$.*

Proof. The substantive direction is established by induction on $|D|$. The result is immediate if $|D|$ is finite. Suppose now that $|D| < \kappa$ is infinite and for each $\subseteq$-directed $E \subseteq X$ with $|E| < |D|$, $\bigcup E \in X$. Enumerate D as $\{x_\alpha \mid \alpha < |D|\}$. Recursively define D_α for $\alpha < |D|$ to be a $\subseteq$-directed subset of D of minimal cardinality including $\{x_\alpha\} \cup \bigcup_{\beta<\alpha} D_\beta$. By a simple closure argument, if α is finite, then D_α is finite, and otherwise $|D_\alpha| = |\bigcup_{\beta<\alpha} D_\beta| = |\alpha|$. Set $y_\alpha = \bigcup D_\alpha$ for $\alpha < |D|$. Then by induction each $y_\alpha \in X$, and since $y_\alpha \subseteq y_{\overline{\alpha}}$ for $\alpha \leq \overline{\alpha} < |D|$, $\bigcup_{\alpha<|D|} y_\alpha = \bigcup D \in X$. ⊣

The following provides analogues to most of 0.1 and is established in like manner.

25.2 Exercise (Jech [71]).

(a) If $\eta < \kappa$ and $\langle C_\alpha \mid \alpha < \eta \rangle$ is a sequence of sets closed unbounded in $\mathcal{P}_\kappa\gamma$, then $\bigcap_{\alpha<\eta} C_\alpha$ is closed unbounded in $\mathcal{P}_\kappa\gamma$.

(b) If $\langle C_\alpha \mid \alpha < \gamma \rangle$ is a sequence of sets closed unbounded in $\mathcal{P}_\kappa\gamma$, then its diagonal intersection $\Delta_{\alpha<\gamma} C_\alpha = \{x \in \mathcal{P}_\kappa\gamma \mid x \in \bigcap_{\alpha\in x} C_\alpha\}$ is closed unbounded in $\mathcal{P}_\kappa\gamma$.

(c) If S is stationary in $\mathcal{P}_\kappa\gamma$ and f is a choice function of S (i.e. $f(x) \in x$ for every $x \in S - \{\emptyset\}$), then there is a $T \subseteq S$ stationary in $\mathcal{P}_\kappa\gamma$ such that f is constant on T. ⊣

Let $\mathcal{C}_{\kappa,\gamma}$ denote the filter generated by the closed unbounded subsets of $\mathcal{P}_\kappa\gamma$:

$$\mathcal{C}_{\kappa,\gamma} = \{X \subseteq \mathcal{P}_\kappa\gamma \mid \exists C(C \text{ is closed unbounded in } \mathcal{P}_\kappa\gamma \ \wedge \ C \subseteq X)\} .$$

Then by 25.2, $\mathcal{C}_{\kappa,\gamma}$ *is a normal filter over* $\mathcal{P}_\kappa\gamma$, called the *closed unbounded filter* over $\mathcal{P}_\kappa\gamma$.

Menas established a useful basis theorem for $\mathcal{C}_{\kappa,\gamma}$. For $n \in \omega$ and $f\colon [\gamma]^n \to \mathcal{P}_\kappa\gamma$, define

$$C(f) = \{x \in \mathcal{P}_\kappa\gamma \mid x \text{ is infinite} \wedge \forall s \in [x]^n (f(s) \subseteq x)\} .$$

Analogously define $C(f)$ for $f\colon [\gamma]^{<\omega} \to \mathcal{P}_\kappa\gamma$. It is simple to see that these $C(f)$'s are closed unbounded.

25.3 Proposition (Menas [74]). *For any $X \subseteq \mathcal{P}_\kappa\gamma$, $X \in \mathcal{C}_{\kappa,\gamma}$ iff there is an $f\colon [\gamma]^2 \to \mathcal{P}_\kappa\gamma$ such that $C(f) \subseteq X$.*

Proof. In the substantive direction, suppose that C is closed unbounded; an $f\colon [\gamma]^2 \to \mathcal{P}_\kappa\gamma$ must be found so that $C(f) \subseteq C$. To this end, a $g\colon [\gamma]^{<\omega} \to \mathcal{P}_\kappa\gamma$ is first found so that $C(g) \subseteq C$ and then refined later:

Define $g(s)$ for $s \in [\gamma]^{<\omega}$ by recursion on $|s|$ as follows: Let $g(\emptyset)$ be any infinite member of C, and if $|s| = n+1$, let $g(s)$ satisfy $s \cup \{g(t) \mid t \in [s]^n\} \subseteq g(s)$ and $g(s) \in C$. Then $C(g) \subseteq C$: If $x \in C(g)$, then x is infinite and $\bigcup\{g(s) \mid s \in [x]^{<\omega}\} \subseteq x$, and so equality holds since $s \subseteq g(s)$. But clearly, $\{g(s) \mid s \in [x]^{<\omega}\}$ is $\subseteq$-directed, and so by 25.1, $x \in C$.

The following will be established next:

$$(*) \quad \text{Whenever } 1 < n < \omega \text{ and } h\colon \gamma^n \to \mathcal{P}_\kappa\gamma, \\ \text{there is an } \overline{h}\colon [\gamma]^2 \to \mathcal{P}_\kappa\gamma \text{ such that } C(\overline{h}) \subseteq C(h) .$$

Once done, the proof can be completed as follows: For each $n \in \omega$ let $g_n\colon [\gamma]^2 \to \mathcal{P}_\kappa\gamma$ be such that $C(g_n) \subseteq C(g|[\gamma]^n)$. Define $f\colon [\gamma]^2 \to \mathcal{P}_\kappa\gamma$ by: $f(s) = \bigcup_n g_n(s)$. Then $C(f) \subseteq \bigcap_n C(g_n) \subseteq C(g) \subseteq X$.

To establish $(*)$, it will be convenient to show that whenever $n > 1$ and $h\colon [\gamma]^{n+1} \to \mathcal{P}_\kappa\gamma$, there is an $\overline{h}\colon [\gamma]^n \to \mathcal{P}_\kappa\gamma$ such that $C(\overline{h}) \subseteq C(h)$. This suffices by finite descent. To this end, let $p\colon [\gamma]^2 \to \gamma$ be a bijection such that $\max(\alpha, \beta) \leq p(\alpha, \beta)$, and $\pi_0, \pi_1\colon \gamma \to \gamma$ be such that $p(\pi_0(\alpha), \pi_1(\alpha)) = \alpha$ for every $\alpha < \gamma$. Define $\overline{h}\colon [\gamma]^n \to \mathcal{P}_\kappa\gamma$ so that for any $\alpha_1 < \ldots < \alpha_n < \gamma$, $\overline{h}(\alpha_1, \ldots, \alpha_n) \in C(h)$ and

$$\overline{h}(\alpha_1, \ldots, \alpha_n) \supseteq \{p(\alpha_{n-1}, \alpha_n)\} \cup h(\alpha_1, \ldots, \alpha_{n-1}, \pi_0(\alpha_n), \pi_1(\alpha_n)) .$$

Then $C(\overline{h}) \subseteq C(h)$: If $x \in C(\overline{h})$ and $\alpha_1 < \ldots < \alpha_{n+1}$ are all in x, then $p(\alpha_n, \alpha_{n+1}) \in \overline{h}(\alpha_1, \ldots, \alpha_{n-2}, \alpha_n, \alpha_{n+1}) \subseteq x$, and so

$$h(\alpha_1, \ldots, \alpha_{n+1}) \subseteq \overline{h}(\alpha_1, \ldots, \alpha_{n-1}, p(\alpha_n, \alpha_{n+1})) \subseteq x . \qquad \dashv$$

This result shows in particular that for any C closed unbounded in $\mathcal{P}_\kappa\gamma$ there is a $\overline{C} \subseteq C$ closed unbounded in $\mathcal{P}_\kappa\gamma$ such that for any $x, y \in \overline{C}$, $x \cap y \in \overline{C}$. If C is closed unbounded in $\mathcal{P}_\kappa\gamma$ so that for any $x, y \in C$, $x \cup y \in C$, then there is an $f\colon [\gamma]^1 \to \mathcal{P}_\kappa\gamma$ such that $C(f) \subseteq C$: For $\alpha < \lambda$, just let $f(\{\alpha\}) \in C$ satisfy $\alpha \in f(\{\alpha\})$. However, Menas observed that for $\gamma \geq \kappa^+$, the $[\gamma]^2$ in 25.3 cannot be replaced by $[\gamma]^1$. In particular, for such γ there are C closed unbounded in $\mathcal{P}_\kappa\gamma$ such that for any $\overline{C} \subseteq C$ closed unbounded in $\mathcal{P}_\kappa\gamma$, there are $x, y \in \overline{C}$ such that $x \cup y \notin \overline{C}$. Donna Carr [82] investigated closed unbounded sets closed under unions; she also observed the following, almost explicit in Menas [74: 332], which is the analogue of the fact that the closed unbounded filter $\mathcal{C}_\kappa$ is included in every normal filter over κ.

25.4 Proposition (Carr [82], Menas [74]). *If F is a normal filter over $\mathcal{P}_\kappa\gamma$, then $\mathcal{C}_{\kappa,\gamma} \subseteq F$.*

Proof. Assume that this fails; then by the previous proposition there is an $f\colon [\gamma]^2 \to \mathcal{P}_\kappa\gamma$ such that $X = \mathcal{P}_\kappa\gamma - C(f)$ is F-stationary. For each $x \in X$, there are $\alpha_x < \beta_x$ both in x so that $f(\alpha_x, \beta_x) \nsubseteq x$. By applying normality (22.6) twice, there are F-stationary $Y \subseteq X$ and $\alpha < \beta$ such that for every $x \in Y$, $\alpha_x = \alpha$ and $\beta_x = \beta$. But this is a contradiction, since Y is unbounded and so there must be a $y \in Y$ such that $f(\alpha, \beta) \subseteq y$. $\dashv$

Jech [73] had previously observed that if U is a normal ultrafilter over $\mathcal{P}_\kappa\gamma$, then $\mathcal{C}_{\kappa,\gamma} \subseteq U$.

Pursuing analogies, the next problem that naturally arises is to generalize Solovay's result 16.9 that every set stationary in κ can be partitioned into κ disjoint stationary sets. Menas [74: 337] conjectured for $\lambda > \kappa$:

Every stationary subset of $\mathcal{P}_\kappa\lambda$ can be
partitioned into $\lambda^{<\kappa}$ disjoint stationary sets .

Most of the results have been about this assertion with $\lambda^{<\kappa}$ replaced by λ. The first observations can be deduced from consideration of one particular stationary set.

25.5 Proposition. *Suppose that $\kappa \leq \lambda$ and $S = \{x \in \mathcal{P}_\kappa\lambda \mid |x \cap \kappa| = |x|\}$. Then:*

(a) S is stationary in $\mathcal{P}_\kappa\lambda$.

(b) If κ is a successor cardinal, then $S \in \mathcal{C}_{\kappa,\lambda}$.

(c) If $\kappa < \lambda$ and κ is λ-supercompact, then $S \notin \mathcal{C}_{\kappa,\lambda}$.

(d) (Baumgartner) *If $X \subseteq S$ is stationary, then X can be partitioned into λ disjoint stationary subsets.*

Proof. (a) Suppose that C is closed unbounded in $\mathcal{P}_\kappa\lambda$. Then it is easy to find $x_n \in \mathcal{P}_\kappa\lambda$ such that $x_n \subseteq x_{n+1}$ for $n \in \omega$ with $x_{2m} \in S$ and $x_{2m+1} \in C$ for $m \in \omega$. It follows that $\bigcup_{n\in\omega} x_n \in S \cap C$.

(b) If $\kappa = \nu^+$, then $\{x \in \mathcal{P}_\kappa\lambda \mid |x \cap \kappa| = \nu\}$ is closed unbounded in $\mathcal{P}_\kappa\lambda$, and this set is included in S.

(c) If U is a normal ultrafilter over $\mathcal{P}_\kappa\lambda$, then $\kappa < \lambda$ implies by 22.11(a) that $\{x \in \mathcal{P}_\kappa\lambda \mid |x \cap \kappa| < |x|\} \in U$. But $\mathcal{C}_{\kappa,\lambda} \subseteq U$ by the previous proposition, and so $S \notin \mathcal{C}_{\kappa,\lambda}$.

(d) For each $x \in X$, let $f_x\colon x \to x \cap \kappa$ be injective. For $\alpha < \lambda$, define g_α on the stationary set $\{x \in X \mid \alpha \in x\}$ by: $g_\alpha(x) = f_x(\alpha)$. Then by 25.2(c) there is a stationary $X_\alpha \subseteq X$ such that g_α is constant on X_α, say with value $\eta_\alpha < \kappa$.

Now for any $\nu \leq \lambda$ such that $\mathrm{cf}(\nu) > \kappa$, there is an $\eta < \kappa$ such that $|\{\alpha < \nu \mid \eta_\alpha = \eta\}| = \nu$, and so it is simple to see that $\{X_\alpha \mid \alpha < \nu \wedge \eta_\alpha = \eta\}$ is a family of ν disjoint stationary subsets of X. Hence, if $\mathrm{cf}(\lambda) > \kappa$, the result follows. If on the other hand $\mathrm{cf}(\lambda) \leq \kappa$, it can be assumed that $\kappa < \lambda$, else the result follows from Solovay's 16.9 and 22.5(b). So, let $\langle \mu_\alpha \mid \alpha < \mathrm{cf}(\lambda)\rangle$ be an increasing sequence of regular cardinals cofinal in λ with $\kappa < \mu_0$. By taking $\nu = \kappa^+ \leq \lambda$ in the above, first find a partition $X = \bigcup_{\alpha<\kappa^+} Y_\alpha$ into disjoint stationary sets. Then for each $\alpha < \mathrm{cf}(\lambda) \leq \kappa$ the argument with X replaced by Y_α can be applied to get a partition of Y_α into μ_α disjoint stationary sets. Finally, the conglomeration of these partitions is a partition of X into λ disjoint stationary sets as desired. ⊣

25.6 Corollary. *Suppose that $\kappa \leq \lambda$. Then:*

(a) $\mathcal{P}_\kappa\lambda$ can be partitioned into λ disjoint stationary sets.

(b) (Jech [73] for regular λ; Matsubara [87]) *If κ is a successor cardinal, then every stationary subset of $\mathcal{P}_\kappa\lambda$ can be partitioned into λ disjoint stationary sets.*

Di Prisco (see Di Prisco-Marek [82]) showed that for regular $\lambda \geq \kappa$, every closed unbounded subset of $\mathcal{P}_\kappa\lambda$ can be partitioned into λ disjoint stationary sets. For κ weakly inaccessible, he formulated a concept, *superstationary*, intermediate between stationary and closed unbounded, and showed that for regular $\lambda \geq \kappa$ every superstationary subset of $\mathcal{P}_\kappa\lambda$ can be partitioned into λ disjoint superstationary sets. When $S \notin \mathcal{C}_{\kappa,\lambda}$ as in 25.5(c), then S is an example of a stationary but not superstationary subset of $\mathcal{P}_\kappa\lambda$.

In 1981, Baumgartner made the following observation about the set S that reestablished direct connections with large cardinals:

25.7 Theorem (Baumgartner). *Assume the hypothesis of 25.5. If $0^{\#}$ does not exist, then $S \in \mathcal{C}_{\kappa,\lambda}$, and so every stationary subset of $\mathcal{P}_{\kappa}\lambda$ can be partitioned into λ disjoint stationary sets.*

Proof. Assume to the contrary that $S \notin \mathcal{C}_{\kappa,\lambda}$. Then $\kappa < \lambda$. Let $f\colon \lambda \to L_\lambda$ be a bijection which is the identity on $\kappa + 1$. Then

$$C = \{x \in \mathcal{P}_{\kappa}\lambda \mid x \cap \kappa \in \kappa \ \wedge \ \kappa \in x \ \wedge \ f``x \prec L_\lambda\}$$

is readily seen to be closed unbounded. By assumption $\mathcal{P}_{\kappa}\lambda - S$ is stationary, so there is a $y \in C$ such that $|y \cap \kappa| < |y|$. Let L_α be the transitive collapse of $f``y$ and j the inverse of the collapsing isomorphism, so that $j\colon L_\alpha \prec L_\lambda$. Setting $\delta = y \cap \kappa$, since f is the identity on $\kappa + 1$ and $\kappa \in y$, $j(\delta) = \kappa$ and $\mathrm{crit}(j) = \delta$. But $\delta < |y| = |\alpha|$, and so by 21.1, $0^{\#}$ exists. ⊣

Hence, the refutation of Menas's conjecture with $\lambda^{<\kappa}$ replaced by λ necessitates the existence of $0^{\#}$! The following observation made in passing precludes the possibility of developing an inner model theory for γ-supercompactness in a naïve manner:

25.8 Exercise. *Suppose that $\kappa < \lambda$ and U is a normal ultrafilter over $\mathcal{P}_{\kappa}\lambda$. Then:*
(a) $\mathcal{P}_{\kappa}\lambda \cap L \notin U$.
(b) $L[U] = L$.

Hint. For (a), assume to the contrary that $\mathcal{P}_{\kappa}\lambda \cap L \in U$, and noting that $\mathcal{P}_{\kappa}\lambda - S \in U$ as for the proof of 25.5(c), use the proof of 25.7 to find a y as there satisfying $y \in L$ to derive contradiction. (b) follows from (a). ⊣

A contradiction from $\mathcal{P}_{\kappa}\lambda \cap L \in U$ can also be derived by considering the $0^{\#}$ Skolem term generation of members of $\mathcal{P}_{\kappa}\lambda \cap L$ and stabilizing components with normality.

With a surprising result that raised the level of sophistication in this area, Gitik in 1982 showed relative to supercompactness that Menas's conjecture can fail in a strong sense:

25.9 Theorem (Gitik [85]). *Suppose that κ is supercompact and $\lambda > \kappa$. Then there is a p.o. P that preserves cardinals $\geq \kappa$ such that $\Vdash_P$ κ* is inaccessible $\wedge\ \exists S(S$ is stationary in $\mathcal{P}_{\kappa}\lambda \ \wedge\ S$ cannot be partitioned into κ^+ disjoint stationary sets).

Gitik observed that this is optimal in the sense that for any $\lambda \geq \kappa$ every stationary subset of $\mathcal{P}_{\kappa}\lambda$ can be partitioned into κ disjoint stationary sets, using a version of Solovay's argument for 16.9. He also noted that κ can also be made the least inaccessible cardinal, using a construction due to Woodin. The exact consistency strength of the proposition that there is a stationary subset of $\mathcal{P}_{\kappa}\kappa^+$ that cannot be partitioned into κ^+ disjoint stationary sets is unknown; Gitik's proof established as an upper bound the existence of a κ that is κ^{+++}-supercompact. Recently, Shioya [03] improved the upper bound to the existence of a κ that is κ^+-supercompact and moreover provided a much simpler proof of 25.9.

The focal set S of 25.5 occurred at one point in Gitik's argument. Moreover, Baldwin [83] was led to S in the course of generalizing the Mahlo hierarchy for applications to the developing inner model theory; the result of Jech [86] is based on S; and S serves to distinguish between versions of the closed unbounded filter formulated in Matsubara [88a]. Consider the possibility $S \notin \mathcal{C}_{\kappa,\lambda}$:

$$\text{For some } \lambda,\ \{x \in \mathcal{P}_\kappa\lambda \mid |x \cap \kappa| < |x|\} \text{ is stationary in } \mathcal{P}_\kappa\lambda\ .$$

As a nice example of how questions are resolved in terms of large cardinals, the exact consistency strength of this proposition was determined in the linear hierarchy. By 25.5(c) and 25.7 the consistency strength lies between the existence of $0^\#$ and the existence of a κ that is κ^+-supercompact. Baldwin [84] localized it to the lower end by showing that it suffices to take λ satisfying a partition property weaker than Ramsey with respect to a weakly inaccessible $\kappa < \lambda$. Then Donder-Koepke-Levinski [88] pinned down the exact partition property needed and confirmed its necessity using the newly available inner model theory to good effect. As was already evident from Levinski [84: 229] however, for $\kappa \geq \omega_3$ the proposition with the additional requirement that $\lambda = \kappa^+$ implies at least the existence of $0^\dagger$, and the exact consistency strength is unknown.

Menas's original conjecture, partitioning into $\lambda^{<\kappa}$ sets, was refuted in ZFC with $\kappa = \omega_1$:

25.10 Theorem (Baumgartner-Taylor [82]). *Assume* GCH *and* $\nu = \lambda^{\aleph_0}$. *Then for any* ω_1*-c.c. p.o.* P, $\Vdash_P \exists S(S$ is stationary in $\mathcal{P}_{\omega_1}\lambda\ \wedge\ S$ cannot be partitioned into ν^{++} disjoint stationary sets). *In particular,* Con(ZFC) *implies* Con(ZFC + $2^{\aleph_0}$ is large + $\exists S(S$ is stationary in $\mathcal{P}_{\omega_1}\omega_2\ \wedge\ S$ cannot be partitioned into ω_4 disjoint stationary sets)).

On the other hand, Matsubara established the consistency of Menas's conjecture as a corollary to the following case:

25.11 Proposition (Matsubara [90]). *Suppose that* $\kappa^{<\kappa} < \lambda^{<\kappa} = 2^\lambda$. *Then every stationary subset of* $\mathcal{P}_\kappa\lambda$ *can be partitioned into* $\lambda^{<\kappa}$ *disjoint stationary sets.*

Proof. Note first that if Y is unbounded in $\mathcal{P}_\kappa\lambda$, then $\mathcal{P}_\kappa\lambda \subseteq \bigcup_{y\in Y}\mathcal{P}(y)$ so that $\lambda^{<\kappa} \leq \kappa^{<\kappa} \cdot |Y|$, and $\kappa^{<\kappa} < \lambda^{<\kappa}$ thus implies that $|Y| = 2^\lambda$. Next, as $\lambda^{<\kappa} = 2^\lambda$, let $\langle f_\alpha \mid \alpha < 2^\lambda\rangle$ enumerate the collection of functions: $[\lambda]^2 \to \mathcal{P}_\kappa\lambda$ so that each function appears cofinally often. In terms of 25.3, each $C(f_\alpha)$ is closed unbounded, and for any $Z \in \mathcal{C}_{\kappa,\lambda}$ and $\xi < 2^\lambda$, there is an α such that $\xi < \alpha < 2^\lambda$ and $C(f_\alpha) \subseteq Z$.

Suppose now that X is stationary in $\mathcal{P}_\kappa\lambda$. For each $\alpha < 2^\lambda$ choose a sequence $\langle \eta^\alpha_\xi \mid \xi < \alpha\rangle$ of distinct members of X by recursion on α as follows: Having chosen such sequences for $\beta < \alpha$, first note that $X \cap C(f_\alpha)$ is stationary, hence unbounded, and therefore has cardinality 2^λ as noted above. Let $\langle \eta^\alpha_\xi \mid \xi < \alpha\rangle$ be a sequence of distinct members drawn from $(X \cap C(f_\alpha)) - \{\eta^\beta_\xi \mid \beta < \alpha \wedge \xi < \beta\}$.

For $\xi < 2^\lambda$, set $X_\xi = \{\eta^\alpha_\xi \mid \xi < \alpha < 2^\lambda\}$. Then $\{X_\xi \mid \xi < 2^\lambda\}$ is a collection of pairwise disjoint subsets of X. The proof is completed by showing that each X_ξ is stationary: If C is closed unbounded, then as noted before there is an α such that $\xi < \alpha < 2^\lambda$ and $C(f_\alpha) \subseteq C$. But then, $\eta^\alpha_\xi \in X_\xi \cap C(f_\alpha) \subseteq X_\xi \cap C$. ⊣

25.12 Corollary. *Assume* GCH *and* $\kappa \leq \lambda$. *Then:*

(a) If $0^\#$ *does not exist, then every stationary subset of* $\mathcal{P}_\kappa\lambda$ *can be partitioned into* $\lambda^{<\kappa}$ *disjoint stationary sets.*

(b) $\mathcal{P}_\kappa\lambda$ *can be partitioned into* $\lambda^{<\kappa}$ *disjoint stationary sets.*

Proof. If $\lambda^{<\kappa} = \lambda$, the results follow from 25.6(a) and 25.7, and if $\lambda^{<\kappa} > \lambda$, from 25.11. ⊣

(b) also follows from the work on combinatorial principles in Donder-Matet [93], which showed that if $2^{<\kappa} < \lambda$, then $\mathcal{P}_\kappa\lambda$ can be partitioned into $\lambda^{<\kappa}$ stationary sets, improving a result of Matsubara [88]. Both papers include results showing that some such hypothesis is necessary.

A major development from Jech [73] was the investigation of partition properties for $\mathcal{P}_\kappa\gamma$. Note that for $|\gamma| > \kappa$ every unbounded subset of $\mathcal{P}_\kappa\gamma$ has members $x_1 \subset y_1$ and $x_2 \not\subset y_2$. Hence, for $n \in \omega$ it is natural to consider partitions of

$$[\mathcal{P}_\kappa\gamma]^n_\subset = \{\{x_1, \ldots, x_n\} \mid x_1 \subset \ldots \subset x_n \in \mathcal{P}_\kappa\gamma\} .$$

For functions f on such sets, $f(x_1, \ldots, x_n)$ is written for $f(\{x_1, \ldots, x_n\})$ with the understanding that $x_1 \subset \ldots \subset x_n$. For such an f,

$H \subseteq \mathcal{P}_\kappa\gamma$ is *homogeneous for* f *iff* there is an i such that $x_1 \subset \ldots \subset x_n$ all in H implies that $f(x_1, \ldots, x_n) = i$.

The following are generalizations (from $n = 2$) of propositions formulated by Jech.

$\text{Part}(\kappa, \gamma)^n$ *iff* for any $f\colon [\mathcal{P}_\kappa\gamma]^n_\subset \to 2$ there is an unbounded $H \subseteq \mathcal{P}_\kappa\gamma$ homogeneous for f .

$\text{Part}^*(\kappa, \gamma)^n$ *iff* for any $f\colon [\mathcal{P}_\kappa\gamma]^n_\subset \to 2$ there is a stationary $H \subseteq \mathcal{P}_\kappa\gamma$ homogeneous for f .

Jech observed that $\text{Part}(\kappa, \gamma)^2$ implies that κ is weakly compact, and raised several questions.

The early work along these lines was done around 1971 by Solovay and Menas, who considered the generalization of Rowbottom's 7.17 to normal ultrafilters over $\mathcal{P}_\kappa\gamma$. For U such an ultrafilter,

U has the *partition property iff* for any $f\colon [\mathcal{P}_\kappa\gamma]^2_\subset \to 2$ there is a set in U homogeneous for f .

This property implies $\text{Part}^*(\kappa, \gamma)^2$ by 25.4. Except for the case $\gamma = \kappa$ handled by 7.17 and 22.5(b), that every normal ultrafilter over $\mathcal{P}_\kappa\gamma$ has the partition property was not evident. Menas found a simple characterization:

25.13 Theorem (Menas [76]). *For U a normal ultrafilter over $\mathcal{P}_\kappa\gamma$, U has the partition property iff there is an $A \in U$ such that $x \subset y$ both in A implies that $|x| < |y \cap \kappa|$.*

Proof. In the forward direction, define $f\colon [\mathcal{P}_\kappa\gamma]^2_\subset \to 2$ by: $f(x, y) = 0$ *iff* $|x| < |y \cap \kappa|$. Let $H \in U$ be homogeneous for f; since there are $x_0, y_0 \in H$ such that $|x_0| < |y_0 \cap \kappa|$, the desired result follows.

The converse is established by adapting the argument for Rowbottom's 7.17: Let A be as hypothesized, and suppose that $f\colon [\mathcal{P}_\kappa\gamma]^2_\subset \to 2$. For each $x \in \mathcal{P}_\kappa\gamma$ define $f_x\colon \{y \in \mathcal{P}_\kappa\gamma \mid x \subset y\} \to 2$ by $f_x(y) = f(x, y)$. Then for each $x \in \mathcal{P}_\kappa\gamma$ there is an $H_x \in U$ and an $i_x < 2$ such that $f_x``H_x = \{i_x\}$. It follows that there is a $B \subseteq A$ with $B \in U$ and an $i < 2$ such that for any $x \in B$, $i_x = i$. Set $H = \{x \in B \mid x \in \bigcap\{H_y \mid y \subset x \ \wedge \ y \in B\}\}$. Then $y \subset x$ both in H implies that $f(x, y) = i$, so it suffices to show that $H \in U$:

Assume that this fails, so that $B - H \in U$. Then for any x in this set, there is an $h(x) \in B$ so that $h(x) \subset x$ and $x \notin H_{h(x)}$. Let $j\colon V \prec M \cong \mathrm{Ult}(V, U)$. Then $[h]_U \subset j``\lambda$, and as $B \subseteq A$ so that $|h(x)| < |x \cap \kappa|$ for $x \in B - H$, $|[h]_U| < \kappa$ by 22.11(a). Hence, a simple argument using κ-completeness shows that there is an $s \in \mathcal{P}_\kappa\gamma$ such that $[h]_U = j(s)$. But for any $x \in B - H$ satisfying $h(x) = s$ we have $x \notin H_s$, and this contradicts $H_s \in U$. ⊣

Proceeding inductively as for 7.17, this argument shows that U has the partition property *iff* for any $n \geq 2$, whenever $f\colon [\mathcal{P}_\kappa\gamma]^n_\subset \to 2$, there is a set in U homogeneous for f.

Menas moreover found a useful property that implies the partition property. For U a normal ultrafilter over $\mathcal{P}_\kappa\gamma$,

$$\chi(U) \ \textit{iff} \ \text{ for some } f \in {}^\kappa\kappa,\ \{x \in \mathcal{P}_\kappa\gamma \mid f(|x \cap \kappa|) = |x|\} \in U\ .$$

With $j_U\colon V \prec M \cong \mathrm{Ult}(V, U)$, $\chi(U)$ translates by 22.11(a) into the assertion that for some $f \in {}^\kappa\kappa$, $j_U(f)(\kappa) = |\gamma|$. That $\chi(U)$ implies that U has the partition property depends on the following lemma, a derivative of Kunen's proof of his 23.12:

25.14 Lemma (Solovay [74: 371]). *For U a normal ultrafilter over $\mathcal{P}_\kappa\gamma$, there is an $E \in U$ such that $x \subset y$ both in E implies that $|x| < |y|$.*

Proof. Let f be ω-Jónsson for γ, and set

$$E = \{x \in \mathcal{P}_\kappa\gamma \mid f|[x]^\omega \text{ is } \omega\text{-Jónsson for } x\}\ .$$

Then for $x \subseteq y$ both in E, $|x| = |y|$ implies that $x = f``[x]^\omega = y$. Hence, it suffices to show that $E \in U$:

Let $j\colon V \prec M \cong \mathrm{Ult}(V, U)$. Then by Łoś's Theorem, it suffices to show that $j(f)|[j``\gamma]^\omega$ is $\omega$-Jónsson for $j``\gamma$, noting that we do not need to distinguish between V and M as ${}^\gamma M \subseteq M$. We indeed have $j(f)``[j``\gamma]^\omega \subseteq j``\gamma$ (as in

Kunen's original proof of his 23.12). Suppose now that $y \subseteq j``\gamma$ with $|y| = |j``\gamma| = |\gamma|$. Setting $z = j^{-1}(y)$, $f``[z]^\omega = \gamma$ since $f$ is $\omega$-Jónsson for $\gamma$. It follows that for any $\alpha < \gamma$, there is an $s \in [z]^\omega$ satisfying $\alpha = f(s)$, so that $j(\alpha) = j(f(s)) = j(f)(j(s))$ where $j(s) \in [y]^\omega$. Hence $j(f)``[y]^\omega = j``\gamma$. ⊣

25.15 Proposition (Menas [76]). *For U a normal ultrafilter over $\mathcal{P}_\kappa\gamma$, $\chi(U)$ implies that U has the partition property.*

Proof. Let $f \in {}^\kappa\kappa$ be such that $X = \{x \in \mathcal{P}_\kappa\gamma \mid f(|x \cap \kappa|) = |x|\} \in U$. With $j\colon V \prec M \cong \mathrm{Ult}(V, U)$, clearly $j(f)``\kappa = f``\kappa \subseteq \kappa$ as $\mathrm{crit}(j) = \kappa$. It follows by 22.11(a) that $Y = \{x \in \mathcal{P}_\kappa\gamma \mid f``|x \cap \kappa| \subseteq |x \cap \kappa|\} \in U$. Finally, by 25.14 let $E \in U$ be such that $x \subset y$ both in E implies that $|x| < |y|$. Now for $x \subset y$ both in $X \cap Y \cap E \in U$, since $|x| < |y|$, $|x \cap \kappa| < |y \cap \kappa|$ by definition of X, and moreover $f(|x \cap \kappa|) = |x| < |y \cap \kappa|$ by definition of Y. Consequently, U has the partition property by 25.13. ⊣

It follows that for γ easily definable from κ, like $\gamma = \kappa^+$ or $\gamma = 2^\kappa$, every normal ultrafilter over $\mathcal{P}_\kappa\gamma$ has the partition property. The initial observations about the partition property were along these lines, made by Solovay assuming GCH. If $\gamma = \kappa^+$ for example, then the function $f \in {}^\kappa\kappa$ given by $f(\alpha) = \alpha^+$ verifies $\chi(U)$ for any normal ultrafilter over $\mathcal{P}_\kappa\kappa^+$. Donald Pelletier (see Kunen-Pelletier [83]) later established that $\chi(U)$ is strictly stronger than the partition property for U.

Menas used the property χ to show that if κ is supercompact, then for every $\gamma \geq \kappa$ there is a normal ultrafilter over $\mathcal{P}_\kappa\gamma$, in fact the maximal possible number of such ultrafilters, with the partition property. The argument is as for 22.14, with a further twist loaded into Solovay's for 22.13:

25.16 Exercise (Menas [76]). *Suppose that κ is $2^{|\gamma|^{<\kappa}}$-supercompact. Then for any $A \in \mathcal{P}(\mathcal{P}(\mathcal{P}_\kappa\gamma))$ there is a normal ultrafilter W over $\mathcal{P}_\kappa\gamma$ with $\chi(W)$ such that $A \in M_W$. Hence, there are $2^{2^{|\gamma|^{<\kappa}}}$ normal ultrafilters over $\mathcal{P}_\kappa\gamma$ with the partition property.*

Hint. Assume that this fails for some $\gamma \geq \kappa$, and let γ_0 be the least such. Following the proof of 22.13, let U be a normal ultrafilter over $\mathcal{P}_\kappa(2^{|\gamma_0|^{<\kappa}})$ with corresponding $j\colon V \prec M \cong \mathrm{Ult}(V, U)$, and let $\overline{\varphi}(v_0, v_1, v_2)$ be such that $\overline{\varphi}[A, \kappa, \gamma]$ *iff* $A \in \mathcal{P}(\mathcal{P}(\mathcal{P}_\kappa\gamma))$ and for any normal ultrafilter W over $\mathcal{P}_\kappa\gamma$ with $\chi(W)$, $A \notin M_W$. It follows as before that if $j_0\colon V \prec M_0 \cong \mathrm{Ult}(V, U|\gamma_0)$, then by the choice of γ_0, $M_0 \models \gamma_0$ is the least ordinal such that $\exists v_0\overline{\varphi}[\kappa, \gamma_0]$. A contradiction results, once it is verified that $\chi(U|\gamma_0)$. But if $f \in {}^\kappa\kappa$ is defined by $f(\alpha) = |\beta|$ for the least β such that $\exists v_0\overline{\varphi}[\alpha, \beta]$ if one exists, and $= 0$ otherwise, then $j_{U|\gamma_0}(f)(\kappa) = |\gamma_0|$. ⊣

What then heightened interest in the partition property was Solovay's discovery that it does not always hold:

25.17 Proposition (Solovay – Menas [76]). *Suppose that $\kappa < \lambda \leq \delta$ and both κ and λ are δ-supercompact. Then there is a normal ultrafilter $\mathcal{P}_\kappa\delta$ without the partition property.*

Proof. Let D be a normal ultrafilter over $\mathcal{P}_\kappa\delta$, and for each $x \in \mathcal{P}_\lambda\delta$ with $x \supseteq \kappa$ set

$$D|x = \{\{s \cap x \mid s \in Z\} \mid Z \in D\} .$$

Then $D|x$ is a normal ultrafilter over $\mathcal{P}_\kappa x$ (cf. after 22.12). Now let U be a normal ultrafilter over $\mathcal{P}_\lambda\delta$, and define W by:

$$X \in W \text{ iff } X \subseteq \mathcal{P}_\kappa\delta \wedge \{x \in \mathcal{P}_\lambda\delta \mid x \supseteq \kappa \wedge X \cap \mathcal{P}_\kappa x \in D|x\} \in U .$$

Then W is a normal ultrafilter over $\mathcal{P}_\kappa\delta$ (cf. the hint for 22.9), one that glues together approximations to D using U. D may or may not have the partition property, but W does not:

To show this, it suffices by 25.13 to take an arbitrary $X \in W$ and find $x \subset y$ both in X such that $|x| \geq |y \cap \kappa|$. For such an X, set

$$Y = \{x \in \mathcal{P}_\kappa\delta \mid x \supseteq \kappa \wedge X \cap \mathcal{P}_\kappa x \in D|x\} ,$$

and for each $x \in Y$ set

$$Z_x = \{s \in \mathcal{P}_\kappa\delta \mid s \cap x \in X\} .$$

Then by definition of W, $Y \in U$ and $Z_x \in D$ for every $x \in Y$. Now let $\overline{x} \subset \overline{y}$ be both in Y, and by unboundedness let $s \in Z_{\overline{x}} \cap Z_{\overline{y}}$ be such that $s \cap (\overline{y} - \overline{x}) \neq \emptyset$. Set $x = \overline{x} \cap s$ and $y = \overline{y} \cap s$. Then $x \subset y$ and both are in X, yet as $\overline{x} \cap \overline{y} \supseteq \kappa$

$$|x| = |\overline{x} \cap s| \geq |\kappa \cap s| = |\overline{y} \cap \kappa \cap s| = |y \cap \kappa| . \quad \dashv$$

Taking $\delta = \lambda$, if $\kappa < \lambda$, κ is λ-supercompact, and λ is measurable, then there is a normal ultrafilter over $\mathcal{P}_\kappa\lambda$ without the partition property. Just after Solovay established this result, Kunen (see Kunen-Pelletier [83]) improved it by showing that if $\kappa < \lambda$, κ is λ-supercompact, and λ is ineffable (a property consistent with $V = L$ discussed in volume II), then there is a $\gamma < \lambda$ and a normal ultrafilter over $\mathcal{P}_\kappa\gamma$ without the partition property. Moreover, he showed that if κ is supercompact and there is a $\gamma > \kappa$ so that some normal ultrafilter over $\mathcal{P}_\kappa\gamma$ does not have the partition property, then γ is Π^2_1-indescribable. Hence, it is consistent that there is no such γ.

Further results about the partition property for normal ultrafilters were established in Menas [76], Di Prisco [77], Pelletier [81], Kunen-Pelletier [83], and Barbanel [86, 92, 93, 93a].

Turning to the more general propositions $\text{Part}(\kappa, \gamma)^n$ and $\text{Part}^*(\kappa, \gamma)^n$, by Menas's 25.16, if κ is $2^{|\gamma|^{<\kappa}}$-supercompact, then $\text{Part}^*(\kappa, \gamma)^2$ (and in fact $\text{Part}^*(\kappa, \gamma)^n$ for every $n \in \omega$ by the comment after 25.13). Citing this for one direction, Magidor established the following combinatorial characterization of supercompactness:

25.18 Theorem (Magidor [74]). *κ supercompact iff* $\text{Part}^*(\kappa, \gamma)^2$ *for every $\gamma \geq \kappa$.*

It was recently observed that the forward direction has a sharp local form, improving what can be deducted from 25.16. Shizuo Kamo [97, 02] established: *If κ is γ-supercompact, then there is a normal ultrafilter over $\mathcal{P}_\kappa\gamma$ with the partition property (and hence* $\text{Part}^*(\kappa, \gamma)^2$ *holds).* See Shioya [∞] for a short proof. There is a quick argument for showing that if κ is γ-supercompact, then $\text{Part}^*(\kappa, \gamma)^2$ holds:

Assume to the contrary that the proposition fails for some $\gamma \geq \kappa$, and let γ_0 be the least such. Let U be a normal ultrafilter over $\mathcal{P}_k\gamma_0$ with corresponding $j\colon V \prec M \cong \text{Ult}(V, U)$. Define $f\colon \kappa \to \kappa$ by: $f(\xi) = 0$ unless ξ is a regular uncountable cardinal and $\text{Part}^*(\xi, \gamma)$ fails for some $\gamma \geq \xi$, in which case $f(\xi) =$ the least such γ. Since M is closed under $|\gamma_0|^{<\kappa}$-sequences (22.11(b)), it follows that $j(f)(\kappa) = \gamma_0$. But then, $\chi(U)$, so that by 25.15, $\text{Part}^*(\kappa, \gamma)$, which is a contradiction.

Magidor also established an intermediate characterization of supercompactness in terms of a concept formulated in Jech [73], γ-ineffability. Di Prisco and William Zwicker in their [80] refined this result, and moreover provided an analogous characterization of strong compactness. Through such means Baumgartner, Carr [87], and Di Prisco independently showed that *if* $\text{Part}(\kappa, \gamma)^3$ *holds for every $\gamma \geq \kappa$, then κ is strongly compact.* The exponent here was recently lowered to 2, but it remains unknown whether strong compactness yields even exponent 2:

25.19 Question. *If κ is strongly compact, does* $\text{Part}(\kappa, \gamma)^3$ *hold for every $\gamma \geq \kappa$?*

25.20 Theorem (Matet). *If* $\text{Part}(\kappa, \gamma)^2$ *holds for every $\gamma \geq \kappa$, then κ is strongly compact.*

The property $\text{Part}(\kappa, \gamma)^2$ generalizes the characterization $\kappa \longrightarrow (\kappa)^2_2$ of the weak compactness of κ. Carr [85] considered generalizations of other characterizations, and in [86, 87, 87a] pursued their study in considerable detail. (See also Johnson [88, 88a].)

Recent work has been directed toward shortcomings in the analogical treatment of $\langle \mathcal{P}_\kappa\gamma, \subset\rangle$ in terms of $\langle \kappa, <\rangle$. Zwicker [84] defined a *stationary coding set for $\mathcal{P}_\kappa\gamma$* to be a set A stationary in $\mathcal{P}_\kappa\gamma$ together with an injective function $c\colon A \to \gamma$ such that $x \subset y$ both in A implies that $c(x) \in y$. He showed that restricting the study of $\langle \mathcal{P}_\kappa\gamma, \subset\rangle$ to such a set A results in more features analogous to $\langle \kappa, <\rangle$. Shelah [86: §§7, 9] then established the existence of stationary coding sets in a variety of situations, e.g. when $\kappa > \omega_2$ is a successor cardinal and $\gamma = \kappa^+$. Zwicker [86] contained related work, and his [89] subsequently used stationary coding sets to begin a study of structural properties of ideals over $\mathcal{P}_\kappa\gamma$ in the spirit of the extensive treatment in Baumgartner-Taylor-Wagon [82] for ideals over κ.

Carr-Pelletier [89] also began such a study, but by proceeding analogously with $\langle \mathcal{P}_\kappa\gamma, <\rangle$ where $<$ on $\mathcal{P}_\kappa\gamma$ is defined by:

$$x < y \quad \textit{iff} \quad x \subset y \ \wedge \ |x| < |y \cap \kappa| \, .$$

This new order is suggested by 25.13, and was first used systematically in Carr [87]. The following exercise provides versions of previous results in terms of $<$. If $\langle X_a \mid a \in \mathcal{P}_\kappa\gamma \rangle \in {}^{\mathcal{P}_\kappa\gamma}\mathcal{P}(\mathcal{P}_\kappa\gamma)$, then its *$<$-diagonal intersection* is $\{x \in \mathcal{P}_\kappa\gamma \mid x \in \bigcap_{a<x} X_a\}$. For a filter F over $\mathcal{P}_\kappa\gamma$,

F is *strongly normal* *iff* F is fine, and for any $\langle X_a \mid a \in \mathcal{P}_\kappa\gamma \rangle \in {}^{\mathcal{P}_\kappa\gamma}F$, its $<$-diagonal intersection is in F.

For $X \subseteq \mathcal{P}_\kappa\gamma$, a function $f \colon X \to \mathcal{P}_\kappa\gamma$ is *$<$-regressive on X iff* $f(x) < x$ for every $x \in X$. For $n \in \omega$, set

$$[\mathcal{P}_\kappa\gamma]^n_< = \{\{x_1, \ldots, x_n\} \mid x_1 < \ldots < x_n \text{ all in } \mathcal{P}_\kappa\gamma\} \, .$$

For functions on this set, that a subset is *homogeneous* has the expected meaning from the $[\mathcal{P}_\kappa\gamma]^n_\subset$ context.

25.21 Exercise (Carr-Pelletier [89]).

(a) A strongly normal filter over $\mathcal{P}_\kappa\gamma$ is normal.

(b) A fine filter F over $\mathcal{P}_\kappa\gamma$ is strongly normal iff whenever X is F-stationary and f is $<$-regressive on X, there is an F-stationary $Y \subseteq X$ such that f is constant on Y.

(c) A normal ultrafilter over $\mathcal{P}_\kappa\gamma$ is strongly normal.

(d) If U is a normal ultrafilter over $\mathcal{P}_\kappa\gamma$ and $f \colon [\mathcal{P}_\kappa\gamma]^n_< \to 2$ for some $n \in \omega$, then there is a set in U homogeneous for f.

Hint. For (c) and (d) recall the argument for 25.13. ⊣

Thus, the partition property for $[\mathcal{P}_\kappa\gamma]^2_<$, unlike for $[\mathcal{P}_\kappa\gamma]^2_\subset$, holds for all normal ultrafilters. On the other hand, the approach with $<$ is less general, for Carr-Levinski-Pelletier [90] showed that for some (every) $\gamma \geq \kappa$ there is a strongly normal filter over $\mathcal{P}_\kappa\gamma$ *iff* κ is Mahlo or $\kappa = \nu^+$ where $\nu^{<\nu} = \nu$.

The investigation of $\mathcal{P}_\kappa\gamma$ is far from exhausted, and both the stationary coding set and $<$ ordering approaches to structural analysis hold promise. Roughly speaking, the first approach seems more appropriate when either γ is small with respect to κ or the property being investigated is invariant under isomorphisms induced by mappings: $\mathcal{P}_\kappa\gamma \to \mathcal{P}_\kappa\gamma$, and the second, when κ is large, and either γ is large with respect to κ or the property being investigated is not invariant under isomorphisms. For ideals over $\mathcal{P}_\kappa\gamma$ a full structure theory has yet to be worked out.

26. Extenders

As a nod to modernity, this section reaches ahead to provide a basic analysis of elementary embeddings and related hypotheses that was to become crucial to both the development of inner model theory and the consistency analysis of strong propositions from the late 1970's onwards. (Volume II proceeds with the orderly elaboration of strong hypotheses, especially with regard to combinatorics and relative consistency, carried out systematically through the 1970's and after.)

As described in §20, if U is a normal ultrafilter over κ, then $L[U]$ is an inner model of ZFC in which κ is measurable. Moreover, these models have strong canonicity properties and establish the relative consistency of minimal principles like GCH, and their structural coherence provide a persuasive argument for the consistency of measurability. A first attempt at relativizing strong hypotheses to develop a comparable inner model theory would be to analogously take a normal ultrafilter U over $\mathcal{P}_\kappa\gamma$ and consider $L[U]$. In the case of measurability, the absoluteness $\kappa \cap L = \kappa$ of the underlying set for measurability is implicitly exploited in the results on canonicity. However, in the presence of a normal ultrafilter U over $\mathcal{P}_\kappa\gamma$ with $|\gamma| > \kappa$, $\mathcal{P}_\kappa\gamma \cap L = \mathcal{P}_\kappa\gamma$ fails, and indeed $\mathcal{P}_\kappa\gamma \cap L \notin U$ so that $L[U] = L$ (25.8).

The absoluteness problem can be dealt with, at the cost of weakening the hypotheses to be relativized, by considering *extenders*, sequences of ultrafilters over absolute sets whose corresponding ultrapowers form a directed system. Extenders had occurred in a general context in Powell [74]. However, they did not enter large cardinal theory until after Mitchell [79] discovered a way to develop a canonical inner model theory for weak versions of γ-supercompactness. Dodd and Jensen (see Dodd [82]) then arrived at the extender concept as an improvement over Mitchell's formulations, and first used the term. The basic theory of extenders and related hypotheses is developed here, these scant remarks alone suggesting their crucial role in the emerging inner model theory. Unlike for γ-supercompactness, it is the method of analysis rather than those hypotheses characterizable by it that is described first, as it provides the main motivation.

The basic discovery is that any elementary embedding between inner models can be approximated arbitrarily closely as direct limits of ultrapowers with constructive features reminiscent of iterated ultrapowers. Proceeding in a general setting that anticipates wide applicability, suppose that

$$N \text{ and } M \text{ are inner models of ZFC and } j \colon N \prec M\ .$$

There is no presumption here that j is a (definable) class of N or that $M \subseteq N$. By 5.1(b), j has a critical point.

$$\text{Let } \beta > \kappa = \operatorname{crit}(j), \text{ and } \zeta \geq \kappa \text{ the least ordinal satisfying } \beta \leq j(\zeta)\ .$$

The basic case is $\zeta = \kappa$, i.e. $\beta \leq j(\kappa)$; the general context is for the study of very strong hypotheses. For each $a \in [\beta]^{<\omega}$, define E_a by:

$$X \in E_a \ \text{ iff } \ X \in \mathcal{P}([\zeta]^{|a|}) \cap N \ \wedge\ a \in j(X)\ .$$

This is yet another version of the now familiar idea of generating ultrafilters from embeddings. If $X \in \mathcal{P}([\zeta]^{|a|}) \cap N$, then $j(X) \in \mathcal{P}([j(\zeta)]^{|a|}) \cap M$, so that querying $a \in j(X)$ is apropos. Although E_a may not be in N, $\langle N, \in, E_a \rangle \models E_a$ is a κ-complete ultrafilter over $[\zeta]^{|a|}$. Strictly speaking, E_a is a principal ultrafilter for $a \in [\kappa]^{<\omega}$, but for notational simplicity conventions and concepts are hereby liberalized to allow such E_a throughout this section. Stipulate that

$$\mathcal{E} = \langle E_a \mid a \in [\beta]^{<\omega} \rangle \text{ is the } (\kappa, \beta)\text{-}\textit{extender derived from } j \ .$$

For $a \in [\beta]^{<\omega}$ let

$$j_a \colon N \prec \mathrm{Ult}(N, E_a)$$

be the usual ultrapower embedding. It is simple to check that $k_a \colon \mathrm{Ult}(N, E_a) \to M$ defined by

$$k_a((f)^0_{E_a}) = j(f)(a)$$

(where $(f)^0_{E_a} \in \mathrm{Ult}(N, E_a)$ is the equivalence class of $f \in {}^{[\zeta]^{|a|}}N \cap N$) is elementary, and that $k_a \circ j_a = j$. In particular, $\mathrm{Ult}(N, E_a)$ is well-founded and so has a transitive collapse M_a.

$$\mathrm{Ult}(N, E_a) \text{ is identified with } M_a \ .$$

Next, for $a \subseteq b$ both in $[\beta]^{<\omega}$, say $b = \{\alpha_1, \ldots, \alpha_n\}$ with the usual understanding that $\alpha_1 < \ldots < \alpha_n$ and $a = \{\alpha_{i_1}, \ldots, \alpha_{i_m}\}$ where $1 \le i_1 < \ldots < i_m \le n$, define $\pi_{ba} \colon [\zeta]^n \to [\zeta]^m$ by:

$$\pi_{ba}(\{\xi_1, \ldots, \xi_n\}) = \{\xi_{i_1}, \ldots, \xi_{i_m}\} \ .$$

Then it is simple to check that $i_{ab} \colon M_a \to M_b$ defined by

$$i_{ab}([f]_{E_a}) = [f \circ \pi_{ba}]_{E_b}$$

is elementary, and that

$$i_{ab} \circ j_a = j_b \ \text{ and } \ k_b \circ i_{ab} = k_a \ .$$

Finally, $\langle \langle M_a \mid a \in [\beta]^{<\omega} \rangle, \langle i_{ab} \mid a \subseteq b \rangle \rangle$ is seen to be a directed system, so stipulate that

$$\langle M_{\mathcal{E}}, \in_{\mathcal{E}} \rangle \text{ is the direct limit ,}$$

and

$$j_{\mathcal{E}} \colon \langle N, \in \rangle \prec \langle M_{\mathcal{E}}, \in_{\mathcal{E}} \rangle \ ,$$
$$k_{a\mathcal{E}} \colon \langle M_a, \in \rangle \prec \langle M_{\mathcal{E}}, \in_{\mathcal{E}} \rangle \ , \text{ and}$$
$$k_{\mathcal{E}} \colon \langle M_{\mathcal{E}}, \in_{\mathcal{E}} \rangle \prec \langle M, \in \rangle$$

the corresponding embeddings, so that for $a \in [\beta]^{<\omega}$,

$$k_{\mathcal{E}} \circ j_{\mathcal{E}} = j \ , \ k_{a\mathcal{E}} \circ j_a = j_{\mathcal{E}} \ , \ \text{and } k_{\mathcal{E}} \circ k_{a\mathcal{E}} = k_a \ .$$

In particular, $\langle M_{\mathcal{E}}, \in_{\mathcal{E}} \rangle$ is well-founded, and so assume that

$$M_{\mathcal{E}} \text{ is transitive and } \in_{\mathcal{E}} = \in \cap (M_{\mathcal{E}} \times M_{\mathcal{E}}) \ .$$

The crucial features of the extender analysis of elementary embeddings are given by the following lemma. Its (a) provides a useful representation of $M_{\mathcal{E}}$ (cf. 19.6), and its (b) and (c) describe to what extent $M_{\mathcal{E}}$ approximates M. Although there are at most $2^{2^{|\zeta|}}$ distinct M_a's, it is how they are interwoven by the i_{ab}'s that leads to these considerable properties.

26.1 Lemma.

(a) $M_{\mathcal{E}} = \{j_{\mathcal{E}}(f)(a) \mid a \in [\beta]^{<\omega} \wedge f \in {}^{[\zeta]^{|a|}}N \cap N\}$.

(b) For any γ *satisfying* $|V_\gamma|^M \leq \beta$: $V_\gamma^M \subseteq \operatorname{ran}(k_{\mathcal{E}})$, $V_\gamma^{M_{\mathcal{E}}} = V_\gamma^M$, *and* $k_{\mathcal{E}}(x) = x$ *for* $x \in V_\gamma^{M_{\mathcal{E}}}$.

(c) $\operatorname{crit}(k_{\mathcal{E}}) \geq \beta$, *and so* $\operatorname{crit}(j_{\mathcal{E}}) = \kappa$ *and* $\beta \leq j_{\mathcal{E}}(\zeta)$. *If* $\beta = j(\zeta)$, *then* $\operatorname{crit}(k_{\mathcal{E}}) > \beta$, *and so* $\beta = j_{\mathcal{E}}(\zeta)$.

Proof. (a) If $x \in M_{\mathcal{E}}$, there is an $a \in [\beta]^{<\omega}$ and an $f \in {}^{[\zeta]^{|a|}}N \cap N$ such that $x = k_{a\mathcal{E}}([f]_{E_a})$, and so

$$k_{\mathcal{E}}(x) = k_{\mathcal{E}}(k_{a\mathcal{E}}([f]_{E_a})) = k_a([f]_{E_a}) = j(f)(a)\ .$$

Consequently,

$$\operatorname{ran}(k_{\mathcal{E}}) = \{j(f)(a) \mid a \in [\beta]^{<\omega} \wedge f \in {}^{[\zeta]^{|a|}}N \cap N\}\ .$$

The result now follows since $k_{\mathcal{E}} \circ j_{\mathcal{E}} = j$.

(b) Let $g \in {}^{[\zeta]^1}N \cap N$ be such that whenever $|V_\alpha|^N \leq \zeta$, $g|[|V_\alpha|^N]^1$ is a bijection between $[|V_\alpha|^N]^1$ and V_α^N. Then for any γ satisfying $|V_\gamma|^M \leq \beta$ and any $x \in V_\gamma^M$, there is a $\xi < \beta$ such that $j(g)(\{\xi\}) = x$. It follows from the characterization of $\operatorname{ran}(k_{\mathcal{E}})$ above that $V_\gamma^M \subseteq \operatorname{ran}(k_{\mathcal{E}})$, and the rest follows since the inverse of $k_{\mathcal{E}}$ is just the collapsing isomorphism: $\operatorname{ran}(k_{\mathcal{E}}) \to M_{\mathcal{E}}$.

(c) This is similar, since it is readily seen that $\beta \subseteq \operatorname{ran}(k_{\mathcal{E}})$. Note that if $\beta = j(\zeta) = k_{\mathcal{E}}(j_{\mathcal{E}}(\zeta))$, then in fact $\beta + 1 \subseteq \operatorname{ran}(k_{\mathcal{E}})$. ⊣

Having formulated extenders in terms of an ambient elementary embedding, the essential features are abstracted in a definition. For N an inner model of ZFC, κ a cardinal in the sense of N, $\beta > \kappa$, and $\mathcal{E} = \langle E_a \mid a \in [\beta]^{<\omega}\rangle$, $\mathcal{E}$ is an *N-(κ,β)-extender* (and simply a *(κ,β)-extender* if $N = V$) *iff* for some $\zeta \geq \kappa$:

(i) For each $a \in [\beta]^{<\omega}$, $\langle N, \in, E_a\rangle \models E_a$ is a κ-complete ultrafilter over $[\zeta]^{|a|}$, and:

(1) For at least one such a, $\langle N, \in, E_a\rangle \models E_a$ is not κ^+-complete.

(2) For each $\xi \in \zeta$, there is such an a so that $\{s \in [\zeta]^{|a|} \mid \xi \in s\} \in E_a$.

(ii) (Coherence) Suppose that $a \subseteq b$ are both in $[\beta]^{<\omega}$ and $\pi_{ba}\colon [\zeta]^{|b|} \to [\zeta]^{|a|}$ is defined as before, i.e. if $b = \{\alpha_1, \ldots, \alpha_n\}$ and $a = \{\alpha_{i_1}, \ldots, \alpha_{i_m}\}$, then $\pi_{ba}(\{\xi_1, \ldots, \xi_n\}) = \{\xi_{i_1}, \ldots, \xi_{i_m}\}$. Then

$$X \in E_a \ \textit{ iff } \ \{s \mid \pi_{ba}(s) \in X\} \in E_b\ .$$

(iii) (Well-foundedness) Whenever $a_m \in [\beta]^{<\omega}$ and $X_m \in E_{a_m}$ for $m \in \omega$, there is an $d\colon \bigcup_m a_m \to \zeta$ such that $d``a_m \in X_m$ for every such m.

(iv) (Normality) Whenever $a \in [\beta]^{<\omega}$, $f \in {}^{[\zeta]^{|a|}}N \cap N$, and

$$\{s \in [\zeta]^{|a|} \mid f(s) \in \max(s)\} \in E_a ,$$

there is a $b \in [\beta]^{<\omega}$ with $a \subseteq b$ such that, with π_{ba} as in (ii),

$$\{s \in [\zeta]^{|b|} \mid f(\pi_{ba}(s)) \in s\} \in E_b .$$

Again, $\zeta = \kappa$ is the basic case, and the general context is for the study of very strong hypotheses. *For* (i), *the* E_a*'s are allowed to be principal.* (i)(1) establishes the role of κ, and as will become clear, (i)(2) specifies the ζ. It is simple to check that *for any* $\alpha < \beta$, *if the* a*'s from* $[\alpha]^{<\omega}$ *satisfy* (i)(1) *and* (i)(2), *then* $\mathcal{E}|[\alpha]^{<\omega}$ *is an* N-(κ, α)*-extender.*

We now proceed to use these clauses to construct an $M_{\mathcal{E}}$ and a $j_{\mathcal{E}}\colon N \prec M_{\mathcal{E}}$ as before; throughout, π_{ba} is as in (ii). First, for each $a \in [\beta]^{<\omega}$ we can by (i) form the ultrapower $\mathrm{Ult}(N, E_a)$. By (iii) applied with each $a_m = a$, E_a is countably complete (as defined before 19.11, i.e. for any $\{X_n \mid n \in \omega\} \subseteq E_a$, $\bigcap_n X_n \neq \emptyset$). It follows that $\mathrm{Ult}(N, E_a)$ is well-founded and so has a transitive collapse M_a. Let

$$j_a\colon N \prec M_a \cong \mathrm{Ult}(N, E_a) .$$

Next, for $a \subseteq b$ both in $[\beta]^{<\omega}$, with (ii), $i_{ab}\colon M_a \prec M_b$ can be defined as before by

$$i_{ab}([f]_{E_a}) = [f \circ \pi_{ba}]_{E_b} ,$$

so that

$$i_{ab} \circ j_a = j_b .$$

Again, $\langle\langle M_a \mid a \in [\beta]^{<\omega}\rangle, \langle i_{ab} \mid a \subseteq b\rangle\rangle$ is a directed system, so stipulate that

$$\langle M_{\mathcal{E}}, \in_{\mathcal{E}}\rangle \text{ is the direct limit,}$$

and

$$j_{\mathcal{E}}\colon \langle N, \in\rangle \prec \langle M_{\mathcal{E}}, \in_{\mathcal{E}}\rangle \text{ and}$$
$$k_a\colon \langle M_a, \in\rangle \prec \langle M_{\mathcal{E}}, \in_{\mathcal{E}}\rangle$$

the corresponding embeddings so that for $a \subseteq b$ both in $[\beta]^{<\omega}$,

$$k_a \circ j_a = j_{\mathcal{E}} \text{ and } k_b \circ i_{ab} = k_a .$$

It is straightforward to see that (iii) implies that $\langle M_{\mathcal{E}}, \in_{\mathcal{E}}\rangle$ is well-founded. (By the argument for 15.7(a), (iii) in fact characterizes well-foundedness.) Moreover, an argument as for 19.3 shows that $\in_{\mathcal{E}}$ is set-like. Hence, we can assume (0.4) that

$$M_{\mathcal{E}} \text{ is transitive and } \in_{\mathcal{E}} = \in \cap (M_{\mathcal{E}} \times M_{\mathcal{E}}).$$

Extending a previous notation, the situation is summed up by

$$j_{\mathcal{E}}\colon N \prec M_{\mathcal{E}} \cong \mathrm{Ult}(N, \mathcal{E})\ ,$$

although a single ultrapower is not being taken. As usual, subscripts will be suppressed when clear from the context.

(a) of the following lemma is based on (iv) and shows why it was called normality, and (b) is based on the subsidiary conditions of (i). For $n \in \omega$, let $\mathrm{id}_n\colon [\zeta]^n \to [\zeta]^n$ be the identity map. Note that for $a \subseteq b$ both in $[\beta]^{<\omega}$, $i_{ab}([\mathrm{id}_{|a|}]_{E_a}) = [\pi_{ba}]_{E_b}$.

26.2 Lemma.

(a) For any $a \in [\beta]^{<\omega}$, $k_a([\mathrm{id}_{|a|}]_{E_a}) = a$.

(b) $\mathrm{crit}(j_{\mathcal{E}}) = \kappa$ *and* ζ *is the least ordinal satisfying* $\beta \leq j_{\mathcal{E}}(\zeta)$.

(c) $M_{\mathcal{E}} = \{j_{\mathcal{E}}(f)(a) \mid a \in [\beta]^{<\omega} \ \wedge \ f \in {}^{[\zeta]^{|a|}}N \cap N\}$.

Proof. (a) This is first established for $a \in [\beta]^1$ by showing that $k_{\{\alpha\}}([\mathrm{id}_1]_{E_{\{\alpha\}}}) = \{\alpha\}$ for $\alpha < \beta$ by induction on α. So, assume that this holds for $\alpha < \gamma$ where $\gamma < \beta$. $k_{\{\gamma\}}([\mathrm{id}_1]_{E_{\{\gamma\}}}) = \{\delta\}$ for some δ, and it must shown that $\delta = \gamma$.

Consider first an $\alpha < \gamma$, and set $a = \{\alpha, \gamma\}$. For each $s \in [\zeta]^2$, $\pi_{a\{\alpha\}}(s) = \{\xi^s_0\}$ and $\pi_{a\{\gamma\}}(s) = \{\xi^s_1\}$ where $s = \{\xi^s_0, \xi^s_1\}$ with $\xi^s_0 < \xi^s_1$. It follows from the definitions that the unique member of $k_{\{\alpha\}}([\mathrm{id}_1]_{E_{\{\alpha\}}})$, which inductively is α, is less than the unique member of $k_{\{\gamma\}}([\mathrm{id}_1]_{E_{\{\gamma\}}})$, which is δ. Hence, $\gamma \leq \delta$.

For the converse, suppose that $\rho < \delta$, say $\rho = k_a([f]_{E_a})$ where $a \in [\beta]^{<\omega}$ with $\gamma \in a$ and $f \in {}^{[\zeta]^{|a|}}N \cap N$. By the definition of δ,

$$(*) \quad \{s \in [\zeta]^{|a|} \mid f(s) \text{ is less than the unique member of } \pi_{a\{\gamma\}}(s)\} \in E_a\ .$$

By condition (iv) there is a $b \in [\beta]^{<\omega}$ with $a \subseteq b$ such that

$$\{s \in [\zeta]^{|b|} \mid f(\pi_{ba}(s)) \in s\} \in E_b\ .$$

It follows that for some i,

$$\{s \in [\zeta]^{|b|} \mid f(\pi_{ba}(s)) = (s)_i\} \in E_b$$

where $(s)_i$ temporarily denotes the ith member in increasing order of a set s of ordinals. If $(b)_i = \alpha$, then

$$\{s \in [\zeta]^{|b|} \mid \{f(\pi_{ba}(s))\} = \pi_{b\{\alpha\}}(s)\} \in E_b\ .$$

But then, condition (ii) applied to $(*)$ implies that $\pi_{b\{\alpha\}}(s) < \pi_{a\{\alpha\}}(\pi_{ba}(s))$ for s in a set in E_b, so that $\alpha < \gamma$. It follows from the definitions and induction that

$$\{\rho\} = \{k_a([f]_{E_a})\} = k_{\{\alpha\}}([\mathrm{id}_1]_{E_{\{\alpha\}}}) = \{\alpha\}\ .$$

Hence $\delta \leq \gamma$, and so $\delta = \gamma$.

The result for general $a \in [\beta]^{<\omega}$ now follows easily: Suppose that $\rho = k_b([f]_{E_b})$, taking $a \subseteq b$. Then as in the above arguments,

$$\begin{aligned}\rho \in k_a([\mathrm{id}_{|a|}]_{E_a}) \ &\textit{iff} \ [f]_{E_b} \in [\pi_{ba}]_{E_b} \\ &\textit{iff} \ \{[f]_{E_b}\} = [\pi_{b\{\alpha\}}]_{E_b} \text{ for some } \alpha \in a \\ &\textit{iff} \ \{\rho\} = k_a([\mathrm{id}_1]_{E_{\{\alpha\}}}) = \{\alpha\} \text{ for some } \alpha \in a\ .\end{aligned}$$

(b) To check that $\mathrm{crit}(j_{\mathcal{E}}) = \kappa$, that $j_{\mathcal{E}}(\alpha) = \alpha$ for $\alpha < \kappa$ follows by a simple inductive argument based on the "N-κ-completeness" of the E_a's, and then (i)(1) implies that for some a, $\mathrm{crit}(j_{\mathcal{E}}) = \mathrm{crit}(j_a) = \kappa$.

That $\beta \leq j_{\mathcal{E}}(\zeta)$ follows from (a): For any $\alpha < \beta$, $[\mathrm{id}_1]_{E_{\{\alpha\}}} \in [j_{\{\alpha\}}(\zeta)]^1$ so that $k_{\{\alpha\}}([\mathrm{id}_1]_{E_{\{\alpha\}}}) \in [j_{\mathcal{E}}(\zeta)]^1$, and $k_{\{\alpha\}}([\mathrm{id}_1]_{E_{\{\alpha\}}}) = \{\alpha\}$ by (a).

Finally, if $\xi < \zeta$, then (i)(2) implies that for some a, $j_a(\xi) \in [\mathrm{id}_{|a|}]_{E_a}$ so that $j_{\mathcal{E}}(\xi) \in a \subseteq \beta$ by (a), and so indeed ζ is the least ordinal satisfying $\beta \leq j_{\mathcal{E}}(\zeta)$.

(c) This is like 26.1(a): If $x \in M_{\mathcal{E}}$, there is an $a \in [\beta]^{<\omega}$ and an $f \in {}^{[\zeta]^{|a|}}N \cap N$ such that $x = k_a([f]_{E_a})$. But $[f]_{E_a} = j_a(f)([\mathrm{id}_{|a|}]_{E_a})$, and so

$$x = k_a([f]_{E_a}) = k_a((j_a(f))(k_a([\mathrm{id}_{|a|}]_{E_a})) = j_{\mathcal{E}}(f)(a)\ ,$$

the last equality being a consequence of (a). ⊣

The correlation of the two kinds of extenders is now at hand:

26.3 Exercise. *Let N be an inner model of* ZFC.

(a) Suppose that $j\colon N \prec M$ with $\mathrm{crit}(j) = \kappa$, *and $\mathcal{E}$ is the (κ, β)-extender derived from j. Then $\mathcal{E}$ is an N-(κ, β)-extender.*

(b) Suppose that $\mathcal{E} = \langle E_a \mid a \in [\beta]^{<\omega}\rangle$ is an N-(κ, β)-extender. Then the (κ, β)-extender derived from $j_{\mathcal{E}}\colon N \prec M_{\mathcal{E}} \cong \mathrm{Ult}(N, \mathcal{E})$ is again $\mathcal{E}$.

Hint. (b) applies 26.2. ⊣

The following observations are as for ordinary ultrapowers (cf. 5.7). Recall from §19 that for N a transitive $\in$-model of ZFC$^-$ and U and S such that $\langle N, \in, U\rangle \models U$ is an ultrafilter over S, $\langle N, U\rangle$ is *weakly amenable iff* whenever $F \in {}^S N \cap N$, $\{i \in S \mid F(i) \in U\} \in N$.

26.4 Exercise. *Suppose that N is an inner model of* ZFC, *$\mathcal{E} = \langle E_a \mid a \in [\beta]^{<\omega}\rangle$ is an N-(κ, β)-extender with each E_a over $[\zeta]^{|a|}$, and $j\colon N \prec M \cong \mathrm{Ult}(N, \mathcal{E})$. Then:*

(a) $j(x) = x$ for every $x \in V_\kappa \cap N$. If $\langle N, E_a\rangle$ is weakly amenable for each $a \in [\beta]^{<\omega}$, then $V_\kappa \cap N = V_\kappa \cap M$, $\mathcal{P}(\kappa) \cap N = \mathcal{P}(\kappa) \cap M$, and $\kappa^{+N} = \kappa^{+M}$.

(b) For any ξ, $j(\xi) < (|{}^\zeta\xi \cap N| \cdot |\beta|)^+$.

(c) If $\theta > \beta$ is a cardinal and $N \models \ulcorner\theta$ is a strong limit of cofinality greater than $\zeta\urcorner$, *then $j(\theta) = \theta$.*

(d) For any set X such that $|X| > \zeta$, $j``X \notin M$.

(e) $\mathcal{E} \notin M$.

Hint. (a) is like 19.1(c), (b) like 19.7(a) because of 26.2(c), (c) like 19.7(c), (d) like 22.4(c), and (e) like 22.4(d). ⊣

The following simple observation precludes the possibility of target models for extender embeddings having global closure properties in general.

26.5 Exercise. *Suppose that* $\kappa < \lambda$, λ *is a strong limit cardinal of cofinality* ω, *and there is a* $j\colon V \prec M$ *with* $\mathrm{crit}(j) = \kappa$ *and* $V_\lambda \subseteq M$. *Let* $\mathcal{E}$ *be the* (κ, λ)*-extender derived from* j *and* $j_\mathcal{E}\colon V \prec M_\mathcal{E} \cong \mathrm{Ult}(V, \mathcal{E})$. *Then* ${}^\omega M_\mathcal{E} \not\subseteq M_\mathcal{E}$.

Hint. $V_\lambda \subseteq M_\mathcal{E}$ by 26.1(b), and so if $\kappa \leq \beta < \lambda$, the (κ, β)-extender $\mathcal{E}|[\beta]^{<\omega} \in M_\mathcal{E}$. But $\mathcal{E} \notin M_\mathcal{E}$ by 26.4(e). ⊣

Strong, Woodin, and Superstrong Cardinals

While extenders may not generate inner models closed under the taking of arbitrary ω-sequences, they suggest appropriately tailored hypotheses:

κ is γ*-strong iff* there is a $j\colon V \prec M$ such that
(a) $\mathrm{crit}(j) = \kappa$ and $\gamma < j(\kappa)$, and
(b) $V_{\kappa+\gamma} \subseteq M$.

κ is *strong iff* κ is γ-strong for every γ.

κ is *superstrong iff* there is a $j\colon V \prec M$ with $\mathrm{crit}(j) = \kappa$ and $V_{j(\kappa)} \subseteq M$.

Thus, *if* $\gamma \leq \delta$ *and* κ *is* δ*-strong, then* κ *is* γ*-strong*; and κ *is measurable iff* κ *is* 0*-strong iff* κ *is* 1*-strong*. These new hypotheses are clearly weak analogues of previous ones: *if* κ *is supercompact, then* κ *is strong*; and *if* κ *is huge, then* κ *is superstrong*. And as before, although the $\exists j$ ranging over classes is not directly formalizable, there will be characterizations that provide *a posteriori* formalizations. By these means the superstrongness of κ will actually be seen to be consistency-wise weaker than its 2^κ-supercompactness (26.11(b)). Gaifman [74: 85ff] had considered strong and superstrong cardinals, but they did not become a focus of attention until Mitchell [79] and Dodd and Jensen (see Dodd [82]) showed that strong cardinals have a canonical inner model theory. Concerning our definition of γ-strong, the specific form of (b) is pertinent only for small γ, since $\gamma \geq \kappa \cdot \omega$ implies that $\kappa + \gamma = \gamma$, and has some nominal advantages. Also the condition $\gamma < j(\kappa)$ is included for convenience as for γ-supercompactness; it is superfluous when γ is a successor and hence for full strongness. (See 26.7(b), which gives Gaifman's original formulation; if $j(\kappa) \leq \gamma$, then κ is superstrong, a consistency-wise stronger assertion than being strong (26.12, 26.13).) Note that $\gamma < j(\kappa)$ implies that $|V_{\kappa+\gamma}| < j(\kappa)$: since $V_{\kappa+\gamma} = V^M_{\kappa+\gamma}$ and $j(\kappa)$ is inaccessible in M, $|V_{\kappa+\gamma}| \leq |V_{\kappa+\gamma}|^M < j(\kappa)$.

Strongness suffices for drawing former conclusions deducible with reflection arguments depending only on having sufficiently large V_α included in the target model:

26.6 Exercise.

(a) If there is a strong cardinal, then $V \neq L(A)$ *for any set* A.

(b) If κ *is 2-strong, there is a normal ultrafilter* U *over* κ *such that*

$$\{\alpha < \kappa \mid \alpha \text{ is measurable}\} \in U \ .$$

(c) If κ *is* γ*-strong and* $2^\alpha = \alpha^+$ *for every* $\alpha < \kappa$*, then* $2^\beta = \beta^+$ *for every* $\beta < \kappa + \gamma$.

(d) If κ *is strong, then* $V_\kappa \prec_2 V$.

Hint. For (a), assume to the contrary that $V = L(A)$ for set A. Let $j\colon V \prec M$ with $\mathrm{crit}(j) = \kappa$ and $A \in M$. Then $M = L(A)$ since M is an inner model of $L(A)$ with $A \in M$. But then $j\colon V \prec V$, contradicting Kunen's 23.12.

(b) follows from the argument for 22.1, (c) from that for 22.2, and (d) from that for 22.3. ⊣

(a) recalls 5.9, but the relationship between strong compactness and strongness is unclear.

These observations may fuel the speculation that strongness can supplant supercompactness, but there are important relative consistency results based on the latter for which the former does not seem to suffice.

The following characterizations are in the spirit of the ultrafilter characterization of γ-supercompactness:

26.7 Exercise.

(a) κ *is* γ*-strong iff there is a* $(\kappa, |V_{\kappa+\gamma}|^+)$*-extender* $\mathcal{E}$ *such that* $V_{\kappa+\gamma} \subseteq M_{\mathcal{E}}$ *and* $\gamma < j_{\mathcal{E}}(\kappa)$.

(b) κ *is* $(\gamma + 1)$*-strong iff there is a* (κ, β)*-extender* $\mathcal{E}$ *for some* $\beta > \kappa$ *such that* $V_{\kappa+\gamma+1} \subseteq M_{\mathcal{E}}$*. Consequently,* κ *is strong iff for any set* x *there is a* $j\colon V \prec M$ *with* $\mathrm{crit}(j) = \kappa$ *and* $x \subseteq M$.

(c) κ *is superstrong iff there is a* (κ, β)*-extender* $\mathcal{E}$ *for some* $\beta > \kappa$ *such that* $V_{j_{\mathcal{E}}(\kappa)} \subseteq M_{\mathcal{E}}$.

Hint. Apply 26.3(b) and 26.1(b)(c). For (a), note that if $j\colon V \prec M$ witnesses the γ-strongness of κ, then $|V_{\kappa+\gamma}| \leq |V_{\kappa+\gamma}|^M < |V_{\kappa+\gamma}|^+$. For (b), argue as for 23.15(a) that the condition $\gamma < j_{\mathcal{E}}(\kappa)$ can be adjoined. ⊣

((b) can be slightly strengthened with occurrences of $\gamma + 1$ replaced by γ, by further developing the theory of iterated embeddings broached in the proof of 23.15(a) to get a $j^{\omega+1}\colon V \prec M_{\omega+1}$ with $\gamma < j^{\omega+1}(\kappa)$.)

These characterizations lead to formalizations in ZFC, since clauses of the sort $V_\alpha \subseteq M_{\mathcal{E}}$ can be rendered in terms of the "extender ultrapower" of a sufficiently large V_ρ. Such clauses can be further embedded by formulating the concept of a (κ, Y)-extender for arbitrary transitive sets Y: Simply replace the β in the definition of (κ, β)-extender by Y, replacing ζ by V_ζ and making notational changes to accommodate Y not necessarily having a natural well-ordering. The argument for 26.2(a) then shows that Y is included in the target model $M_{\mathcal{E}}$ of the resulting extender embedding, so that $V_\alpha \subseteq M_{\mathcal{E}}$ becomes automatic if $V_\alpha \subseteq Y$. Such extenders were used in Martin-Steel [89]. However, the simpler $Y = \beta$ extenders are appropriate for the inner model theory because of the absoluteness of the corresponding ζ, and hence of the underlying $[\zeta]^{|a|}$'s for the E_a's.

The following analogue of 22.8(b) is a consequence of 26.7(a):

26.8 Exercise. *If* Con(ZFC + $\exists\kappa(\kappa$ is strong)), *then* Con(ZFC + $\exists\kappa(\kappa$ is strong $\wedge\ \neg\exists\lambda(\lambda > \kappa\ \wedge\ \lambda$ is inaccessible))). ⊣

In contrast, the argument from 1-extendibility given after 23.3 shows that there are many measurable cardinals above a superstrong cardinal.

We can deduce as for γ-supercompactness that $\ulcorner\kappa$ is γ-strong$\urcorner$ is Δ_2^{ZF} (cf. before 22.9), and as for hugeness that $\ulcorner\kappa$ is superstrong$\urcorner$ is Σ_2^{ZF}. This leads to a direct analogue of 24.10:

26.9 Exercise. *If κ is strong and there is a superstrong cardinal above it, then there are κ superstrong cardinals below κ. In particular, if there are strong and superstrong cardinals, then the least superstrong cardinal is less than the least strong cardinal.* ⊣

This has no bearing on the consistency strength of superstrongness, which is stronger than that of strongness. Before establishing this, a further hypothesis is tucked in:

$$\kappa \text{ is } \textit{Woodin} \textit{ iff } \begin{array}{l} \text{for any } f \in {}^{\kappa}\kappa, \text{ there is an } \alpha < \kappa \\ \text{with } f``\alpha \subseteq \alpha \text{ and a } j\colon V \prec M \text{ with} \\ \mathrm{crit}(j) = \alpha \text{ and } V_{j(f)(\alpha)} \subseteq M\ . \end{array}$$

This concept was isolated in 1984, and through a series of remarkable advances came to play a central role in consistency analysis (§32). A more technical hypothesis, its full significance can only be appreciated after an extended exegesis. Note that as for 26.7, *being Woodin is characterizable in terms of the existence of extenders*; a refined form of this will be established below (26.14).

26.10 Exercise. *Suppose that κ is Woodin. Then κ is regular and* $\{\alpha < \kappa \mid \alpha$ is measurable$\}$ *is stationary in κ, and so κ is κ-Mahlo.*

Hint. For the regularity of κ, note that if κ were singular and $f \in {}^{\kappa}\kappa$ were such that $\langle f(\xi) \mid \xi < \mathrm{cf}(\kappa)\rangle$ is increasing and cofinal in κ with $f(0) > \mathrm{cf}(\kappa)$, there would be no $\alpha < \kappa$ such that $f``\alpha \subseteq \alpha$. ⊣

This section is brought to a close by establishing the hierarchical relationships among the new hypotheses and providing useful characterizations of being Woodin in the process.

26.11 Proposition.

(a) If κ is 1-extendible, then κ is superstrong, and there is a normal ultrafilter U over κ such that

$$\{\alpha < \kappa \mid \alpha \text{ is superstrong}\} \in U .$$

(b) If κ is 2^κ-supercompact, then there is a normal ultrafilter over κ such that

$$\{\alpha < \kappa \mid \alpha \text{ is superstrong}\} \in U .$$

Proof. (a) Suppose that $j\colon V_{\kappa+1} \prec V_{j(\kappa)+1}$ witnesses the 1-extendibility of κ. Let $\mathcal{E}$ be "the $(\kappa, j(\kappa))$-extender derived from j"; we had only defined extenders derived from embeddings when they are between inner models, but in this case we can still define the E_a's, as $\mathcal{P}([\kappa]^{|a|}) \subseteq V_{\kappa+1}$. It is simple to check that $\mathcal{E}$ is indeed a $(\kappa, j(\kappa))$-extender, and so has a corresponding $j_\mathcal{E}\colon V \prec M_\mathcal{E} \cong \mathrm{Ult}(V, \mathcal{E})$. If $g \in V_{\kappa+1}$ is a bijection: $[\kappa]^1 \to V_\kappa$, then $j(g) \in V_{j(\kappa)+1}$ is a bijection: $[j(\kappa)]^1 \to V_{j(\kappa)}$, so that the argument for 26.1(b) shows that $V_{j(\kappa)} \subseteq M$. Also, the argument for 26.1(c) shows that $j_\mathcal{E}(\kappa) = j(\kappa)$. Hence, κ is superstrong by the characterization 26.7(c). Finally, $\mathcal{E} \in V_{j(\kappa)+1}$, and so if U is the normal ultrafilter defined from j as usual, then again by 26.7(c), mindful of how the condition $V_{j_\mathcal{E}(\kappa)} \subseteq M_\mathcal{E}$ can be expressed in terms of the "extender ultrapower" of V_κ, $\{\alpha < \kappa \mid \alpha \text{ is superstrong}\} \in U$.

(b) This follows from (a) and 23.5 for $\eta = 1$. ⊣

(b) is rather striking, in that a $j\colon V \prec M$ with the closure condition ${}^{2^{\mathrm{crit}(j)}}M \subseteq M$ implies the consistency of the existence of a $j'\colon V \prec M'$ with the closure condition $V_{j'(\mathrm{crit}(j'))} \subseteq M'$; $j'(\mathrm{crit}(j'))$ is only determined *a posteriori* by j' and is much larger than $\mathrm{crit}(j')$, so that a large initial segment of the universe is absolute for M'. This serves to mitigate against a heuristic argument given for the strength of n-hugeness just after that concept was defined, for the case $n = 1$.

In contrast, a real strengthening is to go to an $(n + 1)$-huge analogue of superstrongness: There is a $j\colon V \prec M$ with $\mathrm{crit}(j) = \kappa$ and $V_{j^{n+1}(\kappa)} \subseteq M$. This proposition implies the n-hugeness of κ by the argument for 24.8.

26.12 Proposition. *If κ is superstrong, then κ is Woodin and there is a normal ultrafilter U over κ such that*

$$\{\alpha < \kappa \mid \alpha \text{ is Woodin}\} \in U .$$

Proof. Suppose that $j\colon V \prec M$ witnesses the superstrongness of κ. Towards verifying that κ is Woodin, let $f \in {}^\kappa\kappa$. Then $j(f) \in {}^{j(\kappa)}j(\kappa)$, and $j(f)``\kappa = f``\kappa \subseteq \kappa$. Let $\mathcal{E}$ be the $(\kappa, |V_{j(f)(\kappa)}|^M)$-extender derived from j. Since $j(\kappa)$ is inaccessible in M and $j(f)(\kappa) < j(\kappa)$, it follows that $|V_{j(f)(\kappa)}|^M < j(\kappa)$, and so it is simple to see that $\mathcal{E} \in V_{j(\kappa)} \subseteq M$. Moreover, by 26.1(b), $V_{j(f)(\kappa)} \subseteq M_\mathcal{E}$ and $j(f) = k_\mathcal{E}(j_\mathcal{E}(f)) = j_\mathcal{E}(f)$, so that $V_{j_\mathcal{E}(f)(\kappa)} \subseteq M_\mathcal{E}$. It follows by elementarity that there is an $\alpha < \kappa$ with $f``\alpha \subseteq \alpha$ and an (α, β)-extender $\mathcal{E}' \in V_\kappa$ for some β such that $V_{j_{\mathcal{E}'}(f)(\alpha)} \subseteq M_{\mathcal{E}'}$. Hence, κ is Woodin.

The rest of the proposition now follows by the usual reflection argument, since the existence of the requisite extenders in $V_\kappa \subseteq M$ shows that $M \models \kappa$ is Woodin. ⊣

26.13 Proposition. *If κ is Woodin, then*

$$\{\alpha < \kappa \mid \alpha \text{ is } \gamma\text{-strong for every } \gamma < \kappa\}$$

is stationary in κ.

Proof. Suppose that κ is Woodin, and C closed unbounded in κ. A member of C must be found γ-strong for every $\gamma < \kappa$.

Remembering that κ is inaccessible by 26.10, define $g \in {}^\kappa\kappa$ by:

$$g(\xi) = \begin{cases} 0 & \text{if } \xi \text{ is } \gamma\text{-strong for every } \gamma < \kappa \text{ , else} \\ \gamma & \text{where } \gamma \text{ is the least strong limit cardinal} \\ & \text{less than } \kappa \text{ such that } \xi \text{ is not } \gamma\text{-strong .} \end{cases}$$

Then define $f \in {}^\kappa\kappa$ by:

$$f(\xi) = \max(\{g(\xi) + 5, \min(C - \xi)\}) \text{ .}$$

By hypothesis, there is an $\alpha < \kappa$ with $f``\alpha \subseteq \alpha$ and a $j\colon V \prec M$ with $\operatorname{crit}(j) = \alpha$ and $V_{j(f)(\alpha)} \subseteq M$. Now $j(C)$ is closed unbounded in $j(\kappa)$; $j(C) \cap \alpha = C \cap \alpha$; and $f``\alpha \subseteq \alpha$ implies that $C \cap \alpha$ is closed unbounded in α. Consequently, $\alpha \in j(C)$. To complete the proof, it suffices by elementarity to show that $M \models \alpha$ is γ-strong for every $\gamma < j(\kappa)$:

Let

$$\mathcal{E} = \langle E_a \mid a \in [j(g)(\alpha) + 1]^{<\omega} \rangle$$

be the $(\alpha, j(g)(\alpha) + 1)$-extender derived from j. By elementarity,

$$j(g)(\alpha) < j(f)(\alpha) < j(\alpha) \text{ ,}$$

the latter inequality by $f``\alpha \subseteq \alpha$ and $\alpha < j(\alpha)$. Consequently, each E_a is over $[\alpha]^{|a|}$, and $j(g)(\alpha) < j_{\mathcal{E}}(\alpha)$ by 26.1(c) (in its terms, $\operatorname{crit}(k_{\mathcal{E}}) \geq j(g)(\alpha)+1$). Also, $j(g)(\alpha)$ is a strong limit in M and $V_{j(f)(\alpha)} \subseteq M$, so that by 26.1(b), $V_{j(g)(\alpha)} \subseteq M_{\mathcal{E}}$. Finally, $\operatorname{rank}(\mathcal{E}) \leq j(g)(\alpha)+5$, so that $\mathcal{E} \in V_{j(f)(\alpha)} \subseteq M$. Moreover, the foregoing properties of $\mathcal{E}$ also hold in the sense of M:

$\mathcal{E}$ is an extender in M (note that the functions needed for the well-foundedness condition are in $V_{j(f)(\alpha)} \subseteq M$), and $j^M_{\mathcal{E}}\colon M \prec M^M_{\mathcal{E}} \cong \operatorname{Ult}(M, \mathcal{E})$ is a class of M definable from $\mathcal{E}$. $j_{\mathcal{E}}(\alpha)$ is the ordertype of

$$\{k_a([f]_{E_a}) \mid a \in [j(g)(\alpha) + 1]^{<\omega} \wedge f\colon [\alpha]^{|a|} \to \alpha\}$$

and $V^M_{\alpha+1} = V_{\alpha+1}$, and so it is straightforward to check that the computation of $j_{\mathcal{E}}(\alpha)$ is absolute for M, i.e. $j^M_{\mathcal{E}}(\alpha) = j_{\mathcal{E}}(\alpha)$, and consequently $j(g)(\alpha) < j^M_{\mathcal{E}}(\alpha)$. Finally, a similar argument about the "extender ultrapower" of V_α shows that $j^M_{\mathcal{E}}(V_\alpha) = j_{\mathcal{E}}(V_\alpha)$, and so the consequence $V_{j(g)(\alpha)} \subseteq V^M_{j_{\mathcal{E}}(\alpha)} = j_{\mathcal{E}}(V_\alpha)$ of $j(g)(\alpha) < j_{\mathcal{E}}(\alpha)$ implies that $V^M_{j(g)(\alpha)} = V_{j(g)(\alpha)} \subseteq M^M_{\mathcal{E}}$.

All this confirms that $M \models \alpha$ is $j(g)(\alpha)$-strong, and so by the definition of g, $M \models \alpha$ is γ-strong for every $\gamma < j(\kappa)$. ⊣

When κ is inaccessible and $\xi < \kappa$, then through the use of extenders as in the above proof, ξ is γ-strong for every $\gamma < \kappa$ *iff* $V_\kappa \models \xi$ is strong. Hence, 26.13 establishes transcendence of Woodin cardinals over strong cardinals. An extension of its argument leads to a useful characterization of Woodinness in terms of a "boldface" version of strongness. For any set A,

κ is *γ-strong for A* *iff* there is a $j\colon V \prec M$ such that
(a) $\text{crit}(j) = \kappa$ and $\gamma < j(\kappa)$,
(b) $V_{\kappa+\gamma} \subseteq M$, and
(c) $A \cap V_{\kappa+\gamma} = j(A) \cap V_{\kappa+\gamma}$.

There is an obvious analogue of 26.7(a) characterizing γ-strongness for A in terms of the existence of extenders.

26.14 Theorem (Woodin). *The following are equivalent:*

(a) κ is Woodin.

(b) For any $A \subseteq V_\kappa$,

$$\{\alpha < \kappa \mid \alpha \text{ is } \gamma\text{-strong for } A \text{ for every } \gamma < \kappa\}$$

is stationary in κ.

(c) For any $A \subseteq V_\kappa$, there is an $\alpha < \kappa$ γ-strong for A for every $\gamma < \kappa$.

(d) For any $f \in {}^\kappa\kappa$ there is an $\alpha < \kappa$ with $f``\alpha \subseteq \alpha$ and an extender $\mathcal{E} \in V_\kappa$ with $\text{crit}(j_{\mathcal{E}}) = \alpha$, *$j_{\mathcal{E}}(f)(\alpha) = f(\alpha)$, and $V_{j_{\mathcal{E}}(f)(\alpha)} \subseteq M_{\mathcal{E}}$.*

Proof. (a) → (b) κ is inaccessible by 26.10. Following the proof of 26.13, suppose now that κ is Woodin, $A \subseteq V_\kappa$, and C is closed unbounded in κ. Define $g \in {}^\kappa\kappa$ by:

$$g(\xi) = \begin{cases} 0 & \text{if } \xi \text{ is } \gamma\text{-strong for } A \text{ for every } \gamma < \kappa \text{ , else} \\ \gamma & \text{where } \gamma \text{ is the least strong limit cardinal} \\ & \text{less than } \kappa \text{ such that } \xi \text{ is not } \gamma\text{-strong for } A \text{ .} \end{cases}$$

Then define $f \in {}^\kappa\kappa$ by

$$f(\xi) = \max(\{g(\xi) + 5, \min(C - \xi)\}) \text{ .}$$

By hypothesis, there is an $\alpha < \kappa$ with $f``\alpha \subseteq \alpha$ and a $j\colon V \prec M$ with $\text{crit}(j) = \alpha$ and $V_{j(f)(\alpha)} \subseteq M$. As before, $\alpha \in j(C)$, and it suffices to show that $M \models \alpha$ is γ-strong for $j(A)$ for every $\gamma < j(\kappa)$:

Again let $\mathcal{E}$ be the $(\alpha, j(g)(\alpha) + 1)$-extender derived from j. Then the previous results about $\mathcal{E}$ and $j^M_{\mathcal{E}}\colon M \prec M^M_{\mathcal{E}} \cong \text{Ult}(M, \mathcal{E})$ will complete the proof, once we establish the further property

$$j(A) \cap V^M_{j(g)(\alpha)} = j^M_{\mathcal{E}}(j(A)) \cap V^M_{j(g)(\alpha)} \,.$$

To do this, it suffices to show that

$$j(A) \cap V^M_{j(\alpha)} = j^M_{\mathcal{E}}(j(A)) \cap j^M_{\mathcal{E}}(V_\alpha)$$

since as observed before, $j(g)(\alpha) < j(\alpha)$, $j(g)(\alpha) < j^M_{\mathcal{E}}(\alpha)$, and $V^M_{j(g)(\alpha)} = V_{j(g)(\alpha)} \subseteq M^M_{\mathcal{E}}$. But note that

$$j(A) \cap V^M_{j(\alpha)} = j(A \cap V_\alpha) = j_{\mathcal{E}}(A \cap V_\alpha) = j^M_{\mathcal{E}}(A \cap V_\alpha)\,,$$

the second equality from 26.1(b) and the third, by an absoluteness argument as before since $V^M_{\alpha+1} = V_{\alpha+1}$. Furthermore, $A \cap V_\alpha = j(A) \cap V_\alpha$ as $\mathrm{crit}(j) = \alpha$, and so the chain of equalities can be continued with

$$j^M_{\mathcal{E}}(A \cap V_\alpha) = j^M_{\mathcal{E}}(j(A) \cap V_\alpha) = j^M_{\mathcal{E}}(j(A)) \cap j^M_{\mathcal{E}}(V_\alpha)$$

completing the proof.

(b) $\to$ (c) This is immediate.

(c) $\to$ (d) Suppose first that $g \in {}^\kappa\kappa$. Then by hypothesis there is an $\alpha \in \kappa$ γ-strong for $g \subseteq V_\kappa$ for every $\gamma < \kappa$. This implies that $g``\alpha \subseteq \alpha$:

For $\xi \in \kappa$, set $m_g(\xi) = \max(\{\xi, g(\xi)\}) + 3$. Suppose now that $\xi < \alpha$. Let $j \colon V \prec M$ with $\mathrm{crit}(j) = \alpha$, $g(\xi) < j(\alpha)$, and $g \cap V_{m_g(\xi)} = j(g) \cap V_{m_g(\xi)}$. Since $\langle \xi, g(\xi)\rangle \in g \cap V_{m_g(\xi)}$, it follows that $j(g)(\xi) = g(\xi) < j(\alpha)$, and hence that $g(\xi) < \alpha$.

The argument for 26.10 using various g's now shows that κ must be inaccessible.

To complete the proof, suppose that $f \in {}^\kappa\kappa$. As before, let $\alpha < \kappa$ be γ-strong for $f \subseteq V_\kappa$ for every $\gamma < \kappa$, so that in particular $f``\alpha \subseteq \alpha$. As we now know that κ is inaccessible, by characterizability through extenders (cf. 26.7(a)) there is an extender $\mathcal{E} \in V_\kappa$ such that $\mathrm{crit}(j_{\mathcal{E}}) = \alpha$, $V_{f(\alpha)} \subseteq M_{\mathcal{E}}$, and $f \cap V_{m_f(\alpha)} = j_{\mathcal{E}}(f) \cap V_{m_f(\alpha)}$ (where $m_f(\alpha)$ is defined as in the second paragraph). But as before, $j_{\mathcal{E}}(f)(\alpha) = f(\alpha)$, and so the result follows.

(d) $\to$ (a) This is immediate. $\dashv$

Woodin cardinals are best viewed in terms of (c), which captures their essence as a kind of reflection property. (d) shows that κ being Woodin is a Π^1_1 property of $\langle V_\kappa, \in\rangle$, and so *the least Woodin cardinal is not weakly compact.*

26.14(c) is similar to the characterizations 24.19 of Vopěnka's Principle, and this promotes the analogy that Woodin cardinals are to strong cardinals as Vopěnka's Principle is to supercompact cardinals. As with that principle, Woodin cardinals can be further analyzed in terms of corresponding filters: For $X \subseteq \kappa$,

X is *Woodin in* κ *iff* for any $f \in {}^\kappa\kappa$, there is an $\alpha \in X$ with $f``\alpha \subseteq \alpha$ and a $j \colon V \prec M$ with $\mathrm{crit}(j) = \alpha$ and $V_{j(f)(\alpha)} \subseteq M$.

It is simple to check that κ is Woodin *iff*

$$\{X \subseteq \kappa \mid \kappa - X \text{ is not Woodin in } \kappa\}$$

is a (proper) filter, the *Woodin filter* over κ.

The following propositions are analogous to those for Vopěnka's Principle, but the proofs are less straightforward at some points.

26.15 Exercise. *Suppose that κ is Woodin, and F is the Woodin filter over κ.*

(a) F is normal.

(b) For any $A \subseteq V_\kappa$, $\{\alpha < \kappa \mid \alpha$ is γ-strong for A for every $\gamma < \kappa\} \in F$.

(c) If $X \in F$, $\{\alpha < \kappa \mid \alpha$ is measurable and there is a normal ultrafilter over α containing $X \cap \alpha\} \in F$.

Hint. (a) recalls 24.14. For (b), argue as for 26.14(a) $\rightarrow$ (b), but with a j that also satisfies $g(\alpha) = j(g)(\alpha)$; this is possible by the argument for 26.14(c) $\rightarrow$ (d). (c) recalls 24.16 and uses 26.14(c). $\dashv$

In the context of inner model theory Schimmerling [02] provided a further characterization of Woodin cardinals and went on to formulate the *hyperWoodin cardinals* and the *weakly hyperWoodin cardinals*. These cardinals fit into the large cardinal hierarchy in the following increasing order of strength: measurable Woodin, weakly hyperWoodin, hyperWoodin, and superstrong. A cardinal κ is *hyperWoodin iff* it is Woodin and there is a normal ultrafilter over κ extending the Woodin filter.

With this look at extenders and related hypotheses we have peered ahead at trappings of major advances of the 1980's. These were made over a broad front: dramatic progress in inner model theory in conjunction with crowning achievements in the concerted investigation of a hypothesis that arose in the 1960's from an entirely different quarter, the *Axiom of Determinacy*.

Chapter 6

Determinacy

The investigation of the determinacy of games is perhaps the most distinctive and intriguing development of modern set theory, and the correlations eventually established with large cardinals the most remarkable and synthetic. Although focused on sets of reals, the subject was to expand across the breadth of set theory from combinatorics and forcing to large cardinals and inner model theory. As a topical and anticipatory conclusion to the present volume this chapter describes this development. §27 discusses the historical beginnings of the study of infinite games and the early work that led to the formulation of determinacy hypotheses. §28 starts with Solovay's seminal result on the connection with large cardinals and proceeds to develop the combinatorial theory in that direction. §29 and §30 explore the structural consequences of determinacy in descriptive set theory, a direction of investigation first pursued by Moschovakis and Martin. §31 starts the discussion the consistency of determinacy hypotheses, with Martin's groundbreaking work. Finally, §32 gives a panoramic survey of recent relative consistency results of Martin, Steel, and especially Woodin.

27. Infinite Games

With roots in the analysis of actual games, the concepts of infinite game, winning strategy, and determinacy (i.e. having a winning strategy) have become a rich paradigm for the articulation of dichotomies across the breadth of set theory. In light of the first results along these lines, Polish mathematicians in 1962 introduced the Axiom of Determinacy (AD), postulating for every set of reals the determinacy of a corresponding game, largely because it established the regularity properties (§12) for *all* sets of reals. With the conclusion contradicting the Axiom of Choice, AD was initially cast adrift in the sea-change of set theory brought on by the development of forcing and the gathering results about large cardinal hypotheses. On the other hand, after Solovay established in 1965 that the regularity properties for the $\mathbf{\Sigma}^1_2$ sets follow from the existence of a measurable cardinal (§14), there was a predisposition to consider the liberal use of strong hypotheses to settle questions of descriptive set theory.

Two events in 1967 stimulated the further research that soon transformed the investigation of determinacy into a mainstream of set theory. The first was Solovay's demonstration that AD implies the measurability of ω_1, which anchored unexpected consequences for the transfinite in the emerging theory of large cardinals. The second was the appearance of a new and elegant game proof by David Blackwell [67] of a result of classical descriptive set theory, which secured a mooring for a systematic approach to extending the methods and results of that theory based on determinacy.

Acknowledging an apparently strong hypothesis based on a new paradigm and attracted by the possibility of developing more consequent structure about the transfinite cardinals and definable sets of reals, a host of fine mathematicians soon came on board and carried out a concerted effort: Solovay, Martin, and Moschovakis, joined by Kunen and Kechris by the early 1970's. A wide array of techniques ranging from the Recursion Theorem to ultrapowers was brought to bear to systematically substantiate an often strange and striking but internally coherent web of connections imposed by AD and other determinacy hypotheses. That these hypotheses had no particular justification was widely acknowledged, but their consequences as well as the modes of argument, particularly in descriptive set theory, were thought to be considerable motivation for their further study.

By the later 1970's a more or less complete theory for the projective sets was in place, a resilient edifice founded on determinacy with both strong buttresses and fine details. (There was one notable lacuna, the determination of the so-called projective ordinals. This was to be carried out in a culminating feat of technical virtuosity by Steve Jackson.) With the study *per se* thus vindicated the researchers started the Cabal Seminar in the Los Angeles area, and particularly with two fine younger mathematicians, Steel and Woodin, joining the ranks of that seminar, attention began to shift to sets of reals beyond the projective sets, to inner models, and to questions of overall consistency.

From the beginning, the investigation of determinacy not only involved its consequences, but how much determinacy is derivable. Early on Martin had established another synthetic connection between determinacy and large cardinals by showing that the existence of a measurable cardinal implies a significant amount of determinacy. In 1974, complementing early work of Harvey Friedman, Martin essentially established just how much determinacy is derivable in ZFC, and several years later, building on work of Solovay and Harrington he established an exact correlation between amounts of determinacy and the existence of many measurable cardinals.

With its seemingly boundless strength, does determinacy subsume the large cardinal hierarchy? Or are all the determinacy hypotheses derivable, after all, from large cardinals? Are the two somehow orthogonal in the higher reaches? How this central problem was eventually resolved in the later 1980's has elements of an exciting mystery novel: the spurts and lulls, the red herrings and sudden leads, the long-distance phone calls, and once the parts of the puzzle began to fit into place, the orderly denouement. Large cardinals, and relatively modest ones at that, do provide the requisite strength as shown by Martin and Steel, and ultimately by Woodin, in remarkable equiconsistency results. This was a resounding triumph for the modern methods of set theory and an unforseen affirmation of the relevance of large cardinals.

This overview has gone far beyond our historical frame, and to provide an adequate account of the foregoing would go far beyond the scope of the present volume. The approach that will be taken is as follows: This section describes the historical beginnings of determinacy and the early work that led to the formulation of determinacy hypotheses. The next pursues the more combinatorial consequences of determinacy, particularly those of large cardinal character. §29 and §30 explore the structural consequences of determinacy hypotheses in descriptive set theory. §31 discusses the early work on the consistency of determinacy hypotheses, and finally §32 surveys the crowning achievements of the 1980's in this direction. By then, most proofs will have been dispensed with in favor of a coherent account of the overall context and the historical progression of ideas. The text Moschovakis [80] serves as the reference for the decriptive set theory part, and §§28, 31, 32 play roles complementary to it, the first in emphasizing combinatorics and large cardinals and the latter in focusing on relative consistency.

As much of the work in determinacy must proceed without AC,

ZF serves as the ambient theory for this section ,

and uses of AC will be explicitly noted, reversing the usual procedure.

To establish the basic context, let X be a non-empty set. For $A \subseteq {}^{\omega}X$, $G_X(A)$ denotes the following "infinite two-person game with perfect information": There are two players, I and II. I initially chooses an $x(0) \in X$; then II chooses an $x(1) \in X$; then I chooses an $x(2) \in X$; then II chooses an $x(3) \in X$; and so forth:

$$\begin{array}{llllll} I\ : & x(0) & & x(2) & & \dots \\ II : & & x(1) & & x(3) & \dots \end{array}$$

Each choice is a *move* of the game, and each player before making each of his moves is privy to all the previous moves ("perfect information"). The resulting $x \in {}^{\omega}X$ is a *play* of the game, an initial segment of x a *partial play*, and I wins if $x \in A$, and otherwise II wins. A is the *payoff* for the game $G_X(A)$.

A *strategy* for I is a function $\sigma\colon \bigcup_{n\in\omega} {}^{2n}X \to X$ that tells him what move to make given the previous moves, so that a *(partial) play according to* σ is a (partial) play of form

$$\begin{array}{llllll} I : & \sigma(\emptyset) & & \sigma(\langle\sigma(\emptyset), y(0)\rangle) & & \sigma(\langle\sigma(\emptyset), y(0), \sigma(\langle\sigma(\emptyset), y(0)\rangle), y(1)\rangle) \\ II : & & y(0) & & y(1)\ \dots & \end{array}$$

With II's moves thus enumerated by $y \in {}^{\omega}X$, this play is denoted

$$\sigma * y\ .$$

σ is a *winning strategy* for I *iff* $\{\sigma * y \mid y \in {}^{\omega}X\} \subseteq A$, i.e. no matter what moves II makes, plays according to σ always result in a member of A. Analogously, a *strategy* for II is a function $\tau\colon \bigcup_{n\in\omega} {}^{2n+1}X \to X$; and a *(partial) play according to* τ is a (partial) play of form

$$\begin{array}{llllll} I\ : & z(0) & & z(1) & & \dots \\ II : & & \tau(\langle z(0)\rangle) & & \tau(\langle z(0), \tau(\langle z(0)\rangle), z(1)\rangle) & \end{array}$$

for some $z \in {}^{\omega}X$, denoted

$$z * \tau\ .$$

τ is a *winning strategy* for II *iff* $\{z * \tau \mid z \in {}^{\omega}X\} \cap A = \emptyset$. Note that for any strategy σ for I, the function assigning $y \in {}^{\omega}X$ to $\sigma * y \in {}^{\omega}X$ is injective, so that if $|A| < |X|^{\omega}$, then I cannot have a winning strategy; analogous remarks apply to II and $|{}^{\omega}X - A|$. Finally,

$$G_X(A) \text{ is } \textit{determined iff} \text{ a player has a winning strategy}\ .$$

Note that the players cannot both have winning strategies. For brevity in the main case of $X = \omega$ and $A \subseteq {}^{\omega}\omega$,

$$A \text{ is } \textit{determined iff } G_{\omega}(A) \text{ is determined}\ .$$

The following notation is helpful in this setting: For $x \in {}^{\omega}X$, define $x_I, x_{II} \in {}^{\omega}X$ by

$$x_I(i) = x(2i)\ , \text{ and}$$
$$x_{II}(i) = x(2i+1)\ .$$

If σ is a strategy for I as above and $y \in {}^{\omega}X$, then $(\sigma * y)_{II} = y$, and if τ is a strategy for II and $z \in {}^{\omega}X$, then $(z * \tau)_I = z$.

The games $G_X(A)$ are taken to be paradigmatic, but various games will also be considered with more involved rules. The foregoing terminology adapts naturally, and will be applied without further explanation. Sometimes a "payoff scheme" is described without making explicit a corresponding set A, but this will be simple to do in each case.

The determinacy of games having emerged in modern set theory as a major area of investigation, it is notable that important figures in foundational studies first entertained the possibility of an analysis of extant games and game-like contexts. In an early adumbration focusing on chess Zermelo [13: 502-503] argued that if from a position q a player can force a win at all, then there is a $t(q) \in \omega$ such that he can win in at most $t(q)$ moves no matter how his opponent plays. This was later taken up by Hungarian mathematicians: In the paper that introduced his well-known tree lemma König [27] filled a gap in Zermelo's argument using that lemma, following a suggestion of von Neumann. At the end König [27: 130] gave a simple argument communicated to him by Zermelo that also filled gap; Zermelo was arguing in effect that *finite versions of the games $G_X(A)$ are determined.* König considered an infinite chessboard, and László Kalmár [28] pursued a transfinite generalization.

Before this, Borel [21] had described a wide range of finitary game-theoretic problems, introduced a general concept of strategy, and formulated the now familiar minimax strategy. Then von Neumann [28] proved the crucial minimax theorem, the result that really began the mathematical theory of games. A simple consequence of that theorem is again that finite versions of the games $G_X(A)$ are determined. Focusing on chess Max Euwe [29] had also described the minimax strategy; he was, incidentally, the world champion of chess, 1935-1937.

Speculating about possible consequences for sets of reals Polish mathematicians first contemplated infinite games involving reals. Following another anticipation of the theory of games by Hugo Steinhaus [25], Banach and Stanisław Mazur in 1925 (according to Steinhaus [65: 464]) found a non-determined game of the sort $G_X(A)$ (using AC; cf. 27.2). In the Scottish book of problems (see Mauldin [81: 113ff]) Mazur described an infinite game and conjectured its connection to Baire category, a conjecture confirmed by Banach in 1935. Modifying the problem Ulam essentially asked: *For which $A \subseteq {}^{\omega}\omega$ does I (or respectively II) have a winning strategy in $G_{\omega}(A)$?* It was to take set theorists half a century to provide a fair answer to a related question: For which $A \subseteq {}^{\omega}\omega$ does *either* I or II have a winning strategy in $G_{\omega}(A)$, i.e. is $G_{\omega}(A)$ determined?

By the mid-1940's von Neumann and Oskar Morgenstern in their [44] had codified the theory of games and its approach to the analysis of economic behavior, stimulating research in this direction for decades to come. It was against this backdrop that a more systematic investigation of the determinacy of infinite games began.

David Gale and Frank Stewart initiated the study of the games $G_X(A)$. For non-empty X and $s \in {}^{<\omega}X$ let $O(s) = \{x \in {}^{\omega}X \mid s \subseteq x\}$. $A \subseteq {}^{\omega}X$ is *open iff* it is a union of such $O(s)$'s, and is *closed iff* ${}^{\omega}X - A$ is open. With the discrete topology on X, these are just the concepts of open and closed in the product topology, consistent with the case $X = \omega$. The following result is simple yet fundamental for the whole subject; the argument is as for related finite games (essentially given by Zermelo and von Neumann as described above) as membership in open sets is secured at a finite stage.

27.1 Proposition (AC)(Gale-Stewart [53]). *If* $A \subseteq {}^{\omega}X$ *is either open or closed, then* $G_X(A)$ *is determined.*

Proof. For any $B \subseteq {}^{\omega}X$ and $s \in {}^{<\omega}X$, let

$$B/s = \{x \in {}^{\omega}X \mid s^\frown x \in B\}\ .$$

If I has no winning strategy in $G_X(B/s)$, then for any $i \in X$ there is a $j \in X$ such that I has no winning strategy in $G_X(B/s^\frown\langle i, j\rangle)$: Otherwise, let $i \in X$ be such that for every $j \in X$, I has a winning strategy σ_j in $G_X(B/s^\frown\langle i, j\rangle)$. Then I would have a winning strategy in $G_X(B/s)$ after all: Initially make the move i, and after any reply j by II, play according to σ_j.

Suppose now that $A \subseteq {}^{\omega}X$ is open, and assume that I has no winning strategy in $G_X(A)$. Then by the above remark, a strategy τ for II can be defined recursively so that for any partial play $s \in \bigcup_{n\in\omega} {}^{2n}X$ according to τ, I has no winning strategy in $G_X(A/s)$. Now if x were a play according to τ yet $x \in A$, then by openness there would be a $2n \in \omega$ such that $O(x|2n) \subseteq A$. But then, any strategy for I in $G_X(A/x|2n)$ would be a winning one, reaching a contradiction. Hence, τ is a winning strategy for II in $G_X(A)$.

The argument for closed A is analogous, with the roles of I and II interchanged. ⊣

As in this proof, an informal description of a strategy will be typically given without bothering to write out a formal definition. Although the main case is $X = \omega$, proofs of determinacy often turn on the determinacy of auxiliary games corresponding to open $A \subseteq {}^{\omega}X$ for some X, the crux of the matter being that membership in A is secured at a finite stage.

The full exercise of AC delimits the possibilities:

27.2 Proposition (AC)(Gale-Stewart [53]). *There is a set of reals which is not determined.*

Proof. As there are $2^{\aleph_0}$ strategies for games of form $G_\omega(A)$, let $\langle\sigma_\alpha \mid \alpha < 2^{\aleph_0}\rangle$ enumerate the strategies for I and $\langle\tau_\alpha \mid \alpha < 2^{\aleph_0}\rangle$ those for II. Recursively choose $a_\alpha, b_\alpha \in {}^{\omega}\omega$ for $\alpha < 2^{\aleph_0}$ as follows: Having chosen a_β and b_β for $\beta < \alpha$, choose b_α so that $b_\alpha = \sigma_\alpha * y$ for some y yet $b_\alpha \notin \{a_\beta \mid \beta < \alpha\}$. This is possible since $|\{\sigma_\alpha * y \mid y \in {}^{\omega}\omega\}| = 2^{\aleph_0}$, the function taking $y \in {}^{\omega}\omega$ to $\sigma_\alpha * y$ being injective. Similarly, choose a_α so that $a_\alpha = z * \tau_\alpha$ for some z yet $a_\alpha \notin \{b_\beta \mid \beta < \alpha\}$. It is simple to check that the resulting sets $A = \{a_\alpha \mid \alpha < 2^{\aleph_0}\}$ and $B = \{b_\alpha \mid \alpha < 2^{\aleph_0}\}$ are disjoint, and that neither I nor II has a winning strategy for $G_\omega(A)$. ⊣

Gale-Stewart [53] also asked the basic question whether every Borel set of reals determined. By 27.1 open sets are determined; Philip Wolfe [55] proved that

$\boldsymbol{\Sigma}_2^0$ sets are determined; and almost a decade later, Morton Davis [64] proved that $\boldsymbol{\Sigma}_3^0$ sets are determined. The emerging difficulties were to be explained, and further progress to be stimulated, by metamathematical arguments (see after 31.2). Prior developments are discussed next that motivated the formulation of new hypotheses in set theory.

Games and Regularity Properties

Putting the regularity properties for sets of reals in a new light, successively considered are three games closely related to the Baire property, the perfect set property, and Lebesgue measurability.

For $A \subseteq {}^\omega\omega$ Mazur's Scottish book game mentioned earlier is essentially the game $G_\omega^{**}(A)$ formulated as follows: There are two players I and II, but instead of choosing members of ω they choose members of ${}^{<\omega}\omega - \{\emptyset\}$:

$$\begin{array}{lllll} I\ : & s_0 & & s_2 & & \dots \\ II : & & s_1 & & s_3 & \dots \end{array}$$

The previous terminology adapts, and with x the concatenation $s_0{}^\frown s_1{}^\frown s_2{}^\frown\dots$ of the resulting play, I wins if $x \in A$, and otherwise II wins. For the following, Mazur initially noted the forward directions, Banach provided the converse, and John Oxtoby, a generalization to arbitrary topological spaces.

27.3 Proposition (Mazur, Banach – Mauldin [81: 113ff]; Oxtoby [57]). *For $A \subseteq {}^\omega\omega$,*

(a) A is meager iff II has a winning strategy in $G_\omega^{**}(A)$.

(b) $O(s) - A$ is meager for some $s \in {}^{<\omega}\omega$ iff I has a winning strategy in $G_\omega^{**}(A)$.

Proof. (a) Suppose first that A is meager, so that $A \subseteq \bigcup_{n\in\omega} C_n$ where each C_n is closed and nowhere dense. Then a strategy τ for II can be defined so that $\tau(\langle s_0\rangle)$ is some t such that $O(s_0{}^\frown t) \cap C_0 = \emptyset$, and generally $\tau(\langle s_0, s_1, \dots, s_{2n}\rangle)$ is some t such that $O(s_0{}^\frown s_1{}^\frown\dots{}^\frown s_{2n}{}^\frown t) \cap C_n = \emptyset$. Such a τ is winning for II.

Suppose now that II has a winning strategy τ. For each partial play according to τ of form $p = \langle s_0, \dots, s_{2n}\rangle$, let $p_* = s_0{}^\frown\dots{}^\frown s_{2n}$ and set

$$D_p = \{x \in {}^\omega\omega \mid p_* \subseteq x \to \exists t \in {}^{<\omega}\omega - \{\emptyset\}(p_*{}^\frown t^\frown \tau(p^\frown\langle t\rangle) \subseteq x)\}\,.$$

Then each D_p is open (noting in part that $O(p_*)$ is clopen) and dense (for if $u \in {}^{<\omega}\omega$, either $u \not\supseteq p_*$ so that $O(u) \subseteq D_p$, or else there is a $t \in {}^{<\omega}\omega - \{\emptyset\}$ such that $p_*{}^\frown t = u$ and so any $x \in {}^\omega\omega$ with $x \supseteq p_*{}^\frown t^\frown\tau(p^\frown\langle t\rangle)$ satisfies $x \in O(u) \cap D_p$). Moreover, for any $x \in \bigcap_p D_p$ we can recursively define a play $\langle s_i \mid i \in \omega\rangle$ according to τ such that $x = s_0{}^\frown s_1{}^\frown s_2{}^\frown\dots$, and so $x \notin A$. Consequently, $A \subseteq \bigcup_p({}^\omega\omega - D_p)$, a countable union of nowhere dense sets.

(b) Suppose first that $O(s) - A$ is meager for some $s \in {}^{<\omega}\omega$, which can be taken to be non-empty. Then I has a winning strategy beginning with initial move s and then playing to avoid $O(s) - A$ as in the forward direction of (a).

For the converse, if I has a winning strategy σ, say with $\sigma(\emptyset) = s$, then it is simple to see that II has a winning strategy in $G^{**}_{\omega}(O(s) - A)$ derived from σ, and so $O(s) - A$ is meager by (a). ⊣

27.4 Corollary. *For $A \subseteq {}^{\omega}\omega$, setting*

$$O_A = \bigcup\{O(s) \mid s \in {}^{<\omega}\omega \wedge O(s) - A \text{ is meager}\}\ ,$$

if $G^{**}_{\omega}(A - O_A)$ is determined, then A has the Baire property.

Proof. If I had a winning strategy in $G^{**}_{\omega}(A - O_A)$, then for some $t \in {}^{<\omega}\omega$, $O(t) - (A - O_A)$ would be meager. But then, $O(t) - A$ would also be meager and so that $O(t) \subseteq O_A$ and $O(t) - (A - O_A) = O(t)$, which is a contradiction (0.11). Hence, II has a winning strategy in $G^{**}_{\omega}(A - O_A)$, and so $A - O_A$ is meager. But $O_A - A$ is also meager by definition of O_A, and so A has the Baire property. ⊣

The game for the perfect set property is based on $X = 2$. For $A \subseteq {}^{\omega}2$, $G^{*}_{2}(A)$ is the game formulated as follows: I chooses members of ${}^{<\omega}2$ and II, members of 2:

$$\begin{array}{llllll} I: & s_0 & & s_2 & & \ldots \\ II: & & k_1 & & k_3 & \ldots \end{array}$$

The previous terminology adapts, and with x being the concatenation $s_0{}^\frown\langle k_1\rangle{}^\frown s_2{}^\frown\langle k_3\rangle{}^\frown\ldots$ of the resulting play, I wins if $x \in A$, and otherwise II wins. With ${}^{\omega}2$ topologized with basic open sets $O(s) = \{x \in {}^{\omega}2 \mid s \subseteq x\}$ for $s \in {}^{<\omega}2$, a perfect set is as before: non-empty, closed and containing no isolated points.

27.5 Proposition (Davis [64]). *For $A \subseteq {}^{\omega}2$,*
(a) A is countable iff II has a winning strategy in $G^{*}_{2}(A)$.
(b) A has a perfect subset iff I has a winning strategy in $G^{*}_{2}(A)$.

Proof. (a) Suppose first that there is an enumeration $\langle a_i \mid i \in \omega\rangle$ of A. Then II has a simple winning strategy: He plays his ith move to ensure that the concatenation of the resulting play differs from a_i.

Suppose now that II has a winning strategy τ. Proceeding as in the proof of 27.3(a), for a partial play according to τ of form $p = \langle s_0, k_1, \ldots, s_{2n}, k_{2n+1}\rangle$ let $p_* = s_0{}^\frown\langle k_1\rangle{}^\frown\ldots{}^\frown s_{2n}{}^\frown\langle k_{2n+1}\rangle$ and set

$$D_p = \{x \in {}^{\omega}2 \mid p_* \subseteq x \rightarrow \exists t \in {}^{<\omega}2(p_*{}^\frown t^\frown\tau(p^\frown\langle t\rangle) \subseteq x)\}\ .$$

Then as before, $A \subseteq \bigcup_p({}^{\omega}2 - D_p)$.

Now for each p, $x \in {}^{\omega}2 - D_p$ exactly when $p_* \subseteq x$ and $\forall t \in {}^{<\omega}2(p_*{}^\frown t^\frown\tau(p^\frown\langle t\rangle) \nsubseteq x)$. But then, ${}^{\omega}2 - D_p$ has a unique member x_p: With $|p_*| = m$, $x_p|m = p_*$; necessarily $x_p(m) = 1 - \tau(p^\frown\langle\emptyset\rangle)$; and recursively

$$x_p(e) = 1 - \tau(p^\frown\langle x_p(m), \ldots, x_p(e-1)\rangle)$$

for $e > m$. (This argument is possible because of the switch from ${}^{\omega}\omega$ to ${}^{\omega}2$.) Hence, A is countable.

(b) Suppose first that A has a perfect subset P. Setting

$$T = \{x|n \mid x \in P \ \wedge \ n \in \omega\}\,,$$

a strategy for I can be described as follows: Let the initial move be an $s \in T$ such that both $s^\frown\langle 0\rangle$ and $s^\frown\langle 1\rangle$ are in T; such an s exists as P has no isolated points. Generally, in response to a partial play p with concatenation $p_* \in T$, let the move be an s such that both $p_*{}^\frown s^\frown\langle 0\rangle$ and $p_*{}^\frown s^\frown\langle 1\rangle$ are in T; again such an s exists as P has no isolated points. This is a winning strategy for I as P is closed.

For the converse, note that if σ is a winning strategy for I, $\{\sigma * y \mid y \in {}^{\omega}2\}$ is a perfect subset of A. ⊣

It is simple to make the transfer from ${}^{\omega}2$ to ${}^{\omega}\omega$. This can be done on general grounds, or by applying a specific map:

For $n, k \in \omega$ define $b^k_n\colon n+1 \to 2$ as follows: If k is even, set $b^k_n(i) = 1$ for $i < n$ and $b^k_n(n) = 0$; if k is odd, set $b^k_n(i) = 0$ for $i < n$ and $b^k_n(n) = 1$. Then define $\Psi\colon {}^{\omega}\omega \to {}^{\omega}2$ by:

$$\Psi(x) = b^0_{x(0)}{}^\frown b^1_{x(1)}{}^\frown b^2_{x(2)}{}^\frown \ldots$$

It is straightforward to check that the range consists of all but the countably many members of ${}^{\omega}2$ that are eventually constant, and that Ψ is actually a homeomorphism onto its range considered as a subspace of ${}^{\omega}2$. Also, for $B \subseteq {}^{\omega}2$, B is perfect *iff* $\Psi^{-1}(B)$ is perfect, and B is countable *iff* $\Psi^{-1}(B)$ is countable. This leads to the following:

27.6 Corollary. *For $A \subseteq {}^{\omega}\omega$, $G^*_2(\Psi``A)$ is determined iff A has the perfect set property.* ⊣

For Lebesgue measurability, Mycielski-Swierczkowski [64] provided the first game argument. Harrington then found a simpler approach with his *covering game*. This is best cast as a game for ${}^{\omega}2$, but the ${}^{\omega}\omega$ context is maintained through contrivance: For $A \subseteq {}^{\omega}\omega$ and $\epsilon \in \mathbb{R}$ with $\epsilon > 0$, the covering game $G(A, \epsilon)$ is played with integer moves:

I :	$x(0)$		$x(2)$		$\ldots$
II :		$x(1)$		$x(3)$	$\ldots$

However, the payoff scheme is somewhat involved. Let $\Psi\colon {}^{\omega}\omega \to {}^{\omega}2$ be as above, and on its range, let Ψ^{-1} be its inverse function. I must always play 0's or 1's so that the resulting x_I is not eventually constant. Each of II's moves is regarded as coding a finite union of basic open sets; for definiteness, in terms of the fixed recursive enumeration $\langle \mathbf{s}_i \mid i \in \omega\rangle$ of ${}^{<\omega}\omega$ set:

$$N_i = O(\mathbf{s}_{\mathbf{s}_i(0)}) \cup \ldots \cup O(\mathbf{s}_{\mathbf{s}_i(|\mathbf{s}_i|-1)}) .$$

II's move $x(2n+1)$ must then satisfy

$$m_L(N_{x(2n+1)}) < \epsilon/2^{2(n+1)} .$$

Now if either I or II fails to meet these conditions, the first to do so loses. Assuming that this does not happen, the idea is that II tries to cover A with $\bigcup_{n\in\omega} N_{x(2n+1)}$. Accordingly, I wins if $\Psi^{-1}(x_I) \in A - \bigcup_{n\in\omega} N_{x(2n+1)}$, and otherwise II wins.

27.7 Lemma. *In $G(A, \epsilon)$:*

(a) If I has a winning strategy, there is a Lebesgue measurable $B \subseteq A$ such that $m_L(B) > 0$.

(b) If II has a winning strategy, there is an open set $O \supseteq A$ such that $m_L(O) < \epsilon$.

Proof. (a) Suppose that σ is a winning strategy for I. Then the set of I's outputs according to σ,

$$B = \{\Psi^{-1}((\sigma * y)_I) \mid y \in {}^\omega\omega\} ,$$

is seen to be a $\mathbf{\Sigma}^1_1$ set (in a parameter coding σ) and so is Lebesgue measurable. The rules of the game imply that $m_L(B) > 0$: Otherwise, it is simple (using 0.8(d) and 0.9(a)) to find $\langle x(2n+1) \mid n \in \omega\rangle$ so that II wins a play even according to σ.

(b) Suppose that τ is a winning strategy for II. For any $s \in {}^{<\omega}2 - \{\emptyset\}$, say with $|s| = n+1$, let $d(s)$ be II's response (the $(2n+1)$st move of the game) of the play according to τ where I's moves are successively $s(0), \ldots, s(n)$. Then as τ is winning for II, for any $x \in A$ there must be an $n \in \omega$ such that $x \in N_{d(\Psi(x)|(n+1))}$. Consequently, if

$$O = \bigcup\{N_{d(s)} \mid s \in {}^{<\omega}2 - \{\emptyset\}\} ,$$

then $A \subseteq O$. Moreover, according to the rules of the game $m_L(N_{d(s)}) < \epsilon/2^{2(n+1)}$ where $|s| = n+1$, and there are 2^{n+1} such s's. Hence,

$$m_L(O) < \textstyle\sum_{n\in\omega}(\epsilon/2^{2(n+1)}) \cdot 2^{(n+1)} = \sum_{n\in\omega}\epsilon/2^{(n+1)} = \epsilon . \qquad \dashv$$

For $A \subseteq {}^\omega\omega$, $B \subseteq {}^\omega\omega$ a *minimal cover* for A *iff* $A \subseteq B$, B is Lebesgue measurable, and if $Z \subseteq B - A$ is Lebesgue measurable, then $m_L(Z) = 0$. Any $A \subseteq {}^\omega\omega$ has a minimal cover: take any Lebesgue measurable $B \supseteq A$ with $m_L(B)$ minimal. In fact, B can be taken to be G_δ by 0.9(b). Noting that 27.7 can be established in our ambient theory ZF, the argument for 14.4 showing in ZF that every $\mathbf{\Sigma}^1_1$ set is Lebesgue measurable, there is the following corollary:

27.8 Corollary. *Suppose that $A \subseteq {}^\omega\omega$, $B \subseteq {}^\omega\omega$ is a minimal cover for A, and for any $\epsilon \in \mathbb{R}$ with $\epsilon > 0$, $G(B - A, \epsilon)$ is determined. Then A is Lebesgue measurable.*

Proof. By choice of B, II must have a winning strategy in each of the games $G(B - A, \epsilon)$. Hence, $B - A \subseteq G$ where G is a $\mathbf{\Pi}^0_2$ set with $m_L(G) = 0$. ⊣

Determinacy Hypotheses

In Poland, with renewed interest in infinite games (Mycielski-Zieba [55], Mycielski-Swierczkowski-Zieba [56]) Mycielski and Steinhaus in their [62] proposed the following axiom, now known as the *Axiom of Determinacy*:

(AD) Every set of reals is determined .

(Steinhaus [65: 465]) recalled how the axiom came to be formulated. He there used the term "determinacy", but the early papers on the subject used the term "determinateness". By the 1970's "determinacy" came to prevail.) The basic theory of AD was soon set out in Mycielski [64, 66] and Mycielski-Swierczkowski [64].

Mycielski [66] considered a further hypothesis:

($\mathrm{AD}_{\mathbb{R}}$) $G_{^\omega\omega}(A)$ is determined for every $A \subseteq {}^\omega({}^\omega\omega)$.

$\mathrm{AD}_{\mathbb{R}}$ readily implies AD. AD was to be the main preoccupation, but $\mathrm{AD}_{\mathbb{R}}$ was to figure as a suitably strong hypothesis in later developments.

Concerning the regularity properties, the games above can all be recast as games of form $G_\omega(B)$. For $G^{**}_\omega(A)$ for example, in terms of the fixed recursive enumeration $\langle \mathbf{s}_i \mid i \in \omega \rangle$ of ${}^{<\omega}\omega$ with $\mathbf{s}_0 = \emptyset$, define $B \subseteq {}^\omega(\omega - \{0\})$ by

$$x \in B \textit{ iff } \mathbf{s}_{x(0)}{}^\frown\mathbf{s}_{x(1)}{}^\frown\mathbf{s}_{x(2)}{}^\frown \ldots \in A .$$

Then I has a winning strategy in $G^{**}_\omega(A)$ exactly when I has a winning strategy in $G_\omega(B)$, and similarly for II. This leads to the following summarizing result:

27.9 Theorem (Mycielski-Swierczkowski [64]; Mazur, Banach; Davis [64]). *Assume* AD. *Then every set of reals is Lebesgue measurable, has the Baire property, and has the perfect set property.* ⊣

This was the main incentive behind the formulation of AD. AD thus contradicts AC by actually banishing those "pathological" non-measurable sets and the like given by unfettered applications of AC. Nonetheless, accepting AC Mycielski and Steinhaus wrote [62: 2] of their axiom that it could be considered as leading to a nice submodel:

> It is not the purpose of this paper to depreciate the classical mathematics with its fundamental 'absolute' intuitions on the universum of sets (to which belongs the axiom of choice), but only to propose another theory which seems very interesting although its consistency is problematic. Our axiom can be considered as a restriction of the classical notion of a set leading to a smaller universum, say of determined sets, which reflect some physical intuitions which are not fulfilled by the classical sets (e.g. paradoxical decompositions of the sphere are eliminated by [AD]). Our axiom could be considered as an axiom added to the classical set theory claiming the existence of a class of sets satisfying [AD] and the classical axioms without the axiom of choice.

Mycielski [64: 205] hoped for an inner model: "a subclass of the class of all sets with the same membership relation. It would be still more pleasant if such a submodel contains all the real numbers." As Solovay and Gaisi Takeuti pointed out, the natural possibility is $L(\mathbb{R})$, the smallest inner model containing every real. If AD holds, then so does $\mathrm{AD}^{L(\mathbb{R})}$: Every strategy is coded by a real and so belongs to $L(\mathbb{R})$, and the property of being a winning strategy for a game $G_\omega(A)$ with $A \in L(\mathbb{R})$ is absolute for $L(\mathbb{R})$.

In the late 1960's Solovay made the prescient conjecture that $\mathrm{AD}^{L(\mathbb{R})}$ holds relative to some large cardinal hypothesis. (In print Solovay [69: 60] conjectured that if there is a supercompact cardinal, then in $L(\mathbb{R})$ every set of reals has the perfect set property.) ZF + AD was never widely entertained as a serious alternative to ZFC, and increasingly from the early 1970's onward consequences of ZF + AD were regarded as what holds in $L(\mathbb{R})$ assuming $\mathrm{AD}^{L(\mathbb{R})}$. This compromise, if you will, reaffirmed ZFC as the ambient theory while at the same time vastly increasing the sets of reals possessing the regularity properties through a maximal principle.

AD itself provides a countable choice principle, $\mathrm{AC}_\omega({}^\omega\omega)$ of §§12, 13:

27.10 Proposition (Swierczkowski, Mycielski [64], Scott). *Assume* AD. *Then* $\mathrm{AC}_\omega({}^\omega\omega)$, *i.e. every countable set consisting of non-empty sets of reals has a choice function. Consequently,* ω_1 *is regular.*

Proof. Suppose that $\{Z_n \mid n \in \omega\} \subseteq \mathcal{P}({}^\omega\omega) - \{\emptyset\}$. Define $A \subseteq {}^\omega\omega$ by: $x \in A$ *iff* $x_{II} \notin Z_{x(0)}$. Clearly I cannot have a winning strategy in $G_\omega(A)$. But then, with τ a winning strategy for II, $f\colon \omega \to {}^\omega\omega$ defined by $f(n) = (\langle n, 0, 0, \ldots\rangle * \tau)_{II}$ is a choice function as desired. ⊣

As noted in §§12, 13 $\mathrm{AC}_\omega({}^\omega\omega)$ suffices for the results in those sections, and so they can be brought into play when establishing results in ZF + AD. Nonetheless, it soon became *de rigueur* to assume the stronger DC in much of the investigation of AD. DC is needed for those consequences of AD in descriptive set theory depending on the characterization of well-foundedness in terms of the lack of infinite descending chains. Moreover, DC relativizes to $L(\mathbb{R})$ (11.13), the natural inner model for AD.

As described in §11 Cantor initiated a programmatic approach to the Continuum Hypothesis via the perfect set property. Since AD implies the perfect set property for every set of reals, AD realizes that program in the sense that for no set X does $\aleph_0 < |X| < 2^{\aleph_0}$ (where $<$ between "cardinals" in the ZF setting is defined via injective embeddability). However, it should be stressed that Cantor wanted to establish the stronger $2^{\aleph_0} = \aleph_1$, i.e. that there is some well-ordering of the continuum in ordertype ω_1. Observations that were applied in §11, incorporating a deduction by Specker [57], were first made for AD:

27.11 Proposition (Mycielski [64]).

(a) Assume AD. *Then* $\omega_1 \not\leq 2^{\aleph_0}$, *i.e. there is no uncountable well-orderable set of reals.*

(b) Con(ZF + AD) *implies* Con(ZFC + $\exists\kappa(\kappa$ is inaccessible$)$).

Proof. (a) follows from 11.4(b), and (b), from (a), the regularity of ω_1 (27.10), and (c) $\to$ (a) of 11.6. ⊣

(b) was the first connection made with large cardinals. It follows from (a) that an extension of AD in a different direction from $\mathrm{AD}_{\mathbb{R}}$ is already inconsistent with ZF:

27.12 Exercise (Mycielski [64]). *There is an* $A \subseteq {}^{\omega}\omega_1$ *such that* $G_{\omega_1}(A)$ *is not determined.*

Hint. Let $A \subseteq {}^{\omega}\omega_1$ be defined by: $x \in A$ *iff* $x(0) \geq \omega$ and x_{II} is not an injective enumeration of $x(0) = \{\xi \mid \xi < x(0)\}$. If $G_{\omega_1}(A)$ is determined, it must be II who has a winning strategy, say τ. For any $z \in {}^{\omega}\omega_1$ with $z(0) \geq \omega$, $\{\langle m, n\rangle \mid z * \tau(2m+1) < z * \tau(2n+1)\}$ is then a well-ordering of ordertype $z(0)$. It follows that there is an injection: $\omega_1 \to {}^{\omega}\omega$ definable from τ, so that AD fails by 27.11(a). ⊣

This argument does not provide an $A \subseteq {}^{\omega}\omega_1$ such that $G_{\omega_1}(A)$ is not determined, and curiously it may not be possible to do so (see Harrington-Kechris [81: 133-134]).

With full AD confronting AC and raising problematic issues, a natural approach in the ZFC context is simply to consider determinacy as yet another regularity property of sets of reals, one that subsumes the others by local versions of 27.9, and to investigate its possible extent. Consider

(Det(Λ)) Every set of reals in Λ is determined .

Thus, AD is Det($\mathcal{P}({}^{\omega}\omega)$), and $\mathrm{AD}^{L(\mathbb{R})}$ is Det($L(\mathbb{R})$) by absoluteness. Det($\mathbf{\Pi}^1_n$) is known as $\mathbf{\Pi}^1_n$ *Determinacy*, and so forth. Focal became the hypothesis of *Projective Determinacy*:

(PD) $\mathrm{Det}(\bigcup_{n\in\omega}\mathbf{\Pi}^1_n)$.

The consequences of PD will be analyzed at length in §29, and *Borel Determinacy* discussed after 31.2. $\mathbf{\Pi}^1_1$ *Determinacy* was soon seen to transcend ZF, by the first extension of 27.11(b):

27.13 Exercise ($\mathrm{AC}_\omega({}^{\omega}\omega)$). Det($\mathbf{\Pi}^1_1$) *implies* $\forall a \in {}^{\omega}\omega(\omega_1$ is inaccessible in $L[a])$.

Hint. By 11.5 it suffices to show that $\omega_1^{L[a]} < \omega_1$ for every $a \in {}^{\omega}\omega$. But if $\omega_1^{L[a]} = \omega_1$, then by 13.12 relativized there would a be $\Pi^1_1(a)$ set A without the perfect set property. Now apply 27.6, after checking that $G_2^*(\Psi``A)$ can be construed as $G_\omega(B)$ for some $\Pi_1^1(a)$ set B of reals. ⊣

The concept of determinacy having emerged and its basic consequences and connections worked out, two seminal results appeared in 1967 as was mentioned at the outset. With the result of Blackwell [67] to serve as the starting point of §29, the development from the other result is explored in the upcoming section. But before concluding this one, some related results of later vintage are tucked in that will serve subsequent commentary as counterpoint.

A local version of 27.9 is: *If* Det($\mathbf{\Pi}_n^1$), *then every* $\mathbf{\Pi}_n^1$ *set of reals is Lebesgue measurable, has the Baire property, and has the perfect set property.* As anticipated, this merely requires checking for $n > 0$ that for any $\mathbf{\Pi}_n^1$ set A of reals, the games corresponding to the regularity properties for A can each be recast in form $G_\omega(B)$ for some $\mathbf{\Pi}_n^1$ set B of reals.

It was noticed in the early 1970's that a simple way of modifying games due to Solovay called "unfolding" leads to stronger conclusions:

27.14 Exercise ($\mathrm{AC}_\omega({}^\omega\omega)$)(Kechris, Martin). *Assume* Det($\mathbf{\Pi}_n^1$). *Then every* $\mathbf{\Sigma}_{n+1}^1$ *set of reals is Lebesgue measurable, has the Baire property, and has the perfect set property.*

Hint. Suppose that $A \subseteq {}^\omega\omega$ is $\mathbf{\Sigma}_{n+1}^1$. To establish the Baire property for A, apply Solovay's unfolding idea as follows:

Let O_A be the open set as in the statement of 27.4, and set $B = A - O_A$, a $\mathbf{\Sigma}_{n+1}^1$ set. Let $C \subseteq {}^2({}^\omega\omega)$ be a $\mathbf{\Pi}_n^1$ set such that $B = pC$. The unfolded version of the game $G_\omega^{**}(B)$ is played as follows:

$$\begin{array}{lllll} I: & \langle y(0), s_0\rangle & & \langle y(1), s_2\rangle & \dots \\ II: & & s_1 & & s_3 \quad \dots \end{array}$$

As for $G_\omega^{**}(B)$, I and II choose members s_i of ${}^{<\omega}\omega - \{\emptyset\}$, but in addition I chooses $y(i) \in \omega$. With x the concatenation $s_0{}^\frown s_1{}^\frown s_2{}^\frown \dots$ and $y \in {}^\omega\omega$ the result of I's integer choices, I wins if $\langle x, y\rangle \in C$, and otherwise II wins. As C is $\mathbf{\Pi}_n^1$, this unfolded game can be cast as one of form $G_\omega(D)$ for some $\mathbf{\Pi}_n^1$ set D of reals, and so the game is determined by hypothesis.

Suppose first that II has a winning strategy τ. Then argue much as for 27.3(a) that B must be meager: Disregarding I's integer moves define the open dense sets D_p as before, and then show that if $x \in \bigcap_p D_p$, then $x \notin B$. (If to the contrary $x \in B$, then there would be a $y \in {}^\omega\omega$ such that $\langle x, y\rangle \in C$. But then, a play according to τ can be recursively defined with $y(i)$'s as I's integer moves and x as the concatenation of the finite sequences, contradicting τ being a winning strategy.)

Suppose next that I has a winning strategy σ. Then I has a winning strategy in the original game $G_\omega^{**}(B)$: He simply keeps the $y(i)$'s to himself and plays only the $s(2i)$'s. Hence, by 27.3(b), $O(s) - B$ is meager for some $s \in {}^{<\omega}\omega$.

Now conclude by the argument for 27.4 that A has the Baire property.

For the perfect set property, argue similarly with an unfolded version of the game $G_2^*(\Psi``A)$ where I makes extra choices $y(i)$ (cf. 27.5 and 27.6). The resulting "witness" y can be taken to be in ${}^\omega 2$ or just in ${}^\omega\omega$, depending on whether the homeomorphism $\Psi\colon {}^\omega\omega \to {}^\omega 2$ is to be applied first or not.

Finally, establish Lebesgue measurability for the complement $\tilde{A} = {}^\omega\omega - A$ as follows: As described before 27.8 let B be a G_δ minimal cover for $\tilde{A}$, and for rational $\epsilon \in \mathbb{R}$ with $\epsilon > 0$ argue with unfolded versions of $G(B - \tilde{A}, \epsilon)$ where I makes extra choices $y(i)$ (cf. 27.7 and 27.8). Dealing with the complement $\tilde{A}$ instead of A, these games can be recast in form $G_\omega(D)$ for some $\mathbf{\Pi}^1_n$ set D of reals. Again, the resulting "witness" y can be taken to be in ${}^\omega 2$ or just in ${}^\omega\omega$. ⊣

As a nice complement, note that the $n = 0$ case provides a new proof, based on the determinacy of closed games, of the classical result that the analytic sets have the regularity properties.

The next observations concern $\mathrm{AD}_{\mathbb{R}}$, that mightier hypothesis lurking large and inchoate behind AD. In analogy to 27.10, $\mathrm{AD}_{\mathbb{R}}$ implies that every ${}^\omega\omega$-indexed family of nonempty sets of reals has a choice function:

27.15 Exercise. *Assume* $\mathrm{AD}_{\mathbb{R}}$. *Then every subset of* ${}^2({}^\omega\omega)$ *can be uniformized.*

Hint. Suppose that $A \subseteq {}^2({}^\omega\omega)$ and consider the two-move game where I chooses a real x and II wins *iff* he responds with a real y such that $\langle x, y\rangle \in A$. $\mathrm{AD}_{\mathbb{R}}$ implies that II must have a winning strategy, and such a strategy provides a uniformization. ⊣

In contrast, the following delimitation was seen through a diagonalization argument showing that unlike AD, $\mathrm{AD}_{\mathbb{R}}$ does not relativize to $L(\mathbb{R})$ as a natural inner model.

27.16 Proposition (Solovay [78a: 176]). *Assume* $V = L(\mathbb{R})$ *and there is no well-ordering of the reals. Then there is a subset of* ${}^2({}^\omega\omega)$ *that cannot be uniformized.*

Proof. For any $x \in {}^\omega\omega$, let $\mathrm{OD}(x)$ denote the class of sets definable using parameters from $\mathrm{On} \cup \{x\}$. As for the well-known class OD of ordinal definable sets, $\mathrm{OD}(x)$ has a definable well-ordering, and the 2-ary $\ulcorner y \in \mathrm{OD}(x)\urcorner$ is definable via the Reflection Principle for ZF. Let $A \subseteq {}^2({}^\omega\omega)$ be such that

$$\langle x, y\rangle \in A \leftrightarrow y \notin \mathrm{OD}(x)\,.$$

Since by hypothesis there is no well-ordering of the reals, $\forall x \exists y(\langle x, y\rangle \in A)$.

Assume now to the contrary that there is a $B \subseteq A$ uniformizing A. Since by hypothesis $V = L(\mathbb{R})$ and there is a definable surjection: $\mathrm{On} \times {}^\omega\omega \to L(\mathbb{R})$ (cf. the proof of 11.13), $B \in \mathrm{OD}(x_0)$ for some $x_0 \in {}^\omega\omega$. But then, that unique y_0 such that $B(x_0, y_0)$ would be in $\mathrm{OD}(x_0)$. Contradiction! ⊣

For any $S \subseteq {}^{\omega}\omega$, by considering an analogous concept $\mathrm{OD}(\{S, x\})$ for $x \in {}^{\omega}\omega$ and noting that $\{y \in {}^{\omega}\omega \mid y \in \mathrm{OD}(\{S, x\})\}$ is still well-orderable, 27.16 can be extended to $L(\mathbb{R} \cup \{S\})$. This leads to the following:

27.17 Proposition (Solovay [78a: 177]). *Assume* $\mathrm{AD}_{\mathbb{R}}$. *Then for any* $S \subseteq {}^{\omega}\omega$, $V \neq L(\mathbb{R} \cup \{S\})$. ⊣

Solovay [78a] went on to formulate sharps for *sets* of reals and show that $\mathrm{AD}_{\mathbb{R}}$ implies their existence for every such set. In contrast to the situation for AD, for quite some time it was unclear whether there is a natural inner model of $\mathrm{AD}_{\mathbb{R}}$ containing every real. This issue was resolved in the 1980's as broached in the last section (cf. 32.19).

The final remarks concern the Ramsey property discussed at the end of §11; for present purposes the $\mathcal{P}(\omega)$ context there is shifted to ${}^{\omega}\omega$. Being Ramsey is also regarded as a regularity property for sets of reals. Like the others, every set of reals is Ramsey in Solovay's model (11.15), and there was progress through the first levels of the projective hierarchy: Galvin-Prikry [73] showed that every Borel set of reals is Ramsey, and Silver [70a] that every analytic set is Ramsey, and if there is measurable cardinal, that every $\mathbf{\Sigma}^1_2$ set is Ramsey. Erik Ellentuck [74] devised proofs of these results using a topology based on the Mathias forcing conditions and observing that a set of reals is Ramsey exactly when it (through an identification) has the Baire property in that topology. Making the first connection with determinacy, Prikry [76] then appealed to the determinacy of the corresponding **-game with real moves (cf. 27.3) to show that $\mathrm{AD}_{\mathbb{R}}$ *implies that every set of reals is Ramsey.* (Ilias Kastanas [83] established related results.) Consequences of later developments are that PD *implies that every projective set of reals is Ramsey* (Harrington-Kechris [81]), and that AD *and* $V = L(\mathbb{R})$ *implies that every set of reals is Ramsey* (Martin-Steel [83]). However, with the corresponding games naturally based on real moves, the following question remains open:

27.18 Question. *Does* AD *imply that every set of reals is Ramsey?*

28. AD and Combinatorics

Solovay established in 1967 that AD implies that ω_1 is measurable, injecting emerging large cardinal intuitions, techniques, and results into a novel setting without AC. Unexpected was how strong an effect AD has on the transfinite, and intriguing were the possibilities for further large cardinal consequences. Others soon joined in, with Martin making particularly incisive contributions, in a first wave of what was to become a broad combinatorial investigation. This section pursues this development, emphasizing the richness of the game paradigm with a variety of games. As for §27,

$$\text{ZF serves as the ambient theory for this section .}$$

Although the preoccupation is with direct consequences of AD, it is noted again that they were increasingly regarded in ZFC as what holds in $L(\mathbb{R})$ assuming $\mathrm{AD}^{L(\mathbb{R})}$.

The following was an early observation about ultrafilters.

28.1 Proposition. *Assume* AD. *Then there are no (non-principal) ultrafilters over* ω. *Hence, every ultrafilter is* ω_1*-complete.*

Proof. Assume to the contrary that there is an ultrafilter U over ω. Consider a game where the players choose members of $[\omega]^{<\omega}$:

$$\begin{array}{llllll} I: & s_0 & & s_2 & & \dots \\ II: & & s_1 & & s_3 & \dots \end{array}$$

If a move s_n is not disjoint from $\bigcup_{i<n} s_i$, then the player to make the first such move loses. Assuming that this does not happen, I wins if $\bigcup_{i\in\omega} s_{2i} \in U$, and otherwise II wins. A contradiction is derived by showing that a winning strategy for either player can be converted to one for the other:

Suppose first that σ is a winning strategy for I. Define a strategy τ_σ for II by:

$$\tau_\sigma(\langle s_0\rangle) = \sigma(\emptyset), \text{ and for } 0 < i \in \omega,\ \tau_\sigma(\langle s_0, \dots, s_{2i}\rangle) = \sigma(\langle s_1, \dots, s_{2i}\rangle) - s_0\ ,$$

i.e. ignore the first move s_0 and play according to σ, subtracting s_0 to maintain disjointness of moves. For any play $\langle s_0, s_1, s_2, \dots\rangle$ according to τ_σ consisting of disjoint moves, $s_0 \cup \bigcup_{i\in\omega} s_{2i+1} \in U$ as σ is winning for I, and so $\bigcup_{i\in\omega} s_{2i+1} \in U$ as U is presumed non-principal. Hence, τ_σ is winning for II since U is ultra – a contradiction in this case.

Suppose next that τ is a winning strategy for II. First modify τ to a strategy $\overline{\tau}$ for II defined by:

$$\overline{\tau}(\langle s_0, \dots, s_{2i}\rangle) = \begin{cases} \tau(\langle s_0, \dots, s_{2i}\rangle) \cup \{i\} & \text{if } i \notin \bigcup_{j\leq 2i} s_j\ , \text{ and} \\ \tau(\langle s_0, \dots, s_{2i}\rangle) & \text{otherwise .} \end{cases}$$

Then $\overline{\tau}$ is also winning for II, and moreover for any play $\langle s_0, s_1, s_2 \dots\rangle$ according to $\overline{\tau}$ consisting of disjoint moves, $\bigcup_{i\in\omega} s_{2i+1} \in U$ as $\bigcup_{i\in\omega} s_i = \omega$ and U is ultra.

Now define a strategy σ_τ for I by

$$\sigma_\tau(\langle s_0, \ldots, s_{2i-1}\rangle) = \overline{\tau}(\langle \emptyset, s_0, \ldots, s_{2i-1}\rangle) ,$$

i.e. assume there is a first move $\emptyset$ and play according to $\overline{\tau}$. Then as in the previous argument, σ_τ is winning for I – a contradiction in this case also.

For the second assertion, if W were an ultrafilter over a set S which is not ω_1-complete, there would be a function $f\colon S \to \omega$ such that

$$f_*(W) = \{X \subseteq \omega \mid f^{-1}(X) \in W\}$$

is a (non-principal) ultrafilter over ω. ⊣

This result can also be seen, albeit less directly, via 27.9 and the result of Sierpiński [38] that an ultrafilter over ω construed as a set of reals is not Lebesgue measurable. Of course, ω_1-completeness is only of consequence if there are ultrafilters at all under AD, and this was established by Solovay:

28.2 Theorem (Solovay). *Assume* AD. *Then* ω_1 *is measurable.*

Measurable cardinals are regular in ZF because of the completeness of their witnessing ultrafilters. It is AC that renders them weakly inaccessible (and much more) through the enumerability of power sets. However, the absence of AC does not affect consistency strength. $\ulcorner$ZF + ω_1 is measurable$\urcorner$ is equiconsistent with $\ulcorner$ZFC + $\exists\kappa(\kappa$ is measurable)$\urcorner$: Jech [68], Levy, and Takeuti [70] each observed that in ZFC if a measurable cardinal is Levy collapsed to ω_1, then in an inner model of the resulting extension ω_1 is measurable. Conversely, if in ZF, U is a κ-complete ultrafilter over some $\kappa > \omega$, then $L[U] \models$ ZFC + κ is measurable. Moreover, for any $a \in {}^\omega\omega$, $L[U, a] \models$ ZFC + κ is measurable, where $L[U, a]$, defined using predicate symbols for U and a, is the smallest inner model such that $U \cap M \in M$ and $a \in M$. Thus, although AC is not actually needed in §9, a ZFC context is in any case available to develop the theory of $a^\#$, and its absoluteness properties (cf. 14.12(a)) ensure its substance.

The following was an early extension of 27.11(b) about the consistency strength of AD.

28.3 Corollary.

(a) Con(ZF + AD) *implies* Con(ZFC + $\exists\kappa(\kappa$ is measurable)).

(b) Assume AD. *Then* $\forall a \in {}^\omega\omega(a^\#$ exists). ⊣

The measurability of ω_1 also has direct consequences in ZF, e.g. there are no ω_1-Aronszajn trees (cf. 7.8, 7.10).

Starting with his original argument two proofs of Solovay's theorem are given, both of which became paradigmatic for further work.

First Proof of 28.2. Solovay devised a game where the players are required to play codes for well-orderings and used a bounding argument with $\mathbf{\Sigma}^1_1$ sets to show that under AD, the closed unbounded filter $\mathcal{C}_{\omega_1}$ over ω_1 is an ultrafilter. By 28.1 the measurability of ω_1 follows.

To motivate Solovay's argument consider for $Y \subseteq \omega_1$ a game where the players choose countable ordinals:

$$\begin{array}{lllllll} I & : & \xi_0 & & \xi_2 & & \dots \\ II & : & & \xi_1 & & \xi_3 & \dots \end{array}$$

I wins if $(\sup\{\xi_i \mid i \in \omega\}) \in Y$, and otherwise II wins. It is argued that if I has a winning strategy σ, then $Y \in \mathcal{C}_{\omega_1}$ (and the argument will show that if II has a winning strategy, then $\omega_1 - Y \in \mathcal{C}_{\omega_1}$):

As ω_1 is regular (27.10), for each $\eta < \omega_1$ let $w(\eta) < \omega_1$ be least so that $\sigma|^{<\omega}\eta \subseteq w(\eta)$, i.e. for any finite partial play all of whose moves (or equivalently, just II's moves) are less than η, the strategic response according to σ is less than $w(\eta)$. Then

$$C = \{\rho < \omega_1 \mid \rho \text{ is an infinite limit } \wedge \ \forall \eta < \rho(w(\eta) < \rho)\}$$

is closed unbounded in ω_1, and $C \subseteq Y$: Suppose that $\rho \in C$, and let $\{\eta_i \mid i \in \omega\} \subseteq \rho$ be unbounded in ρ. Then in the play $\sigma * \langle \eta_i \mid i \in \omega\rangle$ each move is less than ρ, and since σ is winning, $\rho \in Y$.

Solovay's idea was to simulate this game by one to which AD is applicable. Consider for $Y \subseteq \omega_1$ a game where the players choose integers:

$$\begin{array}{lllllll} I & : & x(0) & & x(2) & & \dots \\ II & : & & x(1) & & x(3) & \dots \end{array}$$

Recall from §13 that each $y \in {}^\omega\omega$ encodes a sequence $\langle (y)_i \mid i \in \omega\rangle \in {}^\omega({}^\omega\omega)$ and each $z \in {}^\omega\omega$ encodes a relation $E_z \subseteq \omega \times \omega$ with $\|z\|$ denoting its ordertype when it is a well-ordering. I is regarded as playing a sequence of relations $\langle E_{(x_I)_i} \mid i \in \omega\rangle$ and II, $\langle E_{(x_{II})_i} \mid i \in \omega\rangle$. Suppose first that for some $i \in \omega$, either $E_{(x_I)_i}$ or $E_{(x_{II})_i}$ is not a well-ordering; then for the least such i, II wins if $E_{(x_I)_i}$ is not a well-ordering, and otherwise I wins. Assuming that this did not happen, I wins if

$$\sup(\{\|(x_I)_i\| \mid i \in \omega\} \cup \{\|(x_{II})_i\| \mid i \in \omega\}) \in Y\ ,$$

and otherwise II wins.

This game can be construed as a game $G_\omega(A)$ for some $A \subseteq {}^\omega\omega$, and so is determined. We argue that, as before, if I has a winning strategy σ, then $Y \in \mathcal{C}_{\omega_1}$ (and the argument will show that if II has a winning strategy, then $\omega_1 - Y \in \mathcal{C}_{\omega_1}$) thus completing the proof by confirming that $\mathcal{C}_{\omega_1}$ is an ultrafilter.

Considering the previous argument, it will suffice to get bounds in the following sense: For each $\eta < \omega_1$, there is a $\beta < \omega_1$ such that: for any $n \in \omega$ and $y \in {}^\omega\omega$, if for each $i < n$, $E_{(y)_i}(= E_{((\sigma*y)_{II})_i})$ is a well-ordering with $\|(y)_i\| < \eta$, then $E_{((\sigma*y)_I)_n}$ is a well-ordering with $\|((\sigma * y)_I)_n\| < \beta$. So, suppose that $\eta \in \omega_1$. Set

$$A_\eta = \{((\sigma * y)_I)_n \mid n \in \omega \ \wedge \ y \in {}^\omega\omega \ \wedge \\ \forall i < n(E_{(y)_i} \text{ is a well-ordering with } \|(y)_i\| < \eta)\} \ .$$

Then A_η is $\mathbf{\Sigma}^1_1$ in a real coding the ordertype η, since it can be defined in terms of embeddability into that ordertype (cf. the second proof of 13.4). Moreover, in terms of 13.6, $A_\eta \subseteq \mathrm{WF}$ as σ is winning, and so $A_\eta \subseteq \mathrm{WF}_\beta$ for some $\beta < \omega_1$. Hence, a β as desired has indeed been found. Taking $w(\eta)$ to be the least such β, the proof can now be completed with the defined function $w \colon \omega_1 \to \omega_1$ as anticipated. ⊣

For the foregoing and the forthcoming, note again for provability in ZF that AD implies $\mathrm{AC}_\omega({}^\omega\omega)$ (27.10), which suffices for the results of §13. That $\mathcal{C}_{\omega_1}$ itself witnesses the measurability of ω_1 turned out to be a substantial hypothesis. In the mid-1970's Martin and Mitchell (see their [79]) applied Kunen's methods for 21.25 to conclude from the hypothesis that for every λ there is an inner model of ZFC with λ measurable cardinals. By the early 1980's the hypothesis was gauged as considerably stronger (see Radin Forcing in volume II).

Martin provided a proof of Solovay's theorem via a nice observation of independent significance. Like Solovay, Martin made crucial contributions in this area as well as in descriptive set theory (§§29, 30), and moreover was to forge the initiative for establishing the consistency of determinacy hypotheses (§31). Some well-known concepts and notation are first affirmed: For $a, b \in {}^\omega\omega$,

$$a \leq_{\mathrm{T}} b \ \ \textit{iff} \ \ a \text{ is recursive in } b \ ,$$
$$a \equiv_{\mathrm{T}} b \ \ \textit{iff} \ \ a \leq_{\mathrm{T}} b \ \wedge \ b \leq_{\mathrm{T}} a \ , \text{ and}$$
$$[a]_{\mathrm{T}} = \{c \in {}^\omega\omega \mid c \equiv_{\mathrm{T}} a\} \ .$$

$\leq_{\mathrm{T}}$ is the well-known relation of *Turing reducibility*, and $\equiv_{\mathrm{T}}$ that of *Turing equivalence*, with the corresponding equivalence classes $[a]_{\mathrm{T}}$ being the *Turing degrees*. Every $a \in {}^\omega\omega$ has countably many $\leq_{\mathrm{T}}$-predecessors; in terms of §12 there are countably many formulas to consider. Letting $\mathcal{D}_{\mathrm{T}}$ be the set of such degrees,

$$\mathcal{D}_{\mathrm{T}} = \{[a]_{\mathrm{T}} \mid a \in {}^\omega\omega\} \ ,$$

$\leq_{\mathrm{T}}$ lifts to an ordering on $\mathcal{D}_{\mathrm{T}}$ well-defined by:

$$d_0 \leq d_1 \ \ \textit{iff} \ \ \exists a \exists b(d_0 = [a]_{\mathrm{T}} \wedge d_1 = [b]_{\mathrm{T}} \wedge a \leq_{\mathrm{T}} b) \ .$$

Martin considered the filter M_{T} over $\mathcal{D}_{\mathrm{T}}$ generated by the "cones" of degrees:

$$X \in M_{\mathrm{T}} \ \ \textit{iff} \ \ \exists d_0 \in \mathcal{D}_{\mathrm{T}}(\{d \in \mathcal{D}_{\mathrm{T}} \mid d_0 \leq d\} \subseteq X) \ .$$

That it is indeed a filter is simple to see, since for any $a, b \in {}^\omega\omega$ there is always a $c \in {}^\omega\omega$ such that $a \leq_{\mathrm{T}} c$ and $b \leq_{\mathrm{T}} c$. Generalizing this, AC_ω implies that M_{T} is ω_1-complete. On the other hand, AC implies that M_{T} is not an ultrafilter (cf. the proof of 11.4(a), noting that every Turing degree is a countable set).

28.4 Proposition (Martin [68]). *Assume* AD. *Then* M_{T} *is an ultrafilter.*

Proof. Suppose that $X \subseteq \mathcal{D}_{\mathrm{T}}$, and consider the game $G_\omega(\bigcup X)$. Assume first that I has a winning strategy σ in this game. Considering σ as a member of ${}^\omega\omega$ (through a recursive coding of ${}^{<\omega}\omega$) set $d_0 = [\sigma]_{\mathrm{T}}$. Then for any $d \in \mathcal{D}_{\mathrm{T}}$ with $d_0 \leq d$ and $y \in d$, clearly $\sigma * y \in d$ and hence $d \in X$. A similar argument shows that if II has a winning strategy τ, then $\{d \in \mathcal{D}_{\mathrm{T}} \mid [\tau]_{\mathrm{T}} \leq d\} \subseteq \mathcal{D}_{\mathrm{T}} - X$. ⊣

With AD, that M_{T} is an ultrafilter already implies that it is ω_1-complete by 28.1.

Second Proof of 28.2. For $a, b \in {}^\omega\omega$ with $a \equiv_{\mathrm{T}} b$, $\omega_1^{L[a]} = \omega_1^{L[b]}$, and this is a countable ordinal (27.13). Thus, $f\colon \mathcal{D}_{\mathrm{T}} \to \omega_1$ given by:

$$f([a]_{\mathrm{T}}) = \omega_1^{L[a]}$$

is well-defined. With 28.4,

$$f_*(M_{\mathrm{T}}) = \{X \subseteq \omega_1 \mid f^{-1}(X) \in M_{\mathrm{T}}\}$$

is then an ω_1-complete ultrafilter over ω_1. (To check that it is non-principal, note that for any $\alpha < \omega_1$ and $a \in {}^\omega\omega$ there is a $b \in {}^\omega\omega$ with $a \leq_{\mathrm{T}} b$ such that $\alpha < \omega_1^{L[b]}$, so that $f^{-1}(\{\alpha\}) \notin M_{\mathrm{T}}$.) ⊣

There is such an f so that $f_*(M_{\mathrm{T}}) = \mathcal{C}_{\omega_1}$; the proof of 28.20 provides a general result along these lines.

Solovay in the meanwhile had established a structure result important for later work, playing the first of what came to be known as a *Solovay game,* a game where the players must again play codes for countable ordinals, but where the rules preclude I from having a winning strategy.

28.5 Theorem (Solovay). *Assume* AD. *Suppose that* $Y \subseteq \omega_1$. *Then there is an* $a \in {}^\omega\omega$ *such that* $Y \in L[a]$.

Proof. Consider a game where the players choose integers:

I :	$x(0)$		$x(2)$		$\ldots$
II :		$x(1)$		$x(3)$	$\ldots$

The rules this time are that II wins if either E_{x_I} is not a well-ordering, or else $E_{(x_{II})_i}$ is a well-ordering for every $i \in {}^\omega\omega$ and $\{\|(x_{II})_i\| \mid i \in \omega\} = Y \cap \beta$ for some $\beta > \|x_I\|$. Otherwise, I wins.

Using a bounding argument as in Solovay's previous proof, it is first established that I cannot have a winning strategy: Assuming that σ is such a strategy, $\{(\sigma * y)_I \mid y \in {}^\omega\omega\}$ is a $\mathbf{\Sigma}^1_1$ subset of WF. Hence, there is a $\beta < \omega_1$ such that for any $y \in {}^\omega\omega$, $\|(\sigma * y)_I\| < \beta$. Consequently, for any $y \in {}^\omega\omega$ with $E_{(y)_i}$ a well-ordering for every $i \in \omega$ and $\{\|(y)_i\| \mid i \in \omega\} = Y \cap \beta$, $\sigma * y$ is a play that II wins – a contradiction.

It follows that II must have a winning strategy, say τ. The proof is completed by showing that $Y \in L[\tau]$ (regarding τ as a real): Note first that there is a simple formula $\varphi(v_0, v_1)$ such that for any $z \in {}^\omega\omega$ with E_z a well-ordering,

$$\|z\| \in Y \text{ iff } \varphi[z, \tau] .$$

As τ is winning, φ can assert that there is an $i \in \omega$ and a real coding an isomorphism between $\langle \omega, E_z \rangle$ and $\langle \omega, E_{((z*\tau)_{II})_i} \rangle$. Now if ${\omega_1}^{L[\tau]} = \omega_1$, then definability via such a φ and absoluteness would imply that $Y \in L[\tau]$. However, ω_1 is inaccessible in $L[\tau]$ by 27.13, but a forcing argument can be applied:

For each $\xi < \omega_1$, let P_ξ be the p.o. for adjoining a surjection from ω onto ξ via finite conditions, and let $\dot{z}_\xi$ be a P_ξ-name, defined uniformly in ξ, for a $z \in {}^\omega\omega$ such that E_z has ordertype ξ in any generic extension via P_ξ. Then for φ as above,

$$\xi \in Y \text{ iff } L[\tau] \models (\Vdash_{P_\xi} \varphi(\dot{z}_\xi, \check{\tau})) .$$

For the right-to-left direction note that $\mathcal{P}(P_\xi) \cap L[\tau]$ is countable by the inaccessibility of ω_1 in $L[\tau]$, and so there are P_ξ-generics over $L[\tau]$. It follows by this definability (and an application of the ZF Reflection Principle) that $Y \in L[\tau]$. ⊣

Kechris [88b] has recent results on subsets of ω_1 so constructible from a real.

Upon seeing Martin's result 28.4 Solovay raised the ante in 1968 with the following result:

28.6 Theorem (Solovay). *Assume* AD. *Then* ω_2 *is measurable.*

Proof. The argument starts out as for the second proof of 28.2 and then has features of the proof of 28.5. For $a, b \in {}^\omega\omega$ with $a \equiv_T b$, $(\omega_1)^{+L[a]} = (\omega_1)^{+L[b]}$, i.e. the respective successors in $L[a]$ and in $L[b]$ of (the real) ω_1 are equal, and this ordinal is less than (the real) ω_2 by the existence of sharps. Thus, $f \colon \mathcal{D}_T \to \omega_2$ given by

$$f([a]_T) = (\omega_1)^{+L[a]}$$

is well-defined. Using 28.4,

$$f_*(M_T) = \{X \subseteq \omega_2 \mid f^{-1}(X) \in M_T\}$$

is then an ω_1-complete ultrafilter over ω_2. (To check that it is non-principal, note that for any $\alpha < \omega_2$ and $a \in {}^\omega\omega$, if $p\colon \alpha \to \omega_1$ is an injection, then by 28.5 there is a $b \in {}^\omega\omega$ such that $p \in L[b]$, and taking $a \leq_T b$ implies that $f^{-1}(\{a\}) \notin f_*(M_T)$.) To complete the proof, it suffices to verify ω_2-completeness:

Suppose then that $\{X_\xi \mid \xi < \omega_1\} \subseteq f_*(M_T)$ where it can assumed that $X_\eta \subseteq X_\xi$ for $\xi < \eta < \omega_1$ by ω_1-completeness. It must be shown that $\bigcap_{\xi<\omega_1} X_\xi \in f_*(M_T)$. To this end, consider the following Solovay game:

$$\begin{array}{lllll} I: & x(0) & & x(2) & & \dots \\ II: & & x(1) & & x(3) & \dots \end{array}$$

II wins if either E_{x_I} is not a well-ordering, or else $\{d \in \mathcal{D}_{\mathrm{T}} \mid [x_{II}]_{\mathrm{T}} \leq d\} \subseteq f^{-1}(X_{\|x_I\|})$.

A bounding argument as for 28.5 shows that I cannot have a winning strategy: Assuming that σ is such a strategy, $\{(\sigma * y)_I \mid y \in {}^\omega\omega\}$ is a $\mathbf{\Sigma}^1_1$ subset of WF, and so there is a $\beta < \omega_1$ such that $\|(\sigma * y)_I\| < \beta$ for every $y \in {}^\omega\omega$. Since $X_\beta \in f_*(M_{\mathrm{T}})$, there is a $b \in {}^\omega\omega$ such that $\{d \in \mathcal{D}_{\mathrm{T}} \mid [b]_{\mathrm{T}} \leq d\} \subseteq f^{-1}(X_\beta)$. But then, as $\xi < \beta$ implies that $f^{-1}(X_\beta) \subseteq f^{-1}(X_\xi)$, it follows that $\sigma * b$ is a play that I loses – a contradiction.

It follows that II must have a winning strategy, say τ. Regarding τ as a real, the proof is completed by showing that $\{d \in \mathcal{D}_{\mathrm{T}} \mid [\tau]_{\mathrm{T}} \leq d\} \subseteq f^{-1}(\bigcap_{\xi<\omega_1} X_\xi)$:

Suppose that $\xi < \omega_1$ and $x \in {}^\omega\omega$ with $\tau \leq_{\mathrm{T}} x$; it must be shown that $f([x]_{\mathrm{T}}) \in X_\xi$. As in the proof of 28.5 let P_ξ be the p.o. for adjoining a surjection from ω onto ξ via finite conditions. Since $\mathcal{P}(P_\xi) \cap L[x]$ is countable by the inaccessibility of ω_1 in $L[x]$, there is a g P_ξ-generic over $L[x]$. g can be regarded as a real such that $\|g\| = \xi$. Let $y \in {}^\omega\omega$ be such that $x \leq_{\mathrm{T}} y$, $g \leq_{\mathrm{T}} y$, and $L[x][g] = L[y]$. Since forcing with P_ξ preserves all cardinals greater than ξ, $f([x]_{\mathrm{T}}) = f([y]_{\mathrm{T}})$. Now τ is winning for II, so if $h = (g * \tau)_{II}$, then $\{d \in \mathcal{D}_{\mathrm{T}} \mid [h]_{\mathrm{T}} \leq d\} \subseteq f^{-1}(X_\xi)$. But it follows from $\tau \leq_{\mathrm{T}} x$ and the choice of y that $h \leq_{\mathrm{T}} y$. Hence, $f([x]_{\mathrm{T}}) = f([y]_{\mathrm{T}}) \in X_\xi$. ⊣

With an apparent trend set, quite unexpected was the next advance: Martin in 1970 established that under AD the ω_n's for $3 \leq n < \omega$ are *singular* with cofinality ω_2. It is a brave new world that has such properties! Martin's result was a corollary to his incisive analysis of $\mathbf{\Sigma}^1_3$ sets under AD. Setting the stage, recall from §14 that under the assumption $\forall a \in {}^\omega\omega(a^\#$ exists), u_ξ denotes the ξth uniform indiscernible, so that $u_1 = \omega_1$ and $u_\xi \leq \omega_\xi$ for $\xi > 0$.

$$\text{AD implies } u_2 = \omega_2 .$$

This follows from the argument for 28.6, which can be viewed as showing that u_2 is a measurable cardinal; put another way, $\omega_2 = \sup(\mathrm{ran}(f)) \leq u_2$, the equality by the ω_2-completeness of $f_*(M_{\mathrm{T}})$ and the inequality by the definition of f.

The following is established, working with $\mathbf{\Sigma}^1_3$ sets and assuming a coding lemma, as 30.18:

28.7 Theorem (Martin). *Assume* AD. *Then* $u_\omega = \omega_\omega$.

With the regularity of ω_2 from 28.6 and the above observation that $u_2 = \omega_2$, Martin's corollary follows:

28.8 Corollary. *Assume* AD. *Then* $\mathrm{cf}(\omega_n) = \omega_2$ *for every* $2 \leq n < \omega$.

Proof. For $2 \leq n < \omega$, $\omega_n = u_k$ for some k, which by the theorem satisfies $1 < k < \omega$. But then by Solovay's 14.18(a), $\mathrm{cf}(\omega_n) = \mathrm{cf}(u_2) = \omega_2$. ⊣

28.7 in fact implies that the first ω infinite cardinals are exactly the first ω uniform indiscernibles:

28.9 Theorem (DC)(Kunen; Solovay). *Assume* AD. *Then for* $1 \leq n \leq \omega$, $u_n = \omega_n$.

Proof. Ultrapowers of sets (regarded as structures for $\mathcal{L}_\in$) via ω_1-complete ultrafilters lack infinite descending chains, and so DC ensures that such ultrapowers are well-founded. In particular, such ultrapowers of ordinals are well-ordered under the induced ordering. The following will be established:

(∗) For $1 \leq n < \omega$, ${}^{\omega_1}u_n/\mathcal{C}_{\omega_1}$ has ordertype u_{n+1} .

Granted this, if for some such n, $|u_n| = |u_{n+1}|$, then

$$|u_{n+2}| = |{}^{\omega_1}u_{n+1}/\mathcal{C}_{\omega_1}| = |{}^{\omega_1}u_n/\mathcal{C}_{\omega_1}| = |u_{n+1}|$$

and recursively $|u_{n+i}| = |u_n|$ for every $i \in \omega$, so that $|u_\omega| = |u_n| \leq \omega_n$, contradicting 28.7. It follows that $|u_n| < |u_{n+1}|$ for $1 \leq n < \omega$, and this together with $u_n \leq \omega_n$ for $1 \leq n < \omega$ imply the desired conclusion.

Toward establishing (∗), it is first observed that for $\xi < u_\omega$, ${}^{\omega_1}\xi \subseteq \tilde{L}$ where $\tilde{L} = \bigcup_{a \in {}^\omega\omega} L[a]$ (as defined before 15.8). For $\xi < \omega_2$ this follows from 28.5. For $\omega_2 \leq \xi < u_\omega$, proceeding by induction on ξ, assume first that for some $a \in {}^\omega\omega$, ξ is not a cardinal in $L[a]$, so that there is a bijection $g\colon \xi \to \zeta$ for some $\zeta < \xi$ with $g \in L[a]$. Then for any $f \in {}^{\omega_1}\xi$, $g \circ f \in L[b]$ for some $b \in {}^\omega\omega$ by induction, and hence $f \in L[c]$ for any $c \in {}^\omega\omega$ coding a and b. Now assume that ξ is a cardinal in $L[a]$ for every $a \in {}^\omega\omega$. Then ξ is a uniform indiscernible: If for some $b \in {}^\omega\omega$, ξ were strictly between two consecutive members of the corresponding class of indiscernibles I_b, then in $L[b^\#]$ these indiscernibles are of the same cardinality, and hence ξ would not be a cardinal there. With $\omega_2 \leq \xi < u_\omega$, $\xi = u_n$ with $2 \leq n < \omega$, and so it follows by 14.18(a) that $\mathrm{cf}(\xi) = \mathrm{cf}(u_2) = \omega_2$. But then, ${}^{\omega_1}\xi = \bigcup_{\zeta<\xi} {}^{\omega_1}\zeta$, and so the result follows by induction.

Finally, to establish (∗) the argument for 15.11 is used. Suppose first that $h \in {}^{\omega_1}u_n$. Then $h \in \tilde{L}$ by our observation. It follows from 14.18(b) that there is an $a \in {}^\omega\omega$ and a Skolem term t_1 such that

$$h = t_1{}^{\langle L[a], \in, a\rangle}(u_1, \ldots, u_n) .$$

A small adjustment leads to a Skolem term t such that for any $\xi < \omega_1$ $(= u_1)$,

$$h(\xi) = t^{\langle L[a], \in, a\rangle}(\xi, u_1, \ldots, u_n) .$$

Now consider

$$\eta(h, a, t) = t^{\langle L[a], \in, a\rangle}(u_1, \ldots, u_{n+1}) < u_{n+1}$$

following the 15.11 proof; as there, if $\overline{h} \in {}^{\omega_1}u_n$ with $\{\xi < \omega_1 \mid \overline{h}(\xi) = h(\xi)\} \in \mathcal{C}_{\omega_1}$ and $\eta(\overline{h}, \overline{a}, \overline{t})$ is derived from some $\overline{a}$ and $\overline{t}$ for $\overline{h}$, then

$$\eta(\overline{h}, \overline{a}, \overline{t}) = \eta(h, a, t) .$$

Hence, the map

$$e: {}^{\omega_1}u_n/\mathcal{C}_{\omega_1} \rightarrow u_{n+1}$$

given by $e([h]) = \eta(h, a, t)$ for some a and t for h is an order-preserving bijection. This completes the proof. ⊣

This synthetic result provided a structural description of the first ω uncountable cardinals under AD through the theory of sharps, and accounted for the apparent pathology of $\mathrm{cf}(\omega_n) = \omega_2$ for $n \geq 3$ through Solovay's 14.18(a).

The investigations of the combinatorial and descriptive set theory consequences of AD were both to provide a context for and be advanced by an initiative from a different quarter, the study of strong partition relations in infinitary combinatorics.

Infinite Exponent Partition Relations

Eugene Kleinberg, a student of Martin, pursued the consequences of infinite exponent partition relations. The Ramsey property for sets of reals, discussed at the ends of §§11, 27, evolved from the possibility $\omega \longrightarrow (\omega)^{\omega}_2$. Although such relations contradict AC (7.1), early on in 1968 Kleinberg argued for their intrinsic interest at uncountable cardinals and showed that versions imply measurability in a strong sense through canonical ultrafilters.

For $\lambda < \kappa$ both regular let $\mathcal{C}^{\lambda}_{\kappa}$ denote the filter over κ generated by the sets λ-closed unbounded in κ. If D is such a set, then letting C consist of the limit points of D other than κ, $C \cap \{\xi < \kappa \mid \mathrm{cf}(\xi) = \lambda\} \subseteq D$. Thus,

$$\mathcal{C}^{\lambda}_{\kappa} = \{X \subseteq \kappa \mid \exists C \in \mathcal{C}_{\kappa}(C \cap \{\xi < \kappa \mid \mathrm{cf}(\xi) = \lambda\} \subseteq X)\}\,.$$

With AC, through arguments that choose for each $X \in \mathcal{C}^{\lambda}_{\kappa}$ a corresponding $C \in \mathcal{C}_{\kappa}$ satisfying $C \cap \{\xi < \kappa \mid \mathrm{cf}(\xi) = \lambda\} \subseteq X$, it can be shown that $\mathcal{C}^{\lambda}_{\kappa}$ is a κ-complete, in fact normal, filter over κ; on the other hand, $\{\xi < \kappa \mid \mathrm{cf}(\xi) = \lambda\}$ can be partitioned into disjoint stationary sets by 16.9, and so $\mathcal{C}^{\lambda}_{\kappa}$ is not an ultrafilter. Infinite exponent partition relations contradict AC, but establish that the $\mathcal{C}^{\lambda}_{\kappa}$ witness the measurability of κ:

28.10 Theorem (Kleinberg [70]). *Suppose that $\lambda < \kappa$, λ is regular, and $\kappa \longrightarrow (\kappa)^{\lambda+\lambda}_2$. Then κ is measurable, and $\mathcal{C}^{\lambda}_{\kappa}$ is a normal ultrafilter over κ.*

Proof. First observe that $\kappa \longrightarrow (\kappa)^2_2$ already implies the regularity of κ (by the proof of 7.6). Next note that $\kappa \longrightarrow (\kappa)^{\lambda}_{\gamma}$ for every $\gamma < \kappa$ by the argument for 7.14(c):

Suppose that $f\colon [\kappa]^{\lambda} \rightarrow \gamma$ where $\gamma < \kappa$. Define $g\colon [\kappa]^{\lambda+\lambda} \rightarrow 2$ by: $g(s) = 0$ *iff* $f(s_0) = f(s_1)$, where $s_0, s_1 \in [s]^{\lambda}$, $s_0 \cup s_1 = s$, and $\bigcup s_0 \leq \bigcap s_1$. Let $H \in [\kappa]^{\kappa}$ be homogeneous for g. Since $\lambda \cdot \kappa = \kappa > \gamma$, there must be $t_0, t_1 \in [H]^{\lambda}$ such that $\bigcup t_0 \leq \bigcap t_1$ and $f(t_0) = f(t_1)$. Thus, $g(t_0 \cup t_1) = 0$, and so by homogeneity $g``[H]^{\lambda+\lambda} = \{0\}$. Now for any $u, v \in [H]^{\lambda}$ there is by the regularity of κ a

$w \in [H]^\lambda$ satisfying $\bigcup u, \bigcup v \leq \bigcap w$, so that $f(u) = f(w) = f(v)$. Hence, H is homogeneous for f as well.

We next show that $\mathcal{C}_\kappa^\lambda$ is a κ-complete ultrafilter. For this purpose, it suffices to assume that $F\colon \kappa \to \gamma$ where $\gamma < \kappa$, and find a $\beta < \gamma$ such that $F^{-1}(\{\beta\}) \in \mathcal{C}_\kappa^\lambda$. To do this, define $G\colon [\kappa]^\lambda \to \gamma$ by: $G(s) = \alpha$ *iff* $F(\bigcup s) = \alpha$. By $\kappa \longrightarrow (\kappa)_\gamma^\lambda$, let $H \in [\kappa]^\kappa$ be homogeneous for G. Set

$$C = \{\xi < \kappa \mid \bigcup(H \cap \xi) = \xi \ \wedge \ \mathrm{cf}(\xi) = \lambda\} .$$

Then C is λ-closed unbounded in κ, and if $G``[H]^\lambda = \{\beta\}$, then $C \subseteq F^{-1}(\{\beta\})$ so that $F^{-1}(\{\beta\}) \in \mathcal{C}_\kappa^\lambda$.

We finally show that $\mathcal{C}_\kappa^\lambda$ is normal. Noting that the regressive function characterization 5.10 holds in ZF, it suffices to assume that $f \in {}^\kappa\kappa$ is regressive and by κ-completeness, to find an $X \in \mathcal{C}_\kappa^\lambda$ and a $\delta \in \kappa$ such that $f``X \subseteq \delta$. To do this, define $g\colon [\kappa]^\lambda \to 2$ by: $g(s) = 0$ *iff* $f(\bigcup s) < \bigcap s$. By $\kappa \longrightarrow (\kappa)_2^\lambda$, let $H \in [\kappa]^\kappa$ be homogeneous for g. Taking any $s \in [H]^\lambda$, let $\overline{s} = s - (f(\bigcup s) + 1)$. Then $\overline{s} \in [H]^\lambda$ as $f(\bigcup s) < \bigcup s$, and $f(\bigcup \overline{s}) = f(\bigcup s) < \bigcap \overline{s}$. Thus, $g(\overline{s}) = 0$, and so by homogeneity $g``[H]^\lambda = \{0\}$. But then, if δ is the least member of H and

$$C = \{\xi < \kappa \mid \bigcup(H \cap \xi) = \xi \ \wedge \ \mathrm{cf}(\xi) = \lambda\}$$

as before, then C is λ-closed unbounded in κ, and it is simple to check that $f``C \subseteq \delta$. ⊣

The $\mathcal{C}_\kappa^\lambda$'s become focal when κ is not too inaccessible:

28.11 Exercise (Kleinberg [70]). *Suppose that κ is not weakly Mahlo and the $\mathcal{C}_\kappa^\lambda$'s for regular $\lambda < \kappa$ are normal ultrafilters. Then they are the only normal ultrafilters over κ.*

Hint. Normal ultrafilters over κ include $\mathcal{C}_\kappa$, and by hypothesis $\{\xi < \kappa \mid \mathrm{cf}(\xi) < \xi\} \in \mathcal{C}_\kappa$. ⊣

These results would have remained curiosities had it not been for a succession of partition theorems that was to span the accessible transfinite. It was established in the early 1970's that under AD + DC both ω_1 and ω_2 satisfy substantial infinite exponent partition relations, and later, that many cardinals satisfy such relations. The first inkling for ω_1 was Martin's groundbreaking result in late 1968 that under AD, $\omega_1 \longrightarrow (\omega_1)_2^\omega$. Kunen then extended the conclusion to $\omega_1 \longrightarrow (\omega_1)_2^\alpha$ for every $\alpha < \omega_1$. Then in 1971, building on Solovay's structure result 28.5, Martin established the following assertion of strong dichotomy about ω_1:

28.12 Theorem (Martin). *Assume* AD. *Then* $\omega_1 \longrightarrow (\omega_1)_2^{\omega_1}$.

Proof. For $H \in [\omega_1]^{\omega_1}$, with $h\colon \omega_1 \to H$ enumerating it in increasing order set

$${}_\omega H = \{\sup(\{h(\omega{\cdot}\xi + i) \mid i \in \omega\}) \mid \xi < \omega_1\} \in [\omega_1]^{\omega_1} ,$$

i.e. ${}_\omega H$ consists of the suprema of ω consecutive members of H.

Suppose now that $f\colon [\omega_1]^{\omega_1} \to 2$. A C closed unbounded in ω_1 will be found so that ${}_\omega C$ is homogeneous for f. C will consist of the closure ordinals provided by a Σ^1_1 bounding argument applied to a winning strategy for a certain game associated with f. By 28.5 every $Y \in [{}_\omega C]^{\omega_1}$ is in some $L[a]$ with $a \in {}^\omega\omega$ and so can be anticipated using sharps and Skolem terms.

Toward defining the game, we first review and tailor the §§9, 14 sharp theory relativized to reals. An EM blueprint for $a \in {}^\omega\omega$ is the theory of some structure $\langle L_\delta[a], \in, a, \zeta_k \rangle_{k\in\omega}$ where δ is a limit ordinal $> \omega$ and $\{\zeta_k \mid k \in \omega\}$ is a set of ordinal indiscernibles for $\langle L_\delta[a], \in, a\rangle$ indexed in increasing order. Here, such a theory is identified with some $z \in {}^\omega\omega$ through some arithmetization (so that a can be recovered from z), and $\mathcal{M}(z, \alpha)$ denotes the model unique up to isomorphism generated by indiscernibles in ordertype α (cf. 9.4). For $z \in {}^\omega\omega$,

$$\exists a \in {}^\omega\omega (z = a^\#) \;\leftrightarrow\; S(z) \;\wedge\; W(z)$$

where S is arithmetical, asserting that z codes an EM blueprint for some a containing sentences for satisfying the §9 conditions (I) and (III), and W is Π^1_2, asserting that $\mathcal{M}(z, \alpha)$ is well-founded for every $\alpha < \omega_1$ (cf. 14.11).

When $S(z)$, whether $\mathcal{M}(z, \alpha)$ is well-founded or not, it has a largest initial segment of its "ordinals" which is actually well-ordered. Identifying members of this initial segment with corresponding *bona fide* ordinals, temporarily write

$$\eta <_{\text{is}} \mathcal{M}(z, \alpha)$$

to indicate that η is a member of the initial segment.

Let $\langle t_n \mid n \in \omega\rangle$ enumerate the Skolem terms (like those described after 9.3, but) for $\mathcal{L}_\in(\dot{A})$ and in one free variable. For $z \in {}^\omega\omega$, let $z^+ \in {}^\omega\omega$ be defined by $z^+(i) = z(i+1)$, so that $z = \langle z(0)\rangle^\frown z^+$. Now define for $\alpha < \omega_1$:

$$A_\alpha = \{z \in {}^\omega\omega \mid S(z^+) \;\wedge\; \alpha <_{\text{is}} \mathcal{M}(z^+, \alpha+\omega) \;\wedge$$
$$\exists \eta \geq \alpha (t_{z(0)}^{\mathcal{M}(z^+,\alpha+\omega)}(\alpha) = \eta <_{\text{is}} \mathcal{M}(z^+, \alpha+\omega))\}\,.$$

The significance of these sets becomes clear through their role in the following game associated with $f\colon [\omega_1]^{\omega_1} \to 2$:

The players choose integers as usual:

$$\begin{array}{llllll} I\ : & x(0) & & x(2) & & \ldots \\ II: & & x(1) & & x(3) & \ldots \end{array}$$

Here, I is regarded as playing an integer $x(0)$ and a real $(x_I)^+$, and II as playing an integer $x(1)$ and a real $(x_{II})^+$. There are two cases in the payoff scheme:

(i) There is an α such that either $x_I \notin A_\alpha$ or $x_{II} \notin A_\alpha$. Then for the least such α, II wins if $x_I \notin A_\alpha$, and otherwise I wins.

This case leads to the crucial bounding argument. Loosely speaking, a player wins if his real satisfies S and generates structures with well-founded parts "more closed" under his Skolem term.

Suppose that $z \in A_\alpha$ for every $\alpha < \omega_1$. Then since $\alpha <_{\text{is}} \mathcal{M}(z^+, \alpha + \omega)$ for such α, all the $\mathcal{M}(z^+, \beta)$'s are well-founded as they are embeddable into each other, and so there is an a such that $z^+ = a^\#$. Moreover, as $\mathcal{M}(z^+, \alpha + \omega) \prec \langle L[a], \in, a\rangle$, $\alpha \leq t_{z(0)}^{\langle L[a], \in, a\rangle}(\alpha) < \omega_1$ for every $\alpha < \omega_1$. These remarks lead to the other case:

(ii) $(x_I)^+ = a^\#$ and $(x_{II})^+ = b^\#$ for some $a, b \in {}^\omega\omega$. Then defining $p_x \colon \omega_1 \to \omega_1$ by

$$p_x(\xi) = \sup(\{t_{x(0)}^{\langle L[a], \in, a\rangle}(\omega{\cdot}\xi + i) \mid i \in \omega\} \cup \{t_{x(1)}^{\langle L[b], \in, b\rangle}(\omega{\cdot}\xi + i) \mid i \in \omega\}) ,$$

I wins if $f(\text{ran}(p_x)) = 0$ and II wins if $f(\text{ran}(p_x)) = 1$.

It will be argued that a winning strategy for I provides a requisite homogeneous set with f value 0 (and the argument will show that a winning strategy for II provides a requisite homogeneous set with f value 1).

Suppose then that σ is a winning strategy for I. As anticipated, a $\mathbf{\Sigma}_1^1$ bounding argument is first applied to define a closed unbounded subset of ω_1 (using the notations E_r, $\|r\|$, and WF as e.g. in the first proof of 28.2). Define for $\eta < \omega_1$:

$$\begin{aligned} B_\eta = \{r \in {}^\omega\omega \mid \exists z \in {}^\omega\omega \exists \alpha \leq \eta(S(z^+) \ &\wedge\ \alpha <_{\text{is}} \mathcal{M}(z^+, \alpha + \omega) \\ &\wedge\ t_{z(0)}^{\mathcal{M}(z^+, \alpha+\omega)}(\alpha) = \eta <_{\text{is}} \mathcal{M}(z^+, \alpha + \omega) \\ &\wedge\ E_r \text{ has ordertype } t_{\sigma(\emptyset)}^{\mathcal{M}(((\sigma * z)_I)^+, \alpha+\omega)}(\alpha)\} . \end{aligned}$$

Then B_η is $\mathbf{\Sigma}_1^1$ in a real coding the ordertype η: The various assertions about ordinals can be cast in terms of reals coding them and embeddings (cf. the second proof of 13.4). Note that since σ is winning for I and the conditions on z ensure that $z \in A_\alpha$, $t_{\sigma(\emptyset)}^{\mathcal{M}(((\sigma * z)_I)^+, \alpha+\omega)}(\alpha)$ is a *bona fide* ordinal by clause (i). This implies that $B_\eta \subseteq \text{WF}$, so that as in previous arguments a function $w \colon \omega_1 \to \omega_1$ can be defined such that if $r \in B_\eta$, then $\|r\| < w(\eta)$. Set

$$C = \{\gamma < \omega_1 \mid \forall \eta < \gamma(w(\eta) < \gamma)\} .$$

Then C is closed unbounded in ω_1, and the proof is completed by showing that ${}_\omega C$ is a homogeneous set for f with value 0.

Suppose that $p \colon \omega_1 \to {}_\omega C$ is increasing; it must be shown that $f(\text{ran}(p)) = 0$. By definition of ${}_\omega C$ there is an increasing function $q \colon \omega_1 \to C$ such that for any $\xi < \omega_1$,

$$p(\xi) = \sup(\{q(\omega{\cdot}\xi + i) \mid i \in \omega\}) .$$

$q \in L[b]$ for some $b \in {}^\omega\omega$ by 28.5, and in fact the last paragraph of its proof provides a specific definition via forcing so that for some $k \in \omega$,

$$\forall \alpha < \omega_1(t_k^{\langle L[b], \in, b\rangle}(\alpha) = q(\alpha)) .$$

Set $z = \langle k \rangle^\frown b^\#$. Then since σ is winning for I, $(\sigma * z)_I = \langle \sigma(\emptyset) \rangle^\frown a^\#$ for some $a \in {}^\omega\omega$. For $p_{\sigma * z}$ as in the statement of (ii) it now suffices to show that $p = p_{\sigma * z}$, for then $f(\mathrm{ran}(p)) = f(\mathrm{ran}(p_{\sigma * z})) = 0$ as σ is winning for I.

Suppose then that $\xi < \omega_1$. For each $i \in \omega$,

$$\omega{\cdot}\xi + i \leq t_k^{\langle L[b], \in, b \rangle}(\omega{\cdot}\xi + i) = t_k^{\mathcal{M}(b^\#, \omega\cdot(\xi+1))}(\omega{\cdot}\xi + i) \ , \ \text{and}$$
$$t_{\sigma(\emptyset)}^{\langle L[a], \in, a \rangle}(\omega{\cdot}\xi + i) = t_{\sigma(\emptyset)}^{\mathcal{M}(a^\#, \omega\cdot(\xi+1))}(\omega{\cdot}\xi + i)$$

as $\mathcal{M}(b^\#, \omega{\cdot}(\xi + 1)) \prec \langle L[b], \in, b \rangle$, and similarly for a. Hence,

$$t_{\sigma(\emptyset)}^{\langle L[a], \in, a \rangle}(\omega{\cdot}\xi + i) < w(q(\omega{\cdot}\xi + i)) < p(\xi) \ ,$$

the first inequality by the choice of k and the definitions of $B_{q(\omega\cdot\xi+i)}$ and w, and the second, as $q(\omega{\cdot}\xi + i) < p(\xi)$ and $p(\xi) \in C$. Thus $p(\xi)$, being the supremum of the $t_k^{\langle L[b], \in, b \rangle}(\omega{\cdot}\xi + i)$'s, equals $p_{\sigma * z}(\xi)$, which is the supremum of the $t_{\sigma(\emptyset)}^{\langle L[a], \in, a \rangle}(\omega{\cdot}\xi + i)$'s and $t_k^{\langle L[b], \in, b \rangle}(\omega{\cdot}\xi + i)$'s. This completes the proof of the theorem. ⊣

This proof was to be generalized to higher cardinals using analogues of the $\mathbf{\Sigma}_1^1$ bounding argument and the coding of functions: $\omega_1 \to \omega_1$ by reals (30.26).

The proof admits a stronger conclusion:

28.13 Exercise. *Assume* AD. *Then* $\omega_1 \longrightarrow (\omega_1)^{\omega_1}_{2^{\aleph_0}}$, *and so* $\omega_1 \longrightarrow (\omega_1)^{\omega_1}_\alpha$ *for every* $\alpha < \omega_1$.

Hint. Suppose that $f \colon [\omega_1]^{\omega_1} \to {}^\omega 2$. For each $n \in \omega$ let $f_n \colon [\omega_1]^{\omega_1} \to 2$ be defined by $f_n(s) = f(s)(n)$, and by the proof of 28.12 let C_n be closed unbounded in ω_1 so that ${}_\omega C_n$ is homogeneous for f_n. Then ${}_\omega(\bigcap_n C_n)$ is homogeneous for f.

For the latter assertion, note that any countable ordinal is injectible into ${}^\omega\omega$. ⊣

The function $f \colon [\omega_1]^{\omega_1} \to \omega_1$ defined by $f(s) = \bigcap s$ is a counterexample to $\omega_1 \longrightarrow (\omega_1)^{\omega_1}_{\omega_1}$, but recall that $\omega_1 \not\leq 2^{\aleph_0}$ under AD.

Martin had also shown by then, using the emerging structure theory for sets of reals (§§29, 30), that under AD + DC there are many cardinals κ satisfying $\forall \alpha < \omega_1 (\kappa \longrightarrow (\kappa)^\alpha_2)$.

In the next advance, Martin and Paris built on 28.12 by taking an ultrapower by $\mathcal{C}_{\omega_1}$ to establish that under AD + DC, $\forall \alpha < \omega_2 (\omega_2 \longrightarrow (\omega_2)^\alpha_2)$. As 28.12 did for ω_1, this result considerably strengthened the result that ω_2 is measurable, and implied through 28.10 and 28.11 that there are exactly two normal ultrafilters over ω_2, $\mathcal{C}^\omega_{\omega_2}$ and $\mathcal{C}^{\omega_1}_{\omega_2}$. On the other hand, Kunen showed that $\omega_2 \not\longrightarrow (\omega_2)^{\omega_2}_2$. (See Kechris [78] for these various results.)

Kleinberg showed in 1975 that many of these results under AD are actually consequences solely of Martin's $\omega_1 \longrightarrow (\omega_1)^{\omega_1}_2$. He emphasized the generality, and the Jónsson and Rowbottom partition properties:

28.14 Theorem (DC)(Kleinberg [77, 77a]). *Suppose that $\kappa > \omega$ and $\kappa \longrightarrow (\kappa)^{\kappa}_{2}$. Then there are cardinals $\kappa = \kappa_1 < \kappa_2 < \kappa_3 < \ldots$ such that:*

(a) κ_2 as well as κ_1 are measurable, and moreover $\kappa_2 \longrightarrow (\kappa_2)^{\alpha}_{2}$ for every $\alpha < \omega_1$.

(b) κ_n for $n \geq 3$ are singular Jónsson cardinals of cofinality κ_2 and $\sup(\{\kappa_n \mid n \in \omega\})$ is a Rowbottom cardinal.

Moreover, if there is a normal ultrafilter U over κ such that ${}^{\kappa}\kappa/U$ has ordertype κ^+, then:

(c) $\kappa_{n+1} = \kappa_n^+$ for $n \geq 1$.
(d) $\kappa_2 \longrightarrow (\kappa_2)^{\alpha}_{2}$ for every $\alpha < \kappa_2$.
(e) Every normal ultrafilter over κ_1 or κ_2 is of form $\mathcal{C}^{\lambda}_{\mu}$.

For $\kappa = \omega_1$ under AD, (c) - (e) apply since ${}^{\omega_1}\omega_1/\mathcal{C}_{\omega_1}$ has ordertype $u_2 = \omega_2$ by (∗) in the proof of 28.9. Some surface remarks are made about the web of connections described by 28.14: (b) and (c) provide a purely combinatorial explication of $\mathrm{cf}(\omega_n) = \omega_2$ for $n \geq 2$ (28.8). The arguments for (a) and (b) are implicit in the Martin-Paris result of the special case $\omega_2 \longrightarrow (\omega_2)^{\alpha}_{2}$ for every $\alpha < \omega_2$. Generalizing $\omega_2 \not\longrightarrow (\omega_2)^{\omega_2}_{2}$, Henle [79] showed that $\kappa_2 \not\longrightarrow (\kappa_2)^{\kappa_2}_{2}$ delimiting (d). As described in Kleinberg [77a] the proof of (b) was simplified by another result of Henle, one that generalizes 28.13: *If $\kappa \longrightarrow (\kappa)^{\kappa}_{2}$, then $\kappa \longrightarrow (\kappa)^{\kappa}_{2^{\aleph_0}}$.* Also, Everett Bull [78: 171] showed that the existence of a U as described implies that κ is a successor cardinal, so that (e) follows from 28.11.

The combinatorial consequences of $\kappa \longrightarrow (\kappa)^{\kappa}_{2}$ and its variants were further studied by Henle [77, 79, 84], Kleinberg [81, 82], and Henle-Kleinberg-Watro [84]. As for consequences of AD along these lines, it was known by 1977 that AD implies that there are many cardinals κ such that $\kappa \longrightarrow (\kappa)^{\kappa}_{2}$. (However, not until the work of Jackson [88] did it become known that there are such cardinals greater than ω_1 that satisfy the "Moreover, . . ." clause of 28.14.) Such results about AD and infinite exponent partition relations are discussed toward the end of §30.

On Θ

With AD seen to cast a long shadow over the transfinite, we next turn to the early analysis of its extent. In gauging the continuum with respect to the ordinals the consequence $\omega_1 \not\leq 2^{\aleph_0}$ of AD implies that the continuum is small with respect to possibilities for injections: $\xi \to {}^{\omega}\omega$. On the other hand, there is always a surjection: ${}^{\omega}\omega \to \omega_1$ (sending reals coding well-ordered sets into their ordertypes), and the length of the continuum can be considered in terms of the possibilities for surjections: ${}^{\omega}\omega \to \xi$. Moschovakis [70] drew attention to the following ordinal:

$$\Theta = \sup(\{\xi \mid \text{there is a surjection: } {}^{\omega}\omega \to \xi\}) .$$

A surjection: ${}^{\omega}\omega \to \xi$ amounts to a stratification of ${}^{\omega}\omega$ into well-ordered layers, and so corresponds to a well-ordering of some subset of $\mathcal{P}({}^{\omega}\omega)$. It follows by

Replacement that in ZF, Θ is indeed an ordinal. Moreover, it is readily seen that in ZF, Θ is a cardinal $\geq \omega_2$, and that in ZFC, $\Theta = (2^{\aleph_0})^+$. The determination of Θ in the sequence of alephs can be regarded as a version of the Continuum Problem in settings without AC.

Θ came to be seen in various ways as the upper bound on the effect of AD on the transfinite, and Θ under AD also turns out to be very large. The first results along these lines were established independently by H. Friedman and Moschovakis in 1968. The latter soon isolated his fundamental Coding Lemma, which figures crucially in many results concerning Θ and its definability analogues (see before 30.16). The proof of the lemma used what came to be known as a *Friedman game*, a game where the players must play codes for countable ordinals but unlike in a Solovay game, either player can have a winning strategy. Such a game was first used in the following striking result, and its anticipations by H. Friedman:

28.15 Theorem (Moschovakis [70]). *Assume* AD. *If there is a surjection:* ${}^{\omega}\omega \to \alpha$, *then there is a surjection:* ${}^{\omega}\omega \to \mathcal{P}(\alpha)$.

Proof. Let $f\colon {}^{\omega}\omega \to \alpha$ be a surjection. A sequence $\langle g_\xi \mid \xi \leq \alpha\rangle$ will be recursively defined so that each g_ξ is a surjection: ${}^{\omega}\omega \to \mathcal{P}(\xi)$. Let $g_0\colon {}^{\omega}\omega \to \mathcal{P}(0)$ be the only possible map. At successor stages, define $g_{\xi+1}$ directly if ξ is finite, and in terms of g_ξ via a bijection: $\xi \to \xi + 1$ defined uniformly in ξ. It remains to define g_δ for limit δ with $0 < \delta \leq \alpha$:

First, for $\xi < \delta$ and $y \subseteq \xi$, temporarily stipulate that $z \in {}^{\omega}\omega$ *codes* y *iff* $f((z)_0) = \xi$ and $g_\xi((z)_1) = y$. Now for $X \subseteq \delta$ consider a game G^X where the players choose integers:

$$\begin{array}{llllll} I: & x(0) & & x(2) & & \ldots \\ II: & & x(1) & & x(3) & \ldots \end{array}$$

I loses unless x_I codes $X \cap \xi$ for some $\xi < \delta$, in which case II loses unless x_{II} codes $X \cap \eta$ for some η with $\xi < \eta < \delta$.

Suppose first that I has a winning strategy σ. Then

$$\forall y \in {}^{\omega}\omega((\sigma * y)_I \text{ codes } X \cap \xi \text{ for some } \xi < \delta)\ ,$$

and X is the unique subset of δ with this property: If $Y \subseteq \delta$ with $Y \neq X$, let $\zeta \in Y \triangle X$ and $y \in {}^{\omega}\omega$ code $X \cap (\zeta+1)$. Then $(\sigma * y)_I$ must code $X \cap \xi$ for some $\xi \geq \zeta$, and $X \cap \xi \neq Y \cap \xi$.

Suppose next that II has a winning strategy τ. Then

$$\begin{aligned} &\forall z \in {}^{\omega}\omega(z \text{ codes } X \cap \xi \text{ for some } \xi < \delta \to \\ &\qquad (z * \tau)_{II} \text{ codes } X \cap \eta \text{ for some } \eta \text{ with } \xi < \eta < \delta) \end{aligned}$$

and X is the unique subset of δ with this property: If $Y \subseteq \delta$ with $Y \neq X$, let ξ be the least member of $Y \triangle X$ and $z \in {}^{\omega}\omega$ code $Y \cap \xi (= X \cap \xi)$. Then $(z * \tau)_{II}$ must code $X \cap \eta$ for some $\eta > \xi$, and $X \cap \eta \neq Y \cap \eta$.

It follows that for any strategy for either player, if it is winning in some G^X, then that X is unique as it can be reconstructed from the strategy. Thus, the function g: $\{\rho \mid \rho$ is a strategy for either I or $II\} \to \mathcal{P}(\delta)$ given by:

$$g(\rho) = \begin{cases} X & \text{if } \rho \text{ is a winning strategy} \\ & \text{for either } I \text{ or } II \text{ in } G^X \text{, and} \\ \emptyset & \text{otherwise.} \end{cases}$$

is well-defined and by AD, surjective. The domain of g can be reconstituted to get a surjective g_δ: ${}^\omega\omega \to \mathcal{P}(\delta)$, completing the recursive construction. ⊣

There is a simple corollary:

28.16 Exercise (H. Friedman). *Assume* AD. *Then Θ is a limit cardinal.*

Hint. Suppose that $\lambda < \Theta$, and show that $\lambda^+ < \Theta$: By the theorem there is a surjection: ${}^\omega\omega \to \mathcal{P}(\lambda)$. First compose it with a bijection: $\mathcal{P}(\lambda) \to \mathcal{P}(\lambda \times \lambda)$ induced say by the Gödel pairing bijection: $\lambda \to \lambda \times \lambda$. Then compose with that function mapping $R \in \mathcal{P}(\lambda \times \lambda)$ to its ordertype if R is a well-ordering, and to 0 otherwise. ⊣

The arguments for 28.15 and 28.16 actually show that from a surjection: ${}^\omega\omega \to \lambda$ a surjection: ${}^\omega\omega \to \lambda^+$ can be defined. A more substantial corollary follows from this observation:

28.17 Exercise (Solovay). *Assume* AD. *Then* $\Theta = \omega_\Theta$.

Hint. Suppose that $\alpha < \Theta$, and show that $\omega_\alpha < \Theta$: Using a surjection: ${}^\omega\omega \to \alpha$, recursively define a sequence $\langle g_\xi \mid \xi \leq \alpha \rangle$ such that each g_ξ is a surjection: ${}^\omega\omega \to \omega_\xi$. ⊣

Applying his Coding Lemma Moschovakis [70] extended further results of Solovay by showing that under AD *there are cardinals κ that are κ-weakly Mahlo below Θ*. However, it turned out that the *cofinality* of Θ may be small without further hypotheses. The following uses the expected argument, depending only on AC_ω:

28.18 Exercise (DC). $\mathrm{cf}(\Theta) > \omega$. ⊣

We had mentioned $L(\mathbb{R})$ as the natural inner model for AD in the ZFC context.

28.19 Exercise (Solovay). *Assume* $V = L(\mathbb{R})$. *Then Θ is regular.*

Hint. It suffices to show (*sans* AC!) that there is a sequence $\langle h_\xi \mid 0 < \xi < \Theta \rangle$ such that each h_ξ is a surjection: ${}^\omega\omega \to \xi$, for then the regularity of Θ can be established in the expected way. To get such a sequence let Φ: On $\times\ ^{\omega}\omega \to V$ be a definable surjection (cf. the proof of 11.13). Define t: $\Theta - \{0\} \to$ On by: $t(\xi)$ is the least ordinal such that for some $y \in {}^{\omega}\omega$, $\Phi(t(\xi), y)$ is a surjection: ${}^{\omega}\omega \to \xi$. Then define h_ξ: ${}^{\omega}\omega \to \xi$ for $0 < \xi < \Theta$ by:

$$h_\xi(x) = \begin{cases} \Phi(t(\xi), (x)_0)((x)_1) & \text{if } \Phi(t(\xi), (x)_0) \text{ is a map into } \xi \text{ , and} \\ 0 & \text{otherwise .} \end{cases}$$

⊣

Thus, although $\Theta^{L(\mathbb{R})} \leq \Theta$ may be a small ordinal, it follows from $\mathrm{AD}^{L(\mathbb{R})}$ that $\Theta^{L(\mathbb{R})}$ is a weakly inaccessible cardinal in $L(\mathbb{R})$. The possibilities for Θ in $L(\mathbb{R})$ were later to be considerably clarified (30.32).

Introducing sharps for *sets* of reals Solovay [78a] showed that

$$\mathrm{ZF} + \mathrm{AD}_{\mathbb{R}} + \mathrm{cf}(\Theta) > \omega \vdash \mathrm{Con}(\mathrm{ZF} + \mathrm{AD}_{\mathbb{R}}) \ ,$$

and so by Gödel's Second Incompleteness Theorem,

$$\mathrm{Con}(\mathrm{ZF} + \mathrm{AD}_{\mathbb{R}}) \text{ implies } \mathrm{Con}(\mathrm{ZF} + \mathrm{AD}_{\mathbb{R}} + \mathrm{cf}(\Theta) = \omega) \ .$$

By 28.18, Con(ZF + $\mathrm{AD}_{\mathbb{R}}$) thus implies Con(AD + ¬DC), the independence of DC from AD. Steel and Robert Van Wesep in their [82] established related results.

Strong and Supercompactness under AD

This section is concluded with combinatorial versions of strong and supercompactness derivable from AD. The equivalences in terms of elementary embeddings of V are no longer available, the proof of the basic Łoś's Theorem 5.2 depending on AC. Thus bereft, the strength of these combinatorial versions is unclear. The results of the late 1980's about the relative consistency of AD (§32) were to indicate that, at least for supercompactness, there is not much consistency strength. Nonetheless, it is interesting that AD has these other ways of affecting the transfinite.

In terms of an early characterization (4.1) of strong compactness, Kunen observed in 1970 that under AD, ω_1 is strongly compact up to the natural bound Θ. The proof uses Martin's ultrafilter over Turing degrees to project down as before, and has a further consequence.

28.20 Theorem (Kunen). *Assume* AD. *Then for any* $\lambda < \Theta$, *any* ω_1*-complete filter over* λ *can be extended to an* ω_1*-complete ultrafilter over* λ.

Proof. By 28.15 let g: ${}^{\omega}\omega \to \mathcal{P}(\lambda)$ be surjective. Suppose now that F is an ω_1-complete filter over λ. Recalling Martin's 28.4 and its terminology, define f: $\mathcal{D}_{\mathrm{T}} \to \lambda$ by:

$$f(d) = \ \text{least ordinal in} \bigcap\{g(a) \in F \mid a \in {}^{\omega}\omega \ \wedge \ [a]_{\mathrm{T}} \leq d\} \ .$$

(Since there are countably many $a \in {}^{\omega}\omega$ such that $[a]_{\mathrm{T}} \leq d$, this intersection is not empty by ω_1-completeness.) Then for any $X \in F$, with $a \in {}^{\omega}\omega$ such that $g(a) = X$, $f``\{d \in \mathcal{D}_{\mathrm{T}} \mid [a]_{\mathrm{T}} \leq d\} \subseteq X$. With this it is simple to check that

$$f_*(M_T) = \{X \subseteq \lambda \mid f^{-1}(X) \in M_T\}$$

is an ω_1-complete ultrafilter over λ extending F. ⊣

28.21 Corollary (DC)(Kunen). *Assume* AD. *Then for any* $\lambda < \Theta$,

$$\beta(\lambda) = \{U \mid U \text{ is an ultrafilter over } \lambda\}$$

is well-orderable.

Proof. By the proof of the theorem, any ultrafilter over λ, being ω_1-complete (28.1), is of form $f_*(M_T)$ for some $f\colon \mathcal{D}_T \to \lambda$. It is simple to see that for any two functions f and g with domain $\mathcal{D}_T$ if $\{d \in \mathcal{D}_T \mid f(d) = g(d)\} \in M_T$, then $f_*(M_T) = g_*(M_T)$. Hence, the map with domain the ultrapower ${}^{\mathcal{D}_T}\lambda/M_T$ sending $[f]_{M_T}$ to $f_*(M_T)$ is a well-defined surjection onto $\beta(\lambda)$. But DC ensures that ultrapowers of ordinals via ω_1-complete ultrafilters are well-ordered under the induced ordering. Hence, the surjective image $\beta(\lambda)$ is also well-orderable. ⊣

Thus, under AD, if $\lambda < \Theta$, then $\mathcal{P}(\lambda)$ is not well-orderable, but $\beta(\lambda)$ is. Under AC, for any κ, that $|\beta(\kappa)| = 2^{2^\kappa}$ is a classical result of Pospíšil [37] and Tarski [39]. Under AD it may happen that $|\beta(\kappa)| = \kappa$. The β function on cardinals less than Θ has been latterly regarded as a power set operation (see Kechris [85]). Assuming DC and a hypothesis intermediate between AD and $\mathrm{AD}_{\mathbb{R}}$, Kechris [88a] established the following result analogous to 28.15: *If there is a surjection:* ${}^\omega\omega \to \lambda$, *then there is a surjection:* ${}^\omega\omega \to \beta(\lambda)$.

Turning to supercompactness, the existence under AD of normal ultrafilters over various $\mathcal{P}_\kappa\gamma$ was established in the latter 1970's. Solovay had proved the following in one of the early uses of the stronger $\mathrm{AD}_{\mathbb{R}}$.

28.22 Theorem (Solovay [78a: §3]). *Assume* $\mathrm{AD}_{\mathbb{R}}$. *Then for any* $\gamma < \Theta$ *there is a normal ultrafilter over* $\mathcal{P}_{\omega_1}\gamma$.

Proof. Solovay actually showed that there is an ω_1-complete ultrafilter U over $\mathcal{P}_{\omega_1}{}^\omega\omega$ that has the following fineness and normality properties:

(i) $\forall a \in {}^\omega\omega(\{X \in \mathcal{P}_{\omega_1}{}^\omega\omega \mid a \in X\} \in U)$.
(ii) If $\langle A_a \mid a \in {}^\omega\omega\rangle \in {}^{{}^\omega\omega}U$, then $\{X \in \mathcal{P}_{\omega_1}{}^\omega\omega \mid X \in \bigcap_{a\in X} A_a\} \in U$.

Such a U plays a role similar to Martin's M_T in Kunen's 28.20: For $\gamma < \Theta$, let $f\colon {}^\omega\omega \to \gamma$ be surjective; then it is straightforward to show that

$$\{Y \subseteq \mathcal{P}_{\omega_1}\gamma \mid \{x \in \mathcal{P}_{\omega_1}{}^\omega\omega \mid f``x \in Y\} \in U\}$$

is a normal ultrafilter over $\mathcal{P}_{\omega_1}\gamma$.

Toward the formulation of U, for $A \subseteq \mathcal{P}_{\omega_1}{}^\omega\omega$ consider a game $G(A)$ where the players choose finite, possibly empty, subsets of ${}^\omega\omega$:

$$\begin{array}{llllll} I\ : & s_0 & & s_2 & & \dots \\ II: & & s_1 & & s_3 & \dots \end{array}$$

II wins if $\bigcup_{i\in\omega} s_i \in A$, and otherwise I wins. Coding finite subsets of reals by reals, $\mathsf{AD}_{\mathbb{R}}$ implies that $G(A)$ is determined. Now define $U \subseteq \mathcal{P}(\mathcal{P}_{\omega_1}{}^{\omega}\omega)$ by:

$$A \in U \textit{ iff } II \text{ has a winning strategy in } G(A)\ .$$

Clearly property (i) obtains, and if $A \in U$ and $A \subseteq B \subseteq \mathcal{P}_{\omega_1}{}^{\omega}\omega$, then $B \in U$. Moreover, if $A \subseteq \mathcal{P}_{\omega_1}{}^{\omega}\omega$ and $A \notin U$, then $\mathcal{P}_{\omega_1}{}^{\omega}\omega - A \in U$, for if σ is a winning strategy for I in $G(A)$, then the following is a winning strategy for II in $G(\mathcal{P}_{\omega_1}{}^{\omega}\omega - A)$: After I's initial move s_0, II moves $\sigma(\emptyset)$, and after I's next move s_2, II moves $\sigma(\langle\sigma(\emptyset), s_0 \cup s_2\rangle)$, and thereafter according to σ.

We next show that U is an ω_1-complete ultrafilter by verifying ω_1-completeness, the argument soon to be extended to establish (ii): Suppose that $\langle A_m \mid m \in \omega\rangle \in {}^{\omega}U$. Consider a game where I's initial move must be an integer, say m, II must respond with $\emptyset$, and then the players' moves and result are as in $G(A_m)$. It is simple to see that it is II who must have a winning strategy in this new game, and that from it a *sequence* $\langle\tau_m \mid m \in \omega\rangle$ can be read off where each τ_m is a winning strategy for II in $G(A_m)$. (This strategem avoids AC_{ω}.) A winning strategy for II in $G(\bigcap_m A_m)$ can now be formulated:

Recall our recursive coding of $\omega \times \omega$ with $(m, n) \in \omega$ coding the ordered pair of m and n; it can be assumed that $(m, n) < (m, n+1)$ for every $m, n \in \omega$. For II's ith move (the $(2i+1)$st move of the game), if $i = (m, n)$, II makes the strategic response according to τ_m to the partial play up until then, but with all the reals played by either player after II's $(m, n-1)$th move (or 0th move, if $n = 0$) taken together as a single move by I. Thus, II's $(m, 0)$th, $(m, 1)$th, $(m, 2)$th, ... moves ensure that the resulting set belongs to A_m.

(ii) is established by building on this argument: Suppose that $\langle A_a \mid a \in {}^{\omega}\omega\rangle \in {}^{{}^{\omega}\omega}U$. Playing an auxiliary game as before but with I's initial move a real a, we get a sequence $\langle\tau_a \mid a \in {}^{\omega}\omega\rangle$ such that each τ_a is a winning strategy for II in $G(A_a)$. A winning strategy for II in $G(\{X \in \mathcal{P}_{\omega_1}{}^{\omega}\omega \mid X \in \bigcap_{a\in X} A_a\})$ can now be formulated as follows: With the countable set of A_a's into which the union of the resulting play must get only determined by the progress of the game, II applies a dovetailing procedure to ensure that whenever a real a occurs in a move of either player, II eventually applies τ_a infinitely many times. As before, each application of τ_a is based on regarding all the reals played after the previous application taken together as a single move of I. ⊣

Whether the conclusion already follows from AD is unknown. Martin observed in 1975 that in ZF, if κ and κ^+ are both measurable, then there is a normal ultrafilter over $\mathcal{P}_{\kappa}\kappa^+$. A normal ultrafilter over κ^+ can be used to "glue together" normal ultrafilters over $\mathcal{P}_{\kappa}\gamma$ for $\kappa \leq \gamma < \kappa^+$ canonically defined from a normal ultrafilter over κ (see Di Prisco-Henle [78]). It thus follows from AD that ω_1 is "ω_2-supercompact", but also it was not difficult to simulate the game of 28.22 for small γ to draw its conclusion for such γ. In a significant advance for the general theory Harrington-Kechris [81] showed that certain games played with reals coding ordinals uniformly bounded below Θ can be simulated with games played with integers, and so are determined under AD. Through this means they simulated Solovay's game to show that under AD there is a normal ultrafilter over $\mathcal{P}_{\omega_1}\delta$ where $\delta < \Theta$ is a definability analogue of Θ, weakly Mahlo but not the least such. Becker [81] applied Martin's ultrafilter M_{T} instead of Solovay's game to get the same result, and showed in [81a] that this approach shows that under AD there is a normal ultrafilter over $\mathcal{P}_{\omega_2}\delta$. Becker [81] also showed that under AD, for several γ there is exactly one normal ultrafilter over $\mathcal{P}_{\omega_1}\gamma$, and Woodin [83] extended this result to all γ covered by the Solovay result under $\mathrm{AD}_{\mathbb{R}}$ and by the Harrington-Kechris result under AD.

These various results as well as 28.7, from which we have drawn conclusions, are very much part of the study of definability under AD, the subject of the succeeding section.

29. Prewellorderings

This and the next sections explore the structural consequences of determinacy hypotheses in descriptive set theory, a direction of investigation inspired by Blackwell [67] and first pursued by Moschovakis and Martin. A theme broached in §27 is broadly developed, that of regarding determinacy itself as a regularity property and investigating the consequences of $\mathrm{Det}(\Lambda)$ for various Λ, particularly those in the projective hierarchy. The Moschovakis text [80] provides an extensive account, and most of the work before the crowning achievements of the later 1980's appears in the proceedings of the Cabal Seminar: Kechris-Moschovakis [78], Kechris-Martin-Moschovakis [81, 83], and Kechris-Martin-Steel [88].

The conventions of §§12, 13 are maintained, and as there

$$\mathrm{ZF} + \mathrm{AC}_\omega({}^\omega\omega) \text{ serves as the ambient theory for this section .}$$

AD implies $\mathrm{AC}_\omega({}^\omega\omega)$ (27.10), and versions of this observation for definable sets of reals give the first equivalence of $(*)$ before 12.4 for definable relations. It will follow that several of the forthcoming results featuring determinacy hypotheses can be established in just ZF, but this is not traced. Uses of DC are explicitly noted however.

The focus will be on the projective sets, and PD, Projective Determinacy, will suffice for the main development of this section. In any case results are stated parsimoniously with determinacy hypotheses only as needed. To cast the discussion in some generality, a *projective class* is some $\mathbf{\Sigma}^1_n$ or $\mathbf{\Pi}^1_n$ for $1 \leq n \in \omega$, or for the effective content, $\Sigma^1_n(a)$ or $\Pi^1_n(a)$ for some $a \in {}^\omega\omega$ *restricted to the traditional relations* $\subseteq {}^k({}^\omega\omega)$, which are again referred to as $\Sigma^1_n(a)$ or $\Pi^1_n(a)$ respectively. For Γ such a class, $\neg\Gamma$ denotes its dual class, i.e.

$$\neg\Gamma = \{A \mid \exists k \in \omega(A \subseteq {}^k({}^\omega\omega) \ \wedge\ {}^k({}^\omega\omega) - A \in \Gamma)\} ,$$

and $\exists^1\Gamma$ and $\forall^1\Gamma$ those classes corresponding to the adjunction of the corresponding quantifier, i.e.

$$\exists^1\Gamma = \{A \mid \exists B \in \Gamma(A = pB)\} , \text{ and}$$
$$\forall^1\Gamma = \neg\exists^1\neg\Gamma .$$

For example, $\exists^1\mathbf{\Pi}^1_3$ is $\mathbf{\Sigma}^1_4$, and $\forall^1\Pi^1_3(a)$ is again $\Pi^1_3(a)$. It will be convenient to have some standing meta-variables for this section; as above,

$$\Gamma \text{ denotes a projective class ,}$$

and

$$\Delta_\Gamma \text{ denotes } \Gamma \cap \neg\Gamma .$$

For example, $\Delta_{\mathbf{\Sigma}^1_3}$ is $\mathbf{\Delta}^1_3$. The Moschovakis text [80] formulates results more abstractly in terms of general classes satisfying closure and other properties, and these generalizations are suggested by the formulations here.

Determinacy hypotheses for projective classes and their duals do not have to be considered separately by a simple observation:

29.1 Lemma. Det(Γ) *iff* Det($\neg\Gamma$).

Proof. For any $A \subseteq {}^{\omega}\omega$, set

$$A^{-} = \{\langle i\rangle^\frown x \mid i \in \omega \ \wedge \ x \in A\} .$$

Then it is readily seen that a winning strategy for I in $G_\omega(A^-)$ converts to a winning strategy for II in $G_\omega({}^{\omega}\omega - A)$, and that a winning strategy for II in $G_\omega(A^-)$ converts to a winning strategy for I in $G_\omega({}^{\omega}\omega - A)$. But $A^- \in \Gamma$ *iff* $A \in \Gamma$. ⊣

Toward giving the seminal Blackwell argument and beginning the appropriate structural approach, two simple properties are discussed that were widely studied in classical descriptive set theory: For any A and B, $\langle A^*, B^*\rangle$ *reduces* $\langle A, B\rangle$ *iff* $A^* \subseteq A$, $B^* \subseteq B$, $A^* \cup B^* = A \cup B$, and $A \cap B = \emptyset$.

Γ has the *reduction property* *iff* for any A, B in Γ there are A^*, B^* in Γ so that $\langle A^*, B^*\rangle$ reduces $\langle A, B\rangle$.

Γ has the *separation property* *iff* for any disjoint $A, B \subseteq {}^{k}({}^{\omega}\omega)$ with A, B in Γ there is a C in Δ_Γ so that $A \subseteq C$ and $B \cap C = \emptyset$.

Simple diagrams illustrate these properties, and the following is an exercise in propositional logic.

29.2 Exercise. *If Γ has the reduction property, then $\neg\Gamma$ has the separation property.* ⊣

An early argument using universal sets was recast for the following:

29.3 Proposition (Novikov [35: 465], Addison [58]). *If Γ has the reduction property, then Γ does not have the separation property.*

Proof. Let $U \subseteq {}^{2}({}^{\omega}\omega)$ be universal for Γ (12.7), and set

$$A = \{\langle x, y\rangle \mid \langle (x)_0, y\rangle \in U\} \quad \text{and} \quad B = \{\langle x, y\rangle \mid \langle (x)_1, y\rangle \in U\} .$$

A and B are in Γ, so let A^* and B^* in Γ be such that $\langle A^*, B^*\rangle$ reduces $\langle A, B\rangle$. Assume to the contrary that there is a C in Δ_Γ so that $A^* \subseteq C$ and $B^* \cap C = \emptyset$. Then $C \subseteq {}^{2}({}^{\omega}\omega)$ is universal for Δ_Γ:

For any $D \subseteq {}^{\omega}\omega$ with D in Δ_Γ, let $a \in {}^{\omega}\omega$ be such that for any $y \in {}^{\omega}\omega$,

$$y \in D \leftrightarrow \langle (a)_0, y\rangle \in U , \quad \text{and} \quad y \notin D \leftrightarrow \langle (a)_1, y\rangle \in U .$$

Then for such a, $\langle a, y\rangle \in A$ *iff* $\langle a, y\rangle \notin B$, so that $\langle a, y\rangle \in A^*$ *iff* $\langle a, y\rangle \notin B^*$, and consequently $y \in D \leftrightarrow \langle a, y\rangle \in C$.

However, there can be no set universal for Δ_Γ by Cantor's diagonal argument (as noted after 12.8). ⊣

29.4 Corollary. *If Γ has the reduction property, then $\neg\Gamma$ does not have the reduction property.*

Proof. By 29.2 and 29.3. ⊣

Luzin [27: 50ff] formulated and established the separation property for $\mathbf{\Sigma}^1_1$ (cf. before 13.5). Kuratowski formulated the reduction property and established the following, which subsumes Luzin's result by 29.2. Addison, as with 29.3, later noted the effectivization. In a move that brought on a conceptual shift, Blackwell provided a proof based on the determinacy of open games:

29.5 Theorem (Kuratowski [36], Addison [58]). *For any $a \in {}^\omega\omega$, $\Pi^1_1(a)$ has the reduction property.*

Proof (Blackwell [67]). Suppose that A and B are $\Pi^1_1(a)$, taking $A, B \subseteq {}^\omega\omega$ for simplicity. Let T and U be trees on $\omega \times \omega$ as provided by the basic 13.1 representation of $\mathbf{\Pi}^1_1$ sets so that for any $z \in {}^\omega\omega$,

$$A(z) \leftrightarrow T_z \text{ is well-founded , and}$$
$$B(z) \leftrightarrow U_z \text{ is well-founded .}$$

For $z \in {}^\omega\omega$, consider a game G^z played with integers:

$$\begin{array}{llllll} I: & x(0) & & x(2) & & \dots \\ II: & & x(1) & & x(3) & \dots \end{array}$$

I wins exactly when there is an $i \in \omega$ such that $x_I|i \in U_z$ and $x_{II}|i \notin T_z$, i.e. I manages to stay on the tree U_z until II falls off of T_z and confirms membership for z in A.

Set

$$A^* = \{z \in A \mid II \text{ has no winning strategy in } G^z\}\ , \text{ and}$$
$$B^* = \{z \in B \mid I \text{ has no winning strategy in } G^z\}\ .$$

If $z \in A - B$, then $z \in A^*$: Letting $y \in [U_z]$, no play where I's moves are $x(2i) = y(i)$ for $i \in \omega$ can be a win for II. Analogously, if $z \in B - A$, then $z \in B^*$. Finally, if $z \in A \cap B$, then $z \in A^* \cup B^*$ as I and II cannot both have winning strategies in G^z. This case-by-case analysis shows that $A \cup B = A^* \cup B^*$.

Next, since in each G^z any win for I is secured at a finite stage, by 27.1 (cf. the remarks afterward) G^z is determined. Hence, $A^* \cap B^* = \emptyset$.

Finally, the 13.1 representation implies that for any $z \in {}^\omega\omega$,

$$A^*(z) \leftrightarrow A(z) \wedge \forall\tau(\tau \text{ is a strategy for } II \to \dots)$$

where ... is arithmetical, so that A^* is $\Pi^1_1(a)$. B^* is similarly $\Pi^1_1(a)$, and so the proof is complete. ⊣

This is a singular example of a new proof of an old result that stimulates a regeneration of the ambient theory. Addison and Martin quickly and independently came to the idea of *assuming* Projective Determinacy (PD) to derive structural conclusions. With $\mathbf{\Sigma}^1_2$ having the reduction property also a classical result (see after 29.9) they each saw how to generalize Blackwell's proof to show that $\mathrm{Det}(\mathbf{\Delta}^1_2)$ implies that $\mathbf{\Pi}^1_3$ has the reduction property. Then Martin and Moschovakis independently propagated the reduction property through the projective classes under PD. With ZFC securing the reduction property only for $\mathbf{\Pi}^1_1$ and $\mathbf{\Sigma}^1_2$ as it turns out, and delimiting it to *at most* one of Γ or $\neg\Gamma$ (29.4), PD is a regularity hypothesis that provided the deductive power to establish reduction for *at least* one of Γ or $\neg\Gamma$ for every Γ by a natural inductive process.

The Prewellordering Property

Although the Blackwell proof can be generalized to propagate reduction through the projective hierarchy under PD, attention soon shifted to a stronger property that could be similarly propagated. That property had already been abstracted by Moschovakis in 1964 from the classical ordinal analysis of $\mathbf{\Pi}^1_1$ and $\mathbf{\Sigma}^1_2$ sets, and being intrinsic to sets rather than relating two sets it soon assumed a greater prominence. The formulation proceeds through several definitions:

A relation $\preceq$ on some A is a *prewellordering (of A) iff* $\preceq$ is reflexive on A, transitive, connected on A (i.e. $\forall x, y \in A(x \preceq y \vee y \preceq x)$), and well-founded. Thus, such a $\preceq$ satisfies all the conditions for a well-ordering of A except possibly anti-symmetry: there may be distinct $x, y \in A$ such that $x \preceq y$ and $y \preceq x$. With a $\preceq$ contextually clear, $\prec$, $\succeq$, and $\succ$ have the expected derived meanings. A *norm* on A is a function $\rho\colon A \to \mathrm{On}$. Such a ρ induces a prewellordering $\preceq_\rho$ of A defined by:

$$x \preceq_\rho y \;\; \textit{iff} \;\; \rho(x) \leq \rho(y) .$$

Conversely, for any prewellordering $\preceq$ of A there is a norm $\rho\colon A \to \mathrm{On}$ such that $\preceq_\rho \,=\, \preceq$. A canonical such norm is the rank function of $\preceq$ (cf. 0.3); it is the one norm whose range is an ordinal.

To illustrate, let U be an ω_1-complete ultrafilter over a set S, and for $f, g\colon S \to \mathrm{On}$ define

$$f \preceq_U g \;\; \textit{iff} \;\; \{i \in S \mid f(i) \leq g(i)\} \in U .$$

Assuming DC, the lack of infinite descending chains implies well-foundedness, so $\preceq_U$ is a prewellordering. The corresponding rank function is just that function taking f to its equivalence class $[f]_U$ in the (transitive collapse of the) ultrapower.

Definability is incorporated next. For $A \subseteq {}^k({}^\omega\omega)$, $\rho\colon A \to \mathrm{On}$ is a *Γ-norm iff* there are R^+ in Γ and R^- in $\neg\Gamma$ with $R^+, R^- \subseteq {}^{2k}({}^\omega\omega)$ such that for any $\mathbf{y} \in A$,

$$\mathbf{x} \in A \ \wedge \ \rho(\mathbf{x}) \leq \rho(\mathbf{y}) \quad \leftrightarrow \quad R^+(\mathbf{x}, \mathbf{y}) \quad \leftrightarrow \quad R^-(\mathbf{x}, \mathbf{y}) \ ,$$

i.e. $\{\mathbf{x} \in A \mid \rho(\mathbf{x}) \leq \rho(\mathbf{y})\}$ is in Δ_Γ "uniformly in $\mathbf{y}$". For A in Γ, this is stronger than requiring that the induced prewellordering $\preceq_\rho$ be in Γ, but weaker than requiring that $\preceq_\rho$ be in Δ_Γ. Finally,

Γ has the *prewellordering property iff* every A in Γ has a Γ-norm .

This definition serves for classes more general than the projective classes. For the latter only a representative set need be considered: *Γ has the prewellordering property iff a set $A \subseteq {}^2({}^\omega\omega)$ universal for Γ has a Γ-norm.* (For $1 \leq k \in \omega$ any subset of ${}^k({}^\omega\omega)$ can be homeomorphically identified with a subset of ${}^\omega\omega$, which can then be endowed with a Γ-norm derived from one for the universal set.)

A useful, if less motivated, characterization of Γ-norm casts a different light on Moschovakis's property:

29.6 Exercise. *Suppose that A is in Γ and ρ: $A \to$ On. Then ρ is a Γ-norm iff the following relations are in Γ:*

$$\mathbf{x} \leq^*_\rho \mathbf{y} \quad \leftrightarrow \quad \mathbf{x} \in A \ \wedge \ (\mathbf{y} \in A \to \rho(\mathbf{x}) \leq \rho(\mathbf{y})) \ .$$
$$\mathbf{x} <^*_\rho \mathbf{y} \quad \leftrightarrow \quad \mathbf{x} \in A \ \wedge \ (\mathbf{y} \in A \to \rho(\mathbf{x}) < \rho(\mathbf{y})) \ .$$

Hint. Note first that if R^+ and R^- are as in the definition of Γ-norm, then

$$\mathbf{x} \leq^*_\rho \mathbf{y} \quad \leftrightarrow \quad \mathbf{x} \in A \ \wedge \ (R^+(\mathbf{x}, \mathbf{y}) \vee \neg R^-(\mathbf{y}, \mathbf{x})) \ , \text{ and}$$
$$\mathbf{x} <^*_\rho \mathbf{y} \quad \leftrightarrow \quad \mathbf{x} \in A \ \wedge \ \neg R^-(\mathbf{y}, \mathbf{x}) \ .$$

For the converse, set

$$R^+(\mathbf{x}, \mathbf{y}) \ \leftrightarrow \ \mathbf{x} \leq^*_\rho \mathbf{y} \ , \text{ and}$$
$$R^-(\mathbf{x}, \mathbf{y}) \ \leftrightarrow \ \neg(\mathbf{y} <^*_\rho \mathbf{x}) \ .$$

⊣

For Γ-norms ρ, $\leq^*_\rho$ and $<^*_\rho$ are taken as basic. They are just what is needed for a simple reduction process:

29.7 Proposition (Kuratowski [36], Addison [58]). *If Γ has the prewellordering property, then Γ has the reduction property.*

Proof. Suppose that $A, B \subseteq {}^k({}^\omega\omega)$ are both in Γ. With the constant functions $\hat{0}$: $\omega \to \{0\}$ and $\hat{1}$: $\omega \to \{1\}$, set

$$R = (A \times \{\hat{0}\}) \cup (B \times \{\hat{1}\}) \ .$$

R is in Γ, so let ρ: $R \to$ On be a Γ-norm. The idea is to see to which set A or B membership is first secured according to the norm. Define $A^*, B^* \subseteq {}^k({}^\omega\omega)$ by

$$\mathbf{x} \in A^* \quad \leftrightarrow \quad \langle \mathbf{x}, \hat{0} \rangle \leq^*_\rho \langle \mathbf{x}, \hat{1} \rangle \text{ , and}$$
$$\mathbf{x} \in B^* \quad \leftrightarrow \quad \langle \mathbf{x}, \hat{1} \rangle <^*_\rho \langle \mathbf{x}, \hat{0} \rangle \text{ .}$$

By the exercise $\langle A^*, B^* \rangle$ reduces $\langle A, B \rangle$ and A^* and B^* are in Γ. ⊣

The following is a simple consequence of the basic 13.1 representation of $\mathbf{\Pi}^1_1$ sets, and indeed harkens back to the original ordinal analysis of $\mathbf{\Pi}^1_1$ sets in Luzin-Sierpiński [18].

29.8 Exercise. *For any $a \in {}^\omega\omega$, $\Pi^1_1(a)$ has the prewellordering property.*

Hint. Suppose that $A \subseteq {}^k({}^\omega\omega)$ is $\Pi^1_1(a)$, with T a tree on ${}^k\omega \times \omega$ such that

$$A(\mathbf{w}) \quad \leftrightarrow \quad T_\mathbf{w} \text{ is well-founded .}$$

Define $\rho\colon A \to \mathrm{On}$ by

$$\rho(\mathbf{w}) = \|T_\mathbf{w}\| \text{ ,}$$

the height of the well-founded tree $T_\mathbf{w}$. That this is a $\Pi^1_1(a)$-norm follows from the arguments for the relations $R^\leq$ and $R^<$ in the proof of 13.17: Fixing the parameter x there, take $\mathbf{y} \leq^*_\rho \mathbf{z}$ to correspond to $R^\leq(0, x, y, z)$ and $\mathbf{y} <^*_\rho \mathbf{z}$ to $\neg R^<(0, x, z, y)$. ⊣

29.7 now implies 29.5, and this is essentially the classical proof of that result.

The early descriptive set theorists did not get to a useful representation of $\mathbf{\Sigma}^1_2$ sets (cf. 13.14), but Novikov [35] did get results about them using an ordinal analysis lifted from the $\mathbf{\Pi}^1_1$ sets. Moschovakis formulated that lifting schematically:

29.9 Proposition (Novikov [35], Moschovakis). *Suppose that $\forall^1\Gamma \subseteq \Gamma$ and Γ has the prewellordering property. Then so does $\exists^1\Gamma$. Consequently, for any $a \in {}^\omega\omega$, $\Sigma^1_2(a)$ has the prewellordering property.*

Proof. Suppose that

$$A(\mathbf{x}) \leftrightarrow \exists^1 w B(\mathbf{x}, w)$$

where B is in Γ. Let ν be a Γ-norm on B, and define $\rho\colon A \to \mathrm{On}$ by:

$$\rho(\mathbf{x}) = \min(\{\nu(\mathbf{x}, w) \mid B(\mathbf{x}, w)\}) \text{ .}$$

Then in terms of 29.6,

$$\mathbf{x} \leq^*_\rho \mathbf{y} \quad \leftrightarrow \quad \exists^1 w \forall^1 z(\langle \mathbf{x}, w \rangle \leq^*_\nu \langle \mathbf{y}, z \rangle) \text{ , and}$$
$$\mathbf{x} <^*_\rho \mathbf{y} \quad \leftrightarrow \quad \exists^1 w \forall^1 z(\langle \mathbf{x}, w \rangle <^*_\nu \langle \mathbf{y}, z \rangle) \text{ ,}$$

and since $\forall^1\Gamma \subseteq \Gamma$, these relations are in $\exists^1\Gamma$. ⊣

This implies the result of Kuratowski [36] that $\mathbf{\Sigma}^1_2$ has the reduction property, and so the results of Novikov [35] that $\mathbf{\Pi}^1_2$ has the separation property but $\mathbf{\Sigma}^1_2$

does not. Note the importance of the well-foundedness of the prewellordering in the proof; well-foundedness was not needed in 29.7, but is a crucial feature in the inductive propagation of the prewellordering property.

What happens at the higher levels? L not only delimited the classical efforts but also provided the first analysis consistent with ZFC. We again start with an abstract definition for a general statement:

Assuming that $\prec$ is a (strict) well-ordering of ${}^{\omega}\omega$ in ordertype ω_1, let $\mathrm{IS}^{\prec} \subseteq {}^{2}({}^{\omega}\omega)$ be given by

$$\mathrm{IS}^{\prec}(x, y) \quad \leftrightarrow \quad \{(x)_i \mid i \in \omega\} = \{z \in {}^{\omega}\omega \mid z \prec y\} \,.$$

Note that if $\prec$ is in Γ, then it is also in $\neg\Gamma$ as $x \prec y \;\leftrightarrow\; x \neq y \;\wedge\; \neg(y \prec x)$, and so

$$\mathrm{IS}^{\prec}(x, y) \quad \leftrightarrow \quad \forall^0 i((x)_i \prec y) \;\wedge\; \forall^1 z(z \prec y \rightarrow \exists^0 i(z = (x)_i))$$

shows that $\mathrm{IS}^{\prec}$ is in $\forall^1\Gamma \cap \forall^1\neg\Gamma$. The following imposes a stronger restriction on $\mathrm{IS}^{\prec}$:

$$\prec \text{ is } \Gamma\text{-}good \textit{ iff } \mathrm{IS}^{\prec} \text{ is in } \Delta_{\Gamma} \,.$$

It is simple to see that if $\prec$ is Γ-good, then it is in Γ, and that $\prec$ is Γ-good *iff* $\prec$ is $\neg\Gamma$-good.

29.10 Proposition. *Suppose that $\neg\Gamma \subseteq \exists^1\Gamma$ and there is a $\exists^1\Gamma$-good well-ordering of ${}^{\omega}\omega$. Then $\exists^1\Gamma$ has the prewellordering property.*

Proof. Suppose that A is $\exists^1\Gamma$, where for simplicity $A \subseteq {}^{\omega}\omega$, say

$$A(x) \;\leftrightarrow\; \exists^1 w B(x, w)$$

where B is in Γ. Let $\prec$ be a Γ-good well-ordering of ${}^{\omega}\omega$, $\prec_2$ the induced lexicographic well-ordering of ${}^{2}({}^{\omega}\omega)$, and ν: ${}^{2}({}^{\omega}\omega) \rightarrow \mathrm{On}$ the corresponding rank function. For each $x \in A$, let w_x by the $\prec$-least w so that $B(x, w)$, and define ρ: $A \rightarrow \mathrm{On}$ by

$$\rho(x) = \nu(w_x, x) \,.$$

($\rho(x)$ could have been taken to be just the $\prec$-rank of w_x, but for a later purpose an injective norm is produced.) Then

$$\begin{aligned} x <^*_{\rho} y \;\leftrightarrow\; & \exists^1 w(B(x, w) \;\wedge\; \exists^1 u(\mathrm{IS}^{\prec}(u, w) \wedge \forall^0 i \neg B(y, (u)_i))) \;\vee \\ & \exists^1 w[B(x, w) \;\wedge\; B(y, w) \;\wedge \\ & \qquad \exists^1 u(\mathrm{IS}^{\prec}(u, w) \wedge \forall^0 i(\neg B(x, (u)_i) \wedge \neg B(y, (u)_i)) \;\wedge\; x \prec y] \,, \end{aligned}$$

and similarly for $\leq^*_{\rho}$. Since $\neg\Gamma \subseteq \exists^1\Gamma$, these relations are thus seen to be in $\exists^1\Gamma$, and so the result follows. ⊣

29.11 Corollary. *If ${}^{\omega}\omega \subseteq L$, then for any $a \in {}^{\omega}\omega$ and $2 \leq n \in \omega$, $\Sigma^1_n(a)$ has the prewellordering property.*

Proof. By 13.11, ${}^2({}^{\omega}\omega) \cap <_L$ is a Σ^1_2-good well-ordering of ${}^{\omega}\omega$. ⊣

It was to become clear from Silver [71b: 440] that the same conclusion can be drawn if the reals are included in an inner model $L[U]$ of measurability; by 20.17 the reals in these models coincide.

29.12 Corollary. *If ${}^{\omega}\omega \subseteq L[U]$, then for any $a \in {}^{\omega}\omega$ and $2 \leq n \in \omega$, $\Sigma^1_n(a)$ has the prewellordering property.*

Proof. The result holds in any case for $n = 2$ by 29.9, and for $n > 2$ there is by 20.20 a Σ^1_3-good well-ordering of ${}^{\omega}\omega$. ⊣

Thus, if $V = L$ or even $V = L[U]$, then the reduction property holds for $\mathbf{\Pi}^1_1$ and then for $\mathbf{\Sigma}^1_n$ for every $n \geq 2$. This was first seen for L by Addison [59]; Novikov [51] had announced the first result of this kind, that if $V = L$, then for sufficiently large n the separation property holds for $\mathbf{\Pi}^1_n$.

This persistence of reduction on the $\mathbf{\Sigma}$ side depended on the external circumstance of having a well-ordering of ${}^{\omega}\omega$ with nice definability properties. Addison [74: 9-10] speculated in 1967 on other possibilities, and his remarks "turned out to be prophetic", for soon afterwards a more natural pattern for reduction through the projective hierarchy was established under PD, and even more to the point, the argument featured a natural inductive propagation. It was for this purpose that Martin introduced his filter over Turing degrees. In another approach Moschovakis explicitly propagated the prewellordering property using strategies for specific games, establishing what he later called the *First Periodicity Theorem*:

29.13 Theorem (DC)(Martin [68]; Moschovakis – Addison-Moschovakis [68]). *Assume* $\mathrm{Det}(\Delta_{\Gamma})$. *Suppose that $\exists^1\Gamma \subseteq \Gamma$ and Γ has the prewellordering property. Then so does $\forall^1\Gamma$.*

Both the Martin and Moschovakis arguments will be given, but first the consequence of this result and 29.9 that PD inductively propagates the prewellordering property in a "zig-zag" pattern:

29.14 Corollary (DC). *Assume* PD. *Then for any $a \in {}^{\omega}\omega$, the following classes have the prewellordering (and hence the reduction) property:*

	$\Sigma^1_2(a)$		$\Sigma^1_4(a)$		...
$\Pi^1_1(a)$		$\Pi^1_3(a)$			...

⊣

In contrast to the stark situation in L and in $L[U]$, this perpetuation of the alternation beyond the classical "zig" from $\mathbf{\Pi}^1_1$ to $\mathbf{\Sigma}^1_2$ of the reduction property suggested that PD may be a hypothesis with the deductive power to settle major questions about the projective sets in ways that naturally extend the classical theory.

The following is Martin's argument with Turing degrees, cast in terms of prewellorderings. The idea of ranking functions via ultrapowers, so crucial in later arguments in descriptive set theory, first appeared here (cf. after 15.4).

First Proof of 29.13. Suppose for simplicity that $A \subseteq {}^{\omega}\omega$ and

$$A(x) \leftrightarrow \forall^1 w B(x, w)$$

where B is in Γ. Let ν be a Γ-norm on B. Recalling the terminology toward 28.4, for $x \in A$ define $s_x\colon \mathcal{D}_{\mathrm{T}} \to \mathrm{On}$ by:

$$s_x(d) = \sup(\{\nu(x, w) + 1 \mid [w]_{\mathrm{T}} \leq d\}) ,$$

and for $y \in A$ also, set

$$Z_{xy} = \{d \in \mathcal{D}_{\mathrm{T}} \mid s_x(d) \leq s_y(d)\} .$$

Note first that for $a, x, y \in {}^{\omega}\omega$,

$$s_x([a]_{\mathrm{T}}) \leq s_y([a]_{\mathrm{T}}) \leftrightarrow \forall w \leq_{\mathrm{T}} a \exists z \leq_{\mathrm{T}} a(\nu(x, w) \leq \nu(y, z)) .$$

It is well-known that the displayed quantifiers are arithmetical, about number codes for "reduction procedures" (with the codes being for formulas and finite sequences if we proceed in terms of 12.4). Consequently, it follows from ν being a Γ-norm and the definitions that

$$\{a \in {}^{\omega}\omega \mid [a]_{\mathrm{T}} \in Z_{xy}\}$$

is in Δ_Γ. Hence, with $\mathrm{Det}(\Delta_\Gamma)$ Martin's argument for 28.4 implies that for any $x, y \in A$,

$$\text{either } Z_{xy} \in M_{\mathrm{T}} \text{ or else } \mathcal{D}_{\mathrm{T}} - Z_{xy} \in M_{\mathrm{T}} .$$

Define an ordering $\preceq$ of A by:

$$x \preceq y \;\; \textit{iff} \;\; Z_{xy} \in M_{\mathrm{T}} .$$

$\preceq$ is a prewellordering, DC and the ω_1-completeness of M_{T} ensuring well-foundedness, so let $\rho\colon A \to \mathrm{On}$ be the corresponding rank function. (ρ ranks the members x of A as per the corresponding s_x's in an ultrapower of On by M_{T}.) To complete the proof, we show that ρ is a $\forall^1\Gamma$-norm:

First, for $y \in A$,

$$x \in A \wedge \rho(x) \leq \rho(y) \leftrightarrow \exists^1 a \forall^1 b(a \leq_{\mathrm{T}} b \to$$
$$\forall w \leq_{\mathrm{T}} b \exists z \leq_{\mathrm{T}} b[B(x, w) \wedge \nu(x, w) \leq \nu(y, z)]) .$$

(Note that for any $w, a \in {}^{\omega}\omega$ there is always a $b \in {}^{\omega}\omega$ such that $w \leq_{\mathrm{T}} b$ and $a \leq_{\mathrm{T}} b$, and so $x \in A$ is indeed a consequence of the right side.) The quantifiers involving $\leq_{\mathrm{T}}$ are arithmetical as observed before, and the relation $[\ldots]$ is in $\neg\Gamma$ as ν is a Γ-norm. Consequently, the overall relation is in $\exists^1\forall^1\neg\Gamma$, or equivalently in $\neg\forall^1\exists^1\Gamma$, and hence in $\neg\forall^1\Gamma$ as $\exists^1\Gamma \subseteq \Gamma$.

We also have for $y \in A$

$$\begin{aligned} x \in A \wedge \rho(x) \leq \rho(y) &\leftrightarrow x \in A \wedge \{d \in \mathcal{D}_{\mathrm{T}} \mid s_y(d) < s_x(d)\} \notin M_{\mathrm{T}} \\ &\leftrightarrow x \in A \wedge \forall^1 a \exists^1 b(a \leq_{\mathrm{T}} b \wedge \\ &\qquad \exists w \leq_{\mathrm{T}} b \forall z \leq_{\mathrm{T}} b[\nu(y,z) \leq \nu(x,w)]) \,. \end{aligned}$$

Since for $y \in A$ the relation $[\ldots]$ is in Γ as ν is a Γ-norm, the overall relation is in $\forall^1\exists^1\Gamma$ and hence in $\forall^1\Gamma$ as $\exists^1\Gamma \subseteq \Gamma$. The definition of $\forall^1\Gamma$-norm thus satisfied, the proof is complete. ⊣

Just as the classical argument for 29.9 lifts the prewellordering property from $\mathbf{\Pi}^1_n$ to $\mathbf{\Sigma}^1_{n+1}$ by the simple expedient of taking minima to define a new norm, Martin had found a way to "integrate" over suprema to lift the prewellordering property from $\mathbf{\Sigma}^1_n$ to $\mathbf{\Pi}^1_{n+1}$. Moschovakis was able to take suprema more concretely with specific strategies, with an argument that was to have an important extension (30.8):

Second Proof of 29.13. Suppose again that $A \subseteq {}^\omega\omega$ and

$$A(x) \leftrightarrow \forall^1 w B(x,w)$$

where B is in Γ.

For $x, y \in {}^\omega\omega$ consider the game $G(x,y)$ played with integers:

$$\begin{array}{llllll} I: & t(0) & & t(2) & & \ldots \\ II: & & t(1) & & t(3) & \ldots \end{array}$$

II wins if either: (a) $\neg B(y, t_{II})$, or else (b) $B(x, t_I)$ and $\nu(x, t_I) \leq \nu(y, t_{II})$. This game is determined: If $y \notin A$, then II can always win by ensuring (a). If $y \in A$, then (a) never occurs, and the definition of Γ-norm ensures that the payoff set according to (b) is in Δ_Γ, and $\mathrm{Det}(\Delta_\Gamma)$ is being assumed.

Define $\preceq$ on A by:

$$x \preceq y \leftrightarrow II \text{ has a winning strategy in } G(x,y) \,.$$

Thus, $x \preceq y$ exactly when $\sup(\{\nu(x,w) \mid w \in {}^\omega\omega\}) \leq \sup(\{\nu(y,z) \mid z \in {}^\omega\omega\})$ uniformly, in the sense that there is a winning strategy τ for II so that for any $w \in {}^\omega\omega$, $\nu(x,w) \leq \nu(y,(w * \tau)_{II})$.

That $\preceq$ is a prewellordering of A is checked clause by clause:

$\preceq$ is reflexive: For $x \in A$, II has a simple winning strategy in $G(x,x)$; just copy I's moves.

$\preceq$ is transitive: Suppose that $x, y, z \in A$, $x \preceq y$, and $y \preceq z$. Letting τ_1 be a winning strategy for II in $G(x,y)$ and τ_2 winning for II in $G(y,z)$, if a strategy τ for II could be devised so that for any $w \in {}^\omega\omega$,

$$(w * \tau)_{II} = ((w * \tau_1)_{II} * \tau_2)_{II}$$

with the consequent

$$\nu(x, w) \leq \nu(y, (w * \tau_1)_{II}) \leq \nu(z, (w * \tau)_{II}) ,$$

then τ would be winning for II in $G(x, z)$ and $x \preceq z$. But this is simple to do:

For the initial response set $\tau(\langle i \rangle) = \tau_2(\langle \tau_1(\langle i \rangle) \rangle)$. Generally, II keeps partial plays of $G(x, y)$ and $G(y, z)$ going, with each of I's moves in $G(x, z)$ viewed as a move in $G(x, y)$, considering the response according to τ_1 as I's next move in $G(y, z)$, and then responding to it according to τ_2.

$\preceq$ is connected: Suppose that $x, y \in Z$ and $\neg(x \preceq y)$. By determinacy, I then has a winning strategy σ in $G(x, y)$, and since $x, y \in A$, $\nu(x, (\sigma * w)_I) > \nu(y, w)$ for any $w \in {}^{\omega}\omega$. If a strategy τ for II could be devised in $G(y, x)$ so that for any $w \in {}^{\omega}\omega$,

$$(\sigma * w)_I = (w * \tau)_{II}$$

with the consequent

$$\nu(y, w) < \nu(x, (\sigma * w)_I) = \nu(x, (w * \tau)_{II}) ,$$

then τ would be winning for II in $G(y, x)$ and $y \preceq x$. But this is simple to do:

Set $\tau(\langle i \rangle) = \sigma(\emptyset)$, and generally

$$\tau(\langle s(0), \ldots, s(2n+2) \rangle) = \sigma(\langle s(1), s(0), s(3), s(2), \ldots, s(2n+1), s(2n) \rangle) .$$

At each turn II temporarily ignores I's previous move $s(2n+2)$, transposing the consecutive odd and even values of the partial play until then, and responds according to σ.

$\preceq$ is well-founded: With DC in hand, assume to the contrary that $x_0 \succ x_1 \succ x_2 \ldots$ are all in A, where with connectedness $x \succ y$ *iff* $\neg(x \preceq y)$. For each $i \in \omega$, let σ_i be a winning strategy for I in $G(x_i, x_{i+1})$, so that for any $w \in {}^{\omega}\omega$, $\nu(x_i, (\sigma_i * w)_I) > \nu(x_{i+1}, w)$. Simultaneously produce plays t_i for $G(x_i, x_{i+1})$ for every $i \in \omega$,where I follows σ_i in $G(x_i, x_{i+1})$, and II's nth move in $G(x_i, x_{i+1})$ (the $(2n+1)$st move of the game) is I's nth move in $G(x_{i+1}, x_{i+2})$ (the $2n$th move of that game). We then have $(t_i)_{II} = (t_{i+1})_I$ for every $i \in \omega$, arriving at the contradiction

$$\nu(x_0, (t_0)_I) > \nu(x_1, (t_0)_{II}) = \nu(x_1, (t_1)_I) > \nu(x_2, (t_1)_{II}) = \nu(x_2, (t_2)_I) > \ \ldots .$$

It remains to verify that the norm ρ corresponding to $\preceq$ is a $\forall^1\Gamma$-norm. Suppose that $y \in A$. Then if II has a winning strategy in $G(x, y)$, it follows that $x \in A$. (Otherwise, for any w such that $\neg B(x, w)$, $w * \tau$ would be a play that loses for II according to the rules.) Hence,

$$\begin{aligned} x \in A \ \wedge \ \rho(x) \leq \rho(y) \ &\leftrightarrow \ II \text{ has a winning strategy in } G(x, y) \\ &\leftrightarrow \ \exists\tau\forall^1 w(\tau \text{ is a strategy for } II \ \wedge \\ &\qquad [B(x, w) \ \wedge \ \nu(x, w) \leq \nu(y, (w * \tau)_{II})]) . \end{aligned}$$

Strategies are coded by reals and the relation [...] is in $\neg\Gamma$ as ν is a Γ-norm. Consequently, the overall relation is in $\exists^1\forall^1\neg\Gamma$, or equivalently in $\neg\forall^1\exists^1\Gamma$, and hence in $\neg\forall^1\Gamma$ as $\exists^1\Gamma \subseteq \Gamma$.

Also,

$$\begin{aligned} x \in A \;\wedge\; \rho(x) \leq \rho(y) &\leftrightarrow I \text{ has no winning strategy in } G(x, y) \\ &\leftrightarrow \forall\sigma\exists^1 z(\sigma \text{ is a strategy for } I \to \\ &\qquad [B(x, (\sigma * z)_I) \wedge \nu(x, (\sigma * z)_I) \leq \nu(y, z)]) \,. \end{aligned}$$

Since the relation [...] is in Γ as ν is a Γ-norm, the relation is in $\forall^1\exists^1\Gamma$ and hence in $\forall^1\Gamma$ as $\exists^1\Gamma \subseteq \Gamma$. The definition of $\forall^1\Gamma$-norm thus satisfied, the proof is complete. ⊣

This section is brought to a close with a discussion of the structural properties thus far considered for more general classes under full AD. These results are part of a general theory, one underpinned by a basic lemma established by William Wadge in 1968 toward his Berkeley dissertation [83]. For $A, B \subseteq {}^\omega\omega$ define

$$A \leq_W B \;\; \textit{iff} \;\; A = f^{-1}(B) \text{ for some continuous } f\colon {}^\omega\omega \to {}^\omega\omega \,,$$

i.e. A is a continuous pre-image of B. The following is *Wadge's Lemma*:

29.15 Lemma (Wadge [72]). Assume AD. *Suppose that $A, B \subseteq {}^\omega\omega$. Then either $A \leq_W B$ or else $B \leq_W {}^\omega\omega - A$.*

Proof. Consider the *Wadge game* WG(A, B) played with integer moves:

$$\begin{array}{llllll} I: & x(0) & & x(2) & & \dots \\ II: & & x(1) & & x(3) & \dots \end{array}$$

II wins exactly when $x_I \in A$ *iff* $x_{II} \in B$. If τ is a winning strategy for II, then $z \in A$ *iff* $(z * \tau)_{II} \in B$ for any $z \in {}^\omega\omega$, and as the map sending z to $(z * \tau)_{II}$ is clearly continuous, $A \leq_W B$. On the other hand, if σ is a winning strategy for I, then $(\sigma * y)_I \notin A$ *iff* $y \in B$ for any $y \in {}^\omega\omega$, and so similarly $B \leq_W {}^\omega\omega - A$. ⊣

This simple but revelatory lemma established a basic connectedness under AD between arbitrary subsets of ${}^\omega\omega$. For boldface Γ and just with a local Γ version of the lemma, note that any $B \subseteq {}^\omega\omega$ in $\Gamma - \neg\Gamma$ is complete in the sense that for any $A \subseteq {}^\omega\omega$ in Γ, $A \leq_W B$. (The alternative $B \leq_W {}^\omega\omega - A$ would imply by the simple definability of continuous functions that B is in $\neg\Gamma$.)

For any $\Lambda \subseteq \bigcup_{k\in\omega} \mathcal{P}({}^k({}^\omega\omega))$,

$$\Lambda \text{ is } \textit{continuously closed} \;\; \textit{iff} \;\; \Lambda \text{ is closed under continuous pre-images}$$

i.e. if $B \in \Lambda$ and $A \leq_W B$, then $A \in \Lambda$. Extending the terminology for projective classes to arbitrary $\Lambda \subseteq \bigcup_{k\in\omega} \mathcal{P}({}^k({}^\omega\omega))$ to get e.g. $\neg\Lambda$ with the expected meaning,

$$\Lambda \text{ is } \textit{nonselfdual} \;\; \textit{iff} \;\; \neg\Lambda \neq \Lambda \,.$$

Next, for $A, B \subseteq {}^{\omega}\omega$ set

$$A \equiv_{\mathrm{W}} B \;\; \textit{iff} \;\; A \leq_{\mathrm{W}} B \wedge B \leq_{\mathrm{W}} A \text{ , and}$$
$$[A]_{\mathrm{W}} = \{C \subseteq {}^{\omega}\omega \mid C \equiv_{\mathrm{W}} A\} \text{ .}$$

The equivalence classes $[A]_{\mathrm{W}}$ are known as the *Wadge degrees*. Wadge in his dissertation [83] latterly appearing carried out a detailed study of Wadge degrees of Borel sets assuming the corresponding determinacy, one that revealed in particular that $\leq_{\mathrm{W}}$ on these sets is well-founded. Leonard Monk found a short argument directly relating winning strategies, and applying this with AD Martin established a basic structure result in 1973:

29.16 Theorem (DC)(Martin and Monk). *Assume* AD. *Then* $\leq_{\mathrm{W}}$ *is a well-founded relation.*

Van Wesep [78a] exposed the basic role of Wadge degrees and provided a proof of this result. It follows from the hypotheses and Wadge's Lemma that if each Wadge degree is coalesced with its dual class, then $\leq_{\mathrm{W}}$ well-orders the resulting classes: Each rank of $\leq_{\mathrm{W}}$ is exactly one Wadge degree, or else the union of a Wadge degree and its dual class when they are nonselfdual. Since each of the classes $\mathbf{\Sigma}^0_\xi$, $\mathbf{\Pi}^0_\xi$, $\mathbf{\Sigma}^1_n$, and $\mathbf{\Sigma}^1_n$ is readily seen by induction to be continuously closed, their restrictions to $\mathcal{P}({}^{\omega}\omega)$ are initial segments of $\leq_{\mathrm{W}}$. Determinacy provides an ultimately fine stratification of $\mathcal{P}({}^{\omega}\omega)$ via continuous reducibility.

Against this backdrop the following results established under AD articulate the pervasiveness of the separation and reduction properties under that hypothesis: Van Wesep [78] initially observed that if Λ is continuously closed and nonselfdual, then *at most* one of Λ and $\neg\Lambda$ can have the separation property. Steel [81] then showed that for such Λ *at least* one of Λ and $\neg\Lambda$ does have the separation property. The arguments for 29.2 and 29.3 relate the reduction and separation properties, so that if there is a set universal for Λ, then *at most* one of Λ and $\neg\Lambda$ can have the reduction property. If Λ is continuously closed and nonselfdual, then it turns out that there is a set universal for Λ; however, Van Wesep [78a: 163] observed that there are such Λ so that neither Λ nor $\neg\Lambda$ has the reduction property. Improving a previous result Steel [81] then showed that if Λ is continuously closed, nonselfdual *and* closed under the taking of finite unions, then *at least* one of Λ and $\neg\Lambda$ does have the reduction property.

Further results about general classes were soon established in ZF + DC + AD: Kechris-Solovay-Steel [81] explored the extent of the prewellordering property and showed for example that *if* $\Lambda \subseteq L(\mathbb{R})$ *is nonselfdual and closed under real quantification, then either* Λ *or* $\neg\Lambda$ *has the prewellordering property.* Steel [81a] established closure properties of nonselfdual Λ based on those for Δ_Λ. Becker [88] established closure properties of various Λ under category and measure quantifiers.

As for finer work in the ZF + DC context, *Borel Wadge Determinacy* is the proposition that for every pair of Borel sets $A, B \subseteq {}^{\omega}\omega$ the Wadge game $\mathrm{WG}(A, B)$ (as in the proof of Wadge's Lemma) is determined. Assuming only this Alain Louveau [83] described the Wadge ordering of Borel sets, recasting work from Wadge's dissertation [83]. Then Louveau and Jean Saint-Raymond (see their [87, 88]) showed rather unexpectedly that Borel Wadge Determinacy is already provable in second-order arithmetic.

30. Scales, Projective Ordinals, and $L(\mathbb{R})$

This section continues the exploration began in §29 of the structural consequences of determinacy hypotheses in descriptive set theory. The references and conventions are as for §29, and as there,

$$\mathrm{ZF} + \mathrm{AC}_\omega({}^\omega\omega) \text{ serves as the ambient theory for this section .}$$

But what serves to distinguish is the increased complexity, and the dispensing of several proofs in favor of a brisk account of the main developments. At the end is a survey of recent results, notably those about the structure of $L(\mathbb{R})$.

A question left unanswered by the prewellordering theory of the previous section is: What about uniformization? Continuing in our general setting,

Γ has the *uniformization property iff*
every subset of ${}^2({}^\omega\omega)$ in Γ can be uniformized by a set in Γ.

Thus, $\mathbf{\Pi}^1_1$ has the uniformization property. Using a coded union as in the argument for 29.7, it is simple to see that *if Γ has the uniformization property, then Γ has the reduction property.*

Martin [68: 689] conjectured that under PD the uniformization property would hold with the same zig-zag pattern as for the reduction property. Analyzing the uniformization arguments of Kondô (13.17) and Martin-Solovay (15.14), Moschovakis in 1971 isolated a strengthening of the prewellordering property appropriate for propagating uniformization. Although somewhat technical, this new property soon assumed a central role, for it abstracted a substantial aspect of determinacy for various classes of sets of reals. The formulation proceeds through several definitions:

For $A \subseteq {}^k({}^\omega\omega)$ and ordinal γ, a *scale* is a sequence $\langle \rho_n \mid n \in \omega \rangle$ of norms: $A \to \gamma$ satisfying the following property: Suppose that

$$\{\mathbf{x}_i \mid i \in \omega\} \subseteq A \text{ with } \lim_{i\to\omega}\mathbf{x}_i = \mathbf{x} ,$$

and there is a $g\colon \omega \to \gamma$ such that

$$\text{for every } n \in \omega \text{ and sufficiently large } i \in \omega,\ \ \rho_n(\mathbf{x}_i) = g(n) .$$

Then

$$\mathbf{x} \in A, \text{ and for every } n \in \omega,\ \ \rho_n(\mathbf{x}) \leq g(n) .$$

Here, $\lim_{i\to\omega}$ is in the usual sense of convergence with respect to the topology of ${}^k({}^\omega\omega)$, so that topological convergence through A together with a norm convergence is to imply membership in A as well as a lower semicontinuity with respect to the norms. Phrases like *scale on A into γ* will make explicit the parameters involved.

Definability is incorporated as for the prewellordering property: First, for $n \in \omega$ let $\hat{n}$ be the constant function: $\omega \to \{n\}$. Then a scale $\langle \rho_n \mid n \in \omega \rangle$ on A

is a *Γ-scale iff* there are R^+ in Γ and R^- in $\neg\Gamma$ with $R^+,\ R^- \subseteq {}^{2k+1}({}^\omega\omega)$ so that for any $\mathbf{y} \in A$ and $n \in \omega$,

$$\mathbf{x} \in A \ \wedge\ \rho_n(\mathbf{x}) \le \rho_n(\mathbf{y}) \ \leftrightarrow\ R^+(\hat{n}, \mathbf{x}, \mathbf{y}) \ \leftrightarrow\ R^-(\hat{n}, \mathbf{x}, \mathbf{y})\ .$$

It is simple to check that the analogue of 29.6 holds:

30.1 Exercise. *Suppose that A is in Γ and $\langle \rho_n \mid n \in \omega\rangle$ is a scale on A. Then it is a Γ-scale iff the following relations are in Γ:*

$$S_{\le}(\hat{n}, \mathbf{x}, \mathbf{y}) \ \leftrightarrow\ \mathbf{x} \le^*_{\rho_n} \mathbf{y}$$
$$S_{<}(\hat{n}, \mathbf{x}, \mathbf{y}) \ \leftrightarrow\ \mathbf{x} <^*_{\rho_n} \mathbf{y} \qquad \dashv$$

Finally,

Γ has the *scale property iff* every set in Γ has a Γ-scale .

The following useful characterization serves to motivate the concept of scale. In the terminology of §13, for a tree T on γ an *honest leftmost branch* is an $f \in [T]$ such that for any $g \in [T]$, $f(i) \le g(i)$ for *every* $i \in \omega$. Such branches occurred in the proof of $\mathbf{\Pi}^1_1$ Uniformization 13.17 and were emphasized in Kechris [81]; trees T with $[T] \ne \emptyset$ always have lexicographically least, or leftmost, branches, but not necessarily honest leftmost branches. For a tree T on ${}^k\omega \times \gamma$, *T has honest leftmost branches iff* for any $\mathbf{x} \in p[T]$, $T_{\mathbf{x}}$ has an honest leftmost branch.

30.2 Proposition. *For any $A \subseteq {}^k({}^\omega\omega)$, there is a scale on A into γ iff there is a tree T on ${}^k\omega \times \gamma$ having honest leftmost branches such that $A = p[T]$.*

Proof. Taking $k = 1$ for simplicity, suppose first that $\langle \rho_n \mid n \in \omega\rangle$ is a scale on A into γ. Set

$$T = \{\langle x|i, \langle \rho_0(x), \ldots, \rho_{i-1}(x)\rangle\rangle \mid x \in A\}\ .$$

Then T is a tree on ${}^k\omega \times \gamma$, and $A \subseteq p[T]$. In fact, $A = p[T]$: Suppose that for some $x \in {}^\omega\omega$, $g \in [T_x]$, so that for some $\{x_i \mid i \in \omega\} \subseteq A$, $\langle x_i|i, \langle \rho_0(x_i), \ldots, \rho_{i-1}(x_i)\rangle\rangle = \langle x|i, g|i\rangle$ for every $i \in \omega$. Since $\lim_{i\to\omega} x_i = x$ and $\rho_n(x_i) = g(n)$ for $n < i \in \omega$, $x \in A$ by the definition of scale. It also follows that $\rho_n(x) \le g(n)$ for every $n \in \omega$, and hence $\langle \rho_n(x) \mid n \in \omega\rangle$ is an honest leftmost branch of $[T_x]$.

For the converse, suppose that $A = p[T]$ for a tree T on $\omega \times \gamma$ such that for each $x \in A$, f_x is an honest leftmost branch of T_x. For each $n \in \omega$ define a norm $\rho_n\colon A \to \gamma$ by:

$$\rho_n(x) = f_x(n)\ .$$

Suppose now that $\{x_i \mid i \in \omega\} \subseteq A$ with $\lim_{i\to\omega} x_i = x$ and $g\colon \omega \to \gamma$ is such that for every n and sufficiently large i, $\rho_n(x_i) = g(n)$. Then it is simple to see that $\langle x, g\rangle \in [T]$. Hence, $x \in A$ and for every n, $\rho_n(x) = f_x(n) \le g(n)$. $\dashv$

Thus, a Γ-scale on A provides a representation of A as $p[T]$ for some T having honest leftmost branches *and* satisfying a Γ definability condition. How that condition is related to uniformization is clarified by the next two results, implicit in the proof of $\mathbf{\Pi}^1_1$ Uniformization.

30.3 Exercise. *For any $a \in {}^\omega\omega$, $\Pi^1_1(a)$ has the scale property.*

Proof. Suppose that $A \subseteq {}^k({}^\omega\omega)$ is $\Pi^1_1(a)$, with T a tree on ${}^k\omega \times \omega$ such that

$$A(\mathbf{w}) \;\leftrightarrow\; T_{\mathbf{w}} \text{ is well-founded}\,.$$

The argument for 13.14 provides a corresponding tree $\hat{T}$ on ${}^k\omega \times \omega_1$ such that

$$A(\mathbf{w}) \;\leftrightarrow\; \exists g \in {}^\omega\omega_1(\mathbf{w}, g) \in [\hat{T}])$$

where g provides an order-preserving map: $T_{\mathbf{w}} \;\to\; \omega_1$ verifying the well-foundedness of $T_{\mathbf{w}}$. As described in the proof of 13.17, $\hat{T}$ has honest leftmost branches defined when $A(\mathbf{w})$ given by:

$$g_{\mathbf{w}}(n) = \begin{cases} \rho_{T_{\mathbf{w}}}(\mathbf{s}_n) & \text{if } \mathbf{s}_n \in T_{\mathbf{w}}\text{ , and} \\ 0 & \text{otherwise}\,. \end{cases}$$

It follows from the proof of 30.2 that $\langle \rho_n \mid n \in \omega\rangle$ defined by $\rho_n(\mathbf{w}) = g_{\mathbf{w}}(n)$ is a scale on A. Finally, that it is a $\Pi^1_1(a)$-scale follows from the arguments for the relations $R^{\leq}$ and $R^{<}$ in the proof of 13.17: In terms of 30.1, take $S_{\leq}(\hat{i}, \mathbf{y}, \mathbf{z})$ to correspond to $R^{\leq}(i, x, y, z)$ and $S_{<}(\hat{i}, \mathbf{y}, \mathbf{z})$ to $R^{<}(i, x, y, z)$. ⊣

30.4 Proposition (Moschovakis [71]). *If $\forall^1\Gamma \subseteq \Gamma$ and Γ has the scale property, then Γ has the uniformization property.*

Proof. Suppose that $A \subseteq {}^2({}^\omega\omega)$ is in Γ, with $\langle \rho_n \mid n \in \omega\rangle$ a Γ-scale on A into γ say, and corresponding relations $R^+ \in \Gamma$ and $R^- \in \neg\Gamma$. The argument for 30.2 shows how the scale leads directly to a tree T on ${}^2\omega \times \gamma$ having honest leftmost branches such that $A = p[T]$. Imposing the definability condition the uniformization can be carried out as in the proof of 13.17:

Define $A_0 \subseteq {}^2({}^\omega\omega)$ by:

$$\begin{aligned} A_0(x, y) \;\leftrightarrow\; & A(x, y) \wedge \forall^1 z \forall^0 m[(\overline{y}(m) = \overline{z}(m) \\ & \wedge \forall^0 n < m(R^-(\hat{n}, x, y, x, z) \wedge R^-(\hat{n}, x, z, x, y)) \\ & \to (y(m) < z(m) \vee (y(m) = z(m) \wedge R^+(\hat{m}, x, y, x, z)))]\,. \end{aligned}$$

This is in Γ by the definability of R^+ and R^-. Moreover, if $A(x, y)$ and $A(x, z)$, then $R^-(\hat{n}, x, y, x, z) \wedge R^-(\hat{n}, x, z, x, y))$ is equivalent to $\rho_n(x, y) = \rho_n(x, z)$, and $R^+(\hat{m}, x, y, x, z)$ to $\rho_m(x, y) \leq \rho_m(x, z)$. By how the tree T was defined, it follows that for any $x \in {}^\omega\omega$, through honest leftmost branches $A_0(x, y)$ does indeed pick out that y of the leftmost pair $\langle y, g\rangle$ such that $\langle x, y, g\rangle \in T$ if there is one, and so the proof is complete. ⊣

Previous arguments lead to the existence of scales in other situations:

30.5 Exercise. *Suppose that $\neg\Gamma \subseteq \exists^1\Gamma$ and there is a $\exists^1\Gamma$-good well-ordering of ${}^\omega\omega$. Then $\exists^1\Gamma$ has the scale property. In particular, if ${}^\omega\omega \subseteq L$, or even ${}^\omega\omega \subseteq L[U]$ for an inner model $L[U]$ of measurability, then for any $a \in {}^\omega\omega$ and $2 \leq n \in \omega$, $\Sigma^1_n(a)$ has the scale property.*

Hint. Consider the proof of 29.10. The norm ρ provided is injective, and so it is simple to see that $\langle \rho \mid n \in \omega\rangle$ is a scale. Furthermore, it is a Γ-scale by 30.1. The second assertion of the exercise follows as for 29.11 and 29.12. ⊣

Note that 13.19 establishes the uniformization properties in L more directly than through 30.4 and 30.5.

For the next result our definitions are temporarily liberalized to allow "$\Delta^1_n(a)$-scales" with the expected meaning.

30.6 Exercise (Mansfield [71], Martin-Solovay [69]). *Assume that* $\forall a \in {}^\omega\omega(a^\#$ exists). *Then for any $b \in {}^\omega\omega$, every $\Pi^1_2(b)$ set has a $\Delta^1_3(b)$-scale into u_ω.*

Hint. Argue as for 30.3, but using 15.14, particularly the parenthetical comment at the end of its proof. ⊣

The scale property propagates in a manner similar to the prewellordering property, and this reinforced the zig-zag pattern under determinacy. The following is analogous to 29.9:

30.7 Exercise (Moschovakis [71]). *Suppose that $\forall^1\Gamma \subseteq \Gamma$ and Γ has the scale property. Then so does $\exists^1\Gamma$. Hence, for any $a \in {}^\omega\omega$, $\Sigma^1_2(a)$ has the scale property.*

Hint. Suppose for simplicity that $A \subseteq {}^\omega\omega$ and

$$A(x) \leftrightarrow \exists^1 w B(x, w)$$

where B is in Γ, and let $\langle \nu_n \mid n \in \omega\rangle$ be a Γ-scale on B. Define new norms through a minimization process:

For convenience, assume by 30.4 that whenever $A(x)$, there is a unique w_x such that $B(x, w_x)$. For each $n \in \omega$, define $\preceq_n$ on ${}^\omega\omega$ by

$$\begin{aligned} x \preceq_n y \ \textit{ iff } \ &\langle \nu_0(x, w_x), w_x(0), \nu_1(x, w_x), w_x(1), \ldots, \nu_n(x, w_x), w_x(n)\rangle \\ &\leq_{\text{lex}} \langle \nu_0(y, w_y), w_y(0), \nu_1(y, w_y), w_y(1), \ldots, \nu_n(y, w_y), w_y(n)\rangle \end{aligned}$$

where $\leq_{\text{lex}}$ is the lexicographic ordering, and let ρ_n be the corresponding rank function. Then $\langle \rho_n \mid n \in \omega\rangle$ is a scale on A:

Suppose that $\{x_i \mid i \in \omega\} \subseteq A$ with $\lim_{i\to\omega} x_i = x$, and there is a $g\colon \omega \to$ On such that for every n and sufficiently large i, $\rho_n(x_i) = g(n)$. Then it is straightforward to see that $\lim_{i\to\omega} w_{x_i} = w$ for some w and that for some $h\colon \omega \to$ On read off from g, $\nu_n(x_i, w_{x_i}) = h(n)$ for every n and sufficiently large i.

Consequently, $B(x, w)$ (and so $w = w_x$), and $\nu_n(x, w) \leq h(n)$ for every n. But then, $A(x)$ and for every n, $\rho_n(x) \leq g(n)$ by definition of the $\preceq_n$'s.

That $\langle \rho_n \mid n \in \omega \rangle$ is a $\exists^1 \Gamma$-scale follows from 30.1 and the argument for 30.4. ⊣

Just as the First Periodicity Theorem 29.13 had complemented the process of taking minima for the prewellordering property, Moschovakis again took suprema to establish the *Second Periodicity Theorem*:

30.8 Theorem (DC)(Moschovakis [71]). *Assume* $\mathrm{Det}(\Delta_\Gamma)$. *Suppose that* $\exists^1 \Gamma \subseteq \Gamma$ *and* Γ *has the scale property. Then so does* $\forall^1 \Gamma$. ⊣

The proof proceeds by elaborating Moschovakis's argument for 29.13, interlacing countably many of the games $G(x, y)$ used there; see Moschovakis [80: 311ff] for details.

30.9 Corollary (DC). *Assume* PD. *Then for any* $a \in {}^\omega\omega$, *the following classes have the scale (and hence the uniformization) property:*

$$\begin{array}{ccccc} & \Sigma^1_2(a) & & \Sigma^1_4(a) & \dots \\ \Pi^1_1(a) & & \Pi^1_3(a) & & \dots \end{array}$$

⊣

Moschovakis [73] provided a third periodicity theorem. Proceeding more generally, Moschovakis [80: 325ff] gave a new explanation for the periodicity of the prewellordering property, one that encompassed both 29.9 and First Periodicity in a single norm propagation theorem: For a class Λ satisfying some weak closure conditions, *if* $\mathrm{Det}(\Lambda)$, *then the class of those* $A \subseteq {}^k({}^\omega\omega)$ *defined by*

$$A(\mathbf{w}) \leftrightarrow \{\forall^0 n_0 \exists^0 n_1 \forall^0 n_2 \exists^0 n_3 \dots\} B(\mathbf{w}, \langle n_0, n_1, n_2, n_3, \dots \rangle) ,$$

where $\{\forall^0 n_0 \exists^0 n_1 \forall^0 n_2 \exists n_3 \dots\}$ *is the game quantifier and* $B \subseteq {}^{k+1}({}^\omega\omega) \in \Lambda$, *has the prewellordering property.* The game quantifier has the expected interpretation, asserting in this instance that $\{x \mid \neg B(\mathbf{w}, x)\}$ is determined with a winning strategy for II.

Moschovakis [80: 335ff] then applied the ideas to scales to derive in ZF + DC the *Third Periodicity Theorem*, which showed that when winning strategies exist, they can be simply defined. Moschovakis [73] had stated the main consequence for the projective classes: *If* $n \in \omega$, $\mathrm{Det}(\mathbf{\Sigma}^1_{2n})$, $A \subseteq {}^\omega\omega$ *is* $\Sigma^1_{2n}(a)$, *and* I *has a winning strategy in* $G_\omega(A)$, *then* I *has a winning strategy which (construed as a real) is* $\Delta^1_{2n+1}(a)$.

From these periodicity theorems flowed results that generalized most of the structure theory for $\mathbf{\Pi}^1_1$ and $\mathbf{\Sigma}^1_2$ to all levels of the projective hierarchy, often with proofs applicable to the first levels and arguably simpler than the original ones there. The elegant Spector-Gandy Theorem was so generalized by Moschovakis [73] (indeed, Martin had conjectured the consequence of Third Periodicity stated above for this purpose); measure and category extensively analyzed by Kechris [73]; and the theory of largest countable sets developed by Kechris-Moschovakis [72] and Kechris [75].

Expanding on this last, there is a Π^1_1 set C_1 without a perfect subset such that any Π^1_1 set without a perfect subset is a subset of C_1 (see after 13.12), and a Σ^1_2 set C_2 with analogous properties (see after 14.9; in fact $C_2 = C^{\emptyset}_2 = {}^{\omega}\omega \cap L$). Under local versions of 27.9, to be bereft of perfect subsets is equivalent to being countable, and one can inquire after the existence of a largest countable Γ set, i.e. one that contains every other countable Γ set. Results in the cited papers show that under PD those Γ so that there is a largest countable Γ set have the same zig-zag pattern as for the prewellordering and scale properties. The structure of these largest countable sets is analyzed in Kechris [75], Guaspari-Harrington [76], and Martin [83].

Moschovakis and Kechris also used scales in the early 1970's to develop a theory of "partially playful universes". Moschovakis (see Kechris-Moschovakis [78a: §5]) showed initially that for $2 \leq n \in \omega$, $\mathrm{Det}(\mathbf{\Delta}^1_k)$ where k is the greatest *even* integer less than n implies in ZF + DC that there is a smallest inner model M^n of ZFC absolute for Σ^1_n relations (as described before 13.15). He verified that M^n satisfies GCH, has a Σ^1_{n+1}-good well-ordering of its reals, and because of the absoluteness, satisfies $\mathrm{Det}(\mathbf{\Delta}^1_{n-1})$ if V does. This in particular established the relative consistency of having finitely many zig-zag's for the scale property and then the property settling on the $\mathbf{\Sigma}$ side thereafter. Moschovakis and Kechris then considered for $1 \leq n \in \omega$ the models $L[C_{2n}]$ where C_{2n} is the largest countable Σ^1_{2n} set as described in the previous paragraph. These models are "higher analogs" of L since $C_2 = {}^{\omega}\omega \cap L$ so that $L[C_2] = L$. Moschovakis observed that $L[C_{2n}]$ is also absolute for Σ^1_{2n} relations and so satisfies $\mathrm{Det}(\mathbf{\Delta}^1_{2n-1})$ if V does, and Kechris showed that in fact it has a Σ^1_{2n}-good well-ordering, and so does not satisfy $\mathrm{Det}(\mathbf{\Sigma}^1_{2n-1})$ (by 27.14 and the argument for 13.10). Martin then established in 1973 a somewhat unexpected result that complemented this:

30.10 Theorem (DC)(Martin). *For $n \in \omega$,* $\mathrm{Det}(\mathbf{\Delta}^1_{2n})$ *implies* $\mathrm{Det}(\mathbf{\Sigma}^1_{2n})$.

Kechris-Solovay [85: 196ff] provided a simple proof of this, and Kechris-Woodin [83: 1784] drew a stronger, contextually optimal determinacy conclusion.

Becker [78] and Moschovakis [80: 8G] provide detailed accounts of the theory of partially playful universes, with both considering models $L[T_{2n+1}]$ naturally defined from a scale and the latter also introducing similarly defined models H_{2n+1}. Discussed is an informative result from Harrington-Kechris [81] on simulating ordinal games that showed, generalizing previous work of Kechris-Martin [78], that $\mathrm{Det}(\mathbf{\Delta}^1_{2n+2})$ implies that ${}^{\omega}\omega \cap L[T_{2n+1}] = C_{2n+2}$, the largest countable Σ^1_{2n+2} set. Moschovakis [81] provided an analogous characterization of $\mathcal{P}(\lambda) \cap H_{2n+1}$ for many λ's; for such λ Becker [80] established an analogue of the perfect set property for $X \subseteq \mathcal{P}(\lambda)$ (proved independently by Ramez Sami), verified $(2^\lambda = \lambda^+)^{H_{2n+1}}$ in some cases, and extended the work of Kechris [75]; and Becker-Moschovakis [81] studied the measurable cardinals in these models.

The T_{2n+1} above is in fact the tree corresponding to a Π^1_{2n+1}-scale (cf. 30.2) on a "complete" Π^1_{2n+1} set. Answering a long-standing question of Moschovakis, Becker-Kechris [84] showed that $L[T_{2n+1}]$ is independent of the choice of the complete set and the scale on that set, and moreover that $L[T_{2n+1}] = H_{2n+1}$. This is a central result that unified the theory of partially playful universes and revealed a strong canonicity for the concept of scale.

Recent results about the extent of scales are surveyed toward the end of this section. Before then a theory of lengths of prewellorderings is developed, one that was buttressed by the results on scales through their tree characterization 30.2.

Projective Ordinals

Moschovakis's groundbreaking work was also to lead to elegant structural analyses of the projective sets, generalizing Suslin's fundamental theorem 12.1 that a set is $\boldsymbol{\Delta}^1_1$ *iff* it is Borel, and Sierpiński's result 13.7 that every $\boldsymbol{\Sigma}^1_2$ set is a union of $\aleph_1$ Borel sets. $\aleph_1$ is an important parameter in these results, and for the higher projective classes new ordinals were to play analogous roles. In his study of prewellorderings Moschovakis [70] had introduced the following; the *length* of a well-founded relation is that ordinal which is the range of the corresponding rank function.

$$\delta_\Gamma = \sup(\{\xi \mid \xi \text{ is the length of a prewellordering } \subseteq {}^2({}^\omega\omega) \text{ in } \Delta_\Gamma\})\,.$$

In particular, for $1 \leq n \in \omega$ the *projective ordinals* are defined by

$$\delta^1_n = \delta_{\boldsymbol{\Sigma}^1_n} = \delta_{\boldsymbol{\Pi}^1_n}\,.$$

These ordinals are evidently definability analogues of Θ discussed at the end of §28. It is consistent with ZFC that they are at most ω_2, as Θ could be ω_2, but under AD things turn out to be dramatically different.

Some simple observations set the stage:

30.11 Exercise.

(a) Suppose that A is in Γ and there is a Γ-norm ρ: $A \to$ On *whose range is an ordinal. Then ρ: $A \to \delta_\Gamma$, and $A = \bigcup\{A_\xi \mid \xi < \delta_\Gamma\}$ for some sets A_ξ all in Δ_Γ.*

(b) $\mathrm{cf}(\delta_\Gamma) > \omega$, *and* $\delta^1_1 = \omega_1$.

Hint. For (b), note that the union of countably may sets in Γ is again in Γ. It remains to verify that $\delta^1_1 \leq \omega_1$:

Suppose that $\preceq\, \subseteq {}^2({}^\omega\omega)$ is a $\boldsymbol{\Sigma}^1_1$ prewellordering. Set

$$X = \{x \in {}^\omega\omega \mid \exists^1 y \forall^0 m \forall^0 n (x(\langle m, n\rangle) = 0 \leftrightarrow (y)_m \preceq (y)_n)\}\,.$$

Then X is $\mathbf{\Sigma}_1^1$ and $X \subseteq \mathrm{WF}$ in the terminology of 13.6, so that by it $X \subseteq \mathrm{WF}_\alpha$ for some $\alpha < \omega_1$. This implies that the length of $\preceq$ is less than ω_1. $\dashv$

Simple conclusions can now be drawn from Second Periodicity 30.8 for the concept of γ-Suslin from 13.13:

30.12 Exercise (DC)(Moschovakis, Kechris). *Assume $n \in \omega$ and* $\mathrm{Det}(\mathbf{\Delta}_{2n}^1)$. *Then:*

(a) Every $\mathbf{\Sigma}_{2n+2}^1$ set is δ_{2n+1}^1-Suslin.

(b) Every $\mathbf{\Sigma}_{2n+1}^1$ set is γ-Suslin for some $\gamma < \delta_{2n+1}^1$.

Hint. For (a), it suffices by 13.13(d) to show that every $\mathbf{\Pi}_{2n+1}^1$ set is δ_{2n+1}^1-Suslin, but this follows from 30.9, 30.2, and 30.11(a).

For (b), it suffices to show that every $\mathbf{\Pi}_{2n}^1$ set is γ-Suslin for some $\gamma < \delta_{2n}^1$. For such a set A, there is by 30.9 a $\mathbf{\Pi}_{2n+1}^1$-scale on A into δ_{2n+1}^1. Since A is in $\mathbf{\Delta}_{2n+1}^1$, each of the prewellorderings corresponding to norms in the scale is actually in $\mathbf{\Delta}_{2n+1}^1$, and so by 30.11(b) there must be a $\gamma < \delta_{2n+1}^1$ bounding all their lengths. $\dashv$

To generalize the classical Suslin and Sierpiński results, call by *η-union* the process of taking the union $\bigcup\{A_\xi \mid \xi < \eta\}$ of a (well-ordered) sequence of sets $\langle A_\xi \mid \xi < \eta\rangle$, and say that $A \subseteq {}^k({}^\omega\omega)$ is *γ-Borel iff* A is in the smallest collection of sets containing the open subsets of ${}^k({}^\omega\omega)$ and closed under complementation and the taking of η-unions for every $\eta < \gamma$. Thus, A is ω_1-Borel *iff* A is Borel. Moreover, for any γ not a cardinal, A is γ-Borel *iff* A is γ^+-Borel.

The following is based on classical arguments going back to Luzin-Sierpiński [18, 23]; Martin saw, particularly through (b), how to make incisive use of the generalizations.

30.13 Exercise. *Suppose that $A \subseteq {}^k({}^\omega\omega)$ is γ-Suslin. Then:*

(a) A is a γ^+-union of $(\gamma+1)$-Borel sets, and so is $(\gamma^+ + 1)$-Borel.

(b) If $\mathrm{cf}(\gamma) > \omega$, then A is γ^+-Borel.

(c) If $B \subseteq {}^k({}^\omega\omega)$ is also γ-Suslin and $A \cap B = \emptyset$, then there is a $(\gamma+1)$-Borel set C that separates A and B, i.e. $B \subseteq C$ and $A \cap C = \emptyset$.

Hint. For (a), suppose that $A = p[T]$ where T is a tree on ${}^k\omega \times \gamma$. Proceed as in the proof of 13.3, first defining B_α^s analogously for $s \in {}^{<\omega}\gamma$ and $\alpha < \gamma^+$ and showing by induction on α that they are $(\gamma+1)$-Borel. Then define C_α analogously for $\alpha < \gamma^+$, and show that they too are $(\gamma+1)$-Borel. Finally, argue as in 13.3 but in terms of complements to show that

$$A = \bigcup_{\alpha<\gamma^+}({}^k({}^\omega\omega) - C_\alpha) .$$

For (b), amend the previous argument as in the second proof of 13.7 (given after 13.14), defining $T^\xi = T \cap ({}^k({}^{<\omega}\omega) \times {}^{<\omega}\xi)$ for $\xi < \gamma$ and observing that $A = \bigcup_{\xi<\gamma} p[T^\xi]$ as $\mathrm{cf}(\gamma) > \omega$.

For (c), suppose that $A = p[T]$ and $B = p[U]$ where T and U are trees on ${}^{k}\omega \times \gamma$. Argue as for 13.4 with ${}^{k}({}^{\omega}\omega) - A$ in the role of the A there, to get an $\alpha < \gamma^{+}$ such that

$$B \subseteq \{\mathbf{w} \in {}^{k}({}^{\omega}\omega) \mid T_{\mathbf{w}} \text{ is well-founded with } \|T_{\mathbf{w}}\| \leq \alpha\} .$$

This set, $B_{\alpha}^{\emptyset}$ in a previous notation, is $(\gamma + 1)$-Borel and is disjoint from A. ⊣

The following is a consequence of (c) and 30.12(b).

30.14 Exercise (DC)(Moschovakis [71]). *Assume $n \in \omega$ and* $\mathrm{Det}(\mathbf{\Delta}^1_{2n})$. *Then every $\mathbf{\Delta}^1_{2n+1}$ set is δ^1_{2n+1}-Borel.*

Martin used his analysis of $\mathbf{\Sigma}^1_3$ sets, discussed in §15, to draw an unexpected conclusion from AD. The following goes much of the way to that result.

30.15 Proposition (Martin). *Assume $\forall a \in {}^{\omega}\omega(a^{\#}$ exists). If $\mathrm{cf}(|u_{\omega}|) > \omega$, then every $\mathbf{\Sigma}^1_3$ set is δ^1_3-Borel.*

Proof. Every $\mathbf{\Sigma}^1_3$ set is u_{ω}-Suslin (15.12), and so is ${u_{\omega}}^{+}$-Borel (30.13(b)). But every cardinal $\leq u_{\omega}$ must have cofinality $\mathrm{cf}(u_2)$ by 14.18(a), and so it is readily seen that every $\mathbf{\Sigma}^1_3$ set is in fact $(u_2 + 1)$-Borel. To complete the proof, it suffices to observe that for $1 \leq n \leq \omega$, $u_n < \delta^1_3$ (and consequently $u_{\omega} < \delta^1_3$ by 30.11(b)). To this end, note that each u_n is the length of a prewellordering in Δ^1_3:

By 14.18(b) any $\gamma < u_n$ is of form $t^{\langle L[a], \in, a\rangle}(u_1, \ldots, u_{n-1})$ for some Skolem term t and $a \in {}^{\omega}\omega$. Coding pairs $\langle t, a\rangle$ by reals, the prewellordering $\preceq$ given by

$$\langle t, a\rangle \preceq \langle \bar{t}, \bar{a}\rangle \quad \textit{iff} \quad t^{\langle L[a], \in, a\rangle}(u_1, \ldots, u_{n-1}) \leq \bar{t}^{\langle L[\bar{a}], \in, \bar{a}\rangle}(u_1, \ldots, u_{n-1})$$

is then Δ^1_3 by 14.12(b) relativized, being equivalent to $c(a, b)^{\#}$ containing (a code for) a sentence $\sigma(c_1, \ldots, c_{n-1})$ depending recursively on t and $\bar{t}$, where $c(a, b) \in {}^{\omega}\omega$ recursively codes a and b. ⊣

We have now arrived at a notable juncture. Further progress cannot be made without the Coding Lemma of Moschovakis [70], a technical assertion for applications that necessarily does not admit a simple statement. Devised in the study of Θ before the concept of scale was isolated, the lemma incorporates definability into antecedents of H. Friedman and Solovay and asserts loosely speaking that bounded subsets of Θ and functions on ordinals less than Θ are definable. The proof requires the full exercise of AD in that winning strategies are needed for games of arbitrary complexity, and for the incorporation of definability, Kleene's Recursion Theorem. This intriguing convergence of recursion theory and a strong hypothesis in set theory was perhaps novel at first, but the lemma steadily gained in importance through various applications so that it came to be recognized as a cornerstone of the theory of determinacy. Moschovakis's first application of his lemma was to establish the following result:

30.16 Theorem (Moschovakis [70: 41]). *Assume* AD. *Then for* $1 \leq m \in \omega$ *and* $\eta < \delta^1_m$*, any* η*-union of* $\mathbf{\Sigma}^1_m$ *subsets of* ${}^k({}^\omega\omega)$ *is also a* $\mathbf{\Sigma}^1_m$ *set.*

See Moschovakis [80: 434ff] for a proof. Incidentally, in ZFC Martin's Axiom MA + $\neg$CH + $\exists a \in {}^\omega\omega(\omega_1^{L[a]} = \omega_1)$ also implies the conclusion for $m = 2$ (Martin-Solovay [70: §3]). Before the emergence of scales Martin had applied 30.16 to establish the converse to 30.14 from AD; the proof is given assuming 30.16, partly to highlight an early use of Wadge's Lemma 29.15:

30.17 Theorem (DC)(Martin). *Assume* $n \in \omega$ *and* AD. *Then every* δ^1_{2n+1}*-Borel set is* $\mathbf{\Delta}^1_{2n+1}$.

Proof. By the definition of δ^1_{2n+1}-Borel set, if the class of $\mathbf{\Delta}^1_{2n+1}$ subsets of ${}^k({}^\omega\omega)$ is closed under the taking of η-unions for every $\eta < \delta^1_{2n+1}$, then every δ^1_{2n+1}-Borel set is $\mathbf{\Delta}^1_{2n+1}$. It thus suffices to assume that the class is not so closed and derive a contradiction.

Suppose then that $\eta < \delta^1_{2n+1}$ is least such that there is an η-union of $\mathbf{\Delta}^1_{2n+1}$ subsets of ${}^k({}^\omega\omega)$ which is not $\mathbf{\Delta}^1_{2n+1}$, and let $A = \bigcup_{\xi<\eta} A_\xi$ be such an η-union. By 30.16, A is $\mathbf{\Sigma}^1_{2n+1}$. We first show that A has a $\mathbf{\Sigma}^1_{2n+1}$-norm:

As $\eta < \delta^1_{2n+1}$, it is simple to see that there is a $\mathbf{\Delta}^1_{2n+1}$ prewellordering of ${}^\omega\omega$ with length η, say with rank function ρ. For each $\xi < \eta$, $R_\xi \subseteq {}^2({}^\omega\omega)$ defined by

$$R_\xi(x, z) \leftrightarrow x \in A_{\rho(z)} \wedge \rho(z) = \xi$$

is seen to be $\mathbf{\Delta}^1_{2n+1}$, so again by 30.16, $R \subseteq {}^2({}^\omega\omega)$ defined by

$$R(x, z) \leftrightarrow x \in A_{\rho(z)}$$

is $\mathbf{\Sigma}^1_{2n+1}$, being the η-union of the R_ξ's.

By the minimality of η, for any $\xi < \eta$ the sets $\bigcup_{\zeta<\xi} A_\zeta$ and $\bigcup_{\zeta\leq\xi} A_\zeta$ are $\mathbf{\Delta}^1_{2n+1}$, and so by a similar argument $S, T \subseteq {}^2({}^\omega\omega)$ defined by

$$S(x, z) \leftrightarrow x \notin \textstyle\bigcup_{\zeta<\rho(z)} A_\zeta\ , \text{ and}$$
$$T(x, z) \leftrightarrow x \notin \textstyle\bigcup_{\zeta\leq\rho(z)} A_\zeta\ .$$

are both $\mathbf{\Sigma}^1_{2n+1}$.

Finally, let ν be that norm on A defined by:

$$\nu(x) = \text{least } \xi \text{ such that } x \in A_\xi\ .$$

Then

$$x \leq^*_\nu y \leftrightarrow \exists^1 z(R(x, z) \wedge S(x, z))$$
$$x <^*_\nu y \leftrightarrow \exists^1 z(R(x, z) \wedge T(x, z))$$

both are evidently $\mathbf{\Sigma}^1_{2n+1}$, and so ν is a $\mathbf{\Sigma}^1_{2n+1}$-norm.

The proof is completed with Wadge's Lemma 29.15. For any $\boldsymbol{\Sigma}^1_{2n+1}$ set B of reals, either $A \leq_{\mathrm{W}} {}^{\omega}\omega - B$ or $B \leq_{\mathrm{W}} A$. But the first alternative readily implies that A is $\boldsymbol{\Pi}^1_{2n+1}$ contrary to assumption. On the other hand, the second alternative readily implies that B has a $\boldsymbol{\Sigma}^1_{2n+1}$ norm induced by ν. Applying homeomorphisms to encompass subsets of ${}^k({}^{\omega}\omega)$, it thus follows that $\boldsymbol{\Sigma}^1_{2n+1}$ has the prewellordering property, contradicting 29.14. $\dashv$

This leads to that unexpected result heralded by 30.15:

30.18 Theorem (Martin). *Assume* AD. *Then* $u_{\omega} = \omega_{\omega}$.

Proof. First observe that for $1 \leq n \in \omega$, $\mathrm{AC}_{\omega}({}^{\omega}\omega)$ together with the existence of a surjection: ${}^{\omega}\omega \to \mathcal{P}(\omega_n)$ (28.15, 28.16) readily imply that $\mathrm{cf}(\omega_n) > \omega$. Hence, the least uncountable cardinal of cofinality ω is ω_{ω}.

If DC were assumed, we could now argue as follows: Since sharps exist under AD and there are $\boldsymbol{\Sigma}^1_3$ sets that are not $\boldsymbol{\Delta}^1_3$, 30.15 and 30.17 imply that $\mathrm{cf}(|u_{\omega}|) = \omega$. But then, the previous observation together with $\omega_1 = u_1 < u_{\omega} \leq \omega_{\omega}$ imply that $u_{\omega} = \omega_{\omega}$.

Finally, the use of DC can be eliminated by analyzing the appeal to 29.14 in the argument for 30.17: First, the prewellordering property for $\boldsymbol{\Sigma}^1_{2n+1}$ is derived. Since the prewellordering property implies the reduction property and the latter cannot hold for both $\boldsymbol{\Sigma}^1_{2n+1}$ and $\boldsymbol{\Pi}^1_{2n+1}$, a contradiction results from $\boldsymbol{\Pi}^1_{2n+1}$ having the prewellordering property. It is only this last appeal to 29.14 that depends on DC. However, in the 29.13 derivation of the prewellordering property for $\boldsymbol{\Pi}^1_3$ from $\mathrm{Det}(\boldsymbol{\Delta}^1_2)$, DC is only used to get the well-foundedness of the prewellordering relation, the classical prewellordering property for $\boldsymbol{\Sigma}^1_2$ not being dependent on DC. The other properties short of well-foundedness suffice for the 29.7 derivation of the reduction property for $\boldsymbol{\Pi}^1_3$ from the prewellordering property, and hence the contradiction. $\dashv$

This is the result from which was derived the remarkable consequence of AD that $\mathrm{cf}(\omega_n) = \omega_2$ for $2 \leq n \in \omega$ (28.8), and indeed that $u_n = \omega_n$ for $1 \leq n \in \omega$ (28.9).

The following are two characterizations depending on 30.16 that generalize classical results. 30.14 and 30.17 provide a generalization of Suslin's result 12.1 to the odd levels of the projective hierarchy:

30.19 Theorem (DC)(Martin, Moschovakis). *Assume* AD. *Then for* $n \in \omega$ *and* $A \subseteq {}^k({}^{\omega}\omega)$,

$$A \text{ is } \boldsymbol{\Delta}^1_{2n+1} \text{ iff } A \text{ is } \delta^1_{2n+1}\text{-Borel}\,.$$

Moschovakis established the following extension of Sierpiński's 13.7 in the wake of scales:

30.20 Theorem (DC)(Moschovakis [71]). *Assume* AD. *Then for* $n \in \omega$ *and* $A \subseteq {}^k({}^{\omega}\omega)$,

$$A \text{ is } \mathbf{\Sigma}^1_{2n+2} \text{ iff } A \text{ is a } \delta^1_{2n+1}\text{-union of } \mathbf{\Delta}^1_{2n+1} \text{ subsets of } {}^k({}^\omega\omega)\ .$$

In particular,

$$A \text{ is } \mathbf{\Sigma}^1_2 \text{ iff } A \text{ is a union of } \aleph_1 \text{ Borel subsets of } {}^k({}^\omega\omega)\ . \qquad \dashv$$

See Moschovakis [80: 435] for a proof. As for the extent of closure of $\mathbf{\Sigma}^1_{2n+2}$ instead of an analysis in terms of simpler sets, Kechris [78a] used this theory to observe in ZF + DC + AD that *for arbitrary η any η-union of $\mathbf{\Sigma}^1_{2n+2}$ subsets of ${}^k({}^\omega\omega)$ is $\mathbf{\Sigma}^1_{2n+2}$.* However, this may be vacuous for large η, with the following question unresolved:

30.21 Question (DC). *Assume* AD. *Is there a sequence of distinct $\mathbf{\Sigma}^1_{2n+2}$ sets of reals of length $(\delta^1_{2n+1})^+$? In particular, is there a sequence of $\mathbf{\Sigma}^1_2$ sets of reals of length ω_2?*

This is a major open question in the area; partial results appear in Jackson-Martin [83] and Jackson [90], e.g. there no such sequences that are increasing under $\subseteq$.

With the elegant structure results 30.19 and 30.20 recasting the projective sets in terms of well-ordered unions of Borel sets, it remained to get an extrinsic understanding of the possible lengths of these unions, i.e. the projective ordinals. Recalling that

$$\delta^1_1 = \omega_1$$

is essentially a classical result (30.11(b)), the following summarizes the first results that situated the further projective ordinals in the transfinite under AD:

Moschovakis [70] had used his Coding Lemma to show under AD that the projective ordinals are *cardinals*, and that for $n \in \omega$, δ^1_{2n+1} is regular and less than δ^1_{2n+2}. Martin showed (without AD) that $\delta^1_2 \leq \omega_2$ and so concluded with AD that

$$\delta^1_2 = \omega_2\ .$$

Using his analysis of $\mathbf{\Sigma}^1_3$ sets, Martin showed (from just the existence of sharps) that $\delta^1_3 \leq \omega_{\omega+1}$, and so concluded with AD that

$$\delta^1_3 = \omega_{\omega+1}\ ,$$

the third regular uncountable cardinal under AD. Martin's results followed from a basic result on the lengths of well-founded relations established by him and Kunen independently:

30.22 Theorem (DC)(Kunen; Martin). *Suppose that $\preceq\ \subseteq {}^2({}^\omega\omega)$ is a well-founded relation which is κ-Suslin. Then the length of $\preceq$ is less than κ^+.*

The following is then a simple consequence of 30.12(a):

30.23 Corollary (DC). *Assume* $n \in \omega$ *and* $\mathrm{Det}(\mathbf{\Delta}^1_{2n})$. *Then* $\delta^1_{2n+2} \leq (\delta^1_{2n+1})^+$.

It follows under AD that

$$\delta^1_4 = \omega_{\omega+2} .$$

Some results generalizing properties of these first projective ordinals were soon established. Kechris [74] showed that for $n \in \omega$, δ^1_{2n+1} is the cardinal successor of a cardinal of cofinality ω and that $\delta^1_{2n+2} < \delta^1_{2n+3}$. Generalizing Solovay's early measurability results, Martin showed that for each $n \in \omega$, δ^1_{2n+1} is measurable, and then Kunen showed that δ^1_{2n+2} is also measurable. The following summarizes what was known by the early 1970's; for proofs see Kechris [78].

30.24 Theorem (DC). *Assume* AD. *Then* $\langle \delta^1_n \mid 1 \leq n \in \omega \rangle$ *is a strictly increasing sequence of measurable cardinals such that for odd* n, δ^1_n *is the successor of a cardinal of cofinality* ω *and* $\delta^1_{n+1} = (\delta^1_n)^+$. *Moreover,* $\delta^1_1 = \omega_1$, $\delta^1_2 = \omega_2$, $\delta^1_3 = \omega_{\omega+1}$, *and* $\delta^1_4 = \omega_{\omega+2}$ *are the first four regular uncountable cardinals.*

But which cardinal is δ^1_5 under AD? This became the focal question, one that was to remain unanswered for well over a decade. It was felt that once δ^1_5 was determined, the techniques developed might suffice to determine all the projective ordinals. Moreover, the question was related to a refined, structural approach to the Continuum Problem: Stepping back into ZFC, we have the classical $\delta^1_1 = \omega_1$, and Martin's $\delta^1_2 \leq \omega_2$. Under the existence of sharps, Martin's AC result 15.13 together with 30.22 imply that

$$\delta^1_3 \leq \omega_3 .$$

(Some assumption is necessary here, for Harrington [77] established the consistency of δ^1_3 being arbitrarily large in ZFC.) Mindful of how the consequences of AD were increasingly viewed as consequences of $\mathrm{AD}^{L(\mathbb{R})}$ in $L(\mathbb{R})$, one more deduction can be made: Note first that the notions having to do with projective sets like the δ^1_n's are readily seen to be absolute for $L(\mathbb{R})$. Assuming $\mathrm{AD}^{L(\mathbb{R})}$ and comparing the fourth regular uncountable cardinals in $L(\mathbb{R})$ and in V,

$$\delta^1_4 = \omega^{L(\mathbb{R})}_{\omega+2} \leq \omega_4 .$$

Here was the new approach to the Continuum Problem. With the projective ordinals being definability analogues of Θ construed as the length of the continuum, it may be that $\mathrm{AD}^{L(\mathbb{R})}$ imposes similar upper bounds on all the projective ordinals. A version of a 1970 conjecture of Martin's is that the hypothesis implies that for every $1 \leq n \in \omega$, $\delta^1_n \leq \omega_n$. But even $\delta^1_n < \omega_\omega$ could not be established.

How these issues came to be resolved in the 1980's is discussed at the end of this section. Although the ground was to lie fallow after 1971 for well over a decade, the remarkable progress made by Kunen in that year should be mentioned. Beyond what was chronicled above, he proceeded to "dualize" the tree representation of $\mathbf{\Sigma}^1_3$ sets to get one for the $\mathbf{\Pi}^1_3$ sets (as suggested at the end of §15), and forging ahead, similarly comprehended the $\mathbf{\Pi}^1_4$ sets (see Kechris [81] and Solovay

[78]). This work stood as the high point of the contextually optimal analysis of projective sets until further progress could be made by the determination of δ^1_5.

The Extent of Scales

Having developed the theory of scales and projective ordinals available in the early 1970's this section is brought to a close with a survey of the further work, starting with the investigation of the extent of scales beyond the projective classes. As Steel [83: 107] put it,

> Scales are important in Descriptive Set Theory because they provide the only known general method which will take arbitrary definitions in a given logical form of sets of reals, and produce definitions of members of those sets.

Using a process known as *inductive definability* Moschovakis [78] formulated the class IND of *inductive* sets as a natural extension of the projective sets. For a succinct characterization, $A \subseteq {}^k({}^\omega\omega)$ is *inductive iff* there is an open set $B \subseteq {}^{k+1}({}^\omega\omega)$ such that

$$A(\mathbf{w}) \leftrightarrow \{\forall^1 x_0 \exists^1 x_1 \forall^1 x_2 \exists^1 x_3 \ldots\} B(\mathbf{w}, \langle x_0, x_1, x_2, x_3, \ldots\rangle) ,$$

where the *real game quantifier* $\{\forall^1 x_0 \exists^1 x_1 \forall^1 x_2 \exists^1 x_3 \ldots\}$ has the expected interpretation and $\langle x_0, x_1, x_2, x_3, \ldots\rangle$ is that real x such that $(x)_i = x_i$ for every $i \in \omega$. Applying (a general form of) his Second Periodicity Theorem 30.8 Moschovakis showed under Det(Δ_{IND}) that the inductive sets, both in boldface and lightface forms, have the scale property. (See Moschovakis [80: 7C].) After a lull of several years, the extent of definable scales was considerably clarified as outlined in Martin-Moschovakis-Steel [82]:

Moschovakis [83] realized that with sufficient determinacy members of the dual class of *coinductive* sets have scales definable not much beyond inductivity in complexity. Martin [83] noted however that this complexity cannot be reduced in general. Building on Moschovakis's construction Martin-Steel [83] then demarcated the extent of scales in $L(\mathbb{R})$:

With *third-order arithmetic* $\mathcal{A}^3$ the expected expansion of $\mathcal{A}^2$ incorporating $\mathcal{P}({}^\omega\omega)$ with new quantifiers $\forall^2$ and $\exists^2$ ranging over that domain, $A \subseteq {}^k({}^\omega\omega)$ is Σ^2_1 *iff*

$$A(\mathbf{w}) \leftrightarrow \mathcal{A}^3 \models \exists^2 X \varphi[\mathbf{w}] ,$$

where φ is a formula with only real and number quantifiers. The dual class Π^2_1 is defined analogously with $\forall^2$ in place of $\exists^2$, and the incorporation of parameters $a \in {}^\omega\omega$ leads to the boldface versions $\mathbf{\Sigma}^2_1$ and $\mathbf{\Pi}^2_1$. Correlating with the inductive sets Martin-Steel [83] first showed that A is $(\mathbf{\Sigma}^2_1)^{L(\mathbb{R})}$ (i.e. $\exists^2$ is restricted to $\mathcal{P}({}^\omega\omega) \cap L(\mathbb{R})$) *iff* there is an $\mathbf{\Pi}^1_1$ set $B \subseteq {}^{k+1}({}^\omega\omega)$ such that

$$A(\mathbf{w}) \leftrightarrow \{\forall^1 x_0 \exists^1 x_1 \forall^1 x_2 \exists^1 x_3 \ldots\} B(\mathbf{w}, \langle x_0, x_1, x_2, x_3, \ldots\rangle) .$$

They then established: *Assume* AD *and* $V = L(\mathbb{R})$. *Then* $\mathbf{\Sigma}^2_1$ *has the scale property.* This is indeed a demarcation, by a 1976 observation of Kechris and Solovay based on an early counterexample:

30.25 Exercise (Kechris and Solovay). *Suppose that* $V = L(\mathbb{R})$, *and there is no well-ordering of the reals. Then there is a* Π^2_1 *subset of* ${}^2({}^{\omega}\omega)$ *that cannot be uniformized.*

Hint. Verify that the set A of 27.16 is a counterexample by checking that $\ulcorner y \in \mathrm{OD}(x)\urcorner$ is Σ^2_1: First use a Löwenheim-Skolem argument to show that definability using parameters from $\mathrm{On}\cup\{x\}$ in $L(\mathbb{R})$ is equivalent to such definability in $L_\gamma(\mathbb{R})$ for some $\gamma < \Theta$. (This recalls the proof of the Reflection Principle for ZF in its role for formalizing the concept of ordinal definability. Remembering that $\mathrm{cf}(\Theta) > \omega$ in $L(\mathbb{R})$, an argument in ω stages minimizing according to a definable surjection: $\mathrm{On} \times {}^{\omega}\omega \to L(\mathbb{R})$ is helpful to get $\gamma < \Theta$.) Then note that any $L_\gamma(\mathbb{R})$ for $\gamma < \Theta$ can be coded by a set of reals. ⊣

In further work Martin [83a] showed that assuming corresponding versions of $\mathrm{AD}_{\mathbb{R}}$, if a reasonably closed class Λ has the scale property, then so does the class consisting of sets defined via the real game quantifier preceding Λ relations, and moreover via analogous quantifiers corresponding to games of any countable length. A high plateau was reached by Steel [83] who developed a "fine structure" theory for $L(\mathbb{R})$ and considerably refined the previous results. Analyzing the minimal complexity of scales in $L(\mathbb{R})$, Steel essentially extended the structure theory for the projective sets provided by determinacy to all sets of reals in $L(\mathbb{R})$.

In the next advance upward, Woodin proved in the mid-1980's that $\mathrm{AD}_{\mathbb{R}}$ *implies that every set of reals has a scale.* This implies by the previous work of Martin that there are nonselfdual classes arbitrarily high in the Wadge ordering $\leq_W$ with the scale property and reasonable closure properties. Woodin also established, relative to large cardinals, the consistency of the existence of an inner model containing every real and satisfying $\mathrm{AD}_{\mathbb{R}}$ – but here we are treading into matters taken up in §32.

The Structure of $L(\mathbb{R})$

With $L(\mathbb{R})$ seen from the beginning as the natural inner model for determinacy, not only were the consequences of AD increasingly regarded as what holds in $L(\mathbb{R})$ assuming $\mathrm{AD}^{L(\mathbb{R})}$, but the structure of the inner model came to be studied for its own sake and as an avenue to relative consistency results. In contrast to the investigation of L which can be regarded as the extended analysis of a minimal principle, this study of $L(\mathbb{R})$ sought clarification of a maximal principle in sharp focus. The sophisticated Steel [83] analysis of scales in $L(\mathbb{R})$ was the high point of the internal analysis, and was to serve as a basis for the wide-ranging assertions about AD in $L(\mathbb{R})$ that resolved various issues raised in the 1970's. Kechris [85] provides a summarizing account. As already with the extent of scales, the short discussion which follows can convey little of the depth and complexity of the results being surveyed.

As described in §28 determinacy has a strong effect on the structure of the transfinite cardinals up to the natural bound Θ. Taking up again the discussion of the infinite exponent partition relations,

$$\kappa \text{ has the } \textit{strong partition property iff } \forall\alpha < \kappa(\kappa \longrightarrow (\kappa)^{\kappa}_{\alpha}) .$$

By Martin's 28.13, AD implies that ω_1 has the strong partition property. For some time there was no other example known, the focal question of whether δ^1_3 has the strong partition property remaining open. Then in late 1977 Kechris bypassed the projective ordinals, and taking features of Martin's proof and using properties of the class IND of inductive sets, established that *under* AD + DC *there is a cardinal* κ *with the strong partition property such that* $\{\lambda < \kappa \mid \lambda$ has the strong partition property $\}$ *is stationary in* κ. The argument was soon extended to establish the following:

30.26 Theorem (DC)(Kechris-Kleinberg-Moschovakis-Woodin [81]). *Assume* AD. *Then* Θ *is a limit of cardinals having the strong partition property.*

These cardinals with the strong partition property are analogous to ω_1 in its role as δ^1_1 in the proof of Martin's 28.12, and a uniform version of the Moschovakis Coding Lemma was used to code functions: $\kappa \to \kappa$ by reals, replacing Martin's coding via sharps. The possibility of a converse in $L(\mathbb{R})$ was raised, and this was established by Kechris and Woodin in 1982:

30.27 Theorem (DC)(Kechris-Woodin [83]). *Assume* $V = L(\mathbb{R})$. *Then the following are equivalent:*

(a) AD.

(b) Θ *is a limit of cardinals having the strong partition property.*

This illuminating characterization reduced AD to a purely combinatorial assertion about the transfinite cardinals, and vindicated the early interest in infinite exponent partition properties. Kechris-Kleinberg-Moschovakis-Woodin [81] had already shown that *if a set* A *of reals is* λ*-Suslin (i.e.* $A = p[T]$ *for some tree* T *on* $\omega\times\lambda$*) and there is a* $\kappa > \lambda$ *with the strong partition property, then* A *is determined.* Loosely speaking, the tree representation was used as in the §15 analysis of $\mathbf{\Sigma}^1_3$ sets, and the strong partition property applied to get an invariance for ordering functions (cf. the end of §15). With this, they had achieved a weak version of 30.27. What Kechris-Woodin [83] established is that *in* $L(\mathbb{R})$, *if every set of reals* λ*-Suslin for some* λ *is determined, then* AD. Proceeding level by level through the fine structure hierarchy of $L(\mathbb{R})$, the argument used the Steel [83] analysis of scales, in their role of witnessing the Suslin property, and applied a technique of Martin [83] for handling quantifier alternations over $\mathbb{R}$ using his filter over Turing degrees. Woodin later confirmed that (b) cannot imply (a) in ZF + DC alone, by showing that (b) is consistent with the existence of an ultrafilter over ω.

Another characterization established around the same time addressed the involvement of Turing degrees in determinacy. Consider *Turing Determinacy,*

$$\mathrm{Det}(\{\bigcup X \mid X \subseteq \mathcal{D}_{\mathrm{T}}\}) ,$$

i.e. the hypothesis that every set of reals closed under Turing equivalence $\equiv_{\mathrm{T}}$ is determined. Turing Determinacy can replace full AD in Martin's result 28.4 that his filter over Turing degrees is ultra and at various junctures (cf. the proofs of First Periodicity 29.13), and the question of its strength was raised early on. In the early 1980's Woodin pushed the techniques that had become available to establish the following:

30.28 Theorem (DC)(Woodin). *Assume* $V = L(\mathbb{R})$. *Then the following are equivalent:*

(a) AD.

(b) Turing Determinacy.

A decade later Woodin was to establish an equivalence in $L(\mathbb{R})$ for AD merely in terms of some of its structural consequences for sets of reals (32.22).

A more basic issue had to do with DC. With much of the consequences of AD appealing to DC, how substantive a role does DC play was a question often raised during the 1970's. Does Con(ZF + AD) imply Con(ZF + AD + DC)? As mentioned after 28.19, Solovay [78a] established that Con(ZF + $\mathrm{AD}_{\mathbb{R}}$) implies Con(ZF + $\mathrm{AD}_{\mathbb{R}}$ + $\neg$DC), and hence the independence of DC from AD relative to $\mathrm{AD}_{\mathbb{R}}$. It was Kechris who eventually settled the consistency question:

30.29 Theorem (Kechris [84]). *Assume* AD *and* $V = L(\mathbb{R})$. *Then* DC.

Thus the role of $L(\mathbb{R})$ as an inner model for ZF + AD became more fully analogous to the role of L for ZF. Just as ZF + $V = L$ procures AC, ZF + AD + $V = L(\mathbb{R})$ procures DC, and so indeed Con(ZF + AD) implies Con(ZF + AD + DC) by absoluteness. The proof also used the Steel [83] analysis of scales, in their role of securing uniformization as a means to establish DC.

Woodin was able to replace $\mathrm{AD}_{\mathbb{R}}$ by AD in Solovay's DC independence result:

30.30 Theorem (Woodin – Kechris [84: §3]). *Assume* AD *and* $V = L(\mathbb{R})$. *Then in a forcing extension there is an inner model satisfying* AD + $\neg\mathrm{AC}_\omega$.

Thus, Con(ZF + AD) implies Con(ZF + AD + $\neg$DC) and the independence of DC relative to AD.

Woodin's counterexample to AC_ω occurs high in the cumulative hierarchy however, and toward the formulation of a refined question consider DC restricted to relations on reals:

$$(\mathrm{DC}({}^{\omega}\omega)) \qquad \forall R(R \subseteq {}^{2}({}^{\omega}\omega) \;\wedge\; \forall^1 x \exists^1 y(\langle x, y\rangle \in R) \\ \to\; \exists f \in {}^{\omega}({}^{\omega}\omega)\forall^0 n(\langle f(n), f(n+1)\rangle \in R))\,.$$

30.31 Question (Woodin). *Does* AD *imply* DC$({}^{\omega}\omega)$*?*

An independence result here would seem to require inner models of AD qualitatively different from the onces concocted thus far.

Yet another issue clarified by $L(\mathbb{R})$ is the size of Θ. Recall from §28 that under AD, there are cardinals κ that are κ-weakly Mahlo below Θ, and under AD $+ V = L(\mathbb{R})$, Θ itself is weakly inaccessible (28.18 and before, and 28.19).

30.32 Theorem (DC)(Kechris and Woodin – Kechris [85]). *Assume* AD *and* $V = L(\mathbb{R})$. *Then:*

(a) Θ *is* Θ*-weakly Mahlo.*

(b) Θ *is not weakly compact, i.e. there is a* Θ*-tree without a* Θ*-branch.*

(c) Θ *is the supremum of the measurable cardinals.*

(a) established that Θ is hierarchically large in $L(\mathbb{R})$, but (b) imposed a limitation. (c) provided a purely combinatorial description of Θ, and was a consequence of the arguments toward (a); a question related to it is stated at the end of the section.

The Determination of the Projective Ordinals

The structural analysis under AD of the projective sets was essentially completed in the mid-1980's with the calibration of the projective ordinals in the sequences of cardinals. This was a veritable *tour de force* of technical virtuosity by Steve Jackson, a student of Martin. More in the fullness of time rather than by the weight of new techniques, Jackson determined δ^1_5 in his 1983 U.C.L.A. dissertation. The starting point was some observations of Martin on normal ultrafilters over δ^1_3 leading to a putative lower bound for δ^1_5. By 1985 Jackson had carried out the determination of all the projective ordinals with the large part of the upper bound calculations provided in his formidable [88]. 30.24 summarizes what had been known before with $\delta^1_{2n+2} = (\delta^1_{2n+1})^+$ reducing the analysis to the δ^1_m's for odd m; define ordinals $E(n)$ for $n \in \omega$ by recursion as follows: $E(0) = 1$, and $E(n+1) = \omega^{E(n)}$ via ordinal exponentiation.

30.33 Theorem (DC)(Jackson [88]). *Assume* AD. *Then for* $n \in \omega$*,*

$$\delta^1_{2n+3} = \omega_{E(2n+1)+1}$$

and δ^1_{2n+3} *has the strong partition property.*

The δ^1_m's for odd m are thus the successors of cardinals indexed by a corresponding tower of ω's, starting with Martin's $\delta^1_3 = \omega_{\omega+1}$ (cf. before 30.22).

That they have the strong partition property is both an outgrowth and a crucial component of the proof, which proceeds by induction. Beyond the Martin result 28.13 that δ^1_1 has the strong partition property, even that δ^1_3 has the strong partition property had not been known. The inductive scheme can be elaborated at the basis:

As motivated through methodological considerations at the end of §15, Kunen had applied Martin's strong partition property for δ^1_1 to get ultrafilters over ω_ω ($= u_\omega$) that provide the necessary homogeneity properties for the contextually optimal tree representation of $\mathbf{\Pi}^1_3$ sets. Forging ahead, Kunen carried out a detailed analysis of these ultrafilters and their ultrapowers (see Solovay [78]). This led to the property $\delta^1_3 \longrightarrow (\delta^1_3)^\alpha_2$ for every $\alpha < \delta^1_3$, which in turn provided the ultrafilters over δ^1_3 (cf. 28.10) necessary for the optimal tree representation of $\mathbf{\Pi}^1_4$ sets.

δ^1_5 is characterizable in terms of a certain supremum of ultrapowers via those ultrafilters over δ^1_3 (see Kechris [81: 62]), and Martin had noted that only certain canonical functions need be considered in the calculation of this supremum. Applying this Jackson was able to complete the calculation by an intricate analysis that further reduced the supremum to one of ultrapowers via certain canonical ultrafilters. In the process, he was able to lift the Kunen situation to the next level: the strong partition property for δ^1_3 (not just the "weak" one mentioned above) and a detailed analysis of the ultrafilters over δ^1_3 and their ultrapowers that leads to the optimal tree representation of $\mathbf{\Pi}^1_5$ sets. With sufficient inductive hypotheses thus in place Jackson completed the calculation of all the projective ordinals, getting the optimal tree representations for all the projective sets.

The Jackson analysis confirmed that under AD the regular cardinals strictly between δ^1_{2n+1} and δ^1_{2n+3} are exactly the ultrapowers of δ^1_{2n+1} via the normal ultrafilters over δ^1_{2n+1} derived from its strong partition property. Moreover, these ultrapowers are distinct, and by 28.11 there are just as many as there are regular infinite cardinals below δ^1_{2n+1}. With this in mind, define $F\colon \omega \to \omega$ by: $F(0) = 1$ and $F(n+1) = 2F(n)+1$. Then *under* AD *there are exactly* $F(n)$ *regular infinite cardinals below* δ^1_{2n+1} *for every* $n \in \omega$:

This is clear for $n = 0$, and for $n = 1$, with the $F(1) = 3$ cardinals below $\delta^1_3 = \omega_{\omega+1}$ being ω, ω_1, and ω_2. Assuming inductively that there are $F(n)$ regular infinite cardinals below δ^1_{2n+1}, it follows that there are $F(n)$ regular cardinals strictly between δ^1_{2n+1} and δ^1_{2n+3}, and so counting the regular infinite cardinals below δ^1_{2n+1} and δ^1_{2n+1} itself, there are exactly $2F(n) + 1 = F(n+1)$ regular cardinals below δ^1_{2n+3}.

The Jackson analysis thus established in ZFC that *under* $\mathrm{AD}^{L(\mathbb{R})}$ *the projective ordinals are all less than* ω_ω. In fact, noting once again that the δ^1_m's are absolute for $L(\mathbb{R})$ and comparing regular cardinals there and in V, the list of inequalities after 30.24 can be continued assuming $\mathrm{AD}^{L(\mathbb{R})}$ with the following for every $n \in \omega$:

$$\delta^1_{2n+1} \leq \omega_{F(n)} \text{ and } \delta^1_{2n+2} \leq \omega_{F(n)+1} .$$

In particular, the projective sets are comprehended in ZFC + $\mathrm{AD}^{L(\mathbb{R})}$ through the characterizations 30.19 and 30.20 as unions of λ Borel sets with the λ's concretely inset among the ω_n's. This is a remarkable advance, mediated by determinacy, beyond the early results of Suslin and Sierpiński.

Taking stock however, the following version of an early conjecture of Martin remains unresolved:

30.34 Question (ZFC). *Does* $\mathrm{AD}^{L(\mathbb{R})}$ *imply that* $\delta^1_n \leq \omega_n$ *for every* $1 \leq n \in \omega$*?*

Jackson's analysis raised the possibility of a similarly detailed analysis of $L(\mathbb{R})$ right up to Θ, with the first excursions made by him in [91, 92]. A question thought to be along these lines was answered in 1993 by Steel and Woodin, in the wake of important advances made by Steel in inner model theory. Moschovakis and Kechris had shown that under ZF + $\mathrm{AD}^{L(\mathbb{R})}$, every regular (in V) uncountable cardinal below $\Theta^{L(\mathbb{R})}$ is measurable in $L(\mathbb{R})$. Jackson's work affirmed that under AD + DC, every regular uncountable cardinal below $\sup(\{\delta^1_n \mid n \in \omega\})$ is measurable.

30.35 Theorem (ZFC) (Steel and Woodin). *Assume* AD + $V = L(\mathbb{R})$. *Then every regular uncountable cardinal below* Θ *is measurable.*

Taken together, the various results through the early 1980's on internal questions about determinacy and the structure of $L(\mathbb{R})$ provide a cumulative impression of systematic progress using more and more sophisticated methods of descriptive set theory. The consistency of AD from the vantage point of ZFC seemed ever remote however, but surprisingly a denouement in this direction was to soon take place, and this notably through new initiatives involving supercompact cardinals as described in §32.

31. Det(α-$\mathbf{\Pi}^1_1$)

The consequences of determinacy having been explored in the previous sections, this one begins the final ascent to the consistency of determinacy hypotheses. Martin forged the initiative in this direction, and starting afresh with his ground-breaking work in 1967 his results through the 1970's are described. They provided the first solid evidence towards Solovay's general conjecture that large cardinals would provide the consistency strength to establish determinacy. The next section describes the definitive work of Martin, Steel, and Woodin in this direction.

Returning to safe haven,

ZFC serves as the ambient theory for this and the next sections .

Proofs are increasingly dispensed with, as in any case they are precluded by their range and sophistication, in favor of providing a driving account of the progression of ideas.

In 1967, Martin struck on the fertile idea of using ω_1-complete ultrafilters, viewed as two-valued measures, to "integrate" over possibilities to get structural results about the projective sets. For this purpose he had introduced his filter over Turing degrees (cf. the first proof of 29.13), and this had led to his proof of Solovay's result from AD that ω_1 is measurable (cf. 28.4). The integration idea, mediated by the sharps, filtered into the Martin-Solovay [69] work on $\mathbf{\Sigma}^1_3$ sets (§15), and was to plainly appear in the next result. That Det($\mathbf{\Pi}^1_1$) cannot be established in ZFC was clear (27.13); Martin established that it does follow from the existence of a measurable cardinal. Obliquely complementing Solovay's deduction of measurability from AD Martin's result opened the door to the possibility of getting amounts of determinacy from large cardinals, a possibility that was to be pursued with remarkable and unexpected consequences in the years to come.

31.1 Theorem (Martin [70]). *Suppose that there is a measurable cardinal. Then* Det($\mathbf{\Pi}^1_1$) *holds.*

Proof. Suppose that $A \subseteq {}^\omega\omega$ is $\mathbf{\Pi}^1_1$, so that there is a tree T on $\omega \times \omega$ such that for any $x \in {}^\omega\omega$,

$$A(x) \;\leftrightarrow\; T_x \text{ is well-founded} .$$

That A is determined will be argued in terms of well-orderings, and to this end, the development toward the beginning of §15 is adapted.

First, it can be assumed that $\langle\emptyset, \emptyset\rangle \in T$. With the fixed recursive enumeration $\langle \mathbf{s}_i \mid i \in \omega\rangle$ of ${}^{<\omega}\omega$ such that $|\mathbf{s}_i| \leq i$, for any $x \in {}^\omega\omega$ let $<_x$ be that strict linear ordering of ω defined by:

$$\begin{aligned} i <_x j \;\leftrightarrow\; & \mathbf{s}_i \notin T_x \;\wedge\; \mathbf{s}_j \notin T_x \;\wedge\; i < j \;;\; \text{else} \\ & \mathbf{s}_i \notin T_x \;\wedge\; \mathbf{s}_j \in T_x \;;\; \text{else} \\ & \mathbf{s}_i \in T_x \;\wedge\; \mathbf{s}_j \in T_x \;\wedge\; \mathbf{s}_i <_{\mathrm{KB}} \mathbf{s}_j \end{aligned}$$

where $<_{\text{KB}}$ is the Kleene-Brouwer ordering. Then for any $x \in {}^{\omega}\omega$,

$$A(x) \leftrightarrow <_x \text{ is a well-ordering} .$$

Next, for $s \in {}^{<\omega}\omega$, setting

$$T_{\langle s \rangle} = \{w \mid \exists m < |s|(\langle s|m, w\rangle \in T)\}$$

let

$$\begin{aligned} <_s = &\text{ that strict linear ordering of } |s| = \{i \mid i < |s|\} \\ &\text{ defined like } <_x \text{ but with } T_x \text{ replaced by } T_{\langle s \rangle} . \end{aligned}$$

$|\mathbf{s}_i| \leq i$ implies that for $s \subseteq t$ both in ${}^{<\omega}\omega$, $<_s \subseteq <_t$ as relations, and for $x \in {}^{\omega}\omega$,

$$<_x = \bigcup_{m\in\omega} <_{x|m} .$$

Finally, suppose that κ is a measurable cardinal, and define a tree T^* on $\omega \times \kappa$ by

$$T^* = \{\langle s, u\rangle \mid \forall i, j < |s|(u(i) < u(j) \leftrightarrow i <_s j)\} .$$

Then for any $x \in {}^{\omega}\omega$, since $<_x$ is a well-ordering exactly when there is an order-preserving injection of $\langle \omega, <_x\rangle$ into $\langle \kappa, <\rangle$,

$$A(x) \leftrightarrow \exists g \in {}^{\omega}\kappa(\langle x, g\rangle \in [T^*]) .$$

Proceeding to the main argument, to establish that $G_\omega(A)$ is determined consider an auxiliary game G^*:

$$\begin{array}{lllll} I \;: & \langle x(0), g(0)\rangle & & \langle x(2), g(1)\rangle & & \ldots \\ II : & & x(1) & & x(3) & \ldots \end{array}$$

I and II both play integers $x(i)$ to form a real x, but in addition, I plays ordinals $g(i) < \kappa$ to form a function $g\colon \omega \to \kappa$. I wins if $\langle x, g\rangle \in [T^*]$, so that g is an order-preserving injection of $\langle \omega, <_x\rangle$ into $\langle \kappa, <\rangle$ and $x \in A$, and otherwise II wins.

If I loses in G^*, then he has lost irrevocably at a finite stage: He will have made a move $\langle x(2i), g(i)\rangle$ for some $i > 0$ such that the choice $g(i)$ does not preserve the order of $<_x$, and this can already be seen in terms of the initial segment $<_{x|2i}$ (as $|\mathbf{s}_i| \leq i < 2i$). Hence, by 27.1 (cf. the comments following) G^* is determined. If it is I who has a winning strategy in this game, then he has a winning strategy in $G_\omega(A)$: he simply keeps the $g(i)$'s of his moves to himself and plays only the $x(2i)$'s. If it is II who has a winning strategy in G^*, it remains to invoke the measurability of κ to show that he has a winning strategy in $G_\omega(A)$:

We first adapt the development after 15.4: Note that T^* has the following homogeneity property: For any $s \in {}^{<\omega}\omega$, set $T^*_s = \{u \mid \langle s, u\rangle \in T^*\}$. Then since $<_s$ is a strict linear ordering, $\text{ran}(u) \in [\kappa]^{|s|}$ for any u such that $u \in T^*_s$. Moreover, for any $w \in [\kappa]^{|s|}$ there is a unique $u \in T^*_s$ such that $\text{ran}(u) = w$.

With this in mind, let U be a normal ultrafilter over κ and for $s \in {}^{<\omega}\omega$ define U_s by:

$$X \in U_s \;\; \textit{iff} \;\; X \subseteq T^*_s \wedge \exists H \in U(|H|^{|s|} \subseteq \text{ran``}X) .$$

Then for $s \neq \emptyset$, by the above described homogeneity and Rowbottom's 7.17, U_s is a κ-complete ultrafilter over T^*_s.

Suppose now that τ is a winning strategy for II in G^*. For $n \in \omega$, $t \in {}^{2n+1}\omega$, and $u \in {}^{n+1}\kappa$, let $\tau(t, u)$ denote the strategic response according to τ to

$$\langle \langle t(0), u(0) \rangle, t(1), \ldots, \langle t(2n), u(n) \rangle \rangle \; .$$

Then for such t and $k \in \omega$, set

$$Z_{t,k} = \{u \in T_{t|(n+1)} \mid \tau(t, u) = k\} \; .$$

Next, define a strategy τ_0: $\bigcup_{n \in \omega} {}^{2n+1}\omega \to \omega$ by:

$$\tau_0(t) = k \;\; \textit{iff} \;\; Z_{t,k} \in U_{t|(n+1)} \; .$$

This is well-defined by the ω_1-completeness of the U_s's. The proof is completed by showing that τ_0 is a winning strategy for II in $G_\omega(A)$:

Assume to the contrary that x is a play according to τ_0, yet $x \in A$. By the ω_1-completeness of U there is an $H_0 \in U$ such that for every $n \in \omega$,

$$[H_0]^{n+1} \subseteq Z_{x|(2n+1), \tau_0(s|(2n+1))} \; .$$

Since H_0 is uncountable, there is an order-preserving injection g: $\langle \omega, <_x \rangle \to \langle H_0, < \rangle$. But then, $\langle x, g \rangle$ is in T^* and corresponds to a play of G^* in which II plays according to τ and still loses. Contradiction! ⊣

As in the analysis of $\mathbf{\Sigma}^1_3$ sets (§15) the hypothesis can be weakened to the existence of sharps:

31.2 Theorem (Martin). *Suppose that $a \in {}^\omega\omega$ and $a^\#$ exists. Then* Det($\Pi^1_1(a)$).

Proof. The previous argument is refined. Construing the A as $\Pi^1_1(a)$, it can be assumed that the corresponding tree T is in $L[a]$ by 13.1. Now define the tree T^* from T as before but with (the real) ω_1 in place of the measurable κ. T^* and so (the payoff set of) the corresponding game G^* are both in $L[a]$, and arguing as before, the game is determined in $L[a]$.

Suppose first that in $L[a]$, I has a winning strategy σ in G^*. Then in V, σ is also a winning strategy for I, since if he moves according to σ and still loses, this would have happened at a finite stage, and so also in $L[a]$. It follows as before that I has winning strategy in $G_\omega(A)$.

Suppose on the other hand that in $L[a]$, II has a winning strategy τ in G^*. Let P_τ consist of those finite partial plays according to τ that are not yet secured for II, i.e. those that are extendible to a play that is a win for I. As I loses at a finite stage when he loses at all, it follows that in $L[a]$, $\langle P_\tau, \supset \rangle$ is well-founded. But then, by absoluteness of well-foundedness (0.3), $\langle P_\tau, \supset \rangle$ is well-founded in V. τ is consequently a winning strategy for II in the sense of V. It remains to derive a winning strategy for II in $G_\omega(A)$:

With I_a the closed unbounded class of ordinal indiscernibles for $L[a]$ given by $a^\#$, let $\gamma < \omega_1$ be sufficiently large so that in some Skolem term rendering of τ, no indiscernible in $C = I_a \cap (\omega_1 - \gamma)$ appears. Now define a strategy τ_0: $\bigcup_{n\in\omega} {}^{2n+1}\omega \to \omega$ by:

$$\tau_0(t) = k \ \ \mathit{iff} \ \ \exists u \in {}^{n+1}C(\langle t|(n+1), u\rangle \in T^* \wedge \tau(t,u) = k) ,$$

where $\tau(t,u)$ is defined as before; τ_0 is well-defined by indiscernibility. Again, τ_0 is a winning strategy for II in $G_\omega(A)$:

Assume to the contrary that x is a play according to τ_0 yet $x \in A$. Since C is uncountable, there is an order-preserving injection g: $\langle \omega, <_x\rangle \to \langle C, <\rangle$. But then, by indiscernibility $\langle x, g\rangle$ corresponds to a play of G^* in which II played according to τ yet lost. Contradiction! ⊣

Note that this result together with 27.14 provides a new proof of the version of Solovay's result 14.3 that if $\forall a \in {}^\omega\omega(a^\#$ exists), then the $\mathbf{\Sigma}^1_2$ sets have the regularity properties.

Martin viewed his result as a plausibility argument in two complementary directions. Looking upward he wrote prophetically ([70: 287]): "We believe that larger cardinals will yield a generalization of our proof to all projective sets." This accorded with Solovay's hopes for the relative consistency of $\mathrm{AD}^{L(\mathbb{R})}$, and was to be borne out by developments in the 1980's. In the other direction, he saw telling evidence for a theorem of ZFC:

Martin [70] actually drew his conclusion from the existence of a cardinal κ satisfying $\kappa \longrightarrow (\omega_1)_2^{<\omega}$. Owing to the boundedness of order-preserving maps (cf. 13.4), he observed that if there is a cardinal κ satisfying $\forall\alpha < \omega_1(\kappa \longrightarrow (\alpha)_2^{<\omega})$ then Borel Determinacy holds. He then noted that the partition property relativizes to L (9.15), and although $\mathrm{Det}(\mathbf{\Pi}^1_1)$ fails in L, Borel Determinacy also relativizes: The assertion that $G_\omega(A)$ is determined,

$$\exists\sigma\forall y(\sigma * y \in A) \ \vee \ \exists\tau\forall z(z * \tau \notin A) ,$$

is $\Sigma^1_2(a)$ when A is $\Delta^1_1(a)$, and hence absolute for L when $a \in L$. Note that Borel Determinacy is itself a Π^1_3 assertion, as "$\forall$ Borel A" can be rendered through real codes for Borel sets (cf. before 11.8) as a real quantifier $\forall^1$.

The best outright determinacy result at the time was the Davis [64] result $\mathrm{Det}(\mathbf{\Sigma}^0_3)$, and its proof could be readily carried out in second-order arithmetic. The apparent impasse in stalking of Borel Determinacy with elementary methods was soon to be explicated by a remarkable metamathematical analysis: Harvey Friedman was an early popularizer of the investigation of determinacy hypotheses (cf. 28.16). Formulating games where the players must code models of fragments of set theory, H. Friedman [71] showed in 1968 that establishing Borel Determinacy would require the existence of the cumulative hierarchy through V_{ω_1}, and so in particular cannot be carried out in Zermelo set theory, i.e. ZF − Replacement. This revealed a remarkable state of affairs, since Borel Determinacy is a Π^1_3 assertion about sets of reals yet would have to be proved using uncountably many iterations of the power set operation. Friedman's arguments lead to a level-by-level analysis showing that determinacy at each new level of the Borel hierarchy would require one more iteration of the power set operation. He himself pointed out that Det($\mathbf{\Sigma}^0_5$) cannot be established in second-order arithmetic.

The stakes having been raised, the progress was initially guided by Martin's Det($\mathbf{\Pi}^1_1$) proof cast in terms of indiscernibles. Davis [64] had shown that determinacy is preserved by the taking of countable unions if the sets involved had sufficiently simple representations, and mixing this with Martin's proof, Baumgartner had found an alternate proof of Det($\mathbf{\Sigma}^0_3$). Martin then saw that if there is a weakly compact cardinal, then there is such a simple representation of $\mathbf{\Pi}^0_3$ sets in terms of certain indiscernibles provided, and so derived Det($\mathbf{\Sigma}^0_4$). Soon afterwards in 1971 Paris [72] procured the requisite indiscernibles in ZFC, establishing Det($\mathbf{\Sigma}^0_4$) outright.

There matters stood for several years, and in retrospect the initiative based on indiscernibles impeded progress toward Borel Determinacy. Although the approach of Martin [70] was to be seminal for the investigation of higher determinacy hypotheses, the large cardinals idea had to be cast aside for another. Blass [75] and independently Mycielski established the interesting result that $\mathrm{AD}_{\mathbb{R}}$ is equivalent to the determinacy of all games of length ω^2 with integer moves, specified in terms of payoff sets included in ${}^{\omega^2}\omega$. The proof featured *moves* of an auxiliary game conveying information corresponding to *strategies* for the given game. Fashioning and elaborating this idea for the Borel sets Martin in 1974 established *Borel Determinacy*:

Theorem 31.3 (Martin [75]). Det($\mathbf{\Delta}^1_1$).

Loosely speaking, the proof proceeded by associating to each Borel set A corresponding sets X and $A^* \subseteq {}^{\omega}X$ such that A^* is open and a player has a winning strategy in $G_\omega(A)$ exactly when he has one in $G_X(A^*)$. Consistent with the Friedman anticipation, if A is $\mathbf{\Sigma}^0_\alpha$, then X has roughly the cardinality of $V_{\omega+\alpha}$. Of later emendations, the proof was reorganized in Martin [85] to subordinate various features to a more directly inductive argument, and extended in Martin [90] to games of form $G_X(A)$ for uncountable X and sets A "$\mathbf{\Delta}^1_1$" in this context. The argument was further streamlined in the book Martin [∞], which moreover exposed Martin's analysis of optimal hypotheses: He established in ZFC − Power Set that if $\alpha < \omega_1$ and $V_{\omega+\alpha}$ exists, then Det($\mathbf{\Delta}^0_{1+\alpha+3}$), and refining Friedman's arguments, that the hypotheses do not suffice to establish Det($\mathbf{\Sigma}^0_{1+\alpha+3}$).

The establishment of Borel Determinacy was a significant advance for set theory in that it both resolved an open problem whose stakes had been raised by metamathematical anticipations, and secured the Borel world for the consequences of determinacy. Among a variety of applications, it now followed from the argument for Martin's 28.4 that any Borel subset B of ${}^{\omega}\omega$ either includes or is disjoint from a Turing cone, i.e. there is $a \in {}^{\omega}\omega$ such that $\{b \in {}^{\omega}\omega \mid a \leq_{\mathrm{T}} b\} \subseteq B$ or else $\{b \in {}^{\omega}\omega \mid a \leq_{\mathrm{T}} b\} \cap B = \emptyset$. One interesting turn of events is that while Wadge in his studies of Wadge degrees of Borel sets (recast in Louveau [83]) had assumed Borel Determinacy in the late 1960's in order to avail himself of *Borel Wadge Determinacy* – the proposition that for every pair of Borel sets $A, B \subseteq {}^{\omega}\omega$ the Wadge game $\mathrm{WG}(A, B)$ (from the proof of Wadge's Lemma 29.15) is determined – Louveau-Saint-Raymond [87, 88] later established that this proposition is already provable in second-order arithmetic. In another reverberating development H. Friedman [81] considerably expanded his metamathematical work of [71], formulating in particular a simple "Borel diagonalization" proposition [81: 235] which follows from Borel Determinacy and also requires uncountably many iterations of the power set operation but is moreover Π^1_2.

It is the direction upward from Martin [70] to stronger hypotheses and conclusions that holds the sustained interest. The early efforts at getting more determinacy were informed by results showing that determinacy hypotheses beyond $\mathrm{Det}(\mathbf{\Pi}^1_1)$ *require* the consistency of measurability and more:

The first break was made, once again, by Solovay. Soon after he showed that full AD implies the measurability of ω_1 and ω_2, he established that $\mathrm{Det}(\mathbf{\Pi}^1_3)$ implies the existence of a transitive $\in$-model of ZFC with a measurable cardinal. After further results along these lines by H. Friedman [71a: §3] and Martin, Solovay then established that $\mathrm{Det}(\mathbf{\Delta}^1_2)$ implies the existence of transitive $\in$-models of ZFC with many measurable cardinals. Subsequently stronger conclusions were derived by Martin and others. These results were based on a way that Solovay had found for exploiting local versions of Martin's result 28.4 about Turing degrees.

Returning the compliment Martin in the meanwhile had gone back to Solovay's original method for establishing the measurability of ω_1 from AD. By that means, he was eventually to derive a series of results that exactly correlated the existence of many measurable cardinals with amounts of determinacy, based on the *difference hierarchy*:

For working with classes closed under the taking of countable intersections, if Λ is such a class and $\alpha < \omega_1$, the class α-Λ is defined by stipulating membership for $A \subseteq {}^{k}({}^{\omega}\omega)$ as follows: A is in α-Λ *iff* there is a sequence $\langle A_\xi \mid \xi < \alpha \rangle$ of subsets of ${}^{k}({}^{\omega}\omega)$ in Λ such that $A_\zeta \supseteq A_\xi$ for $\zeta \leq \xi < \alpha$ and:

$$\mathbf{w} \in A \ \textit{ iff } \text{ the least } \xi \text{ such that either } \mathbf{w} \notin A_\xi \text{ or } \xi = \alpha \text{ is odd} .$$

(As usual, an ordinal is odd *iff* it is of the form $\delta + 2n + 1$ for some limit ordinal δ and $n \in \omega$.) A can be viewed as a union of alternate concentric rings of a $\supseteq$-chain of sets in Λ: A is in 1-Λ *iff* it is in Λ; A is in 2-Λ *iff* $A = A_0 - A_1$ for some sets A_i in Λ; A is in 3-Λ *iff* $A = (A_0 - A_1) \cup A_2$ for some sets A_i in Λ; A is in 4-Λ *iff* $A = (A_0 - A_1) \cup (A_2 - A_3)$ for some sets A_i in Λ; and so forth. If $\alpha \leq \beta < \omega_1$, then α-$\Lambda \subseteq \beta$-Λ. Set

$$\mathrm{Diff}(\Lambda) = \bigcup\nolimits_{\alpha < \omega_1} \alpha\text{-}\Lambda .$$

This cumulative hierarchy is the *difference hierarchy for* Λ.

Hausdorff [14: 460ff] formulated the difference hierarchy for closed sets and showed that

$$\mathrm{Diff}(\mathbf{\Pi}^0_1) = \mathbf{\Delta}^0_2 ,$$

using this "resolvability" to establish properties of this class. Kuratowski [58: §37III] generalized this work through the Borel hierarchy, showing that for each $\xi < \omega_1$,

$$\mathrm{Diff}(\mathbf{\Pi}^0_\xi) = \mathbf{\Delta}^0_{\xi+1} .$$

(The work of Louveau-Saint Raymond [87, 88] on Borel Wadge Determinacy is couched in these terms. An analogous characterization of $\mathbf{\Delta}^0_\delta$ for limit δ is available.) Martin's analysis was based on the α-$\mathbf{\Pi}^1_1$ sets; through its discussion it should be kept in mind that in contrast to the Kuratowski result $\mathrm{Diff}(\mathbf{\Pi}^1_1)$ falls far short of $\mathbf{\Delta}^1_2$.

The sensible formulation of lightface versions is not direct, as the $\supseteq$-sequence should be appropriately definable. For (lightface) α-Π^1_1, first assume that α is recursive, i.e. there is a recursive well-ordering $E \subseteq {}^2({}^\omega\omega)$ with ordertype α, and temporarily let $\|m\|$ for $m \in \omega$ denote the ordertype of the predecessors of m according to E. Then for $A \subseteq {}^k({}^\omega\omega)$, A is α-Π^1_1 *iff* there is a sequence $\langle A_\xi \mid \xi < \alpha\rangle$ of subsets of ${}^k({}^\omega\omega)$ such that $\{\langle m, \mathbf{w}\rangle \mid \mathbf{w} \in A_{\|m\|}\}$ is Π^1_1; $A_\zeta \supseteq A_\xi$ for $\zeta \leq \xi < \alpha$; and

$$\mathbf{w} \in A \text{ \textit{iff} the least } \xi \text{ such that either } \mathbf{w} \notin A_\xi \text{ or } \xi = \alpha \text{ is odd} .$$

For $a \in {}^\omega\omega$, analogously define the class

$$\alpha\text{-}\Pi^1_1(a) \text{ for } \alpha \text{ recursive in } a .$$

Then for any $A \subseteq {}^k({}^\omega\omega)$ and $\alpha < \omega_1$, *A is α-$\mathbf{\Pi}^1_1$ iff there is an $a \in {}^\omega\omega$ such that α is recursive in a and A is α-$\Pi^1_1(a)$.*

In all these various cases, there are universal sets at each level, and so the hierarchies are proper (cf. 12.8).

Martin's results on many measurable cardinals went hand-in-hand with a new elucidation of $0^\#$. The first significant result about sets in $\mathrm{Diff}(\mathbf{\Pi}^1_1)$ was established by Friedman [71a], who extended 31.2 by showing that *if $0^\#$ exists, then* Det(3-Π^1_1). He also found a way to show that Det(Π^1_1) fails in every generic extension of L. Martin then established both the intrinsic necessity of $0^\#$ and the relevance of the difference hierarchy by establishing the following characterization:

31.4 Theorem (Martin). *$0^\#$ exists iff* Det($\bigcup_{\beta<\omega^2} \beta$-$\Pi^1_1$).

(See Martin [∞] or Dubose [90] for a proof.)

In the wake of ideas from Jensen's Covering Theorem (see volume II), Martin in 1975 applied 31.4 to establish the converse of H. Friedman's result: *if* Det(3-Π^1_1), *then $0^\#$ exists.* The gap from 3-Π^1_1 to Π^1_1 loomed larger than the rest, but Harrington closed it shortly afterwards by using ideas of Sami and Steel to establish the converse of Martin's 31.2:

31.5 Theorem (Harrington [78]). *Suppose that $a \in {}^{\omega}\omega$ and* $\mathrm{Det}(\Pi^1_1(a))$. *Then $a^{\#}$ exists.*

For the generality, this (and 31.2) were stated in a form relativized to $a \in {}^{\omega}\omega$. The Martin and Harrington results together provide an elegant characterization for any $a \in {}^{\omega}\omega$:

$$a^{\#} \text{ exists } \textit{ iff } \ \mathrm{Det}(\Pi^1_1(a)) .$$

This is a remarkable equivalence between the existence of the metamathematical sharps and the determinacy of the analytic sets, an unexpected confluence that bolstered both the large cardinal and the determinacy theory and intimated further riches to be unearthed about an underlying unity.

That unity was considerably reinforced by Martin's companion results on many measurable cardinals. He saw that starting with $\alpha+1$ measurable cardinals he could get $\mathrm{Det}((\omega^2{\cdot}\alpha+1)\text{-}\mathbf{\Pi}^1_1)$, and in fact determinacy a bit further along the difference hierarchy. Continuing the pattern of 31.1 and 31.2, he then saw that just having an inner model with α measurable cardinals and indiscernibles for that structure suffices, and moreover, a converse can be achieved!

Suppose that M an inner model of ZFC. Just to clarify, say that *M has α measurable cardinals iff* there is a sequence $\langle \rho_\xi \mid \xi < \alpha \rangle$ of ordinals such that for each $\xi < \alpha$, $M \models \rho_\xi$ is measurable. Suppose next that $M \models \delta$ is the supremum of the measurable cardinals. Eschewing details, say that *M has indiscernibles iff* it has an uncountable set of ordinal indiscernibles whose minimum element is greater than δ. In light of §21, having such indiscernibles is a notion of transcendence over the model, and an $0^{\dagger}$-type theory can be developed. By 21.26 and the comment after, for any $a \in {}^{\omega}\omega$, *$a^{\dagger}$ exists iff there is an inner model containing a that has a measurable cardinal and has indiscernibles.*

The following generalizes 31.4 and 31.5, which together can be regarded as the $\alpha = 0$ case:

31.6 Theorem (Martin). *For any $a \in {}^{\omega}\omega$ and α recursive in a, the following are equivalent:*

(a) $\mathrm{Det}(\bigcup_{\beta<\omega^2}(\omega^2{\cdot}\alpha+\beta)\text{-}\Pi^1_1(a))$.

(b) $\mathrm{Det}((\omega^2{\cdot}\alpha+1)\text{-}\Pi^1_1(a))$.

(c) There is an inner model of ZFC *that contains a and has α measurable cardinals and has indiscernibles.*

In particular, for any $a \in {}^{\omega}\omega$,

$$a^{\dagger} \text{ exists } \textit{ iff } \ \mathrm{Det}((\omega^2+1)\text{-}\Pi^1_1(a)) .$$

John Simms [79], a student of Martin, provided an analogous characterization of the existence of a transitive $\in$-model of ZFC satisfying "there are arbitrarily large measurable cardinals" which has indiscernibles in an appropriate sense: This is equivalent to the determinacy of $\mathbf{\Sigma}^0_1(\Pi^1_1)$, the class of countable unions of

Boolean combinations of Π^1_1 sets. (See Martin [∞: §5.5].) Simms in fact established equivalences analogous to 31.6 through the difference hierarchy for $\mathbf{\Sigma}^0_1(\Pi^1_1)$ in terms of hierarchically high measurable cardinals. Loosely speaking, each new level corresponds to a measurable limit of cardinals corresponding to the previous level.

These various determinacy hypotheses fall far short of Det($\mathbf{\Delta}^1_2$) (which by 30.10 is equivalent to Det($\mathbf{\Pi}^1_2$)). An impressive result at the time was due to John Green [78], a student of Harrington, who was able to get more than measurable limits of measurable cardinals: *If* Det(Δ^1_2), *then there is an inner model of* ZFC *that has a measurable cardinal* κ *with a normal ultrafilter* U *such that* $\{\alpha < \kappa \mid \alpha \text{ is measurable}\} \in U$. Such a κ is high in the "measurable limits of measurables" hierarchy; through the study of inner models Steel [82] was able to determine just how much determinacy is derivable from having such a κ.

There were later emendations having to do with refined issues. 31.6 provides no equivalence for Det(ω^2-$\Pi^1_1(a)$) (taking $\alpha = 1$ as a typical case). Martin showed in the late 1970's that the existence of an inner model of ZFC that contains a and has a measurable cardinal suffices, and eventually improved it in [90] to show that *if there is an inner model of* ZFC *that contains a and has a measurable cardinal, then every set of reals such that both it and its complement are in* (ω^2+1)-$\Pi^1_1(a)$ *is determined.* This is optimal in the sense that by 31.6 indiscernibles for the inner model are needed to get Det((ω^2+1)-$\Pi^1_1(a)$). Martin also established technical equivalences for Det(ω^2-$\Pi^1_1(a)$) in terms of inner models that have a measurable cardinal and closed unbounded sets having special properties for such models. Recently, Philip Welch found an equivalence for Det(ω^2-$\mathbf{\Pi}^1_1$) (boldface only) more analogous to 31.6, but for a weaker set theory.

Derrick Dubose, a student of Martin, has explored in detail how sharps correlate with determinacy hypotheses between Det($\bigcup_{\beta<\omega^2}\beta$-$\Pi^1_1$) and Det($\omega^2$-$\Pi^1_1$). In terms of such hypotheses formulated via hierarchies of sets, he [90] found characterizations for the existence of the iterated sharps $0^\#$, $0^{\#\#}$, $0^{\#\#\#}$, and so forth. He [91, 92] also characterized the existence of the sharp function, that function # on the reals given by $\#(a) = a^\#$, and extended versions of this function. Moreover, he [95] characterized the existence of sharp functions on objects of higher type, sets of reals, sets of sets of reals, and so forth.

Having extensively set out the results on the determinacy of sets in the difference hierarchy for $\mathbf{\Pi}^1_1$, how Martin got measurable cardinals out of such hypotheses is illustrated with his earliest result along these lines established in the early 1970's.

31.7 Theorem (Martin). *Assume* Det((ω^2+1)-Π^1_1). *Then there is an inner model with a measurable cardinal.*

Proof. The proof uses a game which is a sort of ω-ary iteration of the one used by Solovay in his original proof of the measurability of ω_1. The players choose integers as usual:

$$\begin{array}{llllll} I & : & x(0) & & x(2) & & \dots \\ II & : & & x(1) & & x(3) & \dots \end{array}$$

Using a recursive coding scheme, x_I is regarded as encoding a sequence $\langle (x_I)_\xi \mid \xi \leq \omega^2 \rangle \in {}^{\omega^2+1}({}^\omega\omega)$, and x_{II}, a sequence $\langle (x_{II})_\xi \mid \xi < \omega^2 \rangle \in {}^{\omega^2}({}^\omega\omega)$. Recall again (§13) that each $z \in {}^\omega\omega$ encodes a relation $E_z \subseteq \omega \times \omega$ with $\|z\|$ denoting its ordertype when it is a well-ordering.

Suppose first that for some $\xi < \omega^2$, either $E_{(x_I)_\xi}$ or $E_{(x_{II})_\xi}$ is not a well-ordering. Then for minimal such ξ, if $E_{(x_I)_\xi}$ is not a well-ordering, then II wins. Otherwise, I wins.

Suppose now that for each $\xi < \omega^2$ both $E_{(x_I)_\xi}$ and $E_{(x_{II})_\xi}$ are well-orderings. For such ξ and $n \in \omega$, set

$$\begin{aligned} \beta^x_\xi &= \|(x_I)_\xi\| , \\ \gamma^x_\xi &= \|(x_{II})_\xi\| , \\ \rho^x_n &= \sup(\{\beta^x_{\omega\cdot n+i} \mid i \in \omega\} \cup \{\gamma^x_{\omega\cdot n+i} \mid i \in \omega\}) , \text{ and} \\ \delta^x &= \sup(\{\rho^x_n \mid n \in \omega\}) . \end{aligned}$$

Let F^x be the filter over δ^x generated by $\{\rho^x_n \mid n \in \omega\}$, i.e.

$$X \in F^x \text{ iff } X \subseteq \delta^x \wedge \exists m \in \omega(\{\rho^x_n \mid m \leq n \in \omega\} \subseteq X) .$$

The intention when the ρ_n's are increasing is that $L[F^x] \models F^x \cap L[F^x]$ is a normal ultrafilter over δ^x, and through the following rules the burden falls on I to try to defeat this. For their formulation, recall (as before 20.18) that each $z \in {}^\omega\omega$ encodes a structure $N_z = \langle \omega, E_z, A_z \rangle$, and when it is well-founded and extensional, $\mathrm{tr}(N_z)$ denotes its transitive collapse and $\pi_z \colon N_z \to \mathrm{tr}(N_z)$ the collapsing isomorphism.

Stipulate that II wins unless $N_{(x_I)_{\omega^2}}$ is well-founded and extensional, and for some limit ordinal $\theta^x > \omega$,

$$\mathrm{tr}(N_{(x_I)_{\omega^2}}) = \langle L_{\theta^x}[F^x], \in, F^x \cap L_{\theta^x}[F^x] \rangle .$$

Assuming that I has managed to arrange this, II wins unless there is a counterexample to normality: an $f \colon \delta^x \to \delta^x$ such that $f \in L_{\theta^x}[F^x]$ and $\{\rho < \delta^x \mid \rho \leq f(\rho)\} \notin F^x$ yet f is not constant on any set in F^x. If there is such an f, taking f^x to be the least such in the canonical well-ordering of $L[F^x]$, II wins unless either

$$\begin{aligned} x(0) = 0 \wedge f^x(\rho^x_0) \geq f^x(\rho^x_1) , \text{ or} \\ x(0) = 1 \wedge f^x(\rho^x_0) < f^x(\rho^x_1) , \end{aligned}$$

in which case I wins.

To analyze the payoff set of this game, consider for $\xi < \omega^2$ the sets

$$\begin{aligned} Y_{2\xi} &= \{x \in {}^\omega\omega \mid E_{(x_I)_\xi} \text{ is a well-ordering}\} , \text{ and} \\ Y_{2\xi+1} &= \{x \in {}^\omega\omega \mid E_{(x_{II})_\xi} \text{ is a well-ordering}\} . \end{aligned}$$

There is also a set corresponding to the various conditions for I winning the game as described in the previous paragraph:

$$Y_{\omega^2} = \{x \in \bigcap_{\xi<\omega^2} Y_\xi \mid E_{(x_I)_{\omega^2}} \text{ is well-founded } \wedge \ N_{(x_I)_{\omega^2}} \models \sigma_1 \wedge \psi \\ \wedge \ \pi_{(x_I)_{\omega^2}}\text{``}A_{(x_I)_{\omega^2}} = F^x \cap \pi_{(x_I)_{\omega^2}}\text{``}\omega\} \,,$$

where σ_1 is the sentence of 3.2(b), and ψ is a sentence asserting that ω exists and that there is no largest ordinal, and sets out the various conditions for I winning the game with a counterexample to normality. It is straightforward to check that with an appropriate ψ,

$$I \text{ wins a play } x \text{ iff the least } \xi \text{ such that } x \notin Y_\xi \text{ or } \xi = \omega^2+1 \text{ is odd} \,.$$

With the defining clauses for Y_{ω^2} other than well-foundedness being arithmetical as for 13.8(a)(c) it is straightforward to see that the payoff set is (ω^2+1)-Π^1_1 and so the game is determined by hypothesis. As in other Solovay games, I cannot have a winning strategy:

Assume to the contrary that σ is a winning strategy for I. A $\mathbf{\Sigma}^1_1$ bounding argument is first applied as in the first proof of 28.2 to get a closed unbounded subset of ω_1: For $\eta < \omega_1$ set

$$B_\eta = \{((\sigma * y)_I)_\xi \mid \xi < \omega^2 \wedge \ y \in {}^\omega\omega \ \wedge \\ \forall\zeta < \xi(E_{(y)_\zeta} \text{ is a well-ordering with } \|(y)_\zeta\| < \eta)\} \,.$$

B_η is $\mathbf{\Sigma}^1_1$ in a real coding the ordertype η. Moreover, in terms of 13.6, $B_\eta \subseteq \mathrm{WF}$ as σ is winning, and so $B_\eta \subseteq \mathrm{WF}_\beta$ for some $\beta < \omega_1$. Taking $w(\eta)$ to be the least such β, the resulting function $w\colon \omega_1 \to \omega_1$ leads to a set

$$C = \{\rho < \omega_1 \mid \rho \text{ is an infinite limit } \wedge \ \forall\eta < \rho(w(\eta) < \rho)\}$$

closed unbounded in ω_1.

C has a crucial property which will be used several times in what follows: *If* $\langle\rho_n \mid n \in \omega\rangle \in {}^\omega C$, *then there is a play* $x \in \bigcap_{\xi<\omega^2} Y_\xi$ *of the game according to* σ *such that* $\rho^x_n = \rho_n$ *for every* $n \in \omega$. To show this, let $\langle\eta_\xi \mid \xi < \omega^2\rangle$ be such that for each $n \in \omega$, $\{\eta_{\omega\cdot n+i} \mid i \in \omega\} \subseteq \rho_n$ and is cofinal in ρ_n. Then let $y \in {}^\omega\omega$ be such that for each $\xi < \omega^2$, $\|E_{(y)_\xi}\| = \eta_\xi$. Then $\sigma * y \in \bigcap_{\xi<\omega^2} Y_\xi$, and it follows from the definition of w and C that $\rho^{\sigma * y}_n = \rho_n$ for every $n \in \omega$.

A contradiction is now derived from the existence of σ: Fix an increasing sequence $\langle\rho_n \mid n \in \omega\rangle$ of limit points of C such that $C \cap \rho_n$ has ordertype ρ_n for each $n \in \omega$. Let $x \in Y_{\omega^2}$ be a play according to σ such that $\rho^x_n = \rho_n$ for every $n \in \omega$. There is a corresponding F^x and a counterexample f^x to normality, and $x(0)$ must be either 0 or 1.

Suppose first that $x(0) = 0$. Fixing $k \in \omega$ for the moment, let $z \in Y_{\omega^2}$ be a play according to σ such that $\rho^z_n = \rho_{n+k}$ for $n \in \omega$, with corresponding F^z and f^z. Evidently $F^z = F^x$, and so also $f^z = f^x$. But $z(0) = x(0) = 0$, as $x(0)$ is, after all, the first move of the game played according to σ. Hence,

$$f^x(\rho_k) = f^z(\rho_0^z) \geq f^z(\rho_1^z) = f^x(\rho_{k+1}) ,$$

the inequality as σ is winning. Now this holds for every $k \in \omega$, and so $f^x(\rho_k)$ must become constant for sufficiently large $k \in \omega$. But this is a contradiction, as f^x cannot be constant on any set in F^x.

Suppose next that $x(0) = 1$. Fixing $k \in \omega$ and $\rho \in C$ with $\rho < \rho_k$ for the moment, let y be a play according to σ such that $\rho_0^y = \rho$ and $\rho_{n+1}^y = \rho_{n+k}$ for $n \in \omega$. Then $F^y = F^x$, $f^y = f^x$, and $y(0) = x(0) = 1$, and so

$$f^x(\rho) = f^y(\rho_0^y) < f^y(\rho_1^y) = f^x(\rho_k) .$$

Fixing a further $\zeta \in C$ with $\rho < \zeta < \rho_k$, a similar argument with a play z according to σ such that $\rho_1^z = \zeta$ and $\rho_n^z = \rho_n^y$ for every $n \neq 1$ shows that

$$f^x(\rho) < f^x(\zeta) .$$

Varying ρ and ζ, it follows that $f^x|(C \cap \rho_k)$ is an increasing function into $f^x(\rho_k)$. But since $C \cap \rho_k$ has ordertype ρ_k,

$$\rho_k \leq f^x(\rho_k) .$$

Finally, varying k leads to $\{\rho < \delta^x \mid \rho \leq f^x(\rho)\} \in F^x$, which is a contradiction of the choice of f^x.

Having established that I cannot have a winning strategy, it follows that II must have one. Let τ be such a strategy. Following the argument leading from σ to the set C, τ leads to a set D closed unbounded in ω_1 that satisfies the following: *If* $\langle \rho_n \mid n \in \omega \rangle \in {}^{\omega}D$, $\theta > \omega$ *is a countable limit ordinal, and* $i < 2$, *then there is a play* $x \in \bigcap_{\xi < \omega^2} Y_\xi$ *of the game according to* τ *such that* $\rho_n^x = \rho_n$ *for every* $n \in \omega$; θ^x *(is defined and) equals* θ; *and* $x(0) = i$.

Fixing an increasing sequence $\langle \rho_n \mid n \in \omega \rangle \in {}^{\omega}D$, set $\delta = \sup(\{\rho_n \mid n \in \omega\})$ and let F be the filter over δ generated by $\{\rho_n \mid n \in \omega\}$. The proof is completed by showing that $L[F] \models F \cap L[F]$ is a normal ultrafilter. This follows from the following property: *If* $f \in {}^{\delta}\delta \cap L[F]$ *with* $\{\rho < \delta \mid \rho \leq f(\rho)\} \notin F$, *then* f *is constant on a set in* F. Granted this it is simple to show first that $F \cap L[F]$ is an ultrafilter in $L[F]$ using functions in ${}^{\delta}2 \cap L[F]$, and then to derive normality.

To verify the property, assume to the contrary that there is a counterexample which is not constant on any set in F, and let f^* be the least such in the canonical well-ordering of $L[F]$. By the argument for 20.2(a) there is a countable θ^*, which can be taken to be a limit ordinal, such that $f^* \in L_{\theta^*}[F]$. Let $i^* = 0$ if $f^*(\rho_0) \geq f^*(\rho_1)$, and $i^* = 1$ if $f^*(\rho_0) < f^*(\rho_1)$. Then by the condition on D there is a play x according to τ such that $\rho_n^x = \rho_n$ for every $n \in \omega$; $\theta^x = \theta^*$; and $x(0) = i^*$. But this is a contradiction, since τ is a winning strategy. The proof is complete. ⊣

The special properties exhibited by the closed unbounded set C in this proof lend plausibility to the optimal conclusion from $\mathrm{Det}((\omega^2+1)\text{-}\Pi_1^1)$, the existence of

$0^\dagger$. Little more complexity beyond this proof is needed to show that if Det($(\omega^2\cdot\alpha+1)$-Π^1_1), then there is an inner model that has α measurable cardinals.

Through such results a pattern of correlations with large cardinals emerged during the mid-1970's, one agreeably reinforced by the amounts of the rank hierarchy necessary to establish fragments of Borel Determinacy as the beginning cases. The determinacy hypotheses involved, however, fell far short of Det($\mathbf{\Pi}^1_2$) which seemed to loom ever stronger in consistency strength. Through the 1970's a variety of substantial propositions were shown to be consistent relative to a supercompact cardinal. But despite various attempts Det($\mathbf{\Pi}^1_2$) was not among them. Then reaching higher to the hypotheses I0-I3 at the limits of consistency (§24) Martin was able to establish a synthetic connection in 1978:

31.8 Theorem (Martin [80]). *Assume* I2. *Then* Det($\mathbf{\Pi}^1_2$).

In fact, sufficient is the existence of a $j\colon V_\delta \prec V_\delta$ satisfying an iterability condition, a hypothesis strictly intermediate between I2 and I3. Martin applied the basic idea of his Det($\mathbf{\Pi}^1_1$) proof, using the representation of $\mathbf{\Pi}^1_2$ sets cast in terms of an embedding j as above in order to apply its 24.6 properties.

This affirmation of large cardinals raised hopes, but also concerns. Could there be some sort of converse? Any such result would likely have to secure an inner model of some large cardinal hypothesis (cf. 31.7), but into the early 1980's the development of inner model theory had not progressed beyond the models of Mitchell [79]. These were for the "hypermeasurable" cardinals, essentially γ-strong cardinals for small γ (§26), and the amount of determinacy in them was analyzed by Steel [82]. In any case, the models all have Δ^1_3 well-orderings of the reals of ordertype ω_1, and so cannot satisfy Det($\mathbf{\Pi}^1_2$) (by 27.14 and the argument for 13.10).

This apparent obstruction to getting inner models for large cardinals implying Det($\mathbf{\Pi}^1_2$) was a major reason advanced for its considerable strength, and Martin's result ruled out inner models of the sort then known for hypotheses at the level of I0-I3.

The other question was where the higher determinacy hypotheses would fit in if at all; perhaps they veer off in some orthogonal direction from large cardinals? With attention shifted upward to the strongest large cardinal hypotheses, Woodin boldly formulated I0 just at the edge of the Kunen inconsistency and amplified Martin's embeddings proof to establish the following in early 1984:

31.9 Theorem (Woodin). *Assume* I0. *Then* $\mathrm{AD}^{L(\mathbb{R})}$.

So determinacy is subsumed by the large cardinal hierarchy, but only just! However, the hypotheses I0-I3 were little understood and might totter as Kunen's proposition did. And there seemed little hope that the inner model theory could be developed enough to achieve equiconsistencies. But at least a mooring was secured, and there was new anticipation of a grand unification of large cardinals and determinacy. How that came about is the subject of the next section.

32. Consistency of AD

As a fitting conclusion to both this chapter and volume, this section provides a panoramic survey of the crowning achievements concerning the relative consistency of determinacy hypotheses. After setting out the basics of what came to serve as a general framework, proofs are dispensed with, as in any case they are precluded by the range and sophistication of the results discussed. Rather, a scaffolding is provided based on the historical progression of ideas, one that may serve as a guide to expositions to come (Martin [∞] and Woodin-Mathias-Hauser [∞]).

The culminating equiconsistency results are due to Hugh Woodin, whose results have figured in previous sections. Already as an undergraduate Woodin in 1976 had reduced a proposition in the theory of Banach algebras to a set-theoretic one whose consistency was then established by Solovay. Then, as his student at Berkeley, Woodin found a more systematic route to that consistency incorporating Martin's Axiom (see Dales-Woodin [87]), and soon established important consistency results using Radin Forcing and about ideals over ω_1 (see volume II). With strong hypotheses and structural issues in set theory becoming the driving imperatives in his work, Woodin joined the Cabal Seminar in the early 1980's and became deeply involved in the investigation of determinacy. By the mid-1980's his many splendid results had earned him a reputation comparable only to that of Shelah, who has been in the limelight since the 1970's. But unlike the latter Woodin is not one to rush to the presses, and the bulk of his work remains unpublished.

Stepping back, it was Martin and Steel who made a major breakthrough toward the relative consistency of determinacy from large cardinals, and for Martin this was the capstone of nearly twenty years of sustained effort. In retrospect the quest for this synthesis was based on a remarkable insight that principles in ZFC based on elementary embeddings could somehow provide combinatorial principles for sets of reals entailing their regularity properties. Affirming Martin's early successes using tree representations to establish determinacy a general approach to such representations is first provided; it is an elegant abstraction of the essentials of previous constructions involving large cardinals.

Homogeneous Trees

Homogeneous trees are implicit in the work of Martin-Solovay [69] (described in §15) and in early work on determinacy: Martin's Det($\mathbf{\Pi}^1_1$) result 31.1 and Kunen's 1971 analysis of δ^1_3 and δ^1_5. The concept itself was isolated much later by Kechris [81:63] and Martin, and came to be the means by which determinacy was derived from large cardinals. In the paradigmatic case, for any $\kappa > \omega$, $\mathbf{\Pi}^1_1$ sets of reals are κ-Suslin (13.14), i.e. projections of trees on $\omega \times \kappa$. But the infinite paths through such a tree form a closed subset of ${}^\omega\omega \times {}^\omega\kappa$ in the expected sense (cf. 12.10), and so are determined (27.1). Martin derived Det($\mathbf{\Pi}^1_1$) by taking κ to be measurable and using a normal ultrafilter over κ to "integrate" over winning strategies for closed subsets of ${}^{\omega}\omega \times {}^{\omega}\kappa$ to get such strategies for their projections, the $\mathbf{\Pi}^1_1$ subsets of ${}^{\omega}\omega$. Recalling the §15 analysis homogeneous trees are just those trees that carry the requisite ultrafilter structure to push through Martin's argument. Some terminology is first developed toward the definition:

For a tree T on some ${}^{k}\omega \times Y$ (as defined after 12.8) and for $s \in \bigcup_{m\in\omega} {}^{k}({}^{m}\omega)$,

$$T_s = \{u \mid \langle s, u\rangle \in T\} .$$

In this section *by tree is meant some such tree, usually taking* $k = 1$ *for notational simplicity, such that* $T_s \neq \emptyset$ *for every such* s. Also, *principal ultrafilters are allowed*, contravening a convention.

Suppose that for some set Y and $m \leq n \in \omega$, U is an ultrafilter over some $I \subseteq {}^{m}Y$ and W is an ultrafilter over some $J \subseteq {}^{n}Y$. Then U is a *projection* of W *iff* for any $X \subseteq I$,

$$X \in U \quad \textit{iff} \quad \{u \in J \mid u|m \in X\} \in W .$$

If moreover U and W are ω_1-complete, then with

$$j_U\colon V \prec M_U \cong \mathrm{Ult}(V, U) \quad \text{and}$$
$$j_W\colon V \prec M_W \cong \mathrm{Ult}(V, W) ,$$

the map $k\colon M_U \to M_W$ given by

$$k([f]_U) = [\langle f(u|m) \mid u \in J\rangle]_W$$

is elementary, and $k \circ j_U = j_W$.

A sequence $\langle U_m \mid m \in \omega\rangle$ of ω_1-complete ultrafilters is a *tower iff* for some set Y, $U_m \subseteq \mathcal{P}({}^{m}Y)$ for $m \in \omega$, and U_m is a projection of U_n for $m \leq n \in \omega$. For such a tower, let

$$j_m\colon V \prec M_m \cong \mathrm{Ult}(V, U_m) ,$$

and

$$j_{mn}\colon M_m \to M_n$$

be the embeddings corresponding to U_m being a projection of U_n, so that $j_{mn} \circ j_m = j_n$. Then

$$\langle\langle M_m \mid m \in \omega\rangle, \langle j_{mn} \mid m \leq n\rangle\rangle$$

is a directed system.

$\langle U_m \mid m \in \omega\rangle$ is *countably complete iff* whenever $X_m \in U_m$ for $m \in \omega$, there is a $d \in {}^{\omega}Y$ such that $d|m \in X_m$ for each such m. By the argument for 15.7, *a tower is countably complete iff the corresponding directed system has a well-founded direct limit.*

To proceed to the main definitions, for a tree T on some ${}^{k}\omega \times Y$ and $\kappa > \omega$, taking $k = 1$ for notational simplicity, T is *κ-homogeneous iff* for $s \in {}^{<\omega}\omega$ there are κ-complete ultrafilters U_s over T_s such that U_s is a projection of $U_{\bar{s}}$ whenever $s \subseteq \bar{s}$, and: For any $x \in {}^{\omega}\omega$,

$$\text{if } x \in p[T], \text{ then } \langle U_{x|m} \mid m \in \omega \rangle \text{ is a countably complete tower .}$$

The converse is immediate, since if $X_m = T_{x|m}$ for $m \in \omega$, a corresponding $d \in {}^{\omega}Y$ such that $d|m \in X_m$ for such m verifies that $x \in p[T]$.

For $A \subseteq {}^{k}({}^{\omega}\omega)$, A is *κ-homogeneously Suslin iff* $A = p[T]$ for some κ-homogeneous tree T. A is *homogeneously Suslin iff* A is κ-homogeneously Suslin for some $\kappa > \omega$. Note that for such sets to exist in any substantive sense, there must be a measurable cardinal: Otherwise the U_s's would have to be principal (2.2), and so being homogeneously Suslin would be equivalent to being closed (cf. 12.10). Also, note that by the argument for 2.8, *it can be assumed that* $\kappa > 2^{\aleph_0}$ and much more.

As anticipated, the following two exercises factor Martin's Det($\mathbf{\Pi}^1_1$) result 31.1 through the concept of homogeneous tree.

32.1 Exercise. *Suppose that there is a measurable cardinal κ. Then every $\mathbf{\Pi}^1_1$ set is κ-homogeneously Suslin.*

Hint. Show, following the proof of 31.1, that the U_s's defined there are as required. ⊣

32.2 Exercise. *Suppose that a set of reals is homogeneously Suslin. Then it is determined.*

Hint. Suppose that $A = p[T]$ say, where T is a κ-homogeneous tree on $\omega \times Y$ with corresponding κ-complete ultrafilters U_s for $s \in {}^{<\omega}\omega$. To establish that $G_\omega(A)$ is determined, consider the auxiliary game G^* as in the proof of 31.1:

$$\begin{array}{llllll} I: & \langle x(0), g(0)\rangle & & \langle x(2), g(1)\rangle & & \ldots \\ II: & & x(1) & & x(3) & \ldots \end{array}$$

I and II both play integers $x(i)$ to form a real x, but in addition, I plays $g(i) \in Y$ to form a function $g\colon \omega \to Y$. I wins if $\langle x, g\rangle \in [T]$, and otherwise II wins.

As before, G^* is determined, and if I has a winning strategy in this game, he has one in $G_\omega(A)$. It remains show that if II has a winning strategy τ in G^*, then he has one strategy in $G_\omega(A)$:

Proceeding as for 31.1, for $n \in \omega$, $t \in {}^{2n+1}\omega$, and $u \in {}^{n+1}Y$ let $\tau(t, u)$ denote the strategic response according to τ to

$$\langle \langle t(0), u(0)\rangle, t(1), \ldots, \langle t(2n), u(n)\rangle\rangle \; .$$

Then for such t and $k \in \omega$, set

$$Z_{t,k} = \{u \in T_{t|(n+1)} \mid \tau(t, u) = k\} \; .$$

Next, define a strategy $\tau_0\colon \bigcup_{n\in\omega} {}^{2n+1}\omega \to \omega$ by:

$$\tau_0(t) = k \ \textit{ iff } \ Z_{t,k} \in U_{t|(n+1)} \; .$$

This is well-defined by the ω_1-completeness of the U_s's. Finally, argue that τ_0 is a winning strategy for II in $G_\omega(A)$:

Assume to the contrary that x is a play according to τ_0, yet $x \in A$. By the homogeneity of T there is a $d \in {}^\omega Y$ such that for every $n \in \omega$, $d|(n+1) \in Z_{x|(2n+1),\tau_0(x|(2n+1))}$. But then, $\langle x, d\rangle$ is in $[T]$ and corresponds to a play of G^* in which II plays according to τ and still loses. Contradiction! ⊣

Weakly Homogeneous Trees

Weak homogeneity for trees is a technical weakening of homogeneity also considered by Kechris [81: 66] and Martin, which through the work of Woodin became pivotal as the first structural property related to determinacy to be obtained from large cardinals. For a tree T on some ${}^k\omega \times Y$ and $\kappa > \omega$, again taking $k = 1$ for simplicity, T is *κ-weakly homogeneous iff* for $s, t \in {}^{<\omega}\omega$ with $|s| = |t|$ there are κ-complete ultrafilters $U_{s,t}$ over T_s such that $U_{s,t}$ is a projection of $U_{\bar{s},\bar{t}}$ whenever $\bar{s}|m = s$ and $\bar{t}|m = t$ for some $m \in \omega$, and: For any $x \in {}^\omega\omega$,

if $x \in p[T]$, then there is a $y \in {}^\omega\omega$ such that
$\langle U_{x|m,y|m} \mid m \in \omega\rangle$ is a countably complete tower .

As with homogeneous trees, the converse is immediate. Clearly, a κ-homogeneous tree is κ-weakly homogeneous.

For $A \subseteq {}^k({}^\omega\omega)$, A is *κ-weakly homogeneously Suslin iff* $A = p[T]$ for some κ-weakly homogeneous tree T, and A is *weakly homogeneously Suslin iff* A is κ-weakly homogeneously Suslin for some $\kappa > \omega$. As with homogeneity *it can be assumed that* $\kappa > 2^{\aleph_0}$ and much more.

The following two exercises cast weak homogeneity in different lights.

32.3 Exercise. *$A \subseteq {}^k({}^\omega\omega)$ is κ-weakly homogeneously Suslin iff $A = pB$ for some κ-homogeneously Suslin $B \subseteq {}^{k+1}({}^\omega\omega)$.*

Hint. Taking $k = 1$, suppose in the forward direction that $A = p[T]$ where T is a κ-weakly homogeneous tree on $\omega \times Y$ with corresponding κ-complete ultrafilters $U_{s,t}$ for $s, t \in {}^{<\omega}\omega$ with $|s| = |t|$. For any $x, y \in {}^\omega\omega$, if $\langle U_{x|m,y|m} \mid m \in \omega\rangle$ is not countably complete, let $X_m^{x,y} \in U_{x|m,y|m}$ for $m \in \omega$ be such that no $d \in {}^\omega Y$ satisfies $d|m \in X_m^{x,y}$ for every such m; otherwise, set $X_m^{x,y} = T_{x|m}$ for every $m \in \omega$. Next, for $s, t \in {}^{<\omega}\omega$ with $|s| = |t|$ set

$$Z_{s,t} = \bigcap\{X_{|s|}^{x,y} \mid x, y \in {}^\omega\omega \ \wedge \ x||s| = s \ \wedge \ y||t| = t\} \, .$$

Then $Z_{s,t} \in U_{s,t}$ since $U_{s,t}$ is $(2^{\aleph_0})^+$-complete (by the argument for 2.8, as noted before). Finally, define a tree T^+ on ${}^2\omega \times Y$ by

$$T^+ = \{\langle s, t, u\rangle \mid u \in Z_{s,t}\} \, ,$$

and set $\overline{U}_{s,t} = U_{s,t} \cap \mathcal{P}(Z_{s,t})$ for $s, t \in {}^{<\omega}\omega$ with $|s| = |t|$. Now verify that T^+ with the $\overline{U}_{s,t}$'s is κ-homogeneous, and that with $B = p[T^+]$, $A = pB$.

The converse is straightforward. ⊣

If homogeneously Suslin sets are to be viewed as generalizations of closed sets, the verifying ultrafilters ensuring determinacy, then weakly homogeneously Suslin sets can be regarded through 32.3 as generalizations of analytic sets. It now follows from 32.1 that *if there is a measurable cardinal κ, then every $\mathbf{\Sigma}^1_2$ set is κ-weakly homogeneously Suslin.*

32.4 Exercise. *Suppose that T is a tree on some ${}^k\omega \times Y$ and $\kappa > \omega$. Then T is κ-weakly homogeneous iff there is a countable collection $\mathcal{D}$ of κ-complete ultrafilters such that for any $x \in {}^k({}^\omega\omega)$: If $x \in p[T]$, then there is a countably complete tower $\langle U_m \mid m \in \omega\rangle \in {}^\omega\mathcal{D}$ such that $T_{x|m} \in U_m$ for every $m \in \omega$.*

Hint. Taking $k = 1$, for the direction starting from $\mathcal{D}$ it can be assumed that $\mathcal{D}$ is closed under projections: If $W \in \mathcal{D}$ and $W \subseteq \mathcal{P}({}^nY)$ for some $n \in \omega$ and set Y, then for each $m \leq n$

$$\{\{u|m \mid u \in Z\} \mid Z \in W\} \in \mathcal{D}\,.$$

Proceed now by induction on $|s|$ to define $U_{s,t}$ so that for any $i \in \omega$, allowing principal ultrafilters if necessary,

$$\{U_{s^\frown\langle i\rangle, t^\frown\langle e\rangle} \mid e \in \omega\} \supseteq \\ \{W \cap \mathcal{P}(T_{s^\frown\langle i\rangle}) \mid W \in \mathcal{D} \wedge T_{s^\frown\langle i\rangle} \in W \wedge U_{s,t} \text{ is a projection of } W\}\,.$$

⊣

32.3 shows how weak homogeneity is closely tied to homogeneity, but 32.4 suggests that the former may be considerably easier to obtain. One consequence is that being weakly homogeneously Suslin is preserved under Wadge reducibility, i.e. the taking of continuous pre-images:

32.5 Exercise. *Suppose that $B \subseteq {}^l({}^\omega\omega)$ is κ-weakly homogeneously Suslin, $f\colon {}^k({}^\omega\omega) \to {}^l({}^\omega\omega)$ is continuous, and $A = f^{-1}(B)$. Then A is κ-weakly homogeneously Suslin.*

Hint. Taking $k = l = 1$, suppose that $B = p[T]$ where T is a tree on $\omega \times Y$, and let $A = p[S]$ where S is the tree on $\omega \times Y$ defined as in the proof of 13.13(g) from T and a function θ. Show that if $\mathcal{D}$ verifies the weak homogeneity of T as in 32.4, then it also does so for S:

Suppose that $x \in p[S]$. Then $f(x) \in p[T]$, so that there is a countably complete tower $\langle U_m \mid m \in \omega\rangle \in {}^\omega\mathcal{D}$ such that $T_{f(x)|n} \in U_n$ for every $n \in \omega$. To complete the proof, check that $S_{x|m} \in U_m$ for every $m \in \omega$:

$$\begin{aligned} S_{x|m} &= \{u \in {}^mY \mid \langle x|m, u\rangle \in S\} \\ &= \{u \in {}^mY \mid \langle \theta(x|m), u||\theta(x|m)|\rangle \in T\} \\ &= \{u \in {}^mY \mid u|n \in T_{f(x)|n}\} \end{aligned}$$

where $n = |\theta(x|m)|$, so that $\theta(x|m) = f(x)|n$. But since $T_{f(x)|n} \in U_n$ and U_m projects to U_n, it follows that $S_{x|m} \in U_m$. ⊣

A crucial feature of weakly homogeneously Suslin sets is that their complements are also Suslin by a canonical "dualization" process. This was already developed in §15, and with the motivation given there we can proceed forthwith to the formulation:

Suppose that T is a κ-weakly homogeneous tree on $\omega \times Y$ with corresponding κ-complete ultrafilters $U_{s,t}$ for $s, t \in {}^{<\omega}\omega$ with $|s| = |t|$. For $s, t, \overline{s}, \overline{t} \in {}^{<\omega}\omega$ with $s = \overline{s}|m$ and $t = \overline{t}|m$ for some $m \in \omega$, let

$$j_{s,t}\colon V \prec M_{s,t} \cong \mathrm{Ult}(V, U_{s,t})$$

and

$$j_{\langle s,t\rangle,\langle \overline{s},\overline{t}\rangle}\colon M_{s,t} \prec M_{\overline{s},\overline{t}}$$

be the embeddings corresponding to $U_{s,t}$ being a projection of $U_{\overline{s},\overline{t}}$, so that $j_{\langle s,t\rangle,\langle \overline{s},\overline{t}\rangle} \circ j_{s,t} = j_{\overline{s},\overline{t}}$. In terms of the fixed recursive enumeration $\langle \mathbf{s}_i \mid i \in \omega\rangle$ of ${}^{<\omega}\omega$, define a tree on $\omega \times (2^{|Y|})^+$ by:

$$\begin{aligned} \tilde{T} = \{\langle s, \langle \delta_i \mid i < |s|\rangle\rangle \ \in \textstyle\bigcup_{m\in\omega}({}^m\omega \times {}^m((2^{|Y|})^+) \mid \\ \forall i, j < |s|(\mathbf{s}_i \supset \mathbf{s}_j \ \rightarrow\ \delta_i < j_{\langle s||\mathbf{s}_j|,\mathbf{s}_j\rangle,\langle s||\mathbf{s}_i|,\mathbf{s}_i\rangle}(\delta_j))\}\,. \end{aligned}$$

The first part of the following is a representation that recasts 15.5 and 15.7 in terms of countably complete towers; the second shows that that representation is quite robust.

32.6 Proposition.
(a) $p[\tilde{T}] = {}^\omega\omega - p[T]$.
(b) For any p.o. P with $|P| < \kappa$, $\Vdash_P p[\check{\tilde{T}}] = {}^\omega\omega - p[\check{T}]$.

Proof. (a) Suppose first that $x \in p[\tilde{T}]$, say with $\langle x, g\rangle \in [\tilde{T}]$. For $i \in \omega$, let $h_i\colon T_{x||\mathbf{s}_i|} \rightarrow \mathrm{On}$ be such that $g(i) = [h_i]$ in $M_{x||\mathbf{s}_i|,\mathbf{s}_i}$. Then for $\mathbf{s}_i \supset \mathbf{s}_j$,

$$Z_{ij} = \{u \in T_{x||\mathbf{s}_i|} \mid h_i(u) < h_j(u||\mathbf{s}_j|)\} \in U_{x||\mathbf{s}_i|,\mathbf{s}_i}\,.$$

Assume now to the contrary that $x \in p[T]$. Then there would be a $y \in {}^\omega\omega$ such that $\langle U_{x|m,y|m} \mid m \in \omega\rangle$ is countably complete, and so a $d \in {}^\omega Y$ such that for any $\mathbf{s}_i \subset y$, $d||\mathbf{s}_i| \in \bigcap\{Z_{ij} \mid \mathbf{s}_i \supset \mathbf{s}_j\}$. Hence, when $y \supset \mathbf{s}_i \supset \mathbf{s}_j$, $h_i(d||\mathbf{s}_i|) < h_j(d||\mathbf{s}_j|)$. This leads to an infinite descending sequence of ordinals and thus a contradiction.

For the converse, suppose that $x \in {}^\omega\omega - p[T]$. Since T_x is well-founded, let $\rho\colon T_x \rightarrow \|T_x\|$ be its rank function. For $i \in \omega$ set:

$$\delta_i = [\rho | T_{x||\mathbf{s}_i|}]_{U_{x||\mathbf{s}_i|,\mathbf{s}_i}} .$$

Then for $\mathbf{s}_i \supset \mathbf{s}_j$,

$$\{u \in T_{x||\mathbf{s}_i|} \mid \rho(u) < \rho(u||\mathbf{s}_j|)\} \in U_{x||\mathbf{s}_i|,\mathbf{s}_i}$$

so that $\delta_i < j_{\langle s||\mathbf{s}_j|,\mathbf{s}_j\rangle,\langle s||\mathbf{s}_i|,\mathbf{s}_i\rangle}(\delta_j)$. To finish off an inconsequential case, if Y is finite, then the $U_{s,t}$'s are all principal, and so $p[T] = {}^{\omega}\omega$. If Y is infinite, then note that for $i \in \omega$,

$$|\delta_i| \leq |\{f \mid f\colon T_{x||\mathbf{s}_i|} \to \|T_x\|\}| \leq 2^{|Y|} .$$

Hence, all the conditions being satisfied, $x \in p[\tilde{T}]$.

(b) Suppose that P is a p.o. with $|P| < \kappa$ and G is P-generic. It must be shown that in $V[G]$, $p[\tilde{T}] = {}^{\omega}\omega - p[T]$. By the proof of 10.15, each $U_{s,t}$ generates a κ-complete ultrafilter $U^+_{s,t}$ over T_s in $V[G]$. For $s, t, \bar{s}, \bar{t} \in {}^{<\omega}\omega$ with $s = \bar{s}|m$ and $t = \bar{t}|m$ for some $m \in \omega$, let

$$j^+_{s,t}\colon V[G] \prec M^+_{s,t} \cong \mathrm{Ult}(V[G], U^+_{s,t}) ,$$

and as $U^+_{s,t}$ is readily seen to be a projection of $U^+_{\bar{s},\bar{t}}$, let

$$j^+_{\langle s,t\rangle,\langle\bar{s},\bar{t}\rangle}\colon M^+_{s,t} \prec M^+_{\bar{s},\bar{t}}$$

be the corresponding embedding, so that $j^+_{\langle s,t\rangle,\langle\bar{s},\bar{t}\rangle} \circ j^+_{s,t} = j^+_{\bar{s},\bar{t}}$. Then:

(i) If $f\colon T_s \to V$ is in $V[G]$, there is a $g\colon T_s \to V$ in V such that $[f]_{U^+_{s,t}} = [g]_{U^+_{s,t}}$.

(ii) For any $f\colon T_s \to V$ in V, $[f]_{U^+_{s,t}} = [f]_{U_{s,t}}$.

(iii) $j^+_{\langle s,t\rangle,\langle\bar{s},\bar{t}\rangle}|\mathrm{On} = j_{\langle s,t\rangle,\langle\bar{s},\bar{t}\rangle}|\mathrm{On}$.

(i) follows from the claim (∗) of the proof of 10.15. (ii) follows from (i) by induction on rank. Finally, for (iii), note first that for any ordinal ξ, $\xi = [g]_{U^+_{s,t}}$ for some $g\colon T_s \to V$ in V by (i). But then,

$$\begin{aligned} j^+_{\langle s,t\rangle,\langle\bar{s},\bar{t}\rangle}([g]_{U^+_{s,t}}) &= [\langle g(u|m) \mid u \in T_{\bar{s}}\rangle]_{U^+_{\bar{s},\bar{t}}} \\ &= [\langle g(u|m) \mid u \in T_{\bar{s}}\rangle]_{U_{\bar{s},\bar{t}}} \quad \text{(by (ii))} \\ &= j_{\langle s,t\rangle,\langle\bar{s},\bar{t}\rangle}([g]_{U_{s,t}}) \\ &= j_{\langle s,t\rangle,\langle\bar{s},\bar{t}\rangle}([g]_{U^+_{s,t}}) \quad \text{(again by (ii))} \end{aligned}$$

confirming (iii).

Proceeding to the argument that $p[\tilde{T}] = {}^{\omega}\omega - p[T]$ in $V[G]$, note first that $p[T] \cap p[\tilde{T}] = \emptyset$ there: Define a tree S merging T and $\tilde{T}$ by

$$S = \{\langle s, u, \langle\delta_i \mid i < |s|\rangle\rangle \mid \langle s, u\rangle \in T \wedge \langle s, \langle\delta_i \mid i < |s|\rangle\rangle \in \tilde{T}\} .$$

Then in V as well as in $V[G]$, $p[S] = p[T] \cap p[\tilde{T}]$ so that $p[T] \cap p[\tilde{T}] = \emptyset$ *iff* S is well-founded. But then, S is well-founded in V, and hence by absoluteness (0.3) in $V[G]$ as well, and the result follows.

It remains to show that in $V[G]$ if $x \in {}^{\omega}\omega - p[T]$, then $x \in p[\tilde{T}]$: Arguing in $V[G]$ as for (a), since T_x is well-founded, let $\rho\colon T_x \to \|T_x\|$ be its rank function, and for $i \in \omega$ set

$$\delta_i = [\rho | T_{x||\mathbf{s}_i|}]_{U^+_{x||\mathbf{s}_i|,\mathbf{s}_i}} .$$

For $\mathbf{s}_i \supset \mathbf{s}_j$, $\delta_i < j^+_{\langle s||\mathbf{s}_j|,\mathbf{s}_j\rangle,\langle s||\mathbf{s}_i|,\mathbf{s}_i\rangle}(\delta_j)$ as before, so that by (iii), $\delta_i < j_{\langle s||\mathbf{s}_j|,\mathbf{s}_j\rangle,\langle s||\mathbf{s}_i|,\mathbf{s}_i\rangle}(\delta_j)$. Hence, $\langle x, \langle \delta_i \mid i \in \omega\rangle\rangle \in [\tilde{T}]$, and so $x \in p[\tilde{T}]$. ⊣

Soon after their formulation, it was seen that weakly homogeneously Suslin sets possess the regularity properties. The first arguments grew out of the generative game context, and Woodin found new forcing proofs indicative of the new abstract approach. Versions of both are given in the following proof.

32.7 Theorem. *Suppose that a set of reals is weakly homogeneously Suslin. Then it is Lebesgue measurable, has the Baire property, and has the perfect set property.*

Proof. Suppose that $A \subseteq {}^{\omega}\omega$ is κ-weakly homogeneously Suslin. The perfect set property for A is first established in terms of games:

Consider the game $G_2^*(\Psi``A)$ from 27.6 for the perfect set property. By having I not play finite sequences but rather integers coding them the game can be recast as one of form $G_\omega(B)$. Some straightforward checking shows that B can be taken to be a continuous pre-image of A, and thus is κ-weakly homogeneously Suslin by 32.5.

By 32.3 there is a $C \subseteq {}^2({}^{\omega}\omega)$ κ-homogeneously Suslin with $B = pC$. As in the proof of 27.14, consider the unfolded version of the game $G_\omega(B)$ where I makes extra choices $y(i)$ leading to a real y, and I wins if $\langle x, y\rangle \in C$, and otherwise II wins. By a simple modification of the argument for 32.2, this unfolded game is determined. But $G_\omega(B)$ is essentially $G_2^*(\Psi``A)$, and so by the argument for 27.14 A has the perfect set property.

Lebesgue measurability and having the Baire property can be established similarly, although more checking has to done to be able to apply 32.5. Instead, Woodin's forcing argument is given, which builds on Solovay's for 11.11:

Let T be a κ-weakly homogeneous tree such that $A = p[T]$, and $\tilde{T}$ the canonical tree for the complement so that 32.6 is satisfied. To show that A is Lebesgue measurable proceed as for 11.11 in terms of random reals, the corresponding p.o. $\mathcal{B}^*$, and closed and G_δ codes:

With $\Vdash$ temporarily denoting a set restriction of the forcing relation for $\mathcal{B}^*$ sufficient for the coming argument, let μ be a regular cardinal so that H_μ contains $T, \tilde{T}, \mathcal{B}^*$, and $\Vdash$, and let $N \prec H_\mu$ be countable and containing these sets. In what follows, assume the straightforward liberalization of the necessary forcing concepts to such models of just ZFC$^-$. Note that

$$\{x \in {}^{\omega}\omega \mid x \text{ is not random over } N\}$$

is null, being equal by 11.10 to

$$\bigcup\{A_c \mid c \in N \text{ is a } G_\delta \text{ code for a null set}\}\ ,$$

a countable union of null sets since N is countable. Hence, it suffices to find a Borel set X such that $A \triangle X$ consists of reals not random over N:

For $\mathcal{B}^*$ forcing in the sense of N with $\dot{r}$ a canonical name for the random real, let $Q \in N$ be a maximal antichain consisting of closed sets each deciding $\dot{r} \in p[\check{T}]$, all in the sense of N. Set

$$X = \bigcup\{A_c \mid A_c^N \in Q \ \wedge\ A_c^N \Vdash \dot{r} \in p[\check{T}]\}\ .$$

Since Q is countable by the ω_1-c.c., X is Borel (in fact F_σ). To complete the proof it suffices to show that for x random over N,

$$x \in A \textit{ iff } x \in X\ .$$

First, if x is random over N and $x \in X$, then $N[x] \models x \in p[T]$ by the definition of X, and hence $x \in p[T] = A$. For the converse, the crucial property 32.6(b) is used: Since $\mathcal{B}^*$ has a dense subset of cardinality $2^{\aleph_0}$ (the closed non-null sets) and $\kappa > 2^{\aleph_0}$ can always be assumed,

$$\Vdash_{\mathcal{B}^*} p[\check{\tilde{T}}] = {}^{\omega}\omega - p[\check{T}]\ .$$

It follows from this and the definition of X that if x is random over N and $x \notin X$, then $N[x] \models x \in p[\tilde{T}]$, and hence $x \in p[\tilde{T}] = {}^{\omega}\omega - A$.

The argument for the Baire property is analogous, using Cohen reals. ⊣

The Ramsey property was discussed at the ends of §§11, 27. An analogous forcing argument using Mathias reals shows that *being Ramsey can be added to the conclusions of 32.7*. And as for Lebesgue measure and Baire category there is a more involved game argument. This depends on Ramseyness being equivalent to having the Baire property in a topology (as mentioned in §27), and the determinacy of the unfolded version of the corresponding **-game as in the proof of 27.14 (see Kechris [95: 19D, 21D]).

The forcing approach depended on the 32.6 property of both a set of reals and its complement being Suslin with representations that are robust through forcing extensions. In Feng-Magidor-Woodin [92], this property is called "universally Baire" and studied for its own sake as well as in connection with determinacy.

Having set the stage with the basic theory of homogeneous and weakly homogeneous trees, we begin the historical ascent. Martin made the first connection between determinacy and weak homogeneity by showing in ZF that $\mathsf{AD}_{\mathbb{R}}$ *implies that every tree is ω_1-weakly homogeneous.* Inspired by this, Woodin by 1982 had established the first result in the general framework of the subsequent results:

32.8 Theorem (Woodin [86]). *Suppose that κ is a supercompact cardinal and G is* Col(ω, κ)*-generic. Then $L(\mathbb{R})^{V[G]} \models$ Every tree is κ-weakly homogeneous.*

The appeal to $L(\mathbb{R})^{V[G]}$ is as for Solovay's 11.1. Woodin used ultrafilters given by supercompactness to verify weak homogeneity, but in any case 32.8 is a relative consistency result. To describe the progress toward actual implications from large cardinals, the situation is first summarized in the investigation of strong hypotheses in the early 1980's, a subject that is taken up at length in volume II.

The Role of Supercompact Cardinals

After the introduction of the strong large cardinal hypotheses (as described in Chapter 5) the work through the 1970's led to a growing conviction that these hypotheses provide an ultimate scale against which all possible consistency strengths can be gauged. The stronger propositions were usually found to be consistent relative to supercompact cardinals, and also to imply the consistency of having many measurable cardinals in a strong sense. There was one early and notable exception, however, the proposition that there is an ω_2-saturated ideal over ω_1. Kunen [78] in 1972 had established the consistency of this proposition relative to a huge cardinal, and even gave a heuristic argument for why something like it may be necessary.

Kunen collapsed a huge cardinal to ω_1 in such a way that a residue of a huge embedding is retained as an ω_2-saturated ideal over ω_1. Not only are such ideals of general interest as a combinatorial feature low in the cumulative hierarchy but Solovay [71] had shown that they have a intriguing technical property: Starting with such an ideal I there is a forcing extension $V[G]$ (in fact via the p.o. $\mathcal{P}(\omega_1) - I$) in which there is a *generic elementary embedding*, a $j \colon V \prec M$ with critical point ω_1. Kunen observed that ${}^{<j(\omega_1)}M \subseteq M$, so that j is like an embedding witnessing the almost hugeness of ω_1. But the embedding has domain just the ground model V, and only exists in the generic extension $V[G]$. Still, this was a heuristic argument taken to suggest that saturated ideals might have consistency strength on the order of hugeness, that the generic embedding might foreshadow a full-fledged embedding.

Kunen's work emboldened set theorists to use strong large cardinal hypotheses as in the formative work of Foreman [82, 83], and generated a continuing interest in ideals with strong properties low in the cumulative hierarchy. In fact, this interest was not only to trigger the major advances toward the consistency of determinacy, but to continue to interweave ideals into the unifying theory as one of a triad: large cardinals, determinacy, and ideals.

Two synthetic results in the late 1970's injected a sense of anticipation as well as mystery into the growing mosaic. One was the Martin result 31.8 that I2 implies Det($\mathbf{\Pi}^1_2$), and the other was the result of Steel-Van Wesep [82] that in ZF *if* $\mathrm{AD}_{\mathbb{R}}$ *and Θ is regular, then in a forcing extension there is an inner model of* ZFC *in which* NS_{ω_1} *is ω_2-saturated.* The previous work of Martin and others had suggested that Det($\mathbf{\Pi}^1_2$) is an apparently very strong hypothesis, but the upper bound established by Martin raised the specter of having to reach up to the top of the large cardinal hierarchy to pull in determinacy. The Steel-Van Wesep result established the consistency of a basic proposition about stationary subsets of ω_1 that had not been derived from large cardinals; recalling the ideal terminology from §16, Kunen's ideal was not even of form $\mathrm{NS}_{\omega_1}|S$ for any stationary $S \subseteq \omega_1$. This raised the prospect of having to enlist determinacy hypotheses in order to secure the consistency of strong combinatorial propositions in ZFC.

Both the Martin and Steel-Van Wesep results were improved by Woodin. Around 1981 he reduced the hypothesis of the latter result to just AD (see his [83a]), and in early 1984 he improved the conclusion of Martin's result to $\mathrm{AD}^{L(\mathbb{R})}$, albeit starting from the stronger I0 (31.9). This was an important advance in that it did subsume AD into the large cardinal hierarchy, but any possibility of an equiconsistency result remained remote. It turned out that, as sometimes happens in mathematics, a fresh infusion of ideas from a different quarter was to lead to a major reorientation and generate new results.

The break came in the work of three fine mathematicians at the forefront in the investigation of large cardinals, Foreman, Magidor, and Shelah. In a major collaboration in early 1984 in Jerusalem they [88, 88a] implemented a project of larger significance for set theory which led to a new understanding of strong propositions and the possibilities with forcing. Applying the powerful semi-proper forcing techniques of Shelah [82], they [88] collapsed a supercompact cardinal κ to ω_2 via a κ-c.c. forcing so that in the extension a provably maximal form of Martin's Axiom, *Martin's Maximum*, holds. They then established the relative consistency of several propositions by deriving them directly from this new axiom. One such proposition was that NS_{ω_1} is ω_2-saturated. Hence, not only was the upper bound for the consistency strength of having an ω_2-saturated ideal over ω_1 considerably reduced from Kunen's huge cardinal, but for the first time the consistency of NS_{ω_1} itself being ω_2-saturated was established relative to large cardinals rather than determinacy.

With their work Foreman, Magidor, and Shelah had overturned a long-held view about the scaling down of large cardinal properties. In the first flush of new hypotheses and propositions, Kunen had naturally enough collapsed a large cardinal to ω_1 in order to transmute strong properties of the cardinal into an ω_2-saturated ideal over ω_1, and this sort of direct connection had become the rule. The new discovery was that a careful collapse of a large cardinal to ω_2 instead can provide enough structure to secure such an ideal. In particular, the generic elementary embedding given by the further forcing with the saturated ideal does not extend any previous embedding associated with the original large cardinal.

And now for a crucial connection that was made with Lebesgue measurability. As described in Foreman-Magidor-Shelah [88: §3], results of Shelah and Woodin when put together led to a direct implication about $L(\mathbb{R})$ from supercompactness:

Martin's Maximum implies that $2^{\aleph_0} = \aleph_2$. Shelah realized that the technique of S-closed forcing from his [82] can be incorporated into the argument for Martin's Maximum to get a version of that axiom consistent with CH: The supercompact cardinal κ is collapsed by an amended κ-c.c. forcing which moreover is ω_1-Baire and so *does not add any new reals*. Starting in the ground model V with an $S \subseteq \omega_1$ so that both S and $\omega_1 - S$ are stationary, this S-closed forcing arranges in any generic extension $V[G]$ that $\mathrm{NS}_{\omega_1}|(\omega_1 - S)$ is ω_2-saturated.

In long-distance telephone conversation with Foreman in April 1984, Woodin outlined a proof that with the reals unchanging in such a collapse, the existence of a generic elementary embedding of the type given by an ω_2-saturated ideal over ω_1 in a further extension of $V[G]$ implies that *in the ground model V every set of reals in $L(\mathbb{R})$ is Lebesgue measurable.* This depended on the Solovay result 11.1 and the known fact that for large cardinals κ, the reals in any generic collapse of κ to ω_1 via a κ-c.c. forcing coincide with the reals of a $\mathrm{Col}(\omega, \kappa)$ Levy collapse. (Notably, arguments using a generic elementary embedding to secure Lebesgue measurability in the ground model had occurred in previous work of Magidor [80] and Foreman [86].) The Shelah realization and the Woodin observation established: *If there is a supercompact cardinal, then every set of reals in $L(\mathbb{R})$ is Lebesgue measurable.*

Such an implication from a large cardinal hypothesis precludes the possibility of any inner model of it with a reasonable well-ordering of the reals (cf. the remarks after 31.8). With the familiar supercompact cardinals exhibiting such "wildness" the quest was on to reduce the large cardinal hypothesis sufficient for the implication. This was later described in Shelah-Woodin [90], the concepts involved having been explored in §26: Woodin first noticed that a superstrong cardinal suffices. While Woodin was visiting Jerusalem in June 1984, Shelah isolated a weaker concept incorporating functional conditions, later called a *Shelah cardinal*, and implemented his crucial idea of investing one such cardinal to get Lebesgue measurability for each level of the projective hierarchy. But such a cardinal still reflected the requisite technical properties downward, and so less would do. Soon afterward Woodin formulated the crucial concept of Woodin cardinal by inverting quantifiers in Shelah's definition, and showed that it suffices for Shelah's scheme:

32.9 Theorem (Shelah-Woodin [90]).

(a) If $n \in \omega$ and there are n Woodin cardinals with a measurable cardinal above them, then every $\mathbf{\Sigma}^1_{n+2}$ set of reals is Lebesgue measurable.

(b) If there are infinitely many Woodin cardinals with a measurable cardinal above them, then every set of reals in $L(\mathbb{R})$ is Lebesgue measurable.

The $n = 0$ case of (a) is Solovay's 14.3. Woodin cardinals seemed to encapsulate just what is needed to carry out the collapsing argument that led to saturated ideals, and moreover the proofs suggested optimality – recall that Woodin cardinals are not necessarily weakly compact so do not directly reflect substantial properties. The measurable cardinal hovering above was to be a recurring theme, the purpose being loosely speaking to maintain a stable environment with the existence of sharps. Since Woodinness in consistency strength is much stronger than measurability, it is a mild if annoying peculiarity of the theory. Finally, 32.9 was the first instance of a circumstance remarkable granting the optimality the hypotheses: (a) leads from infinitely many Woodin cardinals to the Lebesgue measurability of every projective set of reals, yet the marginally stronger hypothesis of (b) suddenly opens the floodgates to the measurability of every set of reals in $L(\mathbb{R})$.

As part of this development Shelah in early 1985 reduced the consistency strength for saturated ideals to a Woodin cardinal:

32.10 Theorem (Shelah). *Suppose that κ is Woodin. Then there is a forcing extension in which $\kappa = \omega_2$ and NS_{ω_1} is ω_2-saturated.*

As with versions of Martin's Maximum consistent with CH, Shelah was also able to get CH to hold in the extension at the cost of the ideal being of the less canonical form $\mathrm{NS}_{\omega_1}|S$. See Shelah [87] for related results.

Woodin made an important conceptual move in the Autumn of 1984 that streamlined much of the previous work. That work had turned on the process of collapsing a large cardinal to get a saturated ideal and then a generic elementary embedding. As in any case there is a forcing that does all this, Woodin stalked for the essence and formulated his *stationary tower forcing*. This notion of forcing, given sufficient hypotheses, directly provides a generic elementary embedding $j\colon V \prec M$ with $\mathrm{crit}(j) = \omega_1$ and ${}^{\omega}M \subseteq M$ among a multitude of possibilities. It looks strange at first, but can be seen as an outgrowth of the work of Foreman-Magidor-Shelah [88], both in the adoption its generalization of stationariness and the use of a key "semi-properness" property needed to "capture" dense open sets. With stationary tower forcing, Woodin eliminated saturated ideal arguments from the derivation of the regularity properties for sets of reals.

In early 1985 Woodin [88] applied his stationary tower forcing both to improve his earlier 32.8 to a direct implication and to replace Lebesgue measurability by a prior representability for sets of reals: *If there is a supercompact cardinal, then every set of reals in $L(\mathbb{R})$ is weakly homogeneously Suslin.* With Woodin cardinals soon seen to provide just the right technical strength to implement stationary tower forcing Woodin then weakened the hypothesis to an accumulation of many Woodin cardinals. He did not get to the optimal result, however, before a major advance was made from a different quarter.

Woodin Cardinals and Determinacy

With the hypotheses dramatically reduced for getting the Lebesgue measurability of the projective sets, the growing web of connections set the stage for Martin and Steel's breakthrough in September 1985. The new Woodin cardinals held out the hope for getting results along the lines of Martin's correlations of $\mathrm{Det}(\alpha\text{-}\mathbf{\Pi}^1_1)$ with inner models having many measurable cardinals. Alive to the possibility Martin and Steel were able to derive determinacy by converting an obstacle in the developing inner model theory into an asset. They wrote in their systematic exposition [89: 74]:

> ... the authors owe a great debt to the work of Foreman, Magidor, Shelah, and Woodin. Nevertheless, the technical ideas in our proof have no relation to that work. We arrived at our proof by asking what it is about Woodin cardinals that makes their inner models so complicated. In earlier work on inner model theory, there had arisen the worry that superstrong cardinals might generate complicated "iteration" trees .... It turns out that Woodin cardinals generate such trees, and, while this is a problem for inner model theory, it can be used to prove determinacy.

Iteration trees was the means through which Martin and Steel established their results, and Woodin cardinals, complementing their role in stationary tower forcing, provided just the right technical strength to construct such trees. Woodin cardinals thus became firmly established as the central large cardinal concept in the study of the consistency of determinacy.

Martin and Steel's main result was a "transfer" theorem that can be motivated as a means of propagating determinacy through the projective hierarchy via homogeneous trees, just as reduction was propagated via prewellorderings and uniformization via scales. If there is a measurable cardinal κ, then the $\mathbf{\Pi}^1_1$ sets are κ-homogeneously Suslin and so are determined (32.1, 32.2), and $\mathbf{\Sigma}^1_2$ sets are κ-weakly homogeneously Suslin (32.3). A means must now be devised, as for the other regularity properties, to continue to the next level. To a weakly homogeneous tree T corresponds a tree $\tilde{T}$ so that $p[\tilde{T}] = {}^{\omega}\omega - p[T]$ (32.6). The transfer theorem used a Woodin cardinal and the aforementioned iteration trees to derive homogeneity for $\tilde{T}$, and thus determinacy for $p[\tilde{T}]$, from the weak homogeneity of T:

32.11 Theorem (Martin-Steel [88, 89]). *Suppose that λ is a Woodin cardinal and T is a λ^+-weakly homogeneous tree. Then for any $\gamma < \lambda$, $\tilde{T}$ is γ-homogeneous.*

Hence, continuing from $\mathbf{\Sigma}^1_2$ being κ-weakly homogeneously Suslin, if there is a Woodin cardinal $\lambda_0 < \kappa$, then the $\mathbf{\Pi}^1_2$ sets are μ-homogeneously Suslin for any $\mu < \lambda_0$ and $\mathbf{\Sigma}^1_3$ sets are μ-weakly homogeneously Suslin for such μ. And if there is a Woodin cardinal $\lambda_1 < \lambda_0$, the propagation can continue:

32.12 Corollary. *If $n \in \omega$ and there are n Woodin cardinals with a measurable cardinal above them, then* $\mathrm{Det}(\mathbf{\Pi}^1_{n+1})$.

The form of this result had been suggested by the reduction 32.9(a), remembering that $\mathrm{Det}(\mathbf{\Pi}^1_{n+1})$ implies that every $\mathbf{\Sigma}^1_{n+2}$ set of reals is Lebesgue measurable (27.14). By July 1986, Martin and Steel (see their [94]) had advanced the inner model theory enough to get such a model with n Woodin cardinals and a Σ^1_{n+2}-good well-ordering of the reals. Such a well-ordering is not Lebesgue measurable (cf. 13.10), and so in 32.12 (and 32.9(a)) the measurable cardinal hovering above cannot be eliminated.

Within weeks after the Martin-Steel result 32.12 was achieved, Woodin in the autumn of 1985 had established the optimal result for getting weakly homogeneous representations in $L(\mathbb{R})$:

32.13 Theorem (Woodin). *Suppose that there are infinitely many Woodin cardinals with a measurable cardinal above them. Then for any λ less than the supremum of those Woodin cardinals, every set of reals in $L(\mathbb{R})$ is λ-weakly homogeneously Suslin.*

The following is an immediate consequence of 32.11 and 32.13, and as with the latter it is surprising in light of the optimality of 32.12 that just having one measurable cardinal above should encompass all sets of reals in $L(\mathbb{R})$.

32.14 Corollary. *If there are infinitely many Woodin cardinals with a measurable cardinal above them, then* $\mathrm{AD}^{L(\mathbb{R})}$.

Although the pace of narrative is slowed, a component of 32.13 is established to convey something of how large cardinals provide weakly homogeneous representations for sets of reals:

32.15 Proposition (Woodin). *Suppose that κ is inaccessible and there is a generic extension $V[G]$ in which: $\kappa = \omega_1$ and there is a $j \colon V \prec M$ with* $\mathrm{crit}(j) = \omega_1$ *and ${}^{\omega}M \subseteq M$. Suppose also that $A \subseteq {}^{\omega}\omega$ and there are trees T and T' such that $A = p[T]$ and $p[T'] = {}^{\omega}\omega - p[T]$, and this also holds in $V[G]$. Suppose finally that $\lambda < \kappa$ is T-strong. Then A is λ-weakly homogeneously Suslin.*

Remarks. The existence of such a generic extension follows via stationary tower forcing when κ is Woodin. The existence of such tree representations for an arbitrary set A of reals and its complement, robust through generic extensions, follows from the existence of infinitely many Woodin cardinals with a measurable cardinal above them. Finally, after a contextual argument showing that for present purposes $T \subseteq V_\kappa$ can be assumed, the existence of arbitrarily large $\lambda < \kappa$ such that λ is T-strong follows from 26.14 when κ is Woodin. In particular, this result is a weak converse to 32.6 in the presence of a Woodin cardinal.

Proof. It is first deduced that because λ is T-strong, $A = p[T] = p[T \cap V_\lambda]$, and the expected argument is used to generate ultrafilters that motivate the main construction:

Suppose that $x \in A$, say with $\langle x, f \rangle \in [T]$. Let $\gamma = \mathrm{rank}(f)$ and $e \colon V \prec N$ be such that $\mathrm{crit}(e) = \lambda$, $\gamma < e(\lambda)$, $V_{\gamma+1} \subseteq N$, and $T \cap V_{\gamma+1} = e(T) \cap V_{\gamma+1}$. It follows readily that $\langle x, f \rangle \in [e(T)]$, and so by elementarity there is some $g \in V_\lambda$ such that $\langle x, g \rangle \in [T]$. But then, $x \in p[T \cap V_\lambda]$.

Continuing with the $\langle x, f \rangle \in [T]$ and e as above, for $m \in \omega$ define U_m by:

$$X \in U_m \ \text{ iff } \ X \subseteq V_\lambda \ \wedge \ f|m \in e(X) \, .$$

Then U_m is a λ-complete (possibly principal) ultrafilter over V_λ. The definition recalls the formulation of extenders from elementary embeddings (§26), except that the function $f|m$ is used rather than the set $\mathrm{ran}(f|m)$. As for extenders,

$\langle U_m \mid m \in \omega \rangle$ is a countably complete tower: That it is a tower follows from the definition of the U_m's. That it is countably complete can be verified directly from elementarity, since if $X_m \in U_m$ for $m \in \omega$, then as $f|m \in e(X_m)$ for such m and $f \in N$, there must be some $d \in V$ such that $d|m \in X_m$ for such m. Finally, note that for $m \in \omega$,

$$(T \cap V_\lambda)_{x|m} \in U_m \textit{ iff } f|m \in e(T_{x|m}) \textit{ iff } f|m \in e(T)_{x|m} \textit{ iff } \langle x|m, f|m \rangle \in e(T) ,$$

but this last holds, as noted before.

Setting

$$\mathcal{E} = \{U \mid U \text{ is a } \lambda\text{-complete ultrafilter over } V_\lambda\} ,$$

the above argument shows that $\mathcal{E}$ verifies the λ-weak homogeneity of $T \cap V_\lambda$ in the sense of the characterization 32.4, *except* that $\mathcal{E}$ is not countable. It is to remedy this defect that the other hypotheses are used in a further reflection argument. Note first the $\mathcal{E}$ *is* countable in $V[G]$ since κ is inaccessible in V and $\kappa = \omega_1$ in $V[G]$. Since ${}^\omega M \subseteq M$, it readily follows that $j``\mathcal{E} \in M$ and is countable there. Building on this it will be argued that there is enough cohesion between V and M to establish the claim, with that 32.4 sense of "verifies", that

$$M \models\ j``\mathcal{E} \text{ verifies the } j(\lambda)\text{-weak homogeneity of } j(T \cap V_\lambda) .$$

Once this is established the proof would be complete, since by elementarity there would be *some* witness to the λ-weak homogeneity of $T \cap V_\lambda$, and $A = p[T \cap V_\lambda]$.

To establish the claim, suppose that $M \models x \in p[j(T \cap V_\lambda)]$. It must be shown that in M there is a countably complete tower $\langle W_m \mid m \in \omega \rangle \in {}^\omega j``\mathcal{E}$ such that $j(T \cap V_\lambda)_{x|m} \in W_m$ for every $m \in \omega$. First note that since $p[T'] = {}^\omega\omega - p[T]$ in V, $M \models x \notin p[j(T')]$ by elementarity. It follows from this that $x \in p[T]$ in $V[G]$:

Arguing in $V[G]$, assume to the contrary that $x \notin p[T]$. Then since $p[T'] = {}^\omega\omega - p[T]$ persists there by hypothesis, $x \in p[T']$, say with $\langle x, g \rangle \in [T']$. It follows readily that $\langle x, j``g \rangle \in [j(T')]$ where $j``g$ is of course the function $\{\langle n, j(g(n)) \rangle \mid n \in \omega\}$, and moreover $j``g \in M$ as ${}^\omega M \subseteq M$. But then, $M \models x \in p[j(T')]$, which is a contradiction.

With $x \in p[T]$ in $V[G]$ confirmed, let $f \in V[G]$ be such that $\langle x, f \rangle \in [T]$ and set $\text{rank}(f) = \gamma$. In V, let $e \colon V \prec N$ be as at the beginning of this proof, with $\text{crit}(e) = \lambda$, $\gamma < e(\lambda)$, $V_{\gamma+1} \subseteq N$, and $T \cap V_{\gamma+1} = e(T) \cap V_{\gamma+1}$. Although f is not necessarily in V, for each $m \in \omega$ the finite set $f|m$ can be used in V as before to define an ultrafilter $U_m \in \mathcal{E}$.

In $V[G]$, f is available and so also is the sequence $\langle U_m \mid m \in \omega \rangle$, and as argued previously this sequence is a countably complete tower such that $(T \cap V_\lambda)_{x|m} \in U_m$ for each $m \in \omega$. It follows from ${}^\omega M \subseteq M$ that $j``f = \{\langle n, j(f(n)) \rangle \mid n \in \omega\} \in M$ and $\langle j(U_m) \mid m \in \omega \rangle \in M$, and from elementarity that this sequence is a tower drawn from $j``\mathcal{E}$ such that $j(T \cap V_\lambda)_{x|m} \in j(U_m)$ for each $m \in \omega$. To complete the proof, it is checked that this tower is countably complete in M:

In V set $\overline{e} = e|V_{\lambda+1}$; then in the definition of the U_m's, e can be replaced by $\overline{e}$. In its terms the previous argument for the U_m's can be transferred from V: Suppose that in M, $X_m \in j(U_m)$ for $m \in \omega$, so that by elementarity $j(f|m) \in j(\overline{e})(X_m)$ for such m. Since $j``f \in M$, and by elementarity $\text{rank}(j``f) \leq j(\gamma)$ and $V^M_{j(\gamma)+1} \subseteq \text{ran}(j(\overline{e}))$, it follows that $j``f \in \text{ran}(j(\overline{e}))$, and of course $(j``f)|m = j(f|m)$ for $m \in \omega$. But then, the elementarity of $j(\overline{e})$ implies that there must be some $d \in M$ such that $d|m \in X_m$ for every $m \in \omega$. This completes the proof of the claim about $j``\mathcal{E}$ in M, and with it, the theorem. ⊣

This argument highlights the crucial roles played in the theory of the property ${}^{\omega}M \subseteq M$ and of $p[T'] = {}^{\omega}\omega - p[T]$ holding in $V[G]$ as well as in V.

With the consistency strength of AD gauged by 32.14, Woodin soon established the crowning equiconsistency result:

32.16 Theorem (Woodin). *The following are equiconsistent:*
(a) ZFC + There are infinitely many Woodin cardinals.
(b) ZF + AD.

Whether or not this had to be the theorem, the scant remarks made here can convey little of the depth and sophistication of the proof (for which and others see Woodin-Mathias-Hauser [∞]). In the forward direction, infinitely many Woodin cardinals are all made countable and a sequence $\langle G_n \mid n \in \omega\rangle$ of generics for their stationary tower forcings is developed that cohere: In each $V[G_n]$ there is a generic elementary embedding $j_n \colon V \prec M_n$, *and* for $n \leq \overline{n}$ there are maps $j_{n,\overline{n}} \colon M_n \prec M_{\overline{n}}$ that commute with the j_n's. Taking $\mathbb{R}_n$ to be the reals of M_n, the desired model of AD is $L(\bigcup_n \mathbb{R}_n)$. The final argument is one by contradiction based on a constructively least counterexample. For that allegedly undetermined set A of reals a robust tree representation for it and its complement is found, and from the consequent weak homogeneity a contradiction is reached by applying the Martin-Steel result 32.11 to derive the determinacy of A.

For the converse direction, it is first established that *if* AD *and* $V = L(\mathbb{R})$, *then* Θ *is Woodin in* HOD: HOD is of course the class of hereditarily ordinal definable sets. AD is applied most directly in an argument, recalling Solovay's original proof of the measurability of ω_1, to get certain ultrafilters. A crucial normality property is then derived for these ultrafilters through the definability properties available in HOD. Finally, a uniform version of the Moschovakis Coding Lemma is applied to show that these ultrafilters provide embeddings of HOD that witness the strongness properties below Θ needed to affirm its Woodinness.

To get an inner model with infinitely many Woodin cardinals, an infinite $\supseteq$-decreasing chain $\langle M_i \mid i \in \omega\rangle$ of inner models is first constructed based on the above HOD result. Starting with $M_0 = \text{HOD}^{L(\mathbb{R})}$, M_1 is (the transitive collapse of) the ultrapower of M_0 constructed in $L(\mathbb{R})$ using an analogue of the Martin ultrafilter over the Turing degrees. The ultrapower indirectly generates a new set of reals $\mathbb{R}_1$ such that $\mathbb{R}_1 \supseteq \mathbb{R}$, and an embedding: $L(\mathbb{R}) \prec L(\mathbb{R}_1)$. By elementarity

AD holds in $L(\mathbb{R}_1)$ and $M_1 = \mathrm{HOD}^{L(\mathbb{R}_1)}$, and the construction is iterated to get all the M_i's. Recasting the $\mathrm{HOD}^{L(\mathbb{R})} \models \ulcorner\Theta$ is Woodin$\urcorner$ result, an increasing sequence $\langle\theta_i \mid i \in \omega\rangle$ of ordinals is then found with accompanying "witnesses" $W_i \subseteq \theta_i$ defined uniformly so that θ_i is Woodin in $M_i[W_i]$. With $W = \langle W_i \mid i \in \omega\rangle$ and $M = \bigcap_i M_i$ the argument is concluded by showing that in $M[W]$ each θ_i is Woodin.

As described in §31, $\mathrm{Det}(\mathbf{\Pi}^1_2)$ had been a focal hypothesis for the consistency study of determinacy through the 1970's. In 1989 Woodin settled matters with the following elegant characterizations:

32.17 Theorem (Woodin). *The following are equiconsistent:*
(a) ZFC + $\mathrm{Det}(\Pi^1_2)$.
(b) ZFC + There is a Woodin cardinal.

32.18 Theorem (Woodin). *The following are equivalent:*
(a) $\mathrm{Det}(\mathbf{\Pi}^1_2)$.
(b) For any $a \in {}^\omega\omega$ there is a countable ordinal δ such that δ is Woodin in an inner model of ZFC *containing a.*

The (a) to (b) direction figured in the proof of 32.16, and begins from a surprising result of Kechris-Solovay [85] in ZF + DC, that *for any $a \in \omega$, $L[a] \models$* $\mathrm{Det}(\Delta^1_2)$ *implies* Det(OD). That is, in such models determinacy for *lightface Δ^1_2* sets already implies determinacy for the largest class of lightface "definable"sets, the ordinal definable sets.

It is known that there is no direct analogue of 32.17 for $\mathrm{Det}(\Pi^1_n)$ when $n > 2$ in terms of merely having a number of Woodin cardinals. Woodin has explored this terrain, and has tentative large cardinal formulations for equiconsistencies.

What about $\mathrm{AD}_{\mathbb{R}}$? By our surface remarks at the end of §27 no inner model of $\mathrm{AD}_{\mathbb{R}}$ can be of form $L(\mathbb{R} \cup \{S\})$ for any $S \subseteq \mathbb{R}$. Woodin was able to show that if there is an inner model of $\mathrm{AD}_{\mathbb{R}}$ containing every real, then there is a smallest such model (see Steel [88: §3]). Moreover, he was able to get one from a moderate augmentation of having infinitely many Woodin cardinals:

32.19 Theorem (Woodin). *Suppose that there is a δ such that* $\mathrm{cf}(\delta) = \omega$, *$\delta$ is a limit of Woodin cardinals, and δ is also a limit of cardinals γ-strong for every $\gamma < \delta$. Then there is a forcing extension in which there is an inner model containing every real and satisfying* $\mathrm{AD}_{\mathbb{R}}$.

The proof is along the lines of that for the forward direction of 32.16, and recent results suggest that the hypothesis may be the exact consistency strength for $\mathrm{AD}_{\mathbb{R}}$. As mentioned earlier, ZF + $\mathrm{AD}_{\mathbb{R}} + \ulcorner\Theta$ is regular$\urcorner$ is a strong hypothesis that implies the consistency of ZF + $\mathrm{AD}_{\mathbb{R}}$ (Solovay [78a]), and served as the first sufficient hypothesis for getting the relative consistency of NS_{ω_1} is ω_2-saturated (Steel-Van Wesep [82]). There were various indications that $\mathrm{AD}_{\mathbb{R}} + \ulcorner\Theta$ is regular$\urcorner$ has stronger consistency strength than any accumulation of Woodin cardinals, but in 1992 Woodin was able to tame even this hypothesis. Recall from §24 the hypothesis I1 that for some δ, there is a $j\colon V_{\delta+1} \prec V_{\delta+1}$, and that a marginally stronger hypothesis just short of Kunen's inconsistency figured in the first proof of the relative consistency of AD (31.9).

32.20 Theorem (Woodin). *Suppose that there is a* $j\colon V_{\delta+1} \prec V_{\delta+1}$ *for some* δ*, say with* $\mathrm{crit}(j) = \kappa$. *Then the following holds in any generic extension via* $\mathrm{Col}(\omega, \kappa)$*: There is an inner model containing every real and satisfying* $\mathrm{AD}_{\mathbb{R}} + \ulcorner\Theta$ is regular$\urcorner$.

Large cardinals have reasserted themselves once again, but surely the hypothesis here is not optimal.

Among a growing number of synthetic results that have built a large web of connections around determinacy, the more accessible are cursorily described to bring this overview to a close.

Determinacy in Terms of Regularity Properties

In his early investigations toward equivalences for determinacy (cf. 30.28) Woodin considered the possibility that determinacy hypotheses might be equivalent to some of their structural consequences for sets of reals. He posed the following question for the projective sets, since answered in the negative as dicussed below.

32.21 Question (ZF + DC)(Woodin [82]). *Are the following equivalent:*

(a) PD.

(b) Every projective set of reals is Lebesgue measurable and has the Baire property, and every projective subset of ${}^2({}^\omega\omega)$ *can be uniformized by a projective set.*

Lebesgue measurability and having the Baire property, being antithetical to choice principles, are complemented by uniformizability as just such a principle, and these consequences of PD are together to provide the strength to recover PD. If so, a hypothesis with *ad hoc* features would gain considerably in justification and explication through being equated with consequences of structural immediacy. Although Woodin [82] did not answer his question, he was able to derive from (b) that $\forall a \in {}^\omega\omega(a^\dagger$ exists) ($a^\dagger$ was defined at the end of §21).

With the resolution of 32.21 the ultimate goal, Hauser [95] in the wake of considerable advances in inner model theory established that projective absoluteness, the proposition that projective relations are absolute for V in any generic extension (cf. 15.6), is equiconsistent with the existence of infinitely many strong cardinals.

Then somewhat unexpectedly, Steel in the Fall of 1997 answered 32.21 in the negative. With the relevance of strong cardinals seen and building on work of Woodin, Steel established that the consistency of 32.21(b) already follows from the consistency of the rather weak proposition (*): There is an increasing sequence

$\langle \kappa_n \mid n \in \omega \rangle$ of cardinals with λ their supremum such that: For any $n \in \omega$ and $x \in V_{\lambda+1}$ there is a $j\colon V \prec M$ with $\mathrm{crit}(j) = \kappa_n$ and $x \in M$.

Hauser-Schindler [00] then established a relative consistency result that was just short of being the converse to Steel's, using inner model apparatus that necessitated the existence of a measurable cardinal above. Finally, resolving the issue with an exact large cardinal analysis Schindler [02] established the equiconsistency of 32.21(b) with (*).

In the early 1990's Woodin was able to answer the question analogous to 32.21 for full AD in $L(\mathbb{R})$; recall the 30.25 delimitation at Π^2_1 for uniformization there.

32.22 Theorem (ZF + DC)(Woodin). *Assume* $V = L(\mathbb{R})$. *Then the following are equivalent:*

(a) AD.

(b) Every set of reals is Lebesgue measurable and has the Baire property, and every $\boldsymbol{\Sigma}^2_1$ *subset of* ${}^2({}^\omega\omega)$ *can be uniformized.*

The new (b) to (a) direction provided by Woodin used recent results of Steel in inner model theory. The import of 32.21 for PD carries over to 32.22 for AD, and this result, characterizing the regularity properties for the large class of sets in $L(\mathbb{R})$, can be viewed as a culmination of the efforts of the classical descriptive set theorists.

What about $\mathrm{AD}_{\mathbb{R}}$ in terms of regularity properties? Formulations along these lines stirred interest as a possible approach to comprehending this strong hypothesis. $\mathrm{AD}_{\mathbb{R}}$ implies that every subset of ${}^2({}^\omega\omega)$ can be uniformized (27.15). Consider the following hypothesis intermediate between AD and $\mathrm{AD}_{\mathbb{R}}$:

($\mathrm{AD}^{1/2}_{\mathbb{R}}$) Every game where one player plays reals and the other plays integers is determined .

Clearly this hypothesis suffices for the proof of 27.15 (since player I can choose a real x, and forgetting about the rest of his moves player II can be required to choose integers forming a real y such that $\langle x, y\rangle \in A$). In the early 1980's Kechris [88] established the converse in ZF + DC: $\mathrm{AD}^{1/2}_{\mathbb{R}}$ *iff* AD + ⌜Every subset of ${}^2({}^\omega\omega)$ can be uniformized⌝. With uniformizability seemingly a weak choice principle, this result suggested that $\mathrm{AD}^{1/2}_{\mathbb{R}}$ is a weak augmentation of AD.

On the other hand, Woodin showed that in ZF + DC, AD + ⌜Every set of reals has a scale⌝ *implies* $\mathrm{AD}_{\mathbb{R}}$. Scales provide tree representations (30.2), and these in turn lead to representations as γ-Borel sets (30.13(a)) and analogues of the projective ordinals (cf. 30.12). AD + ⌜Every set of reals has a scale⌝ thus has consequences beyond what was known to follow from $\mathrm{AD}_{\mathbb{R}}$: (i) Every set of reals is ∞-Borel, i.e. in the smallest collection of sets of reals containing the open sets and closed under complementation and the taking of well-ordered unions (of arbitrary length). (ii) There are arbitrarily large regular successor cardinals below Θ (those analogues of the projective ordinals).

Having scales seemed to strengthen AD considerably whereas uniformizability seemed a weak augmentation. However, using results of Becker [85], Woodin in the early 1980's established an equiconsistency: Con(ZF + DC + AD + Every subset of ${}^2({}^\omega\omega)$ can be uniformized) *iff* Con(ZF + DC + AD + Every set of reals has a scale). This foreshadowed equivalence, and in the wake of the developments of the late 1980's Woodin duly established this, bringing together the various hypotheses:

32.23 Theorem (ZF + DC)(Woodin). *The following are equivalent:*

(a) $\mathrm{AD}_{\mathbb{R}}$.

(b) AD + Every set of reals has a scale.

(c) AD + Every subset of ${}^2({}^\omega\omega)$ can be uniformized.

It is striking that $\mathrm{AD}_{\mathbb{R}}$ could be equivalent to such an ostensibly weak form as (c).

Determinacy in Terms of Ideals

With the investigation of ideals having led to the major break for determinacy, subsequent developments were eventually to reverberate back to inform on ideals. Mitchell and Steel has pursued the development of inner models for Woodin cardinals, and the latter was able to establish that Shelah's 32.10 is essentially optimal:

32.24 Theorem (Steel). *Suppose that there is an ω_2-saturated ideal over ω_1 and there is a measurable cardinal. Then there is an inner model of* ZFC *in which there is a Woodin cardinal.*

Again there is the minor annoyance of the measurable cardinal, but an equiconsistency result can at least be stated for saturated ideals and Woodin cardinals by incorporating a measurable cardinal hovering over each.

The inner model theory has been further advanced with persistence and energy. In 1992 Steel used an inner model with infinitely many Woodin cardinals to answer a question of Woodin [88: 6591] with the following result: *If every set of reals in $L(\mathbb{R})$ is weakly homogeneously Suslin, then* $\mathrm{AD}^{L(\mathbb{R})}$.

Sprinting to the finish, a pivotal ideal concept is quickly introduced. Building on our terminology from §16, an ideal I over a cardinal κ is *λ-dense iff* there is a family $D \subseteq \mathcal{P}(\kappa) - I$ with $|D| = \lambda$ such that for any $X \in \mathcal{P}(\kappa) - I$, there is a $Y \in D$ with $Y - X \in I$. This is a natural notion of density for the Boolean algebra $\mathcal{P}(\kappa)/I$. Clearly, *if I is λ-dense and $\lambda < \mu$, then I is μ-saturated.* As part of his sustained interest in ideals, Woodin in the early 1980's had improved the Steel-Van Wesep [82] result mentioned earlier by showing in ZF that *if* $\mathrm{AD}_{\mathbb{R}}$ *and Θ is regular, then in a forcing extension there is an inner model of* ZFC *in which* $\mathrm{NS}_{\omega_1}|S$ *is ω_1-dense for some stationary $S \subseteq \omega_1$.* He later drew the same conclusion from *the existence of an almost huge cardinal.* Shelah [86: 247ff] showed that ZFC imposes an intriguing limitation: *If* $2^{\aleph_0} < 2^{\aleph_1}$, *then* NS_{ω_1} *is not* ω_1*-dense.* In particular, CH implies the conclusion. With the continuum being ω_2 in 32.10, an important open question is whether CH is consistent with NS_{ω_1} being ω_2-saturated.

Pushing the connections between ideals and determinacy, Woodin established the following, which like 32.22 used Steel's recent work in inner model theory.

32.25 Theorem (Woodin). *Suppose that there is an* ω_1*-dense ideal over* ω_1*. Then* $\mathrm{AD}^{L(\mathbb{R})}$.

Woodin [99] would go on to develop a vast and brillant body of work on canonical models based on forcing and large cardinals for the failure of the Continuum Hypothesis, confronting the fundamental problem from the beginnings of set theory. Pivoting toward this new context Woodin in late 1992 established the following converse to 32.25 for consistency strength:

32.26 Theorem (Woodin [99: 366]). *Assume* $\mathrm{AD}^{L(\mathbb{R})}$. *Then there is a generic extension of* $L(\mathbb{R})$ *satisfying* ZFC + $\ulcorner \mathrm{NS}_{\omega_1}$ is ω_1-dense$\urcorner$.

Putting in the AD result and stepping back we conclude this book with a breath-taking synthesis of large cardinals, determinacy, and infinite combinatorics, one that speaks to the great achievements that have been made, but is only a prologue to Woodin [99].

32.27 Theorem (Woodin). *The following are equiconsistent:*

(a) ZF + AD.

(b) ZFC + There are infinitely many Woodin cardinals.

(c) ZFC + The non-stationary ideal over ω_1 is ω_1-dense.

Chart of Cardinals

The arrows indicates direct implications or relative consistency implications, often both.

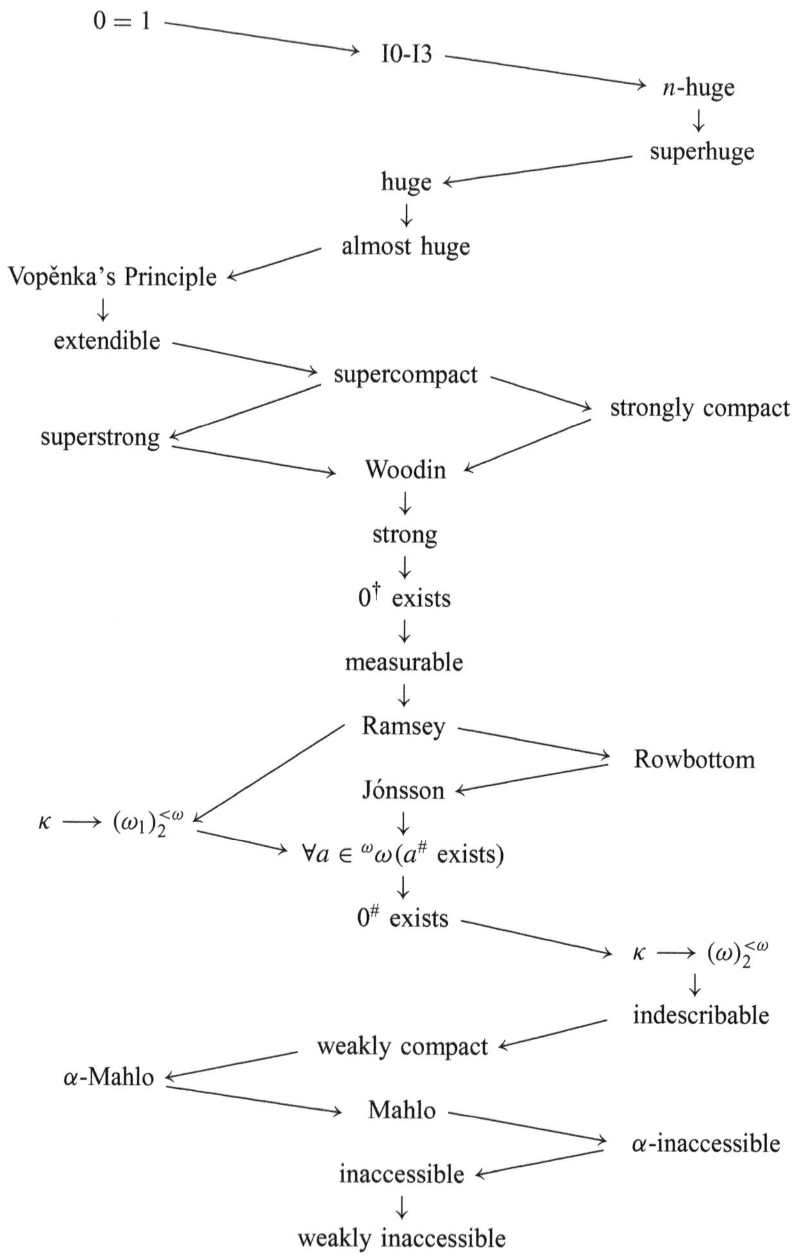

Appendix

(With apologies to Burton Dreben)

This appendix addresses several larger and more discursive issues that may be raised by set theory and made more acute by the theory of large cardinals. An attempt is made to deflate their significance, and thereby to forestall moves that end up in metaphysics. The history and practice of mathematics in general and set theory in particular affirms that they have achieved an evident autonomy, one that should resist external explanations, extrapolations, or prescriptions.

Although there have been strong metaphysical motivations for doing mathematics, it has steadily emerged through a thinning process to achieve a degree of invariance that distinguishes it from other human endeavors. As Laplace replied to Napoleon when asked why God did not appear in his work, "I have no need of that hypothesis!" This invariance is evident in the universality of mathematical language and acceptance of a body of results and procedures, and is arguably part of the sense of being mathematical. Moreover, it is this invariance that has led to the cumulative progress of mathematics, one more sustained than in science and very different from the dialectical to and fro of philosophy.

On the other hand, the broader history of mathematics undercuts any suggestion of orderly progress and moreover features sudden spurts, the major ones brought about by successively new contexts – complexes of terms, approaches, and procedures. These contexts provided new senses that transformed or refined older concepts and resolved previous questions thus recast, eventually stabilizing at new levels of invariance. To frame a discussion of set theory, we shall review (with interpretive twists) the salient features in the evolution of the continuum to the real numbers, i.e. the so-called "arithmetization of the continuum".

The most momentous mathematical discovery made by the classical Greeks was that the side and the diagonal of a square are incommensurable, in our terms that $\sqrt{2}$ is irrational. Filtering through the mysteries of the Pythagorean school in the 5th Century B.C., this first impossibility result, reputedly due to one Hippasus of Metapontum, overturned their belief that geometry can be investigated in terms of ratios of (the natural) numbers. With the situation further accentuated by Zeno's paradoxes, collectively an attack on the intelligibility of infinitary dynamic processes, the response was one of the great achievements of Greek mathematics: Eudoxus's theory of geometrical proportions as described in Book V of Euclid's *Elements*. This context provided for the ordering of possibly incommensurable magnitudes, and was thus the first mathematization of the continuum. But geometry was dominant to the extent that numbers had the sense of geometrical magnitudes, the product of two numbers *qua* lengths being an area and so forth, and multiplication in practice was not iterated more than a few times.

Almost two millenia later, the emergence of mercantile arithmetic and quantitative scientific investigation proceeded hand in hand with the introduction of basic arithmetical notation. By the early 17th Century $+$, $\times$, $=$, $<$ were in common usage, and also $\sqrt{}$ and variables x, y. This amounted to a new context in which long calculations became surveyable, particular irrational numbers gained an operative legitimacy, and algebra could develop systematically. Even before this standardization, the study of equations in streamlined notations had suggested new combinations like $\sqrt{-1}$ that outstripped sense but pointed to new possibilities.

In the mid-17th Century the great philosopher René Descartes set the stage for the subsequent expansionary period in mathematics with his advocacy of a new framework, analytic geometry, for the study of the continuum. Establishing the primacy of algebra over geometry, not only did Descartes make the conceptual move to the familiar coordinate system, but he loosened the connection of multiplication to dimension so that polynomials could be investigated without e.g. x^2 being regarded as an area, and he shifted attention from closed curves to those given by functional variation.

Surely the greatest advance in mathematics since antiquity was the independent creation of the calculus by Newton and Leibniz in the late 17th Century. Newton viewed curves as representing physical motion and made liberal use of infinite series. Leibniz articulated curves with infinitesimals and emphasized the larger possibilities for symbolic manipulation. Both arrived at the fundamental connection between tangents to curves and areas underneath, Newton from the former and Leibniz from the latter. But even in this very multiplicity genius is contextual, not ineffable, and there is something to Newton's remark that he saw farther because he had "stood on the shoulders of giants". What Newton and Leibniz had done was to forge a general approach that subsumed previous piecemeal results, an approach that not only resolved, in the new terms, a host of problems inherited from the Greeks, but suggested new problems and possibilities for the emerging field of mathematical analysis.

With the legacies of Newtonian mechanics and Leibnizian generative symbolism, mathematics expanded tremendously in the 18th Century, especially into the new domains of functions, infinite series, and differential equations. The latter provided a language to describe physical phenomena, and there was overwhelming empirical reinforcement, particularly in celestial mechanics. The century was epitomized by Euler, whose staggering output featured great strides in inductive mathematics bolstered by appeals to empirical evidence, as in physics today, and remarkable computational powers, foreshadowing recent trends in mathematics.

In the 19th Century mathematics not only continued to expand at a tremendous rate, but it also underwent a transformation based on new structural initiatives, a transformation beginning in analysis. In the previous century the vibrating string had been much discussed; the physics suggested the superposition of many frequencies, leading symbolically to infinite trigonometric series, as well as the possibility of an arbitrary initial configuration, which in turn led to the extension of the concept of function beyond those given by analytic expressions. The tension thus created by this juxtaposition of "infinite series" and "function" in new expanded senses was to lead to the arithmetization of analysis initiated by Cauchy and eventually to the creation of set theory by Cantor.

In order to pursue his study of series of functions, Cauchy in the 1820's articulated the concept of limit, and in its terms, convergence of series and continuity of functions. This amounted to a reorientation of mathematical analysis in that divergent series were excised and the ground laid for the deniability of a property of functions that had been implicit in their geometric sense. Indeed, a discontinuous counterexample to one of Cauchy's own assertions, that the sum of a convergent series of continuous functions is continuous, spurred a more careful analysis.

Karl Weierstrass in the 1850's introduced the familiar $\forall\epsilon$-$\exists\delta$ style of formulating limits, and eliminated once and for all the justificatory procedures in terms of infinitesimals. With mathematicians in the process of making their subject free from appeals to physical or geometric intuition, this new language was quickly accepted because of its ability to draw finer distinctions. By this means Weierstrass was able to rectify Cauchy's assertion by incorporating the concept of uniform continuity, and moreover to formulate a continuous yet nowhere differentiable function, driving a wedge between continuity and differentiability.

With infinitesimals replaced by the concept of limit and that cast in the ϵ-δ language, a level of deductive rigor was incorporated into mathematics that had been absent for two millenia. This may be surprising, but the concept of proof as first advanced by the Greeks had not remained crucial to mathematics. With the creation of the calculus, the available calculational procedures and the steady reinforcement of empirical evidence had been sufficient to propel mathematics inductively forward. But with the function concept steadily outgrowing the Newtonian basis in physical motion, a new calculus became imperative. Sense for the new functions given in terms of infinite series could only be developed through carefully specified deductive procedures, and these amounted to a new calculational technique. Just as proof for the Greeks had implicitly been the vehicle for demonstrating geometric constructibility, proof reemerged as an extension of algebraic calculation and soon became intrinsic as the basis for mathematics in general.

With the new articulations to be secured by proof and proof in turn to be based on prior principles, the regress lead in the early 1870's to the appearance of several independent formulations of the real numbers, of which Cantor's and Richard Dedekind's are the best known. It is at first quite striking that the real numbers as a class came to be developed so late, but this can be viewed against the background of the foregoing account as part of a larger conceptual shift from intensional to extensional mathematics, that is from rules to objects:

The geometric investigation of the continuum as transformed by the calculus had made functional variation central to mathematics. But functions were initially identified with analytic expressions, so that they were viewed intensionally as rules, whether generating geometric figures or representing physical motion. Although infinitesimal change served to motivate differentiation and integration as transformations of these rules, there was no mathematical need to analyze the continuum itself. Mathematics has consistently maintained invariance with such minimal commitments, but the 19th Century articulation of limits and continuity to be demonstrated by proof brought to bear new pressures toward an extensional view of functions as acting on points. With geometric assumptions made more explicit and infinite series outstripping sense, it became necessary to adopt an arithmetical view of the continuum given extensionally as a collection of points.

Cantor's formulation of the real numbers appeared in his seminal paper [72] on Fourier series; proceeding in terms of fundamental sequences, he laid the basis for his theorems on sequential convergence. Dedekind [72] formulated the real numbers in terms of his cuts to express the completeness of the continuum; deriving the least upper bound principle as a simple consequence, he thereby secured the basic properties of continuous functions. In the use of arbitrary sequences and infinite collections, both Cantor's and Dedekind's objectifications of the continuum helped set the stage for the subsequent development of that extensional mathematics *par excellence*, set theory. Cantor was led to his formulation rather pragmatically to secure specific results, but they were also the results that suggested the enumerations leading to the investigation of the transfinite. Dedekind [72] describes how he came to his formulation much earlier, but also acknowledges Cantor's work.

Neither Cantor nor Dedekind regarded their respective formulations and its correlation with an antecedent continuum as automatic. Cantor [72: 97] wrote:

> In order to complete the connection ... with the geometry of the straight line, one must only add an *axiom* which simply says that conversely every numerical quantity also has a determined point on the straight line, whose coordinate is equal to that quantity ... I call this proposition an *axiom* because by its nature it cannot be universally proved. A certain objectivity is then subsequently gained thereby for the quantities although they are quite independent of this.

Dedekind [72] (see [63: 11ff]) wrote:

> If all points on the straight line fall into two classes such that every point of the first class lies to the left of every point of the second class, then there exists one and only one point that produces this division ...
>
> ... I am glad if everyone finds the above principle so obvious and so in harmony with his own ideas of a line; for I am utterly unable to adduce any proof of its correctness, nor has anyone the power. The assumption of this property of the line is nothing else than an axiom by which we attribute to the line its continuity, by which we find continuity in the line.

Dedekind was not above "proving" basic principles (cf. his (in)famous proof of Theorem 66 in his [88], that there is an infinite set), but here he advocates an axiomatic correlation as does Cantor. Dedekind's circumlocutions in terms of "creation" via cuts led to Russell's [03: 279ff] criticism. However, Russell's extensional solution of simply defining the real numbers as the cuts, while consistent with his reductionism, obscures an antecedent sense of the continuum that both Cantor and Dedekind were trying to accommodate. Neither theft nor honest toil sufficed; both Cantor and Dedekind recognized the need for a sort of Church's Thesis, a thesis of adequacy for the new construal of the continuum as a collection of points.

Set theory emerged out of this setting, with the larger backdrop of mathematics also featuring new extensional initiatives in the development of algebraic and geometric structures. The birth of set theory was attended by the metaphysics of Cantor's Absolute; it was raised on the more subtle metaphysical initiatives from logic; and throughout, the mathematization of the infinite confronted concerns about its very possibility.

The numerical infinite of indefinite progression had long been held to be incoherent as a completed totality. With Aristotle's potential infinite vs. actual infinite a traditional demarcation, the occasional excursions into the latter had only been morality tales clinched by apparently paradoxical one-to-one correspondences. In particular, the only possibility for an answer to the question "how many points are there on a line?" had been "potentially infinite", with "infinite" in the decidedly negative, etymological sense of "not finite". It was Cantor's incisive work that made of the infinite a positive concept and provided a structured sense to the question with the answer, $2^{\aleph_0}$. Cantor charted out the transfinite with simple generative and arithmetical rules, and thereby provided a mathematical context for the investigation of infinite collections. The infinite, thus cast, was after all mathematically coherent.

As before in mathematics the new language suggested a host of new possibilities, most notably $2^{\aleph_0} = \aleph_1$, but the whole transfinite landscape was slow in gaining acceptance. Philosophical skepticism about the actual infinite may have played an inhibitory role, just as Kant's dogma of the *a priori* of Euclidean geometry may have done for non-Euclidean geometry, but the main factor may have simply been the mathematical reluctance to contemplate a proliferation of new entities. Whereas the natural numbers were as old as time and at the heart of the intuitive underpinnings of mathematics, the transfinite numbers seemed at first to be contrived and of little mathematical use. While Cauchy's formulation of continuity had been quickly accepted as the articulation of geometric assumptions, there was no comparable backing to the transfinite numbers. That reinforcement was to be secured by the steadily increasing use of the transfinite leading eventually to the full-blown theory of large cardinals, and the explicit extensional casting of set theory through axiomatization.

As described in the introduction, Zermelo axiomatized set theory in order to make explicit some underlying set existence assumptions and thereby advanced a combinatorial view of sets structured solely by $\in$ and simple operations. The vagueness of the *definit* property in the Separation Axiom invited Skolem's proposal to base it on first-order logic, and the addition of the Replacement Axiom figured in von Neumann's reformulation of the transfinite numbers as the ordinals, transitive sets well-ordered by $\in$.

Beginning as a mathematical theory of extensions, axiomatic set theory has carried the weight of a larger significance having to do generally with the existence of mathematical objects. However, just as Euclid's axioms for geometry had set out the permissible geometric constructions, the axioms of set theory set out the specific conditions for set generation, and this in a new calculus based on first-order logic. Unlike the emergence of mathematics from marketplace arithmetic and Greek geometry, sets and transfinite numbers were neither laden nor buttressed with substantial antecedence. Like strangers in a strange land, stalwarts developed a familiarity with them guided hand in hand by their axiomatic scaffolding, which served as new rules of procedure. In particular, existential propositions of set theory are closely tied, even in practice, to the $\exists$ of first-order logic whose sense is determined by its logical rules, and do not call for correlation with "existence" in some larger sense.

As for the contextualized existence of infinite sets, the Axiom of Infinity in its usual formulation is just the extensional counterpart to the principle of mathematical induction as a rule for deriving the universal $\forall n\varphi(n)$. In so far as the natural numbers do have an antecedent sense, $\forall n\varphi(n)$ should be correlated with the informal counterparts to $\varphi(0)$, $\varphi(1)$, $\varphi(2)$, ... However, the correlation here is less direct than that for the continuum with formulations of the real numbers, since number-theoretic assertions are being made. *Contra* Poincaré, Hilbert distinguished between *contentual* (inhaltlich) induction proceeding recursively from one number to the next and *formal* induction by which $\forall n\varphi(n)$ follows immediately from the $\varphi(0) \wedge \forall n(\varphi(n) \rightarrow \varphi(n+1))$ and "through which alone the mathematical variable can begin to play its role in the formal system" (Hilbert [28]; see van Heijenoort [67: 473]). With Cantorian metaphysics thinned out by axiomatization, there is no larger sense of existence beyond the use of formal induction in which the infinite has been domesticated; a telling observation is that if there is some doubt about some $\forall n\varphi(n)$ purportedly established by the rule, no search is undertaken for a particular natural number a so that $\varphi[a]$ fails, but rather the putative proof is carefully scrutinized much as an arithmetical calculation is checked. There is no larger sense to the Axiom of Infinity other than providing the extension of formal induction; one consequence is that the Cantorian move against the natural numbers as having no end in the traditional "after" sense is neatly rendered by extensionalizing induction itself with the ordinal ω, with "after" recast as "$\in$".

The transfinite is similarly contextualized by Replacement and the principle of transfinite induction. Recalling Cantor's unitary view of the finite and the transfinite, the principle is a simple extension of induction through limit points, and the seeming exacerbation of the breach into the actual infinite amounts to just contextual deductions from this new rule. From this viewpoint, the Axiom of Choice can be regarded as a similarly necessary principle for infusing the contextualized transfinite with the order already inherent in the finite.

As described in the introduction, the Foundation Axiom and the iterative conception of set converted set theory into a study of well-foundedness. After the infusion of model-theoretic techniques, the development of forcing and inner model theory established set theory as a sophisticated and distinctive field of mathematics. And to repeat, formalized versions of truth and consistency became matters for combinatorial manipulation as in algebra, with large cardinals providing an elegant and fully sufficient superstructure for the study of consistency strength.

Large cardinal hypotheses as existential propositions have a distinctive status in so far as relative consistency results have made of them a gauge, the interplay between forcing and inner models becoming part of their collective sense in set theory. Just as large finite numbers became surveyable through arithmetical notation, so also did large cardinals through their set-theoretic formulations as part of the calculus of Zermelian set theory. And just as large finite numbers seem hopelessly inaccessible in terms of counting one by one, so also do large cardinals in terms of hierarchically simpler processes. But just as we can work with $10^{1,000,000}$ in the proper context, so we can work with a measurable cardinal.

What about the role of set theory as a foundation for mathematics? As a mathematical theory of extension, a small part of Zermelian set theory can serve as an ambient framework for most of ongoing mathematics. Recapitulating the history above, rules can be extensionalized and then further recast in terms of sets, $\in$, and $=$. Be this as it may, this reduction does not necessarily clarify, for generally speaking, mathematics operates at various levels of organization and articulation: Number theory thrives first at the level of arithmetical rules and then at the level of analytic superstructures, far above Frege and even Peano. Mathematical analysis thrives first at the level of Weierstrass and then at the level of functional superstructures, with only a faint nod to Cantor and Dedekind. Like the relation of organic chemistry to particle physics, reducibility is acknowledged, but the emphasis is rather on the possibilities afforded by specific conceptual schemes at different levels of organization.

On the other hand, as a study of well-foundedness ZFC together with the spectrum of large cardinals serves as a court of adjudication, in terms of relative consistency, for mathematical propositions that can be informatively contextualized in set theory by letting their variables range over the set theoretic universe. Thus, set theory is more of an open-ended framework for mathematics rather than an elucidating foundation. It is as a field *of* mathematics that both proceeds with its own internal questions and is capable of contextualizing over a broad range which makes of set theory an intriguing and highly distinctive subject.

What then is left for philosophy? Existence, truth, and knowledge, when taken in the large, are ultimately contentious subjects for debate rather than concepts for explication. The subjects of much philosophizing, these as well as the whole package of dichotomies like objective/subjective, realist/anti-realist, and contingent/necessary are of unlimited fluidity and variability. For example, a table, the moon, and unicorns exist in such plainly different ways that existence *per se* cannot have sense as a prior category. It is only through communication and learning that we become familiar with the various uses of the term as part of our language. This being the case, such general terms cannot serve, unreflectively or with short "definitions", as the beginning of some analysis, but must themselves be subject to contextualized description. Mathematics has long been held up as a paragon of clarity and knowledge, but even then, assertions like "mathematics is true" or "numbers exist" are without antecedent sense and must be developed and argued for *in toto*, each like a new sentence formed from familiar words.

But where to begin, and how far to go? Considerations of assertions like "a sea-battle will take place in the Aegean tomorrow", "the evening star is the morning star", or even "$7 + 5 = 12$" should begin with their evidently invariant feature, the words themselves. And such considerations can arguably proceed at most to a description of their interplay as part of how we use language, if they are not to overleap the bounds of invariance and become enmeshed in metaphysics. According to Wittgenstein in *Philosophical Investigations* (§109):

> . . . We must do away with all *explanation*, and description alone must take its place. And this description gets its light, that is to say its purpose, from the philosophical problems. These are, of course, not empirical problems; they are solved, rather, by looking into the workings of our language, and that in such a way as to make us recognize those workings; *in despite of* an urge to misunderstand them. The problems are solved, not by giving new information, but by arranging what we have always known. Philosophy is a battle against the bewitchment of our intelligence by means of language.

A historical account was given above "arranging what we have always known" to describe existence as contextualized in set theory. As for truth, the attitude is similar. To be bold, *mathematical truth is what we have come to make of it.* As for knowledge, description ultimately provides no insights beyond description. To be bolder still, it may be that we cannot say anything other than that *the acquisition of mathematical knowledge may be just what happens.* Like morning lilies opening at dawn or squirrels saving up nuts in trees, *we just do it.*

To pursue an analogy, the world of mathematics is like a great cathedral. The thick stone walls along the stately aisles still show the lines of the ancient church that predated the grand edifice. The central dome is supported by high arches of vaulting stone, resilient reminders of the anonymous master masons. Whatever their design, the arches have easily supported the elegant latter-day spires reaching high into the sky. The first adornments can still be seen in the oldest chapels; there in continuing communion with the past steady additions are made, each new age imparting its own distinctive style. In recent memory large new side chapels have been constructed, and new flying buttresses for extra support. Every day the curious enter through the great door of polished wood with the attractive inset figures. Several venture down the long nave seeking instruction, and a few even initiation, quite taken by the the order and beauty of the altar. And the work continues: The architects attempt to chart out large parts of the cathedral, some even proposing vast renovations. The craftsman continue the steady work on the new wood paneling, the restoration of the sculpture, and the mortaring of the cracks that appear with age. And supported by high scaffolding, the artisans continue to work on the fine stained glass. They try to coordinate with their colleagues in the adjoining frames, but sometimes the heady heights inspire them to produce new gems. Those who step back see a larger scheme, but they cannot see across the whole breadth. And they are so high up that they can no longer see their supports. Nevertheless, they are sustained as a community, as part of the ongoing human adventure.

To append an apocryphal tale: A host of industrious spiders started to build an elaborate network of webs in newly excavated vaults beneath the cathedral. It quickly grew so thick and complex that no one could venture across without getting enmeshed. One day, a fearful wind came howling in and blew a gaping hole through the network, and in desperate response the spiders worked frantically to reestablish the connections. For you see, the spiders had become convinced that their carefully constructed webbing was the foundation without which the entire cathedral would totter. Of course, the craftsmen above hardly raised an eyebrow.

There are some remaining possibilities for metaphysical appropriation that should be forestalled. Mathematicians themselves have often described a feeling of dealing with autonomous objects, some professing an avowedly realist view of mathematics. The reply is that this objectification is part of the practice of mathematics, the sense of existence here to be described as in any other concerted human activity. Among many examples of the sort, one should remember how roundly Edward Gibbon was criticized by the faithful for having presented the emergence of Christianity as part of history. And there is the analogy with the blindfold chess player who can play out entire games; he may visualize the pieces on a particular chessboard, but in the end what remains is the structure of the game as communicated by him through the notation. Finally, Gödel's realist arguments in [47] have been much discussed, no doubt in part because of the significance of his mathematical results. But again, it is the invariance of those results that lies at the heart of the matter. Gödel himself much admired the work of Robinson and wanted him as his (Gödel's) successor at the Institute for Advanced Study; that Robinson [65] was a committed formalist was never of mathematical consequence.

But even with the metaphysics thinned out, there is still the recurrent feeling, familiar to the working mathematician, of questions being induced by a context, and once resolved, a gripping sense of inevitability about their solutions. This suggests the possibility of a new metaphysical appropriation, as the mathematician seems to be impelled to extend the boundaries of order against a chaos of possibilities. But this feeling is also familiar to the artist, who can proceed straightforwardly at various junctures once the context has largely precipitated; as James Joyce in *Ulysses* (17: 1012-15) describes Dedalus,

> He affirmed his significance as a conscious animal proceeding syllogistically from the known to the unknown and a conscious rational reagent between a micro and a macrocosm ineluctably constructed upon the incertitude of the void.

Even with this said, there may still be lurking some inchoate feeling that the fact of mathematics still calls for some kind of explanation. Beyond any concerns about its unreasonable effectiveness in the natural sciences, there may remain a larger feeling of *mystery* about how the world of mathematics has come about and fits together into a coherent whole. But this final possibility for metaphysical appropriation, along with the more traditional musings about the starry heaven above or the moral law within, are not in the world but of the mystical, part of the feeling for the unity of experience in the large. According to Wittgenstein in the *Tractatus* (6.41):

> The sense of the world must lie outside the world. In the world everything is as it is and happens as it does happen. *In* it there is no value – and if there were, it would be of no value.
>
> If there is a value which is of value, it must lie outside all happening and being-so. For all happening and being-so is accidental.
>
> What makes it non-accidental cannot lie *in* in the world, for otherwise this would again be accidental.
>
> It must lie outside the world.

Again from the *Tractatus* (6.44):

> Not *how* the world is, is the mystical, but *that* it is.

And this in Wittgenstein's dialectical distinction can at most be *shown*, not *said*, leading in one direction to his admonition to silence at the end of the *Tractatus*. Neither metaphysics nor solipsism, it is that part of human experience beyond human discourse. The *Tao Te Ching* of Lao-Tzu begins:

> The Tao that can be told is not the eternal Tao.
> The name that can be named is not the eternal name.

Indexed References

The following abbreviations are used for long titled or frequently cited journals:

AAMS Abstracts of papers presented to the American Mathematical Society

ALS Acta Litterarum ac Scientiarum Regiae Universitatis Hungaricae Francisco-Josephinae, Sectio Scientiarum Mathematicarum (from 1946: Acta Scientiarum Mathematicarum, Szeged).

AdM Advances in Mathematics

AM Annals of Mathematics

AMAH Acta Mathematica Academiae Scientiarum Hungaricae

AMM American Mathematical Monthly

AML Annals of Mathematical Logic (continued from 1983 by Annals of Pure and Applied Logic)

APAL Annals of Pure and Applied Logic (continues Annals of Mathematical Logic from 1983)

BAMS Bulletin of the American Mathematical Society

BAPS Bulletin de l'Académie Polonaise des Sciences, Série des Sciences Mathématiques, Astronomiques et Physiques (continued from 1983 by Bulletin of the Polish Academy of Sciences. Mathematics.)

BKSG Berichte über die Verhandlungen der Königlich Sächsischen Gesellschaft der Wissenschaften zu Leipzig, Mathematisch-Physikalische Klasse

BLMS Bulletin of the London Mathematical Society

CMUC Commentationes Mathematicae Universitatis Carolinae

CRP Comptes Rendus Hebdomadaires des Séances de l'Académie des Sciences, Paris.

FM Fundamenta Mathematicae

IJM Israel Journal of Mathematics

JAMS Journal of the American Mathematical Society

JSL The Journal of Symbolic Logic

MA Mathematische Annalen

NAMS Notices of the American Mathematical Society

PAMS Proceedings of the American Mathematical Society

PJM Pacific Journal of Mathematics

PLMS Proceedings of the London Mathematical Society

PNAS Proceedings of the National Academy of Sciences U.S.A.

RMS Russian Mathematical Surveys

TAMS Transactions of the American Mathematical Society

ZML Zeitschrift für Mathematische Logik und Grundlagen der Mathematik

The italicized numbers after a publication refer to those pages in the text where it is cited. The italicized numbers directly after a person refer to those pages in the text where he or she is cited, but not only in connection with a publication or specific result.

Aanderaa, Stål O.

74 Inductive definitions and their closure ordinals. In: Fenstad-Hinman [74], 207–220. *66*

Abe, Yoshihiro

84 Strongly compact cardinals, elementary embeddings and fixed points. JSL **49** (1984), 808–812. *306*

85 Some results concerning strongly compact cardinals. JSL **50** (1985), 874–880. *309*

86 Notes on $\mathcal{P}_\kappa\lambda$ and $[\lambda]^\kappa$. Tsukuba Journal of Mathematics **10** (1986), 155–163. *334*

Abel, Niels H. *148*

Aczel, Peter

77 An introduction to inductive definitions. In: Barwise [77], 739–782. *66*

See also Richter, Wayne H., and Peter Aczel.

Addison Jr., John W. *32, 151, 157, 176, 406*

58 Separation principles in the hierarchies of classical and effective descriptive set theory. FM **46** (1958), 123–135. *146, 152, 157, 404, 405, 407*

59 Some consequences of the axiom of constructibility. FM **46** (1959), 337–357. *151, 169, 177, 410*

74 Current problems in descriptive set theory. In: Jech [74], 1–10. *410*

Addison Jr., John W., and Yiannis N. Moschovakis

68 Some consequences of the axiom of definable determinateness. PNAS **59** (1968), 708–712. *410*

Addison Jr., John W., Leon Henkin, and Alfred Tarski

65 *The Theory of Models.* Proceedings of the 1963 International Symposium at Berkeley. Amsterdam, North-Holland 1965.

Albers, Donald J., Gerald L. Alexanderson, and Constance Reid

90 (eds.) *More Mathematical People.* Boston, Harcourt Brace Jovanovich 1990. *116*

Aleksandrov, Pavel S.

16 Sur la puissance des ensembles mesurables B. CRP **162** (1916), 323–325. *147, 148*

79 Pages from an autobiography. RMS **34**(6) (1979), 267–302. *148*

Alexanderson, Gerald L.

See Albers, Donald J., Gerald L. Alexanderson, and Constance Reid.

Apter, Arthur W., and James M. Henle

91 Relative consistency results via strong compactness. FM **139** (1991), 133–149. *238, 243*

92 On box, weak box, and strong compactness. BLMS **24** (1992), 513–518. *238, 243*

Aristotle *XIII, 477*

Aronszajn, Nathan (Nachman) *78, 79*

52 Applied functional analysis. In: *Proceedings of the International Congress of Mathematicians*, Cambridge, Massachusetts 1950, vol. 2. Providence, American Mathematical Society 1952, 123–127. *78*

52a Characterization of types of order satisfying $\alpha_0 + \alpha_1 = \alpha_1 + \alpha_0$. FM **39** (1952), 65–96. *78*

Awerbuch-Friedlander, Tamara E.

See Kanamori, Akihiro, and Tamara E. Awerbuch Friedlander.

Baire, René *145*

99 Sur les fonctions de variables réelles. Annali di Matematica Pura ed Applicata **(3)3** (1899), 1–123. *13, 145*

Baeten, Josephus C.M.

86 *Filters and Ultrafilters over Definable Subsets of Admissible Ordinals.* CWI Tract #24. Amsterdam, Centrum voor Wiskunde en Informatica 1986. *66*

Baldwin, Stewart

83 Generalizing the Mahlo hierarchy, with applications to the Mitchell models. APAL **25** (1983), 103–127. *345*

84 The consistency strength of certain stationary subsets $P_\kappa\lambda$. PAMS **92** (1984), 90–92. *345*

85 The $\triangleleft$-ordering on normal ultrafilters. JSL **51** (1985), 936–952. *305*

Banach, Stefan *22, 371, 373, 377*

30 Über additive Massfunktionen in abstrakten Mengen. FM **15** (1930), 97–101. Reprinted in [67] below, 200–203. *23, 24*

67 Hartman, Stanisław, and Edward Marczewski (eds.) *Oeuvres.* Vol. 1. Warsaw, Państwowe Wydawnictwo Naukowe, 1967.

Banach, Stefan, and Kazimierz Kuratowski

29 Sur une généralisation du problème de la mesure. FM **14** (1929), 127–131. Reprinted in Banach [67], 182–186, and in Kuratowski [88], 327–331. *23*

Barbanel, Julius B.

82 Supercompact cardinals, elementary embeddings, and fixed points. JSL **47** (1982), 84–88. *306*

82a Supercompact cardinals and trees of normal ultrafilters. JSL **47** (1982), 89–109. *306*

85 An ordering of normal ultrafilters. FM **125** (1985), 155–165. *306*

86 Supercompact cardinals, trees of normal ultrafilters, and the partition property. JSL **51** (1986), 701–708. *349*

89 Flipping properties and huge cardinals. FM **132** (1989), 171–188. *334*

91 Almost hugeness and a related notion. The Notre Dame Journal of Formal Logic **32** (1991), 255–265. *334*

91a Making the hugeness of κ resurrectable after κ-directed closed forcing. FM **137** (1991), 9–24. *334*

92 A note on a result of Kunen and Pelletier. JSL **57** (1992), 461–465. *349*

93 On the relationship between the partition property and the weak partition property for normal ultrafilters on $\mathcal{P}_\kappa\lambda$. JSL **58** (1993), 119–127. *349*

93a Some variations on the partition property for normal ultrafilters on $\mathrm{P}_\kappa\lambda$. FM **142** (1993), 163–171. *349*

Barbanel, Julius B., Carlos A. Di Prisco, and It Beng Tan

84 Many times huge and superhuge cardinals. JSL **49** (1984), 112–122. *334*

Bar-Hillel, Yehoshua

65 (ed.) *Logic, Methodology and Philosophy of Science.* Proceedings of the 1964 International Congress, Jerusalem. Amsterdam, North-Holland 1965.

70 (ed.) *Mathematical Logic and Foundations of Set Theory.* Amsterdam, North-Holland 1970.

See also Fraenkel, Abraham A., Yehoshua Bar-Hillel, and Azriel Levy.

Bar-Hillel, Yehoshua, E.I.J. Poznanski, Michael O. Rabin, and Abraham Robinson

61 (eds.) *Essays on the Foundations of Mathematics.* Jerusalem, Magnes Press 1961.

Barua, Rana

84 R-sets and category. TAMS **286** (1984), 125–158. *180*

Barwise, K. Jon

75 *Admissible Sets and Structures.* Perspectives in Mathematical Logic. Berlin, Springer-Verlag 1975. *20*

77 (ed.) *Handbook of Mathematical Logic.* Amsterdam, North-Holland 1977.

Baumgartner, James E. *117, 343, 344, 350, 441*

75 Ineffability properties of cardinals I. In: Hajnal-Rado-Sós [75] vol. 1, 109–130. *77*

76 Almost-disjoint sets, the dense set problem and the partition calculus. AML **9** (1976), 401–439. *213*

83 Iterated forcing. In: Mathias [83], 1–59.

Baumgartner, James E., and Fred Galvin

78 Generalized Erdős cardinals and $0^{\#}$. AML **15** (1978), 289–313. *82, 108*

Baumgartner, James E., Donald A. Martin, and Saharon Shelah

84 (eds.) *Axiomatic Set Theory.* Contemporary Mathematics vol. 31. Providence, American Mathematical Society 1984.

Baumgartner, James E., and Alan D. Taylor

82 Saturation properties of ideals in generic extensions. I. TAMS **270** (1982), 557–574. *213, 345*

82a Saturation properties of ideals in generic extensions. II. TAMS **271** (1982), 587–609. *221*

Baumgartner, James E., Alan D. Taylor, and Stanley Wagon

82 Structural properties of ideals. Dissertationes Mathematicae (Rozprawy Matematyczne) **197** (1982), 1–95. *350*

Becker, Howard S.

78 Partially playful universes. In: Kechris-Moschovakis [78], 55–90. *422*

80 Thin collections of sets of projective ordinals and analogs of L. AML **19** (1980), 205–241. *422*

81 AD and the supercompactness of $\aleph_1$. JSL **46** (1981), 822–842. *402*

81a Determinacy implies that $\aleph_2$ is supercompact. IJM **40** (1981), 229–234. *402*

85 A property equivalent to the existence of scales. TAMS **287** (1985), 591–612. *470*

88 More closure properties of pointclasses. In: Kechris-Martin-Steel [88], 31–36. *415*

92 Descriptive set theoretic phenomena in analysis and topology. In: Judah-Just-Woodin [92], 1–25. *166*

Becker, Howard S., and Alexander S. Kechris

84 Sets of ordinals constructible from trees and the Third Victoria Delfino Problem. In: Baumgartner-Martin-Shelah [84], 13–29. *423*

Becker, Howard S., and Yiannis N. Moschovakis

81 Measurable cardinals in playful models. In: Kechris-Martin-Moschovakis [81], 203–214. *423*

Bell, John L.

85 *Boolean-Valued Models and Independence Proofs in Set Theory.* Second edition. Oxford Logic Guides #12. Oxford, Oxford University Press 1985. *114, 117*

Beller, Aaron, Ronald B. Jensen, and Philip Welch

82 *Coding the Universe.* London Mathematical Society Lecture Note Series #47. Cambridge, Cambridge University Press 1982. *186, 187*

Bendixson, Ivar

83 Quelques théorèmes de la théorie des ensembles de points. Acta Mathematica **2** (1883), 415–429. *133*

Bernays, Paul

37 A system of axiomatic set theory. Part I. JSL **2** (1937), 65–77. *30*

42 A system of axiomatic set theory. Part III. JSL **7** (1942), 65–89. *132*

61 Zur Frage der Unendlichkeitsschemata in der axiomatischen Mengenlehre. In: Bar-Hillel-Poznanski-Rabin-Robinson [61], 3–49. See also [76] below. *59, 71*

76 On the problem of schema of infinity in axiomatic set theory. In: Müller [76], 121–172. English translation of a revised version of Bernays [61].

Bernstein, Felix *16, 134*

08 Zur Theorie der trigonometrischen Reihen. BKSG **60** (1908), 325–338. *134*

Blackwell, David *404, 406*

67 Infinite games and analytic sets. PNAS **58** (1967), 1836–1837. *XXI, 368, 380, 403, 405*

Blass, Andreas R.

75 Equivalence of two strong forms of determinacy. PAMS **52** (1975), 373–376. *441*

88 Selective ultrafilters and homogeneity. APAL **38** (1988), 215–255. *238, 243*

Bloch, Gérard

53 Sur les ensembles stationnaires de nombres ordinaux et les suites distinguées de fonctions régressives. CRP **236** (1953), 265–268. *17*

Boos, William

74 Boolean extensions which efface the Mahlo property. JSL **39** (1974), 254–268. *233*

76 Infinitary compactness without strong inaccessibility. JSL **41** (1976), 33–38. *37*

Borel, Emile F. *11,145*

98 *Leçons sur la Théorie des Fonctions.* Paris, Gauthier-Villars 1898. *12, 145*

05 *Leçons sur les fonctions de variables réelles et les développements en series de polynomes.* Paris, Gauthier-Villars 1905. *157*

21 La théorie du jeu et les équations intégrales á noyau symétrique. CRP **173** (1921), 1304–1308. Reprinted in [72] below, 901–904. Translated in Econometrica **21** (1953), 97–100. *371*

72 *Oeuvres de Emile Borel.* Paris, Editions de C.N.R.S. 1972.

Brouwer, Luitzen E.J. *XIII*

Bukovský, Lev

73 Changing cofinality of a measurable cardinal. CMUC **14** (1973), 689–697. *259*

Bull Jr., Everett L.

78 Successive large cardinals. AML **15** (1978), 161–191. *396*

Bulloff, Jack J., Thomas C. Holyoke, and Samuel W. Hahn

69 (eds.) *Foundations of Mathematics.* Symposium papers commemorating the sixtieth birthday of Kurt Gödel. Berlin, Springer-Verlag 1969.

Burgess, John P.

82 What are R-sets? In: Metakides [82], 307–324. *180*

83 Classical hierarchies from a modern standpoint. II. R-sets. FM **115** (1983), 97–105. *180*

Burke, Maxim, and Menachem Magidor

90 Shelah's pcf theory and its applications. APAL **50** (1990), 207–254. *97, 98*

Cantor, Georg *XI, XII, XIV–XVIII, XXI, 13, 16, 20, 32, 42, 119, 133, 134, 145, 146, 313, 378, 474–479*

72 Über die Ausdehnung eines Satzes aus der Theorie der trignometrischen Reihen, MA **5** (1872), 123–132. Reprinted in [80] below, 92–102. *476*

83 Über unendliche, lineare Punktmannichfaltigkeiten. V. MA **21** (1883), 545–591. Reprinted in [80] below, 165–209. *133*

84 Über unendliche, lineare Punktmannichfaltigkeiten. VI. MA **23** (1884), 453–488. Reprinted in [80] below, 210–246. *133*

84a De la puissance des ensembles parfaits de points. Acta Mathematica **4** (1884), 381–392. Reprinted in [80] below, 252–260. *133*

80 Zermelo, Ernst (ed.) *Gesammelte Abhandlungen mathematischen und philosophischen Inhalts.* Berlin, Springer-Verlag 1980. Reprint of the original 1932 edition, Berlin, Verlag von Julius Springer.

Carr, Donna M.

82 The minimal normal filter on $P_\kappa\lambda$. PAMS **86** (1982), 316–320. *342*

85 $P_\kappa\lambda$ generalizations of weak compactness. ZML **31** (1985), 393–401. *350*

86 The structure of ineffability properties of $\mathcal{P}_\kappa\lambda$. AMAH **47** (1986), 325–332. *350*

87 $P_\kappa\lambda$ partition relations. FM **128** (1987), 181–195. *350*

87a A note on the λ-Shelah property. FM **128** (1987), 197–198. *350*

Carr, Donna M., Jean-Pierre Levinski, and Donald H. Pelletier

90 On the existence of strongly normal ideals over $P_\kappa\lambda$. Archive for Mathematical Logic **30** (1990), 59–72. *351*

Carr, Donna M., and Donald H. Pelletier

89 Towards a structure theory for ideals on $P_\kappa\lambda$. In: Steprāns-Watson [89], 41–54. *350, 351*

Cauchy, Augustin-Louis *148, 474, 476*

Chang, Chen-Chung *85, 94*

71 Sets constructible using $L_{\kappa\kappa}$. In: Scott [71], 1–8. *257*

Chang, Chen-Chung, and H. Jerome Keisler

90 *Model Theory.* Third edition. Amsterdam, North-Holland 1990. *3, 8, 9, 253*

Chuaqui, Rolando

78 Bernays' class theory. In: Arruda, Ayda I., Newton C.A. da Costa, and Rolando Chuaqui (eds.) *Mathematical Logic.* Proceedings of the First Brazilian Conference. New York, Marcel Dekker 1978, 31–55. *59*

81 *Axiomatic Set Theory. Impredicative Theories of Classes.* Amsterdam, North-Holland 1981. *59*

Cohen, Paul J. *XVIII, XIX, 32, 35, 44, 113–117, 119, 126, 132, 134, 295*

63 The independence of the Continuum Hypothesis. I. PNAS **50** (1963), 1143–1148. *XVII, 114, 115*

64 The independence of the Continuum Hypothesis. II. PNAS **51** (1964), 105–110. *XVII, 114, 115, 125*

65 Independence results in set theory. In: Addison-Henkin-Tarski [65], 39–54. *115*

66 *Set Theory and the Continuum Hypothesis.* New York, Benjamin 1966. *115*

71 Comments on the foundations of set theory. In: Scott [71], 9–15. *115*

Cohn, Paul M.

65 *Universal Algebra.* New York, Harper & Row 1965. *340*

Comfort, W. Wistar, and Stylianos Negrepontis

74 *The Theory of Ultrafilters.* Berlin, Springer-Verlag 1974. *39*

Dales, H. Garth, and W. Hugh Woodin

87 *An Introduction to Independence for Analysts.* London Mathematical Society Lecture Note Series #115. Cambridge, Cambridge University Press 1987. *450*

David, René

82 A very absolute Π^1_2 real singleton. AML **23** (1982), 101–120. *188*

84 Generic reals close to $0^{\#}$. In: Baumgartner-Martin-Shelah [84], 63–70. *188*

89 A functorial Π^1_2 singleton. AdM **74** (1989), 258–268. *188*

Davis, Morton

64 Infinite games of perfect information. In: Dresher, Melvin, Lloyd S. Shapley, and Alan W. Tucker (eds.) *Advances in Game Theory.* Annals of Mathematical Studies #52. Princeton, Princeton University Press 1964, 85–101. *373, 374, 377, 440, 441*

Dedekind, Richard *475, 476, 479*

72 *Stetigkeit und irrationale Zahlen.* Braunschweig, F. Vieweg 1872. Translated in [63] below, 1–27. *476*

88 *Was sind und was sollen die Zahlen?* Braunschweig, F. Vieweg 1888. Translated in [63] below, 29–115. *476*

63 *Essays on the Theory of Numbers.* Translations by Wooster W. Beman. New York, Dover 1963. Reprint of original edition, Chicago, Open Court 1901. *476*

Dehornoy, Patrick *238*

75 Solution d'une conjecture de Bukovský. CRP **281** (1975), 821–824. *259*

76 Intersections d'ultrapuissances iterées de modèles de la théorie des ensembles. CRP **282** (1976), 935–938. *260*

78 Iterated ultrapowers and Prikry forcing. AML **15** (1978), 109–160. *238, 240, 242, 259, 260*

83 An application of ultrapowers to changing cofinality. JSL **48** (1983), 225–235. *260*

88 Π^1_1-complete families of elementary sequences. APAL **38** (1988), 257–287. *329*

89 Free distributive groupoids. Journal of Pure and Applied Algebra **61** (1989), 123–146. *329*

89a Sur la structure des gerbes libres. CRP **309** (1989), 143–148. *329*

89b Algebraic properties of the shift mapping. PAMS **106** (1989), 617–623. *329*

92 An alternative proof of Laver's results on the algebra generated by an elementary embedding. In: Judah-Just-Woodin [92], 27–33. *330*

92a Preuve de la conjecture d'irreflexivité pour les structures distributives libres. CRP **314** (1992), 333–336. *330*

94 Braid groups and left distributive operations. TAMS **345** (1994), 115–151. *330*

00 *Braid groups and self-distributivity*. Progress in Mathematics #192. Basel, Birkhäuser Verlag 2000. *330*

Descartes, René *474*

Devlin, Keith J.

73 Some weak versions of large cardinal axioms. AML **5** (1973), 291–325. *94, 96, 104*

74 Some remarks on changing cofinalities. JSL **39** (1974), 27–30. *236*

75 Indescribability properties and small large cardinals. In: Müller-Oberschelp-Potthoff [75], 89–114. *39, 63, 66*

84 *Constructibility*. Perspectives in Mathematical Logic. Berlin, Springer-Verlag 1984. *3, 6, 28, 33, 35, 72, 104, 168, 281, 282*

Devlin, Keith J., and Jeffrey B. Paris

73 More on the free subset problem. AML **5** (1973), 327–336. *108*

Dickmann, Máximo A.

75 *Large Infinitary Languages*. Amsterdam, North-Holland 1975. *36, 39*

85 Larger infinitary languages. In: Barwise, K. Jon, and Solomon Feferman (eds.) *Model-Theoretic Logics*. Berlin, Springer-Verlag 1985, 317–363. *36*

Di Prisco, Carlos A.

77 Supercompact cardinals and a partition property. AdM **25** (1977), 46–55. *349*

See also Barbanel, Julius B., Carlos A. Di Prisco, and It Beng Tan.

Di Prisco, Carlos A., and James M. Henle

78 On the compactness of $\aleph_1$ and $\aleph_2$. JSL **43** (1978), 394–401. *401*

85 Sorts of huge cardinals. In: Caicedo, Xavier, Newton C.A. da Costa, and Rolando Chuaqui (eds.) *Proceedings of the Fifth Latin American Symposium on Mathematical Logic*. Revista Colombiana de Matemáticas **19** (1985), 69–75. *334*

Di Prisco, Carlos A., and Wiktor Marek

82 Some properties of stationary sets. Dissertationes Mathematicae (Rozprawy Matematyczne) **198** (1982). *343*

84 A filter on $[\lambda]^\kappa$. PAMS **90** (1984), 591–598. *334*

85 On the space $(\lambda)^\kappa$. In: Di Prisco, Carlos A. (ed.) *Methods in Mathematical Logic*. Lecture Notes in Mathematics #1130. Berlin, Springer-Verlag 1985, 151–156. *334*

85a Some aspects of the theory of large cardinals. In: Luiz Paulo de Alcantara (ed.) *Mathematical Logic and Formal Systems*. Lecture Notes in Pure and Applied Mathematics #94. New York, Marcel Dekker 1985, 87–139. *298*

88 Reflection properties induced by some large cardinal axioms. In: Carnielli, Walter A., and Luiz Paulo de Alcantara (eds.) *Methods and Applications of Mathematical Logic*. Contemporary Mathematics vol. 69. Providence, American Mathematical Society 1988, 19–25. *334*

Di Prisco, Carlos A., and William S. Zwicker

80 Flipping properties and supercompact cardinals. FM **109** (1980), 31–36. *350*

Dodd, Anthony J. *276, 295*

82 *The Core Model*. London Mathematical Society Lecture Note Series #61. Cambridge, Cambridge University Press 1982. *254, 257, 352, 358*

Donder, Hans-Dieter, and Peter G. Koepke

83 On the consistency strength of 'accessible' Jonsson cardinals and of the weak Chang conjecture. APAL **25** (1983), 233–261. *96, 292*

Donder, Hans-Dieter, Peter G. Koepke, and Jean-Pierre Levinski

88 Some stationary subsets of $P(\lambda)$. PAMS **102** (1988), 1000–1004. *345*

Donder, Hans-Dieter, and Pierre Matet

93 Two cardinal versions of Diamond. IJM **83** (1993), 1–43. *346*

Dougherty, Randall *329*

93 Critical points in an algebra of elementary embeddings. APAL **65** (1993), 211–241. *330, 331*

96 Critical points in an algebra of elementary embeddings II. In: Wilfrid Hodges (ed.) *Logic: From Foundations to Applications*, Staffordshire 1993. New York, Oxford University Press 1996, 103–136. *330*

Dougherty, Randall, and Thomas J. Jech

97 Finite left-distributive algebras and embedding algebras. AdM **130** (1997), 201–241. *331*

Drake, Frank R.

74 *Set Theory*. Amsterdam, North-Holland 1974. *1, 3, 6, 28, 34, 59, 63, 83, 168*

Dreben, Burton, and Warren D. Goldfarb

79 *The Decision Problem: Solvable Classes of Quantificational Formulas*. Reading, Addison-Wesley 1979. *70*

Dubose, Derrick A.

90 The equivalence of determinacy and iterated sharps. JSL **55** (1990), 502–525. *443, 445*

91 Determinacy and the sharp function on the reals. APAL **54** (1991), 59–85. Corrected version **55** (1992), 237–263. *445*

92 Determinacy and extended sharp functions on the reals, Part II: Obtaining sharps from determinacy. APAL **58** (1992), 1–28. *445*

95 Determinacy and the sharp function on objects of type k. JSL **60** (1995), 1025–1053. *445*

Dummett, Michael

77 *Elements of Intuitionism*. Oxford Logic Guides #2. Oxford, Oxford University Press 1977. *114*

Easton, William B.

64 Powers of regular cardinals. Ph.D. thesis, Princeton University 1964. Abstracted as: Proper classes of generic sets. NAMS **11** (1964), 205. Published in abridged form as [70] below. *114, 122*

70 Powers of regular cardinals. AML **1** (1970), 139–178. *114, 122*

Egorov, Dmitry *147*

Ehrenfeucht, Andrzej, and Jerzy Łoś

54 Sur les produits cartésiens des groupes cyclique infinis. BAPS **2** (1954), 261–263. *27*

Ehrenfeucht, Andrzej, and Andrzej M. Mostowski

56 Models of axiomatic theories admitting automorphisms. FM **43** (1956), 50–68. Reprinted in Mostowski [79], 494–512. *100*

Ellentuck, Erik

74 A new proof that analytic sets are Ramsey. JSL **39** (1974), 163–165. *382*

Erdős, Paul *XVIII, 69, 70, 75*

42 Some set-theoretical properties of graphs. Revista, Universidad Nacional de Tucumán, Serie A, Matemáticas y Física Teórica **3** (1942), 363–367. *74, 75*

73 Spencer, Joel H. (ed.) *Paul Erdős: The Art of Counting*. Selected Writings. Cambridge, The MIT Press 1973.

Erdős, Paul, and András Hajnal

58 On the structure of set mappings. AMAH **9** (1958), 111–131. *80, 83, 84*

62 Some remarks concerning our paper "On the structure of set mappings". AMAH **13** (1962), 223–226. *83*

66 On a problem of B. Jónsson. BAPS **14** (1966), 19–23. *85, 93, 94, 319*

71 Unsolved problems in set theory. In: Scott [71], 17–48. *97*

74 Unsolved and solved problems in set theory. In: Henkin *et al.* [74], 269–287. *87*

Erdős, Paul, András Hajnal, Attila Máté, and Richard Rado

84 *Combinatorial Set Theory: Partition Relations for Cardinals*. Amsterdam, North-Holland 1984. *72, 77, 85*

Erdős, Paul, András Hajnal, and Richard Rado

65 Partition relations for cardinal numbers. AMAH **16** (1965), 93–196. *72, 74, 82, 85, 86, 94*

Erdős, Paul, and Shizuo Kakutani

43 On non-denumerable graphs. BAMS **49** (1943), 457–461. *74*

Erdős, Paul, and Richard Rado

52 Combinatorial theorems on classifications of subsets of a given set. PLMS **(3)2** (1952), 417–439. Reprinted in Erdős [73], 383–405. *71, 81*

56 A partition calculus in set theory. BAMS **62** (1956), 427–489. Reprinted in Gessel-Rota [87], 179–241. *71, 72, 74*

Erdős, Paul, and György Szekeres

35 A combinatorial problem in geometry. Compositio Mathematica **2** (1935), 463–470. Reprinted in Erdős [73], 5–12, and in Gessel-Rota [87], 49–56. *70*

Erdős, Paul, and Alfred Tarski

43 On families of mutually exclusive sets. AM **44** (1943), 315–329. Reprinted in Tarski [86] vol. 2, 591–605. *36, 38, 70, 71, 75, 76, 78, 83, 211, 219*

61 On some problems involving inaccessible cardinals. In: Bar-Hillel-Poznanski-Rabin-Robinson [61], 50–82. Reprinted in Tarski [86] vol. 4, 79–111. *36, 76*

Euclid *473, 477*

Eudoxus *473*

Euler, Leonhard *XII, 474*

Euwe, Max

29 Mengentheoretische Betrachtungen über das Schachspiel. Koninklijke Akademi van Wetenschappen, Afdeeling Natuurkundige **32** (1929), 633–642. *371*

Farrington, C. Patrick

82 The first-order theory of the c-degrees with the # operation. ZML **28** (1982), 487–493. *186*

Feferman, Solomon *114*

Feng, Qi, Menachem Magidor, and W. Hugh Woodin

92 Universally Baire sets of reals. In: Judah-Just-Woodin [92], 203–242. *458*

Fenstad, Jens E., and Peter G. Hinman

74 (eds.) *Generalized Recursion Theory*. Amsterdam, North-Holland 1974.

Fenstad, Jens E., and Dag Normann

74 On absolutely measurable sets. FM **81** (1974), 91–98. *180*

Firestone, C.D., and J. Barkley Rosser

49 The consistency of the hypothesis of accessibility (abstract). JSL **14** (1949), 79. *30*

Fisher, Edward R.

77 Vopěnka's Principle, category theory and universal algebra (abstract). NAMS **24** (1977), A-44. *339*

Fleissner, William G.

75 Lemma on measurable cardinals. PAMS **49** (1975), 517–518. *258*

Fodor, Géza *83*

56 Eine Bemerking zur Theorie der regressiven Funktionen. Acta Scientiarum Mathematicarum, Szeged **17** (1956), 139–142. *3*

Foreman, Matthew *XXII, 461, 463*

82 Large cardinals and strong model-theoretic transfer properties. TAMS **272** (1982), 427–463. *459*

83 More saturated ideals. In: Kechris-Martin-Moschovakis [83], 1–27. *459*

86 Potent axioms. TAMS **294** (1986), 1–28. *461*

Foreman, Matthew, Menachem Magidor, and Saharon Shelah

88 Martin's Maximum, saturated ideals and non-regular ultrafilters. Part I. AM **127** (1988), 1–47. *332, 460, 462*

88a Martin's Maximum, saturated ideals and non-regular ultrafilters. Part II. AM **127** (1988), 521–545. *460*

Fraenkel, Abraham A.

21 Über die Zermelosche Begründung der Mengenlehre. Jahresbericht der Deutschen Mathematiker-Vereinigung **30**II (1921), 97–98. *XIV*

22 Zu den Grundlagen der Cantor-Zermeloschen Mengenlehre. MA **86** (1922), 230–237. *XIV*

Fraenkel, Abraham A., Yehoshua Bar-Hillel, and Azriel Levy

73 *Foundations of Set Theory*. Second edition. Amsterdam, North-Holland 1973. *XII*

Frayne, Thomas E., Ann C. Morel, and Dana S. Scott

62 Reduced direct products. FM **51** (1962), 195–228. Abstracted in NAMS **5** (1958), 673–675. *37, 253*

Frege, Gottlob *XIII, 477*

Fremlin, David H. *216*

84 *Consequences of Martin's Axiom*. Cambridge Tracts in Mathematics #84. Cambridge, Cambridge University Press 1984. *124*

93 Real-valued-measurable cardinals. In: Judah, Haim (ed.) *Set Theory of the Reals*. Israel Mathematical Conference Proceedings vol. 6. Providence, American Mathematical Society 1993, 151–304. *26, 61*

Friedman, Harvey M. *369, 397, 398, 425, 440, 441*
- 71 Higher set theory and mathematical practice. AML **2** (1971), 325–357. *440, 442*
- 71a Determinacy in the low projective hierarchy. FM **72** (1971), 79–84. *442, 443*
- 81 On the necessary use of abstract set theory. AdM **41** (1981), 209–280. *442*

Friedman, Sy D.
- 87 Strong coding. APAL **35** (1987), 1–98. *187*
- 87a A guide to "Strong coding". APAL **35** (1987), 99–122. *187*
- 90 The Π^1_2-Singleton Conjecture. JAMS **3** (1990), 771–791. *188*
- 94 A simpler proof of the Coding Theorem. APAL **70** (1994), 1–16. *187*
- 97 Coding without fine structure. JSL **62** (1997), 808–815. *187*

Gaifman, Haim *43, 88, 99, 108, 184, 244, 254, 257*
- 64 Measurable cardinals and constructible sets (abstract). NAMS **11** (1964), 771. *XX, 99*
- 67 A generalization of Mahlo's method for obtaining large cardinal numbers. IJM **5** (1967), 188–200. *17*
- 74 Elementary embeddings of models of set theory and certain subtheories. In: Jech [74], 33–101. *46, 325–327, 358*

Gale, David, and Frank M. Stewart
- 53 Infinite games with perfect information. In: Kuhn, Harold W., and Alan W. Tucker (eds.) *Contributions to the Theory of Games*, vol. 2. Annals of Mathematical Studies #28. Princeton, Princeton University Press 1953, 245–266. *371, 372*

Galvin, Fred
- 65 Problem 5348. AMM **72** (1965), 1136. *319*

See also Baumgartner, James E., and Fred Galvin.

Galvin, Fred, and Karel L. Prikry
- 73 Borel sets and Ramsey's Theorem. JSL **38** (1973), 193–198. *382*
- 76 Infinitary Jonsson algebras and partition relations. Algebra Universalis **6** (1976), 367–376. *319, 321*

Gandy, Robin O., and Charles E.M. Yates
- 71 (eds.) *Logic Colloquium '69*. Amsterdam, North-Holland 1971.

Gentzen, Gerhard *XX*
- 36 Die Widerspruchsfreiheit der reinen Zahlentheorie. MA **112** (1936), 493–565. Translated in [69] below, 132–213. *XIX*
- 43 Beweisbarkeit und Unbeweisbarkeit van Anfangsfällen der transfiniten Induktion in der reinen Zahlentheorie. MA **119** (1943), 140–161. Translated in [69] below, 287–308. *XIX*
- 69 Szabo, M.E. (ed.) *The Collected Papers of Gerhard Gentzen*. Amsterdam, North-Holland 1969.

Gessel, Ira, and Gian-Carlo Rota

87 (eds.) *Classic Papers in Combinatorics.* Boston, Birkhäuser 1987.

Gibbon, Edward *481*

Gitik, Moti *243, 345*

80 All uncountable cardinals can be singular. IJM **35** (1980), 61–88. *238, 243*

85 Nonsplitting subset of $\mathcal{P}_\kappa(\kappa^+)$. JSL **50** (1985), 881–894. *344*

86 Changing cofinalities and the non-stationary ideal. IJM **56** (1986), 280–314. *238, 243*

Gitik, Moti, and Menachem Magidor

92 The Singular Cardinals Hypothesis revisited. In: Judah-Just-Woodin [92], 243–279. *243*

Gloede, Klaus

72 Ordinals with partition properties and the constructible hierarchy. ZML **18** (1972), 135–164. *109*

Gödel, Kurt F. *XIV–XX, 15, 20, 21, 27–32, 35, 44, 70, 74, 101, 113, 119, 125, 145, 167, 169, 178, 179, 261, 274*

30 Die Vollständigkeit der Axiome des logischen Funktionenkalküls. Monatshefte für Mathematik und Physik **37** (1930), 349–360. Reprinted and translated in [86] below, 102–123. *XV*

31 Über formal unentscheidbare Sätze der *Principia Mathematica* und verwandter Systeme I. Monatshefte für Mathematik und Physik **38** (1931), 173–198. Reprinted and translated with minor emendations by the author in [86] below, 144–195. *XV, 19, 28, 29*

38 The consistency of the Axiom of Choice and of the Generalized Continuum-Hypothesis. PNAS **24** (1938), 556–557. Reprinted in [90] below, 26–27. *XV, 28, 29, 33, 150, 169*

39 Consistency-proof for the Generalized Continuum-Hypothesis. PNAS **25** (1939), 220–224. Reprinted in [90] below, 28–32. *XV, 28–30*

40 *The Consistency of the Axiom of Choice and of the Generalized Continuum Hypothesis with the Axioms of Set Theory.* Annals of Mathematics Studies #3. Princeton, Princeton University Press 1940. Reprinted in [90] below, 33–101. See also [51] below. *30*

44 Russell's mathematical logic. In: Schilpp, Paul A. (ed.) *The Philosophy of Bertrand Russell.* Library of Living Philosophers vol. 5. Evanston, Northwestern University 1944, 123–153. Third edition, New York, Tudor 1951. Reprinted in [90] below, 119–141. *29*

47 What is Cantor's Continuum Problem? AMM **54** (1947), 515–525. Errata **55** (1948), 151. Reprinted in [90] below, 176–187. Revised and expanded version in: Benacerraf, Paul, and Hilary Putnam (eds.) *Philosophy of Mathematics. Selected Readings.* Englewood Cliffs, N.J., Prentice Hall 1964, 258–273. This version reprinted with minor emendations by the author in [90] below, 254–270. *31, 32, 115, 481*

51 Second printing (1951) of [40]. *33, 151, 169*

65 Remarks before the Princeton Bicentennial Conference on problems in mathematics. In: Davis, Martin (ed.) *The Undecidable. Basic Papers on Undecidable Propositions, Unsolvable Problems and Computable Functions.* Hewlett, N.Y., Raven Press 1965, 84–88. Reprinted in [90] below, 150–153. *31*

86 Feferman, Solomon, *et al.* (eds.) *Collected Works*, vol. 1. New York, Oxford University Press 1986. *28*

90 Feferman, Solomon, *et al.* (eds.) *Collected Works*, vol. 2. New York, Oxford University Press 1990. *XVII, 30–32, 58, 151*

Goldfarb, Warren D.

See Dreben, Burton, and Warren D. Goldfarb.

Gottwald, Siegfried, and Lothar Kreiser

84 Paul Mahlo – Leben und Werk. NTM. Schriftenreihe für Geschichte der Naturwissenschaften, Technik, und Medizin **21** (1984), 1–22. *16*

Graham, Ronald L., Bruce L. Rothschild, and Joel H. Spencer

90 *Ramsey Theory*. Second edition. New York, Wiley & Sons 1990. *70*

Green, John

78 Determinacy and the existence of large measurable cardinals. Ph.D. thesis, University of California at Berkeley 1978. *445*

Grigorieff, Serge

75 Intermediate submodels and generic extensions in set theory. AM **101** (1975), 447–490. *124*

Guaspari, David

73 Thin and well ordered analytical sets. Ph.D. thesis, Cambridge University 1973. *171*

Guaspari, David, and Leo A. Harrington

76 Characterizing C_3 (the largest countable Π^1_3 set). PAMS **57** (1976), 127–129. *422*

Hájek, Petr

71 Sets, semisets, models. In: Scott [71], 67–81. *116*

Hajnal, András *35, 71*

56 On a consistency theorem connected with the generalized continuum problem. ZML **2** (1956), 131–136. *34*

61 On a consistency theorem connected with the generalized continuum problem. AMAH **12** (1961), 321–376. *34*

See also Erdős, Paul, and András Hajnal.

See also Erdős, Paul, András Hajnal, Attila Máté, and Richard Rado.

Hajnal, András, Richard Rado, and Vera T. Sós

75 (eds.) *Infinite and Finite Sets*. Colloquia Mathematica Societatis Janos Bolyai vol. 10. Amsterdam, North-Holland 1975.

Hallett, Michael
84 *Cantorian Set Theory and Limitation of Size.* Oxford Logic Guides #10. Oxford, Clarendon Press 1984. *XII*

Halmos, Paul R.
50 *Measure Theory.* Princeton, Van Nostrand 1950. *12*

Hanf, William P. *15, 58, 60, 70, 83*
64 Incompactness in languages with infinitely long expressions. FM **53** (1964), 309–324. *XVII, 39, 41, 42*
64a On a problem of Erdős and Tarski. FM **53** (1964), 325–334. Abstracted in NAMS **9** (1962), 229. *76*

Hanf, William P., and Dana S. Scott
61 Classifying inaccessible cardinals (abstract). NAMS **8** (1961), 445. *XVIII, 39, 55, 57, 59, 61, 64*

Harada, Mikio *321, 322*

Harrington, Leo A. *369, 375, 443*
77 Long projective wellorderings. AML **12** (1977), 1–24. *429*
78 Analytic determinacy and $0^{\#}$. JSL **43** (1978), 685–693. *444*
See also Guaspari, David, and Leo A. Harrington.

Harrington, Leo A., and Alexander S. Kechris
77 Π^1_2 singletons and $0^{\#}$. FM **95** (1977), 167–171. *186, 188*
81 On the determinacy of games on ordinals. AML **20** (1981), 109–154. *379, 382, 402, 422*

Hartogs, Friedrich
15 Über das Problem der Wohlordnung. MA **76** (1915), 436–443. *XIV*

Hausdorff, Felix
08 Grundzüge einer Theorie der geordneten Mengen. MA **65** (1908), 435–505. *XVI, 16*
14 *Grundzüge der Mengenlehre.* Leipzig, de Gruyter 1914. Reprinted in New York, Chelsea 1965. *12, 16, 146, 147, 443*
16 Die Mächtigkeit der Borelschen Mengen. MA **77** (1916), 430–437. *147*

Hauser, Kai
91 Indescribable cardinals and elementary embeddings. JSL **56** (1991), 439–457. *67*
92 The indescribability of the order of the indescribable cardinals. APAL **57** (1992), 45–91. *67*
95 The consistency strength of projective absoluteness. APAL **74** (1995), 245–295. *468*
See also Woodin, W. Hugh, Adrian R.D. Mathias, and Kai Hauser.

Hauser, Kai, and Ralf-Dieter Schindler
00 Projective uniformization revisited. APAL **103** (2000), 109–153. *469*

Hawkins, Thomas W.

75 *Lebesgue's Theory of Integration. Its Origins and Development.* Second edition. New York, Chelsea 1975. *22, 145*

Henkin, Leon, *et al.*

74 (eds.) *Proceedings of the Tarski Symposium.* Proceedings of Symposia in Pure Mathematics vol. 25. Providence, American Mathematical Society, 1974.

Henle, James M.

77 Some consequences of an infinite exponent partition relation. JSL **42** (1977), 523–526. *396*

79 Researches into the world of $\kappa \longrightarrow (\kappa)^{\kappa}$. AML **17** (1979), 151–169. *396*

84 Weak strong partition relations. JSL **49** (1984), 555–557. *396*

90 Partition properties and Prikry Forcing on simple spaces. JSL **55** (1990), 938–947. *238*

See also Apter, Arthur W., and James M. Henle.

See also Di Prisco, Carlos A., and James M. Henle.

Henle, James M., Eugene M. Kleinberg, and Ronald J. Watro

84 On the ultrafilters and ultrapowers of strong partition cardinals. JSL **49** (1984), 1268–1272. *396*

Hewitt, Edwin

48 Rings of real-valued continuous functions. TAMS **64** (1948), 45–99. *27*

Hilbert, David *XIII, XX*

28 Die Grundlagen der Mathematik. Abhandlungen aus dem mathematischen Seminar der Hamburgischen Universität **6** (1928), 65–92. Translated in van Heijenoort [67], 464-479. *478*

Hinman, Peter G.

78 *Recursion-Theoretic Hierarchies.* Perspectives in Mathematical Logic. Berlin, Springer-Verlag 1978. *145, 157*

Hippasus of Metapontum *473*

Hrbáček, Karel *51*

See Vopěnka, Petr, and Karel Hrbáček.

Ihoda, Jaime (Haim Judah), and Saharon Shelah

89 Δ^1_2-sets of reals. APAL **42** (1989), 207–223. *181*

See also Judah, Haim (Jaime Ihoda), and Saharon Shelah.

Isbell, John R.

66 Directed unions and chains. PAMS **17** (1966), 1467–1468. *340*

Jackson, Steve *368, 434–436*

88 AD and the projective ordinals. In: Kechris-Martin-Steel [88], 117–220. *396, 434*

90 Partition properties and well-ordered unions. APAL **48** (1990), 81–101. *428*

91 Admissible Suslin cardinals in $L(\mathbb{R})$. JSL **56** (1991), 260–275. *436*

92 Admissibility and Mahloness in $L(\mathbb{R})$. In: Judah-Just-Woodin [92], 63–74. *436*

Jackson, Steve and Donald A. Martin

83 Pointclasses and well-ordered unions. In: Kechris-Martin-Moschovakis [83], 56–66. *428*

Janiszewski, Zygmunt *22*

Jech, Thomas J. *234*

68 ω_1 can be measurable. IJM **6** (1968), 363–367. *384*

71 The closed unbounded filter over $P_\kappa(\lambda)$. NAMS **18** (1971), 663. *340, 341*

73 Some combinatorial problems concerning uncountable cardinals. AML **5** (1973), 165–198. *340, 342, 343, 346, 350*

74 (ed.) *Axiomatic Set Theory*. Proceedings of Symposia in Pure Mathematics vol. 13, part 2. Providence, American Mathematical Society 1974.

86 A theorem on $P_\kappa(\lambda)$. Journal of the Mathematical Society of Japan **38** (1986), 421–425. *345*

92 Singular Cardinal Problem: Shelah's theorem on $2^{\aleph_\omega}$. BLMS **24** (1992), 127–139. *97*

03 *Set Theory*. Heidelberg, Springer-Verlag 2003. Third millenium edition. *1, 4, 6, 117, 120–124, 145, 182, 217, 225, 260*

See also Dougherty, Randall, and Thomas J. Jech.

Jensen, Ronald B. *59, 111, 186–188, 216, 254, 256, 261, 276, 295, 352, 358, 443*

66 Independence of the axiom of dependent choices from the countable axiom of choice (abstract). JSL **31** (1966), 294. *132*

70 Definable sets of minimal degree. In: Bar-Hillel [70], 122–128. *188*

72 The fine structure of the constructible hierarchy. AML **4** (1972), 229–308. *30*

74 Measurable cardinals and the GCH. In: Jech [74], 175–178. *263*

See also Beller, Aaron, Ronald B. Jensen, and Philip Welch.

Jensen, Ronald B. and Håvard Johnsbråten

74 A new construction of a non-constructible Δ^1_3 subset of ω. FM **81** (1974), 279–290. *188*

Jensen, Ronald B., and Robert M. Solovay

70 Some applications of almost disjoint sets. In: Bar-Hillel [70], 84–104. *188*

Johnsbråten, Håvard

See Jensen, Ronald B. and Håvard Johnsbråten.

Johnson, Chris A.

88 Seminormal λ-generated ideals on $\mathcal{P}_\kappa\lambda$. JSL **53** (1988), 92–102. *350*

88a On saturated ideals and $\mathcal{P}_\kappa\lambda$. FM **29** (1988), 215–221. *350*

Jónsson, Bjarni

72 *Topics in Universal Algebra.* Lecture Notes in Mathematics #250. Berlin, Springer-Verlag 1972. *92*

Joyce, James *481*

Judah, Haim (Jaime Ihoda), Winfried Just, and W. Hugh Woodin

92 (eds.) *Set Theory of the Continuum.* Mathematical Sciences Research Institute publication #26. Berlin, Springer-Verlag 1992.

Judah, Haim (Jaime Ihoda), and Saharon Shelah

91 Forcing minimal degree of constructibility. JSL **56** (1991), 769–782. *243*

See also Ihoda, Jaime (Haim Judah), and Saharon Shelah.

Just, Winfried

See Judah, Haim (Jaime Ihoda), Winfried Just, and W. Hugh Woodin.

Kakuda, Yuzuru

72 Saturated ideals in Boolean extensions. Nagoya Mathematical Journal **48** (1972), 159–168. *220*

Kakutani, Shizuo

See Erdős, Paul, and Shizuo Kakutani.

Kalmár, László

28 Zur Theorie der abstrakten Spiele. ALS **4** (1928), 65–85. *371*

Kamo, Shizuo

97 Ineffability and partition property on $\mathcal{P}_\kappa\lambda$. Journal of the Mathematical Society of Japan **49** (1997), 125–143. *350*

02 Partition properties for $\mathcal{P}_\kappa\lambda$. Journal of the Mathematical Society of Japan **54** (2002), 123–133. *350*

Kanamori, Akihiro

78 On Vopěnka's and related principles. In: Macintyre, Angus, Leszek Pacholski, and Jeffrey B. Paris (eds.) *Logic Colloquium '77.* Amsterdam, North-Holland 1978, 145–153. *336, 337*

See also Solovay, Robert M., William N. Reinhardt, and Akihiro Kanamori.

Kanamori, Akihiro, and Tamara E. Awerbuch-Friedlander

90 The compleat $0^\dagger$. ZML **36** (1990), 133–141. *283*

Kanamori, Akihiro, and Menachem Magidor

78 The evolution of large cardinal axioms in set theory. In: Müller, Gert H., and Dana S. Scott (eds.) *Higher Set Theory.* Lecture Notes in Mathematics #669. Berlin, Springer-Verlag 1978, 99–275. *96, 282*

Kanamori, Akihiro, and Saharon Shelah

95 Complete quotient Boolean algebras. TAMS **347** (1995), 1963–1979. *214*

Kanovei, Vladimir G.

85 The development of the descriptive theory of sets under the influence of the work of Luzin. RMS **40**(3) (1985), 135–180. *146*

Kant, Immanuel *477*

Karp, Carol R.

64 *Languages with Expressions of Infinite Length.* Amsterdam, North-Holland 1964. *36*

Kastanas, Ilias G.

83 On the Ramsey property for sets of reals. JSL **48** (1983), 1035–1045. *382*

Kaufman, Matthew J., and Evangelos Kranakis

84 Definable ultrapowers and ultrafilters over admissible ordinals. ZML **30** (1984), 97–118. *66*

Kechris, Alexander S. *XXI, 368, 380, 422, 424, 431, 432, 436*

73 Measure and category in effective descriptive set theory. AML **5** (1973), 337–384. *422*

74 On projective ordinals. JSL **39** (1974), 269–282. *429*

75 The theory of countable analytical sets. TAMS **202** (1975), 259–297. *171, 422*

78 AD and projective ordinals. In: Kechris-Moschovakis [78], 91–132. *395, 429*

78a On transfinite sequences of projective sets with an application to Σ^1_2 equivalence relations. In: Macintyre, Angus, Leszek Pacholski, and Jeffrey B. Paris (eds.) *Logic Colloquium '77.* Amsterdam, North-Holland 1978, 155–160. *428*

81 Homogeneous trees and projective scales. In: Kechris-Martin-Moschovakis [81], 33–73. *192, 418, 429, 435, 450, 453*

84 The Axiom of Determinacy implies dependent choices in $L(\mathbb{R})$. JSL **49** (1984), 161–173. *433*

85 Determinacy and the structure of $L(\mathbb{R})$. In: Nerode-Shore [85], 271–283. *400, 431, 434*

88 "AD + UNIFORMIZATION" is equivalent to "HALF $\mathrm{AD}_{\mathbb{R}}$". In: Kechris-Martin-Steel [88], 98–102. *469*

88a A coding theorem for measures. In: Kechris-Martin-Steel [88], 103–109. *400*

88b Subsets of $\aleph_1$ constructible from a real. In: Kechris-Martin-Steel [88], 110–116. *388*

95 *Classical Descriptive Set Theory.* New York, Springer-Verlag 1995. *145, 458*

See also Becker, Howard S., and Alexander S. Kechris.

Kechris, Alexander S., Eugene M. Kleinberg, Yiannis N. Moschovakis and W. Hugh Woodin

81 The Axiom of Determinacy, strong partition properties and nonsingular measures. In: Kechris-Martin-Moschovakis [81], 77–99. *432*

Kechris, Alexander S., and Donald A. Martin

78 On the theory of Π^1_3 sets of reals. BAMS **84** (1978), 149–151. *422*

Kechris, Alexander S., Donald A. Martin, and Yiannis N. Moschovakis

81 (eds.) *Cabal Seminar 77–79*. Proceedings, Caltech-UCLA Logic Seminar 1977–79. Lecture Notes in Mathematics #839. Berlin, Springer-Verlag 1981. *403*

83 (eds.) *Cabal Seminar 79–81*. Proceedings, Caltech-UCLA Logic Seminar 1979–81. Lecture Notes in Mathematics #1019. Berlin, Springer-Verlag 1983. *403*

Kechris, Alexander S., Donald A. Martin, and John R. Steel

88 (eds.) *Cabal Seminar 81–85*. Proceedings, Caltech-UCLA Logic Seminar 1981–85. Lecture Notes in Mathematics #1333. Berlin, Springer-Verlag 1988. *403*

Kechris, Alexander S., and Yiannis N. Moschovakis

72 Two theorems about projective sets. IJM **72** (1972), 391–399. *422*

78 (eds.) *Cabal Seminar 76–77*. Proceedings, Caltech-UCLA Logic Seminar 1976–77. Lecture Notes in Mathematics #689. Berlin, Springer-Verlag 1978. *403*

78a Notes on the theory of scales. In: Kechris-Moschovakis [78], 1–53. *422*

Kechris, Alexander S., and Robert M. Solovay

85 On the relative consistency strength of determinacy hypotheses. TAMS **290** (1985), 179–211. *422, 467*

Kechris, Alexander S., Robert M. Solovay, and John R. Steel

81 The Axiom of Determinacy and the prewellordering property. In: Kechris-Martin-Moschovakis [81], 101–125. *415*

Kechris, Alexander S., and W. Hugh Woodin

83 Equivalence of partition properties and determinacy. PNAS **80** (1983), 1783–1786. *422, 432*

Keisler, H. Jerome *50, 85*

62 Some applications of the theory of models to set theory. In: Nagel-Suppes-Tarski [62], 80–86. *39, 42, 44, 49*

62a The equivalence of certain problems in set theory with problems in the theory of models (abstract). NAMS **9** (1962), 339. *39, 49*

71 *Model Theory for Infinitary Logic*. Amsterdam, North-Holland 1971. *36*

See also Chang, Chen-Chung, and H. Jerome Keisler.

Keisler, H. Jerome, and Frederick Rowbottom

65 Constructible sets and weakly compact cardinals (abstract). NAMS **12** (1965), 373–4. *93*

Keisler, H. Jerome, and Alfred Tarski

64 From accessible to inaccessible cardinals. FM **53** (1964), 225–308. Corrections **57** (1965), 119. Reprinted in Tarski [86] vol. 4, 129–213. *27, 37, 38, 42–44, 53, 54, 64, 78*

Keldysh, Ljudmila V.
74 The ideas of N.N. Luzin in descriptive set theory. RMS **29**(5) (1974), 179–193. *146, 148*

Ketonen, Jussi A.
72 Strong compactness and other cardinal sins. AML **5** (1972), 47–76. *308*

Kleene, Stephen C. *114, 145, 151, 152, 154, 157*
43 Recursive predicates and quantifiers. TAMS **53** (1943), 41–73. *151*
55 On the forms of predicates in the theory of constructive ordinals (second paper). American Journal of Mathematics **77** (1955), 405–428. *151, 162*
55a Arithmetical predicates and function quantifiers. TAMS **79** (1955), 312–340. *151*
55b Hierarchies of number-theoretic predicates. BAMS **61** (1955), 193–213. *151*

Kleinberg, Eugene M. *391, 395*
70 Strong partition properties for infinite cardinals. JSL **35** (1970), 410–428. *391, 392*
72 The equiconsistency of two large cardinal axioms (abstract). NAMS **16** (1972), 329. *96, 128*
73 Rowbottom cardinals and Jonsson cardinals are almost the same. JSL **38** (1973), 423–427. *95*
77 AD $\vdash$ "The $\aleph_n$ are Jonsson cardinals and $\aleph_\omega$ is a Rowbottom cardinal". AML **12** (1977), 229–248. *396*
77a *Infinitary Combinatorics and the Axiom of Determinateness.* Lecture Notes in Mathematics #612. Berlin, Springer-Verlag 1977. *396*
78 A combinatorial characterization of normal M-ultrafilters. AdM **30** (1978), 77–84. *245*
79 The equiconsistency of two large cardinal axioms. FM **102** (1979), 81–85. *96, 128*
81 Producing measurable cardinals beyond κ. JSL **46** (1981), 643–648. *396*
82 A measure representation theorem for strong partition cardinals. JSL **47** (1982), 161–168. *396*
See also Henle, James M., Eugene M. Kleinberg, and Ronald J. Watro.
See also Kechris, Alexander S., Eugene M. Kleinberg, Yiannis N. Moschovakis, and W. Hugh Woodin.

Kochen, Simon
61 Ultraproducts in the theory of models. AM **74** (1961), 221–261. *37*

Koepke, Peter G.
88 Some applications of short core models. APAL **37** (1988), 179–204. *96*
89 An introduction to extenders and core models for extender sequences. In: Ebbinghaus, Hans-Dieter, *et al.* (eds.) *Logic Colloquium '87.* Amsterdam, North-Holland 1989, 137–182. *96*
See also Donder, Hans-Dieter, and Peter G. Koepke.
See also Donder, Hans-Dieter, Peter G. Koepke, and Jean-Pierre Levinski.

Kolmogorov, Andrei N.
28 Operations on sets (in Russian). Matematicheskij Sbornik **35** (1928), 414–422. *180*

Kondô, Motokiti *167*
37 L'uniformisation des complémentaires analytiques. Proceedings of the Imperial Academy of Japan **13** (1937), 287–291. *150, 176*
39 Sur l'uniformisation des complémentaires analytiques et les ensembles projectifs de la seconde classe. Japanese Journal of Mathematics **15** (1939), 197–230. *150, 176, 177*

König, Dénes
27 Über eine Schlussweise aus dem Endlichen ins Unendliche: Punktmengen. Kartenfärben. Verwandtschaftsbeziehungen. Schachspiel. ALS **3** (1927), 121–130. *72, 75, 371*

Kranakis, Evangelos
82 Reflection and partition properties of admissible ordinals. AML **22** (1982), 213–242. *66*
82a Invisible ordinals and inductive definitions. ZML **28** (1982), 137–148. *66*
83 Definable Ramsey and definable Erdős ordinals. Archiv für Mathematische Logik und Grundlagenforschung **23** (1983), 115–128. *66*
See also Kaufman, Matthew J., and Evangelos Kranakis.

Kranakis, Evangelos, and Iain Phillips
84 Partitions and homogeneous sets for admissible ordinals. In: Müller, Gert H., and Michael M. Richter (eds.) *Models and Sets*. Lecture Notes in Mathematics #1103. Berlin, Springer-Verlag 1984, 235–260. *66*

Kreisel, Georg
80 Kurt Gödel, 28 April 1906 – 14 January 1978. Biographical Memoirs of the Fellows of the Royal Society **26** (1980), 149–224. Corrections **27** (1981), 697 and **28** (1982), 718. *29, 32, 151*

Kreiser, Lothar
See Gottwald, Siegfried, and Lothar Kreiser.

Kueker, David
72 Löwenheim-Skolem and interpolation theorems in infinitary languages. BAMS **78** (1972), 211–215. *340*
77 Countable approximations and Löwenheim-Skolem theorems. AML **11** (1977), 57–103. *340*

Kunen, Kenneth *XXI, XXII, 43, 61, 94, 99, 108, 110, 207, 209, 219, 244, 252, 254, 257, 259, 261, 264, 275, 277, 282, 294, 297, 298, 311, 319–322, 324, 325, 328, 339, 349, 368, 386, 390, 392, 395, 399, 400, 428, 429, 435, 449, 450, 459–461*
68 Inaccessibility properties of cardinals. Ph.D. thesis, Stanford University 1968. Published in part in [70] below. *244, 250, 253, 256, 265, 267, 276*

70 Some applications of iterated ultrapowers in set theory. AML **1** (1970), 179–227. *XX, 220, 233, 244, 245, 250, 253, 256, 265, 267–270, 275, 292*
71 Indescribability and the continuum. In: Scott [71], 199–203. *61*
71a On the GCH at measurable cardinals. In: Gandy-Yates [71], 107–110. *293*
71b Elementary embeddings and infinitary combinatorics. JSL **36** (1971), 407–413. *XX, 319, 320*
73 A model for the negation of the Axiom of Choice. In: Mathias-Rogers [73], 489–494. *258*
78 Saturated ideals. JSL **43** (1978), 65–76. *212, 233, 331, 459*
80 *Set Theory. An Introduction to Independence Proofs*. Amsterdam, North-Holland 1980. *1, 3, 28, 80, 117, 119–121, 123, 124, 139*
84 Random and Cohen reals. In: Kunen-Vaughan [84], 887–911. *14, 223*
03 Compact spaces, compact cardinals, and elementary submodels. Topology and its Applications **130** (2003), 99–109. *303*

Kunen, Kenneth, and Jeffrey B. Paris
71 Boolean extensions and measurable cardinals. AML **2** (1971), 359–377. *226, 228, 230, 231*

Kunen, Kenneth, and Donald H. Pelletier
83 On a combinatorial property of Menas related to the partition property for measures on supercompact cardinals. JSL **48** (1983), 475–481. *348, 349*

Kunen, Kenneth, and Jerry E. Vaughan
84 (eds.) *Handbook of Set-Theoretic Topology*. Amsterdam, North-Holland 1984.

Kuratowski, Kazimierz *16, 22*
21 Sur la notion de l'ordre dans la théorie des ensembles. FM **2** (1921), 161–171. Reprinted in [88] below, 1–11. *XIV*
24 Sur l'état actuel de l'axiomatique de la théorie des ensembles. Annales de la Société Polonaise de Mathématique **3** (1924), 146–147. Reprinted in [88] below, 179. *18*
31 Evaluation de la classe Borélienne ou projective d'un ensemble de points à l'aide des symboles logiques. FM **17** (1931), 249–272. Reprinted in [88] below, 376–399. *152*
36 Sur les théorèmes de séparation dans la théorie des ensembles. FM **26** (1936), 183–191. Reprinted in [88] below, 461–469. *405, 407, 408*
58 *Topologie.* Vol. 1. Warsaw, Państwowe Wydawnictwo Naukowe 1958. English translation, 1966. *443*
80 *A Half Century of Polish Mathematics. Remembrances and Reflections.* Oxford, Pergamon Press 1980. *22, 148*
88 Borsuk, Karol, *et al.* (eds.) *Selected Papers.* Warsaw, Państwowe Wydawnictwo Naukowe 1988.
See also Banach, Stefan, and Kazimierz Kuratowski.

Kuratowski, Kazimierz, and Alfred Tarski

31 Les opérations logiques et les ensembles projectifs. FM **17** (1931), 240–248. Reprinted in Tarski [86] vol. 1, 551–559, and in Kuratowski [88], 367–375. Translated in Tarski [83], 143–151. *152*

Kuratowski, Kazimierz, and Stanisław M. Ulam

32 Quelques propriétés topologiques du produit combinatoire. FM **19** (1932), 247–251. Reprinted in Ulam [74], 32–36. *14*

Kurepa, Djuro R.

35 Ensembles ordonnés et ramifiés. Thèse, Paris. Published as: Publications mathématiques de l'Université de Belgrade **4** (1935), 1–138. *72, 75, 78, 79*

41 Transformations monotones des ensembles partiellement ordonnés (continuation). Revista de Ciencias de la Universidad Mayor de San Marcos (Lima) **43**(437) (1941), 483–500. *74*

59 On the cardinal number of ordered sets and of symmetrical structures in dependence on the cardinal numbers of its chains and antichains. Glasnik Matematičko-fizički i astronomski, Periodicum mathematico-physicum et astronomicum **14** (1959), 183–203. *72, 74*

Kuzawa, Mary Grace

68 *Modern Mathematics. The Genesis of a School in Poland.* New Haven, College & University Press 1968. *22*

70 Fundamenta Mathematicae: an examination of its founding and significance. AMM **77** (1970), 485–492. *22*

Lao-Tzu *482*

Laplace, Pierre Simon *473*

Laver, Richard *117*

82 Saturated ideals and non-regular ultrafilters. In: Metakides [82], 297–305. *221*

86 Elementary embeddings of a rank into itself. AAMS **7** (1986), 6. *329*

92 The left distributive law and the freeness of an algebra of elementary embeddings. AdM **91** (1992), 209–231. *329, 330*

94 A division algorithm for the free left distributive algebra. In: Juha Oikkonen and Jouko A. Väänänen (eds.) *Logic Colloquium '90.* Lecture Notes in Logic #2. New York, Springer-Verlag 1994, 155–162. *330*

95 On the algebra of elementary embeddings of a rank into itself. AdM **110** (1995), 334–346. *330*

97 Implications between strong large cardinal axioms. APAL **90** (1997), 79–90. *328*

01 Reflection of elementary embedding axioms on the $L[V_{\lambda+1}]$ hierarchy. APAL **107** (2001), 227–238. *329*

02 On very large cardinals. In: Gábor Halász, László Lóvasz, Miklós Simonovits, and Vera T. Sós (eds.) *Paul Erdős and his Mathematics II,*

Bolyai Society Mathematical Studies 11. Berlin, Springer-Verlag (2002), 453–469. *329*

Lavine, Shaughan M.

94 *Understanding the Infinite.* Cambridge, Harvard University Press 1994. *XII*

Lebesgue, Henri *132, 145, 146, 151*

02 Intégrale, longueur, aire. Annali di Matematica Pura ed Applicata **(3)7** (1902), 231–359. Reprinted in [72] below, vol. 1, 203–331. *12, 22, 145*

05 Sur les fonctions représentables analytiquement. Journal de Mathématiques Pures et Appliquées **(6)1** (1905), 139–216. Reprinted in [72] below, vol. 3, 103–180. *XVI, 145–147, 148, 150, 157, 159, 166*

07 Contributions à l'etude des correspondances de M. Zermelo. Bulletin de la Société Mathématique de France **35** (1907), 202–212. Reprinted in [72] below, vol. 3, 227–237. *22*

72 *Oeuvres Scientifiques.* Geneva, Kundig 1972.

Leibniz, Gottfried W. *XII, 474*

Lerman, Manuel

83 *Degrees of Unsolvability.* Perspectives in Mathematical Logic. Berlin, Springer-Verlag 1983. *114*

Levinski, Jean-Pierre

84 Instances of the conjecture of Chang. IJM **48** (1984), 225–243. *345*

See also Carr, Donna M., Jean-Pierre Levinski, and Donald H. Pelletier.

See also Donder, Hans-Dieter, Peter G. Koepke, and Jean-Pierre Levinski.

Levy, Azriel *21, 35, 114, 125, 128, 129, 135, 169, 212, 311, 384*

57 Indépendance conditionelle de $V = L$ et d'axiomes qui se rattachent au système de M. Gödel. CRP **245** (1957), 1582–1583. *34*

60 A generalization of Gödel's notion of constructibility. JSL **25** (1960), 147–155. *34*

60a Axiom schemata of strong infinity in axiomatic set theory. PJM **10** (1960), 223–238. *XVIII, 57–59*

63 Independence results in set theory by Cohen's method IV (abstract). NAMS **10** (1963), 593. *126, 136*

64 Measurable cardinals and the continuum hypothesis (abstract). NAMS **11** (1964), 769–770. *126*

65 A hierarchy of formulas in set theory. Memoirs of the American Mathematical Society **57** (1965). *6, 299*

65a Definability in axiomatic set theory I. In: Bar-Hillel [65], 127–151. *177*

70 Definability in axiomatic set theory II. In: Bar-Hillel [70], 129–145. *126, 136, 177*

71 The sizes of the indescribable cardinals. In: Scott [71], 205–218. *XVIII, 57, 63–66*

79 *Basic Set Theory.* Perspectives in Mathematical Logic. Berlin, Springer-Verlag 1979. *1, 11, 72, 223*

See also Fraenkel, Abraham A., Yehoshua Bar-Hillel, and Azriel Levy.

Levy, Azriel, and Robert M. Solovay

67 Measurable cardinals and the continuum hypothesis. IJM **5** (1967), 234–248. *126*

72 On the decomposition of sets of reals to Borel sets. AML **5** (1972), 1–19. *167*

Łoś, Jerzy

55 Quelques remarques, théorèmes et problèmes sur les classes définissables d'algèbres. In: Skolem, Thoralf, *et al.* (eds.) *Mathematical Interpretation of Formal Systems*. Amsterdam, North-Holland 1955, 98–113. *9, 37*

See also Ehrenfeucht, Andrzej, and Jerzy Łoś.

Louveau, Alain

83 Some results in the Wadge hierarchy of Borel sets. In: Kechris-Martin-Moschovakis [83], 28–55. *416, 442*

Louveau, Alain, and Jean Saint-Raymond

87 Borel classes and closed games: Wadge-type and Hurewicz-type results. TAMS **304** (1987), 431–467. *416, 442, 443*

88 The strength of Borel Wadge Determinacy. In: Kechris-Martin-Steel [88], 1–30. *416, 442, 443*

Luzin, Nikolai N. *XVI, 146, 147, 151*

17 Sur la classification de M. Baire. CRP **164** (1917), 91–94. *148, 149, 183*

25 Sur un problème de M. Emile Borel et les ensembles projectifs de M. Henri Lebesgue; les ensembles analytiques. CRP **180** (1925), 1318–1320. *148*

25a Sur les ensembles projectifs de M. Henri Lebesgue. CRP **180** (1925), 1572–1574. *148, 149*

25b Les propriétés des ensembles projectifs. CRP **180** (1925), 1817–1819. *149*

25c Sur les ensembles non mesurables *B* et l'emploi de la diagonale Cantor. CRP **181** (1925), 95–96. *149, 158*

27 Sur les ensembles analytiques. FM **10** (1927), 1–95. *148, 165, 405*

30 *Leçons sur Les Ensembles Analytiques et Leurs Applications*. Paris, Gauthier-Villars 1930. Reprinted with corrections in New York, Chelsea 1972. *148–150*

30a Sur le problème de M.J. Hadamard d'uniformisation des ensembles. CRP **190** (1930), 349–351. *150*

35 Sur les ensembles analytiques nuls. FM **25** (1935), 109–131. *167*

Luzin, Nikolai N., and Petr S. Novikov

35 Choix éffectif d'un point dans un complémentaire analytique arbitraire, donné par un crible. FM **25** (1935), 559–560. *150*

Luzin, Nikolai N., and Wacław Sierpiński

18 Sur quelques propriétés des ensembles (A). Bulletin de l'Académie des Sciences Cracovie, Classe des Sciences Mathématiques et Naturelles,

Série A (1918), 35–48. Reprinted in Sierpiński [75], 192–204. *159, 163–165, 408, 424*

23 Sur un ensemble non mesurable B. Journal de Mathématiques Pures et Appliquées **(9)2** (1923), 53–72. Reprinted in Sierpiński [75], 504–519. *159, 162–166, 424*

Lyubeckij, V.A.

70 The existence of a nonmeasurable set of type A_2 implies the existence of an uncountable set of type CA which does not contain a perfect subset. Doklady Akademii Nauk SSSR **195** (1971), 548–550. Translated in Soviet Mathematics, Doklady **11** (1970), 1513–1515. *184*

71 Independence of certain propositions of set theory from Zermelo-Fraenkel set theory. Vestnik Moskovskogo Universiteta Seriia I, Matematika Mehanika **26**(2) (1971), 78–82. Translated in Moscow University Mathematics Bulletin **26**(2) (1971), 116–119. *184*

Mackey, George W.

44 Equivalence of a problem in measure theory to a problem in the theory of vector lattices. BAMS **50** (1944), 719–722. *27*

Maddy, Penelope

88 Believing the Axioms. I. JSL **53** (1988), 481–511. *XVII*

88a Believing the Axioms. II. JSL **53** (1988), 736–764. *XVII*

Magidor, Menachem *XXII, 295, 309, 325, 463*

71 There are many normal ultrafilters corresponding to a supercompact cardinal. IJM **9** (1971), 186–192. *306, 309*

71a On the role of supercompact and extendible cardinals in logic. IJM **10** (1971), 147–157. *302, 303, 309, 315*

74 Combinatorial characterization of supercompact cardinals. PAMS **42** (1974), 279–285. *350*

76 How large is the first strongly compact cardinal? or: A study on identity crises. AML **10** (1976), 33–57. *309*

80 Precipitous ideals and Σ^1_4 sets. IJM **35** (1980), 109–134. *461*

See also Burke, Maxim, and Menachem Magidor.

See also Feng, Qi, Menachem Magidor, and W. Hugh Woodin.

See also Foreman, Matthew, Menachem Magidor, and Saharon Shelah.

See also Gitik, Moti, and Menachem Magidor.

See also Kanamori, Akihiro, and Menachem Magidor.

Mahlo, Paul *58*

11 Über lineare transfinite Mengen. BKSG **63** (1911), 187–225. *XVI, 16, 17*

12 Zur Theorie und Anwendung der ρ_0-Zahlen. BKSG **64** (1912), 108–112. *XVI, 16*

13 Zur Theorie und Anwendung der ρ_0-Zahlen II. BKSG **65** (1913), 268–282. *XVI, 16*

13a Über Teilmengen des Kontinuums von dessen Mächtigkeit. BKSG **65** (1913), 283–315. *16*

Makowsky, Johann A.
85 Vopěnka's Principle and compact logics. JSL **50** (1985), 42–48. *339*

Mansfield, Richard B.
70 Perfect subsets of definable sets of real numbers. PJM **35** (1970), 451–457. *182*
71 A Souslin operation for Π^1_2. IJM **9** (1971), 367–379. *192, 198, 201, 204, 420*

Mansfield, Richard B., and Galen Weitkamp
85 *Recursive Aspects of Descriptive Set Theory.* Oxford Logic Guides #11. New York, Oxford University Press 1985. *145*

Marek, Wiktor
See Di Prisco, Carlos A., and Wiktor Marek.

Marshall, M. Victoria
89 Higher order reflection principles. JSL **54** (1989), 474–489. *331*

Martin, Donald A. *XXI, 113, 145, 171, 192, 193, 203, 204, 367–369, 380, 383, 386, 389, 391, 392, 395, 401, 403, 406, 411, 412, 415, 417, 422, 424–429, 431–436, 439–445, 450, 453, 458–460, 462, 463*
68 The axiom of determinateness and reduction principles in the analytical hierarchy. BAMS **74** (1968), 687–689. *387, 410*
70 Measurable cardinals and analytic games. FM **66** (1970), 287–291. *XXII, 437, 440–442*
75 Borel determinacy. AM **102** (1975), 363–371. *441*
76 Hilbert's first problem: The Continuum Hypothesis. In: Browder, Felix (ed.) *Mathematical Developments Arising from Hilbert's Problems.* Proceedings of Symposia in Pure Mathematics vol. 28. Providence, American Mathematical Society 1976, 81–92. *31*
80 Infinite games. In: Lehto, Olli (ed.) *Proceedings of the International Congress of Mathematicians*, Helsinki 1978, vol. 1. Helsinki, Academia Scientiarum Fennica 1980, 269–273. *329, 449*
83 The largest countable this, that, and the other. In: Kechris-Martin-Moschovakis [83], 97–106. *422, 430, 432*
83a The real game quantifier propagates scales. In: Kechris-Martin-Moschovakis [83], 157–171. *431*
85 A purely inductive proof of Borel determinacy. In: Nerode-Shore [85], 303–308. *441*
90 An extension of Borel determinacy. APAL **49** (1990), 279–293. *441, 445*
∞ Forthcoming book on determinacy. *441, 443, 445, 450*
See also Baumgartner, James E., Donald A. Martin, and Saharon Shelah.
See also Jackson, Steve, and Donald A. Martin.
See also Kechris, Alexander S., and Donald A. Martin.
See also Kechris, Alexander S., Donald A. Martin, and Yiannis N. Moschovakis.
See also Kechris, Alexander S., Donald A. Martin, and John R. Steel.

Martin, Donald A., and William J. Mitchell

79 On the ultrafilter of closed, unbounded sets. JSL **44** (1979), 503–506. *386*

Martin, Donald A., Yiannis N. Moschovakis, and John R. Steel.

82 The extent of definable scales. BAMS (New Series) **6** (1982), 435–440. *430*

Martin, Donald A., and Robert M. Solovay

69 A basis theorem for Σ_3^1 sets of reals. AM **89** (1969), 138–159. *XIX, 192, 196, 198, 204, 420, 437, 450*

70 Internal Cohen extensions. AML **2** (1970), 143–178. *31, 167, 179, 426*

Martin, Donald A., and John R. Steel

83 The extent of scales in $L(\mathbb{R})$. In: Kechris-Martin-Moschovakis [83], 86–96. *382, 430*

88 Projective determinacy. PNAS **85** (1988), 6582–6586. *463*

89 A proof of Projective Determinacy. JAMS **2** (1989), 71–125. *360, 462, 463*

94 Iteration trees. JAMS **7** (1994), 1–73. *463*

Máté, Attila

See Erdős, Paul, András Hajnal, Attila Máté, and Richard Rado.

Mathias, Adrian R.D. *238*

68 On a generalization of Ramsey's theorem (abstract). NAMS **15** (1968), 931. *143, 144*

73 On sequences generic in the sense of Prikry. Journal of the Australian Mathematical Society **15** (1973), 409–414. *237*

77 Happy families. AML **12** (1977), 59–111. *143, 144, 238, 243*

79 Surrealist landscape with figures (a survey of recent results in set theory). Periodica Mathematica Hungarica **10** (1979), 109–175. *117*

83 (ed.) *Surveys in Set Theory*. London Mathematical Society Lecture Note Series #87. Cambridge, Cambridge University Press, 1983.

See also Woodin, W. Hugh, Robert M. Solovay, and Adrian R.D. Mathias.

Mathias, Adrian R.D., and Hartley Rogers Jr.

73 (eds.) *Cambridge Summer School in Mathematical Logic*. Lecture Notes in Mathematics #337. Berlin, Springer-Verlag 1973.

Matsubara, Yo

87 Menas' conjecture and generic ultrapowers. APAL **36** (1987), 225–234. *343*

88 Splitting $\mathcal{P}_\kappa\lambda$ into stationary subsets. JSL **53** (1988), 385–389. *346*

88a Variations of cub filter on $\mathcal{P}_\kappa\lambda$. PAMS **102** (1988), 1009–1017. *345*

90 Consistency of Menas' conjecture. Journal of the Mathematical Society of Japan **42** (1990), 259–263. *345*

Mauldin, R. Daniel

81 (ed.) *The Scottish Book*. Boston, Birkhäuser 1981. *371, 373*

Mazur, Stanisław *371, 373, 377*

Mazurkiewicz, Stefan

36 Über die Menge der differenzierbaren Funktionen. FM **27** (1936), 244–249. *166*

Menas, Telis K. *344–346*

74 On strong compactness and supercompactness. AML **7** (1974), 327–359. *308, 309, 341, 342*

76 A combinatorial property of $p_\kappa\lambda$. JSL **41** (1976), 225–234. *309, 347–349*

76a Consistency results concerning supercompactness. TAMS **223** (1976), 61–91. *309*

Metakides, George

82 (ed.) *Patras Logic Symposion.* Amsterdam, North-Holland 1982.

Mignone, Robert J.

84 The ultrafilter characterization of huge cardinals. PAMS **90** (1984), 585–590. *334*

Mirimanov, Dmitry

17 Les antinomies de Russell et de Burali-Forti et le problème fondamental de la théorie des ensembles. L'Enseignment Mathematique **19** (1917), 37–52. *XIV, 20*

Mitchell, William J. *117, 276, 470*

74 Sets constructible from sequences of ultrafilters. JSL **39** (1974), 57–66. *305*

79 Hypermeasurable cardinals. In: Boffa, Maurice, Dirk van Dalen, and Kenneth McAloon (eds.) *Logic Colloquium '78.* Amsterdam, North-Holland 1979, 303–316. *352, 358, 449*

See also Martin, Donald A., and William J. Mitchell.

Monk, J. Donald, and Dana S. Scott

64 Additions to some results of Erdős and Tarski. FM **53** (1964), 335–343. *76*

Monk, Leonard *415*

Montague, Richard M.

55 Non-finitizable and essentially non-finitizable theories (abstract). BAMS **61** (1955), 172–173. *57*

61 Fraenkel's addition to the axioms of Zermelo. In: Bar-Hillel-Poznanski-Rabin-Robinson [61], 91–114. *57, 58*

Montague, Richard M., and Robert L. Vaught

59 Natural models of set theories. FM **47** (1959), 219–242. *19*

Moore, Gregory H.

82 *Zermelo's Axiom of Choice. Its Origins, Development and Influence.* New York, Springer-Verlag 1982. *XII, 146*

88 The origins of forcing. In: Drake, Frank R., and John K. Truss (eds.) *Logic Colloquium '86*. Amsterdam, North-Holland 1988, 143–173. *32, 114*

Morel, Ann C.
See Frayne, Thomas E., Ann C. Morel, and Dana S. Scott.

Morgenstern, Carl F.
79 On the ordering of certain large cardinals. JSL **44** (1979), 563–565. *333*

Morgenstern, Oskar
See von Neumann, John, and Oskar Morgenstern.

Morley, Michael D.
65 Categoricity in power. TAMS **114** (1965), 514–538. *100*
65a Omitting classes of elements. In: Addison-Henkin-Tarski [65], 265–273. *100*

Moschovakis, Yiannis N. *XXI, 367, 368, 397, 403, 406, 408, 410, 412, 417, 422–424, 427, 436*
70 Determinacy and prewellorderings of the continuum. In: Bar-Hillel [70], 24–62. *396–398, 423, 425, 426, 428*
71 Uniformization in a playful universe. BAMS **77** (1971), 731–736. *419–421, 425, 427*
73 Analytical definability in a playful universe. In: Suppes, Patrick, *et al.* (eds.) *Logic, Methodology and Philosophy of Science IV*. Proceedings of the 1971 International Congress, Bucharest. Amsterdam, North-Holland 1973, 77–83. *421, 422*
76 Indescribable cardinals in L (abstract). JSL **41** (1976), 554–555. *66*
78 Inductive scales on inductive sets. In: Kechris-Moschovakis [78], 185–192. *430*
80 *Descriptive Set Theory*. Amsterdam, North-Holland 1980. *3, 28, 34, 35, 145, 179, 203, 369, 403, 421, 422, 426, 428, 430*
81 Ordinal games and playful models. In: Kechris-Martin-Moschovakis [81], 169–201. *422*
83 Scales on coinductive sets. In: Kechris-Martin-Moschovakis [83], 77–85. *430*
See also Addison Jr., John W., and Yiannis N. Moschovakis.
See also Becker, Howard S., and Yiannis N. Moschovakis.
See also Kechris, Alexander S., Eugene M. Kleinberg, Yiannis N. Moschovakis, and W. Hugh Woodin.
See also Kechris, Alexander S., Donald A. Martin, and Yiannis N. Moschovakis.
See also Kechris, Alexander S., and Yiannis N. Moschovakis.
See also Martin, Donald A., Yiannis N. Moschovakis, and John R. Steel.

Mostowski, Andrzej M. *36, 76, 99, 151*
47 On definable sets of positive integers. FM **34** (1947), 81–112. Reprinted in [79] below, vol. 1, 339–370. *151*

49 An undecidable arithmetical statement. FM **36** (1949), 143–164. Reprinted in [79] below, vol. 1, 531–552. *7, 18, 19*

79 Kuratowski, Kazimierz, *et al.* (eds.) *Foundational Studies. Selected Works.* Amsterdam, North-Holland 1979.

See also Ehrenfeucht, Andrzej, and Andrzej M. Mostowski.

Müller, Gert H.

76 (ed.) *Sets and Classes.* Amsterdam, North-Holland 1976.

Müller, Gert H., Arnold Oberschelp, and Klaus Potthoff

75 (eds.) *Logic Conference, Kiel 1974.* Lecture Notes in Mathematics #499. Berlin, Springer-Verlag 1975.

Mycielski, Jan *441*

64 On the axiom of determinateness. FM **53** (1964), 205–224. *135, 169, 377–379*

66 On the axiom of determinateness II. FM **59** (1966), 203–212. *377*

Mycielski, Jan, and Hugo Steinhaus

62 A mathematical axiom contradicting the axiom of choice. BAPS **10** (1962), 1–3. Reprinted in Steinhaus [85], 778–781. *XXI, 377*

Mycielski, Jan, and Stanisław Swierczkowski

64 On the Lebesgue measurability and the axiom of determinateness. FM **54** (1964), 67–71. *375, 377*

Mycielski, Jan, Stanisław Swierczkowski, and A. Zieba

56 On infinite positional games. BAPS **4** (1956), 485–488. *377*

Mycielski, Jan, and A. Zieba

55 On infinite games. BAPS **3** (1955), 133–136. *377*

Nagel, Ernest, Patrick Suppes, and Alfred Tarski

62 (eds.) *Logic, Methodology and Philosophy of Science.* Proceedings of the 1960 International Congress, Stanford. Stanford, Stanford University Press 1962.

Namba, Kanji *305*

Napoleon Bonaparte *473*

Negrepontis, Stylianos

See Comfort, W. Wistar, and Stylianos Negrepontis.

Nerode, Anil, and Richard A. Shore

85 (eds.) *Recursion Theory.* Proceedings of Symposia in Pure Mathematics vol. 42. Providence, American Mathematical Society 1985.

Newton, Isaac *XII, 474*

Normann, Dag

See Fenstad, Jens E., and Dag Normann.

Novikov, Petr S.

31 Sur les fonctions implicites mesurables *B*. FM **17** (1931), 8–25. *150*

35 Sur la séparabilité des ensembles projectifs de seconde class. FM **25** (1935), 459–466. *404, 408*

51 On the consistency of some propositions of the descriptive theory of sets (in Russian). Trudy Matematičeskogo Instituat imeni V.A. Steklova **38** (1951), 279–316. Translated in American Mathematical Society Translations **29** (1963), 51–89. *151, 410*

See also Luzin, Nikolai N., and Petr S. Novikov.

Oxtoby, John C.

57 The Banach-Mazur game and Banach Category Theorem. In: Dresher, Melvin, Alan W. Tucker, and Philip Wolfe (eds.) *Contributions to the Theory of Games.* Vol. 3. Annals of Mathematics Studies #39. Princeton, Princeton University Press 1957, 159–163. *373*

71 *Measure and Category. A Survey of the Analogies between Topological and Measure Spaces.* New York, Springer-Verlag 1971. *14*

Paris, Jeffrey B. *236, 282, 395*

69 Boolean extensions and large cardinals. Ph.D. thesis, Manchester University 1969. *270*

72 ZF $\vdash \Sigma^0_4$ determinateness. JSL **37** (1972), 661–667. *441*

74 Patterns of indiscernibles. BLMS **6** (1974), 183–188. *186, 188*

See also Devlin, Keith J., and Jeffrey B. Paris.

See also Kunen, Kenneth, and Jeffrey B. Paris.

Peano, Giuseppe *XIII, 479*

Pelletier, Donald H.

81 The partition property for certain extendible measures on supercompact cardinals. PAMS **81** (1981), 607–612. *349*

See also Carr, Donna M., Jean-Pierre Levinski, and Donald H. Pelletier.

See also Carr, Donna M., and Donald H. Pelletier.

See also Kunen, Kenneth, and Donald H. Pelletier.

Phillips, Esther R.

78 Nicolai Nicolaevich Luzin and the Moscow school of the theory of functions. Historia Mathematica **5** (1978), 275–305. *146*

Phillips, Iain

See Kranakis, Evangelos, and Iain Phillips.

Poincaré, Henri *478*

Pospíšil, Bedřich

37 Remark on bicompact spaces. AM **38** (1937), 845–846. *400*

Post, Emil L. *114*

Powell, William C.

72 Almost huge cardinals and Vopěnka's principle (abstract). NAMS **19** (1972), A-616. *338*

74 Variations of Keisler's theorem for complete embeddings. FM **81** (1974), 121–132. *326, 352*

Prikry, Karel L. *91, 143, 234*

66 Measurable cardinals and saturated ideals (abstract). NAMS **13** (1966), 720–721. *220, 221*

70 Changing measurable into accessible cardinals. Dissertationes Mathematicae (Rozprawy Matematyczne) **68** (1970), 5–52. *90, 220–222, 235, 236, 238, 240, 242*

75 Ideals and powers of cardinals. BAMS **81** (1975), 907–909. *91, 219*

76 Determinateness and partitions. PAMS **54** (1976), 303–306. *382*

See also Galvin, Fred, and Karel L. Prikry.

Rado, Richard *71*

See Erdős, Paul, András Hajnal, Attila Máté, and Richard Rado.

See also Erdős, Paul, András Hajnal, and Richard Rado.

See also Erdős, Paul, and Richard Rado.

See also Hajnal, András, Richard Rado, and Vera T. Sós.

Ramsey, Frank P.

30 On a problem of formal logic. PLMS **(2)30** (1930), 264–286. Reprinted in Gessel-Rota [87], 2–24. *XVIII, 70, 75, 100*

Rasiowa, Helena, and Roman Sikorski

63 *The Mathematics of Metamathematics.* Monografie Matematyczne #41. Warsaw, Państwowe Wydawnictwo Naukowe 1963. *114*

Reid, Constance

See Albers, Donald J., Gerald L. Alexanderson, and Constance Reid.

Reinhardt, William N. *XX, 297, 298, 302, 307, 311–314, 339*

70 Ackermann's set theory equals ZF. AML **2** (1970), 189–249. *311*

74 Set existence principles of Shoenfield, Ackermann, and Powell. FM **84** (1974), 5–34. *312, 313*

74a Remarks on reflection principles, large cardinals, and elementary embeddings. In: Jech [74], 189–205. *313, 318*

80 Satisfaction definitions and axioms of infinity in a theory of properties with necessity operator. In: Arruda, Ayda I., Rolando Chuaqui, and Newton C.A. da Costa (eds.) *Mathematical Logic in Latin America.* Amsterdam, North-Holland 1980, 267–303. *314*

See also Solovay, Robert M., William N. Reinhardt, and Akihiro Kanamori.

Reinhardt, William N., and Jack H. Silver

65 On some problems of Erdős-Hajnal (abstract). NAMS **12** (1965), 723–724. *83, 109*

Richter, Wayne H., and Peter Aczel

74 Inductive definitions and reflecting properties of admissible ordinals. In: Fenstad-Hinman [74], 301–381. *66*

Robinson, Abraham

62 Recent developments in model theory. In: Nagel-Suppes-Tarski [62], 60–79. Reprinted in [79] below, vol. 1, 12–31. *39*

65 Formalism 64. In: Bar-Hillel [65], 228–246. Reprinted in [79] below, vol. 2, 505–523. *115, 481*

79 Keisler, H. Jerome *et al.* (eds.) *Selected Papers of Abraham Robinson.* New Haven, Yale University Press 1979.

Rogers, Claude A., *et al.*

80 (eds.) *Analytic Sets.* London, Academic Press 1980. *145*

Rosser, J. Barkley

69 *Simplified Independence Proofs.* New York, Academic Press 1969. *117*

See also Firestone, C.D., and J. Barkley Rosser.

Rothschild, Bruce L.

See Graham, Ronald L., Bruce L. Rothschild, and Joel H. Spencer.

Rowbottom, Frederick *43, 69, 87, 88, 94, 98, 99, 184*

64 Large cardinals and small constructible sets. Ph.D. thesis, University of Wisconsin at Madison 1964. Published as [71] below. *XVIII, 83, 85, 86, 88, 89*

71 Some strong axioms of infinity incompatible with the axiom of constructibility. AML **3** (1971), 1–44. *XVIII, 83, 86, 88, 89*

See also Keisler, H. Jerome, and Frederick Rowbottom.

Russell, Bertrand *XIII, XIV, 2, 29, 70*

03 *The Principles of Mathematics.* Cambridge, Cambridge University Press 1903. *477*

Sacks, Gerald E.

76 Countable admissible ordinals and hyperdegrees. AdM **19** (1976), 213–262. *171*

90 *Higher Recursion Theory.* Perspectives in Mathematical Logic. Berlin, Springer-Verlag 1990. *157, 166*

Saint-Raymond, Jean

See Louveau, Alain, and Jean Saint-Raymond.

Sami, Ramez *423, 443*

Šanin, Nikolai A.

46 A theorem from the general theory of sets (in Russian). Doklady Academii Nauk S.S.S.R. **53** (1946), 399–400. *121*

Schimmerling, Ernest

02 Woodin cardinals, Shelah cardinals, and the Mitchell-Steel Core Model. PAMS **130** (2002), 3385–3391. *365*

Schimmerling, Ernest, and John R. Steel

96 Fine structure for tame inner models. JSL **61** (1996), 621–639. *472*

Schindler, Ralf-Dieter

02 The core model for almost linear iterations. APAL **116** (2002), 205–272. *469*

See also Hauser, Kai, and Ralf-Dieter Schindler.

Scott, Dana S. *XVIII, 15, 27, 32, 44, 51–53, 55, 58, 60, 88, 114, 116, 244, 257, 261, 295, 378*

55 Definitions by abstraction in axiomatic set theory. BAMS **61** (1955), 442. *47*

61 Measurable cardinals and constructible sets. BAPS **9** (1961), 521–524. *XVII, 44, 49*

67 A proof of the independence of the Continuum Hypothesis. Mathematical Systems Theory **1** (1967), 89–111. *116*

71 (ed.) *Axiomatic Set Theory.* Proceedings of Symposia in Pure Mathematics vol. 13, part 1. Providence, American Mathematical Society 1971.

See also Frayne, Thomas E., Ann C. Morel, and Dana S. Scott.

See also Hanf, William P., and Dana S. Scott.

See also Monk, J. Donald, and Dana S. Scott.

Shelah, Saharon *XXII, 97, 98, 179, 331, 450, 460–463*

75 Notes on partition calculus. In: Hajnal-Rado-Sós [75] vol. 3, 1257–1276. *72*

78 Jonsson algebras in successor cardinals. IJM **30** (1978), 57–64. *96, 97*

79 Weakly compact cardinals: a combinatorial proof. JSL **44** (1979), 559–562. *77*

80 A problem of Kurosh, Jónsson groups and applications. In: Adian, Sergei I., William W. Boone, and Graham Higman (eds.) *Word Problems II.* Amsterdam, North-Holland 1980, 373–394. *95*

82 *Proper Forcing.* Lecture Notes in Mathematics #940. Berlin, Springer-Verlag, 1982. *460*

84 Can you take Solovay's inaccessible away? IJM **48** (1984), 1–47. *XIX, 136, 140, 179*

86 *Around Classification Theory of Models.* Lecture Notes in Mathematics #1182. Berlin, Springer-Verlag 1986. *350, 470*

87 Iterated forcing and normal ideals on ω_1. IJM **60** (1987), 345–380. *462*

See also Kanamori, Akihiro, and Saharon Shelah.

See also Baumgartner, James E., Donald A. Martin, and Saharon Shelah.

See also Foreman, Matthew, Menachem Magidor, and Saharon Shelah.

Shelah, Saharon, and Lee J. Stanley

90 Coding and reshaping when there are no sharps. In: Judah-Just-Woodin [92], 407–416. *187*

95 A combinatorial forcing for coding the universe by a real when there are no sharps. JSL **60** (1995), 1–35. *187*

95a The combinatorics of combinatorial coding by a real. JSL **60** (1995), 36–57. *187*

Shelah, Saharon, and W. Hugh Woodin

90 Large cardinals imply that every reasonably definable set of reals is Lebesgue measurable. IJM **70** (1990), 381–394. *461*

Shepherdson, John C.

51 Inner models for set theory – Part I. JSL **16** (1951), 161–190. *7, 33*

52 Inner models for set theory – Part II. JSL **17** (1952), 225–237. *20, 33*

53 Inner models for set theory – Part III. JSL **18** (1953), 145–167. *33*

Shioya, Masahiro

93 Infinitary Jónsson functions and elementary embeddings. Archive for Mathematical Logic **33** (1994), 81–86. *325*

03 A saturated stationary subset of $\mathcal{P}_\kappa\kappa^+$. Mathematical Research Letters **10** (2003), 493–500. *344*

∞ An ultrafilter with property σ. PAMS. *350*

Shoenfield, Joseph R. *174, 312, 313*

59 On the independence of the axiom of constructibility. American Journal of Mathematics **81** (1959), 537–540. *35*

61 The problem of predicativity. In: Bar-Hillel-Poznanski-Rabin-Robinson [61], 132–139. *171, 173, 175*

71 Unramified forcing. In: Scott [71], 357–381. *117*

Sierpiński, Wacław *XVI, 22, 147, 148, 151, 167*

25 Sur une classe d'ensembles. FM **7** (1925), 237–243. Reprinted in [75] below, 571–576. *148, 149, 158, 166*

28 Sur une décomposition d'ensembles. Monatshefte für Mathematik und Physik **35** (1928), 239–242. Reprinted in [75] below, 719–722. *211*

30 Sur l'uniformisation des ensembles mesurables (B). FM **16** (1930), 136–139. Reprinted in [76] below, 44–46. *150*

33 Sur un problème de la théorie des relations. Annali della Scuola Normale Superiore de Pisa **(2)2** (1933), 285–287. Reprinted in [76] below, 120–122. *74*

38 Fonctions additives non complètement additives et fonctions non mesurables. FM **30** (1938), 96–99. Reprinted in [76] below, 380–382. *384*

50 *Les ensembles projectifs et analytiques*. Mémorial des Sciences Mathématiques #112. Paris, Gauthier-Villars 1950. *147*

75 Hartman, Stanisław, *et al.* (eds.) *Oeuvres Choisies*, vol. 2. Warsaw, Państwowe Wydawnictwo Naukowe 1975.

76 Hartman, Stanisław, *et al.* (eds.) *Ouevres Choisies*, vol. 3. Warsaw, Państwowe Wydawnictwo Naukowe 1976.

See also Luzin, Nikolai N., and Wacław Sierpiński.

Sierpiński, Wacław, and Alfred Tarski

30 Sur une propriété caractéristique des nombres inaccessibles. FM **15** (1930), 292–300. Reprinted in Sierpiński [76], 29–35, and in Tarski [86] vol. 1, 289–297. *XVI, 16, 18*

Sikorski, Roman

See Rasiowa, Helena, and Roman Sikorski.

Silver, Jack H. *43, 69, 87, 88, 96, 98, 99, 101, 107, 110, 184, 209, 212, 217, 234, 244, 261, 264, 274, 293, 295, 311–313, 330*

66 Some applications of model theory in set theory. Ph.D. thesis, University of California at Berkeley 1966. Published in abridged form as [71] below. *XVIII, 80, 82, 99, 100, 107, 108*

66a The consistency of the generalized continuum hypothesis with the existence of a measurable cardinal (abstract). NAMS **13** (1966), 721. *263*

70 A large cardinal in the constructible universe. FM **69** (1970), 93–100. *108*

70a Every analytic set is Ramsey. JSL **35** (1970), 60–64. *382*

71 Some applications of model theory in set theory. AML **3** (1971), 45–110. *XVIII, 39, 80, 100, 104, 107, 184*

71a The consistency of the GCH with the existence of a measurable cardinal. In: Scott [71], 391–395. *263*

71b Measurable cardinals and Δ^1_3 well-orderings. AM **94** (1971), 414–446. *170, 272, 273, 276, 410*

See also Reinhardt, William N., and Jack H. Silver.

Simms, John C. *445*

79 Semihypermeasurables and $\Pi^0_1(\Pi^1_1)$ games. Ph.D. thesis, Rockefeller University 1979. *444*

Simpson, Stephen G.

77 First-order theory of the degrees of unsolvability. AM **105** (1977), 121–139. *186*

Skolem, Thoralf *XV, XVIII, 21, 477*

23 Einige Bemerkungen zur axiomatischen Begründung der Mengenlehre. In: *Matematikerkongressen i Helsingfors den 4–7 Juli 1922, Den femte skandinaviska matematikerkongressen, Redogörelse.* Helsinki, Akademiska-Bokhandeln 1923, 217–232. Reprinted in [70] below, 137–152. Translated in van Heijenoort [67], 290–301. *XIII, XIV, 20, 28, 86*

33 Ein kombinatorischer Satz mit Anwendung auf ein logisches Entscheidungsproblem. FM **20** (1933), 254–261. Reprinted in [70] below, 337–344. *70*

70 Fenstad, Jens E. (ed.) *Selected Works in Logic.* Oslo, Univesitetsforlaget 1970.

Smith Jr., Edgar C., and Alfred Tarski

57 Higher degrees of distributivity and completeness in Boolean algebras. TAMS **84** (1957), 230–257. Reprinted in Tarski [86] vol. 3, 623–650. *213*

Solovay, Robert M. *XX, XXI, 25, 43, 44, 54, 55, 99, 110, 113, 116, 125, 129, 132, 135–139, 141–145, 167, 170, 178–180, 182, 184, 186–188, 190, 192, 209, 212, 220, 222, 223, 259, 261, 262, 273, 274, 277, 283, 288, 297, 298, 302, 305, 307, 308, 321, 327, 340, 346, 348, 349, 367–369, 378, 383–388, 390, 398, 425, 431, 437, 442, 445, 450, 466*

63 Independence results in the theory of cardinals. I, II (abstracts). NAMS **10** (1963), 595. *116*

65 $2^{\aleph_0}$ can be anything it ought to be. In: Addison-Henkin-Tarski [65], 435. *116*

65a Measurable cardinals and the continuum hypothesis (abstract). NAMS **12** (1965), 132. *126*

65b The measure problem (abstract). NAMS **12** (1965), 217. *XIX, 114, 132, 136–139*

66 Real-valued measurable cardinals (abstract). NAMS **13** (1966), 721. *XX, 210, 224*

67 A nonconstructible Δ^1_3 set of integers. TAMS **127** (1967), 50–75. *XVIII, 107, 184*

69 The cardinality of Σ^1_2 sets of reals. In: Bulloff-Holyoke-Hahn [69], 58–73. *XIX, 183, 184, 310, 378*

70 A model of set theory in which every set of reals is Lebesgue measurable. AM **92** (1970), 1–56. *XIX, 114, 116, 121, 130, 132, 136–139*

71 Real-valued measurable cardinals. In: Scott [71], 397–428. *XX, 61, 210, 214, 216–220, 224, 264, 459*

74 Strongly compact cardinals and the GCH. In: Henkin *et al.* [74], 365–372. *347*

78 A Δ^1_3 coding of the subsets of ω_ω. In: Kechris-Moschovakis [78], 133–150. *430, 435*

78a The independence of DC from AD. In: Kechris-Moschovakis [78], 171–184. *381, 382, 399, 400, 433, 467*

See also Jensen, Ronald B., and Robert M. Solovay.

See also Kechris, Alexander S., and Robert M. Solovay.

See also Kechris, Alexander S., Robert M. Solovay, and John R. Steel.

See also Levy, Azriel, and Robert M. Solovay.

See also Martin, Donald A., and Robert M. Solovay.

Solovay, Robert M., William N. Reinhardt, and Akihiro Kanamori

78 Strong axioms of infinity and elementary embeddings. AML **13** (1978), 73–116. *298, 303, 307, 314, 318, 325, 328, 331, 334, 336, 339*

Solovay, Robert M., and Stanley Tennenbaum

71 Iterated Cohen extensions and Souslin's problem. AM **94** (1971), 201–245. *116, 117, 123, 124*

Sós, Vera T.

See Hajnal, András, Richard Rado, and Vera T. Sós.

Specker, Ernst

49 Sur un problème de Sikorski. Colloquium Mathematicum **2** (1949), 9–12. *79*

57 Zur Axiomatik der Mengenlehre (Fundierungs- und Auswahlaxiome). ZML **3** (1957), 173–210. *XIX, 135, 136, 378*

Spector, Clifford *114*

55 Recursive well-orderings. JSL **20** (1955), 151–163. *166*

Spencer, Joel H.

See Graham, Ronald L., Bruce L. Rothschild, and Joel H. Spencer.

Stanley, Lee J.

See Shelah, Saharon, and Lee J. Stanley.

Stanley, Mack C.

94 A Π^1_2 singleton incompatible with $0^\#$. APAL **66** (1994), 27–88. *187*

Steel, John R. *XXII, 276, 330, 367–369, 415, 437, 443, 450, 462, 463, 468, 470*

81 Determinateness and the separation property. JSL **46** (1981), 41–44. *415*

81a Closure properties of pointclasses. In: Kechris-Martin-Moschovakis [81], 147–163.

82 Determinacy in the Mitchell models. AML **22** (1982), 109–125. *445, 449*

83 Scales in $L(\mathbb{R})$. In: Kechris-Martin-Moschovakis [83], 107–156. *430–433, 436*

88 Long games. In: Kechris-Martin-Steel [88], 56–97. *467*

See also Kechris, Alexander S., Robert M. Solovay, and John R. Steel.

See also Martin, Donald A., Yiannis N. Moschovakis, and John R. Steel.

See also Martin, Donald A., and John R. Steel.

Steel, John R., and Robert A. Van Wesep

82 Two consequences of determinacy consistent with choice. TAMS **272** (1982), 67–85. *399, 459, 467, 470*

Steinhaus, Hugo

25 Definicje potrzebne do teorji gry i pościgu. Myśl Akademicka (Lwów) **1** (1925), 13–14. Translated as: Definitions for a theory of games and pursuit. Naval Research Logistics Quarterly **7** (1960), 105–108. Translation reprinted in [85] below, 332–336. *371*

65 Games, an informal talk. AMM **72** (1965), 457–468. Reprinted in [85] below, 805–817. *371, 377*

85 Edward Marczewski *et al.* (eds.) *Selected Papers*. Warsaw, Państwowe Wydawnictwo Naukowe 1985.

See also Mycielski, Jan, and Hugo Steinhaus.

Steprāns, Juris, and Stephen Watson

89 (eds.) *Set Theory and its Applications*. Lecture Notes in Mathematics #1401. Berlin, Springer-Verlag 1989.

Stewart, Frank M.

See Gale, David, and Frank M. Stewart.

Sureson, Claude

86 ω_1-constructible universe and measurable cardinals. APAL **30** (1986), 293–320. *258*

89 About Prikry generic extensions. APAL **51** (1989), 247–278. *238*

Suslin, Mikhail Ya. *145–149, 151, 166, 183*
17 Sur une définition des ensembles mesurables B sans nombres transfinis. CRP **164** (1917), 88–91. *XVI, 147, 148, 165, 180*
20 Problème 3. FM **1** (1920), 223. *22*

Swierczkowski, Stanisław *378*
See Mycielski, Jan, and Stanisław Swierczkowski.
See also Mycielski, Jan, Stanisław Swierczkowski, and A. Zieba.

Szekeres, György
See Erdős, Paul, and György Szekeres.

Takeuti, Gaisi *378*
70 A relativization of axioms of strong infinity to ω_1. Annals of the Japan Association for the Philosophy of Science **3** (1970), 191–204. *384*

Tan, It Beng
81 Sequentially large cardinals. In: Chong, Chi Tat, and Malcolm J. Wicks (eds.) *Southeast Asian Conference on Logic.* Amsterdam, North-Holland 1981, 197–210. *334*
See also Barbanel, Julius B., Carlos A. Di Prisco, and It Beng Tan.

Tarski, Alfred *XIV, 15, 22, 31, 36, 37, 39, 44, 70, 83, 92, 99, 307, 308*
29 Sur les fonctions additives dans les classes abstraites et leur application au problème de la mesure. Comptes Rendus des Séances de la Société des Sciences et des Lettres de Varsovie, Classe III **22** (1929), 114–117. Reprinted in [86] below, vol. 1, 243–248. *24*
33 Pojęcie prawdy w językach nauk dedukcyjnych. (The concept of truth in the languages of deductive sciences.) Prace Towarzystwa Naukowego Warszawskiego, Wydział III (Travaux de la Société des Sciences et des Lettres de Varsovie, Classe III) #34 (1933). See also [35] below. *XV*
35 Der Wahrheitsbegriff in den formalisierten Sprachen. (German translation of [33] with an appendix) Studia Philosophica **1** (1935), 261–405. Reprinted in [86] below, vol. 2, 51–198. Translated in [83] below, 152–278. *XV*
38 Über unerreichbare Kardinalzahlen. FM **30** (1938), 68–89. Reprinted in [86] below, vol. 2, 359–380. *19, 20*
39 Ideale in vollständigen Mengenkörpern. I. FM **32** (1939), 45–63. Reprinted in [86] below, vol. 2, 509–527. *26, 400*
39a On well-ordered subsets of any set. FM **32** (1939), 176–183. Reprinted in [86] below, vol. 2, 551–558. *20*
45 Ideale in vollständigen Mengenkörpern. II. FM **33** (1945), 51–65. Reprinted in [86] below, vol. 3, 3–17. *26, 210, 211*
58 Remarks on predicate logic with infinitely long expressions. Colloquium Mathematicum **6** (1958), 171–176. Reprinted in [86] below, vol. 4, 11–16. *36*

62 Some problems and results relevant to the foundations of set theory. In: Nagel-Suppes-Tarski [62], 125–135. Reprinted in [86] below, vol. 4, 115–125. *XVII, 36, 39, 42*

83 *Logic, Semantics, Metamathematics. Papers from 1923 to 1938.* Translations by J.H. Woodger. Second edition. Indianapolis, Hackett Publishing Company 1983.

86 Givant, Steven R., and Ralph N. McKenzie (eds.) *Collected Papers.* Basel, Birkhäuser 1986.

See also Addison Jr., John W., Leon Henkin, and Alfred Tarski.
See also Erdős, Paul, and Alfred Tarski.
See also Keisler, H. Jerome, and Alfred Tarski.
See also Kuratowski, Kazimierz, and Alfred Tarski.
See also Nagel, Ernest, Patrick Suppes, and Alfred Tarski.
See also Sierpiński, Wacław, and Alfred Tarski.
See also Smith Jr., Edgar C., and Alfred Tarski.

Tarski, Alfred, and Robert L. Vaught

57 Arithmetical extensions of relational systems. Compositio Mathematica **13** (1957), 81–102. Reprinted in Tarski [86], 653–674. *8*

Taylor, Alan D.

See Baumgartner, James E., and Alan D. Taylor.
See also Baumgartner, James E., Alan D. Taylor, and Stanley Wagon.

Teichmüller, Oswald

39 Braucht der Algebraiker das Auswahlaxiom? Deutsche Mathematik **4** (1939), 567–577. Reprinted in [82] below, 323–333. *132*

82 Ahlfors, Lars V., and Frederick W. Gehring (eds.) *Gesammelte Abhandlungen. Collected Papers.* Berlin, Springer-Verlag 1982.

Tennenbaum, Stanley *116*

See Solovay, Robert M., and Stanley Tennenbaum.

Thompson, S. *83*

Todorčević, Stevo B. *80, 87, 98, 320*

84 Trees and linearly ordered sets. In: Kunen-Vaughan [84], 235–293. *72*

87 Partitioning pairs of countable ordinals. Acta Mathematica **159** (1987), 261–294. *78, 79, 85, 97*

94 Some partitions of three-dimensional combinatorial cubes. Journal of Combinatorial Theory, Series A **68** (1994), 410–437. *87*

Tryba, Jan

81 A few remarks on Rowbottom cardinals. IJM **40** (1981), 193–196. *92*

84 On Jónsson cardinals with uncountable cofinality. IJM **49** (1984), 315–324. *96*

86 Rowbottom-type properties and a cardinal arithmetic. PAMS **96** (1986), 661–667. *92, 97*

87 No Jónsson filters over $\aleph_\omega$. JSL **52** (1987), 51–53. *97*

Ulam, Stanisław M. *15, 22, 39, 56, 87, 151, 294, 371*
29 Concerning functions on sets. FM **14** (1929), 231–233. Reprinted in [74] below, 6–8. *24*
30 Zur Masstheorie in der allgemeinen Mengenlehre. FM **16** (1930), 140–150. Reprinted in [74] below, 9–19. *XVI, 24, 26, 210, 211*
74 Beyer, W.A., Jan Mycielski, and Gian-Carlo Rota (eds.) *Sets, Numbers, and Universes*. Selected Works. Cambridge, The MIT Press 1974.
See also Kuratowski, Kazimierz, and Stanisław M. Ulam.

Uspenskii, Vladimir A.
85 Luzin's contribution to the descriptive theory of sets and functions: concepts, problems, predictions. RMS **40**(3) (1985), 97–134. *146*

van Heijenoort, Jean
67 (ed.) *From Frege to Gödel: A Source Book in Mathematical Logic, 1879–1931*. Cambridge, Harvard University Press 1967. *478*

Van Wesep, Robert A.
78 Separation principles and the axiom of determinateness. JSL **43** (1978), 77–81. *415*
78a Wadge degrees and descriptive set theory. In: Kechris-Moschovakis [78], 151–170. *415*
See also Steel, John R., and Robert A. Van Wesep.

Vaughan, Jerry E.
See Kunen, Kenneth, and Jerry E. Vaughan.

Vaught, Robert L. *311*
63 Models of complete theories. BAMS **69** (1963), 299–313. *85*
63a Indescribable cardinals (abstract). NAMS **10** (1963), 126. *61*
See also Montague, Richard M., and Robert L. Vaught.
See also Tarski, Alfred, and Robert L. Vaught.

Veblen, Oswald
08 Continuous increasing functions of finite and transfinite ordinals. TAMS **9** (1908), 280–292. *53*

Vitali, Giuseppe *132*
05 *Sul problema della misura dei gruppi di punti di una retta.* Bologna, Tip. Gamberini e Parmeggiani 1905. *22*

von Neumann, John *28, 71, 371, 372, 477*
23 Zur Einführung der transfiniten Zahlen. ALS **1** (1923), 199–208. Reprinted in [61] below, vol. 1, 24–33. Translated in van Heijenoort [67], 346–354. *XIV*
25 Eine Axiomatisierung der Mengenlehre. Journal für die reine und angewandte Mathematik **154** (1925), 219–240. Reprinted in [61] below, vol. 1, 34–56. Translated in van Heijenoort [67], 393–413. *XIV, 20, 30*
28 Zur Theorie der Gesellschaftsspiele. MA **100** (1928), 295–320. Reprinted in [61] below, vol. 6, 1–26. *371*

49 On rings of operators. Reduction theory. AM **50** (1949), 401–485. Reprinted in [61] below, vol. 3, 400–484. *150*

61 Taub, Abraham H. (ed.) *John von Neumann. Collected Works*. New York, Pergamon Press 1961.

von Neumann, John, and Oskar Morgenstern

44 *Theory of Games and Economic Behavior*. Princeton, Princeton University Press 1944. Second edition, 1947. Third edition, 1953. *371*

Vopěnka, Petr *116, 234*

62 Construction of models of set theory by the method of ultraproducts (in Russian). ZML **8** (1962), 293–304. *51*

64 The independence of the Continuum Hypothesis (in Russian). CMUC **5** Supplement I (1964), 1–48. Translated in American Mathematical Society Translations **57** (1966), 85–112. *116*

65 The first measurable cardinal and the Generalized Continuum Hypothesis. CMUC **6** (1965), 367–370. *56*

67 The general theory of ∇-models. CMUC **8** (1967), 145–170. *116*

Vopěnka, Petr, and Karel Hrbáček

66 On strongly measurable cardinals. BAPS **14** (1966), 587–591. *51*

Wadge, William W. *442*

72 Degrees of complexity of subsets of the Baire space. NAMS **19** (1972), A-714. *414*

83 Reducibility and determinateness on the Baire space. Ph.D. thesis, University of California at Berkeley 1983. *414–416*

Wagon, Stanley

See Baumgartner, James E., Alan D. Taylor, and Stanley Wagon.

Wang, Hao

74 *From Mathematics to Philosophy*. New York, Humanities Press 1974. *29, 31*

Watro, Ronald J.

See Henle, James M., Eugene M. Kleinberg, and Ronald J. Watro.

Wehrung, Friedrich

89 Nonabsoluteness of elementary embeddings. JSL **54** (1989), 774–778. *259*

Weierstrass, Karl *475, 479*

Weitkamp, Galen

See Mansfield, Richard B., and Galen Weitkamp.

Welch, Philip *445*

See Beller, Aaron, Ronald B. Jensen, and Philip Welch.

Weyl, Hermann *XIII*

Whitehead, Alfred N. *XIII, 2*

Wittgenstein, Ludwig *70, 480, 482*

Wolfe, Philip

55 The strict determinateness of certain infinite games. PJM **5** (1955), 841–847. *372*

Woodin, W. Hugh *XXII, 96, 243, 310, 320, 322, 329, 344, 363, 367–369, 431–434, 437, 449, 450, 453, 457–464, 466–471*

82 On the consistency strength of projective uniformization. In: Jacques Stern (ed.) *Proceedings of Herbrand Symposium. Logic Colloquium '81.* Amsterdam, North-Holland 1982, 365–384. *468*

83 AD and the uniqueness of the supercompact measures on $P_{\omega_1}(\lambda)$. In: Kechris-Martin-Moschovakis [83], 67–72. *402*

83a Some consistency results in ZFC using AD. In: Kechris-Martin-Moschovakis [83], 172–198. *460*

86 Aspects of determinacy. In: Marcus, Ruth B., Georg J. Dorn, and Paul Weingartner (eds.) *Logic, Methodology and Philosophy of Science VII.* Proceedings of the 1983 International Congress, Salzburg. Amsterdam, North-Holland 1986, 171–181. *459*

88 Supercompact cardinals, sets of reals, and weakly homogeneous trees. PNAS **85** (1988), 6587–6591. *462, 470*

99 *The Axiom of Determinacy, Forcing Axioms, and the Nonstationary Ideal.* De Gruyter Series in Logic and Applications 1. Berlin, Walter de Gruyter 1999. *471*

See also Dales, H. Garth, and W. Hugh Woodin.

See also Feng, Qi, Menachem Magidor, and W. Hugh Woodin.

See also Judah, Haim (Jaime Ihoda), Winfried Just, and W. Hugh Woodin.

See also Kechris, Alexander S., Eugene M. Kleinberg, Yiannis N. Moschovakis, and W. Hugh Woodin.

See also Kechris, Alexander S., and W. Hugh Woodin.

See also Shelah, Saharon, and W. Hugh Woodin.

Woodin, W. Hugh, Adrian R.D. Mathias, and Kai Hauser

∞ *The Axiom of Determinacy.* To appear. *450, 466*

Yankov, Vladimir

41 Sur l'uniformisation des ensembles *A*. Doklady Akademiia Nauk SSSR **30** (1941), 597–598. *150*

Young, William H.

03 Zur Lehre der nicht abgeschlossenen Punktmengen. BKSG **55** (1903), 287–293. *147*

Zeno *473*

Zermelo, Ernst *XIV, 21, 371, 372*

04 Beweis, dass jede Menge wohlgeordnet werden kann (Aus einem an Herrn Hilbert gerichteten Briefe). MA **59** (1904), 514–516. Translated in van Heijenoort [67], 139–141. *XIII, 146*

08 Neuer Beweis für die Möglichkeit einer Wohlordnung. MA **65** (1908), 107–128. Translated in van Heijenoort [67], 183–198. *XIII, XIV*

08a Untersuchungen über die Grundlagen der Mengenlehre I. MA **65** (1908), 261–281. Translated in van Heijenoort [67], 199–215. *XIII, XIV, 16*

13 Über eine Anwendung der Mengenlehre auf die Theorie des Schachspiels. In: Hobson, E.W., and A.E.H. Love (eds.) *Proceedings of the Fifth International Congress of Mathematicians*, Cambridge 1912, vol. 2. Cambridge, Cambridge University Press 1913, 501–504. *XXI, 371*

29 Über den Begriff der Definitheit in der Axiomatik. FM **14** (1929), 339–344. *19*

30 Über Grenzzahlen und Mengenbereiche: Neue Untersuchungen über die Grundlagen der Mengenlehre. FM **16** (1930), 29–47. *XIII, XIV, XVI, XVII, 18–20, 29, 31, 34, 115*

Zieba, A.

See Mycielski, Jan, Stanisław Swierczkowski, and A. Zieba.

See also Mycielski, Jan, and A. Zieba.

Zwicker, William S.

84 $P_\kappa\lambda$ combinatorics I: Stationary coding sets rationalize the club filter. In: Baumgartner-Martin-Shelah [84], 243–259. *350*

86 $\mathcal{P}_\kappa\lambda$ combinatorics II: The RK ordering beneath a supercompact measure. JSL **51** (1986), 604–616. *350*

89 A beginning for structural properties of ideals on $P_\kappa\lambda$. In: Steprāns-Watson [89], 201–217. *350*

See also Di Prisco, Carlos A., and William S. Zwicker.

Subject Index

This index concentrates on definitions (occurring on the italicized pages), basic properties, and those significant but singular occurrences of larger subjects that might otherwise be difficult to find. Greek letters and other symbols are alphabetized according to their English transliterations.

Absolute, Cantor's XII, XV, XVII, 31, 42, 313, 477
Absolute relation *174*
AC (*see* Axiom of Choice)
AC_ω; $AC_\omega({}^\omega\omega)$ (*see* Countable Axiom of Choice)
Ackermann's set theory (**A**) 311–313, 318
AD (*see* Axiom of Determinacy)
$AD_{\mathbb{R}}$ 329, *377*, 399–402, 441, 458, 467–471
$AD_{\mathbb{R}}^{1/2}$ *469*
Algebra *93*
Almost huge cardinal *331*, 333, 334, 338
α-branch (of a tree) *72*
α-inaccessible cardinal *21*, 33
α-Mahlo cardinal *21*, 33, 41, 58
α-$\Pi^1_1(a)$ set *443*–449
αth iterate (of a structure) *250*
αth level (of a tree) *72*
α-weakly inaccessible cardinal *16*, 17
α-weakly Mahlo cardinal *17*, 216, 217
Amenable pair *187, 244*
Amenable-generic set *187*
Analytic set *147*, 148
Analytical hierarchy *155*
Analytical relation 151, *155*
– in a *156*
Antichain
– of a tree *72*
– of conditions *118*
– of conditions below p *118*
Application (ap) *152*
A-set *147*, 148
$a^\dagger$ (*see also* Daggers) *294*
$a^\#$ (*see also* Sharps) *110*
Arithmetical hierarchy *154*
Arithmetical relation 151, *154*
– in a *156*
Aronszajn tree (*see* κ-Aronszajn tree)
Atom (for a measure) *24*
Atomless measure *24*
Axiom of Choice (AC) XIII, XV, 1, 21, 22, 28, 32, 34, 71, 101, 114, 132, 167, 257, 324, 329, 372, 377, 478
Axiom of Constructibility ($V = L$) (*see also* L) *28*, 44, 66
Axiom of Determinacy (AD) 31, 167, 192, 204, *377*–379, 382, §28, 414, 415, 425–436, 449, §32
Axiom of Infinity XIV, 58, 478
Axiom of Reducibility XIII, 29

Baire Category Theorem *13*
Baire function *145*
Baire property *13*, 14, 136
– for all sets from inaccessibility 132, 136
– for all sets under AD 377, 468
– for $\boldsymbol{\Delta}^1_2$ sets 181
– for provably $\boldsymbol{\Delta}^1_2$ sets 180
– for $\boldsymbol{\Sigma}^1_1$ (analytic) sets 148
– for $\boldsymbol{\Sigma}^1_2$ sets 179
– for $\boldsymbol{\Sigma}^1_{n+1}$ sets under Det($\boldsymbol{\Pi}^1_n$) 380
– for weakly homogeneously Suslin sets 457
– in L 169
– in $L[U]$ 273
– via games 374
Baire space *11*
Banach algebra 450
Begriffsschrift XIII
Berkeley (*see* University of California at Berkeley)
Boolean algebra 26, 116, 120, 129, 213, 214, 223
Boolean-valued model 114, 116, 117, 223
Borel Determinacy 372, 373, *379*, 440–442
Borel hierarchy *146*, 149, 151, 152
Borel measure ($m_{\mathcal{B}}$) *12*
Borel set (*see also* γ-Borel set, ∞-Borel set, Borel Determinacy) *12*, 145–149, 165
Borel Wadge Determinacy *415*, 416, *442*
Bounded quantifier *5, 153*
Branch (of a tree) *72*
Burali-Forti Paradox XIV

Cabal Seminar 368, 403, 450
CH (*see* Continuum Hypothesis)
Chain (of a tree) *72*
Chang's Conjecture *85*, 87, 88, 282, 292
Closed code *137*
Closed set
– in $\mathcal{P}_\kappa\gamma$ *340*, 341
– of ${}^\omega X$ *371*, 372
Closed unbounded filter
– over λ ($\mathcal{C}_\lambda$) *3*
– over $\mathcal{P}_\kappa\gamma$ ($\mathcal{C}_{\kappa,\gamma}$) *341*, 342
Closed unbounded set (*see also* ν-closed unbounded set)
– in λ *2*, 3
– in $\mathcal{P}_\kappa\gamma$ *340*
Coding Lemma 397, 398, 425, 428
Cofinal branch (of a tree) *72*

Cohen real *127*, 130, 177, 179–181
Coinductive set *430*
Collapsing Lemma *7*, 48, 62
Compactness Theorem XV, 36, 39
Comparable structures *265*
Comparison Lemma (for ZFC$^-$ premice) *266*
Compatible conditions ($p \parallel q$) *118*
Complete Boolean algebra 116, 120, *213*, 214, 223
Complete embedding *119*, 120
Complete set (of Skolem functions) *8*
Completeness Theorem XV, 102, 185
Condition *117*
Constructibility degree *186*
Constructible closure (*see* $L(A)$)
Constructible hierarchy (*see* L)
Constructible set (*see also* L) *28*, 29
Continuously closed class *414*, 415
Continuum Hypothesis (CH) *XII*, 1, 28–32, 114–116, 133, 134, 378
Countable Axiom of Choice (AC_ω) *132*, 433
– restricted to sets of reals ($AC_\omega({}^\omega\omega)$) *152*, 159, 162, 378, 403, 417
Countably complete tower *451*
Countably complete ultrafilter *254*, 256
Countably iterable structure *254*, 255
Covering Theorem *111*, 443
Critical point (crit(j)) *47*
Cumulative hierarchy XIII, XIV, XVI, XVII, 1, 20

Daggers *294*, 444, 445, 468
DC (*see* Principle of Dependent Choices)
DC(${}^\omega\omega$) *434*
definit property XIII, 19
$\boldsymbol{\Delta}^1_2$ sets
– Baire property 169, 181
– Determinacy 467
– Lebesgue measurability 150, 169, 181
Δ-system *121*
Δ-system Lemma *121*
Dense embedding *119*, 120
Dense open set (of conditions) *118*
Dense set (of conditions) *118*
Derived set (A') *133*
Det(Λ) *379*
Determinacy
– of Borel sets (see Borel Determinacy)
– of $\boldsymbol{\Pi}^1_1$ sets 437–440
– of Π^1_2 sets 467
– of $\boldsymbol{\Pi}^1_2$ sets 329, 449, 467
– of projective sets (see Projective Determinacy)
Determined game *370*
Determined set *370*
Diagonal intersection
– of sets indexed by λ *2*
– of sets indexed by $[\kappa]^{<\omega}$ *237, 242*
– of subsets of $\mathcal{P}(S)$ *301*

Difference hierarchy (Diff(Λ)) (*see also* α-$\Pi^1_1(a)$ set) *442*
Direct limit 9, *10*
Directed set *9*
Directed system *9, 10*
Dual filter (of an ideal) *210*

Easton product *122*, 228, 230, 232
Elementary embedding (*see also* Σ_n-elementary embedding) *44, 46*
Elementary substructure (*see also* Σ_n-elementary substructure) 39, *44*
EM blueprint *101*–107
– for a *110*
EM^2 blueprint *284*–288
– for a *294*
$\in$-model *4*
Equiconsistency result XIX, *4*, 135
Equivalent ultrafilters *269*
Erdős cardinal ($\kappa(\alpha)$) *80*–84
Erdős-Rado Theorem *74*
η-extendible cardinal *311*–318
η-extendible for A cardinal *339*
η-supercompact for A cardinal *339*
Extender *353–355*, §26
Extendible cardinal *311*–312, 315–318, 322
Extension property *39*, 76
Extensional relation *7*

Filter *3*
Fine filter *301*
Fine structure theory 30, 431
First Periodicity Theorem *410*
Foundation Axiom XIV, 20, 478
Friedman game 397
F_σ set *12*
F-stationary set *53*
Function quantifier *153*
Fundamenta Mathematicae 22

Game *369, 370*
Game quantifier *421*
Γ-good well-ordering 169, 273, *409*, 422, 463
Γ-norm *406*, 407
Γ-scale *418*
γ-Borel set *424*–427, 469
γ-compact cardinal *307*
γ-strong cardinal *358*, 359
γ-strong for A cardinal *363*
γ-supercompact cardinal (*see also* η-supercompact for A cardinal) *298*, §22, 316, 317, 348, 349, 358
GCH (*see* Generalized Continuum Hypothesis)
G_δ code *137*
G_δ set *12*
Generalization (argument) XVII, 32
Generalized Continuum Hypothesis (GCH) 1, 16, 28, 263
Generic elementary embedding 459, 460

Generic extension *118*
Generic ultrapower 210
Genericity Conjecture *186*
Global Axiom of Choice *59*
Good well-ordering (*see* Γ-good well-ordering)
Ground model *118*

Harvard University 115
Hebrew University 309
Height (of a tree) *72*
Hereditarily ${}^{\omega}$On-definable set *141*
Hereditarily ordinal definable set *466*
HOD *466*
Homogeneous Boolean algebra 129
Homogeneous set (for a function)
– on an ordinal *71*
– on $[\mathcal{P}_\kappa \gamma]^n_{\subset}$ *346*
Homogeneous tree 195, 451, 452
Homogeneously Suslin set *452*
Honest leftmost branch 176, 204, *418*
Huge cardinal 233, *331*, 334, 358, 449
Hyperarithmetic hierarchy 157
Hyperarithmetic set 157, 166
HyperWoodin cardinal *365*

I0 325, *329*, 449
I1 *325*–329, 468
I2 *325*–329, 449
I3 *325*–329
Ideal *3*
Ideal over κ *211*
Inaccessible cardinal (*see also* α-inaccessible cardinal, α-weakly inaccessible cardinal, Weakly inaccessible cardinal) *18*, 26, 38, 57, 83, 115, 125, 127, 132, 136, 144
– ω_1 inaccessible from reals *135*, 379
Incompatible conditions ($p \perp q$) *118*
Incompleteness Theorem XV, 28, 70, 151
– Second 19, 399
Indescribable cardinal *58*, §6
Indescribable filter *65*, 66
Indescribable in κ set *64*
Indiscernibles, set of *99*, 253
– double *284*
Inductive definability 430
Inductive set (of reals) 430
∞-Borel set *469*
Inner model *33*
– of measurability §20
Institute of Advanced Study 28, 30, 114
Isolated point (of a set of reals) *133*
Iterable M-ultrafilter *250*
Iterable structure *250*, 255–257
Iterate (of a structure) *250*
Iterated forcing 121, *123*, 124
Iterated ultrapower §19
Iteration (of a structure) *250*
Iteration tree *463*

Iterative conception (of set) (*see also* Cumulative hierarchy) XV

Jónsson algebra *93*
Jónsson cardinal *93*–98, 128, 275, 282, 292, 396

κ-Aronszajn tree *78*–80, 384
κ-Baire partial order *121*
κ-chain condition (for a partial order) *121*
κ-closed partial order *121*
κ-homogeneous tree *451*, 452
κ-homogeneously Suslin set *452*
κ-model (*see also* $L[U]$) *261*
κ-product (of partial orders) *122*
κ-tree *75*
κ-weakly homogeneous tree *453*–459, 463
κ-weakly homogeneously Suslin set *453*–459, 464
Kleene-Brouwer ordering ($\leq_{KB}$) *162*, 193, 194
– $\leq_{KB^2}$ version *194*
Kunen-Paris forcing *226, 227*

L *28*, §3, 44, 49, 62, 66, 88, 89, 93, §9, 150, 151, 167–170, 277–283
$\tilde{L}$ *199*
$L[A]$ (*see also* $L[U]$) *34*
$L(A)$ (*see also* $L(\mathbb{R})$) *34*, 51, 359
$L(\mathbb{R})$ *142*, 378, 381, 398, 431–434, 461, 464, 466, 467
$L[U]$ *261*, §20, 283–295
$\mathcal{L}_\in$ *1*
$\mathcal{L}^*_\in$ *101*, 102
$\mathcal{L}_\in(\dot{A})$ *245*
$\mathcal{L}_\in(\dot{A})^{**}$ *284*
$L_{\lambda,\mu}$ *36*, 257
$L^n_{\lambda,\mu}$ *315*
$L^n_{\lambda,\mu}$-compact cardinal *315*
λ-additive measure *23*
λ-complete filter *3*
λ-complete ideal *3*
λ-dense ideal *470*, 471
λ-saturated ideal *211*, §16, §17
Lebesgue measurability *12*, 13, 61, 133
– from Woodin cardinals 461
– in L 169
– in $L[U]$ 273
– of all sets from inaccessibility 132, 136, 179
– of all sets under AD 377, 468
– of $\boldsymbol{\Delta}^1_2$ sets 181
– of provably $\boldsymbol{\Delta}^1_2$ sets 180
– of $\boldsymbol{\Sigma}^1_1$ (analytic) sets 148
– of $\boldsymbol{\Sigma}^1_2$ sets 149, 178
– of $\boldsymbol{\Sigma}^1_{n+1}$ sets under Det($\boldsymbol{\Pi}^1_n$) 380
– of weakly homogeneously Suslin sets 457
– via games 376
Lebesgue measure (m_L) *12*, 22, 24
– algebra *136*

Left distributive law *329*
Length (of an iteration) *250*
Levy Collapse (Col(λ, κ)) *128*, 135, 139, 221
Levy hierarchy *5*
Limit point
– of a set of ordinals *2*
– of a set of reals *133*
Łoś's Theorem *9*, 47, 52, 245
Löwenheim-Skolem Theorem XIV, *8*, 19
– for $L_{\lambda\mu}$ *40*

Mahlo cardinal (*see also* α-Mahlo cardinal, α-weakly Mahlo cardinal, Weakly Mahlo cardinal) *21*, 57, 58, 125
Mahlo's Operation *18*, 54, 66, 216
Martin's Axiom (MA) *124*, 167, 179, 219, 426, 450, 460
– for κ (MA(κ)) *124*
Martin's Maximum *460*
Mathematische Annalen 16
Mathias real *143*, 144, 238
Maximal Principle *119*
meager set *13*, 373
Measurable cardinal *26*, 27, 38, 39, 44, 49–56, 61, 83, 90, 126, 179, 195, 210–212, 221, 224, 226–237, §20, 283–295, 299, 312, 359, 384–389, 434, 444, 445
measure
– on a σ-algebra *12*
– over S *23*
Measure Problem *22*, 23, 132
Mild extension *125*, 126
Minimal cover (of a set of reals) *376*
Move (of a game) *370*
M-ultrafilter *244–257*

Name *118*
Natural sequence (of structures) *336–339*
n-huge cardinal *331–333*
N-λ-saturated ideal *227*
N-normal ultrafilter *227*
Non-rigidity (of an inner model) *277*
Nonselfdual class *414*, 415
Non-stationary ideal over λ (NS_λ) *3*
Norm *406*
Normal filter (*see also* Strongly normal filter)
– over λ *52–56*
– over $\mathcal{P}_\kappa S$ *301*
– over $\mathcal{P}(\lambda)$ *332*
Normal ideal over κ *214*
Nowhere dense set *13*
ν-closed unbounded set *2*, 3, 391
ν-Rowbottom cardinal *88–92*
ν-Rowbottom filter *218*
ν-satisfiable set
– of $L_{\lambda\mu}$ sentences *36*
– of $L^n_{\lambda\mu}$ sentences *315*
Null set (of reals) *12*
Number quantifier *153*

ω-Jónsson function *319*–321, 325, 347
${}^\omega$On-definable set *139*
ω_1-dense ideal 470, 471
Operation (A) 147, 148
Open code *137*
Open set (of ${}^\omega X$) *371*, 372
Order-preserving map (for trees) *160*
Ordinal definable set 139, 381, 466

Partial order (p.o.) *117*
Partial play (of a game) *370*
– according to σ (resp. τ) *370*
Partition property (for ultrafilters) *346*–349
Partition relation
– infinite exponent 71, 391–396
– of all finite subsets *80*–84
– ordinary *71–77*
– square brackets *85*, §8
Payoff (of a game) *370*, 371
pcf theory 97
Perfect set *133*
Perfect set property 133, *134*, 310
– for all sets from inaccessibility 132, 136
– for all sets under AD 377
– for Borel sets 147
– for $\boldsymbol{\Pi}^1_1$ sets 149, 170, 184
– for $\boldsymbol{\Sigma}^1_1$ (analytic) sets 148, 183
– for $\boldsymbol{\Sigma}^1_2$ sets 184
– for $\boldsymbol{\Sigma}^1_{n+1}$ sets under Det($\boldsymbol{\Pi}^1_n$) 380
– for weakly homogeneously Suslin sets *457*
– in L 150, 169, 170
– in $L[U]$ 273
– via games 375
P-generic *119*
– over M *118*
Philosophical Investigations 480
$\boldsymbol{\Pi}^1_1$ sets
– Determinacy 437–440
– Perfect set property 149, 150, 170, 184
– Tree representation 162
– Uniformization *150*, 175–177
– Union of $\aleph_1$ Borel sets 163
$\boldsymbol{\Pi}^1_2$ sets
– Determinacy 329, 449, 467
– Perfect set property 273
– Tree representation 192, 196, 201
– Uniformization 150, 177, 204
Π^1_2 Singleton Conjecture *187*
Π^1_n singleton *187*
Play (of a game) *370*
– according to σ (resp. τ) *370*
P-name *118*
p.o. *117*
Predense set (of conditions) *118*
Premouse, ZFC$^-$ *264–266*
Prewellordering 178, *406*
Prewellordering property *407*
– for $\Pi^1_1(a)$ sets 408

– for $\Sigma^1_2(a)$ sets 408
– in L 410
– in $L[U]$ 410
– under PD 410
Prikry forcing *234*, §18, 259–260
Prikry sequence *235*
Princeton University 30, 31, 114
Principia Mathematica XIII
Principle of Dependent Choices (DC) *132*, 142, 152, 378, 433
Proceedings of the National Academy of Sciences U.S.A. 28
Product forcing *121*–123
Product measure (on ${}^\kappa 2$ (m_κ)) *222*
Product measure forcing *222, 223*
Projection (of a set (pB)) *147*
Projective class *403*
Projective Determinacy (PD) *379*, 406, 410, 421–425, 468
Projective hierarchy *148, 149*
Projective ordinal *423*, 428, 429, 434–436, 469
Projective set 148
Provably $\boldsymbol{\Delta}^1_2$ set *179, 180*

Quotient Boolean algebra *213*

Radon-Nikodym Theorem 225
Ramification Lemma 72
Ramsey cardinal *81*, 84, 89, 126, 263, 275
Ramsey property *144*, 382, 458
Ramsey's Theorem
– finite version 70, 100
– infinite version 70, 75, 76
Random real *138*, 178, 180, 181, 223
Rank function *7*
– on a tree *160*
Real game quantifier *430*
Real-valued measurable cardinal *24*, 80, 210, 219, 224
Recursion Theorem 151, 368, 425
Recursive relation *154*
– in a *156*
Reduction property *404*, 405–407, 415
– for $\boldsymbol{\Pi}^1_1$ sets 405
– for $\boldsymbol{\Sigma}^1_2$ sets 406
– under PD 410
Reflection (argument) XVII–XVIII, 30, 31
Reflection Principle 31, *58*
Regressive function *2*
Relative constructibility (*see* $L[A]$)
Replacement Axiom (Schema) XIV, XV, 2, 58, 115, 440, 477, 478
Root (of a tree) *72*
Rowbottom cardinal *88*, 91, 92, 96, 128, 396
Rowbottom filter *218*, 236
R-set 180
Russell's Paradox XIII, XIV

Satisfiable set
– of $L_{\lambda\mu}$ sentences *36*
– of $L^n_{\lambda\mu}$ sentences *315*
Saturated ideal *211*, §16, §17
Scale 177, *417*, 418, 430, 431
Scale property *418*, 419
– for inductive sets 430
– for $\boldsymbol{\Pi}^1_1$ sets 419
– for $\boldsymbol{\Sigma}^1_2$ sets 420
– for $\boldsymbol{\Sigma}^2_1$ sets 431
– under PD 421
Schröder-Bernstein Theorem *134*
S-closed forcing 460
Second-order arithmetic *152*, 416
– in a *156*
Second Periodicity Theorem *421*
Semi-proper forcing 460
Separation Axiom (Schema) XIV, 19, 30, 245, 477
Separation property *404*, 415
Separative partial order *120*, 129, 130
Separative quotient (of a partial order) *120*, 121
Set-generic set *186*
Sharps *110*, 188–191, 198–207, 420, 425, 439, 444, 445
Shelah cardinal 461
σ-algebra *11*
Σ_n-elementary embedding *45*, 46, 281
Σ_n-elementary substructure *45*
Σ_1 Skolem hull *279*
$\boldsymbol{\Sigma}^1_1$ (analytic) sets
– Bounding principle 164
– Lebesgue measurability 148
– Perfect set property 148, 183
– Union of $\aleph_1$ Borel sets 163
$\boldsymbol{\Sigma}^1_2$ sets
– Absoluteness 174, 175
– Baire property 179
– Lebesgue measurability 178, 179
– Perfect set property 183, 184
– Tree representation 173, 174
– Uniformization 177
– Union of $\aleph_1$ Borel sets 149, 166, 167, 174
$\boldsymbol{\Sigma}^1_3$ sets
– Absoluteness 197
– Tree representation 192, 201, 204
– Uniformization 206
– Union of $\aleph_2$ Borel sets 204
Singular Cardinals Problem *122*, 217, 308
Skolem function *8*
Skolem hull (and argument) (*see also* Σ_1 Skolem hull) *8, 101, 266, 284*
Skolem paradox XIV
Skolem term *8*
– for $0^\#$ *101*
– for $L[U]$ *266*
– for $0^\dagger$ *284*
Special κ-tree *79*, 80
Solovay game 387

Spector-Gandy Theorem 422
Stanford University 114, 244
Stationary coding set *350*
Stationary set
– in λ *2*, 3, 17, 217
– in $\mathcal{P}_\kappa\gamma$ *340*, 341
Stationary tower forcing 462, 464, 466
Strategy *370*
Strong cardinal (*see also* γ-strong cardinal, γ-strong for A cardinal) *358*–360, 468
Strong limit cardinal *1*, 18
Strong partition property *432*, 434, 435
Strongly compact cardinal (*see also* γ-compact cardinal) *37*, 38, 51, 298, 307–310, 350, 399
Strongly inaccessible cardinal (*see* Inaccessible cardinal)
Strongly normal filter *350*, 351
Subalgebra *93*
Subtree *72*
Supercompact cardinal (*see also* γ-supercompact cardinal, η-supercompact for A cardinal) *298*, §22, 316, 317, 322, 344, 350, 358, 400–402, 459–462
Superhuge cardinal *334*
Superstrong cardinal *358*–361
Support (of a condition (supt(p)) *122*
Suslin set *171*, 424, 428
Suslin tree (and problem) 22, 72, 116

Tao Te Ching 482
Θ *396*–400, 432, 434, 467, 468
Third-order arithmetic 430
Third Periodicity Theorem 421, 422
Totally indescribable cardinal *59*, 61, 83, 109
Tower (of ultrafilters) *451*
Tractatus (Logico-Philosophicus) 482
Transitive collapse *7*
Tree *72*
– on ${}^\kappa\omega \times Y$ *159*–161
– on Y *159*
Tree property *75*, 76, 78, 212
Tree representation
– for $\boldsymbol{\Pi}^1_1$ sets 162
– for $\boldsymbol{\Pi}^1_2$ sets 192, 196, 201
– for $\boldsymbol{\Sigma}^1_2$ sets 173, 174, 192
– for $\boldsymbol{\Sigma}^1_3$ sets 192, 201
– for $\boldsymbol{\Sigma}^1_4$ sets 207
Turing degree ($[a]_\mathrm{T}$) 114, 186, *386*
Turing Determinacy *433*
Turing reducibility ($\leq_\mathrm{T}$ and $\equiv_\mathrm{T}$) *386*

Ulam matrix *24*
Ultrafilter *3*
– over ω 24, 26, 383, 384
Ultrapower *9*
Ultraproduct *9*, 37
Ulysses 481
Unbounded set (in $\mathcal{P}_\kappa\gamma$) *340*
Uniform filter *3*
Uniform indiscernible 188, *189*–191, 202–204, 389, 427
Uniformization *150*, 151, 468
– of $\boldsymbol{\Pi}^1_1$ sets 150, 175–177
– of $\boldsymbol{\Pi}^1_2$ sets 204
– of $\boldsymbol{\Sigma}^1_2$ sets 177
– of $\boldsymbol{\Sigma}^1_3$ sets 206
– in L 177
– under $\mathrm{AD}_\mathbb{R}$ 381, 469
Uniformization property *417*, 419, 421
Universal set *158*, 159, 404
University of California at Berkeley 36, 39, 76, 99, 311, 450
University of California at Los Angeles (U.C.L.A) 115–117, 434
University of Moscow 146
University of Warsaw 22
University of Wisconsin 85, 244
Urelement XIV, 20, 34

Vopěnka cardinal *336*, 338, 339
Vopěnka filter *336*–338
Vopěnka in κ set *336*
Vopěnka's Principle *335*

Wadge degree ($[A]_\mathrm{W}$) *415*
Wadge game 414, 415, 442
Wadge reducibility ($\leq_\mathrm{W}$ and $\equiv_\mathrm{W}$) *414*, *415*, 431, 454
Wadge's Lemma *414*, 415
Weakly amenable pair *244*
Weakly compact cardinal *37*–43, 59, 62, 76, 126, 364
Weakly compact filter *65*
Weakly homogeneous partial order *129*
Weakly homogeneous tree 453–459, 464
Weakly homogeneously Suslin set *453*–459, 464
Weakly inaccessible cardinal *16*
Weakly Mahlo cardinal *17*, 392, 402
Well-founded relation *7*
Well-founded tree (on ${}^k\omega \times Y$) *159*
Well-Ordering Theorem *XIII*, 146
Winning strategy *370*
Woodin cardinal *360*–365, 461–468, 470, 471
Woodin filter *364*, 365
Woodin in κ set *364*

Y-Suslin set *171*, 424, 428

Zermelo-Fraenkel set theory (ZF) 1, 369, 383
Zermelo-Fraenkel set theory with Choice (ZFC) XIV, 1, 437
$0^\dagger$ 283, *288*–295
$0^\#$ 99, *107*–111, 184–188, 277, 282, 344, 346
Zig-zag pattern 410, 420–422

Springer Monographs in Mathematics

This series publishes advanced monographs giving well-written presentations of the "state-of-the-art" in fields of mathematical research that have acquired the maturity needed for such a treatment. They are sufficiently self-contained to be accessible to more than just the intimate specialists of the subject, and sufficiently comprehensive to remain valuable references for many years. Besides the current state of knowledge in its field, an SMM volume should also describe its relevance to and interaction with neighbouring fields of mathematics, and give pointers to future directions of research.

Abhyankar, S.S. **Resolution of Singularities of Embedded Algebraic Surfaces** 2nd enlarged ed. 1998
Alexandrov, A.D. **Convex Polyhedra** 2005
Andrievskii, V.V.; Blatt, H.-P. **Discrepancy of Signed Measures and Polynomial Approximation** 2002
Angell, T.S.; Kirsch, A. **Optimization Methods in Electromagnetic Radiation** 2004
Ara, P.; Mathieu, M. **Local Multipliers of C*-Algebras** 2003
Armitage, D.H.; Gardiner, S.J. **Classical Potential Theory** 2001
Arnold, L. **Random Dynamical Systems** corr. 2nd printing 2003 (1st ed. 1998)
Arveson, W. **Noncommutative Dynamics and E-Semigroups** 2003
Aubin, T. **Some Nonlinear Problems in Riemannian Geometry** 1998
Auslender, A.; Teboulle, M. **Asymptotic Cones and Functions in Optimization and Variational Inequalities** 2003
Banagl, M. **Topological Invariants of Stratified Spaces** 2006
Banasiak, J.; Arlotti, L. **Perturbations of Positive Semigroups with Applications** 2006
Bang-Jensen, J.; Gutin, G. **Digraphs** 2001
Baues, H.-J. **Combinatorial Foundation of Homology and Homotopy** 1999
Bonfiglioli, A.; Lanconelli, E.; Uguzzoni, F. **Stratified Lie Groups and Potential Theory for Their Sub-Laplacians** 2007
Böttcher, A.; Silbermann, B. **Analysis of Toeplitz Operators** 2006
Brown, K.S. **Buildings** 3rd printing 2000 (1st ed. 1998)
Chang, K.-C. **Methods in Nonlinear Analysis** 2005
Cherry, W.; Ye, Z. **Nevanlinna's Theory of Value Distribution** 2001
Ching, W.K. **Iterative Methods for Queuing and Manufacturing Systems** 2001
Chudinovich, I. **Variational and Potential Methods for a Class of Linear Hyperbolic Evolutionary Processes** 2005
Coates, J.; Sujatha, R. **Cyclotomic Fields and Zeta Values** 2006
Crabb, M.C.; James, I.M. **Fibrewise Homotopy Theory** 1998
Dineen, S. **Complex Analysis on Infinite Dimensional Spaces** 1999
Dugundji, J.; Granas, A. **Fixed Point Theory** 2003
Ebbinghaus, H.-D.; Flum, J. **Finite Model Theory** 2006
Edmunds, D.E.; Evans, W.D. **Hardy Operators, Function Spaces and Embeddings** 2004
Elstrodt, J.; Grunewald, F.; Mennicke, J. **Groups Acting on Hyperbolic Space** 1998
Engler, A.J.; Prestel, A. **Valued Fields** 2005
Fadell, E.R.; Husseini, S.Y. **Geometry and Topology of Configuration Spaces** 2001
Fedorov, Yu. N.; Kozlov, V.V. **A Memoir on Integrable Systems** 2001
Flenner, H.; O'Carroll, L.; Vogel, W. **Joins and Intersections** 1999
Gelfand, S.I.; Manin, Y.I. **Methods of Homological Algebra** 2nd ed. 2003 (1st ed. 1996)
Griess, R.L. Jr. **Twelve Sporadic Groups** 1998
Gras, G. **Class Field Theory** corr. 2nd printing 2005
Greuel, G.-M.; Lossen, C.; Shustin, E. **Introduction to Singularities and Deformations** 2007
Hida, H. ***p*-Adic Automorphic Forms on Shimura Varieties** 2004
Ischebeck, E; Rao, R.A. **Ideals and Reality** 2005
Ivrii, V. **Microlocal Analysis and Precise Spectral Asymptotics** 1998
Jakimovski, A.; Sharma, A.; Szabados, J. **Walsh Equiconvergence of Complex Interpolating Polynomials** 2006
Jech, T. **Set Theory** (3rd revised edition 2002)
Jorgenson, J.; Lang, S. **Spherical Inversion on SLn (R)** 2001
Kanamori, A. **The Higher Infinite** corr. 2nd printing 2005 (2nd ed. 2003)
Kanovei, V. **Nonstandard Analysis, Axiomatically** 2005

Khoshnevisan, D. **Multiparameter Processes** 2002
Koch, H. **Galois Theory of *p*-Extensions** 2002
Komornik, V. **Fourier Series in Control Theory** 2005
Kozlov, V.; Maz'ya, V. **Differential Equations with Operator Coefficients** 1999
Lam, T.Y. **Serre's Problem on Projective Modules** 2006
Landsman, N.P. **Mathematical Topics between Classical & Quantum Mechanics** 1998
Lau, D. **Function Algebras on Finite Sets: Basic Course on Many-Valued Logic and Clone Theory** 2006
Leach, J.A.; Needham, D.J. **Matched Asymptotic Expansions in Reaction-Diffusion Theory** 2004
Lebedev, L.P.; Vorovich, I.I. **Functional Analysis in Mechanics** 2002
Lemmermeyer, F. **Reciprocity Laws: From Euler to Eisenstein** 2000
Levine, M.; Morel, F. **Algebraic Cobordism** 2007
Malle, G.; Matzat, B.H. **Inverse Galois Theory** 1999
Mardesic, S. **Strong Shape and Homology** 2000
Margulis, G.A. **On Some Aspects of the Theory of Anosov Systems** 2004
Miyake, T. **Modular Forms** 2006
Murdock, J. **Normal Forms and Unfoldings for Local Dynamical Systems** 2002
Narkiewicz, W. **Elementary and Analytic Theory of Algebraic Numbers** 3rd ed. 2004
Narkiewicz, W. **The Development of Prime Number Theory** 2000
Neittaanmaki, P.; Sprekels, J.; Tiba, D. **Optimization of Elliptic Systems. Theory and Applications** 2006
Onishchik, A.L. **Projective and Cayley–Klein Geometries** 2006
Parker, C.; Rowley, P. **Symplectic Amalgams** 2002
Peller, V. (Ed.) **Hankel Operators and Their Applications** 2003
Prestel, A.; Delzell, C.N. **Positive Polynomials** 2001
Puig, L. **Blocks of Finite Groups** 2002
Ranicki, A. **High-dimensional Knot Theory** 1998
Ribenboim, P. **The Theory of Classical Valuations** 1999
Rowe, E.G.P. **Geometrical Physics in Minkowski Spacetime** 2001
Rudyak, Y.B. **On Thorn Spectra, Orientability and Cobordism** 1998
Ryan, R.A. **Introduction to Tensor Products of Banach Spaces** 2002
Saranen, J.; Vainikko, G. **Periodic Integral and Pseudodifferential Equations with Numerical Approximation** 2002
Schneider, P. **Nonarchimedean Functional Analysis** 2002
Serre, J-P. **Complex Semisimple Lie Algebras** 2001 (reprint of first ed. 1987)
Serre, J-P. **Galois Cohomology** corr. 2nd printing 2002 (1st ed. 1997)
Serre, J-P. **Local Algebra** 2000
Serre, J-P. **Trees** corr. 2nd printing 2003 (1st ed. 1980)
Smirnov, E. **Hausdorff Spectra in Functional Analysis** 2002
Springer, T.A.; Veldkamp, F.D. **Octonions, Jordan Algebras, and Exceptional Groups** 2000
Székelyhidi, L. **Discrete Spectral Synthesis and Its Applications** 2006
Sznitman, A.-S. **Brownian Motion, Obstacles and Random Media** 1998
Taira, K. **Semigroups, Boundary Value Problems and Markov Processes** 2003
Talagrand, M. **The Generic Chaining** 2005
Tauvel, P.; Yu, R.W.T. **Lie Algebras and Algebraic Groups** 2005
Tits, J.; Weiss, R.M. **Moufang Polygons** 2002
Uchiyama, A. **Hardy Spaces on the Euclidean Space** 2001
Üstünel, A.-S.; Zakai, M. **Transformation of Measure on Wiener Space** 2000
Vasconcelos, W. **Integral Closure. Rees Algebras, Multiplicities, Algorithms** 2005
Yang, Y. **Solitons in Field Theory and Nonlinear Analysis** 2001
Zieschang, P.-H. **Theory of Association Schemes** 2005

GPSR Compliance
The European Union's (EU) General Product Safety Regulation (GPSR) is a set of rules that requires consumer products to be safe and our obligations to ensure this.

If you have any concerns about our products, you can contact us on

ProductSafety@springernature.com

In case Publisher is established outside the EU, the EU authorized representative is:

Springer Nature Customer Service Center GmbH
Europaplatz 3
69115 Heidelberg, Germany

www.ingramcontent.com/pod-product-compliance
Ingram Content Group UK Ltd.
Pitfield, Milton Keynes, MK11 3LW, UK
UKHW020437080426
469735UK00006B/194
9783540888666